Jerzy Leszczynski • Manoj K

Editors

Practical Aspects of Computational Chemistry I

An Overview of the Last Two Decades
and Current Trends

Springer

Editors
Prof. Jerzy Leszczynski
Department of Chemistry
Jackson State University
P.O. Box 17910
1400 Lynch Street
Jackson, MS 39217
USA
jerzy@icnanotox.org

Prof. Manoj K. Shukla
Department of Chemistry
Jackson State University
P.O. Box 17910
1400 Lynch Street
Jackson, MS 39217
USA

Present affiliation:
Environmental Laboratory
US Army Engineer Research
and Development Center
3909 Halls Ferry Road
Vicksburg, MS 39180
USA
mshukla@icnanotox.org

ISBN 978-94-007-9771-0   ISBN 978-94-007-0919-5 (eBook)
DOI 10.1007/978-94-007-0919-5
Springer Dordrecht Heidelberg London New York

Springer is part of Springer Science+Business Media (www.springer.com)

Practical Aspects of Computational Chemistry I

Preface

It is a rare event that the impressive group of leading experts is willing to share their views and reflections on development of their research areas in the last few decades. The editors of this book have been very fortunate to attract such contributions, and as an effect two volumes of "Practical Aspects of Computational Chemistry: Overview of the Last Two Decades and Current Trends" are being published. Astonishingly, we found that this task was not so difficult since the pool of authors was derived from a large gathering of speakers who during the last 20 years have participated in the series of meetings "Conferences on Current Trends in Computational Chemistry" (CCTCC) organized by us in Jackson, Mississippi. We asked this group to prepare for the 20th CCTCC that was hold in October 2011 the reviews of the last 20 years of the progress in their research disciplines. Their response to our request was overwhelming. This initiative was conveyed to Springer who in collaboration with the European Academy of Sciences (EAS) invited as to edit such a book.

The current volume presents the compilation of splendid contributions distributed over 21 chapters. The very first chapter contributed by Istvan Hargittai presents the historical account of development of structural chemistry. It also depicts some historical memories of scientists presented in the form of their pictures. This historical description covers a vast period of time. Intruder states pose serious problem in the multireference formulation based on Rayleigh-Schrodinger expansion. Ivan Hubac and Stephen Wilson discuss the current development and future prospects of Many-Body Brillouin-Wigner theories to avoid the problem of intruder states in the next chapter. The third chapter written by Vladimir Ivanov and collaborators reveals the development of multireference state-specific coupled cluster theory. The next chapter from Maria Barysz discusses the development and application of relativistic effects in chemical problems while the fifth chapter contributed by Manthos Papadopoulos and coworkers describes electronic, vibrational and relativistic contributions to the linear and nonlinear optical properties of molecules.

James Chelikowsky and collaborators discuss use of Chebyshen-filtered subspace iteration and windowing methods to solve the Kohn-Sham problem in the sixth chapter. Next chapter contributed by Karlheinz Schwarz and Peter Blaha provides a detailed account of applications of WIEN2K program to determination of electronic structure of solids and surfaces. The recent development of model core potentials during the first decade of the current century is discussed by Tao Zeng and Mariusz Klobukowski in the Chap. 8. Next two chapters discuss Monte Carlo method. Chapter 9 written by William Lester and coworkers describes practicality of Monte Carlo method to study electronic structure of molecules and Chap. 10 describes the relativistic quantum Monte Carlo method and is written by Takahito Nakajima and Yutaka Nakatsuka.

There are two chapters presenting discussion on the various important aspects of nanoscience. Chapter 11 is written by Kwang Kim and coworkers and presents description of computer aided nanomaterial design techniques applying to nanooptics, molecular electronics, spintronics and DNA sequencing. Jorge Seminario and coworkers describe application of computational methods to design nanodevices and other nanosystems in the Chap. 12. The problem of DNA photodimerization has always been very attractive to research communities. Martin McCullagh and George Schatz discuss the application of ground state dynamics to model the thymine-thymine photodimerization reaction in the Chap. 13. In the next chapter A. Luzanov and O. Zhikol review the excited state structural analysis using the time dependent Density Functional Theory approach.

The next four chapters deal with molecular interactions. In the Chap. 15 Joanna Sadlej and coworkers reveal the application of VCD chirality transfer to study the intermolecular interactions. Peter Politzer and Jane Murray review different aspects of non-hydrogen bonding intramolecular interactions in the Chap. 16. The next chapter by Slawomir Grabowski describes characterization of X-$H \ldots \pi$ and X-$H \ldots \sigma$ interactions. Chapter 18 deals with role of cation-π, π-π and hydrogen bonding interaction towards modeling of finite molecular assemblies and is written by A.S. Mahadevi and G.N. Sastry. In the Chap. 19, Oleg Shishkin and Svitlana Shishkina discuss the conformational analysis of cyclohexene, its derivatives and heterocyclic analogues. The stabilization of bivalent metal cations in zeolite catalysts is reviewed by G. Zhidomirov in the Chap. 20. The last chapter of the current volume written by Andrea Michalkova and Jerzy Leszczynski deals with the interaction of nucleic acid bases with minerals that could shed a light on the understanding of origin of life.

With great pleasure, we take this opportunity to thank all authors for devoting their time and hard work enabling us to complete the current volume "Practical Aspects of Computational Chemistry I: Overview of the Last Two Decades and Current Trends". We are grateful to excellent support from the President of the EAS as well as Editors at the Springer. Many thanks go to our families and friends without whom the realization of the book would be not possible.

Jackson, Mississippi, USA   Jerzy Leszczynski
                                                                                              Manoj K. Shukla

Contents

1   **Models—Experiment—Computation: A History of Ideas
    in Structural Chemistry** 1
    Istvan Hargittai

2   **Many-Body Brillouin-Wigner Theories: Development
    and Prospects** 33
    Ivan Hubač and Stephen Wilson

3   **Multireference State–Specific Coupled Cluster Theory
    with a Complete Active Space Reference** 69
    Vladimir V. Ivanov, Dmitry I. Lyakh, Tatyana A. Klimenko,
    and Ludwik Adamowicz

4   **Relativistic Effects in Chemistry and a Two-Component Theory** 103
    Maria Barysz

5   **On the Electronic, Vibrational and Relativistic
    Contributions to the Linear and Nonlinear Optical
    Properties of Molecules** 129
    Aggelos Avramopoulos, Heribert Reis,
    and Manthos G. Papadopoulos

6   **Using Chebyshev-Filtered Subspace Iteration and
    Windowing Methods to Solve the Kohn-Sham Problem** 167
    Grady Schofield, James R. Chelikowsky, and Yousef Saad

7   **Electronic Structure of Solids and Surfaces with WIEN2k** 191
    Karlheinz Schwarz and Peter Blaha

8   **Model Core Potentials in the First Decade of the XXI Century** 209
    Tao Zeng and Mariusz Klobukowski

9   **Practical Aspects of Quantum Monte Carlo for the**
     Electronic Structure of Molecules 255
     Dmitry Yu. Zubarev, Brian M. Austin,
     and William A. Lester Jr.

10   **Relativistic Quantum Monte Carlo Method** 293
     Takahito Nakajima and Yutaka Nakatsuka

11   **Computer Aided Nanomaterials Design – Self-assembly,**
     Nanooptics, Molecular Electronics/Spintronics,
     and Fast DNA Sequencing 319
     Yeonchoo Cho, Seung Kyu Min, Ju Young Lee,
     Woo Youn Kim, and Kwang S. Kim

12   **Computational Molecular Engineering for Nanodevices**
     and Nanosystems 347
     Norma L. Rangel, Paola A. Leon-Plata,
     and Jorge M. Seminario

13   **Theoretical Studies of Thymine–Thymine**
     Photodimerization: Using Ground State Dynamics
     to Model Photoreaction 385
     Martin McCullagh and George C. Schatz

14   **Excited State Structural Analysis: TDDFT and Related Models** 415
     A.V. Luzanov and O.A. Zhikol

15   **VCD Chirality Transfer: A New Insight**
     into the Intermolecular Interactions 451
     Jan Cz. Dobrowolski, Joanna E. Rode, and Joanna Sadlej

16   **Non-hydrogen-Bonding Intramolecular Interactions:**
     Important but Often Overlooked 479
     Peter Politzer and Jane S. Murray

17   **X –H$\cdots\pi$ and X –H$\cdots\sigma$ Interactions – Hydrogen Bonds**
     with Multicenter Proton Acceptors 497
     Sławomir J. Grabowski

18   **Computational Approaches Towards Modeling Finite**
     Molecular Assemblies: Role of Cation-π, $\pi -\pi$
     and Hydrogen Bonding Interactions 517
     A. Subha Mahadevi and G. Narahari Sastry

19   **Unusual Properties of Usual Molecules. Conformational**
     Analysis of Cyclohexene, Its Derivatives and Heterocyclic
     Analogues 557
     Oleg V. Shishkin and Svitlana V. Shishkina

20 Molecular Models of the Stabilization of Bivalent Metal Cations in Zeolite Catalysts 579
G.M. Zhidomirov, A.A. Shubin, A.V. Larin, S.E. Malykhin, and A.A. Rybakov

21 Towards Involvement of Interactions of Nucleic Acid Bases with Minerals in the Origin of Life: Quantum Chemical Approach 645
Andrea Michalkova and Jerzy Leszczynski

Index 673

Contributors

Ludwik Adamowicz University of Arizona, Tucson, AZ, USA, ludwik@u.arizona.edu

Brian M. Austin Kenneth S. Pitzer Center for Theoretical Chemistry, Department of Chemistry, University of California, Berkeley, CA 94720-1460, USA

National Energy Research Scientific Computing, Lawrence Berkeley National Laboratory, Berkeley, CA 94720, USA, baustin@lbl.gov

Aggelos Avramopoulos Institute of Organic and Pharmaceutical Chemistry, National Hellenic Research Foundation, 48 Vas. Constantinou Ave, Athens 116 35, Greece, aavram@eie.gr

Maria Barysz Institute of Chemistry, N. Copernicus University, Gagarina 7, Toruń 87 100, Poland, teomjb@chem.uni.torun.pl

Peter Blaha Institute of Materials Chemistry, Vienna University of Technology, Getreidemarkt 9/165-TC, A-1060 Vienna, Austria, pblaha@theochem.tuwien.ac.at

James R. Chelikowsky Institute for Computational Engineering and Sciences, University of Texas, Austin, TX, 78712, USA

Departments of Physics and Chemical Engineering, University of Texas, Austin, TX 78712, USA, jrc@ices.utexas.edu

Yeonchoo Cho Center for Superfunctional Materials, Department of Chemistry and Department of Physics, Pohang University of Science and Technology, Hyojadong, Namgu, Pohang 790-784, South Korea

Jan Cz. Dobrowolski National Medicines Institute, 30/34 Chełmska Street, 00-725 Warsaw, Poland

Industrial Chemistry Research Institute, 8 Rydygiera Street, 01-793 Warsaw, Poland

Sławomir J. Grabowski Kimika Fakultatea, Euskal Herriko Unibertsitatea and Donostia International Physics Center (DIPC), P.K. 1072, 20080 Donostia, Euskadi, Spain

Ikerbasque, Basque Foundation for Science, 48011 Bilbao, Spain, s.grabowski@ikerbasque.org

Istvan Hargittai Materials Structure and Modeling of the Hungarian Academy of Sciences at Budapest, University of Technology and Economics, POBox 91, 1521 Budapest, Hungary, istvan.hargittai@gmail.com

Ivan Hubač Department of Nuclear Physics and Biophysics, Division of Chemical Physics, Faculty of Mathematics, Physics and Informatics, Comenius University, Bratislava, 84248, Slovakia

Institute of Physics, Silesian University, P.O. Box 74601, Opava, Czech Republic, belaxx@gmail.com

Vladimir V. Ivanov V. N. Karazin Kharkiv National University, Kharkiv, Ukraine, vivanov@univer.kharkov.ua

Kwang S. Kim Center for Superfunctional Materials, Department of Chemistry and Department of Physics, Pohang University of Science and Technology, Hyojadong, Namgu, Pohang 790-784, South Korea, kim@postech.ac.kr

Woo Youn Kim Department of Chemistry, KAIST, Daejeon 305-701, South Korea

Tatyana A. Klimenko V. N. Karazin Kharkiv National University, Kharkiv, Ukraine, generalchem@mail.ru

Mariusz Klobukowski Department of Chemistry, University of Alberta, Edmonton, AL, Canada T6G 2G2, mariusz.klobukowski@ualberta.ca

A.V. Larin Chemistry Department, Lomonosov Moscow State University, Leninskiye Gory 1-3, Moscow GSP-2, 119992, Russia, nasgo@yandex.ru

Ju Young Lee Center for Superfunctional Materials, Department of Chemistry and Department of Physics, Pohang University of Science and Technology, Hyojadong, Namgu, Pohang 790-784, Korea

Paola A. Leon-Plata Department of Chemical Engineering, Texas A&M University, College Station, TX, USA, paola.leon@tamu.com

William A. Lester Jr. Chemical Sciences Division, Lawrence Berkeley National Laboratory, Berkeley, CA 94720, USA

Kenneth S. Pitzer Center for Theoretical Chemistry, Department of Chemistry, University of California, Berkeley, CA 94720-1460, USA, walester@lbl.gov

Jerzy Leszczynski Interdisciplinary Nanotoxicity Center, Jackson State University, Jackson, MS 39217, USA, jerzy@icnanotox.org

A.V. Luzanov STC "Institute for Single Crystals" of National Academy of Sciences of Ukraine, 60 Lenin ave, Kharkiv 61001, Ukraine, luzanov@xray.isc.kharkov.com

Dmitry I. Lyakh V. N. Karazin Kharkiv National University, Kharkiv, Ukraine, lyakh@univer.kharkov.ua

A. Subha Mahadevi Molecular Modeling Group, Indian Institute of Chemical Technology, Tarnaka, Hyderabad, 500607, India

S.E. Malykhin Boreskov Institute of Catalysis, Siberian Branch of the Russian Academy of Sciences, Pr. Akad. Lavrentieva 5, Novosibirsk 630090, Russia, s.e.malykhin@gmail.com

Martin McCullagh Department of Chemistry, Northwestern University, Evanston, IL 60208-3113, United States

Andrea Michalkova Interdisciplinary Nanotoxicity Center, Jackson State University, Jackson, MS, 39217, USA

Seung Kyu Min Center for Superfunctional Materials, Department of Chemistry and Department of Physics, Pohang University of Science and Technology, Hyojadong, Namgu, Pohang 790-784, South Korea

Jane S. Murray CleveTheoComp, 1951 W. 26th Street, Cleveland, OH 44113, USA

Takahito Nakajima Computational Molecular Science Research Team, Advanced Institute for Computational Science, RIKEN, 7-1-26, Minatojima-minami, Cyuo, Kobe, Hyogo 650-0047, Japan, nakajima@riken.jp

Yutaka Nakatsuka Computational Molecular Science Research Team, Advanced Institute for Computational Science, RIKEN, 7-1-26, Minatojima-minami, Cyuo, Kobe, Hyogo 650-0047, Japan, yutakana@riken.jp

Manthos G. Papadopoulos Institute of Organic and Pharmaceutical Chemistry, National Hellenic Research Foundation, 48 Vas. Constantinou Ave, Athens 116 35, Greece, mpapad@eie.gr

Peter Politzer CleveTheoComp, 1951 W. 26th Street, Cleveland, OH 44113, USA, ppolitze@uno.edu

Norma L. Rangel Department of Chemical Engineering, Texas A&M University, College Station, TX, USA

Materials Science and Engineering, Texas A&M University, College Station, TX, USA, normalucre@gmail.com

Heribert Reis Institute of Organic and Pharmaceutical Chemistry, National Hellenic Research Foundation, 48 Vas. Constantinou Ave, Athens 116 35, Greece, hreis@eie.gr

Joanna E. Rode Industrial Chemistry Research Institute, 8 Rydygiera Street, 01-793 Warsaw, Poland

A.A. Rybakov Chemistry Department, Lomonosov Moscow State University, Leninskiye Gory 1-3, Moscow GSP-2, 119992, Russia, rybakovy@mail.ru

Yousef Saad Department of Computer Science and Engineering, University of Minnesota, Minneapolis, MN, 55455, USA, saad@cs.umn.edu

Joanna Sadlej National Medicines Institute, 30/34 Chełmska Street, 00-725 Warsaw, Poland

Faculty of Chemistry, University of Warsaw, 1 Pasteura Street, 02-093 Warsaw, Poland, sadlej@chem.uw.edu.pl

G. Narahari Sastry Molecular Modeling Group, Indian Institute of Chemical Technology, Tarnaka, Hyderabad, 500607, India, gnsastry@gmail.com

George C. Schatz Department of Chemistry, Northwestern University, Evanston, IL, 60208-3113, United States, schatz@chem.northwestern.edu

Grady Schofield Institute for Computational Engineering and Sciences, University of Texas, Austin, TX, 78712, USA, grady@ices.utexas.edu

Karlheinz Schwarz Institute of Materials Chemistry, Vienna University of Technology, Getreidemarkt 9/165-TC, A-1060 Vienna, Austria, kschwarz@theochem.tuwien.ac.at

Jorge M. Seminario Department of Chemical Engineering, Texas A&M University, College Station, TX, USA

Materials Science and Engineering, Texas A&M University, College Station, TX, USA

Department of Electrical and Computer Engineering, Texas A&M University, College Station, TX, USA, seminario@tamu.edu

Oleg V. Shishkin Division of Functional Materials Chemistry, SSI "Institute for Single Crystals", National Academy of Science of Ukraine, 60 Lenina ave, Kharkiv 61001, Ukraine, shishkin@xray.isc.kharkov.com

Svitlana V. Shishkina Division of Functional Materials Chemistry, SSI "Institute for Single Crystals", National Academy of Science of Ukraine, 60 Lenina ave, Kharkiv 61001, Ukraine

A.A. Shubin Boreskov Institute of Catalysis, Siberian Branch of the Russian Academy of Sciences, Pr. Akad. Lavrentieva 5, Novosibirsk 630090, Russia, A.A.Shubin@catalysis.ru

Stephen Wilson Theoretical Chemistry Group, Physical and Theoretical Chemistry Laboratory, University of Oxford, Oxford, OX1 3QZ, UK

Division of Chemical Physics, Faculty of Mathematics, Physics and Informatics Comenius University, Bratislava 84248, Slovakia, quantumsystems@gmail.com

Tao Zeng Department of Chemistry, University of Alberta, Edmonton, AL, Canada T6G 2G2

G.M. Zhidomirov Boreskov Institute of Catalysis, Siberian Branch of the Russian Academy of Sciences, Pr. Akad. Lavrentieva 5, Novosibirsk 630090, Russia

Chemistry Department, Lomonosov Moscow State University, Leninskiye Gory 1-3, Moscow GSP-2, 119992, Russia, zhidomirov@mail.ru

O.A. Zhikol STC "Institute for Single Crystals" of National Academy of Sciences of Ukraine, 60 Lenin ave, Kharkiv 61001, Ukraine, zhikol@xray.isc.kharkov.com

Dmitry Yu. Zubarev Kenneth S. Pitzer Center for Theoretical Chemistry, Department of Chemistry, University of California, Berkeley, CA 94720-1460, USA, dmitry.zubarev@berkeley.edu

Chapter 1
Models—Experiment—Computation: A History of Ideas in Structural Chemistry

Istvan Hargittai

Abstract Ideas about chemical structures have developed over hundreds of years, but the pace has greatly accelerated during the twentieth century. The mechanical interactions among building blocks of structures were taken into account in the computational models by Frank Westheimer and by Terrel Hill, and Lou Allinger's programs made them especially popular. G. N. Lewis provided models of bonding in molecules that served as starting points for later models, among them for Ron Gillespie's immensely popular VSEPR model. Accounting for non-bonded interactions has conveniently augmented the considerations for bond configurations. The emergence of X-ray crystallography almost 100 years ago, followed by other diffraction techniques and a plethora of spectroscopic techniques provided tremendous headway for experimental information of ever increasing precision. The next step was attaining comparable accuracy that helped the meaningful comparison and ultimately the combination of structural information from the most diverse experimental and computational sources. Linus Pauling's valence bond theory and Friedrich Hund's and Robert Mulliken's molecular orbital approach had their preeminence at different times, the latter finally prevailing due to its better suitability for computation. Not only did John Pople build a whole systematics of computations; he understood that if computation was to become a tool on a par with experiment, error estimation had to be handled in a compatible way. Today, qualitative models, experiments, and computations all have their own niches in the realm of structure research, all contributing to our goal of uncovering "coherence and regularities"—in the words of Michael Polanyi and Eugene Wigner—for our understanding and utilization of the molecular world.

I. Hargittai (✉)
Materials Structure and Modeling of the Hungarian Academy of Sciences at Budapest,
University of Technology and Economics, POBox 91, 1521 Budapest, Hungary
e-mail: istvan.hargittai@gmail.com

J. Leszczynski and M.K. Shukla (eds.), *Practical Aspects of Computational Chemistry I:* 1
An Overview of the Last Two Decades and Current Trends,
DOI 10.1007/978-94-007-0919-5_1, © Springer Science+Business Media B.V. 2012

Keywords Structural chemistry • Molecular mechanics • Gilbert N. Lewis • Non-bonded interactions • Molecular structure • Molecular biology • Theory of resonance • Alpha-helix • Geometrical parameters • John Pople • Eugene P. Wigner

1.1   Introduction

Philosophically, Democritos's maxim that "Nothing exists except atoms and empty space; everything else is opinion" has been around for millennia [1]. Modern atomistic approach dates only back a few hundred years. Johannes Kepler is credited with being the first to build a model in which he packed equal spheres representing in modern terms water molecules. He published his treatise in Latin in 1611, De nive sexangula (The Six-cornered Snowflake) [2]. He tried to figure out why the snowflakes have hexagonal shapes and in this connection discussed the structure of the honeycomb. His drawings of closely packed spheres were forward-pointing (Fig. 1.1a). It preceded another model of close packing of spheres which Dalton produced almost two hundred years later, in 1805 with which he illustrated his studies of the absorption of gases (Fig. 1.1b) [3].

There may be different considerations of what the beginning of modern chemistry was. To me, it was the recognition that the building blocks—atoms—of the same or different elements link up for different substances. Somehow—and for a long time it was not clear how—in such a linkage the atoms must undergo

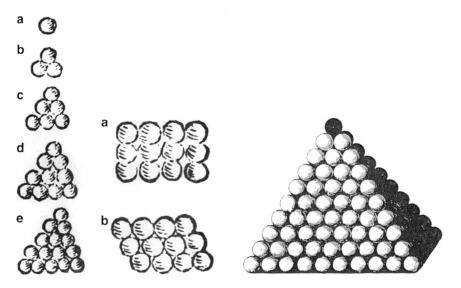

Fig. 1.1 (**a**) Packing of water "molecules" according to Johannes Kepler in 1611 (Ref. [2]); (**b**) Packing of gaseous "molecules" in absorption according to John Dalton's packing in 1805 (Ref. [3])

some change which could only be consistent with throwing out the dogma of the indivisibility of the atom. By advancing this concept chemistry anticipated—even if only tacitly—the three major discoveries at the end of the nineteenth century. They included the discoveries of radioactivity, the electron, and X-rays. This is also why it is proper to say that the science of the twentieth century had begun at the end of the previous century. These experimental discoveries created also the possibilities of testing the various models that have been advanced to describe the structure of matter.

Kepler used modeling not only in his studies of snowflakes but in his investigation of celestial conditions. Curiously though, his three-dimensional planetary model appears to be closer to modern models in structural chemistry than to astronomy. Albert Einstein referred to the significance of modeling in scientific research on the occasion of the 300th anniversary of Kepler's death in 1930 in his article published in Frankfurter Zeitung: "It seems that the human mind has first to construct forms independently before we can find them in things. Kepler's marvelous achievement is a particularly fine example of the truth that knowledge cannot spring from experience alone but only from the comparison of the inventions of the intellect with observed facts" [4]. In much of the success of structural chemistry models have played a ubiquitous role.

1.2 Frank Westheimer and the Origin of Molecular Mechanics

At one point in the history of structural chemistry molecular mechanics calculations dominated the computational work for relatively large molecules. The origins of these calculations were intimately connected to another modeling approach that one of its initiators vividly described. Frank Westheimer (Fig. 1.2a) had participated in the American defense efforts during WWII and when the war had ended, he returned to the University of Chicago to resume his teaching and research. He had to start anew and had time to think about basic problems. This is how half a century later he remembered the birth of molecular mechanics [5]:

> I thought through the idea of calculating the energy of steric effects from first principles and classical physics, relying on known values of force constants for bond stretching and bending, and known values of van der Waals constants for interatomic repulsion. I applied this idea to the calculation of the energy of activation for the racemization of optically active biphenyls. Minimizing the energy of a model for the transition state leads to a set of n equations in n unknowns, one for each stretch or bend of a bond in the molecule. It seemed to me that, to solve these equations, one needed to solve a huge $n \times n$ determinant.
>
> Fortunately for me, Joe Mayer came to the University of Chicago at the end of WWII. Joe was an outstanding physical chemist; he and his wife Maria [Goeppert Mayer] wrote the outstanding text in statistical mechanics. During the war, he had been working at Aberdeen, Maryland, using the world's first digital computer to calculate artillery trajectories. Perhaps Joe could have access to that computer, and could show me how to solve my determinant on it. So I went to him and asked him to help me. He didn't know about optically active biphenyls, so I made some molecular models and explained the stereochemistry to him, and

Fig. 1.2 (**a**) Frank Westheimer in the laboratory (Photograph by MINOT, courtesy of the late Frank Westheimer); (**b**) Norman (Lou) Allinger (Photograph and © by I. Hargittai)

showed him my mathematical development, up to the determinant. Then, in something like half an hour, he found a mathematical trick that we used to solve my equations without needing the determinant. That's how the solution of real problems in molecular mechanics got started. It has become big business since. Furthermore, it turns out that my instinct for computerizing was correct, since that is the way in which the field has since been developed.

The history of molecular mechanics must include—in fact perhaps begins with—a publication by Terrell Hill that presented the same general method I had invented for expressing the energy of molecules in terms of bond stretching, bond bending, and van der Waals interactions, and then minimizing that energy. Hill published the method, but with no application, no "reduction to practice" [6]. I hadn't known that we had a competitor, or that one could publish a bare research idea. After Hill published, I immediately wrote up the work that Mayer and I had already done, theory *and* successful application to determining the activation energy for the racemization of an optically active biphenyl, and submitted it for publication [7].

Eventually, Norman ("Lou") Allinger's (Fig. 1.2b) programs made molecular mechanics accessible for many chemists and he kept expanding the scope of these calculations toward further classes of compounds [8].

1.3   Gilbert N. Lewis's Models of Atoms and Bonding

As for modeling and advancement in the description of chemical bonding prior to quantum chemistry, the importance of Gilbert N. Lewis's (Fig. 1.3a) contributions could hardly be overestimated. They were trend-setters in the first half of twentieth century chemistry and his missing Nobel Prize has been rightly lamented about a great deal.

The quantum chemical description of the covalent bond was given by Walter Heitler and Fritz W. London, but their rigorous treatment severely limited their approach to be utilized directly in chemistry. Heitler himself appreciated Lewis's forward-pointing contribution when he referred to it in his 1945 book Wave Mechanics: "Long before wave mechanics was known Lewis put forward a semi-empirical theory according to which the covalent bond between atoms was effected by the formation of pairs of electrons shared by each pair of atoms. We see now that wave mechanics affords a full justification of this picture, and, moreover, gives a precise meaning to these electron pairs: they are pairs of electrons with antiparallel spins" [9]. Figure 1.3b illustrates Lewis's cubical atoms and some molecules built from such atoms with his original sketches at the bottom [10].

Another testimony for the advanced nature of Lewis's theory was given by Robert S. Mulliken in his Nobel lecture. He described the relation of Lewis's theory to molecular orbital (MO) theory using chemical orbitals. Mulliken emphasized that "Lewis resolved the long-standing conflict between, on the one hand, ionic and charge-transfer theories of chemical bonding and, on the other hand, the kind of bonding which is in evidence in bonds between equal atoms . . . " [11]. Further, in the same lecture, Mulliken writes, "for individual atoms, Lewis' electron shells were three-dimensional, in contrast to Bohr's planar electron orbits, in this respect being closer to the present quantum mechanics than the Bohr theory." Nonetheless, of course, Lewis's theory was "empirical, schematic, and purely qualitative," as Mulliken pointed this out as well. Mulliken appreciated Lewis's contribution so much that he mentioned as a merit of the MO theory that it best approximates Lewis's theory. He writes, " . . . These localized MO's I like to call chemical MO's (or just chemical orbitals because of the fact that some of the orbitals used are now really AO's [atomic orbitals]). In simple molecules, electrons in chemical MO's usually represent the closest possible quantum-mechanical counterpart to Lewis' beautiful pre-quantum valence theory . . . "

1.4   VSEPRing an Efficient Model

The name of the model, VSEPR stands for Valence Shell Electron Pair Repulsion and usually pronounced as "vesper," almost like "whisper," and I have used it as a verb [12] to imply that its principal creator, Ron Gillespie often appeared

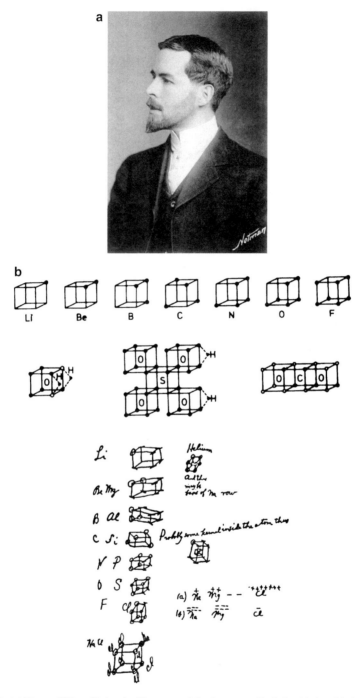

Fig. 1.3 (a) Young Gilbert N. Lewis (Courtesy of the Lawrence Berkeley National Laboratory); (b) G. N. Lewis's cubical atoms and some molecules built from such atoms, first proposed in 1916; his original sketches are at the lower part of the Figure (Ref. [10])

embarrassed by its great success on the background of its rudimentary nature, and would have liked to lend it "respectability" by linking it directly to quantum mechanical considerations.

The origin of the model goes back to N. V. Sidgwick and H. M. Powell who correlated the number of electron pairs in the valence shell of the central atom and its bond configuration in a molecule [13]. Then Ronald J. Gillespie and Ronald S. Nyholm introduced allowances for the differences between the effects of bonding pairs and lone pairs, and applied the model to large classes of inorganic compounds [14]. With coining the VSEPR name the model was ready in its initial formulation. It has since gone through improvements mainly by introducing additional sub-rules and defining its scope of validity. A plethora of examples of VSEPR geometries and geometrical variations through the compounds of main group elements have been presented [15].

The attempts to provide a quantum-mechanical foundation for the VSEPR model have occurred in two directions. One has been to understand better the reason why the model works so well in large classes of compounds, and its basic tenets have been interpreted by the Pauli exclusion principle. Another direction has been to encourage comparisons between sophisticated computations and the application of the model. It could have been expected that calculations of the total electron density distribution should mimic the relative space requirements of the various electron pairs. This was though not too successful—apparently due to the core electron densities suppressing the minute variations in the valence shell. Closer scrutiny, however, revealed that the spatial distributions of the various electron pairs—modeled by electron densities assigned to molecular orbitals—indeed showed distinguishing features in accordance with the expectations of the VSEPR model. A set of examples are shown in Fig. 1.4 [16]. Here, close to the sulfur core, the lone pair of electrons has the largest space requirement; next to it is that of the SO double bond; the SH bonding pair follows; and the smallest space requirement in this series characterizes that of the bonding pair linking the very electronegative fluorine to sulfur.

There have been other approaches to enhance the relative contributions of the valence shell electron density distributions. Thus, visualizing the second derivative of the electron density distribution led to success and the emerging patterns paralleled some important features predicted by the VSEPR model [17].

Some structures, however, have resisted persistently an unambiguous classification of their geometries. The XeF_6 structure was originally considered a success story for the VSEPR model when—contrary to the then available experimental evidence—Gillespie predicted a distorted octahedral arrangement of the six fluorines about the central xenon atom. Recent computational work, however, has suggested that the disturbing lone pair is so much beneath the xenon valence shell that it is hardly expected to distort the regular octahedral arrangement of the ligands. Thus the best that could be said about this molecular shape is that we still don't know it but today we don't know it on a much more sophisticated basis than before [18].

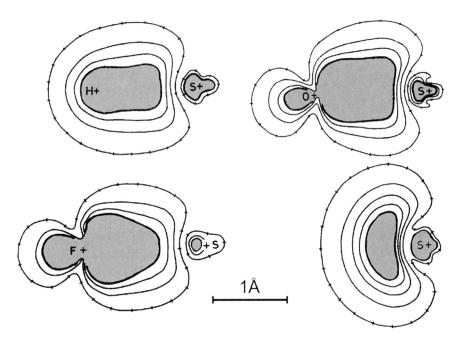

Fig. 1.4 Localized molecular orbitals represented by *contour lines* denoting electron densities of 0.02, 0.04, 0.06, etc. electron/bohr^3 from theoretical calculations for the S–H, S–F, and S=O bonds and the lone pair on sulfur; the *pluses* indicate the positions of the atomic nuclei (After Ref. [16])

1.5 Non-bonded Interactions

A model usually singles out one or a few effects that it takes into consideration and ignores the rest. Hence a reliable application of any model requires the delineation of its applicability. Since the VSEPR model considers the interactions of the electron pairs—even better to say, electron domains as the bonds may correspond to multiple bonds—ligand–ligand interactions are ignored. Accordingly, the applicability of the VSEPR model is enhanced with increasing central atom size with respect to ligand sizes. Conversely, increasing ligand sizes with respect to the size of the central atom diminishes the applicability of the VSEPR model.

In some molecular geometries of fairly large series of compounds, the distances between atoms separated by another atom between them remain remarkably constant, which points to the importance of non-bonded interactions. Thus, for example, the O...O nonbonded distances in XSO_2Y sulfones have been found to hardly deviate from 2.48 Å while the lengths of the S=O bonds vary up to 0.05 Å and the bond angles O=S=O up to 5°, depending on the nature of the ligands X and Y. This is depicted in Fig. 1.5 [19].

These geometrical variations and constancies could be visualized as if the two oxygen ligands were firmly attached to two of the four vertices of the tetrahedron

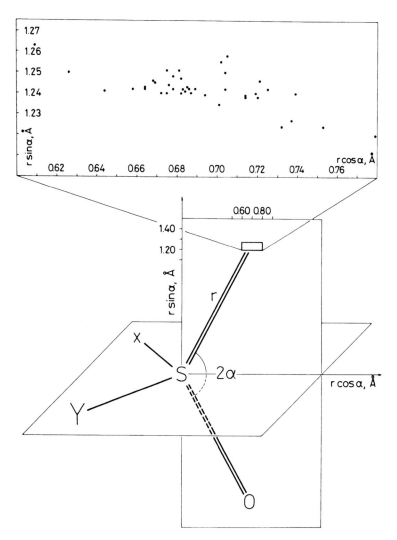

Fig. 1.5 Illustration for the constancy of the O . . . O nonbonded distances in an extended class of sulfones (After Ref. [19])

formed by the four ligands about the central sulfur, and this central sulfur moved along the bisector of the OSO angle depending on the nature of the ligands X and Y. The molecule of sulfuric acid, H_2SO_4 or $(HO)SO_2(OH)$, has its four oxygens about the central sulfur at the vertices of a nearly regular tetrahedron (Fig. 1.6). The six O . . . O distances in this case show a relatively large variation of 0.07 Å , which is still much smaller than the variations of the lengths of the four SO bonds, viz. 0.15 Å and the bond angles OSO vary up to 20° [20].

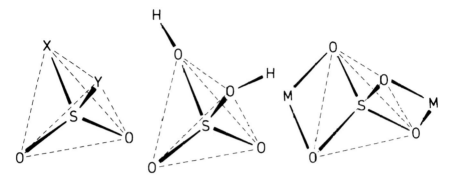

Fig. 1.6 Tetrahedral sulfur bond configurations in (from the *left*) sulfones, sulfuric acid, and alkali sulfates (After Ref. [20])

The importance of intramolecular nonbonded interactions were first recognized by Lawrence S. Bartell when he observed that the three outer carbon atoms in $H_2C=C(CH_3)_2$ were arranged as if they were at the corners of an approximately equilateral triangle [21]. Obviously, the central carbon in this arrangement is not in the center of the triangle and the bond angle between the two bulky methyl groups is smaller than the ideal 120°. For the same reason, the length of the single C–C bonds increases in the series of molecules in which the adjacent CC bonds to the single bond change from triple bond to double bond and to single bond. The single bond under consideration lengthens from 1.47 Å by 0.03 Å at every step. Bartell originally pronounced his considerations as early as 1968, but it caught more attention when, decades later, Gillespie arranged the considerations of intra-ligand interactions into a system and gave it a name, viz. LCP or Ligand Close Packing model (see, Ref. [21]).

1.6 Origins of Experimental Molecular Structure Determination

The determination of molecular structures by X-ray crystallography was one of the great success stories in twentieth century science. It is typical that its roots go back to considerations of theoretical physicists. The possibility of crystals scattering X-rays was raised in Paul P. Ewald's doctoral dissertation in theoretical physics in early 1912 at the Theoretical Physics Institute headed by Arnold Sommerfeld at the University of Munich. Ewald considered the propagation of electromagnetic radiation in a medium having a regular arrangement of resonators and he thought that crystals could be such resonators. The distances between the resonators in a crystal would be much shorter than the wavelength of light. Ewald consulted Max Laue, also a member of Sommerfeld's institute and it was Laue's idea to use X-rays rather than visible light after Ewald had assured him that his theory

Fig. 1.7 John A. Pople (*second from left*) and Herbert A. Hauptman (*second from right*) at the 10th CCTCC in the company of the author on the *left* and Peter Pulay, the initiator of the force method, on the *right* (Photograph by Jerzy Leszczynski, 2001)

was independent on the wavelength of the electromagnetic radiation. Two junior members of the institute, Walter Friedrich and Paul Knipping carried out the experiment, which was then interpreted by Laue. They communicated the results in June of 1912 [22]. In Britain, W. H. Bragg and W. L. Bragg, father and son, initiated the necessary theoretical and experimental work to utilize the new experiments for crystal structure determination. Their papers started appearing in late 1912, and thus X-ray crystallography was launched [23].

The 100-year history of X-ray crystallography is a history of ever improving techniques and crumbling of dogmas, such as the one about the impossibility of determining the phase of the scattering. Two of the principal workers who brought down this dogma, Herbert Hauptman [24] and David Sayre [25] reflected on this history with lessons that point to a broader scope than just one of the techniques of structure determination. Herbert Hauptman and Jerome Karle were awarded the Nobel Prize in Chemistry for 1985 "for their outstanding achievements in the development of direct methods for the determination of crystal structures." Hauptman is a mathematician and so is John Pople, yet both became Nobel laureates in chemistry. They are seen in Fig. 1.7 at the 10th Conference on Current Trends in Computational Chemistry (CCTCC).

At this point I would like to single out one important consideration, namely, that in the application of modern techniques it is no longer a requirement to possess crystals for the X-ray diffraction structure determination; rather, noncrystalline specimens are also possible to use that may be as small as an individual protein molecule or as large as a whole cell. This extends the possibilities of this technique. The suitability of noncrystalline specimens has special significance for the determination of biologically important substances as the bottleneck of such studies used to be the preparation of single crystals from such materials. There remain difficulties to be sure because the experiments with noncrystalline specimens necessitate higher exposure than crystals. This is how one of the pioneers, David Sayre summarized the situation less than a decade ago: "... the problems of crystallization and phasing promise to disappear in the newer technique, while the problem of damage, due to the increased exposure, will become more important" (Ref. [25], p 81).

X-ray crystallography has been a widespread technique and has expanded its scope toward even the largest biologically important systems. As if a younger brother, the gas-phase electron diffraction technique has remained applicable to a limited circle of substances, but is capable of providing unique information, often augmenting the information from X-ray crystallography. It is amazing how much information may be extracted from the diffraction pattern—a set of concentric interference rings—of a gaseous sample [26, 27]. However, the primary information is scarce that is directly obtainable from such a pattern. It may be just about the magnitude of the principal internuclear distances in the molecule and about the relative rigidity of the molecule. The same is true from the visual inspection of the intensity distribution that comes directly from the experimental pattern. The sine Fourier transform of the intensity distribution is related to the probability density distribution of the internuclear distances in the molecule, and thus it provides a considerable amount of information graphically, in a visually perceivable way. However, since it is obtained via certain mathematical manipulations, it is used for general orientation rather than for quantitative elucidation of parameters.

It is the intensity distribution, referred to above, that is the primary source of the reliable quantitative structural information, and most often the analysis utilizes a least-squares procedure. Such a procedure, however—it being based on a non-linear relationship—necessitates suitable initial sets of parameters for best results. Here is where model building comes into the structure analysis for which the sources include already existing structural information, intuition, information directly read off from the Fourier transform of the intensity data, and—increasingly—quantum chemical calculations. A poor initial model may result in reaching a local minimum in the structure refinement yielding a false structure for which there have been plenty of examples in the literature. We refer here only to one such example in which the previously reported erroneous structure [28] was corrected in a reanalysis involving quantum chemical calculations in addition to the electron diffraction data [29]. The situation may be remedied by careful compilation of the model, by testing the results against all available other evidence, and by employing more than one technique simultaneously in the structure determination.

The other experimental techniques include high-resolution rotational spectroscopy and other kinds of spectroscopy. The microwave region is where the pure rotational spectra may be obtained, but the other regions that are used for various spectroscopies also have rotational structure at sufficiently high resolution. This is for the determination of metrical aspects of structure. The scope of techniques that yield information on molecular shape and symmetry is much broader. One of the "other techniques" is computations that have become popular not only as being applied on their own, but also as part of such concerted structure analysis.

There is no doubt that combined application of different techniques is the most promising approach in modern structure analysis. However, it also necessitates a careful consideration of the meaning of structural information derived from different techniques before they would be combined. In this, the concepts of precision and accuracy are to be distinguished. Precision expresses the internal consistency of the data while accuracy refers to the differences between the results obtained for the same parameter by different techniques with the same precision. With the increasing precision achieved by the various physical techniques, the question of accuracy has come into the forefront in demanding studies. From the point of view of accuracy, one of the sources of differences may be the difference in the nature of the physical phenomena used in different techniques. (As an example, X-rays are scattered by the electron density distribution and electrons by the total charge density distribution. Hence, the interatomic distance from X-ray diffraction is the distance between charge centroids of the electron density distribution whereas from electron diffraction it is the internuclear distance.) Further differences originate from the differences in the averaging in different techniques over intramolecular motion. All experimental structure determinations refer to some sort of averages whereas all quantum chemical calculations to the hypothetically motionless structures— equilibrium structures—in the minimum position of the potential energy surface. Averaging over motion depends on the relationship between the lifetime of the structure and the interaction time in a given physical technique. Further consideration should be given to the environment of the structures in the experiment. A crystalline molecular structure carries the impact of neighboring structures that are absent for the isolated molecule in the gas phase. The reference given here discusses the considerations for bond length determination from the points of view mentioned above [30].

1.7   Structural Chemistry in Molecular Biology

It is remarkable how soon following the discovery of the diffraction techniques for structure determination—and while it still required a major effort to elucidate reliable information even for simple molecules—the quest had already begun for the structure determination of large biological systems. Some quotations succinctly illuminate the thrust of structural chemistry in this area. In 1945 Oswald T. Avery was awarded the most prestigious distinction of the Royal Society (London), the

Copley Medal for his and his two associates' discovery that DNA was the substance of heredity. On this occasion, Sir Henry Dale, the President of the Royal Society said, "Here surely is a change to which ... we should accord the status of a genetic variation; and the substance inducing it—the gene in solution, one is tempted to call it—appears to be a nucleic acid of the desoxyribose type. Whatever it be, it is something which should be capable of complete description in terms of *structural chemistry*" (italics added) [31].

The application of rigorous physical techniques to biological systems was in its infancy. Warren Weaver coined the term "molecular biology" only a few years before, and defined it as a new field "in which delicate modern techniques are being used to investigate ever more minute details of certain life processes" [32]. Weaver occupied a crucial position at the Rockefeller Foundation and was responsible for supporting many projects worldwide that laid the foundation of the new science. It was only a few years before Weaver's semantic innovation that in 1934 J. Desmond Bernal subjected protein samples to X-ray diffraction. His experiments gave hope that maybe one day structural information would throw light on the essence of life. The Royal Society distinguished Bernal, at the same ceremony as it did Avery, with a Royal Medal. The citation referred to Bernal's achievements with the following words, among others, "With admirable enterprise he, with his pupils and associates, proceeded to apply the methods of X-ray crystallography to crystals of some of the simpler proteins, as these became available, such as crystalline pepsin and, later, insulin" (Ref. [31]).

In the first decades of the application of X-ray crystallography to biological systems there were only a few scientists in fewer laboratories that could take up the challenge of such work. It appeared possible for two British scientists, Bernal and William Astbury—both pioneers to be sure—to come to a gentleman's agreement to divide the whole area between the two of them. Astbury and Bernal early on recognized the significance of difference between the fully and partially crystalline structures. According to their agreement, Bernal took the crystalline substances and Astbury the more amorphous ones, and the former seemed initially more advantageous than the latter. The diffraction patterns of the regular three-dimensional crystalline substances contained more spots than those of the less regular systems. Thus, it was more hopeful to deduce detailed information on atomic positions for the regular systems. Eventually, however, it turned out that the seemingly information-poor diffraction patters on the less regular systems were easier to interpret as far as the overall structure was concerned. Following Watson's and Crick's spectacular success in proposing a double-helix model for deoxyribonucleic acid, Bernal noted: "It may be paradoxal that the more information-carrying methods should be deemed the less useful to examine a really complex molecule but this is so as a matter of analytical strategy rather than accuracy." To this, Bernal added that "A strategic mistake may be as bad as a factual error" [33]. He meant that he had made the wrong choice when Astbury and he had divided the area and he took the more regular systems. In our reference to Sayre's studies above we have already alluded to the question of ever increasing precision in structure elucidation of biological systems. This progress in precision concerns the regular systems as well as the less regular, more amorphous ones [34]. With the elucidation of the physiological importance of even minute structural effects, the maxim that Max Perutz quoted from Linus Pauling might only be expected gaining significance in the future: "To understand the properties of molecules, not only must you know their structures, but you must know them accurately" [35].

Astbury and Bernal were mistaken when they thought that only the two of them figured in the quest for biological structures. In the US, Linus Pauling led the efforts in attacking the frontier problems of ever larger structures with broad-scale efforts. In the present account we merely single out one feature in the long quest of structural chemistry in understanding biological systems. We do this in order to demonstrate how crucial simple theoretical considerations of the molecular electronic structure could be in such development. Above we have alluded to the fact that protein X-ray diffraction experiments commenced in the early 1930s. Linus Pauling who championed the application of quantum mechanics to a plethora of chemical problems played a pioneering role in the first attempts in the structure elucidation of biological molecules.

1.8 The Theory of Resonance and the Discovery of Alpha-Helix

Linus Pauling (1901–1994) did his PhD work under Roscoe Dickenson who had earned the first PhD degree in X-ray crystallography at the fledgling California Institute of Technology. Like many other young, aspiring American researchers at the time, Pauling also gained postdoctoral experience in European laboratories and spent some time with Arnold Sommerfeld in Munich and Erwin Schrödinger in Zurich. Pauling's goal was to apply the latest discoveries in physics to chemistry. He became very successful and bridged the gap between the rigorous treatment of the covalent bond in hydrogen by Heitler and London and the larger systems chemists were interested in. His theoretical approach was called the valence-bond or VB theory, which built the molecules from atoms linked by electron-pair bonds. It was one of the two major theories, the other being the molecular orbital or MO theory. The latter started from a given arrangement of the atomic nuclei and added all the electrons to this framework. For a long time the VB theory dominated the field because it appealed to chemists as if it were more straightforward of the two, alas, it did not stand well the test of time. The MO theory has proved more amenable to computations, which itself has become a major thrust in modern structural chemistry.

The VB theory described molecular structure by a set of "resonating" structures. This did not mean that each structure in such a set would be considered as present individually, but that the sum of these would correspond to the set. The resonance theory provided merely a model, an approach, rather than a unique reflection of reality. There were proponents and opponents of the theory as is the case with most theories. George Wheland (Fig. 1.8a) published a book in 1944 about the theory of resonance [36]. Linus Pauling (Fig. 1.8b) also contributed to the theory especially

a   b

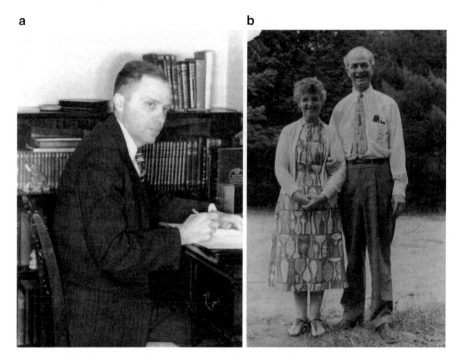

Fig. 1.8 (**a**) George Wheland in 1949 (Photograph courtesy of Betty C. Wheland); (**b**) Ava and
Linus Pauling (Photograph by and courtesy of Karl Maramorosch)

with applications that gained widespread popularity. The theory proved eminently
useful for Pauling in his quest for the protein structure [37].

Pauling was advancing in a systematic manner in his quest for building up
structural chemistry, first in inorganic chemistry and then in organic chemistry. From
the mid-1930s, his attention turned toward proteins and he hoped that understanding
their structure would lead him toward the understanding of biological processes.
Hemoglobin was the first protein that attracted his attention, and he formulated a
theory about the oxygen uptake of hemoglobin and about the structural features of
this molecule related to its function of disposing of and taking up oxygen.

Pauling and Alfred Mirsky recognized the importance of folding in protein
structures whose stability was provided by hydrogen bonds. Hydrogen bonding was
a pivotal discovery, but its significance emerged only gradually over the years. For
many biological molecules it is the hydrogen bonds that keep their different parts
together. Pauling postulated that the subsequent amino acid units are linked to each
other in the folded protein molecule not only by the normal peptide bond but also
by hydrogen bonding that is facilitated by the folding of the protein, which brings
the participating atoms sufficiently close to each other for such interactions.

By the time Pauling became engaged in this research it had been established
from rudimentary X-ray diffraction patterns that there might be two principal
types of protein structure. Keratin fibers, such as hair, horn, porcupine quill, and

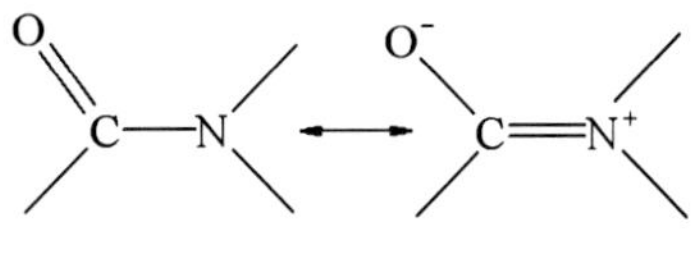

Fig. 1.9 Peptide bond
resonance

fingernail belonged to one, and silk to the other. William Astbury showed in the early 1930s that the X-ray diffraction pattern of hair underwent changes when it was stretched. He called the one producing the normal pattern, alpha keratin and the other, which was similar to the pattern from silk, beta keratin. In 1937, Pauling set out to determine the structure of alpha keratin. He mobilized all his accumulated knowledge in structural chemistry to find the best model that would make sense on this background and would be compatible with the X-ray diffraction pattern.

There seemed to be a good point of reference from X-ray diffraction that the structural unit—whatever it would be—along the axis of the protein molecules repeated at the distance of 5.1 Å. From the studies of smaller molecules, Pauling knew the dimensions of the peptide group, that is, the characteristic sizes of the group linking the amino acids to each other in the protein chain. The C–N bond in the peptide linkage was not a single bond, but it was not a double bond either. Pauling represented the emerging structure by two resonating structures as shown in Fig. 1.9.

Thus, his theory suggested that the C–N bond in the peptide linkage had a partial double bond character, and he knew that the bonds around a double bond were all in the same plane. This allowed him to greatly reduce the number of possible models that had to be considered for alpha keratin. Still, Pauling was unable at this time to find a model that would fit the X-ray diffraction pattern.

When Robert Corey, an expert in X-ray crystallography, joined Pauling at Caltech, they expanded the experimental work determining the structures of individual amino acids and simple peptides. The work was interrupted by World War II, but resumed immediately after the war, and Pauling returned to the structure of alpha keratin in 1948 while he was a visiting professor at Oxford University in England.

The single most important difference in Pauling's renewed approach to the problem was that in 1948 he decided to ignore the differences among the amino acid units in alpha keratin, and assumed all of them to be equivalent for the purpose of his model. Further, he remembered a theorem in mathematics that proved to be most helpful for his purpose. According to this theorem, the most general operation to convert an asymmetric object—an amino acid in this case—into an equivalent asymmetric object—that is, another amino acid—is a rotation–translation. Here the rotation takes place about the molecular axis of the protein; and the translation is the movement ahead along the chain. The repeated application of this rotation–translation produces a helix. The amount of rotation was such that took the chain from one amino acid to the next while the peptide group was kept planar, and this operation was being repeated and repeated all the time. An additional restriction was keeping the adjacent peptide groups apart at a distance that corresponded to hydrogen bonding. In Pauling's model the turn of the protein chain did not involve

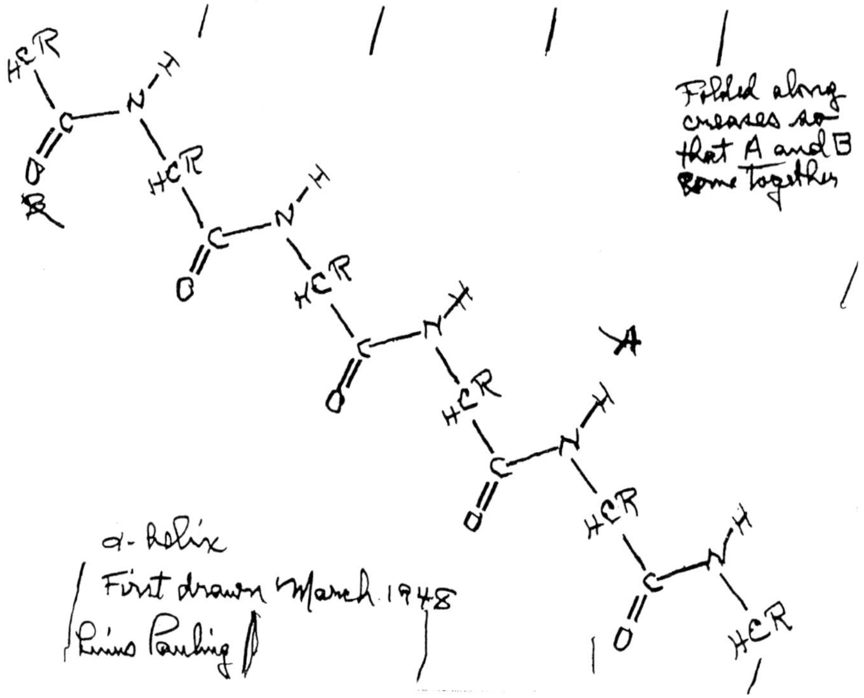

Fig. 1.10 Alpha-helix, as first drawn by Linus Pauling in March 1948. Reproduced from Ref. [38] with kind permission of © Springer

an integral number of amino acids—he did not consider this a requirement. This was yet another relaxed feature of the structure that served him well in finding the best model.

Pauling sketched a protein chain on a piece of paper and folded it along the creases that he had marked, and looked for structures that would satisfy the assumptions he had made (Fig. 1.10) [38]. He found two such structures and called one the alpha helix and the other the gamma helix. The latter appeared to be much less probable than the former.

When Pauling turned around the chain in order to form a helix the fact that non-integer number of amino acid units occurred at any given turn gained added significance: the intramolecular hydrogen bonds did not link identical parts in the chain (Fig. 1.11) [39].

For this model, Pauling determined the distance between repeating units in the protein chain and there was still a marked difference between the distance in the model and from the diffraction experiment. However, Pauling liked the model so much that he thought it had to be correct. Soon afterwards, he saw the diffraction patterns of the British group that was also involved in the structure elucidation of proteins, and it was a much improved diagram compared with the one from

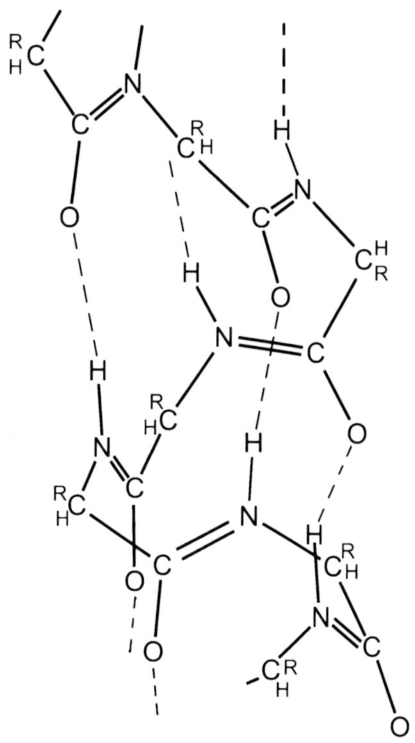

Fig. 1.11 Projection representation of the three-dimensional alpha-helix model. The peptide linkage CN bonds are shown exaggeratedly as *double bonds*. The hydrogen bonds are depicted by *dashed lines* (see text)

Pauling's own laboratory. From this point, Pauling had no longer any doubt that the structure was what he called alpha helix.

The most common example of the resonance theory is the description of the benzene structure. The experimentally precisely determined and accurately known carbon–carbon bond length is consistent with the model as average of the resonance structures. When Pauling's resonance description of the benzene structure was criticized, the physicist Edward Teller and his colleagues provided spectroscopic evidence to support it [40]. The Nobel laureate physicist Philip Anderson was oblivious of Teller's and his co-workers paper (Private communication from Philip Anderson to the author by e-mail in 2009), and 68 years after Teller's contribution, in 2008, Anderson communicated another supportive paper for Pauling's model [41].

1.9 Some Major Contributors to the MO Approach

Above, we narrated the story of alpha helix to stress the merits of utilizing the theory of resonance in this pivotal discovery. There were other theories developed at the same time. Erich Hückel's (Fig. 1.12) studies of the double bond and of aromatic

systems need special mention. He was a physicist who contributed a great deal to the development of theoretical organic chemistry. His approaches to calculations of the electronic structures were in general use for a long time after he had stopped doing creative science before World War II and after he had died many years later. The recognition accorded to him while he was alive was not in proportion with his oeuvre. Later on, however, it has been increasingly appreciated and not only in accolades, but also in the utilization of his research achievements in the studies of such prominent scientists as William N. Lipscomb and Roald Hoffmann, and others. There is a new book, which properly appreciates Hückel's science [42].

As far as recognition during one's lifetime is concerned, Robert S. Mulliken (1896–1986) was the most successful. He was both an experimentalist and a theoretician [43]. For quite some time it seemed that Pauling's VB theory will emerge the winner from the competition of the two strongest theories. Whereas Pauling was flamboyant and an excellent presenter, Mulliken's withdrawn demeanor was the opposite, he was quiet and inclined to reflecting. He is much less known today than Pauling. Mulliken considered both philosophy and science for his career in high school, and he chose science. He received his undergraduate degree in chemistry at MIT. During World War I he did research for the US Chemical Warfare Service. He did his doctoral work at the University of Chicago between 1919 and 1922 where one of his interests was in isotope separation. He did war service during World War II in the framework of the Metallurgical Laboratory in Chicago. Figure 1.13 shows Mulliken in the company of another theoretician, Charles Coulson of Oxford University, who did a great deal for the application of the MO theory in chemistry.

Mulliken spent a few postdoctoral years in Europe where the closest interaction developed between him and Friedrich Hund (Fig. 1.14). They first met in 1925 in

Fig. 1.13 Robert Mulliken and Charles Coulson in 1953 in Oxford (Photograph by and © 1953 of John D. Roberts, used with permission)

Fig. 1.14 Friedrich Hund (*left*) in the company of Max Born (*right*) and Werner Heisenberg in 1966 (Courtesy of Gerhard Hund)

Göttingen at the time when Hund was Max Born's assistant in the mid-1920s. They shared interest in science, developed a fruitful interaction, but never published a joint paper. In Göttingen, in 1927, Hund and Mulliken generalized the ideas of atomic orbitals, and the concept of molecular orbitals was born about which each started publishing in 1928. Mulliken was critical not so much of the valence bond theory, but more of the way Pauling publicized it. He stated that "Pauling made a special point in making everything sound as simple as possible and in that way making it [the VB theory] very popular with chemists but delaying their understanding of the true [complexity of molecular structure]" (Ref. [43]). The merits of the MO theory and Mulliken's contributions were recognized in 1966 by the Nobel Prize in Chemistry "for his fundamental work concerning chemical bonds and the electronic structure of molecules by the molecular orbital method." As up to three persons may share a Nobel Prize in any given category, Hund's omission from this award has been a puzzle. Mulliken gave ample exposure in his Nobel lecture to Hund's contributions.

He described, among others, Hund's works on applying quantum mechanics to the understanding of atoms and their spectra and molecules and their spectra (Ref. [11], p 141): "Using quantum mechanics, he [Hund] quickly clarified our understanding of diatomic molecular spectra, as well as important aspects of the relations between atoms and molecules, and of chemical bonding. It was Hund who in 1928 proposed the now familiar Greek symbols Σ, Π, Δ, for the diatomic molecular electronic states which I had been calling S, P, and D. Molecular orbitals also began to appear in a fairly clear light as suitable homes for electrons in molecules in the same way as atomic orbitals for electrons in atoms. MO theory has long been known as the Hund–Mulliken theory in recognition of the major contribution of Professor Hund in its early development." Mulliken wrote series of articles throughout his career and through them he influenced the development of chemical science and the spreading of his molecular orbitals approach. Steven Berry's Mulliken obituary was concluded with the following evaluation, "He was ready for the unexpected, but he was in tune with nature, and knew inside himself what was real and deserving his acute thought. He set a style and a standard that are as much his legacy as the body of scientific understanding he created" (Ref. [11]).

1.10 Physical Content of Metric Aspects

X-ray diffraction first both in gases and crystals, then increasingly in crystals pioneered the determination of metric aspects of structures. It was eventually joined by other diffraction and various spectroscopy techniques on the experimental side, and quantum chemical computations of ever increasing sophistication. Comparisons and later combined and concerted analysis of metric aspects has proved instructive and useful, and could be done without critical examination of the meaning of information from different sources only within certain precision levels. With increasing precision, the question arose to better understand the physical meanings of geometrical characteristics originating from the different physical techniques and computational techniques as well.

With the availability of improved computed bond lengths, for example, their comparison with experimental information must take into account the physical meaning of the experimentally determined bond lengths. The computed equilibrium distance (r_e) corresponds to the minimum energy position on the potential energy surface and thus should be smaller than the experimental average-distance bond length (r_g). With increasing temperature and enhanced floppiness, the differences may amount to a few hundredths of an angstrom. Hence, for accurate comparison, experimental bond lengths should be compared with computed ones only following necessary corrections that reduce all information involved in the comparison to a common denominator (Ref. [30]). Note that the energy requirements of changes of bond lengths are the largest among changing various geometrical characteristics in molecular structure. Accordingly, similar considerations for bond angles and torsional angles have yet greater consequences than those for bond lengths [44]. A comprehensive discussion extending to most experimental and computational techniques has been available [45].

1.11   John A. Pople's Comprehensive Program

The citation for John A. Pople's (1925–2004) Nobel Prize in Chemistry for 1998 stated, "for his development of computational methods in quantum chemistry." It was a shared prize; the other recipient was Walter Kohn whose citation read "for his development of the density-functional theory." The two together are shown in Fig. 1.15.

In Pople's career in computational chemistry he was first instrumental in the introduction and dissemination of semiempirical techniques, and contributed to the development of a whole set of successful methods that gained broad acceptance and applications. Here we mention only two of the several major figures in addition to Pople who contributed greatly to the development and spreading of semiempirical methods, Robert G. Parr (Fig. 1.16a) and Michael Dewar (Fig. 1.16b).

The semiempirical approach was meant to overcome the barriers represented by the difficulties in calculating integrals. For the more difficult ones approximations were introduced, while for others parameters were adjusted by empirically fitting the experimental data. In time, the semiempirical methods were superseded by more modern approaches, but they had had a pioneering contribution not only by providing a plethora of results, but also by educating the community of chemists to the possibilities of quantum chemical computations and wetted their appetites for more. The approximate methods of Vladimir A. Fock (Fig. 1.17a) and Douglas R. Hartree (Fig. 1.17b) [46] pointed the way toward the more objective non-empirical or ab initio techniques [47].

Eventually Pople embarked on developing ever improving approaches to non-empirical, called *ab initio*, computations. There is a tremendous literature about the plethora of his contributions that have remained essential in current research. Rather than surveying them, I would present a selection of his views based on a recording of

Fig. 1.15 Walter Kohn (*left*) and John Pople in 2001 in Stockholm (Photograph by and © I Hargittai)

Fig. 1.16 (**a**) Robert Parr in 1997 in Wilmington, NC; (**b**) Michael Dewar in 1988 in Austin, TX (Both photographs and © by I. Hargittai)

a   b

Fig. 1.17 (**a**) Vladimir A. Fock (Courtesy of the late Lev Vilkov); (**b**) Douglas R. Hartree (Fischer CF (2004) Douglas Rayner Hartree: His life in science and computing. World Scientific, Singapore)

a conversation we had in 1995, because it seems to me that this would best represent his concerted approach to theoretical chemistry in which he constantly looked out for the needs and interest not only of computational workers, but of experimental chemists as well [48].

Pople formulated the essence of computational chemistry "as the implementation of the existing theory, in a practical sense, to studying particular chemical problems by means of computer programs." He stressed not to draw a distinction between computational chemistry and the underlying theory, because the computer only enabled the theory to be applied more broadly than was possible before.

At the time of our conversation, Pople was developing theories to include the density functional theory and he aimed to treat quantum mechanical problems more efficiently than before. He emphasized the importance of the possibility to make comparisons with experimental information. From this point of view, the density distribution of electrons is the same thing what X-ray diffraction provides, that is, the electron density distribution. In reality, when plots of the total electron density are calculated or measured the features of bonding (or the features of nonbonding electron pairs) are not directly discernable because the total electron density distribution suppresses the fine information related to them. There have been techniques that help us make the bonding features (as well as nonbonding electron

a   b

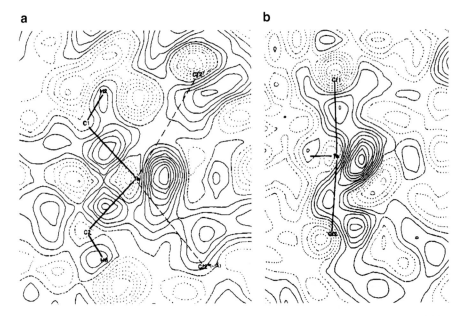

Fig. 1.18 Electron density difference maps for the $(CH_3)_2TeCl_2$ molecule. *On the left* (**a**), in the equatorial plane defined by the atoms Te, S1, and C2, showing the Te–C bond and Te(IV) lone-pair densities. Contours are at 0.03e/A3 with negative contours broken. *On the right* (**b**), in the plane perpendicular to the plane of the left, showing the Te–Cl (chlorine) bond and the Te(IV) lone-pair densities (Ref. [53], © 1983 American Chemical Society)

pairs) visible and one such approach is mentioned here. When the total electron density of the molecule is measured or computed, the measured or computed electron densities of the atoms constituting the molecule may be subtracted from it yielding the features sought. The difference maps in two different cross sections of the $(CH_3)_2TeCl_2$ molecule are demonstrated from X-ray diffraction measurements [49]. In this case both the total electron density and the electron densities of the atoms constituting the molecule were obtained from the X-ray diffraction measurements. The latter was achieved by excluding data from low-angle scattering thereby minimizing the impact of bonding features and those of the nonbonding electron pairs. The emerging patterns were consistent with the expected trigonal bipyramidal tellurium electron pair configuration with the lone electron pair in equatorial position (Fig. 1.18).

Pople saw the advantage in density functional theory versus the quantum chemical methods in that the former dealt with a function of three dimensions whereas to get the full wave function of the electrons, a problem in $3n$ dimensions had to be considered (with n being the total number of electrons).

I found Pople very sensitive to the question of experimental error in computational work, which is a cornerstone issue for a meaningful comparison of experimental and computational results. Because of its importance, I communicate his response to my question about it in full (Ref. [48], pp 182–183):

> This is a good question. The way I like to do this is to set up a theoretical model. You apply one theoretical model essentially to all molecules. This model is one level of approximation. Then you apply this one level of calculation to a very large number of different molecules. In fact, one level of approximation is applied to all molecules, giving you an entire chemistry corresponding to that approximation. That chemistry, of course, would not be the same as real chemistry but it would approach that chemistry and if it is a good model, it will approach real chemistry well. What I try to do is to take a given model and then to use that model to try to reproduce a lot of well-known facts of experimental chemistry. For example you try to reproduce the bond lengths in a number of simple organic molecules, or the heats of formation for that set of molecules, in a situation where the experiment is beyond question. Then you can actually do statistics and say that this theory reproduces all known heats of formation to the root-mean-square accuracy of 2 kcal/mol. When you've done that you build some confidence in the level of theory. If you then apply the same theory in a situation where experiment may not exist, you know the level of confidence of your calculations.

Pople fully agreed that it should always be an objective in computational work to indicate experimental errors. He envisioned an ideal relationship between experimental and computational work in which any chemist could use computations. He was not shy in admitting that he, more than anybody else, was responsible for transforming chemistry and making it a computational science, as far as electronic structure was concerned (he added this qualifying expression). Our conversation was recorded three years before his Nobel Prize.

More than 150 years ago, Joseph Louis Gay-Lussac prophesized that the time was not far when everything would be possible to calculate in chemistry. In 1966 Mulliken concluded his Nobel lecture with the following words: "I would like to emphasize strongly my belief that the era of computing chemists, when hundreds if not thousands of chemists will go to the computing machine instead of the laboratory for increasingly many facets of chemical information, is already at hand" (Ref. [11]). If we look around today, Mulliken's evaluation has been proved correct many times over.

1.12   Final Thoughts

Model building, experimental, and computational techniques, all have contributed to the collection of structural information that has become enormous. Linus Pauling could only have dreamt about such a data base when he constructed the building of his maxims about structural chemistry. It is to his credit that he created principles of long-lasting value based on data that were a miniscule fraction of one percent of the structural information we possess collectively today. The availability of this much structural information today calls for emulating Pauling even if not in his comprehensibility but even in specific domains of knowledge. For science such activities should be appreciated.

Fig. 1.19 Eugene P Wigner and the author on the campus of the University of Texas at Austin in 1969 (By unknown photographer, © I Hargittai)

In this connection I would like to refer to Eugene P. Wigner's characterization of scientific research. He expressed it eloquently in 1963 in his Nobel lecture when he quoted his teacher, Michael Polanyi in that "...science begins when a body of phenomena is available which shows some coherence and regularities, [that] science consists in assimilating these regularities and in creating concepts which permit expressing these regularities in a natural way." Wigner (and Polanyi) saw in this the real transferability of the scientific approach, and more so than in transferring concepts, such as energy, for example, "to other fields of learning" [50]. The beginning of my interactions with Wigner dated back to 1964 when he wrote me a long letter in response to an article I had published in a Hungarian literary magazine in reference to his essay on the limits of science. This article was my first ever publication and it was in the senior year of my university studies. Our interactions culminated in our meeting in person and extended conversations in 1969 at the University of Texas at Austin (Fig. 1.19).

Wigner made fundamental contributions to structural science, and in 1931 he summarized them in his German-language monograph about the applications of group theory to the quantum mechanics of atomic spectra. The book was not considered dated even three decades later when it was published in English translation [51]. Wigner's studies included considerations for chemical reactions and his and Witmer's paper in 1928 contained the first application of symmetry considerations to chemical reactions. The Wigner–Witmer rules referred to the conservation of spin and orbital angular momentum in the reactions of diatomic molecules [52].

In our extended conversations—they were in fact tutorials for me—Wigner introduced me to the intricacies and broad applications of the symmetry concept [53]. We then remained in on–and–off correspondence throughout the years. The utilization of the symmetry concept has become an all-embracing feature of our work in structural chemistry [54]. It was probably Wigner's influence that I have avoided falling into narrowly focused research directions and pursued my interest in the determination and modeling of molecular structure utilizing both experimental and theoretical approaches.

References

1. Democritos [of Abdera], quoted in Mackay AL (1991) A dictionary of scientific quotations. Adam Hilger, Bristol
2. Kepler J (1611) Strena seu de nive sexangula. Godefridum Tampach, Francofurti ad Moenum (English translation 1966) The six-cornered snowflake. Oxford, Clarendon Press
3. Dalton J (1805) Memoirs and proceedings of the manchester literary and philosophical society, vol. 6, Manchester, p 271 (Alembic Club Reprints 1961, Edinburgh, no 2, p 15)
4. Einstein A (1954) Ideas and opinions. Crown Publishers, New York, p 266
5. Hargittai I (2000) Frank H Westheimer. In: Hargittai M (ed) Candid science: conversations with famous chemists. Imperial College Press, London, pp 38–53 (the actual quotation is on pp 41–42)
6. Hill TL (1946) On steric effects J Chem Phys 14:465 (one page)
7. Westheimer FH, Mayer JE (1946) The theory of the racemization of optically active derivatives of diphenyl. J Chem Phys 14:733–738
8. Allinger NL (1992) Molecular mechanics. In: Domenicano A, Hargittai I (eds) Accurate molecular structures: their determination and importance. Oxford University Press, New York, pp 336–354
9. Heitler W, Wave Mechanics as quoted by Laidler KJ (1995) Lessons from the history of chemistry. Acc Chem Res 28:187–192 (actual quote on p 190)
10. Lewis GN (1916) The atom and the molecule. J Am Chem Soc 38:762–785
11. Mulliken RS (1999) Spectroscopy, molecular orbitals, and chemical bonding. In: Nobel lectures chemistry 1963–1970. World Scientific, Singapore, pp 131–160
12. Hargittai I (2000) Echoes of our VSEPRing. Coordination Chem Rev 197:21–35
13. Sidgwick NV, Powell HM (1940) Bakerian lecture. Stereochemical types and valency groups. Proc R Soc Lon Ser A 176:153–180
14. Gillespie RJ, Nyholm RS (1957) Inorganic stereochemistry. Quart Rev Chem Soc 11:339–380
15. Gillespie RJ, Hargittai I (1991) The VSEPR model of molecular geometry. Allyn & Bacon, Boston. Reprint edition: (2012) Dover Publications, Mineola, New York
16. Schmiedekamp A, Cruickshank DWJ, Skaarup S, Pulay P, Hargittai I, Boggs JE (1979) Investigation of the basis of the valence shell electron pair repulsion model by ab initio calculation of geometry variations in a series of tetrahedral and related molecules. J Am Chem Soc 101:2002–2010
17. Bader RFW, MacDougall PJ, Lau CDH (1984) Bonded and nonbonded charge concentrations and their relation to molecular geometry and reactivity. J Am Chem Soc 106:1594–1605
18. Hargittai I, Menyhard DK (2010) Further VSEPRing about molecular geometries. J Mol Struct 978:136–140
19. Hargittai M, Hargittai I (1987) Gas-solid molecular structure differences. Phys Chem Miner 14:413–425
20. Hargittai I (1985) The structure of volatile sulphur compounds. Reidel Publ Co, Dordrecht

21. See eg, Bartell LS (2011) A personal reminiscence about theories used and misused in structural chemistry. Struct Chem 22:247–251. doi:10.1007/s11224-010-9693-8

22. Ewald PP (ed) (1962) Fifty years of X-ray diffraction: dedicated to the International Union of Crystallography on the occasion of the commemoration meeting in Munich, July 1962. A. Oosthoek's Uitgeversmaatschappij N. V, Utrecht

23. Bijvoet JM, Burgers WG, Hägg G (eds) (1969) Early papers on diffraction of X-rays by crystals. A. Oosthoek's Uitgeversmaatschappij N. V, Utrecht

24. Hauptman HA (1990) History of x-ray crystallography. Struct Chem 1:617–620

25. Sayre D (2002) X-ray crystallography: past and present of the phase problem. Struct Chem 13:81–96

26. Hargittai I, Hargittai M (2010) Electron diffraction theory and methods. In: John L, George T, David K (eds) Encyclopedia of spectroscopy and spectrometry, vol 1, 2nd edn. Elsevier, Oxford, pp 461–465

27. Hargittai M, Hargittai M (2010) Electron diffraction applications. In: John L, George T, David K (eds) Encyclopedia of spectroscopy and spectrometry, vol 1, 2nd edn. Elsevier, Oxford, pp 456–460

28. Hargittai M, Dotofeeva OV, Tremmel J (1985) Molecular structure of vanadium dichloride and chromium dichloride from electron diffraction. Inorg Chem 24:3963–3965

29. Varga Z, Vest B, Schwerdtferger P, Hargittai M (2010) Molecular geometry of vanadium dichloride and vanadium trichloride: a gas-phase electron diffraction and computational study. Inorg Chem 49:2816–2821

30. Hargittai M, Hargittai I (1992) Experimental and computed bond lengths: the importance of their differences. Int J Quant Chem 44:1057–1067

31. Dale H (1945) Anniversary address to the Royal Society, London, pp 1–17 (actual quotation, p 2)

32. Weaver W (1970) Molecular biology: origin of the term. Science 170:581–582

33. Bernal JD (1968) The material theory of life. Labour Monthly, July, pp 323–326

34. Hargittai M, Hargittai I (2002) Aspects of structural chemistry in molecular biology. In: Domenicano A, Hargittai I (eds) Strength from weakness: structural consequences of weak interactions in molecules, supermolecules, and crystals. Kluwer, Dordrecht, pp 91–119

35. Perutz M (1998) I wish I'd made you angry earlier: essays on science and scientists. Oxford University Press, Oxford/New York, p 167

36. Wheland GW (1944) The theory of resonance and its applications to organic chemistry. Chapman and Hall, London

37. Hargittai I (2010) Linus Pauling's quest for the structure of proteins. Struct Chem 21:1–7

38. Pauling L (1996) The discovery of the alpha helix. Chem Intell 2(1):32–38, This was a posthumous publication communicated by Dorothy Munro, Linus Pauling's former secretary/assistant between 1973 and 1994

39. Our representation of the alpha-helix follows how Francis Crick depicted it. Crick's involvement in the alpha-helix story is given Olby R (2009) Francis Crick: hunter of life's secrets. Cold Spring Harbor Laboratory Press, Cold Spring Harbor, pp 99–101

40. Nordheim G, Sponer H, Teller E (1940) Note on the ultraviolet absorption systems of benzene vapor. J Chem Phys 8:455–458

41. Anderson P (2008) Who or what is RVB? Physics Today 8–9 April 2008 (RVB = Resonance Valence Bond)

42. Karachalios A (2010) Erich Hückel (1896–1980): from physics to quantum chemistry. Springer Science + Business Media, Dordrecht

43. Berry RS (2000) Robert Sanderson Mulliken 1896–1986. Biographical memoirs, vol 78. The National Academy Press, Washington, DC, pp 146–165

44. Hargittai I, Levy JB (1999) Accessible geometrical changes. Struct Chem 10:387–389

45. Domenicano A, Hargittai I (eds) (1992) Accurate molecular structures. Oxford University Press, Oxford

46. Fischer CF (2004) Douglas Rayner Hertree: his life in science and computing. World Scientific, Singapore

47. Faddeev LD, Khalfin LA, Komarov IV (2004) V. A. Fock—Selected Works: Quantum Mechanics and Quantum Field Theory. Chapman and Hall/CRC Press, Boca Raton
48. Hargittai I (2000) John Pople. In: Hargittai M (ed) Candid science: conversations with famous chemists. Imperial College Press, London, pp 178–189
49. Ziolo RF, Troup JM (1983) Experimental observation of the tellurium(IV) bonding and lone-pair electron density in dimethyltellurium dichloride by x-ray diffraction techniques. J Am Chem Soc 105:229–235
50. Wigner EP (1967) City Hall speech—Stockholm, 1963. In: Wigner EP (ed) Symmetries and reflections: scientific essays of Eugene P Wigner. Indiana University Press, Bloomington and London, pp 262–263
51. Wigner E (1931) Gruppentheorie und ihre Anwendung auf die Quantenmechanik der Atom-spektren. Vieweg, Braunschweig, Germany. English translation: Wigner EP (1959) Group Theory and Its Application to Quantum Mechanics of Atomic Spectra. Academic Press, New York
52. Wigner E, Witmer EE (1928) Über die Struktur der zweiatomigen Molekelspektren nach der Quantenmechanik. Z Phys 51:859–886
53. Hargittai I (2002) Learning symmetry from Eugene P. Wigner. In: Marx G (ed) Eugene Paul Wigner Centennial. Roland Eötvös Physical Society, Budapest, pp 124–143
54. Hargittai M, Hargittai I (2009, 2010) Symmetry through the eyes of a chemist, 3rd edn. Springer, Dordrecht

Chapter 2
Many-Body Brillouin-Wigner Theories: Development and Prospects

Ivan Hubač and Stephen Wilson

Abstract We describe a quantum chemical project focussed on the development of state-specific many-body Brillouin–Wigner methods which was undertaken during the period 1994 to the present day. The Brillouin–Wigner methodology has been shown to provide an approach to the many-body problem which is especially useful when a multireference formulation is required since it completely avoids the 'intruder state' problem that often plagues the traditional Rayleigh–Schrödinger expansion. The many-body Brillouin–Wigner approach provides the basis for robust methods which can be applied routinely in situations where the more familiar single-reference formalism is not adequate in Coupled Cluster (CC) and Perturbation Theory (PT) formalisms. It can also be employed in Configuration Interaction (CI) studies. Although the Brillouin–Wigner expansion is not itself a 'many-body' theory, it can be subjected to a posteriori adjustments which removes unphysical terms, which in the diagrammatic formalism correspond to unlinked diagrams, and recover a 'many-body' method.

As well as reviewing progress and prospects of the state-specific Brillouin–Wigner approach, we describe the new methods of communication that were deployed to facilitate effective collaboration between researchers located as geographically distributed sites. A web-based collaborative virtual environment (CVE) was designed for research on molecular electronic structure theory which

I. Hubač (✉)
Department of Nuclear Physics and Biophysics, Division of Chemical Physics, Faculty of Mathematics, Physics and Informatics, Comenius University, Bratislava, 84248, Slovakia

Institute of Physics, Silesian University, P.O. Box 74601, Opava, Czech Republic
e-mail: belaxx@gmail.com

S. Wilson
Theoretical Chemistry Group, Physical and Theoretical Chemistry Laboratory, University of Oxford, Oxford, OX1 3QZ, UK

Division of Chemical Physics, Faculty of Mathematics, Physics and Informatics Comenius University, Bratislava, 84248, Slovakia
e-mail: quantumsystems@gmail.com

J. Leszczynski and M.K. Shukla (eds.), *Practical Aspects of Computational Chemistry I: An Overview of the Last Two Decades and Current Trends,*
DOI 10.1007/978-94-007-0919-5_2, © Springer Science+Business Media B.V. 2012

will support the development of quantum chemical methodology for challenging applications. This CVE was developed whilst actually carrying out a significant but specific, 'real life' quantum chemical project so that those features which were found to be useful in facilitating remote collaboration could be evaluated and incorporated in an emerging framework.

Keywords Many-body theory • Brillouin–Wigner theory • State-specific multireference correlation problem • Many-body perturbation theory • Coupled cluster theory • Configuration interaction • Collaborative virtual enviroments

Preamble

Almost 175 years ago in 1837, Charles Babbage, "the father of the computer" [39, 60], wrote, with prophetic insight [1]

> The whole of chemistry, and with it crystallography, would become a branch of mathematical analysis which, like astronomy, taking its constants from observation, would enable us to predict the character of any new compound.

At that time, the molecular theory of matter was not universally accepted. This had to wait for Einstein's 1905 explanation [32] of a study of the irregular movement of microscopic particles suspended in a liquid by the botanist Robert Brown [12] in 1828 – Brownian motion.[1] Rutherford's 1911 paper [101] establishing the nuclear model of the atom was followed Bohr's quantum theory of the atom [8] and the flowering of Quantum Mechanics during the 1920s.[2] Quantum Chemistry, now mature field of scientific investigation[3] and one of the central pillars of Computational Chemistry, was established in two seminal papers published in 1927 by Heitler and London [40] and Born and Oppenheimer [9].

During the second half of the twentieth century, electronic digital computers of increasing power facilitated the computational study of chemistry and chemical processes [108, 109]. They allowed the development of practical approximation schemes that Dirac [28] had seen the need for in 1929 when he stated that

> underlying physical laws necessary for the mathematical theory of [. . .] the whole of chemistry are [. . .] completely known

but

> the exact application of these laws leads to equations much too complicated to be soluble.

---

[1] For a recent account of Einstein's explanation of Brownian motion see Renn [99].

[2] For a recent account of the development of quantum theory see Kumar's *Quantum – Einstein, Bohr and the Great Debate about the Nature of Reality* [73]. The six volumes by Mehra and Rechenberg on *The Historical Development of Quantum Theory* [86] provide a definitive account.

[3] See, for example, the *Handbook of Molecular Physics and Quantum Chemistry* [123] for a recent major reference work.

Of these schemes, quantum many-body theories implemented within the algebraic approximation (using finite basis set expansions) have come to dominate quantum chemistry over the past 50 years underpinning a systematic approach to the theoretical study of molecular electronic structure.

It is against this broad historical background that, in this contribution, we review the development of a methodology suitable for handling the multi-reference correlation problem in molecules. Such problems frequently arise when dissociative processes or electronic excited states are investigated and have attracted a great deal of attention over recent years. The present review is focussed on the state-specific many-body Brillouin–Wigner methodology[4] which has been developed and applied to the molecular electronic structure problem over the period 1994 to the present day.

Calculations based on ab initio methods of molecular electronic structure theory are very demanding computationally and any review of progress over the past two decades must be made against the background of the continually increasing power of contemporary computers. The past 20 years have witnessed a relentless increase in the power of computing machines. It has been observed[5] that the processing power of computers seems to double every 18 months. As the historian Roberts [100] points out in his book *Twentieth Century*

> No other technology has ever improved so rapidly for so long.

This ever-increasing computing power has led to both higher accuracy in molecular electronic structure calculations, often because larger basis sets can be utilized, and has opened up the possibility of applications to larger molecules and molecular systems. In Fig. 2.1, we show the increasing speed of the most powerful computers in the period 1993–2010 [42]. During the early 1990s, high performance computers achieved rates of computation of the order of a few gigaflops.[6] Such performance was realized in many-body quantum chemical algorithms [2]. By 2010, rates of computation of a few petaflops[7] had been realized and machines capable of delivering rate at the exaflops[8] level are anticipated.

During the period under review, there were other developments in computer and communications technology which had and indeed continue to have a profound influence on the way that scientific research is carried out. The growth of the internet has been a world-wide phenomenon. This growth is illustrated in Fig. 2.2 where we display the growth in the number of internet users in the authors' home countries, the United Kingdom (left-hand scale) and Slovakia (right-hand scale), during the period 1998–2008 [4]. The trends observed are similar in both countries. In this

---

[4]Brillouin–Wigner perturbation theory was originally introduced in independent publications by Lennard-Jones [74] in 1930, by Brillouin [11] in 1932 and by Wigner [107] in 1935.

[5]By Moore [87] of Intel. This has been dubbed "Moore's Law".

[6]gigaflops: 10^9 floating-point operations per second.

[7]petaflops: 10^{15} floating-point operations per second.

[8]exaflops: 10^{18} floating-point operations per second.

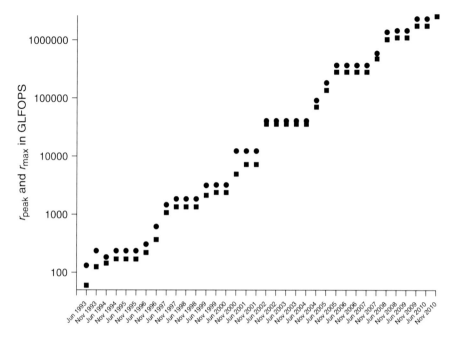

Fig. 2.1 Increasing speed of the most powerful computers in the period 1993–2010. The figure shows values of r_{peak}, the theoretical peak performance, and r_{max}, the maximal LINPACK performance, achieved by the most powerful machine in the '*Top500*' list. (LINPACK is a program that solves a dense system of linear equations that is often used in assessing the performance of computers) (Data taken from the Top500 website [42] www.top500.org)

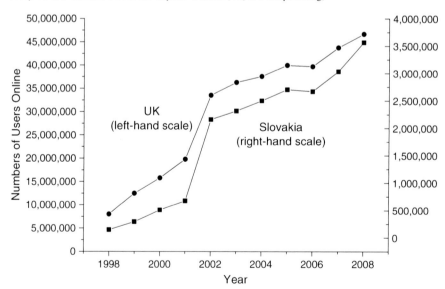

Fig. 2.2 Growth in the Number of Internet Users in the United Kingdom (*left-hand scale*) and Slovakia (*right-hand scale*) during the period 1998–2008. (Data taken from the BBC News website[4])

contribution, we describe the impact of these developments in the collaborative study of quantum chemical methodology and of state-specific many-body Brillouin–Wigner methods in particular.

The internet offers new ways for communication between scientists especially when they are situated in geographically distributed locations. As Ziman [125] has pointed out:

> the communication system is [...] vital to science; it lies at the very heart of the 'scientific method'.

Over the past 20 years, most learned journals have become available online. Some new journals are only available online. Online versions of journals can offer significant advantages over the traditional printed version: fast, economical distribution, hypertext links to cited literature, easy (low cost) incorporation of colour figures and illustrations, in-text movies, and so on.

However, as well as changing the way in which scientists formally present their completed work to their peers, the internet is also changing the way that researchers collaborate whilst carrying out their projects. This paper describes some of our recent work aimed at the development of methods for collaborative research in molecular physics and quantum chemistry via the internet. Collaborative virtual environments have been developed for quantum chemistry whilst actually carrying out significant, 'real life' projects. The first of these was entitled *European Metalaboratory for multireference quantum chemical methods*. Features which were found to be useful in facilitating the collaboration could be evaluated and incorporated in the emerging environment. A collaborative virtual environment[9] actively supports human–human communication in addition to human–machine communication and uses a virtual environment as the interface. The specific quantum chemical projects considered involved the development of Brillouin–Wigner methods for handling the many-body problem which arises in molecular electronic structure theory especially when a multireference formulation is required.

2.1 Introduction

In this article,[10] we describe a quantum chemical project focussed on the development of state-specific many-body Brillouin–Wigner methods which was undertaken during the period 1994–2010. We outline the motivation for this project, survey the progress made and consider the prospects for future progress. The Brillouin–Wigner methodology has been shown [49,51,124] to provide an approach to the many-body

---

[9]For an introduction to the field of collaborative virtual environments see, for example, Churchill et al. [19].

[10]Parts of this article are taken from the authors' book entitled *Brillouin–Wigner Methods for Many-Body Systems*.

problem which is especially useful when a multireference formulation is required. For multireference problems, the more widely used Rayleigh–Schrödinger-based formalism can often suffer from the 'intruder state' problem. The many-body Brillouin–Wigner approach provides the basis for robust methods which can be applied routinely in situations where the more familiar single-reference formalism is not adequate in Configuration Interaction (CI), Coupled Cluster (SC), and Perturbation Theory (PT) formalisms. Although the Brillouin–Wigner expansion is not itself a 'many-body' theory, it can be subjected to a posteriori adjustments which removes unphysical terms, which scale nonlinearly with the number of electrons in the molecular system under consideration and in the diagrammatic formalism correspond to unlinked diagrams. In this way, we recover a 'many-body' method.

As well as outlining the motivation for this project, survey the progress made during the period 1994–2010 and consider the prospects for future progress, we describe the new methods of internet-based communication that were deployed to facilitate effective collaboration between scientists working at geographically distributed sites. A web-based collaborative virtual environment (CVE) was designed for research on molecular electronic structure theory which will support the development of quantum chemical methodology for challenging applications [120]. This CVE was developed whilst actually carrying out a significant but specific, real life' quantum chemical project so that those features which were found to be useful in facilitating remote collaboration could be evaluated and incorporated in an emerging framework. A CVE of this type actively supports human–human communication as well as human–machine communication and uses a web-based virtual environment as the interface.

The second half of the twentieth century saw a sustained attack on the correlation problem in atoms and molecules. By the mid-1950s the basic structure of correlated wavefunctions was understood, thanks mainly to developments in solid state physics and nuclear physics. The linked diagram theorem of the many-body perturbation theory and the connected cluster structure of the exact wavefunctions were firmly established. Goldstone [38] exploited diagrammatic techniques developed in quantum field theory [30, 31, 36, 37], to complete Brueckner's work [13] on the scaling of energies and other expectation values with the number of electrons in the system. Hugenholtz [58] provided an alternative approach. The exponential ansatz for the wave operator suggested by Hubbard [45] in 1957 was first exploited in nuclear physics by Coester [22] and by Coester and Kümmel [23].

In applications to the atomic and molecular electronic structure problem – the problem of describing the motion of the electrons in the field of clamped nuclei – there was, as Paldus [90] describes, an

> initial hope that the configuration interaction approach limited to doubly excited configurations, originating from a single reference state, [would] provide a satisfactory description of correlation effects.

However, this approach, which exploits the variation theorem to determine the correlated wavefunction for the non-relativistic problem, was soon thwarted by the slow convergence of the configuration interaction expansion.

The 1960s saw the applications of the many-body perturbation theory developed during the 1950s by Brueckner [13], Goldstone [38] and others to the atomic structure problem by Kelly [63–72].[11] These applications used the numerical solutions to the Hartree–Fock equations which are available for atoms because of the special coordinate system. Kelly also reported applications to some simple hydrides in which the hydrogen atom nucleus is treated as an additional perturbation.

At about the same time, Čížek [20] developed the formalism of the coupled cluster approach for use in the context of molecular electronic structure theory.

By the early 1970s, both the many-body perturbation theory and the coupled cluster theory had been implemented in the algebraic approximation and applications to arbitrary molecular systems became a reality. Schulman and Kaufman [104] reported many-body perturbation theory calculations for the hydrogen molecule using a basis set of uncontracted Gaussian functions in 1970. In 1972, Paldus, Čížek and Shavitt [93] initiated applications of ab initio coupled cluster theory using finite basis set expansions, whilst in the following year Kaldor [61] first invoked the algebraic approximation in an application of the many-body perturbation theory to the hydrogen molecule ground state. In 1976, Wilson and Silver [122] compared finite order many-body perturbation theory with limited configuration interaction calculations when both methods are formulated in the algebraic approximation.

Progress was particularly rapid during the late 1970s and early 1980s with the introduction of a new generation of 'high performance' computing machines which enabled the realization of practical schemes of calculation which in turn gave new levels of understanding of the nature of the electron correlation problem in atoms and, more particularly, in molecules. It became widely recognized that a successful description of correlation effects in molecules must have two key ingredients:

1. It must be based either directly or indirectly on the linked diagram theorem of many-body perturbation theory so as to ensure that the calculated energies and other expectation values scale linearly with particle number
2. It must be based on a careful and systematic realization of the algebraic approximation (i.e., the use of finite basis set expansions), since this can often be the dominant source of error in calculations which aim to achieve high precision. Over the past 20 years a great deal of progress has been made in the systematic and accurate implementation of the algebraic approximation in many-body quantum chemical studies. This will not be reviewed here and the interested reader is referred elsewhere for detailed reviews [89, 110].

Today, there remain a number of problems in molecular electronic structure theory. The most outstanding of these is undoubtedly the development of a robust theoretical apparatus for the accurate description of dissociative processes which usually demand the use of multi-reference functions. This requirement has recently kindled a renewal of interest in the Brillouin–Wigner perturbation theory and its application to such problems. This contribution describes the application of

---

[11] See also the work of Das and his coworkers [29].

Brillouin–Wigner methods to many-body systems and, in particular, to molecular systems requiring a multi-reference formalism.

During the period under review in this contribution, progress has been made in the development of state-specific Rayleigh-Schrödinger formalism for the multi-reference correlation problem. Mukherjee and his collaborators [15–18, 24–26, 41, 77–80] have developed a theoretical framework for the multi-reference problem which scales linearly with the number of electrons in the target system. Schaefer and his colleagues [33–35] have also explored state-specific Rayleigh-Schrödinger formalism. The recent work of Pittner and his coworkers [5, 6, 27] provides an alternative perspective on the state-specific Rayleigh-Schrödinger formalism of Mukherjee and his collaborators. In this review, we shall concentrate on the state-specific Brillouin-Wigner approach.

2.2   Brillouin–Wigner Theories

The use of Brillouin–Wigner perturbation theory in describing many-body systems has been critically re-examined in recent years. It has been shown that under certain well-defined circumstances, the Brillouin–Wigner expansion can be regarded as a valid many-body theory. The primary purpose of this contribution is to provide an overview of the Brillouin–Wigner methods in the study of the 'many-body' problem in atomic and molecular quantum mechanics. A detailed and coherent account can be found in the volume *Brillouin–Wigner Methods for Many-Body Systems* [51]. We also summarize the progress made to date and briefly describe the internet-based collaborative tools deployed in executing this some of this work.

The renewal of interest in Brillouin–Wigner perturbation theory for many-body systems seen in recent years, is driven by the need to develop a robust multi-reference theory. Multi-reference formalisms are an important prerequisite for theoretical descriptions of dissociative phenomena and of many electronically excited states. Brillouin–Wigner perturbation theory is seen as a remedy to a problem which plagues multi-reference Rayleigh-Schrödinger perturbation theory; the so-called intruder state problem.

Multi-reference Rayleigh–Schrödinger perturbation theory is designed to describe a manifold of states. However, as the perturbation is 'switched on' the relative disposition of these states and those states outside the reference space may change in such a way that convergence of the perturbation series is impaired or even destroyed. States from outside the reference space which assume an energy below that of any of the states among the reference set when the perturbation is switched on are termed intruder states. The situation is illustrated schematically in Fig. 2.3 which provides a representation of the intruder state problem. In this figure, the reference space $\mathcal{P}$ consists of three states with energies E_0, E_1 and E_2 which are represented on the left-hand side. The exact energies $\mathcal{E}_0$, $\mathcal{E}_1$, ... obtained when the perturbation is turned on, are represented on the right-hand side of the figure. But the exact energy $\mathcal{E}_2$ corresponds to E_3 in the reference space. This is an intruder state.

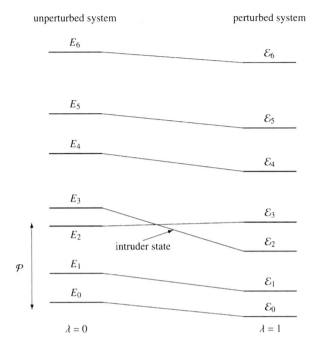

unperturbed system   perturbed system

$\mathcal{P}$

$\lambda = 0$ $\lambda = 1$

Fig. 2.3 In multireference Rayleigh-Schrödinger perturbation theory, states from outside the reference space, $\mathcal{P}$, which assume an energy below that of any state among the reference set when the perturbation is switched on, are termed intruder states

Intruder states arising when $0 < \lambda \leq +1$ often have a physical origin. The so-called 'backdoor' intruder states, which arise for $-1 \leq \lambda < 0$ are frequently unphysical. The occurence of 'backdoor' intruder states is illustrated schematically in Fig. 2.4. In this figure, the reference space $\mathcal{P}$ again consists of three states with energies E_0, E_1 and E_2 which are represented in the central column. The perturbed system with energies $\mathcal{E}_0$, $\mathcal{E}_1$, ... is represented on the right-hand side and corresponds to $\lambda = +1$. The 'backdoor' spectrum corresponding to $\lambda = -1$ is shown on the left-hand side. 'Backdoor' intruder states arise in this 'backdoor' spectrum.

Multi-reference Brillouin–Wigner theory overcomes the intruder state problem because the exact energy is contained in the denominator factors. Calculations are therefore 'state specific', that is they are performed for one state at a time. This is in contrast to multi-reference Rayleigh–Schrödinger perturbation theory which is applied to a manifold of states simultaneously. Multi-reference Brillouin–Wigner perturbation theory is applied to a single state. Wenzel and Steiner [105] write (see also [106]):

> ... the reference energy in Brillouin–Wigner perturbation theory is the fully dressed energy ... This feature guarantees the existence of a natural gap and thereby rapid convergence of the perturbation series.

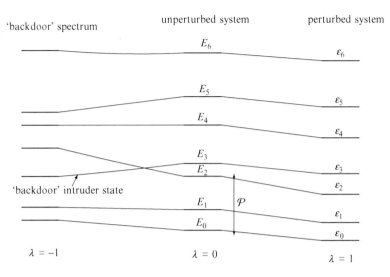

Fig. 2.4 In multireference Rayleigh-Schrödinger perturbation theory, states from outside the reference space, $\mathcal{P}$, which assume an energy below that of any state among the reference set for $-1 \leq \lambda < 0$ are termed 'backdoor' intruder states. Unlike the intruder states corresponding to $0 < \lambda \leq +1$, which often have a physical origin, 'backdoor' intruder states are frequently unphysical

Multi-reference Brillouin–Wigner perturbation theory overcomes the intruder state problem which has plagued multi-reference Rayleigh–Schrödinger perturbation theory for many years. However, in general, Brillouin–Wigner expansions do not scale linearly with the number of electrons; they are not extensive. These insights have led to the critical re-examination of Brillouin–Wigner perturbation theory in describing many-body systems in recent years [5, 6, 14, 46–56, 76, 81–85, 94–96, 103, 104, 124].

When we consider the application of multi-reference Brillouin–Wigner methods to many-body systems, two distinct approaches can be taken which we consider now in turn:

1. The Brillouin–Wigner perturbation theory can be employed to solve the equations associated with an explicit 'many-body' method.

   For example, the full configuration interaction problem can be solved by making a Brillouin–Wigner perturbation theory expansion through infinite order as can the equations of the coupled cluster theory. In the case of the full configuration interaction, it is obvious that provided the Brillouin–Wigner perturbation series is summed through all orders, then a result will be obtained which is entirely equivalent, at least as far as the results are concerned,[12] to any other technique for solving the full configuration interaction problem. We might designate the Brillouin–Wigner-*based* approach to the Full Configuration Interaction (FCI) model "Brillouin–Wigner-Full Configuration Interaction" or (BW-FCI).

---

[12] We are not concerned here with questions of computational efficiency.

Hubač and Neogrády [53] have explored the use of Brillouin–Wigner perturbation theory in solving the equations of coupled cluster theory. In a paper published in *The Physical Review* in 1994 entitled *Size-consistent Brillouin–Wigner perturbation theory with an exponentially parametrized wave function: Brillouin–Wigner coupled cluster theory*, they write

> Size consistency of the Brillouin–Wigner perturbation theory is studied using the Lippmann–Schwinger equation and an exponential ansatz for the wave function. Relation of this theory to the coupled cluster method is studied and a comparison through the effective Hamiltonian method is also provided.

By adopting an exponential expansion for the wave operator, they ensure that their method is extensive. Hubač and his coworkers, Neogrády and Mášik, obtained the "Brillouin–Wigner" coupled cluster theory [52, 53, 81, 82] which is entirely equivalent to other many-body formulations of coupled cluster theory for the case of a single reference function, since the Brillouin–Wigner perturbation expansion is summed through all orders. We designate the Brillouin–Wigner-*based* approach to the Coupled Cluster (CC) model 'Brillouin–Wigner Coupled Cluster' or (BW-CC). If, for example, the equations of a limited coupled cluster expansion, such as that usually designated CCSD (in which all single and double excitation cluster operators with respect to a single reference function are considered), are solved by means of Brillouin–Wigner perturbation theory, then a method, designated 'Brillouin–Wigner-Coupled Cluster Singles Doubles' or (BW-CCSD) theory, is obtained which is entirely equivalent to 'standard' CCSD theory.

2. A posteriori corrections can be developed for calculations performed by using the Brillouin–Wigner perturbation expansion. These a posteriori corrections can be obtained for the Brillouin–Wigner perturbation theory itself and, more importantly, for methods, such as limited configuration interaction or multireference coupled cluster theory, which can be formulated within the framework of a Brillouin–Wigner perturbation expansion.

These a posteriori corrections are based on a very simple idea which is suggested by the work of Brandow [10]. Brandow used the Brillouin–Wigner perturbation theory as a starting point for a derivation of the Goldstone "linked diagram" expansion "by elementary time-independent methods". At a NATO Advanced Study Institute held in 1991, Wilson wrote [112]:

> The Rayleigh–Schrödinger perturbation theory can be derived from the Lennard–Jones–Brillouin–Wigner perturbation theory by expanding the energy-dependent denominators which occur in the latter.

In the work of Brandow [10], Brillouin–Wigner perturbation theory is used as a step in the theoretical development of first Rayleigh–Schrödinger perturbation theory and then the many-body perturbation theory. In the a posteriori correction developed by the present authors in a paper entitled *On the use of Brillouin–Wigner perturbation theory for many-body systems* and published in the *Journal of Physics B: Atomic, Molecular and Optical Physics* in 2000,

> The use of Brillouin–Wigner perturbation theory in describing many-body systems is critically re-examined.

Brillouin–Wigner perturbation theory is employed as a computational technique – a technique which avoids the intruder state problem – and then the relation between the Brillouin–Wigner and Rayleigh–Schrödinger propagators is used to correct the calculation for lack of extensivity.

2.2.1   Brillouin–Wigner Expansions

The time-independent Schrödinger equation for an N-particle system described by the hamiltonian $\hat{\mathcal{H}}$ can be written as the eigenvalue equation

$$\hat{\mathcal{H}}\Psi_\mu = \mathcal{E}_\mu \Psi_\mu, \qquad \mu = 0, 1, \ldots \tag{2.1}$$

in which the eigenfunction, the wave function Ψ_μ, depends on the space and spin co-ordinates of the constituent particles and the eigenvalue $\mathcal{E}_\mu$ is the energy of the μth state. For a many-body system solution of Eq. 2.1 is impossible and approximations must be developed. The perturbation method is ubiquitous in quantum mechanics. A series expansion for the solution of the Schrödinger equation (2.1) is developed with respect to some suitably chosen model problem. We can write the time-independent Schrödinger equation for this model or reference problem with hamiltonian $\hat{\mathcal{H}}_0$ in the form

$$\hat{\mathcal{H}}_0 \Phi_\mu = E_\mu \Phi_\mu, \qquad \mu = 0, 1, \ldots \tag{2.2}$$

where the eigenfunction Φ_μ is the reference wave function and the eigenvalue E_μ is the corresponding energy. The difference between the hamiltonian $\hat{\mathcal{H}}$ and the unperturbed hamiltonian $\hat{\mathcal{H}}_0$ is the perturbation $\hat{\mathcal{H}}_1$

$$\hat{\mathcal{H}}_1 = \hat{\mathcal{H}} - \hat{\mathcal{H}}_0 \tag{2.3}$$

By introducing the perturbation parameter λ, we can define the hamiltonian $\hat{\mathcal{H}}(\lambda)$ as

$$\hat{\mathcal{H}}(\lambda) = \hat{\mathcal{H}}_0 + \lambda \hat{\mathcal{H}}_1 \tag{2.4}$$

so that

$$\lim_{\lambda \to 0} \hat{\mathcal{H}}(\lambda) = \hat{\mathcal{H}}_0 \tag{2.5}$$

and

$$\lim_{\lambda \to 1} \hat{\mathcal{H}}(\lambda) = \hat{\mathcal{H}}. \tag{2.6}$$

Similarly, we can define the energy $\mathcal{E}_\mu(\lambda)$ and the wave function $\Psi_\mu(\lambda)$ as follows

$$\hat{\mathcal{H}}(\lambda)\,\Psi_\mu(\lambda) = \mathcal{E}_\mu(\lambda)\,\Psi_\mu(\lambda)\,, \qquad \mu = 0, 1, \ldots \tag{2.7}$$

such that

$$\lim_{\lambda \to 0} \mathcal{E}_\mu(\lambda) = E_\mu, \tag{2.8}$$

$$\lim_{\lambda \to 1} \mathcal{E}_\mu(\lambda) = \mathcal{E}_\mu, \tag{2.9}$$

$$\lim_{\lambda \to 0} \Psi_\mu(\lambda) = \Phi_\mu, \tag{2.10}$$

and

$$\lim_{\lambda \to 1} \Psi_\mu(\lambda) = \Psi_\mu. \tag{2.11}$$

The familiar Rayleigh–Schrödinger perturbation theory is obtained by making power series expansions in the perturbation parameter λ for the exact energy

$$\mathcal{E}_\mu = E_\mu + \lambda\epsilon_1^{(\mu)} + \lambda^2\epsilon_2^{(\mu)} + \ldots \tag{2.12}$$

and the corresponding wavefunction

$$\Psi_\mu = \Phi_\mu + \lambda\chi_1^{(\mu)} + \lambda^2\chi_2^{(\mu)} + \ldots \tag{2.13}$$

Substituting these two expansions in Eq. 2.7, using (2.4) and equating powers of λ leads to the working equations of the Rayleigh–Schrödinger perturbation theory.[13]

Here we shall concentrate on the Brillouin–Wigner form of perturbation theory. We begin in Sect. 2.2.2 by considering the single reference formalism. The multi-reference formalism is considered in Sect. 2.2.3.

2.2.2 Single-Reference Brillouin–Wigner Expansions

By combining Eq. 2.4 and the full Schrödinger equation (2.1), we can write

$$\left(\mathcal{E}_\mu - \hat{\mathcal{H}}_0\right)\Psi_\mu = \lambda\hat{\mathcal{H}}_1\Psi_\mu, \qquad \mu = 0, 1, \ldots \tag{2.14}$$

This equation can be rewritten in matrix form as

$$\left(\mathcal{E}_\mu - E_m\right)\langle\Phi_m|\Psi_\mu\rangle = \lambda\langle\Phi_m|\hat{\mathcal{H}}_1|\Psi_\mu\rangle \tag{2.15}$$

---

[13]Further details of the Rayleigh–Schrödinger perturbation expansion can be found elsewhere [110].

or

$$\langle \Phi_m | \Psi_\mu \rangle = \frac{\lambda \langle \Phi_m | \hat{\mathcal{H}}_1 | \Psi_\mu \rangle}{\left(\mathcal{E}_\mu - E_m \right)} \tag{2.16}$$

In Eqs. 2.15 and 2.16, the state Φ_m is the unperturbed eigenket of $\hat{\mathcal{H}}_0$ with eigenenergy E_m; that is Eq. 2.2.

It is convenient to put

$$\langle \Phi_\mu | \Psi_\mu \rangle = 1 \tag{2.17}$$

to simplify the following formulae. $|\Psi_\mu\rangle$ is not normalized to unity, but satisfies the *intermediate normalization condition* (2.17).

Putting $\mu = m$ in Eq. 2.15 and using (2.17) leads to the following expression for the exact energy eigenvalue, $\mathcal{E}_\mu$

$$\mathcal{E}_\mu = E_\mu + \lambda \langle \Phi_\mu | \hat{\mathcal{H}}_1 | \Psi_\mu \rangle \tag{2.18}$$

The eigenstate Ψ_μ is constructed by exploiting the completeness of the unperturbed basis, Φ_μ, so that

$$|\Psi_\mu\rangle = \sum_m |\Phi_m\rangle \langle \Phi_m | \Psi_\mu \rangle \tag{2.19}$$

or

$$|\Psi_\mu\rangle = |\Phi_\mu\rangle \langle \Phi_\mu | \Psi_\mu \rangle + \sum_{m \neq \mu}' |\Phi_m\rangle \langle \Phi_m | \Psi_\mu \rangle \tag{2.20}$$

where the prime on the summation emphasizes that the term for which $m = \mu$ is excluded. Using Eq. 2.16 for $\langle \Phi_m | \Psi_\mu \rangle$ together with the intermediate normalization condition 2.17, we can rewrite Eq. 2.20 in the form

$$|\Psi_\mu\rangle = |\Phi_\mu\rangle + \lambda \sum_m' |\Phi_m\rangle \frac{1}{\left(\mathcal{E}_\mu - E_m \right)} \langle \Phi_m | \hat{\mathcal{H}}_1 | \Psi_\mu \rangle \tag{2.21}$$

where the prime again indicates that the case $m = \mu$ is excluded from the summation. Equation 2.21 is the basic formula of Brillouin–Wigner perturbation theory for the case of a single reference function.

We can develop a series expansion for Ψ_μ in powers of λ. The coefficients in this expansion depend on the exact perturbed energy $\mathcal{E}_\mu$ rather than, as is the case in the more familiar Rayleigh–Schrödinger expansion, on the unperturbed energy E_m. Iterating the basic formula (2.21), we obtain

$$|\Psi_\mu\rangle = |\Phi_\mu\rangle + \lambda \sum_{m \neq \mu} |\Phi_m\rangle \frac{1}{\left(\mathcal{E}_\mu - E_m \right)} \langle \Phi_m | \hat{\mathcal{H}}_1 | \Phi_\mu \rangle$$

$$+ \lambda^2 \sum_{j,m \neq \mu} |\Phi_j\rangle \frac{1}{\left(\mathcal{E}_\mu - E_j \right)} \langle \Phi_j | \hat{\mathcal{H}}_1 | \Phi_m \rangle \frac{1}{\left(\mathcal{E}_\mu - E_m \right)} \langle \Phi_m | \hat{\mathcal{H}}_1 | \Phi_\mu \rangle + \dots$$

$$\tag{2.22}$$

By introducing the Brillouin–Wigner resolvent

$$\hat{B}_\mu = \sum_{m \neq \mu} |\Phi_m\rangle \frac{1}{\left(\mathcal{E}_\mu - E_m\right)} \langle\Phi_m| , \qquad (2.23)$$

we can write Eq. 2.23 in the form

$$|\Psi_\mu\rangle = \left(1 + \lambda\hat{B}_\mu\hat{\mathcal{H}}_1 + \lambda^2\hat{B}_\mu\hat{\mathcal{H}}_1\hat{B}_\mu\hat{\mathcal{H}}_1 + \ldots\right)|\Phi_\mu\rangle \qquad (2.24)$$

By substituting the expansion for the exact wavefunction (2.22) into the expression (2.18) for the exact eigenvalue, we can develop the following expansion for the perturbed energy values

$$\mathcal{E}_\mu = E_\mu + \lambda\langle\Phi_\mu|\hat{\mathcal{H}}_1|\Phi_\mu\rangle + \lambda^2 \sum_{m \neq \mu} \langle\Phi_\mu|\hat{\mathcal{H}}_1|\Phi_m\rangle \frac{1}{\left(\mathcal{E}_\mu - E_m\right)} \langle\Phi_m|\hat{\mathcal{H}}_1|\Phi_\mu\rangle + \ldots$$

$$(2.25)$$

or, in terms of the Brillouin–Wigner resolvent, $\hat{B}_\mu$,

$$\mathcal{E}_\mu = E_\mu + \lambda\langle\Phi_\mu|\hat{\mathcal{H}}_1|\Phi_\mu\rangle + \lambda^2\langle\Phi_\mu|\hat{\mathcal{H}}_1\hat{B}_\mu\hat{\mathcal{H}}_1|\Phi_\mu\rangle + \ldots \qquad (2.26)$$

Let us make three comments about Eq. 2.26. First, the expansion (2.26) is not a simple power series expansion in the perturbation parameter λ. The exact energy $\mathcal{E}_\mu$ which appears in the denominators of the resolvent $\hat{B}_\mu$ is also dependent on λ. Secondly, we note that if the denominator terms $\frac{1}{\mathcal{E}_\mu - E_m}$ are expanded in powers of λ then the usual Rayleigh–Schrödinger expansion for the exact energy $\mathcal{E}_\mu$ is obtained. Thirdly, because of the restriction on the indices in the summations in Brillouin–Wigner expansion, it terminates in no more than k steps for a finite k dimensional space. For example, it is easy to see that the Brillouin–Wigner expansion for a two-dimensional space is exact when taken to second order in λ.

We are now ready to introduce the Brillouin–Wigner *wave operator* for the μth state. The *state specific* wave operator $\hat{\Omega}_\mu(\lambda)$ is defined by the relation

$$|\Psi_\mu\rangle = \hat{\Omega}_\mu(\lambda)|\Phi_\mu\rangle. \qquad (2.27)$$

The application of the wave operator to the reference function $|\Phi_\mu\rangle$ gives the exact wavefunction $|\Psi_\mu\rangle$. Comparing the expansion (2.24) with (2.27), we can write the wave operator as an expansion in $\lambda\hat{B}_\mu\hat{\mathcal{H}}_1$

$$\hat{\Omega}_\mu(\lambda) = 1 + \lambda\hat{B}_\mu\hat{\mathcal{H}}_1 + \lambda^2\hat{B}_\mu\hat{\mathcal{H}}_1\hat{B}_\mu\hat{\mathcal{H}}_1 + \ldots \qquad (2.28)$$

We then note that this expansion can also be written as a recursion

$$\hat{\Omega}_\mu(\lambda) = 1 + \lambda\hat{B}_\mu\hat{\mathcal{H}}_1\hat{\Omega}_\mu(\lambda) \qquad (2.29)$$

We emphasize that the wave operator $\hat{\Omega}_\mu(\lambda)$ is applied to the μth state only. It is *state specific*. Equation 2.29 is termed the Bloch equation [7] in Brillouin–Wigner form. Having introduced the Brillouin–Wigner wave operator, we turn our attention now to the corresponding reaction operator $\hat{V}_\mu(\lambda)$.

To introduce the reaction operator, let us begin by using the expansion (2.28) for the wave operator to rewrite Eq. 2.26 for the exact energy $\mathcal{E}_\mu$ as

$$\mathcal{E}_\mu = E_\mu + \lambda \left\langle \Phi_\mu \middle| \hat{\mathcal{H}}_1 \hat{\Omega}_\mu(\lambda) \middle| \Phi_\mu \right\rangle. \tag{2.30}$$

Defining the Brillouin–Wigner *reaction operator* $\hat{V}_\mu(\lambda)$ as

$$\hat{V}_\mu(\lambda) = \hat{\mathcal{H}}_1 \hat{\Omega}_\mu(\lambda) \tag{2.31}$$

we can then write Eq. 2.30 as

$$\mathcal{E}_\mu = E_\mu + \lambda \left\langle \Phi_\mu \middle| \hat{V}_\mu(\lambda) \middle| \Phi_\mu \right\rangle. \tag{2.32}$$

so that the exact energy $\mathcal{E}_\mu$ is given by the sum of the reference energy E_μ and the expectation value of the reaction operator $\hat{V}_\mu(\lambda)$. Note that the reaction operator, like the wave operator, has a subscript μ because it is *state specific*. We can combine the Bloch equation in Brillouin–Wigner form, Eq. 2.29, with the definition of the reaction operator, Eq. 2.31 to obtain the following recursion for $\hat{V}_\mu$:

$$\hat{V}_\mu(\lambda) = \hat{\mathcal{H}}_1 \lambda \hat{\mathcal{H}}_1 \hat{B}_\mu \hat{V}_\mu(\lambda) \tag{2.33}$$

This is the Brillouin–Wigner form of the Lippmann–Schwinger equation [75]. Equations 2.32 and 2.33 together define the exact energy eigenvalue. The solution for these two equations is entirely equivalent to the solution of the Schrödinger equation (2.1).

2.2.3   Multi-Reference Brillouin-Wigner Expansions

We turn now to the Brillouin–Wigner perturbation theory for a system described in zero order by a multi-reference function. The multi-reference formalism closely parallels that given in the previous section for the case of a single-reference function.

Let us begin by defining a *reference space* $\mathcal{P}$. Let

$$\{\Phi_v : v = 0, 1, 2, \ldots, p - 1\} \tag{2.34}$$

be a set of linearly independent eigenfunctions of the zero-order Hamiltonian which constitute the chosen reference space. (It is not our purpose here to discuss how this space should be chosen. This will be determined by the nature of the particular system and physical processes under investigation. We should, however, emphasize that the choice of reference space is crucial to the successful application of the method in practical applications.)

Let $\hat{P}$ be the projection operator on to the chosen reference space $\mathcal{P}$; that is

$$\hat{P} = \sum_{v \in \mathcal{P}} |\Phi_v\rangle \langle \Phi_v|, \tag{2.35}$$

and let $\hat{Q}$ be its orthogonal complement so that

$$\hat{P} + \hat{Q} = \hat{I} \tag{2.36}$$

and therefore

$$\hat{Q} = \hat{I} - \hat{P} \tag{2.37}$$

where $\hat{I}$ is the identity operator. In sum-over-states form, the operator Q can be written

$$\hat{Q} = \sum_{v \notin \mathcal{P}} |\Phi_v\rangle \langle \Phi_v|. \tag{2.38}$$

The projection operators (2.35) and (2.38) satisfy the relations

$$\hat{P}^2 = \hat{P}$$
$$\hat{Q}^2 = \hat{P}$$
$$\hat{P}\hat{Q} = 0. \tag{2.39}$$

Let $\left|\Psi_\mu^{\mathcal{P}}\right\rangle$ by the projection of the exact wavefunction $|\Psi_\mu\rangle$ onto the reference space $\mathcal{P}$; that is,

$$\left|\Psi_\mu^{\mathcal{P}}\right\rangle = \hat{P}|\Psi_\mu\rangle. \tag{2.40}$$

The function $\left|\Psi_\mu^{\mathcal{P}}\right\rangle$ is usually termed the *model function*. It can be written as a linear combination of the set $\{\Phi_v : v = 0, 1, 2, \ldots, p - 1\}$:

$$\left|\Psi_\mu^{\mathcal{P}}\right\rangle = \sum_{v \in \mathcal{P}} |\Phi_v\rangle C_{v,\mu} \tag{2.41}$$

where the coefficients $C_{v,\mu}$ are, at this stage, unknown. In general, the eigenfunctions, $\left|\Psi_\mu^{\mathcal{P}}\right\rangle$ are non-orthogonal, but are assumed to be linearly independent.

As in the single-reference formulation of the theory, the exact eigenstate is constructed by exploiting the completeness of the unperturbed basis. For a single state (but multi-reference function), Eq. 2.19 is now written as

$$|\Psi_\mu\rangle = \sum_{\nu \in \mathcal{P}} |\Phi_\nu\rangle \langle \Phi_\nu | \Psi_\mu\rangle + \sum_{m \notin \mathcal{P}} |\Phi_m\rangle \langle \Phi_m | \Psi_\mu\rangle \tag{2.42}$$

which can be rewritten as

$$|\Psi_\mu\rangle = |\Psi_\mu^{\mathcal{P}}\rangle + \sum_{m \notin \mathcal{P}} |\Phi_m\rangle \langle \Phi_m | \Psi_\mu\rangle \tag{2.43}$$

Using (2.15) and (2.16), we can put (2.43) in the form

$$|\Psi_\mu\rangle = |\Psi_\mu^{\mathcal{P}}\rangle + \lambda \sum_{m \notin \mathcal{P}} |\Phi_m\rangle \frac{1}{(\mathcal{E}_\mu - E_m)} \langle \Phi_m | \Psi_\mu\rangle \tag{2.44}$$

The exact wavefunction is then given by the following recursion

$$|\Psi_\mu\rangle = |\Psi_\mu^{\mathcal{P}}\rangle + \lambda \hat{\mathcal{B}}_\mu^{\mathcal{P}} \hat{H}_1 |\Psi_\mu\rangle \tag{2.45}$$

in which the resolvent is defined as

$$\hat{\mathcal{B}}_\mu^{\mathcal{P}} = \sum_{m \notin \mathcal{P}} |\Phi_m\rangle \frac{1}{(\mathcal{E}_\mu - E_m)} \langle \Phi_m | . \tag{2.46}$$

From Eq. 2.45, we can write down the following expansion for the exact wavefunction:

$$|\Psi_\mu\rangle = \left(1 + \lambda \hat{\mathcal{B}}_\mu^{\mathcal{P}} \hat{\mathcal{H}}_1 + \lambda^2 \hat{\mathcal{B}}_\mu^{\mathcal{P}} \hat{\mathcal{H}}_1 \hat{\mathcal{B}}_\mu^{\mathcal{P}} \hat{\mathcal{H}}_1 + \dots \right) \tag{2.47}$$

It should be noted that Eq. 2.47 for the multi-reference formalism is analogous to Eq. 2.24 in the single-reference case.

We are now ready to define the wave operator for the multi-reference formalism. By analogy with the single-reference wave operator which we defined in Eq. 2.27, the multi-reference Brillouin–Wigner wave operator is defined as

$$|\Psi_\mu\rangle = \hat{\Omega}_\mu^{\mathcal{P}} (\lambda) |\Psi_\mu^{\mathcal{P}}\rangle \tag{2.48}$$

Comparing Eqs. 2.47 and 2.48, we can write the Brilloun–Wigner wave operator as the expansion

$$\hat{\Omega}_\mu^{\mathcal{P}} (\lambda) = 1 + \lambda \hat{\mathcal{B}}_\mu^{\mathcal{P}} \hat{\mathcal{H}}_1 + \lambda^2 \hat{\mathcal{B}}_\mu^{\mathcal{P}} \hat{\mathcal{H}}_1 \hat{\mathcal{B}}_\mu^{\mathcal{P}} \hat{\mathcal{H}}_1 + \dots \tag{2.49}$$

or as the recursion

$$\hat{\Omega}_\mu^{\mathcal{P}} (\lambda) = 1 + \lambda \hat{\mathcal{B}}_\mu^{\mathcal{P}} \hat{\mathcal{H}}_1 \hat{\Omega}_\mu^{\mathcal{P}} (\lambda) . \tag{2.50}$$

Equation 2.50 is the Bloch equation for multi-reference Brillouin–Wigner theory. It should be emphasized that the wave operator $\hat{\Omega}_\mu^P (\lambda)$ has the subscript μ indicating that it is state-specific.

The multi-reference Brillouin–Wigner reaction operator is defined as

$$\hat{\mathcal{V}}_\mu^P (\lambda) = \hat{\mathcal{H}}_1 \hat{\Omega}_\mu^P (\lambda) \tag{2.51}$$

Like the wave operator, the reaction operator $\hat{\mathcal{V}}_\mu^P (\lambda)$ is state-specific. It satisfies a Lippmann–Schwinger-like equation for multi-reference Brillouin–Wigner theory:

$$\hat{\mathcal{V}}_\mu^P (\lambda) = \hat{\mathcal{H}}_1 + \lambda \hat{\mathcal{H}}_1 \hat{\mathcal{B}}_\mu^P \hat{\mathcal{V}}_\mu^P (\lambda) \tag{2.52}$$

2.2.4 Rayleigh–Schrödinger and Brillouin–Wigner Perturbation Theories and A Posteriori 'Many-Body' Corrections

If we compare the Brillouin-Wigner resolvent for the ground state given in Eq. 2.23

$$\mathcal{B} = \sum_{m \neq 0} |\Phi_m\rangle \frac{1}{(\mathcal{E}_0 - E_m)} \langle \Phi_m| \tag{1.15}$$

with the corresponding Rayleigh-Schrödinger resolvent

$$\mathcal{R} = \sum_{m \neq 0} |\Phi_m\rangle \frac{1}{E_0 - E_m} \langle \Phi_m| \tag{1.47}$$

then we see that they differ only in the denominator factors. Using identity relation [46]

$$(\mathcal{E} - E_k)^{-1} = (E_0 - E_k)^{-1} + (E_0 - E_k)^{-1} (-\Delta E) (\mathcal{E} - E_k)^{-1} \tag{2.53}$$

where the exact ground state energy is written as

$$\mathcal{E}_0 = E_0 + \Delta E_0 \tag{2.54}$$

in which E_0 is the ground state eigenvalue of $\mathcal{H}_0$ and ΔE_0 is termed the level shift, we can relate the Brillouin–Wigner resolvent to the Rayleigh–Schrödinger resolvent. In this way, we can find a posteriori extensivity corrections to any Brillouin–Wigner perturbation series.

We know that the Rayleigh–Schrödinger perturbation theory series leads directly to the many-body perturbation theory by employing the linked diagram theorem. This theory uses factors of the form $(E_0 - E_k)^{-1}$ as denominators. Furthermore, this theory is fully extensive; it scales linearly with electron number. The second term on the right-hand side of Eq. 2.53 can be viewed as an "extensivity correction term" for the Brillouin–Wigner series: a correction term which recovers the Rayleigh–Schrödinger and many-body perturbation theoretic formulations. This simple idea has been used to find a posteriori correction for limited configuration interaction method [55], as well as for state specific multi-reference Brillouin–Wigner coupled cluster theory [54, 102]. Indeed, a posteriori corrections for a lack of extensivity can be obtained for any ab initio quantum chemical method provided that the method can be formulated within Brillouin–Wigner perturbation theory.

Whereas the multi-reference Rayleigh–Schrödinger perturbation theory approximates a manifold of states simultaneously, the multi-reference Brillouin–Wigner perturbation theory approach is applied to a single state – it is said to be 'state-specific'. The multi-reference Brillouin–Wigner perturbation theory avoids the intruder state problem. If a particular Brillouin–Wigner-based formulation is not a valid many-body method, then a posteriori correction can be applied. This correction is designed to restore the extensivity of the method. This extensivity may be restored approximately or exactly, depending on the formulation. The situation is illustrated schematically in Fig. 2.5. Pittner [95] has explored the continuous transition between BrillouinWigner and RayleighSchrödinger perturbation theory and the Hilbert space multireference coupled cluster expansion.

2.3   Digression: Collaborative Virtual Environments for Many-Body Brillouin–Wigner Theories

In this section, we digress to review the use of collaborative virtual environments in prosecuting research on Brillouin–Wigner methods.[14]

Initially, this project was carried out under the auspices of the EU COST programme – Action D23 – METACHEM (*Metalaboratories for Complex Computational Applications in Chemistry*).[15] The establishment of a European Metalaboratory[16] for ab initio multireference quantum chemical methods had two main objectives:

---

[14]Parts of this digression are taken from the authors' article entitled *A Collaborative Virtual Environment for Molecular Electronic Structure Theory: Prototype for the Study of Many-Body Methods* [121] which was published in the volume *Frontiers in Quantum Systems in Chemistry and Physics*.

[15]Project number: D23/0001/01: European Metalaboratory for multireference quantum chemical methods (01/02/2001–18/07/2005). Participants: P. Čársky, J. Pittner (*J. Heyrovsky Institute, Prague, Czech Republic*), I. Hubač (*Comenius University, Slovakia*), S. Wilson (*Rutherford Appleton Laboratory, UK*), W. Wenzel (*Universität Dortmund, Germany*), L. Meissner (*Nicholas Copernicus University, Poland*), V. Staemmler (*Ruhr Universität Bochum Germany*), C. Tsipis (*Aristotle University of Thessaloniki, Greece*), A. Mavridis (*National and Kapodistrian University of Athens, Greece*).

[16]Loosely speaking, a *metalaboratory* may be defined as a cluster of geographically distributed resources.

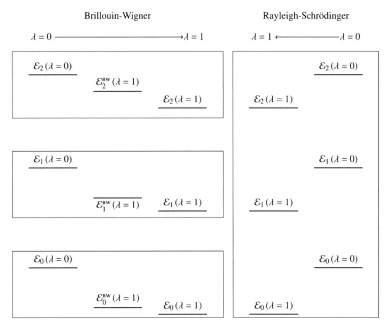

Fig. 2.5 This figure compares schematically the application of Rayleigh-Schrödinger perturbation theory and Brillouin-Wigner perturbation theory to the multireference electron correlation problem. We indicate states which are considered in a single calculation by enclosing them in a box. The Rayleigh-Schrödinger perturbation theory approach approximates the energy expectation values for a manifold of states in a single calculation. In the multireference Rayleigh-Schrödinger perturbation theory, the states with energies $\mathcal{E}_0$, $\mathcal{E}_1$, $\mathcal{E}_2$, ... are considered in a single calculation as represented on the *right-hand side* of the figure. The Brillouin-Wigner perturbation theory approach is 'state specific'; that is, we consider a single state in a given calculation and, if the resulting theory is not a valid many-body theory, then we apply a suitable *a posteriori* correction based on the relation between the Brillouin-Wigner and the Rayleigh-Schrödinger denominators. In principle, this *a posteriori* correction can be rendered exact. In the multireference Brillouin-Wigner perturbation theory, the states with energies $\mathcal{E}_0$, $\mathcal{E}_1$, $\mathcal{E}_2$, ... are considered in separate calculations as represented on the *left-hand side* of the figure. The resulting energies are denoted by $\mathcal{E}_0^{bw}$, $\mathcal{E}_1^{bw}$, $\mathcal{E}_2^{bw}$, ... To each of these energies an *a posteriori* correction to restore linear scaling can be introduced, if necessary, to yield the energies $\mathcal{E}_0$, $\mathcal{E}_1$, $\mathcal{E}_2$, ...

1. The development of a comprehensive suite of capabilities for remote scientific collaboration between geographically distributed sites, creating a prototypical environment tailored to the needs of the quantum chemistry community. This environment is supposed to provide a cross-platform suite of tools for data exchange and sharing and sharing of computer resources. The tools should help to easily build and share our common knowledge base and allow the exchange of draft manuscripts, documents, preprints and reprints, data preparation and analysis, program execution, training, etc. The use of a web-based communication for the Metalaboratory should ensure that all of the data, notes, sketches, molecular structures under consideration, etc., are always available from any desktop to all participants.

2. The Metalaboratory focuses on a specific problem at the cutting-edge of modern ab initio quantum chemical methodology – the development of multireference quantum chemical methods together with the associated algorithms. Such methods are essential for the description of the breaking of bonds, a process which might be regarded as the very essence of chemistry. By establishing a European Metalaboratory directed towards the solution of a specific and challenging scientific problem, a rapid evaluation of facilities for remote collaboration will be achieved as well as a coordinated problem-solving programme directed towards the development of robust, reliable and cost-effective ab initio multireference quantum chemical methods. Such methods are still far from being routine since their widespread use in practical applications is frequently hampered by their complexity and by the problem of intruder states.

The need to establish more effective mechanisms for collaboration has become evident in our previous work under EU COST Action D9.[17] Therefore, in parallel with further development of accurate multireference quantum chemical methods, we decided to investigate the use of web-based and internet tools for remote scientific collaboration. In the past, some of the participants have exchanged data, draft manuscripts and the like by *e*-mail, supplemented by occasional correspondence by post and face-to-face meetings. By exploring alternative mechanisms, in particular, real-time collaboration procedures, we aim to establish a more productive environment. This includes remote execution of computer programs, the use of web-based interfaces for remote collaboration. The intention has been to use only a publicly accessible and user-friendly software for making the expertise accumulated in this project profitable for national and international collaborations in other domains of chemistry and physics.

The development of collaborative virtual environments is a key element of what is becoming known as "*e*-science."[18] In 2000, Sir John Taylor, Director General of Research Councils, Office of Science and Technology, UK, wrote

> *e*-Science is about global collaboration in key areas of science, and the next generation of infrastructure that will enable it.

He predicted that

> *e*-Science will change the dynamic of the way science is undertaken.

We submit that key elements of *e*-science such as collaborative virtual environments will evolve most rapidly and deliver a functionality required by practicing scientists if they are developed as part of a research project in the target discipline. This is the

---

[17] EU COST Action D9 "Advanced computational chemistry of increasingly complex systems".

[18] For some details see, for example the Wikipedia entry on *e*-science at http://en.wikipedia.org/wiki/E-Science

Currently the largest focus in *e*-science is in the United Kingdom. In the United States similar initiatives are termed cyberinfrastructure projects.

approach that we have followed in this work. We have developed a collaborative virtual environment for molecular electronic structure theory whilst undertaking research into aspects of the Brillouin–Wigner theory for many-body systems.

A collaborative virtual environment for molecular electronic structure theory evolved during research into the use of Brillouin–Wigner methodology in handling the electron correlation problem in molecules. We describe all of the elements of a collaborative virtual environment that we have considered and give a brief assessment of how useful each element turned out to be in practice. Others may find different elements to be more or less useful in their own collaborative work. Others may find elements that we have not listed here to be useful. We believe that collaborative virtual environments will become increasing important in the years ahead. They will undoubtedly be at their most useful when tailored to participants' research projects. We are not aware of any previous work which has examined the use of collaborative virtual environments in molecular electronic structure theory or quantum chemistry.

The environment is web-based. This has the advantage that it can be accessed from any machine with an internet connection from home or office or mobile computer. Also the web-pages associated with the environment can be distributed, that is different pages can be hosted by different machines in different locations.

Our collaborative virtual environment consists of three main elements:

1. Details of the participating scientists, i.e., the human resources for the collaboration.
2. A knowledge base of key information of use to scientists involved in the project, i.e., the intellectual foundations upon which the research is built and the intellectual products of the collaborative research.
3. A set of tools for collaboration, i.e., mechanisms for exchanging ideas and for criticising proposals, sharing information and pooling resources.

We have developed a prototype collaborative virtual environment, which actively supports human–human communication in addition to human–machine communication, for molecular electronic structure theory. We submit that key elements of e-science, such as collaborative virtual environments, will evolve most rapidly and deliver a functionality required by practicing scientists if they are developed as part of a research project in the target discipline. This is the approach that we have followed in this work. We have developed a collaborative virtual environment for molecular electronic structure theory whilst undertaking research into aspects of the Brillouin–Wigner theory for many-body systems. Others working in molecular physics and quantum chemistry may find this a useful starting point for the development of improved environments. Such environments will undoubtedly evolve with time as higher bandwidths and new tools become available.

In more recent work we have found Google Sites [43] to provide a convenient platform for collaborative virtual environments. (See, for example, Fig. 2.6).

Google Sites is the easiest way to make information accessible to people who need quick, up-to-date access. People can work together on a Site to add file attachments, information from other Google applications (like Google Docs, Google Calendar, YouTube and Picasa), and new free-form content. Creating a site together is as easy as editing a document, and

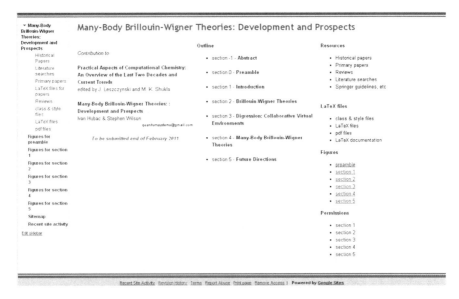

Fig. 2.6 Collaborative Virtual Environment used in preparing this contribution. The home page of this Google Sites platform at https://sites.google.com/site/virtualenvironment25/

you always control who has access, whether it's just yourself, your team, or your whole organization. You can even publish Sites to the world. The Google Sites web application is accessible from any internet connected computer. [44]

Advances in information and communications technology are facilitating widespread cooperation between groups and individuals, who may be physically located at geographically distributed sites (– sites in different laboratories, perhaps in different countries or even different continents), in a way that may disrupt and challenge the traditional structures and institutions of science (as well bringing change to society as a whole). Collaborative virtual environments, such as the one described here, have the potential to transform the 'scientific method' itself by fuelling the genesis, dissemination and accumulation of new ideas and concepts, and the exchange of alternative perspectives on current problems and strategies for their solution. Because of their openness and their global reach, as well as their emergent and thus agile nature, such environments may transform the practice of science over the next decades.

"[T]he communication system is [...] vital to science; it lies at the very heart of the 'scientific method'" [125]. But, at the same time as technology is facilitating radical improvements in communications which can only serve to fuel scientific progress, there are factors in the contemporary structures and institutions governing science which have to potential to seriously inhibit scientific communication and thereby scientific progress. These factors are potentially as limiting as the 'iron curtain'. The science policies of governments and universities increasingly link funding for science with economic return. This can foster practices, such as the proprietary capture of genetic databases, which inhibit the free and open exchange of information. With scientific results cloaked in a veil of commercial confidentiality and/or vested interests, the communication which is the engine of progress in science is curtailed. Even in those areas of science which are of no direct economic benefit there is the outdated 'copyright economy' which forces scientists to assign all rights to a major commercial journal publisher for no remuneration and then buy back their work through monopolistic subscriptions.

2.4 Applications of Many-Body Brillouin–Wigner Theories

In this section, we shall consider the application of Brillouin–Wigner-based methods to the three most commonly employed ab initio methods of approximation in molecular electronic structure theory:

- Coupled Cluster (CC) expansions Approximations based on CC expansions [3, 91, 92] include CCD, CCSD, CCSDT, etc., and their multireference variants MR-CCD, MR-CCSD, MR-CCSDT, etc. The CCSD approximation is also known as the coupled pair approximation (CPA) [59] or the coupled pair many electron theory (CPMET) [20, 21, 93].
- Configuration Interaction (CI) expansions Approximations based on CI expansions include CI D, CI SD, etc., and their multireference variants MR-CI D, MR-CI SD, etc.
- Perturbation Theory (PT) expansions Approximations based on PT expansions include MBPT2, MBPT3, MBPT4, etc.[19] and their multireference variants MR-MBPT2, MR-MBPT3, MR-MBPT4, etc[20] [113–118].

We shall provide an overview of the applications that have been made over the period being review which demonstrate the many-body Brillouin–Wigner approach for each of these methods. By using Brillouin–Wigner methods, any problems associated with intruder states can be avoided. A posteriori corrections can be introduced to remove terms which scale in a non linear fashion with particle number. We shall not, for example, consider in any detail hybrid methods such as the widely used CCSD(T) which employs ccsd theory together with a perturbative estimate of the triple excitation component of the correlation energy.

We note that, provided the respective expansions are convergent, each of the methods listed above should, when all terms are included, lead to the same result,

---

[19]These methods are often written as MP2, MP3, MP4, etc., that is Møller-Plesset perturbation theory [88] in second, third, fourth order.

[20]These methods are often written as MR-MP2, MR-MP3, MR-MP4, etc., that is multireference Møller-Plesset perturbation theory in second, third, fourth order.

i.e., the exact solution of the appropriate Schrödinger equation within the chosen basis set. Futhermore, if, for any reference function, all terms are included in the respective expansions, then, again subject to the proviso that the expansions converge, they will lead to the same result – the exact solution of the relevant Schrödinger equation within the basis set chosen. In the presence of strong quasi-degeneracy effects, an appropriately chosen multireference function may support a convergent expansion whereas another multireference function or a single reference function may not. The appropriate choice of multireference function for quasi-degenerate problems is a significant problem and one which we do not address here. The use of a multireference formalism is required for problems as simple as the dissociation of the ground state of the hydrogen molecule. The choice of mul-tireference function is dictated by the physics and chemistry of the systems under study. For more complicated problems the choice of reference requires considerable care. This choice certainly represents a significant barrier to the development of black box quantum chemical software packages for problems demanding the use of a multireference formalism. In practice, the different expansions for the correlation energy afforded by coupled cluster theory, configuration interaction and perturbation theory have to be truncated in order to render computations tractable. The different methods differ only in the way in which this truncation is carried out. However,

> the method of truncation can significantly affect not only the theoretical properties of a particular approach but also, to some extent, its computational feasibility [111, 119].

Furthermore,

> The fact that one method may include more terms in an expansion than another method does not necessarily imply that it is superior. The terms which are left out of an expansion of the wavefunction or expectation value are, in fact, often just as important as the ones which are actually included [111, 119].

It is well-known, for example, that in a perturbation theory analysis of the method of configuration interaction when restricted to single- and double-excitations with respect to a single determinant reference function includes many terms, which correspond to unlinked diagrams, which are exactly canceled by terms involving higher order excitations.

We leave it as an exercise for the interested reader to consider the application of Brillouin–Wigner methods, together with an a posteriori extensivity correction where appropriate, to other approximation techniques in molecular structure theory. The Brillouin–Wigner approach is a very general technique which has been largely overlooked in many-body studies of molecules and which shows considerable promise for situations in which a single reference formalism is inadequate.

We do not proposed to describe here the theoretical details of many-body Brillouin–Wigner methodology. They can be found in our book *Brillouin–Wigner Methods for Many-Body Systems*. Here we concentrate on applications of Brillouin–Wigner methods to many-body systems in chemistry and physics. Previous reviews can be found in our article in the *Encyclopedia of Computational Chemistry* [49] and in our review entitled *Brillouin–Wigner expansions in quantum chemistry: Bloch-*

like and Lippmann–Schwinger-like equations [124]. Another review [50] can be found in the volume published as a tribute to Professor P.-O. Löwdin. We have established a website at

quantumsystems.googlepages.com/Brillouin-Wigner

where further details of the development and application of Brillouin–Wigner methodology are given. Figure 2.7 shows the list of publications on many-body Brillouin–Wigner methods from this website. The collected results clearly establish the value of the Brillouin–Wigner method in the study of problems requiring a multireference formalism.

2.5 Future Directions

In this review, we have described the development and application of many-body Brillouin–Wigner methods for the molecular electronic structure problem. In our recent monograph [51], we suggest that "such methods will have a broad range of applications in studies of the solid state, in condensed matter theory, in material science and in nuclear physics. In this final section, we shall concentrate on two areas of molecular electronic structure research for which many-body Brillouin–Wigner methodology shows promise.

2.5.1 Relativistic Many-Body Brillouin–Wigner Theories

The theoretical apparatus presented in this review is formulated entirely within the framework of non-relativistic quantum mechanics. This may be adequate for studies of most properties of molecules containing light atoms. However, it is now well established that a fully relativistic formulation is essential for the treatment of heavy and superheavy elements and molecules containing them [62, 97]. A fully relativistic, four component Brillouin-Wigner-based treatment of hydrogen-like systems by Quiney and the present authors has demonstrated [98] the feasibility of general four component Brillouin-Wigner-based theories. There is no problem of principle in extending this approach to arbitrary atomic and molecular systems.

2.5.2 Fock Space Brillouin–Wigner Methods

The Brillouin-Wigner-based methods described in this review can also be employed in the calculation of molecular properties other than the energy. Because of the central role play by the energy expectation value in studies of molecular structure, we have concentrated exclusively on the calculation of approximations to energy.

PUBLICATIONS ON MANY-BODY BRILLOUIN-WIGNER METHODS

Hubac, I., & Wilson, S.
Many-Body Brillouin-Wigner Theories: Development and Prospects
in *Practical Aspects of Computational Chemistry: An Overview of the Last Two Decades and Current Trends*, ed. J. Leszczynski and M. K. Shukla (in press) Springer

Pittner, J., & Piecuch, P.
Method of moments for the continuous transition between the Brillouin–Wigner-type and Rayleigh–Schrodinger-type multireference coupled cluster theories
Molecular Physics **107**, 1209 (2009)

Papp, P., Neogrady, P., Mach, P., Pittner, J., Hubac, I., & Wilson, S.
Many-body Brillouin-Wigner second-order perturbation theory: an application to the autoaromatisation of hex-3-ene-1,5-diyne (the   Bergman reaction)
Molecular Physics **106**, 57 (2008)

Demel, O., & Pittner, J.
Multireference Brillouin-Wigner coupled cluster method with singles, doubles, and triples: efficient implementation and comparison with approximate approaches
Journal of Chemical Physics **128**, 104108 (2008)

Wilson, S, & Hubac, I.
A Collaborative Virtual Environment for Molecular Electronic Structure Theory: A Prototype for the Study of Many-Body Methods
in *Frontiers in Quantum Systems in Chemistry & Physics*, ed. S. Wilson, P.J. Grout, J. Maruani, G. Delgado-Barrio and P. Piecuch, Progress in Theoretical Chemistry & Physics **18**, Springer (2008)

Pittner, J., & Smydke, J.
Analytic gradient for the multireference Brillouin-Wigner coupled cluster method and for the state-universal multireference coupled cluster method
Journal of Chemical Physics **127**, 114103 (2007)

Papp, P., Mach, P., Hubac, I., & Wilson, S.
Many-body Brillouin-Wigner second-order perturbation theory: A robust and efficient approach to the multireference correlation problem
International Journal of Quantum Chemistry **107**, 2622 (2007)

Demel, O., & Pittner, J.
Multireference Brillouin-Wigner coupled clusters method with noniterative perturbative connected triples
Journal of Chemical Physics **124**, 144112 (2006)

Papp, P., Mach, P., Pittner, J., Hubac, I., & Wilson, S.
Many-body Brillouin-Wigner second-order perturbation theory using a multireference formulation: an application to bond breaking in the diatomic hydrides BH and FH
Molecular Physics **104**, 2367 (2006)

Hubac, I., Mach, P., & Wilson, S.
Multireference Brillouin-Wigner coupled cluster (MR-BWCC) theory applied to the H8 model:

Fig. 2.7 The webpages entitled PUBLICATIONS ON MANY-BODY BRILLOUIN-WIGNER METHODS (Taken from the webpage quantumsystems.googlepages.com/Brillouin-Wigner)

Comparison with CCSD(T) theory
International Journal of Quantum Chemistry **104**, 387 (2005)

Pittner, J., & Demel, O.,
Towards the multireference Brillouin-Wigner coupled-clusters method with iterative connected triples: MR BWCCSDT-alpha approximation
Journal of Chemical Physics **122**, 181101 (2005)

Pittner, J., Li, X.Z., & Paldus, J.
Multi-reference Brillouin-Wigner coupled-cluster method with a general model space
Molecular Physics **103**, 2239 (2005)

Teberekidis, V.I., Kerkines, I.S.K., Tsipis, C.A., Carsky, P., & Mavridis, A.
Ground states of BeC and MgC: A comparative multireference Brillouin-Wigner coupled cluster and configuration interaction study
International Journal of Quantum Chemistry **102**, 762 (2005)

Demel, O., Pittner, J., Carsky, P., & Hubac, I.
Multireference Brillouin-Wigner coupled cluster singles and doubles study of the singlet-triplet separation in alkylcarbenes
Journal of Physical Chemistry A **108**, 3125 (2004)

Hubac, I., Mach, P., Papp, P., & Wilson, S.
Multireference second-order Brillouin-Wigner perturbation theory
Molecular Physics **102**, 701 (2004)

Pittner, J., Gonzalez, H.V., Gdanitz, R.J., & Carsky, P.
The performance of the multireference Brillouin-Wigner coupled cluster singles and doubles method on the insertion of Be into H-2
Chemical Physics Letters **386**, 211 (2004)

Hubac, I., Mach, P., & Wilson, S.
A posteriori Brillouin-Wigner correction of limited multireference configuration interaction: analysis for an (H-2)(4) cluster model
Molecular Physics **101**, 3493 (2003)

Pittner, J.
Continuous transition between Brillouin-Wigner and Rayleigh-Schrodinger perturbation theory, generalized Bloch equation, and Hilbert space multireference coupled cluster
Journal of Chemical Physics **118**, 10876 (2003)

Puiggros, O.R., Pittner, J., Carsky, P., Stampfuss, P., & Wenzel, W.
Multireference Brillouin-Wigner coupled cluster singles and doubles (MRBWCCSD) and multireference doubles configuration interaction (MRD-CI) calculations for the Bergman cyclization reaction
Collection of Czechoslovak Chemical Communications **68**, 2309 (2003)

Wilson, S., Hubac, I., Mach, P., Pittner, J., & Carsky, P.
Brillouin-Wigner expansions in quantum chemistry: Bloch-like and Lippmann-Schwinger-like equations
in *Advanced Topics in Theoretical Chemical Physics*, ed. J. Maruani, R. LeFebvre & E. Brandas, Progress in Theoretical Chemistry & Physics **12**, Springer (2003)

Fig. 2.7 (continued)

Hubac, I., Mach, P., & Wilson, S.
A posteriori corrections to multireference limited configuration interaction based on a Brillouin-Wigner perturbative analysis
International Journal of Quantum Chemistry **89**, 198 (2002)

Hubac, I., Mach, P., & Wilson, S.
On the application of Brillouin-Wigner perturbation theory to multireference configuration mixing
Molecular Physics **100**, 859 (2002)

Kerkines, I.S.K., Pittner, J., Carsky, P., Mavridis, A., & Hubac, I.
On the ground states of CaC and ZnC: A multireference Brillouin-Wigner coupled cluster study
Journal of Chemical Physics **117**, 9733 (2002)

Petraco, N.D.K., Horny, L., Schaefer, H.F., & Hubac, I.
Brillouin-Wigner coupled cluster theory: Fock-space approach
Journal of Chemical Physics **117**, 9580 (2002)

Pittner, J., Carsky, P., & Hubac, I.
Four- and 8-reference state-specific Brillouin-Wigner coupled-cluster method: Study of the singlet oxygen
International Journal of Quantum Chemistry **90**, 1031 (2002)

Hubac, I., & Wilson, S.
On the generalized multi-reference Brillouin-Wigner coupled cluster theory
Journal of Physics B: Atomic, Molecular & Optical Physics **34** 4259 (2001)

Pittner, J., Smydke, J., Carsky, P., & Hubac, I.
State-specific Brillouin-Wigner multireference coupled cluster study of the F-2 molecule: assessment of the a posteriori size-extensivity correction
Journal of Molecular Structure - THEOCHEM **547**, 239 (2001)

Pittner, J., Nachtigall, P., Carsky, P., & Hubac, I.
State-specific Brillouin-Wigner multireference coupled cluster study of the singlet-triplet separation in the tetramethyleneethane diradical
Journal of Physical Chemistry A **105**, 1354 (2001)

Quiney, H.M., Hubac, I., & Wilson, S.
On the application of Brillouin-Wigner perturbation theory to a relativistic and non-relativistic hydrogenic model problem
Journal of Physics B: Atomic, Molecular & Optical Physics **34**, 4323 (2001)

Wilson, S., & Hubac, I.
On the use of MP2 theory for electron correlation in atoms and molecules
Molecular Physics **99**, 1813 (2001)

Hubac, I., Mach, P. & Wilson, S.
On the use of limited configuration interaction for many-body systems
Journal of Physics B: Atomic, Molecular & Optical Physics **33** 4735 (2000)

Hubac, I., & Wilson, S.
On the use of Brillouin-Wigner perturbation theory for many-body systems
Journal of Physics B: Atomic, Molecular & Optical Physics **33** 365 (2000)

Fig. 2.7 (continued)

Hubac, I., Pittner, J., & Carsky, P.
Size-extensivity correction for the state-specific multireference Brillouin-Wigner coupled-cluster theory
Journal of Chemical Physics **112**, 8779 (2000)

Sancho-Garcia, J.C., Pittner, J., Carsky, P, & Hubac, I.
Multireference coupled-cluster calculations on the energy of activation in the automerization of cyclobutadiene: Assessment of the state-specific multireference Brillouin-Wigner theory
Journal of Chemical Physics **112**, 8785 (2000)

Masik, J., & Hubac, I.
Multireference Brillouin-Wigner coupled-cluster theory. Single-root approach
Advances in Quantum Chemistry 31, 75 (1999)

Pittner, J., Nachtigall, P., Carsky, P., Masik, J., & Hubac, I.
Assessment of the single-root multireference Brillouin-Wigner   coupled-cluster method: Test calculations on CH2, SiH2, and twisted ethylene
Journal of Chemical Physics **110**, 10275 (1999)

Mach, P., Masik, J., Urban, J., & Hubac, I.
Single-root multireference Brillouin-Wigner coupled-cluster theory. Rotational barrier of the N2H2 molecule
Molecular Physics 94, 173 (1998)

Masik, J., Mach, P., Urban, J., Polasek, M., Babinec, P., & Hubac, I.
Single-root multireference Brillouin-Wigner coupled-cluster theory.   Rotational barrier of the ethylene molecule
Collection of Czechoslovak Chemical Communications **63**, 1213 (1998)

Masik, J., Hubac, I., & Mach, P.
Single-root multireference Brillouin-Wigner coupled-cluster theory: Applicability to the F-2 molecule
Journal of Chemical Physics **108**, 6571 (1998)

Wenzel, W.
Excitation energies in Brillouin-Wigner-based multireference perturbation theory
International Journal of Quantum Chemistry 70, 613 (1998)

Wenzel, W., & Steiner, M.M.
Brillouin-Wigner based multi-reference perturbation theory for electronic correlation effects
Journal of Chemical Physics **108**, 4714 (1998)

Masik, J., & Hubac, I.
Comparison of the Brillouin-Wigner coupled cluster theory with the standard coupled cluster theory. Cancellation of disconnected terms in the Brillouin-Wigner coupled cluster theory
Czechoslovak Chemical Communications **62**, 829 (1997)

Hubac, I.
Size-extensive Brillouin-Wigner perturbation theory. Size-extensive Brillouin-Wigner coupled cluster theory
in *NATO Advanced Research Workshop on New Methods in Quantum Theory*, ed. C.A. Tsipis, V.S. Popov, D.R. Herschbach & J.S. Avery, p. 183 (1996)

Fig. 2.7 (continued)

Carsky, P., Hrouda, V., Sychrovsky, V., Hubac, I., Babinec, P., Mach, P., Urban, J., & Masik, J.
Brillouin-Wigner Perturbation-Theory as a possible more effective alternative to many-body Rayleigh-Schrodinger perturbation theory and coupled cluster theory
Collection of Czechoslovak Chemical Communications **60**, 1419 (1995)

Hubac, I., & Neogrady, P.
Size-consistent Brillouin-Wigner Perturbation-Theory with an Exponentially Parametrized Wave-Function- Brillouin-Wigner Coupled-Cluster Theory
The Physical Review A **50**, 4558 (1994)

Fig. 2.7 (continued)

However, the Brillouin-Wigner-based techniques can be applied to the calculation of first order properties, such as dipole moments and multipole moments, as well as second order properties, such as polarizabilities and hyperpolarizabilities. In calculations of properties such as ionization potentials and electron affinities to use of a Fock space formulation is more appropriate that the Hilbert space formulation that we have followed in this monograph. Some progress has been made [94, 103] in formulating the necessary Fock space BrillouinWigner methodology. The interested reader can find further details in the original publications [94, 103].

Research over the past two decades has demonstrated that, after being abandoned for almost half a century in favor of Rayleigh–Schrödinger-based many-body formalism, the Brillouin–Wigner approach has much to offer in studies of the quantum many-body problem. This short review should convince the reader that the formalism of Brillouin and Wigner, and of Lennard-Jones,[21] has much to contribute to modern many-body theory.

Acknowledgements This work was supported in part by the APVV Grant Agency, Slovakia, under project numbers -0420-10 and -0442-07 and also VEGA Grant Agency, Slovakia, under project number -1/0762/11. IH thanks the Grant Agency of the Czech Republic under project number MSM 4781305903.

Note added in proof: Recent research has shown how the many-body Brillouin-Wigner formalism can be deployed in the study of electron-molecule scattering [57].

References

1. Babbage C (1837) On the great law which regulates matter. Ninth bridgewater treatise. p 167. See Knight DM (1968) Classical scientific papers: chemistry. Mills and Boon, London
2. Baker DJ, Moncrieff D, Saunders VR, Wilson S (1991) Comput Phys Commun 62:25
3. Bartlett RJ (ed) (1997) Recent advances in coupled cluster methods. Recent advances in computational chemistry, vol 3. World Scientific, Singapore

---

[21] See Chap. 1 of reference [51] for a brief discussion of the contribution of Lennard-Jones to Brillouin-Wigner theory.

4. BBC News, http://news.bbc.co.uk/1/hi/8552410.stm Accessed on 26 Jan 2011. Source International Telecommunication Union (ITU), http://www.itu.int
5. Bhaskaran-Nair K, Demel O, Pittner J (2008) J Chem Phys 129:184105
6. Bhaskaran-Nair K, Demel O, Pittner J (2010) J Chem Phys 132:154105
7. Bloch C (1958) Nucl Phys 6:329
8. Bohr N (1913) Phil Mag 26:1, 476, 857
9. Born M, Oppenheimer JR (1927) Ann Physik 84:457 An English translation of the paper by Blinder SM with emendations by Sutcliffe BT, Geppert W (2003) In: Wilson S, Bernath PF, McWeeny R (eds) Handbook of molecular physics and quantum chemistry, vol 1. Wiley, Chichester
10. Brandow BH (1967) Rev Mod Phys 39:771
11. Brillouin L (1932) J Physique 7:373
12. Brown R (1828) Edinb New Philos J 5:358-371
13. Brueckner KA (1955) Phys Rev 100:36
14. Čársky P, Hrouda V, Sychrovsky V, Hubač I Babinec P, Mach P, Urban J, Mášik J (1995) Collect Czech Chem Commun 60:1419
15. Chattopadhyay S, Mahapatra US, Mukherjee D (1999) J Chem Phys 111:3820
16. Chattopadhyay S, Mahapatra US, Mukherjee D (2000) J Chem Phys 112:7939
17. Chattopadhyay S, Ghosh P, Mahapatra US (2004) J Phys B: At Mol Opt Phys 37:495
18. Chattopadhyay S, Pahari D, Mukherjee D, Mahapatra US (2004) J Chem Phys 120:5968
19. Churchill EF, Snowdon DN, Munro AJ (2001) Collaborative virtual environments: digital places and spaces for interaction (Computer supported cooperative work). Springer, London/ New York
20. Čížek J (1966) J Chem Phys 45:4256
21. Čížek J (1969) Adv Chem Phys 45:35
22. Coester F (1958) Nucl Phys 7:421
23. Coester F, Kümmel H (1960) Nucl Phys 17:4777
24. Das S, Mukherjee D, Kállay M (2010) J Chem Phys 132:074103
25. Das S, Kállay M, Mukherjee D (2010) J Chem Phys 133:234110
26. Datta D, Mukherjee D (2011) J Chem Phys 134:054122
27. Demel O, Bhaskaran-Nair K, Pittner J (2010) J Chem Phys 133:134106
28. Dirac PAM (1929) Proc R Soc (London) A 123:714
29. Dutta NC, Matsubara C, Pu RT, Das TP (1969) Phys Rev 177:33
30. Dyson F (1949) Phys Rev 75:486
31. Dyson F (1949) Phys Rev 75:1736
32. Einstein A (1905) Ann Phys (Leipzig) 17:549-560
33. Evangelista FA, Allen WD, Schaefer HF III (2006) J Chem Phys 125:154113
34. Evangelista FA, Allen WD, Schaefer HF III (2007) J Chem Phys 127:024102
35. Evangelista FA, Simmonett AC, Allen WD, Schaefer HF III (2008) J Gauss, J Chem Phys 128:124104 (E)
36. Feynman RP (1949) Phys Rev 76:749
37. Feynman RP (1949) Phys Rev 76:769
38. Goldstone J (1957) Proc R Soc (London) A 239:267
39. Halacy DS (1970) Charles Babbage, father of the computer. Macmillan, New York
40. Heitler W, London F (1927) Zeit f Physik 44:455
41. Hoffmann MR, Datta D, Das S, Mukherjee D, Szabados A, Rolik Z, Surján PR (2009) J Chem Phys 131:204104
42. http://www.top500.org Accessed on 25 Jan 2011
43. http://www.google.com/sites/ Accessed on 25 Jan 2011
44. http://www.google.com/sites/overview.html Accessed 25 Jan 2011
45. Hubbard J (1957) Proc R Soc (London) A 240:539
46. Hubač I, Wilson S (2000) J Phys B: At Mol Opt Phys 33:365
47. Hubač I, Wilson S (2001) Adv Quantum Chem 39:209
48. Hubač I, Wilson S (2001) J Phys B: At Mol Opt Phys 34:4259

49. Hubač I, Wilson S (2003) In: von Ragué Schleyer P, Allinger NL, Schaefer HF III, Clark T, Gasteiger J, Kollman P, Schreiner P (eds) Encyclopedia of computational chemistry, electronic edn. Wiley, Chichester

50. Hubač I, Wilson S (2003) In: Brändas EJ, Kryachko ES (eds) Fundamental world of quantum chemistry – a tribute to the memory of Per-Olov Löwdin. Kluwer, Dordrecht

51. Hubač I, Wilson S (2010) Brillouin-Wigner methods for many-body systems. Progress in theoretical chemistry & physics, vol 21. Springer, Dordrecht/New York

52. Hubač I (1996) In: Tsipis CA, Popov VS, Herschbach DR, Avery JS (eds) New methods in quantum theory. NATO ASI series. Kluwer, Dordrecht, p 183

53. Hubač I, Neogrády P (1994) Phys Rev A50:4558

54. Hubač I, Pittner J, Čársky P (2000) J Chem Phys 112:8779

55. Hubač I, Mach P, Wilson S (2000) J Phys B: At Mol Opt Phys 33:4735

56. Hubač I, Mach P, Wilson S (2001) Adv Quantum Chem 39:225

57. Hubač I, Masarik J and Wilson S (2011) J Phys B: At Mol Opt Phys. 44:205201

58. Hugenholtz N (1957) Physica 23:481

59. Hurley AC (1976) Electron correlation in small molecules. Academic, London

60. Hyman RA (1982) Charles Babbage: pioneer of the computer. Oxford University Press, Oxford

61. Kaldor U (1973) Phys Rev A 7:427

62. Kaldor U, Wilson S (eds) (2003) Theoretical chemistry and physics of heavy and superheavy elements. Progress in theoretical chemistry and physics. Kluwer, Dordrecht

63. Kelly HP (1963) Phys Rev 131:684

64. Kelly HP (1964) Phys Rev A134:1450

65. Kelly HP (1964) Phys Rev B136:896

66. Kelly HP (1966) Phys Rev 144:39

67. Kelly HP (1968) Adv Theor Phys 2:75

68. Kelly HP (1969) Adv Chem Phys 14:129

69. Kelly HP (1969) Phys Rev Lett 23:255

70. Kelly HP (1970) Int J Quantum Chem Symp 3:349

71. Kelly HP (1970) Phys Rev A1:274

72. Kelly HP, Ron A (1971) Phys Rev A4:11

73. Kumar M (2008) Quantum – Einstein, Bohr and the great debate about the nature of reality. Icon, London

74. Lennard-Jones JE (1930) Proc R Soc (London) A 129:598

75. Lippmann BP, Schwinger J (1950) Phys Rev 79:469

76. Mach P, Mášik J, Urban J, Hubač I (1998) Mol Phys 94:173

77. Mahapatra US, Datta B, Bandyopadhyay B, Mukherjee D (1998) Adv Quantum Chem 30:163

78. Mahapatra US, Datta B, Mukherjee D (1999) J Phys Chem A 103:1822

79. Mahapatra US, Datta B, Mukherjee D (1999) Chem Phys Lett 299:42

80. Mahapatra US, Datta B, Mukherjee D (1999) J Chem Phys 110:6171

81. Mášik J, Hubač I (1997) Coll Czech Chem Commun 62:829

82. Mášik J, Hubač I (1997) In: McWeeny R, Maruani J, Smeyers YG, Wilson S (eds) Quantum systems in chemistry and physics: trends in methods and applications. Kluwer, Dordrecht, p 283

83. Mášik J, Hubač I (1998) Adv Quantum Chem 31:75

84. Mášik J, Mach P, Hubač I (1998) J Chem Phys 108:6571

85. Mášik J, Mach P, Urban J, Polasek M, Babinec P, Hubač I (1998) Collect Czech Chem Commun 63:1213

86. Mehra J, Rechenberg H (1982, 1987, 2000, 2001) The historical development of quantum theory, vol 6. Springer, Berlin

87. Moore GE (1965, April 19) Electronics 38(8):114–117

88. Møller C, Plesset MS (1934) Phys Rev 46:618

89. Moncrieff D, Wilson S (2012) Universal Gaussian basis sets for molecular electronic structure theory. Progress in theoretical chemistry & physics, Springer, Dordrecht/New York

90. Paldus J (1992) In: Wilson S, Diercksen GHF (eds) Methods in computational molecular physics. NATO ASI series. Plenum Press, New York, p 99*ff*
91. Paldus J (1994) In: Malli GL (ed) Relativistic and correlation effects in molecules and solids, NATO ASI series. Plenum, New York, p 207
92. Paldus J (2002) In: Wilson S, Bernath PF, McWeeny R (eds) Handbook of molecular physics and quantum chemistry, vol 2. Wiley, Chichester
93. Paldus J, Čížek J, Shavitt I (1972) Phys Rev A 5:50
94. Petraco NDK, Hornyý L, Schaefer HF, Hubač I (2002) J Chem Phys 117:9580
95. Pittner J (2003) J Chem Phys 118:10876
96. Pittner J, Nechtigall P, Čársky P, Mášik J, Hubač I (1999) J Chem Phys 110:10275
97. Quiney HM, Grant IP, Wilson S (1989) In: Kaldor U (ed) Many-body methods in quantum chemistry. Lecture notes in chemistry, vol 52. Springer, Berlin, p 307
98. Quiney HM, Hubač I, Wilson S (2001) J Phys B: At Mol Opt Phys 34:4323
99. Renn J (2005) Ann Phys 14(Supplement):23
100. Roberts JM (1999) Twentieth century: the history of the world – 1901 to present. Allen Lane/The Penguin, London
101. Rutherford E (1911) Phil Mag 21:669
102. Sancho-García J, Pittner J, Čársky P, Hubač I (2000) J Chem Phys 112:8785
103. Schaefer HF, Hornyý L, Hubač I, Pal S (2005) Chem Phys 315:240
104. Schulman JM, Kaufman DN (1970) J Chem Phys 53:477
105. Wenzel W, Steiner MM (1998) J Chem Phys 108:4714
106. Wenzel W (1998) Int J Quantum Chem 70:613
107. Wigner EP (1935) Math naturw Anz ungar Akad Wiss 53:475
108. Wilson S (1987) Chemistry by computer. An overview of the applications of computers in chemistry. Plenum, New York
109. Wilson S (ed) (1987, 1989, 1990, 1992, 1993) Methods in computational chemistry, vol 1, 2, 3, 4, 5. Plenum, New York
110. Wilson S (2003) In: Wilson S, Bernath PF, McWeeny R (eds) Handbook of molecular physics and quantum chemistry, vol 2. Wiley, Chichester
111. Wilson S (1983) Electron correlation in molecules. Clarendon, Oxford
112. Wilson S (1992) In: Methods in computational molecular physics. NATO ASI series B, vol 293. Plenum, New York, p 222
113. Wilson S (2000) In: Chemical modelling: applications and theory. Senior reporter: A. Hinchliffe, Specialist periodical reports, vol 1. The Royal Society of Chemistry, London, p 364
114. Wilson S (2001) J Mol Struct Theochem 547:279
115. Wilson S (2002) In: Hinchliffe A (ed) Chemical modelling: applications and theory. Specialist periodical reports, vol 2. The Royal Society of Chemistry, London, p 329
116. Wilson S (2002) In: Wilson S, Bernath PF, McWeeny R (ed) Handbook of molecular physics and quantum chemistry, vol 2. Wiley, Chichester
117. Wilson S (2004) In: Hinchliffe A (ed) Chemical modelling: applications and theory. Specialist periodical reports, vol 3. The Royal Society of Chemistry, London, p 379
118. Wilson S (2006) In: Hinchliffe A (ed) Chemical modelling: applications and theory. Specialist periodical reports, vol 4. The Royal Society of Chemistry, London, p 470
119. Wilson S (2007) Electron correlation in molecules. Dover, New York
120. Wilson S (2008) In: Chemical modelling: applications and theory. Hinchliffe A (ed) Specialist periodical reports, vol 5. The Royal Society of Chemistry, London, p 208
121. Wilson S, Hubač I (2008) In: Wilson S, Grout PJ, Maruani J, Delgado-Barrio G, Piecuch P (eds) Frontiers in quantum systems in chemistry and physics. Progress in theoretical chemistry and physics, vol 18. p 561
122. Wilson S, Silver DM (1976) Phys Rev A 14:1949
123. Wilson S, Bernath PF, McWeeny R (eds) (2003) In: Handbook of molecular physics and quantum chemistry, vol 3. Wiley, Chichester

124. Wilson S, Hubač I, Mach P, Pittner J, Čársky P (2003) In: Maruani J, Lefebvre R, Brändas EJ (eds) Advanced topics in theoretical chemical physics. Progress in theoretical chemistry & physics, vol 12. p 71
125. Ziman J (1976) The force of knowledge. The scientific dimension of society. Cambridge University Press, Cambridge, pp 90–91

Chapter 3
Multireference State–Specific Coupled Cluster Theory with a Complete Active Space Reference

Vladimir V. Ivanov, Dmitry I. Lyakh, Tatyana A. Klimenko,
and Ludwik Adamowicz

Abstract The multireference state specific coupled cluster theory with a complete active space reference (CASCCSD) is described and its application to calculate electronic ground states is discussed. The working algorithm for the CASCCSD method was derived with a computer-based automated approach that generates the coupled-cluster diagrams and the corresponding amplitude equations. The method has been used to calculate the potential energy curves, spectroscopic parameters and vibrational levels of diatomic molecules. The test calculations have demonstrated high efficiency of the approach in comparison with other accurate approaches and in comparison with experimental data.

Keywords State-specific coupled-cluster theory • Electron correlation • Configuration interaction • Many-body perturbation theory • Single and double excitation

3.1 Introduction

The electron correlation problem [1, 2] is a key issue in contemporary quantum chemistry calculations. There are numerous examples demonstrating the importance of the correlation effects in the description of various properties of atoms, molecules, and complexes. Hence, a significant part of the development of the contemporary quantum chemistry methods have been devoted to accounting for those effects (see for instance [1, 3, 4]).

V.V. Ivanov (✉) • D.I. Lyakh • T.A. Klimenko
V. N. Karazin Kharkiv National University, Kharkiv, Ukraine
e-mail: vivanov@univer.kharkov.ua; lyakh@univer.kharkov.ua; generalchem@mail.ru

L. Adamowicz (✉)
University of Arizona, Tucson, Arizona, USA
e-mail: ludwik@u.arizona.edu

J. Leszczynski and M.K. Shukla (eds.), *Practical Aspects of Computational Chemistry I:*   69
An Overview of the Last Two Decades and Current Trends,
DOI 10.1007/978-94-007-0919-5_3, © Springer Science+Business Media B.V. 2012

One can divide all ab initio quantum chemistry methods into three major groups – the variational methods which include the configuration interaction (CI) methods, the methods based on the perturbation theory which include the many–body perturbation theory (MBPT) methods, and the coupled cluster (CC) methods (see, for example, [1, 4]). Among the mentioned methods, the CC methods [5–13] play an especial role, as they are capable of providing the most accurate description of the electron correlation effects in atoms and molecules. In particular, the standard single-reference CC theory for singlet states has established itself as a method which can very accurately describe the ground–state electronic structures of closed-shell molecular systems at equilibrium geometries. The size–extensivity of the CC theory has been very important in this respect. High accuracy and size–extensivity of the standard CC method is achieved due to so–called exponential *ansatz* of the CC wave function:

$$|\Psi_{CC}\rangle = e^{\hat{T}}|0\rangle, \tag{3.1}$$

where the cluster operator, $\hat{T}$, generates a superposition of different types of electronic excitations (i.e. single, double, etc. excitations) from the determinant $|0\rangle$. $|0\rangle$ in most CC calculations is the Hartree-Fock single-determinant wave function. $\hat{T}$ can be expressed as a sum of cluster operators representing different excitation levels:

$$\hat{T} = \hat{T}_1 + \hat{T}_2 + \cdots + \hat{T}_n. \tag{3.2}$$

The determinant $|0\rangle$ plays key role in the CC theory. It is called the reference determinant and in the standard version of the CC theory this is the determinant used to generate all necessary excitations in the wave function. For this reason the standard CC theory is called the single reference theory. Also, in the standard implementation of the theory, the $\hat{T}$ operator includes only single and double excitations from $|0\rangle$ (the CCSD theory),

$$\hat{T} \approx \hat{T}_1 + \hat{T}_2. \tag{3.3}$$

The cluster operators acting on $|0\rangle$ produce linear combinations the determinants:

$$\hat{T}_1|0\rangle = \sum_{i,a} t_i^a |{}_i^a\rangle, \tag{3.4}$$

$$\hat{T}_2|0\rangle = \sum_{i>j,a>b} t_{ij}^{ab} |{}_{i\,j}^{ab}\rangle, \tag{3.5}$$

where the cluster amplitudes, t_i^a and t_{ij}^{ab}, determine the contributions of the corresponding excitations to the CC operator. Indices i, j, etc. denote the spin–orbitals occupied in the reference determinant, $|0\rangle$, while indices a, b, etc. denote the vacant spin–orbitals. The determinants $|{}_i^a\rangle$ and $|{}_{i\,j}^{ab}\rangle$ are singly and double excited determinants, respectively. Expanding the exponent in (3.1) produces nonlinear components in the CC wave function. This leads to an effective, but approximate, accounting for higher (triple, quadruple, etc.) excitations. For example:

$$\hat{T}_1 \hat{T}_2 |0\rangle = \sum_{\substack{i,j>k \\ a,b>c}} t_i^a t_{jk}^{bc} |_{ijk}^{abc}\rangle, \tag{3.6}$$

$$\hat{T}_2^2 |0\rangle = \sum_{\substack{i>j,k>l \\ a>b,c>d}} t_{ij}^{ab} t_{kl}^{cd} |_{ijkl}^{abcd}\rangle. \tag{3.7}$$

The configuration coefficients of those higher excitations are products of amplitudes of lower level excitations. For example, the main contributions of the triple excitations in the CCSD theory are given as products of amplitudes of single excitations and double excitations, $t_i^a t_{jk}^{bc}$ (see Eq. 3.6), and the main contributions of the quadruple excitations are given as products of two double-excitation amplitudes, $t_{ij}^{ab} t_{kl}^{cd}$ (see Eq. 3.7).

The problem of describing the electron correlation effects becomes more complicated when quasidegenerate states are involved. Such states appear when the system has high symmetry, undergoes dissociation of a covalent chemical bond in the ground electronic state, it is in an electronically excited state and undergoes a photochemical transformation, etc. The problem is to describe the electron correlation effects in such situations in a size–extensive manner and as accurately as it is done for systems where the single-reference CC approach is adequate. Thus the challenge is to extend the CC theory to problems that cannot be, even in the zero approximation, described by wave functions being single closed-shell determinants for the singlet states. The multireference CC approach we have developed is an answer to this challenge. It provides a way to systematically generalize the single–reference CC approach to describe quasidegenerate states which arise in bond breaking and electron excitation processes.

3.2 Multireference State–Specific Generalization of CCSD Theory

Multireference coupled cluster (MRCC) models provide a generalization of the single-reference CC approach (3.1) for applications where several reference determinants contribute with similarly large weights to the wave function of the molecular system under consideration. The MRCC models can be divided into state-specific (SS) and multi-state approaches. One of the state–specific MRCC (SSMRCC) approaches is the active space CC method proposed in the works of Adamowicz and co-workers [14–18]. This method established the foundation for the approach developed and implemented by the authors of this article [19–27].

The key feature of our approach is the concept of a formal reference determinant, $|0\rangle$. This determinant is used to generate the complete MRCC wave function by acting on it with an appropriate CC excitation operator. The formal reference determinant, in the case of the ground electronic state, can be the Hartree-Fock determinant. In most cases this determinant has the largest weight in the wave function of the considered state. Moreover, $|0\rangle$ defines the partition of spin–orbital space into occupied and unoccupied orbital subspaces. These orbitals are often referred to as holes and particles, respectively. Hence, the holes correspond to the spin–orbitals occupied in $|0\rangle$ and the particles correspond to the unoccupied orbitals. The formal reference determinant is used to generate all necessary electronically excited configurations in the wave function in the Oliphant–Adamowicz coupled cluster model. The simplest case discussed in the early works of Oliphant and Adamowicz [14–16] is the case of two determinantal reference space:

$$P = span\{|0\rangle, |1\rangle\}. \tag{3.8}$$

where $|0\rangle$ is the formal reference determinant and $|1\rangle$ is the other reference determinant. This model enables description of the dissociation of a single bond in a homonuclear diatomic molecule. Such a dissociation can be described with the following CC wave function:

$$|\Psi_{SSMRCC}\rangle = e^{\hat{T}^{(ext)}}(|0\rangle + c|1\rangle), \tag{3.9}$$

where operator $\hat{T}^{(ext)}$ generates all single and double electronic excitations with respect to both reference determinants, $|0\rangle$ and $|1\rangle$. In the model that includes all single and double excitations from the reference determinants $\hat{T}^{(ext)}$ can be partitioned as:

$$\hat{T}^{(ext)} = \hat{T}_1^{(ext)} + \hat{T}_2^{(ext)}. \tag{3.10}$$

The coefficient c in (3.9) defines the ratio of the two reference determinants in the total wave function (3.9).

The second reference determinant, $|1\rangle$, can be expressed as the following two–electron excitation from $|0\rangle$:

$$|1\rangle = |^{AB}_{IJ}\rangle = a_A^+ a_I a_B^+ a_J |0\rangle. \tag{3.11}$$

In the above equation, operators a_A^+ and a_I are the usual creation and annihilation operators of the second quantization formalism. As common in the multireference approaches, the SSMRCC Oliphant–Adamowicz theory involves the concept of the *model space*. In a single-reference case there is one reference function, $|0\rangle$, and the model space is one–dimensional. $|0\rangle$ defines the Fermi vacuum and, as mentioned, the corresponding orbital-space partitioning into the occupied (with indices $i,j,k,...$) and unoccupied (with indices $a,b,c,...$) orbital subspaces. In a quasidegenerate situation the model space usually contains a larger (suitably chosen) set of configurations $(\{\Phi_\mu\}, \mu - 1, M)$. The model space comprising these N-electron configurations is

Fig. 3.1 Partitioning of the orbital space in the SSMRCC theory

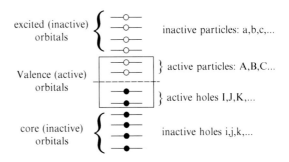

denoted as P. Typically configurations $\{\Phi_\mu\}$ differ by only a few orbitals from each other. These orbitals (called active orbitals) usually lie closely to the Fermi level.

The reference space (reference determinants) define the partition of the spin–orbital space into three groups (Fig. 3.1). These groups are the core, active (or valence), and virtual (excited) spin-orbital subspaces. The core spin-orbitals are those which are occupied in all determinants in the reference (model) space; the virtual (excited) spin–orbitals are not occupied in any reference determinant (these are empty spin-orbitals), and the active (valence) spin-orbitals are those which are occupied in at least one reference determinant and unoccupied in at least one other determinant. All possible different distributions of the valence (active) electrons among the active (valence) spin–orbitals generate all determinants of the *complete* model (reference) space.

We use capital letters (as in (3.11)) for active spin–orbitals. Among those, the letters I, J. etc. are used for active spin-orbitals which are occupied in $|0\rangle$ and letters A, B, etc. are used for active spin-orbitals which are unoccupied in $|0\rangle$. Lowercase indices (i, j, k, l,...) denote the occupied orbitals and (a, b, c, d,...) denote orbitals which are unoccupied in all the CAS determinants. These orbitals are "inactive". Also we use italic–style letters, *a, b, c, d,...* and *i, j, k, l,...*, to label orbitals in the cluster amplitudes and indicate that the coupled–cluster operators corresponding to these amplitudes can involve both active and inactive orbitals.

In the case of complete model space, the more general form of the SSMRCC exponential wave function is:

$$|\Psi_{SSMRCC}\rangle = e^{\hat{T}^{(ext)}} e^{\hat{T}^{(int)}} |0\rangle, \tag{3.12}$$

where the $e^{\hat{T}^{(int)}}$ generates the reference function, i.e. the wave function consisting of determinants from the reference space. In (3.9) and (3.12), the intermediate normalization is assumed:

$$\langle 0|\Psi_{SSMRCC}\rangle = 1. \tag{3.13}$$

It is very important to stress that in SSMRCC both operators, $e^{\hat{T}^{(int)}}$ and $e^{\hat{T}^{(ext)}}$, commute and act on the same "formal-reference" state $|0\rangle$ to generate the part of the wave function which comprises the reference determinants and the part comprising single, double, etc. excitations from the reference determinants, respectively. For example, the double excitation with respect to the 'second' determinant, $|1\rangle \equiv |_{IJ}^{AB}\rangle$:

$$a_a^+ a_i a_b^+ a_j |_{IJ}^{AB}\rangle = |_{IJij}^{ABab}\rangle. \tag{3.14}$$

is expressed as the following quadruple excitation with respect to the "formal-reference" determinant, $|0\rangle$:

$$a_a^+ a_i a_b^+ a_j a_A^+ a_I a_B^+ a_J |0\rangle = |_{IJij}^{ABab}\rangle. \tag{3.15}$$

In SSMRCC all excitations of the type shown in (3.15) are generated by using the following cluster operator include in $\hat{T}^{(ext)}$:

$$\hat{T}_4 \left(_{IJij}^{ABab}\right) = \sum t_{IJij}^{ABab} a_a^+ a_i a_b^+ a_j a_A^+ a_I a_B^+ a_J. \tag{3.16}$$

The test calculations [14, 15] showed that SSMRCC is a very effective approach for describing the dissociation of a single bond. The important feature of the SSMRCC method is that it is a direct generalization of the single-reference CC approach to describe MR problems.

3.2.1   The CASCCD Method

In this section we will describe the semi-linear (SL) SSMRCC method [19, 20]. The wave function of this method is a special case of the wave function represented by *ansatz* (3.12) as proposed in [14–16]. Contrary to the standard CC single-reference (SR) theory, we need to consider more than one reference determinant. This set of reference determinants has to provide an adequate zero-order description of the electronic wave function involved in the bond breaking/formation process or the excited state under consideration. A convenient and well establish way to generate the reference space is by performing a complete active space, self consistent field (CASSCF) calculation. The determinants used in expanding the CASSCF wave function are obtained by considering all possible distributions of active electrons among the active orbitals. Thus, the CASSCF wave function comprises all single, double etc. excitations within the active space. These excitations can be represented as excitations from the formal reference determinant, $|0\rangle$. Thus $|\Psi_{CASSCF}\rangle$ can be written as:

$$|\Psi_{CASSCF}\rangle = (1 + \hat{C}_1 + \hat{C}_2 + \ldots)|0\rangle. \tag{3.17}$$

Obviously, it is appropriate to choose $|0\rangle$ as the most contributing determinant to $|\Psi_{CASSCF}\rangle$.

The $\hat{C}_1$, $\hat{C}_2$, etc. operators can be represented using the standard creation/annihilation operators as:

$$\hat{C}_1 = \sum_{I,A} c_I^A a_A^+ a_I, \tag{3.18}$$

$$\hat{C}_2 = \sum_{I>J,A>B} c_{IJ}^{AB} a_A^+ a_I a_B^+ a_J, \tag{3.19}$$

etc. Thus, $|0\rangle$ defines the Fermi's vacuum for the problem.

In the first implementation of the semi–linear SSMRCC method (the method is termed the complete active space CC method, CASCC) we limited the excitation level of the CI-operators, $\hat{C}_n$, to quadruples:

$$|\Psi_{CASCC}\rangle = e^{\hat{T}^{(ext)}} (1 + \hat{C}_1 + \hat{C}_2 + \hat{C}_3 + \hat{C}_4)|0\rangle \tag{3.20}$$

This allowed us to describe single– and double–bond dissociation (or dissociation of two single bonds) processes.

In the first application of the CASCC method, the **external** part of the CC operator, $\hat{T}^{(ext)}$, in the CASCC wave function contained only doubles excitations with respect to Fermi's vacuum. We should note that $\hat{T}^{(ext)}$ generates external and semi–internal excitations with respect to $|0\rangle$. In the semi–internal excitations electrons are distributed among active (occupied in $|0\rangle$) and inactive (unoccupied in $|0\rangle$) orbitals or between inactive (occupied) and active (unoccupied) orbitals. In general, $\hat{T}^{(ext)}$ has the following form:

$$\hat{T}^{(ext)} = \sum_{\substack{i>j>... \\ a>b>...}} t_{ij...}^{ab...} a_a^+ a_i a_b^+ a_j ..., \tag{3.21}$$

where spinorbitals denoted by italic indices can be either active or inactive, at least one index has to be inactive.

Single excitations can also be included in $\hat{T}^{(ext)}$ (3.20). There are three types of these excitations with the following amplitudes: t_i^a, t_I^a, and t_i^A. The first type are *external* excitations and the second and third types are *semi–internal* excitations.

There are eight types of double external and semi–internal excitations included in operator $\hat{T}_2^{(ext)}$. Their amplitudes are:

$$t_{ij}^{ab}, \quad t_{Ij}^{ab}, \quad t_{ij}^{Ab}, \quad t_{IJ}^{ab}, \quad t_{ij}^{AB}, \quad t_{Ij}^{Ab}, \quad t_{Ij}^{AB}, \quad t_{IJ}^{Ab}. \tag{3.22}$$

Only the first type represents the fully *external* excitations. One can notice that among the amplitudes in (3.22) at least one index is inactive.

The number of the CI-type and t-type amplitudes in the CASCC wave function is exactly equal to number of determinants which the $\hat{C}_n$ and $\hat{T}^{(ext)}$ operators in (3.20) generate. As the amplitude equations are obtained by projecting the Schrödinger equation against those determinants, the number of equations is equal to the number of the amplitudes. This lack of ambiguity is an important and desirable feature of the CASCC method. Another desirable feature is the generation of all (excited) determinants from a single formal-reference determinant, $|0\rangle$. Moreover, the operators that appear in the CASCC wave function:

$$|\Psi_{CASCC}\rangle = e^{\hat{T}^{(ext)}}(1 + \hat{C})|0\rangle = (1 + \hat{C})e^{T^{(ext)}}|0\rangle, \qquad (3.23)$$

mutually commute.

The amplitude *redundancy* problem is present in the SS MRCC theories developed by Mukherjee et al. [28–30] and Bartlett et al. [31, 32]. Its absence in our CASCCD theory distinguishes this approach from those other approaches. Another important characteristic of the CASCCD approach is its size–extensivity which results from the size-extensivity of both internal and external parts of the CASCC wave function.

The simplest version of the CASCCD method includes all double excitations from all reference determinants. This necessitates the inclusion of higher excitations in the $\hat{T}^{(ext)}$ operator, as all excitations are generated as excitations from $|0\rangle$. For example, in the case of two active electrons distributed among two active orbitals (the 2×2 CAS), the CASCCD wave function has the following form:

$$|\Psi_{CASCCD}\rangle = e^{\hat{T}_2 + \hat{T}_4(^{ABab}_{IJij})}(1 + \hat{C}_2)|0\rangle, \qquad (3.24)$$

where we assumed that the excitation generated by $\hat{C}_1$ do not contribute to the wave function (this can happen, for example, due to symmetry constraints). Here, the operator $\hat{T}_2$ generates all necessary double excitations with respect to $|0\rangle$ and the quadruple excitation operator $\hat{T}_4(^{ABab}_{IJij})$ generates the remaining double excitations from the second reference determinant (generated by $\hat{C}_2|0\rangle$) which were not generated by $\hat{T}_2$. Those quadruple excitations with respect to $|0\rangle$ are double excitations from $\hat{C}_2|0\rangle$ because:

$$\hat{T}_4(^{ABab}_{IJij})|0\rangle = \hat{T}_2|^{AB}_{IJ}\rangle. \qquad (3.25)$$

Let us provide more explanation on this point. The CCSD wave function in the simplest model case can be approximately written as:

$$|\Psi_{CCSD}\rangle \approx |0\rangle + x|^{AB}_{IJ}\rangle + y|^{ab}_{ij}\rangle + x \cdot y|^{ABab}_{IJij}\rangle, \qquad (3.26)$$

where $|^{AB}_{IJ}\rangle$ and $|^{ab}_{ij}\rangle$ are the two most important doubly excited configurations relative to $|0\rangle$. The numbers x and y are the corresponding configuration coefficients. The coefficient of the quadruply excited configuration in $|\Psi_{CCSD}\rangle$ does not have an independent variable. This coefficient is fixed as $x \cdot y$. In the absence of quasi-degeneracy the magnitudes of x and y are small. This is a typical situation where the single-reference CCSD model is adequate (see Fig. 3.2). When quasi-degeneracy appears, the energy gap between HOMO and LUMO decreases. This leads to an increase of the weights of some doubly excited configurations (in our example, the configurations with coefficients x and y). Say, the weight of the determinant

Fig. 3.2 A model description of the CCSD virtual excitations

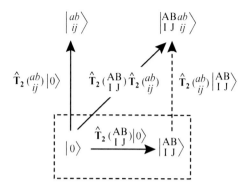

Fig. 3.3 A model description of CASCC virtual excitations. There is a T_4 operator in addition to $T_2 T_2$ product

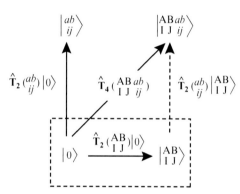

$|{}^{AB}_{IJ}\rangle$ substantially increases and becomes ≈ 1. As a result, the weight of quadruply excited configuration, $|{}^{ABab}_{IJij}\rangle$, also increases. In such a situation the weight of this configuration determined as $x \cdot y$ may not be sufficiently accurate. One may notice that the doubly–excited configurations relative to two (reference) determinants are included in a not completely equivalent way in the CCSD wave function. In the SSMRCC and CASCC approaches the equivalent treatment is achieved by including an operator representing the quadruple excitation (see Fig. 3.3). With that, the CASCC wave function, by analogy with (3.26), is:

$$|\Psi_{CASCC}\rangle \approx |0\rangle + x|{}^{AB}_{IJ}\rangle + y|{}^{ab}_{ij}\rangle + (z + x \cdot y)|{}^{ABab}_{IJij}\rangle, \qquad (3.27)$$

where an additional cluster coefficient z is introduced.

We will now consider the equations for the energy and the amplitudes in the CASCCD method in the most general form. For simplicity let us consider a system which can be described with two active electrons and two active orbitals, i.e. (2,2). As usual in the CC theory, to derive the equations for the energy and the amplitudes we project the Schrödinger equation,

$$(\hat{H} - E_{CASCCD})|\Psi_{CASCCD}\rangle = 0, \qquad (3.28)$$

on the appropriate set of configurations. In (3.28) $\hat{H}$ is the electronic Hamiltonian of the system and E_{CASCCD} is the CASCCD energy. In order to generate the sufficient number of equations for all c (CI) amplitudes (3.28) is projected against all determinants from the CAS space. The projection against $\langle 0|$ gives the equation for the energy:

$$\langle 0|(\hat{H} - E_{CASCCD})|\Psi_{CASCCD}\rangle = 0. \tag{3.29}$$

The remaining projections against the CAS determinants produce equations for the CAS CI amplitudes:

$$\langle^{A\cdots}_{I\cdots}|(\hat{H} - E_{CASCCD})|\Psi_{CASCCD}\rangle = 0. \tag{3.30}$$

To generate equations for the t amplitudes involved in the $\hat{T}^{(ext)}$ operator, the (3.28) equation is projected against the determinants generated by the action of $\hat{T}^{(ext)}$ on the formal reference $|0\rangle$. For example, the equations for the amplitudes corresponding to the double excitations within $\hat{T}^{(ext)}$ are obtained as:

$$\langle^{ab\cdots}_{i\ j\cdots}|(\hat{H} - E_{CASCCD})|\Psi_{CASCCD}\rangle = 0. \tag{3.31}$$

In the last equation the orbital indices correspond to both active and inactive spin–orbitals, but again at least one index has to correspond to an inactive orbital. However, at least one of the indices has to be inactive. In our approach, amplitudes C^A_I, C^{AB}_{IJ}, etc. and the t amplitudes of $\hat{T}^{(ext)}$ are determined simultaneously by iteratively solving the coupled set of Eqs. 3.29–3.31. The detailed algebraic forms of the equations are generated using a computerized automated technique we have developed for this purpose (see Sect. 3.4).

3.3 Multireference State–Specific Coupled Cluster Theory with Complete Account of Single and Double Excitations: The CASCCSD Method

In a more general form the CASCC wave function must include single excitations. This is accomplished by adding operator $\hat{T}_1$ to the exponential CC operator (the approach is termed CASCCSD):

$$|\Psi_{CASCCSD}\rangle = e^{\hat{T}^{(ext)}_1 + \hat{T}^{(ext)}_2}(1 + \hat{C})|0\rangle, \tag{3.32}$$

where the cluster operators $\hat{T}^{(ext)}_1$ and $\hat{T}^{(ext)}_2$ generate all single and double excitations, respectively, from the reference determinants produced by the linear part of the CASCC wave function (3.32). As all those excitations are generated as excitations from $|0\rangle$, some selective higher excitations need to be included in $\hat{T}^{(ext)}_1$ and $\hat{T}^{(ext)}_2$ (for a detailed description see below).

As the SSMRCC approach is based on the standard CC theory, it assumes the intermediate normalization of the wave function (3.13). Using this condition the equation for the energy can be written as:

$$\langle 0|[\hat{H}_N e^{\hat{T}^{(ext)}} e^{\hat{T}^{(int)}}]|0\rangle = \Delta E. \tag{3.33}$$

The equations for the CC amplitudes are derived from projecting the Schrödinger equation against excited determinants:

$$\langle^{ab...}_{i\,j...}|[\hat{H}_N e^{\hat{T}^{(ext)}} e^{\hat{T}^{(int)}}]_c|0\rangle = 0, \tag{3.34}$$

where $\langle^{ab...}_{i\,j...}|$ are those determinants and $\hat{H}_N = \hat{H} - \langle 0|H|0\rangle$ is the Hamiltonian of the system in the "normal" form. As in the case of FCI, the disconnected components on the left side of the Eq. 3.34 cancel out.

In the semi-linear form (3.32) of the CASCC wave function, the CI part, $((1+\hat{C})|0\rangle)$, includes the complete set of determinants $(\Phi^{(int)})$, which are generated by all possible distributions of the active electrons among the active orbitals. In the semilinear CASCCSD method the Eq. 3.34 can be written in the form of the following two coupled equations :

$$\langle\Phi^{(int)}|[(\hat{H}_N - E)e^{\hat{T}^{(ext)}}(1 + \hat{C}^{(int)})]|0\rangle = 0, \tag{3.35}$$

$$\langle\Phi^{(ext)}|[(\hat{H}_N - E)e^{\hat{T}^{(ext)}}(1 + \hat{C}^{(int)})]|0\rangle = 0. \tag{3.36}$$

The first Eq. 3.35, is equivalent to the equation one solves to determine the configuration amplitudes in the CI method and the second Eq. 3.36, is what one solves to determine the CC amplitudes in the CC method.

To obtain an explicit cancelation of the disconnected terms in the Eqs. 3.35 and 3.36 it is necessary to separate those terms out when they appear due the action of the CI operator, $\hat{C}$, on $|0\rangle$. We should note that the use of the semi-linear form of the CASCCSD wave function (3.32) leads to equations which are explicitly dependent on the energy the system, while the fully exponential form of the wave function, (3.12), leads to energy independent equations. We should also add that the CASCC wave function can be alternatively expressed in the following pure linear form:

$$|\Psi_{CASCC}\rangle = (1 + \hat{C}^{(ext)})(1 + \hat{C}^{(int)})|0\rangle = (1 + \hat{C})|0\rangle. \tag{3.37}$$

The linear, semi–linear, and fully exponential forms of the CASCC wave function are equivalent, if the following relations between the CI configuration coefficients and the CC amplitudes are imposed:

$$c^{a...}_{i...} = t^{a...}_{i...} + \sum_{n=2}^{n}\left(\prod_{j=1}^{n} t^{a...}_{i...}\right). \tag{3.38}$$

In the lowest level of the CASCC approach termed CASCCSD(T) (which is equivalent to SSMRCCSD(T)), the wave function includes all singles and doubles from $|0\rangle$, as well as some selected triples excitations. Those selected triples generate some double excitations from the reference determinants which are single and double excitations from $|0\rangle$, as well as some single excitations from the reference determinants which are double excitations from $|0\rangle$. The CASCCSD(T) wave function has the following form:

$$|\Psi_{CASCCSD(T)}\rangle = e^{\hat{T}_1+\hat{T}_2+\hat{T}_3(^{Abc}_{Ijk})}(1 + \hat{C}_1(^A_I) + \hat{C}_2(^{AB}_{IJ}) + \ldots)|0\rangle. \qquad (3.39)$$

In a more balanced approach one needs to include all singles and doubles from all reference determinants. Thus, in the case of the CAS that includes two active electrons distributed among two active orbitals (four spin–orbitals), the CASCC wave function (called in this case CAS(2,2)CCSD) must also include, in addition to the selected triples, also some selected quadruples. The CAS(2,2)CCSD method can be used to describe, for example, dissociation of a single covalent bond. The CAS(2,2)CCSD wave function has the following form:

$$|\Psi_{CAS(2,2)CCSD}\rangle = e^{\hat{T}_1+\hat{T}_2+\hat{T}_3(^{Abc}_{Ijk})+\hat{T}_4(^{ABab}_{IJij})}(1 + \hat{C}_1(^A_I) + \hat{C}_2(^{AB}_{IJ}))|0\rangle. \qquad (3.40)$$

In (3.40), the cluster operators representing triples, $\hat{T}_3(^{Abc}_{Ijk})$, generate all doubly excited configurations from determinant $|^A_I\rangle$. All necessary single excitations from that determinant are generated by operators $\hat{T}_1$ and $\hat{T}_2$. The operators $\hat{T}_4(^{ABab}_{IJij})$ generate double excitations relative to the $|^{AB}_{IJ}\rangle$ reference determinant. All necessary single excited configurations relative to this determinant are generated by the $\hat{T}_3(^{ABc}_{IJk})$ operator. With that, all single and double excitations with respect to all reference CAS(2,2) determinants are included in the CAS(2,2)CCSD wave function.

To describe simultaneous dissociation of two single bonds (or single double bond) one needs to use an active space with four electrons and four orbitals (eight spin–orbitals). Using this active space in the CASCCSD method gives rise to the CAS(4,4)CCSD approach. There are the following five types of reference determinants in the CAS(4,4)CCSD wave function:

$$|0\rangle, \quad |^A_I\rangle, \quad |^{AB}_{IJ}\rangle, \quad |^{ABC}_{IJK}\rangle, \quad |^{ABCD}_{IJKL}\rangle. \qquad (3.41)$$

To generate all single and double excitations from the above reference determinants represented as excitations from the formal reference determinant, $|0\rangle$, one needs the cluster operators which are listed in Table 3.1. As one can see, for example, some double excitations relative to determinant $|^{ABCD}_{IJKL}\rangle)$ are expressed as some selected sextuple excitations from $|0\rangle$.

Thus, the complete CAS(4,4)CCSD wave function has the following form:

$$|\Psi_{CAS(4,4)CCSD}\rangle = e^{\hat{T}_1+\hat{T}_2+\hat{T}_3(^{Abc}_{Ijk})+\hat{T}_4(^{ABab}_{IJij})+\hat{T}_5(^{ABCab}_{IJKij})+\hat{T}_6(^{ABCDab}_{IJKLij})}\left(1 + \sum_{k=1}^{4}\hat{C}_k\right)|0\rangle.$$
$$(3.42)$$

Table 3.1 The cluster operators which generate all singles and doubles excitations relative to reference determinants in the CAS(4,4)CCSD wave function

Level of excitation	$|0\rangle$	$|_1^A\rangle$	$|_{1J}^{AB}\rangle$	$|_{1JK}^{ABC}\rangle$	$|_{1JKL}^{ABCD}\rangle$
Singles	$\hat{T}_1(_i^a)$	$\hat{T}_2(_{1j}^{Ab})$	$\hat{T}_3(_{1Jk}^{ABc})$	$\hat{T}_4(_{1JK1}^{ABCd})$	$\hat{T}_5(_{1JKLm}^{ABCDd})$
Doubles	$\hat{T}_2(_{ij}^{ab})$	$\hat{T}_3(_{1jk}^{Abc})$	$\hat{T}_4(_{1Jk1}^{ABcd})$	$\hat{T}_5(_{1JKlm}^{ABCcd})$	$\hat{T}_6(_{1JKLmn}^{ABCDcd})$

Table 3.2 Four points of the CAS(2,2)CCSD PEC relative to FCI (reported in 10^{-3} au) in comparison with the CASSCF(2,2), CCSD, and CCSD(T) PECes for the hydrogen fluoride molecule (Re = 1.733 au) In the last row the FCI energies (in au) are shown. The calculations have been performed with the DZV basis set

Method	Re	2Re	3Re	5Re
CASSCF(2,2)	99.73	87.82	78.33	77.79
CCSD	1.60	5.97	11.42	12.17
CCSD(T)	0.32	0.08	−23.23	−48.85
CAS(2,2)CCSD	0.77	0.98	1.19	1.21
FCI	−100.146457	−100.007886	−99.972218	−99.970266

Dissociation of a triple covalent bond (or simultaneous dissociation of three single bonds, or simultaneous dissociation of a double bond and a single bond) gives rise to an even more complicated form of the CASCCSD wave function. Such a wave function one needs to use, for example, in describing the bond dissociation in the nitrogen molecule. In this case the active space must comprise six electrons and six orbitals (12 spin-orbitals). Thus, in the reference space there are excitations as high as sextuples relative to the formal reference determinant, $|0\rangle$. Some of the double excitations with respect to those sextuples, which need to be included in the cluster operator in the CASCCSD wave function to have a balanced treatment of the electron correlation effects, are octuples relative to $|0\rangle$. Hence, the complete CAS(6,6)CCSD wave function has the following form:

$$|\Psi_{CAS(6,6)CCSD}\rangle = e^{\hat{T}_1 + \hat{T}_2 + \sum_{k=3}^{8} \hat{T}_k'} \left(1 + \sum_{k=1}^{6} \hat{C}_k\right)|0\rangle. \qquad (3.43)$$

Where " $/$ " in the operator $\hat{T}_k'$ indicates that only a selected subset of the k excitations are included in the operator.

As demonstrated in our recent works [33–35], the CASCCSD wave function constructed in the above-described way provides in all studied cases a very accurate description of the potential energy curve (PEC) in comparison with FCI. Let us, for example, show in more details how the method performs in the case of the dissociation of the hydrogen fluoride (FH) molecule. The calculations have been performed with the standard valence double–zeta (DZV) basis set (implemented in GAMESS package [36]). The results are presented in Table 3.2. As one can see in Table 3.2, the differences between the CAS(2,2)CCSD and FCI energies are almost constant and small for all the PEC points shown in the table. The worst results

are those obtained in the CCSD(T) calculations of PEC. At the bond stretched to
R = 5Re the energy of the FH molecule is below the FCI result by as much as
48.85 mH !

It should be emphasized that the CASCCSD wave function includes only a small
set of all possible (FCI) excitations in the $\hat{C}$ and $\hat{T}$ operators. The $\hat{C}$ excitations
constitute only 0.007% of the total number of the FCI excitations, while for the $\hat{T}$
excitations (i.e. the number of t amplitudes) the corresponding percentage in 2.2%.
Furthermore, the computational costs of CAS(2,2)CCSD and CCSD calculations are
comparable. They both scale as $O(M^6)$, where M is the number of basis functions.

More results of CASCCSD test calculations and their comparison with FCI
results can be found in several articles our group has published in recent years. These
tests have demonstrated high accuracy of the CASCCSD approach in describing
dissociation of a single bond [21,23], a double bond [27,37], and a triple bond [33].
Some application of the CASCCSD approach to describe electronic excited states
have also been presented in our recent works [24,34,35].

3.4 Automated Derivation of the CC Equations and Generation of the Computer Code for Solving Them

In the conventional approach, the CC equations have been derived using diagrammatic techniques which offer a concise way of expressing the individual terms in
the equations in a graphical form. A description of the technique can be found
in several articles (see, for example, [38–40]). As the CC theory becomes more
complex, the number of diagrams increases rapidly. This is particularly true in the
multireference coupled cluster methods. Deriving and coding the corresponding
formulae by hand is error-prone and very time consuming. Consequently, attempts
to automate the derivation of diagrams have been made since 1970s [41]. The
interest in the computer-aided derivation of the CC equations has increased in the
last decade mostly because of the need to handle higher-rank tensors representing
higher excitations. As mentioned, explicit inclusion of higher excitations is needed
in considering quasi-degenerate (multireference) phenomena such as chemical
bond-breaking processes, photo-chemical reactions, electronic excited states, etc.
The presence of higher-rank tensors in the CC equations leads to a large number of
diagrams. To handle these diagrams it is convenient to use an automated computer
approach. Such an approach should be capable of deriving the diagrams and the
corresponding formulae, factorizing them, and producing a computer code for
calculating them. With such a tool, the amount of time spent on implementing,
testing, and debugging of a particular version of the CC method (or any other
electronic-structure method) is significantly reduced.

In the CC theory one of the first attempts to automate the CC diagram derivation
were undertaken by Janssen and Schaefer [42], Li and Paldus [43], and Harris [44].
Since then several rather sophisticated automated schemes have been developed

by different groups. Among them one should mention the general-order scheme of Kállay et al. [45, 46], the tensor contraction engine of Hirata [47], the general-order scheme of Lyakh et al. [25–27], the technique developed by Bochevarov and Sherrill [48], the automated formula derivation tool of Hanrath [49], and the general-order approach for explicitly correlated methods of Shiozaki et al. Nowadays almost every group developing multireference or higher-order methods has its own automated computer code for the formula derivation and implementation.

The CASCCSD approach explicitly involves higher-rank excitation operators because the entire wave function *ansatz* is defined with respect to a single determinant, i.e. the formal reference determinant (Fermi vacuum). The excitation rank of the CC operator depends on the size of the active space. The other complication in deriving the CASCCSD equations appears because there are two kinds of indices involved in the higher-rank tensors, i.e. the general hole/particle indices running over the entire hole/particle set and the active hole/particle indices running over the active hole/particle subsets. This permits splitting the automated derivation of the coupled-cluster diagrams into two subproblems:

1. Automated derivation of the general-order coupled cluster diagrams
2. Automated derivation of diagrams with restrictions imposed on some summations over indices ranging only over the active spin–orbitals

We will now briefly describe the key elements involved in the diagrammatic technique used in the electronic structure theory:

- **Task**: given a product of operators, express the required matrix elements in terms of the products of underlying tensor amplitudes and their contractions.
- **Arguments**: normal-ordered operators anti-symmetrized with respect to indices of the same type. Diagram technique can be easily extended to include operator self-contractions, but this is not needed in the CASCC theory.
- **Operator representation**: in general, a normal-ordered operator splits into different terms having different excitation/de-excitation, hole/particle ranks. For example, the normal-ordered Hamiltonian:

$$\hat{H}_N = \sum_{p,q} f_q^p \, \hat{p}^+ \hat{q} + \frac{1}{4} \sum_{pqsr} v_{rs}^{pq} \, \hat{p}^+ \hat{q}^+ \hat{s}\hat{r}, \tag{3.44}$$

where

$$f_q^p = h_q^p + \sum_i \langle pi \| qi \rangle, \tag{3.45}$$

is a matrix element of the Fock operator,

$$v_{rs}^{pq} = \langle pq \| rs \rangle = \left\langle pq \left| \frac{1}{r_{12}} \right| rs \right\rangle - \left\langle pq \left| \frac{1}{r_{12}} \right| sr \right\rangle \tag{3.46}$$

can be decomposed into the following parts: (a) four one-electron (1e) terms (see Fig. 3.4); and (b) nine two-electron terms (see Fig. 3.5). In Figs. 3.4

Fig. 3.4 Diagrammatic
representations of
one-electron terms

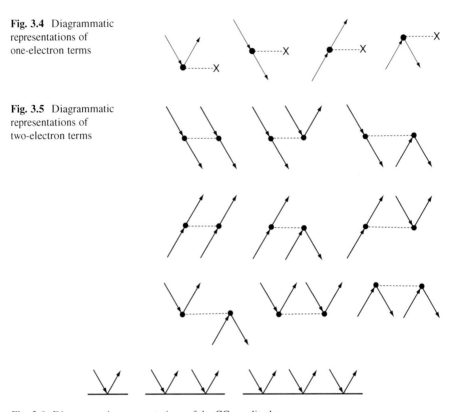

Fig. 3.5 Diagrammatic
representations of
two-electron terms

Fig. 3.6 Diagrammatic representations of the CC amplitudes

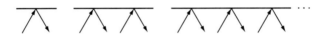

Fig. 3.7 Diagrammatic representations of the Λ amplitudes

and 3.5, down-pointing arrows represent holes, up-pointing arrows represent
particles, incoming arrows are annihilators, and out-going arrows are creators.
The excitation operator $\hat{T}$ consists of different-rank excitation parts (see Eq. 3.9).
In the diagrammatical form it can be represented as shown in Fig. 3.6. The
definition of the "left" state in the CC response theory [50], the Λ operator, can
be expanded in a series of de-excitation terms (show in Fig. 3.7.)

The above graphical constructs represent individual normal-ordered operators.
Using these graphical representations one can derive all unique combinations
(operator contractions) which contribute to the considered matrix element of an
operator product. The rules for the diagram manipulations are standard and can be
found elsewhere [13, 51]. General-order coupled cluster equations can be derived
from the general-order coupled cluster functional:

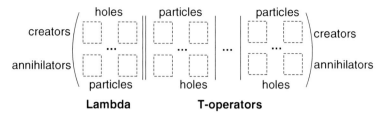

Fig. 3.8 Matrix representation of CC diagram

$$E = \langle 0 | 1 + \hat{\Lambda} | e^{-\hat{T}} \hat{H}_N e^{\hat{T}} | 0 \rangle. \tag{3.47}$$

The graphical representation of the components of this equation was given above. In particular, the T-equations and the energy equation are:

$$\forall \, 0 < k \leq n \quad \langle_{i_1 \, i_2 \, \dots \, i_k}^{a_1 a_2 \dots a_k} | (\hat{H}_N e^{\hat{T}})_c | 0 \rangle = 0, \tag{3.48}$$

$$\Delta E = \langle 0 | (\hat{H}_N e^{\hat{T}})_c | 0 \rangle = 0, \tag{3.49}$$

where n is the highest rank tensor included in the CC *ansatz*. Now, the main task is to encode the graphical objects representing the above operators and establish rules which are sufficient to construct all required diagrams without repetitions.

In the following we present a slightly extended version of our original approach to accomplish this task. The diagrams are represented in a machine-readable form using special two-row matrices called D-matrices. The upper row of the D-matrix corresponds to the creators while the lower row corresponds to the annihilators. The columns of the D-matrix are grouped into submatrices with each submatrix representing an individual operator (T-submatrix or L-submatrix). The number of columns in each submatrix multiplied by two is equal to the rank of the corresponding tensor such that each D-matrix element corresponds to an index of the tensor. The Hamiltonian operator enters the D-matrix only implicitly by storing the corresponding integral-type number. For our purpose, the general form of the D-matrix can be schematically represented as it is shown in Fig. 3.8, where the dotted rectangles designate individual D-matrix elements corresponding to the indices of the tensors. The general-order coupled-cluster equations may involve up to four $\hat{T}$ excitation operators and only one $\hat{\Lambda}$ de-excitation operator (in the standard CC approach). All tensor indices (D-matrix elements) can be classified as either *open* or *free (closed)*. Open indices are not contracted and represent the calculated matrix element of the considered operator product. In this case, the corresponding D-matrix elements contain zeros. Free indices are the indices involved in contractions (summation indices). In this case, the corresponding D-matrix elements contain positive integers representing the free indices. Figure 3.9 illustrates the correspondence between the different diagrams and the D-matrices.

Diagram	D-matrix	Formula (Einstein notation)
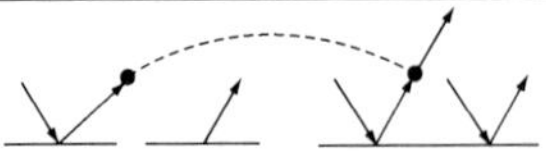	$\begin{pmatrix} 1 & 0 & 1 & 0 \\ 0 & 1 & 0 & 0 \end{pmatrix}$	$-H^{k_1 a_2}_{c_1 c_2} T^{c_1}_{i_1} T^{a_1}_{k_1} T^{c_2 a_3}_{i_2 i_3}$
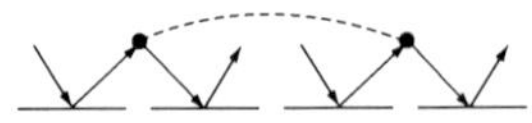	$\begin{pmatrix} 1 & 0 & 1 & 0 \\ 0 & 1 & 0 & 1 \end{pmatrix}$	$H^{k_1 k_2}_{c_1 c_2} T^{c_1}_{i_1} T^{a_1}_{k_1} T^{c_2}_{i_2} T^{a_2}_{k_2}$
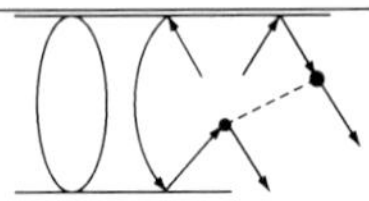	$\begin{pmatrix} 3 & 3 & 2 & 3 & 1 \\ 3 & 0 & 0 & 3 & 3 \end{pmatrix}$	$\Lambda^{k_1 k_2 k_3}_{c_1 a_2 a_2} H^{i_1 i_2}_{c_2 k_3} T^{c_1 c_2}_{k_1 k_2}$

Fig. 3.9 Examples of the correspondence between algebraic formulae, diagrams, and D-matrices

The algorithm for the diagram derivation is the following:

- loop over the products of operators involved in the CC approach under consideration
- loop over the 2nd–quantized structure of the bra/ket (for instance, $\langle {}^{a_1 a_2}_{i_1 i_2}|$, $|{}^{a_1}_{i_1}\rangle$, ...). This determines the required matrix elements for the operator product
- loop over the components of the Hamiltonian (integral types are given in Figs. 3.4 and 3.5 in the diagrammatic representation)
- loop over the number of $\hat{i}^+$ and $\hat{a}$ in the Hamiltonian to be contracted with the $\hat{T}$ operators. These $\hat{i}^+$ and $\hat{a}$ spawn free indices of the 1st class (the so-called H-T contractions). The following restrictions need to be imposed: (a) the numbers of $\hat{i}^+$ and $\hat{a}$, which remain open in the Hamiltonian, must be compatible with the structure of the ket. In the $\hat{\Lambda}$-equations at least one of $\hat{i}^+$ and $\hat{a}$ must remain open. (b) the number of open indices which still remain in the T-submatrices must be compatible with the structure of the bra
- Distribute free indices of the 1st class among the T-submatrices (holes and particles are independent) under the following restrictions: (a) if T-submatrices of the same size are present, the number of the H-T hole contractions of these T-submatrices must constitute a non-descending series; (b) if T-submatrices of the same size are present and they have the same number of the H-T hole contractions, the number of the H-T particle contractions provided to these T-submatrices must constitute a non-descending series; (c) in the "connected" coupled cluster formulation each $\hat{T}$ operator must be involved in at least one contraction
- loop over $\hat{i}$ and $\hat{a}^+$ of the Hamiltonian involved in contractions with $\hat{\Lambda}$ (if present). These $\hat{i}$ and $\hat{a}^+$ spawn free indices of the 2nd class (the so-called L-H contractions). Restrictions: (a) the number of $\hat{i}$ and $\hat{a}^+$ that remains open in the Hamiltonian must be compatible with the structure of the bra; (b) the number

of open indices that remain in $\hat{\Lambda}$ must be compatible with the structure of the ket (taking into account the possible contraction with $\hat{T}$)

- Adjust the total number of open indices commensurate with the bra/ket structure. The adjustment is accomplished by declaring the excessive open indices of the L-submatrix as free and of the 3rd class (the so-called L-T contractions)
- Distribute free indices of the 3rd class among the T-submatrices (holes and particles are independent) with the following restrictions: (a) if T-submatrices of the same size and with the same H-T contraction pattern are present, the number of the L-T hole contractions provided to these T-submatrices must form a non-descending series; (b) if T-submatrices of the same size and with the same H-T contraction pattern are present and they have the same amounts of L-T hole contractions, the number of the L-T particle contractions provided to these T-submatrices must form a non-descending series

These rules enable to generate a complete set of topologically unique diagrams for the coupled cluster approach of interest. If one is also interested in deriving the diagrams for a generalized density matrix using the above approach, the Hamiltonian in the formalism has to be replaced with the appropriate density operator.

In the CASCC approach we have to impose restrictions on the summations if they involve active orbitals. The general-order CC equations generated with the above algorithm are suitable for both the standard general-order coupled-cluster formalism (where there are no active indices) and the active-space CC formalism such as the CASCCSD approach. The CASCCSD approach restricts both $\hat{T}$ and $\hat{\Lambda}$ to contain no more than two general hole/particle indices regardless of the tensor rank. The remaining indices are assumed active. One way to deal with such restrictions is to directly examine each general-order CC diagram and to derive all possible forms of that diagram where the restrictions are explicitly implemented. Such an approach was adopted by Kállay et al. [45, 46]. The resulting number of diagrams with such restrictions quickly reaches tens of thousands when higher tensor ranks are considered. In our work we adopted a different approach where the restrictions on each general-order CC diagram are imposed on the fly during its evaluation. The readers interested in computationally efficient automated implementations of general-order CC codes are referred to [45, 46, 49].

3.5   Numerical Results

We will now examine the effectiveness of the CASCCSD method in calculations concerning vibrational spectra of some diatomic molecules (O_2, LiH, BH, and FH). As a criterion for the effectiveness we use the comparison of the calculated vibrational frequencies and other spectroscopic parameters we calculate with the experimental values. The comparison also includes the results obtained in the calculations performed with the following high-level methods implemented in the GAMESS package [36]: the standard coupled-cluster singles and doubles method (CCSD) [9, 52], the equations-of-motion CCSD(EOM-CCSD) method [52–55], and the multireference configuration interaction method with single and double excitations(CASCISD) that includes the Davidson correction for quadruples (CASCISD[+Q]) [56]. As these methods are analogues with our approach, the comparison provides an assessment of where CASCCSD method should be placed on the relative accuracy scale. To further examine the efficiency of the CASCCSD method, we also compare its results with more recently developed MRCC methods and the full configuration interaction (FCI) method.

The calculation of the spectroscopic parameters for a diatomic molecule requires solving the following one-dimension radial Schrödinger equation with the appropriate potential $U(R)$:

$$\left\{ \frac{\hbar^2}{2\mu} \frac{d^2}{dR^2} - U(R) - \frac{\hbar^2}{2\mu R^2} [J(J+1) - \Omega^2] + G(v, J) \right\} \psi(R; v, J) = 0, \quad (3.50)$$

where $G(v, J)$ denotes the energy of the particular vibrational-rotational level, $\psi(R; v, J)$ is the corresponding wave function, v is the vibrational quantum number, J is the rotational quantum number, Ω is a projection of the electron angular momentum on the bond axis, and μ is the reduced mass. In the **HERZBERG** package we developed the potential energy function, $U(R)$, is approximated with the following generalized Morse function:

$$U_{GM}(R) = \sum_{m=2}^{q} \alpha_m \left\{ 1 - exp\left[\beta_m(R - R_e) \right] \right\}^m, \quad q = 2, 3, 4 \ldots, \quad (3.51)$$

where α_m and β_m are fitting parameters and R_e is the equilibrium internuclear distance. The $U_{GM}(R)$ function with $q = 2$ (the standard Morse potential) usually gives a qualitatively correct form of the curve, but it is still too rough for more accurate calculations. In our calculations we have been using q≥6 which we found optimal to correctly represent the whole range of the potential energy curve. We also use the standard deviation parameter, SD, to examine the accuracy of the vibrational energy calculations:

$$SD = \sqrt{\sum_{v=0}^{n_k} \left[G(v) - G'(v) \right]^2 / (n_k + 1)}, \quad (3.52)$$

where $G(v)$ and $G'(v)$ are the experimental and the calculated energy values, respectively, and n_k is the number of the vibrational levels.

The analytic form of the potential function, $U(R)$, allows one to estimate the spectroscopic parameters [57], which can be compared with the experimental values by using the Dunham expansion:

$$G(v, J) = \sum_i \sum_j Y_{ij} h \left(v + \frac{1}{2} \right)^i J^j (J + 1)^j. \tag{3.53}$$

The Dunham coefficients Y_{ij} are related to the spectroscopical parameters as follows: $Y_{10} \equiv \omega_e$ to the fundamental vibrational frequency, $Y_{20} \equiv \omega_e \chi_e$ to the anharmonicity constant, $Y_{02} \equiv \overline{D}$ to the centrifugal distortion constant, $Y_{11} \equiv \alpha_e$ to the vibrational-rotational interaction constant, and $Y_{01} \equiv \beta_e$ to the rotational constant. These coefficients can be expressed in terms of different derivatives of $U(R)$ at the equilibrium point, $r = R_e$. The derivatives can be either calculated analytically or by using numerical differentiation applied to the PEC points. The numerical differentiation of the total energy of the system, $E_{CASCCSD}$, "point by point" is the simplest way to obtain the parameters. In our works we have used the standard five-point numerical differentiation formula. In the comparison of the calculated values with the experimental results we utilize the "experimental" PECs obtained with the Rydberg-Klein-Rees (RKR) approach [58–60] and with the inverted perturbation approach (IPA) [61,62]. The IPA is method originally intended to improve the RKR potentials.

Oxygen molecule (O_2). The ground-state molecular–orbitals electronic configuration of O_2 is:

$$[\sigma]^8 5\sigma^2 1\pi_x^2 1\pi_y^2 2\pi_x^1 2\pi_y^1 6\sigma^0.$$

The CASCISD/cc-pVTZ wave function for ground state with O–O bond fixed at its equilibrium value is:

$$|\psi(X^3\Sigma_g^-)_{R_e}\rangle \approx 0.67|2\pi_x^1 2\overline{\pi}_y^1\rangle - 0.67|2\overline{\pi}_x^1 2\pi_y^1\rangle$$
$$-0.059|1\overline{\pi}_x^1 1\pi_y^1 2\pi_x^2 2\pi_y^2\rangle + 0.059|1\pi_x^1 1\overline{\pi}_y^1 2\pi_x^2 2\pi_y^2\rangle. \tag{3.54}$$

The wave function is dominated by the following open-shell configurations: $|2\pi_x^1 2\overline{\pi}_y^1\rangle \equiv |\uparrow - - - \downarrow\rangle$ and $|2\overline{\pi}_x^1 2\pi_y^1\rangle \equiv |- - \downarrow \uparrow -\rangle$. One of these configurations could be chosen as the formal-reference determinant, $|0\rangle$, in the CASCCSD calculation. In the case of $|0\rangle \equiv |\uparrow - - - \downarrow\rangle$, the (2,2) active space comprises two π-orbitals and include four singly and doubly exited reference determinants generated by the $(1 + \widehat{C}_1 + \widehat{C}_2)$ operator acting on $|0\rangle$. For the zero value of the S_z quantum number, the active space CAS(2,2) is:

$$(1 + \widehat{C}_1 + \widehat{C}_2)|0\rangle \rightarrow |\uparrow - - - \downarrow\rangle \pm |- - \downarrow \uparrow -\rangle \pm |\uparrow \downarrow$$
$$- -\rangle \pm |- - \uparrow \downarrow\rangle.$$

At the equilibrium distance, besides the configurations that include the $2\pi_2$ orbitals, the only configurations that contribute more noticeably to the wave function $|\psi(X^3\Sigma_g^-)_{R_e}\rangle$, are the two configurations in (3.54) which include the $1\pi_x$ and $1\pi_y$ orbitals. Thus, to increase the accuracy of the calculation, we add the $1\pi_x$ and $1\pi_y$ orbital to the active space. With this addition, the active space becomes (4,6)

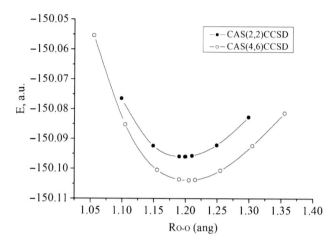

Fig. 3.10 CAS(2,2)CCSD/cc-pVTZ and CAS(4,6)CCSD/cc-pVTZ ground state PECes of O_2 molecule

and contains four π-orbitals. The corresponding reference space contains eight determinants, which are generated by acting with $(1 + \widehat{C}_1 + \widehat{C}_2)$ on $|0\rangle = |{}^{\uparrow--\downarrow}_{\uparrow\downarrow\uparrow\downarrow}\rangle$:

$$(1 + \widehat{C}_1 + \widehat{C}_2)|0\rangle \rightarrow$$

$$|{}^{\uparrow--\downarrow}_{\uparrow\downarrow\uparrow\downarrow}\rangle \pm |{}^{-\downarrow\uparrow-}_{\uparrow\downarrow\uparrow\downarrow}\rangle \pm |{}^{\uparrow\downarrow--}_{\uparrow\downarrow\uparrow\downarrow}\rangle \pm |{}^{-\uparrow\downarrow}_{\uparrow\downarrow\uparrow\downarrow}\rangle \pm |{}^{\uparrow\downarrow\uparrow\downarrow}_{\uparrow--\downarrow}\rangle \pm |{}^{\uparrow\downarrow\uparrow\downarrow}_{-\downarrow\uparrow-}\rangle \pm |{}^{\uparrow\downarrow\uparrow\downarrow}_{\uparrow\downarrow--}\rangle \pm |{}^{\uparrow\downarrow\uparrow\downarrow}_{--\uparrow\downarrow}\rangle.$$

Thus the CAS(4,6)CCSD wave function with the $|{}^{\uparrow--\downarrow}_{\uparrow\downarrow\uparrow\downarrow}\rangle$ reference is:

$$|\psi_{CAS(4,6)CCSD}(X^3\Sigma_g^-)_{R_e}\rangle \approx 1|{}^{\uparrow--\downarrow}_{\uparrow\downarrow\uparrow\downarrow}\rangle + 0.13|{}^{\uparrow\downarrow\uparrow\downarrow}_{\uparrow--\downarrow}\rangle + 0.13|{}^{\uparrow\downarrow\uparrow\downarrow}_{-\downarrow\uparrow-}\rangle + ...$$

In the above we only explicitly show the dominant configurations contributing to the wave function. Numerous more configurations are generated by the cluster operator $\exp(\widehat{T}^{(ext)})$.

The enlarging of the active space has a significant effect on the ground state $(|\psi(X^3\Sigma_g^-)_{R_e}\rangle)$ PEC calculation. The energy at the equilibrium decreased by 0.008 au (or $1,750\,cm^{-1}$; see Fig. 3.10). In Table 3.3 we present the calculated vibration frequencies and the energy of the first vibrational level. The results are compared with those obtained using the method developed by Nooijen [63] and with the experiment values. The experiment value $G(\nu_0)$ is calculated from O_2 experiment values of energy dissociation $D_0 = 41,260\,cm^{-1}$ and $D_e = 42,047\,cm^{-1}$ [64].

Lithium hydride (LiH). For the singly-bonded LiH molecule multi-configuration effects appear close to the dissociation region and result from the degeneracy between the 2σ bonding orbital and the 3σ antibonding orbital. A ground state Hartree-Fock calculation at the LiH equilibrium distance gives the following energy ordering of the molecular orbitals: $1\sigma^2 2\sigma^2 3\sigma^0$.

Table 3.3 Some spectroscopic parameters calculated for the ground electronic state ($X^3\Sigma_g^-$) of the O_2 molecule with different methods including our CAS(n,m)CCSD method. The calculations were performed in the cc-pVTZ basis set

	E_{min}, au	ω_e, cm^{-1}	R_e, Å	$G(\nu_0)$, cm^{-1}
CAS(2,2)CCSD	−150.096015	1,688	1.199	880.7
CAS(4,6)CCSD	−150.103875	1,578	1.205	799.8
CSS-EOM-CCSD	–	1,661	1.199	–
DIP-EOM-CCSD	–	1,634	1.202	–
DIP-EOM-CCSD	–	1,606	1.207	–
RSS-EOM-CCSD	–	1,629	1.205	–
Experiment	–	1,580.19	1.207	(787)

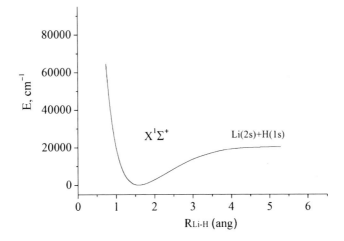

Fig. 3.11 ^{7}LiH CASCCSD/aug-cc-pVQZ ground state energy curve

The reference space in our CASCCSD calculations was taken from a CAS-CISD calculation. The correlation-consistent aug-cc-pVQZ basis-set has been used [65, 66]. The ground-state CASCISD wave function at the equilibrium is:

$$|\psi(X^1\Sigma^+)_{R_e}\rangle \approx 0.98|2\sigma^2\rangle - 0.122|2\sigma^0 3\sigma^2\rangle$$
$$+0.017|2\sigma^1 3\overline{\sigma}^1\rangle + 0.017|2\overline{\sigma}^1 3\sigma^1\rangle + \ldots . \quad (3.55)$$

This shows that the wave function near the equilibrium geometry is dominated by a single closed-shell configuration (see Fig. 3.11). However, as the internuclear distance increases and approaches ~ 4.2 Å, the ground state becomes dominated by a singlet-coupled pair of singly excited open-shell determinants (see Fig. 3.12). In Table 3.4 we compare our calculated CASCCSD spectroscopic data with the 4R-RMR and CCSD results of Li and Paldus [67] and with the experimental IPA results [68]. As one can see, while our results are similar to the results of Li and Paldus

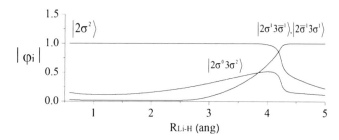

Fig. 3.12 Weights of the reference determinants in the CASCCSD/aug-cc-pVQZ wave function of the ground-state ($X^1\Sigma^+$) state of the LiH molecule

Table 3.4 Experimental IPA vibrational energies in cm^{-1} and the CCSD, 4R-RMR, and CAS(2,2)CCSD absolute deviations from those energies for LiH ground $X^1\Sigma^+$ state (aug-cc-pVQZ basis set)

ν	Exp. (IPA)	CCSD	4R-RMR	CAS(2,2)CCSD
0	697.88	−6.6	−6.7	0.82
1	2,057.59	−14.8	−15.1	9.79
2	3,372.48	−23.5	−23.8	19.14
3	4,643.37	−25.8	−26.3	23.48
4	5,871.14	−22.4	−23.2	22.93
5	7,056.58	−15.0	−16.0	18.41
6	8,200.35	−4.9	−6.1	11.72
7	9,302.95	7.2	5.6	3.79
8	10,364.73	20.5	18.5	−4.61
9	11,385.90	34.6	32.1	−12.96
10	12,366.42	49.2	46.2	−21.26
11	13,306.04	64.2	60.4	−29.87
12	14,204.13	79.3	74.6	−38.89
13	15,059.61	94.5	88.6	−47.02
14	15,870.80	109.7	102.3	−51.61
15	16,635.24	125.1	115.6	−51.47
16	17,349.46	140.7	128.2	−50.42
17	18,008.73	156.5	140.0	−56.23
18	18,606.62	172.8	150.4	−74.46
19	19,134.53	190.0	158.8	−89.21
SD	–	90.44	80.83	39.88

for lower vibrational states, they are significantly better for higher states. In both, our and Li and Paldus calculations, the **LEVEL** package [69] was use to determine the vibrational energies. The systematic increase of the absolute error of the CCSD results, which can be seen in Table 3.4, indicates that the CCSD PEC does not have an adequate accuracy at larger internuclear distance. In contrary, the error in the CAS(2,2)CCSD vibrational energies does not vary much from state to state and is nearly constant for all states in comparison with the IPA experimental values [68]. Also CAS(2,2)CCSD approach performs noticeably better than 4R-RMR.

Table 3.5 Spectroscopic constants ($J = 0$) in cm^{-1} for the LiH ground $X^1\Sigma^+$ state. Experimental values Y_{ij} are obtained from the IPA data. Calculations performed with the aug-cc-pVQZ basis set

$G(v)$	Y_{10}	Y_{20}	Y_{30}	Y_{40}
	$n(v) = 4$			
CAS(2,2)CCSD	1,411.3	−36.42	4.44	−0.42
exp. IPA	1,408.53	−26.17	1.30	−0.14
	$n(v) > 4$			
CAS(2,2)CCSD	1,395.5	−22.24	0.27	−0.011
exp. IPA	1,412.35	−25.89	0.50	−0.016

Table 3.6 LiH dissociation energies and internuclear distances (aug-cc-pVQZ basis set)

Method	D_e, cm^{-1}	D_0, cm^{-1}	R_e, Å
		(ground state) $X^1\Sigma^+$	
CAS(2,2)CCSD	20,314.8	19,617.8	1.588
Experiment	20,287.7 ± 0.3	19,589.8 ± 0.3	1.5956
		(excited state) $A^1\Sigma^+$	
CAS(2,2)CCSD	7,995.6	7,846.2	2.415
Experiment	8,681.6 ± 0.3	8,550.3 ± 0.3	2.5963

The standard deviation in the vibrational energies is more than two times larger for 4R-RMR than for CAS(2,2)CCSD.

The Dunham coefficient for the LiH ground state are reported in Table 3.5. One notices strong dependency of the fundamental vibration frequency and the anharmonicity constant on the quality of the PEC (either calculated or experimentally derived) used in determining them. Especially the results obtained using the experimentally derived PEC strongly depend on the number of the vibrational levels used. It is well known, that the fundamental vibrational frequency depends on the energy of the lowest vibrational level. So the errors in the $\omega_e = Y_{10}$ value for the numbers of levels equal four and greater than four are fairly constant. The first order anharmonicity coefficient, $\omega_e \chi_e = -Y_{20}$, and higher order coefficients depend on the whole PEC and their values vary a lot with the number of the vibrational levels used in calculating Y_{20}. But, as increasingly higher vibrational levels are used ($v > 4$), the differences become smaller and agree better with the experimental results.

The CASCCSD dissociation energy obtained for LiH in the ground ($X^1\Sigma^+$) and excited ($A^1\Sigma^+$) electronic states are reported in Table 3.6. A comparison is made with the recommended values found in [62,68]. D_e was calculated as the difference between the energy at equilibrium (R_e) and the energy at $5R_e$, $E_{CASCCSD}(R_e) - E_{CASCCSD}(5R_e)$. The D_0 value was calculated as: $D_0 = D_e - G(0)$.

Boron hydride (BH). A ground state Hartree-Fock calculation of the BH molecule performed at the equilibrium distance gives the following energy ordering of the

Table 3.7 Spectroscopic constants of BH in the ground electronic state, in cm^{-1}, with different functions used to fit the calculated CASCCSD/aug-cc-pVDZ PEC

Potential function $N_{\alpha_m,\beta_m} = 6$	R_e, Å	ω_e	$\omega_e\chi_e$	$\bar{D}, 10^{-3}$	B_e	α_e	D_e, eV
Lennard-Jones	1.2541	2,343.5	49.26	1.140	11.61	0.445	3.45
Generalized Morse	1.2545	2,346.8	50.75	1.134	11.60	0.412	3.47
James-Coolidge-Vernon	1.2548	2,346.1	49.30	1.134	11.60	0.401	3.44
Numerically	1.2550	2,348.9	43.20	1.127	11.58	0.396	–
FCI [70]	1.2559	2,340.0	48.80	1.100	11.57	0.396	–
Experiment	1.2324	2,366.9	49.39	1.242	12.02	0.412	3.57±0.40

molecular orbitals: $1\sigma^2 2\sigma^2 3\sigma^2 1\pi_x^0 1\pi_y^0 4\sigma^0$. The reference space for the CASCCSD calculations of BH was taken from a CASCISD calculation. The correlation-consistent aug-cc-pVDZ basis-set [65, 66] has been used. A CAS(2,2)CISD calculation performed with the CASSCF orbitals shows that the BH ground-state wave function near the equilibrium geometry is dominated by a single closed-shell configuration:

$$|\psi(X^1\Sigma^+)_{R_e}\rangle \approx 0.95|3\sigma^2\rangle - 0.13|3\sigma^0 4\sigma^2\rangle$$
$$+0.0003|3\sigma^1 4\bar{\sigma}^1\rangle + 0.0003|3\bar{\sigma}^1 4\sigma^1\rangle + ... \qquad (3.56)$$

Table 3.7 presents the results of the CAS(2,2)CCSD calculations of the spectroscopic parameters for the ground state of BH. The calculated PEC was fitted the generalized Morse function mentioned above. For comparison, we also used the James-Coolidge-Vernon and Lennard-Jones potential functions containing the same number of the fitting parameters, α and β, as the generalized Morse function. In the comparison we also show the results obtained using numerical differentiation of the calculated PEC, and the experimental results. The data shown in Table 3.7 exhibit some discrepancies caused by both, the form of the fitting function and the methods used to calculate the derivatives (analytically for the analytical fits or numerically using the five-point procedure). However, regardless of the fit the CASCCSD results agree very well with the FCI results.

Figure 3.13 presents the ground $X^1\Sigma^+$ state PEC of BH and a curve showing the accuracy of the generalized Morse function fit calculated as the difference between the PEC points and the fit ($\Delta U(R)$). One notices that the largest deviation reaches about $500\,cm^{-1}$. Such a deviation can introduce a substantial error in the calculation of the energies of the vibrational levels (see Table 3.8). More detailed description of the calculations of Boron hydride molecule can be found in [71].

Fluorine hydride (FH). The FH ground state $(X^1\Sigma^+)$ PEC calculations have been performed in the range of the internuclear distances $R \in [0.64, 4.0]$ Å. The correlation-consistent aug-cc-pVTZ basis set [65, 66] has been used in the calculations.

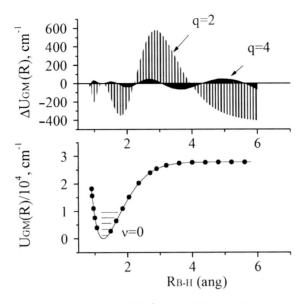

Fig. 3.13 CASCCSD/cc-pVDZ ground $X^1\Sigma^+$ state PEC of BH and the absolute error $(\Delta U_{GM}(R))$ between the generalized Morse fit and the CASCCSD energies

Table 3.8 Experimental vibrational energies compared with the energies obtained from the CAS(2,2)CCSD/aug-cc-pVDZ PEC fitted with different generalized Morse (GM) potential functions and without using a fit (numerically), in cm^{-1}

v	$U_{GM}(R)$ $q = 2$	$U_{GM}(R)$ $q = 3$	$U_{GM}(R)$ $q = 4$	Numerically	$G(v)$ Experiment
0	1,192.2 (21.2)	1,185.3 (14.2)	1,183.8 (12.7)	1,162.5 (−8.6)	1,171.1
1	3,504.3 (63.9)	3,470.4 (30.0)	3,474.0 (33.6)	3,412.6 (−27.8)	3,440.4
2	5,726.8 (112.6)	5,634.6 (20.4)	5,651.1 (36.9)	5,569.3 (−44.9)	5,614.2
3	7,868.1 (173.4)	7,711.9 (17.2)	7,710.7 (16.0)	7,637.7 (−57.0)	7,694.7
4	9,939.4 (255.4)	9,772.8 (88.8)	9,722.9 (38.9)	9,622.8 (−61.2)	9,684.0

The reference space for the CASCCSD calculations was taken from a CASSCF calculation. The ground-state Hartree-Fock calculation at the FH equilibrium distance gives the following ordering of the molecular orbitals: $1_\sigma^2 2\sigma^2 3\sigma^2 1\pi_x^2 1\pi_y^2 4\sigma^0$. From the analysis of the orbital occupation numbers, the bonding 3σ and antibonding 4σ orbitals were chosen as the minimal active space. The CAS(2,2)CISD calculations performed with the CASSCF orbitals showed that the FH ground-state wave function near the equilibrium geometry is dominated by the single closed-shell configuration:

$$|\psi(X^1\Sigma^+)_{R_e}\rangle \approx 0.97|3\sigma^2\rangle + 0.024|3\sigma^1 4\bar{\sigma}^1\rangle$$
$$-0.024|3\bar{\sigma}^1 4\sigma^1\rangle + 0.013|3\sigma^0 4\sigma^2\rangle. \tag{3.57}$$

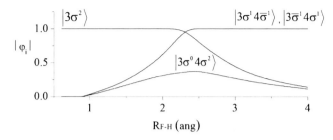

Fig. 3.14 Weights of the reference determinants in the CASCCSD wave function of the ground $X^1\Sigma^+$ state of the FH molecule as a function of the internuclear distance

Table 3.9 Calculated and experimental vibrational energies, $G(v)$, in cm^{-1} for the FH molecule in the ground state. Differences with respect to the experimental RKR values are shown in parentheses. The correlation-consistent aug-cc-pVTZ basis set has been used in the calculations

Gv	RKR	CCSD	CAS(2,2)CISD[+Q]	CAS(2,2)CCSD
0	2,050.761	2,066.4 (15.6)	2,036.8 (-14)	2,042.8 (-8.0)
1	6,012.184	6,066.9 (54.7)	5,974.9 (-37.3)	5,993.9 (-18.3)
2	9,801.555	9,903.5 (101.9)	9,745.6 (-55.9)	9,777.7 (-23.9)
3	13,423.565	13,578.6 (155.0)	13,353.1 (-70.5)	13,396.7 (-26.9)
4	16,882.403	17,095.2 (212.8)	16,802.4 (-80.0)	16,854.9 (-27.5)
5	20,181.700	20,456.8 (275.1)	20,098.7 (-83.0)	20,156.9 (-24.8)
6	23,324.465	23,667.1 (342.7)	23,246.3 (78.1)	23,306.9 (-17.5)
7	26,312.990	26,729.5 (416.5)	26,247.9 (-65.1)	26,307.7 (-5.3)
8	29,148.739	29,646.8 (498.1)	29,103.8 (-44.9)	29,159.9 (-11.3)
9	31,832.203	32,421.4 (589.2)	31,811.9 (-20.3)	31,862.2 (30.0)
10	34,362.711	35,054.7 (691.9)	34,367.6 (-4.9)	34,410.4 (47.7)
SD	–	372.2	57.0	24.7

However, as the internuclear distance increases and approaches $\sim$2.6 Å, the wave function becomes dominated by the following singlet-coupled pair of singly excited open-shell determinants $|3\sigma^1 4\overline{\sigma}^1\rangle - |3\overline{\sigma}^1 4\sigma^1\rangle$ (see Fig. 3.14). To circumvent the possible PEC discontinuity problem in the CASCCSD calculations we have used the same formal reference determinant, the $|3\sigma^2\rangle$ determinant, for all PEC points. Such an approach has generated sufficiently accurate energies to calculate the low-lying vibrational states for FH in the ground state as it can be seen in Table 3.9. As mentioned, the FH ground-state wave function near the equilibrium geometry is dominated by the $|3\sigma^2\rangle$ configuration. This makes the single-reference CCSD method sufficiently accurate to describe this state at this geometry. Indeed, as can be seen in Fig. 3.15, the CCSD and CAS(2,2)CISD results are very similar in this region. However, as the FH bond stretches, the CCSD method starts to show significant differences (see Fig. 3.15) with respect to other more accurate methods.

To evaluate the accuracy of the CAS(2,2)CCSD and CAS(2,2)CISD[+Q] PECes we compared them with the PEC reconstructed from the experimental data us-

Fig. 3.15 The potential energy curve of the FH molecule in the ground state obtained with the CASCCSD, CASCISD, CCSD and CASCISD[+Q] methods (aug-cc-pVTZ basis set) (Reproduced from Ref. [72] with kind permission of Taylor and Francis)

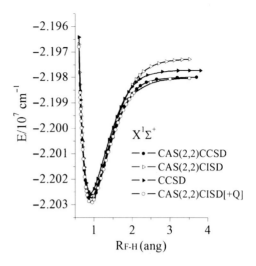

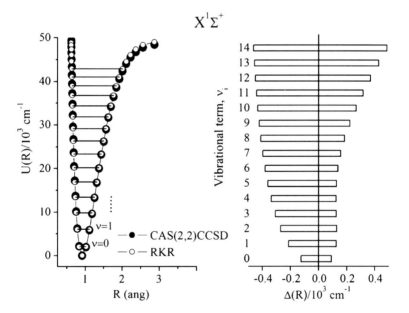

Fig. 3.16 Potential energy functions for the ground state of the FH molecule determined using the CAS(2,2)CCSD/aug-cc-pVTZ and RKR energies. ν_i denotes the vibrational quantum number. Also shown are the errors, $\Delta(R)$, determined with respect to RKR energies (Reproduced from Ref. [72] with kind permission of Taylor and Francis)

ing the Rydberg-Klein-Reese (RKR) approach [73]. The comparison shown in Figs. 3.16 and 3.17 includes curves representing fits obtained with the generalized Morse functions (3.51) for the PECes generated in the calculations. To make the comparison more direct, all plots are renormalized by setting the energy value at

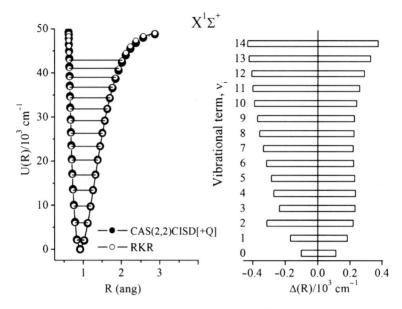

Fig. 3.17 Potential functions for ground state of FH molecule determined with CAS(2,2)CISD[+Q]/aug-cc-pVTZ and RKR energies. v_i denotes the number of terms. Also shown are the errors $\Delta(R)$ with respect to the differences between CASCISD[+Q] and RKR energies (Reproduced from Ref. [72] with kind permission of Taylor and Francis)

the minimum to zero for each curve. The figures include histograms showing the error determined with respect to the RKR PEC, $\Delta(R)$, calculated as: $\Delta(R) = [V(R_{RKR}) - V(R_{method})]$.

Figures 3.16 and 3.17 show that the errors in the ground state calculations using the CAS(2,2)CCSD and CAS(2,2)CISD[+Q] methods with respect to the RKR energies are relatively small, particularly at the energy values of low lying vibrational states ($v \leq 10$). The error does not exceed 1%. The "parallelism" indices calculated for the two methods as:

$$I_p = \int_{R_{min}}^{R_{max}} \Delta^2(R) dR. \tag{3.58}$$

demonstrates high accuracy of our method (see Table 3.10).

For the vibrational frequencies, the CAS(2,2)CCSD method performs noticeably better than the CAS(2,2)CISD[+Q]. This can be seen by examining the standard deviations shown in Table 3.9. In Table 3.11 some selected spectroscopic constants calculated for FH in the ground electronic state ($X^1\Sigma^+$) are shown. The results are compared with the experimental values taken from Huber and Herzberg [64], with the exception of the dissociation energy, D_e, which was taken from Lonardo and Douglas [74]. To calculate the spectroscopic constants we used the numerical differentiation formulas. As one can expect, the values of the spectroscopic constants

Table 3.10 Index of "parallelism" in cm^{-1} determined with respect to the RKR spectroscopic curve

Method	I_p
CAS(2,2)CCSD	0.716
CAS(2,2)CISD[+Q]	0.658
CCSD	14.253

Table 3.11 Calculated and experimental molecular constants the FH molecule in the ground $X^1\Sigma^+$ state. The calculations were performed in the aug-cc-pVTZ basis set

	Experiment[64, 74]	CCSD	CAS(2,2)CISD[+Q]	CAS(2,2)CCSD
D_e, eV	6.12	6.85	6.06	6.07
D_0, eV	5.869	6.643	5.807	5.789
R_e, Å	0.9168	0.9168	0.9194	0.9194
B_e, cm^{-1}	20.96	20.96	20.84	20.84
ω_e, cm^{-1}	4,138.32	4,163.1	4,127.4	4,138.7
$\omega_e\chi_e$, cm^{-1}	89.88	82.74	92.11	89.96
α_e, cm^{-1}	0.798	0.755	0.805	0.791
$\overline{D}_e/10^{-3}$, cm^{-1}	2.15	2.12	2.12	2.11

calculated for the ground state are quite similar for the CCSD and CAS(2,2)CCSD methods. However, the dissociation energy, which strongly depends on the quality of the energy calculated at a large internuclear distance shows significant difference between the CCSD and CAS(2,2)CCSD calculations. A more detailed description of the FH calculations can be found in Ref. [72].

3.6 Conclusions

In this article we have reviewed our state-specific coupled-cluster approach with the CASSCF reference for calculating ground and excited states of molecular systems. The computer program **CLUSTER** we have developed enables automated generation of the CC diagrams, the CC energy and amplitude equations, and the corresponding computer code for solving those equations. This can be done for any version of the CC method including CASCCSD. The approach has been implemented and used to perform several model calculations of some small molecular systems. The results clearly demonstrate an excellent performance of the CASCCSD method in various situations where configurational quasidegeneracy occurs. It is shown that the method enables very accurate and reliable ab initio determination of key spectroscopic constants. The comparison of CASCCSD with FCI demonstrate that the CASCCSD error does not exceed the range of the chemical accuracy which is 10^{-3} au. When a moderate-size basis set is used in the calculations, the errors for individual vibrational levels are in the range of 10–30 cm^{-1} in comparison with the experiment. The discrepancy between the experimental and computed vibrational energies do not increase with the increasing vibrational quantum number, as it does for some other methods. In general, the CASCCSD calculated rovibrational spectra match very well the experimental ones.

Thus we can conclude that the CASCCSD approach is clearly superior to other CC methods. In particular, it represents a significant improvement over the standard single reference CCSD approaches. It also favorably compares with other multireference CC theories and the EOMCC approach.

References

1. Schleyer PvR et al (ed) (2004) Encyclopedia of computational chemistry. Wiley, Chichester
2. Wilson S (1984) Electron correlation in molecules. Clarendon, Oxford
3. Head-Gordon M (1996) J Phys Chem 100:13213
4. Knowles P, Schütz M, Werner HJ (2000) *Ab initio* methods for electron correlation in molecules. In: Grotendorst J (ed) Modern methods and algorithms of quantum chemistry, proceedings, vol 3. John von Neumann institute for computing, Jülich, p 97
5. Coester F (1958) Nucl Phys 7:421
6. Kummel H (1960) Nucl Phys 17:477
7. Čížek J (1966) J Chem Phys 45(11):4256
8. Paldus J, Čížek J, Shavitt I (1972) Phys Rev A5:50
9. Purvis III GD, Bartlett RJ (1982) J Chem Phys 76(4):1910
10. Li X, Paldus J (1994) J Chem Phys 101:8812
11. Bartlett RJ (1981) Ann Rev Phys Chem 32:359
12. Bartlett RJ, Stanton JF (1994) In: Lipkowitz KB, Boyd DB (eds) Reviews in computational chemistry, vol 5. Wiley, New York, p 65
13. Bartlett RJ, Musiał M (2007) Rev Mod Phys 79:291
14. Oliphant N, Adamowicz L, (1991) J Chem Phys 94(2):1229
15. Oliphant N, Adamowicz L (1992) J Chem Phys 96:3739
16. Oliphant N, Adamowicz L (1993) Int Rev Phys Chem 12:339
17. Piecuch P, Oliphant N, Adamowicz L (1993) J Chem Phys 99(3):1875
18. Piecuch P, Adamowicz L (1994) J Chem Phys 100(8):5792
19. Ivanov VV, Adamowicz L (2000) J Chem Phys 112:9258
20. Ivanov VV, Adamowicz L (2000) J Chem Phys 113:8503
21. Ivanov VV, Lyakh DI, Adamowicz L (2005) Collect Czech Chem Commun 70:1017
22. Ivanov VV, Lyakh DI, Adamowicz L (2005) Mol Phys 103:2131
23. Ivanov VV, Adamowicz L, (2006) Lyakh DI Int J Quant Chem 106(14):2875
24. Ivanov VV, Adamowicz L, Lyakh DI (2006) J Chem Phys 124(18):184302
25. Ivanov VV, Lyakh DI (2002) Kharkiv Univ Bull Chem Ser 549:15
26. Lyakh DI, Ivanov VV, Adamowicz L (2003) Kharkiv Univ Bull Chem Ser 596:9
27. Lyakh DI, Ivanov VV, Adamowicz L (2005) J Chem Phys 122:024108
28. Mahapatra US, Datta B, Bandyopadhyay B, Mukherjee D (1998) State–specific muli–reference coupled cluster formulations: two paradigms. In: Löwdin PO (ed) Advances in Quantum Chemistry, vol 30. Academic, San Diego, p 163
29. Mahapatra US, Datta B, Mukherjee D (1998) Mol Phys 94:157
30. Mahapatra US, Datta B, Mukherjee D (1999) J Chem Phys 110:6171
31. Laidig WD, Bartlett RJ (1984) Chem Phys Lett 104(5):424
32. Laidig WD, Saxe P, Bartlett RJ (1987) J Chem Phys 86(2):887
33. Lyakh DI, Ivanov VV, Adamowicz L (2007) Mol Phys 105:1335
34. Lyakh DI, Ivanov VV, Adamowicz L (2008) J Chem Phys 128:074101
35. Ivanov VV, Lyakh DI, Adamowicz L (2009) Phys Chem Chem Phys 11:2355

36. Schmidt MW, Baldridge KK, Boatz JA, Elbert ST, Gordon MS, Jensen JH, Koseki S, Mastunada N, Nguyen KA, Su S, Windus TL, Dupuis M, Montgomery JA (1993) J Comput Chem 14:1347
37. Lyakh DI, Ivanov VV, Adamowicz L (2006) Theor Chem Acc 116:427
38. Crawford TD, Schaefer-III HF (2000) In: Lipkowitz KB, Boyd DB (eds) Reviews in computational chemistry, vol 14. Wiley, New York, p 33
39. Zaitsevsky AV (1993) Methods of many-body theory in quantum chemistry. Moscow University Press, Moscow (in Russian)
40. Harris FE, Monkhorst HJ, Freeman DL (1992) Algebraic and diagrammatic methods in many-fermion theory. Oxford University Press, New York
41. Paldus J, Wong HC (1973) Comput Phys Commun 6:1
42. Janssen CL, Schaefer III HF (1991) Theor Chem Acc 79:1
43. Li X, Paldus J (1994) J Chem Phy 101:8812
44. Harris FE (1999) Int J Quantum Chem 75:593
45. Kállay M, Surján P (2001) J Chem Phys 115:2945
46. Kállay M, Szalay PG, Surján PR (2002) J Chem Phys 117:980
47. Hirata S (2003) J Phys Chem A107:9887
48. Bochevarov AD, Sherrill CD (2004) J Chem Phys 121(8):3374
49. Hanrath M, Engels-Putzka A (2010) J Chem Phys 133:064108
50. Scheiner AC, Scuseria GE, Rice JE, Lee TJ, Schaefer HF (1987) J Chem Phys 87:5361
51. Musiał M, Kucharski SA, Bartlett RJ (2002) Mol Phys 100:1867
52. Piecuch P, Kucharski SA, Kowalski K, Musiał M (2002) Comput Phys Commun 149:71
53. Geertsen J, Rittby M, Bartlett RJ (1989) Chem Phys Lett 164:57
54. Sekino H, Bartlett RJ (1984) Int J Quantum Chem Symp 18:255
55. Kowalski K, Piecuch P (2004) J Chem Phys 120:1715
56. Ivanic J (2003) J Chem Phys 119:9364
57. Levine IN (1975) Molecular Spectroscopy. Wiley, New York, p 491
58. Rydberg R (1931) Z Phys 73:373
59. Klein O (1932) Z Phys 76:226
60. Rees ALG (1947) Proc Phys Soc 59:998
61. Kosman WM, Hinze J (1975) J Phys Chem Ref Data 56:93
62. Stwalley WC, Zemke WT (1993) J Phys Chem Ref Data 22(1):87
63. Nooijen M (2002) Int J Mol Sci 3:656
64. Huber KP, Herzberg G (1979) Molecular spectra and molecular structure. IV: constants of diatomic molecules. Van Nostrand and Reinhold, New York
65. Dunning TH (1989) J Chem Phys 90:1007
66. Kendall R, Dunning T Jr, Harrison R (1992) J Chem Phys 96:6796
67. Li X, Paldus J (2003) J Chem Phys 118:2470
68. Stwalley WC, Zemke WT, Yang SC (1991) J Phys Chem Ref Data 20:153
69. Le Roy RJ (2007) Level 8.0: A Computer program for solving the radial Schrodinger Equation for Bound and Quasibound Levels, University of Waterloo Chemical Physics research report CP-663. See http://leroy.uwaterloo.ca/programs/
70. Abrams ML, Sherrill CD (2003) J Chem Phys 118:1604
71. Klimenko TA, Ivanov VV, Lyakh DI (2010) Ukr J Phys 55(6):657
72. Klimenko TA, Ivanov VV, Lyakh DI, Adamowicz L (2010) Chem Phys Lett 493:173
73. Coxon JA, Hajigeorgiou PG (2006) J Phys Chem A 110:6261
74. Di-Lonardo G, Douglas AE (1973) Can J Phys 51:434

Chapter 4
Relativistic Effects in Chemistry and a Two-Component Theory

Maria Barysz

Abstract In this chapter I demonstrate a series of examples showing the importance of relativistic quantum chemistry to the proper description of variety of molecular and atomic properties including valence and core ionization potentials, electron affinities, chemical reactions, dissociation energies, spectroscopic parameters and other properties. An overview of basic principles of the relativistic quantum chemistry and the reduction of relativistic quantum chemistry to two–component form is also presented. I discuss the transition of the four–component Dirac theory to the infinite–order two–component (IOTC) formalism through the unitary transformation which decouples exactly the Hamiltonian.

Keywords Relativistic effects • Infinite–order Two–component theory • Change of picture • Back unitary transformation • Ionization potentials • Electron affinities • Dissociation energies • Hydrolysis • Electric properties • CEBE • Strong bonds to gold

4.1 Introduction

The two basic theories of modern physics are the theory of relativity and quantum mechanics. For a long time the idea of bringing the theory of relativity into chemistry was considered to be unnecessary. Although the basic equation of relativistic quantum mechanics was presented shortly after the nonrelativistic theory was completed, its use in quantum chemistry is relatively recent. Dirac himself was quite sceptical considering it rather unlikely that his equation may be of any use for chemistry. While today it is generally recognized that a rigorous theoretical

M. Barysz (✉)
Institute of Chemistry, N. Copernicus University, Gagarina 7, Toruń, 87 100, Poland
e-mail: teomjb@chem.uni.torun.pl

J. Leszczynski and M.K. Shukla (eds.), *Practical Aspects of Computational Chemistry I:* 103
An Overview of the Last Two Decades and Current Trends,
DOI 10.1007/978-94-007-0919-5_4, © Springer Science+Business Media B.V. 2012

foundation for chemistry would rest on relativistic quantum theory, the great majority of quantum mechanical calculations of quantities of chemical interest are still based on the nonrelativistic Schrödinger equation, or its equivalent. However, the research and development in the relativistic filed have increased greatly over the decades.

The relativistic effects are important for both light and heavy elements. For very precise calculations, while searching the limit of accuracy of quantum mechanics or quantum electrodynamics the relativistic energy contributions are already needed for H or He atoms. For heavy elements, relativistic effects are important in atomic and in chemical calculations when one search for a chemical accuracy of about 0.1 eV.

Semi-quantitative arguments on the role of relativistic effects in the chemical applications can be already presented within the very simply model. One of the principle axiom of the theory of relativity is the assumption that the speed of light is the finite and constant number ($c \approx 137.036$ au) in all inertial frames. The quantity which characterizes the limit of validity of classical mechanics is the ratio of velocity of particle to the velocity of light:

$$\beta = \frac{v}{c}. \tag{4.1}$$

Similar quantitative estimation of the role of the relativistic effects can be done on the basis of the quantum mechanics by the substitution of the velocity v by its expectation value in the given state. With the virial theorem one obtains for the hydrogen-like ions with the atomic number Z:

$$\beta = \frac{Z}{nc}. \tag{4.2}$$

where n is the principal quantum number. It is evident from Eq. 4.2 that the inner shells will be the most important. Similarly one may expect that in the heavy many–electron elements the core electrons will be mostly modified by the relativistic effects.

The relativistic effects are responsible for radial contraction and energetic stabilization of the $s_{1/2}$ and $p_{1/2}$ shells and for the spin–orbit splitting of shells with $l > 0$ into sub-shells with $j = l - 1/2$ and $j = l + 1/2$. The indirect relativistic effects are consequences of a screening of the d and f electrons by inner electrons occupying the contracted $s_{1/2}$ and $p_{1/2}$ orbitals and lead to the radial expansion of the d and f shells accompanied by their energetic destabilization [1,2].

The relativistic stabilization and splitting of hydrogen–like systems were already found by Sommerfeld [3] using the old quantum theory. The result, incidentally, fully agrees with that for the Dirac equation [4].

Throughout the Periodic Table there are irregularities which were difficult to explain without the relativistic quantum mechanics and chemistry. The unusual properties of gold and mercury are one of the most spectacular manifestation of relativity.

Table 4.1 Nonrelativistic (NR), DKH2[a] relativistic (R) and experimental ionization potentials and electron affinities of the Group 11 and 12 elements[b]

Metal	CCSD(T) NR	CCSD(T) R	Experiment[c]
Ionization potential			
Cu[d]	7.555	7.781	7.733
Ag[d]	6.934	7.461	7.576
Au[d]	7.035	9.123	9.225
Zn[e]	9.167	9.363	9.3942
Cd[e]	8.351	8.851	8.9938
Hg[e]	8.326	10.285	10.4375
Electron affinity			
Cu[d]	1.165	1.236	1.235
Ag[d]	1.054	1.254	1.302
Au[d]	1.161	2.229	2.309

[a]Douglas–Kroll–Hess calculations
[b]All values in eV
[c]Ref. [5]
[d]Ref. [6]
[e]Ref. [7]

It is well known that atomic ionization potentials vary within the Periodic Table and for a group of atoms characterized by the same valence electrons one expects that first ionization potentials will decrease down of the group. This means, the expected sequence of IP's should be $Cu > Ag > Au$ for the coinage elements and $Zn > Cd > Hg$ for the Group 12 elements. The nonrelativistic Schrödinger calculations confirm this expectations. However, on the contrary to our expectations, experimental data show that IPs of gold and mercury are the largest within the group. This observation can be explained by the relativistic theory, only. Similar anomalies are observed for the electron affinities. The numerical data are shown in the Table 4.1.

Many examples of relativistic effects in chemistry have been published in a number of papers. Pekka Pyykkö has published a complete data base of references in a series of books [8–10] which is also available on-line on the Internet [11].

The nature is always relativistic. Non-relativistic objects do not exists. The only way to see the role of relativistic effects in chemistry is through the comparison between relativistic and nonrelativistic theory and results. The values of the relativistic effects for a given quantity X are then usually calculated as a difference between the relativistic and nonrelativistic values of this quantity, i.e. $\Delta X = X(rel) - X(nrel)$ calculated at the same level of theory. Therefore, from both fundamental and practical points of view there is a necessity for relativistic quantum chemistry; theory and applications. While the Hamiltonian of a molecule is exactly known in nonrelativistic quantum mechanics, this is no longer the case for the relativistic formulation. Many relativistic Hamiltonians have been derived over the past decades. It goes beyond the goal of this chapter to review all of the theories. Details can be found in recent excellent reviews and books [12–15]. This chapter has two parts: The first part is an illustration of the role of the relativistic effects. Some selected examples of atomic and molecular relativistic calculations are shown and their comparison with the nonrelativistic data. In the second part, an overview is given over the recently defined, exact relativistic Infinite Order Two– Component approach (IOTC) and related phenomena [16–23]. The IOTC method seems to be a complete relativistic theory ready for the atomic and molecular calculations.

4.2   Relativistic Effects in Atoms and Molecules

4.2.1   Dissociation Energies and Strong Chemical Bonds to Gold

Energy is central to the understanding of chemistry. The potential energy surface of a molecule defines chemical structure and reactivity. A diatomic dissociation curve has two ends, one at the equilibrium distance R_e and another one at the dissociation limit $R \to \infty$. For determining a dissociation energy D_e both counts.

The role of the relativistic effects on the dissociation curve at R_e can be explain already on the level of the Pauli approximation [12]. Scalar Mass Velocity (MV) and Darwin (MD) corrections to the energy have opposite signs. The Mass Velocity is negatively defined while Darwin correction has always a positive value. The Mass Velocity correction lowers the energy while the Darwin correction increase the energy at the minimum of the dissociation curve. Two cases are possible, the absolute value of the Mass Velocity term can be higher or lower then the Darwin correction ($MV > MD$ or $MD > MV$). In the first case one observes the lowering of the minimum at the dissociation curve in the second case the increasing of the minimum at the dissociation curve.

Lowering or increasing values of energies can also be observed at the second end of the dissociation curve ($R \to \infty$). The example is the dissociation curve of the diatomic $AuCl$ structure which may lead to both covalent ($Au^0 + Cl^0$) or ionic ($Au^+ + Cl^-$) structures. The ionization will increase the energy at the dissociation limit ($R \to \infty$). The size of the increasing will depend of the value of the ionization potential of the gold atom. Since the nonrelativistic and relativistic values of the ionization potentials of gold differ significantly, the nonrelativistic and relativistic theory will predict different stability's of the diatomic structure $AuCl$.

A numerical example of the importance of the relativistic effects on the dissociation energies and spectroscopic constants can be illustrated on a series of the diatomic species of gold. Gold forms unusually strong chemical bonds due to relativistic effects. The two main effects are the stabilization of the $6s$ shell and the destabilization of the $5d$ shell. Depending on the system, both trends can lead to stronger bonding. The diatomic $AuBe^+$, AuC^+, $AuMg^+$, and $AuSi^+$ cations have been studied by Pykkö et al. [24]. It was found that three of the species,

Table 4.2 Calculated bond lengths, R_e (in pm), dissociation energies (in eV) and spectroscopic constants[a] (in cm^{-1})

Ion	Property	CASPT2 NR[b]	DKH CASPT2
$AuBe^+$	R_e	236.8	198.3
	D_e	1.253	3.636
	D_o	1.232	3.594
	ω_e	340.4	667.7
$AuC^{+,a}$	R_e	227.6	176.6
	D_e	1.489	5.075
	D_o	1.469	5.021
	ω_e	334.3	873.1
$AuMg^+$	R_e	264.7	233.5
	D_e	1.720	2.316
	D_o	1.707	2.294
	ω_e	208.4	359.6
$AuSi^+$	R_e	253.0	218.8
	D_e	2.808	3.688
	D_o	2.874	3.688
	ω_e	425.6	459.0

[a]Ref. [24, 24a]
[b]On the singlet surface, dissociating to $C^0(^1D) + Au^+(^1S) C^0(^1D)$
$+ Au^+(^1S)$

$AuBe^+$, AuC^+, and $AuSi^+$ have larger calculated dissociation energies than any previously studied gold species. The nonrelativistic results of the calculations did not show the similar trends. The calculated bond lengths, dissociation energies and spectroscopic constants are presented in the Table 4.2. The relativistic bond length and population analysis of AuC^+ suggested that it may approach the triple bond to gold [24].

There are also cases when relativistic effects make the diatomic bond weaker. The diatomic Tl_2 molecule was probably the first showcase molecule for a substantial relativistic bond weakening [25]. Most of this bond-weakening for Tl_2 can be attributed to spin–orbit interactions. Experimental studies have also confirmed the weakness of the bond in Tl_2. For the $^3\Sigma_g$ ground state the experimental D_o is about 0.43 eV [26].

4.2.2 Electric Properties

The electric properties are deduced from the response of the system to an electric external field **E**, which is related to the scalar potential ϕ by

$$\mathbf{E} = -\Delta\phi \qquad (4.3)$$

The simplest electric property of a neutral molecule is the dipole moment, which may be defined in terms of the derivative of the energy E at $\mathbf{E} = \mathbf{0}$ as

$$\mu = -\frac{dE}{d\mathbf{E}} \tag{4.4}$$

According to the Hellman–Feynman theorem the calculations of dipole moment can be performed as an expectation value of the Hamiltonian

$$\mu = -\left\langle \frac{dH}{d\mathbf{E}} \right\rangle = \langle e\mathbf{r} \rangle \tag{4.5}$$

To apply the method based on the expectation value we need a wave function which satisfies the Hellman–Feynman theorem. This is fulfilled for variational wave functions in which all parameters are fully optimized. Such calculations are possible for small molecules only. For the molecular case, generally applicable and straightforward is the so called finite field method. In the usual experimental situation a molecule is somehow placed in the external field, and the external field will influence the internal field. We can take this into account by expressing the energy of the system as a function of the filed. The finite filed method exploits the expansion of the energy with respect to an electric filed $\mathbf{E}$ (or for some properties, also other external fields) along the i-direction,

$$E(\mathbf{E}) = E_0 - \mu_i \mathbf{E}_\mathbf{i}^{\text{ext}} - \frac{1}{2}\alpha_{ij} \mathbf{E}_\mathbf{i}^{\text{ext}} \mathbf{E}_\mathbf{j}^{\text{ext}} - \frac{1}{6}\beta_{ijk} \mathbf{E}_\mathbf{i}^{\text{ext}} \mathbf{E}_\mathbf{j}^{\text{ext}} \mathbf{E}_\mathbf{k}^{\text{ext}} - \ldots \tag{4.6}$$

The first order properties like dipole and quadrupole moments, electric field gradient, etc. are calculated as the first order derivatives:

$$\mu_i = -\left(\frac{dE}{d\mathbf{E}_\mathbf{i}^{\text{ext}}} \right)_{E_i^{ext}=0} \tag{4.7}$$

The second order properties like dipole, quadrupole polarizabilities, force constants, etc. are calculated as the second derivatives:

$$\alpha_{ii} = -\left(\frac{d^2 E}{d\mathbf{E}_\mathbf{i}^{\text{ext}} d\mathbf{E}_\mathbf{i}^{\text{ext}}} \right)_{E_i^{ext}=0} \tag{4.8}$$

Similarly we can calculate also first and second hyperpolarizabilities. More details can be found in the review published recently by M. Ilias et al. [15].

The numerical examples of the role of relativistic effects in the calculations of dipole moments of CuH, AgH and AuH diatomic species and dipole polarizabilities of GeO, SnO and PbO molecules are shown in Tables 4.3 and 4.4.

From the Tables one can see that both, the type of Hamiltonian (relativity) and the wave function (correlation) are important for the accurate calculations of the electric properties of molecular species.

Table 4.3 Dipole moments of CuH, AgH and AuH [a.u.][a]

Method	CuH	AgH	AuH
NR CCSD(T)	1.136	1.362	1.223
DKH2 SCF[b]	1.526	1.618	0.980
DKH2 CCSD(T)[b]	1.052	1.135	0.522

[a]Data from Ref. [15]
[b]Douglas–Kroll–Hess calculations

Table 4.4 Dipole polarizabilities of GeO, SnO, and PbO [a.u.][a]

Method	α_{zz}			α_{xx}		
	GeO	SnO	PbO	GeO	SnO	PbO
NR CCSD(T)	40.52	57.18	65.76	28.23	40.34	45.85
DKH2 SCF[b]	35.02	47.89	53.39	28.19	37.19	34.97
DKH2 CCSD(T)[b]	40.33	56.29	62.55	27.62	37.75	37.46

[a]Data from Ref. [15]
[b]Douglas–Kroll–Hess calculations

Table 4.5 Equilibrium SCF energies[a] ΔE for the reaction $XH_4 \rightarrow XH_2 + H_2$ in kcal/mol

	SiH_2	GeH_2	SnH_2	PbH_2
NR	62.9	45.5	30.9	21.2
DHF	62.3	42.4	23.2	−6.2

[a]Ref. [27, 27a]

4.2.3 Chemical Reactions

4.2.3.1 Equilibrium Reaction Energies

The relativistic effects play an important role in a chemical reaction and influence its energy and the reaction path. It has been observed that if the dominant relativistic effect comes through the s populations, relativistic effects on reaction energies should be large when there is a significant change in s population [12]. This is the case in the reaction studied by Dyall [27]

$$XH_4 \rightarrow XH_2 + H_2 \qquad (4.9)$$

where X is the Group 14 element ($X = Si, Ge, Sn$ and Pb). Much of the relativistic destabilization comes from the change in s−orbital occupation. The XH_4 molecules have sp^3 hybridization, which leads to an effective population of $s^{1.5}p^{2.5}$ in the dihydrides XH_2 molecule [27]. Energies for this reaction calculated at the SCF nonrelativistic and the Dirac-Hartree-Fock (DHF) relativistic level are given in Table 4.5.

The importance of relativity increases as we move towards the heavier elements of the series. It becomes dramatic for Pb element, which yields an endothermic reaction from nonrelativistic data, $\Delta E_{NR} = 21.2$ kcal/mol, but which is actually an

Table 4.6 The pK_1, r_{max}(pm) [29] for the Group 11 and 12 cations

Cation	r_{max}	pK_1	Cation	r_{max}	pK_1
Cu^+	32.3	–	Zn^{2+}	30.5	8.96
Ag^+	54.5	12	Cd^{2+}	52.1	10.08
Au^+	64.0	3.8	Hg^{2+}	61.5	3.404

exothermic reaction when relativity is taken into account, $\Delta E_R = -6.2$ kcal/mol. The tetrahydrides become less stable with respect to the dihydrides and H_2 down to the Group. Relativistic effects make them even less stable, so that for Pb, the 'relativistic' tetrahydride is predicted to be unstable at the SCF level. This is because the divalent species become increasingly more stable as atomic charge Z increases. It has been shown by Dyall [27] that for Sn and Pb there is also a significant spin-orbit effect on the bonding. Interaction with the low-lying triplet state pushes the ground state of the dihydrides down.

4.2.3.2 Hydrolysis of Group 11 and 12 Cations

The hydrolysis reactions occur for nearly all the metal cations and profoundly affect their chemistry. These reactions determine the form of ions to be found in solutions as well as their charge and stability. What is referred to as the hydrolysis of a metal cation Me^{m+} of the charge $m+$ is the process proceeding according to the following reaction:

$$Me(H_2O)_x^{m+} \rightarrow Me(H_2O)_{x-1}(OH)^{(m-1)+} + H^+. \qquad (4.10)$$

Generally, hydrolysis is related to the interaction between oxygen in the water molecule and the metal cation [28]. Therefore, the ability of the cation to hydrolyze is expected to increase with its increasing charge and decreasing radius. With the increase of m and the accompanying reduction of the cationic radius, the Me...O distance is expected to decrease. Hence, the polarizing effect of the cation on the O–H bond in the water ligand will increase. As a consequence less energy will be required to break the O–H bond(s) and to facilitate the proton release. According to this model, the ability to hydrolyse should decrease (pK_1 increases) down the given Group of the Periodic Table because of increasing radius of the outermost shell of the ion. The model works very well for light metals and fails when applied to the hydrolysis of heavier elements. This is illustrated by the data of Table 4.6.

To find the explanation for the observed trends in the hydrolysis of the Group 11 and Group 12 metals Bilewicz et al. [29] investigated a simple model in which the hydrolysis is linked to the changes in the strength of the Me...O and O–H bonds. The nonrelativistic and relativistic calculations on complexes between water and the corresponding cations were carried out. It was found that the relativistic effects in heavy metal ions lead to a significant strengthening of the Me...O bond and simultaneously weaken the O–H bonds, Table 4.7.

Table 4.7 The stretching force constants (in a.u.) of the O–H and Me...O bonds in $Me(H_2O)^+$ and $Me(H_2O)^{2+}$ cations from MP2[a] nonrelativistic (NR) and DKH2 MP2 relativistic (R) calculations[b,c]

	k_{MeO}		k_{OH}	
$Me(H_2O)^+$	NR	R	NR	R
$Cu(H_2O)^+$	0.0948	0.1066	1.0267	1.0177
$Ag(H_2O)^+$	0.0527	0.0608	1.0360	1.0291
$Au(H_2O)^+$	0.0520	0.1084	1.0371	0.9913
$Zn(H_2O)^{2+}$	0.1744	0.1816	0.9200	0.9128
$Cd(H_2O)^{2+}$	0.1215	0.1315	0.9561	0.9439
$Hg(H_2O)^{2+}$	0.1129	0.1544	0.9649	0.8809

[a]Møller Plesset theory
[b]All values in a.u.
[c]Ref. [29]

This explains why a simple extrapolation of the data obtained for lighter elements must fail. The decrease of the pK_1 constant for Au^+ and Hg^{2+} as compared with the pK_1 values for Ag^+ and Cd^{2+}, respectively, is the manifestation of the relativistic effect in heavy elements. The nonrelativistic calculations do not explain the experimental observation.

4.2.4 Atomic Core Ionization Potentials

A very important example of the chemical quantities which are strongly affected by the relativistic effects are the core electron binding energies (CEBE) also known as the core electron ionization potentials (IP). The X-ray absorption structures near the ionization threshold are closely related to the local electronic structure at the atom on which the localized excitation, i.e., the creation of the hole state, takes place. The spectra provide local probes of the charge distribution in molecules and of the ability of a molecule to accept charge. They can be related to and can provide insight into such chemical concepts and properties as electronegativity, acidity, basicity, proton affinities, reactivity and regioselectivity of reactions [30–32]. Already in 1960s K. Siegbahn and his co-workers [33] showed that inner-shell ionization energies depend on the chemical state (environment) of the atom from which the electron is ionized. Recent developments in the high-resolution soft x-ray photoelectron spectroscopy have enabled accurate and detailed experimental characterization of the core-electron ionization processes. As the improved X-ray data have become available, computational methods that can give reliable IP values of different molecular species became an important issue. The X–ray spectra combined with the results of theoretical calculations are expected to lead to the understanding of the underlying electronic aspects of the structure–property relationships. Depending on the ionized state there is various interplay between relativistic and electron correlation contributions to the core ionization potentials. For holes of low energy, e.g., for singly ionized valence states, the proper treatment of the electron correlation contribution is of vital importance for the accuracy of the calculated IP data. On the contrary, the deep core ionized states require that the relativistic effects are taken into account. Since most of the electron correlation contribution is (relatively) important only for valence (and perhaps next–to–valence) ionized states, the main problem in accurate calculations of the core ionization potentials becomes the evaluation of the relativistic contribution to energy of the deep core states. This does not seem to be an easy task if one has in view methods which can be used for large molecules involving heavy atoms.

The Infinite–Order Two–Component (IOTC) method has been recently applied to the calculations of the valence and inner shell ionization potentials for the Ne, Ar, Kr, and Xe elements [34].

In Table 4.8 the IOTC CASSCF (complete active space SCF) and IOTC CASPT2 (CAS second-order perturbation theory) results have been compared with the corresponding nonrelativistic data. The ionization potential corresponding to the creation of a hole in the i–the orbital of a closed shell $2N$–electron system was obtained from:

$$IP(i) = E_i(2N - 1) - E(2N),\qquad (4.11)$$

where $E(2N)$ and $E_i(2N - 1)$ are the energies of the $2N$ and $(2N - 1)$–electron systems, respectively.

One should note that the relativistic effect on the core IPs can already be seen for Ne and progressively increases for heavier atoms. Thus, to compare theoretical results with the experimental data for the ionization core levels one needs to account for relativistic contributions even for relatively light systems. The still existing disagreement between the experimental data and theoretical IOTC results for the inner 1s core IPs can be attributed to the neglection of other higher–order relativistic quantum electrodynamic contributions such as Breit, self energy and vacuum polarization terms [34].

4.2.5 Molecular Core Electron Binding Energies

The core ionization potentials, more frequently called core electron binding energies (CEBEs) when molecular systems are studied, have been also recently calculated for the tautomeric structures of thio- and seleno-cytosine [35]. The role of relativistic effects in 1s ionization have been studied for selenocytosine by the comparison of the nonrelativistic and relativistic SCF and MP2 results of the calculations.

The data in Table 4.9 show the importance of the relativistic and electron correlation effects on the calculated ionization potentials. It is worth to notice

Table 4.8 Comparison of nonrelativistic (NR) and IOTC relativistic results for ionization potentials obtained from CASCF and CASPT2 calculations[a,b]

Atom	Orbital	Exp.	CASSCF		CASPT2	
			NR	IOTC	NR	IOTC
Ne	$2s$	23.12	19.82	19.80	21.48	21.46
Z = 10	$2p$	48.5	49.31	49.44	48.60	48.70
	$1s$	870.1	868.62	869.62	869.76	870.77
Ar	$3p$	15.75	14.76	14.74	15.56	15.54
Z = 18	$3s$	29.20	33.19	33.42	30.41	30.62
	$2p$	249.00	248.91	248.95	249.26	249.31
	$2s$	326.30	324.85	327.14	327.47	327.86
	$1s$	3,206.0	3,195.34	3,208.07	3,196.07	3,208.77
Kr	$4p$	13.99	13.21	13.18	14.03	14.01
Z = 36	$4s$	27.5	30.12	31.01	27.96	28.80
	$3d$	94.28	94.68	92.57	96.24	94.13
	$3p$	216.93	217.69	219.74	218.04	219.25
	$3s$	292.1	286.29	296.12	280.80	296.05
	$2p$	1,692.23	1,685.00	1,698.32	1,682.38	1,698.51
	$2s$	1,921.1	1,875.00	1,932.75	1,869.70	1,911.89
	$1s$	14,325.6	14,101.23	14,349.81	14,106.23	14,351.20
Xe	$5p$	12.13	11.68	11.62	12.15	12.23
Z = 54	$5s$	23.4	24.76	26.46	22.11	23.68
	$4d$	68.3	70.27	67.18	70.64	68.05
	$4p$	145.50	157.66	160.81	161.61	153.88
	$4s$	213.3	207.70	222.95	200.51	213.37
	$3d$	681.8	692.48	681.80	691.37	681.97
	$3p$	961.1	941.97	964.26	943.82	960.95
	$3s$	1,148.7	1,077.25	1,153.14	1,042.30	1,140.43
	$2p$	4,889.37	4,799.07	4,897.13	4,798.64	4,894.08
	$2s$	5,452.8	5,116.71	5,468.83	5,116.12	5,462.78
	$1s$	34,564.4	33,252.26	34,656.64	33,259.61	34,656.51

[a]Ref. [34, 34a]
[b]All values in eV

that already the IOTC SCF calculations, which do not take into account an electron correlation contribution to CEBEs, give the right order of magnitude of the relativistic effects.

The relativistic contributions become increasingly important for the heaviest atom, selenium. For carbon and nitrogen atoms the core electron binding energies are small (about 0.1–0.2 eV) but not negligible if one search for the chemical accuracy. The nonrealtivistic and the relativistic MP2 results, which include the electron correlation effects, show that the latter are not as important as the relativistic contributions.

Table 4.9 Theoretical values of nonrelativistic and relativistic $1s$ core electron binding energies in tautomer A of selenocytosine[a,b]

	NR HF	IOTC HF	NR MP2	IOTC MP2
Se_7	12,484.30	12,679.45	12,495.31	12,682.19
N_1	406.36	406.54	406.81	406.99
N_3	403.90	404.08	404.63	404.81
N_8	405.99	406.18	406.36	406.55
C_2	293.62	293.68	292.85	292.92
C_4	294.22	294.29	293.52	293.59
C_6	293.22	293.30	292.86	292.95
C_5	290.79	290.87	291.05	291.16

[a]All values in eV
[a]Barysz et al. [35]

4.3 Basis of Relativistic Theory

4.3.1 One–Electron Dirac Equation

All important features of the relativistic theory are already present in the one-electron Dirac equation [12, 14, 36].

$$h_D \Psi_D = \epsilon_D \Psi_D \tag{4.12}$$

where

$$h_D = c\boldsymbol{\alpha}\boldsymbol{p} + \beta mc^2 + (V - mc^2)I = \begin{pmatrix} V & c\boldsymbol{\sigma}\boldsymbol{p} \\ c\boldsymbol{\sigma}\boldsymbol{p} & V - 2mc^2 \end{pmatrix}, \tag{4.13}$$

where V is the external potential of the Coulomb type, or alternatively a potential derived from an extended nuclear charge distribution.

$$\boldsymbol{\alpha} = \begin{pmatrix} \mathbf{0} & \boldsymbol{\sigma} \\ \boldsymbol{\sigma} & \mathbf{0} \end{pmatrix}, \beta = \begin{pmatrix} \mathbf{I} & \mathbf{0} \\ \mathbf{0} & -\mathbf{I} \end{pmatrix}, \tag{4.14}$$

with

$$\mathbf{0} = \begin{pmatrix} 0 & 0 \\ 0 & 0 \end{pmatrix}, \mathbf{I} = \begin{pmatrix} 1 & 0 \\ 0 & 1 \end{pmatrix} \tag{4.15}$$

and

$$\alpha_x = \begin{pmatrix} 0 & 1 \\ 1 & 0 \end{pmatrix}, \alpha_y = \begin{pmatrix} 0 & -i \\ i & 0 \end{pmatrix}, \alpha_z = \begin{pmatrix} 1 & 0 \\ 0 & -1 \end{pmatrix} \tag{4.16}$$

are the Pauli matrices. In order to get electronic binding energies from Eq. 4.13 directly comparable to the nonrelativistic theory the energy scale is shifted by the rest mass energy mc^2 of the electron.

The 4×4 matrix form of the Dirac Eq. 4.13 gives for each energy state four solutions which are arranged into a four–component column vector referred to as the 4–spinor. Each of these solutions can be written as

$$\Psi_D = \begin{pmatrix} \Psi^L \\ \Psi^S \end{pmatrix}, \tag{4.17}$$

where the upper and lower two components Ψ^L and Ψ^S are spinors originating from the electronic and positronic degrees of freedom.

In the limit of $c \to \infty$ the lower component vanishes and Ψ^L terms into non–relativistic solutions of the Schrödinger equation. Simultaneously the Dirac energy for these solutions becomes equal to the non–relativistic energy. For this reason the part of the Dirac spectrum which has the Schrödinger non–relativistic limit for $c \to \infty$ is referred to as the positive 'electronic' energy spectrum and is associated with the dominant contribution of Ψ^L in the Ψ_D–spinor (4.17).

One should note, that in the majority of cases in the relativistic quantum chemistry, only the 'electronic' part of the energy spectrum is explicitly considered. The so–called negative energy solutions are simple neglected. Thus, although the underlying theory is expressed in terms of 4–spinors, it can be, at least in principle, exactly transformed into a two–component representation. This feature of most of the practical four–component methods of quantum chemistry, though hardly stated explicitly, makes the background for the development of the two–component representation.

4.3.2 Relativistic Theory of Many–Particle Systems

Once we know the solution of the Dirac equation for a single electron moving in the external potential it is tempting to build the relativistic theory of many–electron systems in a similar way as the non–relativistic theory is built, i.e., by combining the one–electron Dirac Hamiltonian for each electron with the interaction between electrons.

One of the major fundamental difference between nonrelativistic and relativistic many–electron problems is that while in the former case the Hamiltonian is explicitly known from the very beginning, the many–electron relativistic Hamiltonian has only an implicit form given by electrodynamics [13, 37]. The simplest relativistic 'model' Hamiltonian is considered to be given by a sum of relativistic (Dirac) one–electron Hamiltonians h_D and the usual Coulomb interaction term:

$$H^{DC}(1, 2, \ldots n) = \sum_{i=1}^{n} h_D(i) + \sum_{i<j} \frac{1}{r_{ij}} \tag{4.18}$$

where the first part $h_D(i)$ is the one–electron part of the Hamiltonian given by the Eq. 4.13. The second part of the Hamiltonian is the Coulomb interaction, and the operator (4.18) is refereed as Dirac-Coulomb Hamiltonian and represents the lowest order approximation to the electron-electron interaction. When we consider the Coulomb interaction, what is the simplest case, this term is not Lorentz invariant. The Lorenz invariant interaction operator can be approximately derived from quantum electrodynamics which is relativistically correct through $\frac{1}{c^2}$ and is known as the Breit interaction operator $V_B(i, j)$:

$$V^B(i, j) = \frac{1}{r_{ij}} - \frac{1}{2}\left[\frac{\alpha_i\alpha_j}{r_{ij}} + \frac{(\alpha_i r_{ij})(\alpha_j r_{ij})}{r_{ij}^3}\right] \tag{4.19}$$

By substituting in Eq. 4.18 the Coulomb interaction operator by its relativistic extension (4.19) one obtains the so-called Dirac–Breit many electron Hamiltonian H^{DB}:

$$H^{DB} = \sum_{i=1}^{n} h_D(i) + \sum_{i<j} V^B(ij) \tag{4.20}$$

In the context of Breit operator (4.19) one should also mention its approximate form known as the Gaunt interaction V^G:

$$V^G(i, j) = \frac{1}{r_{ij}} - \frac{\alpha_i\alpha_j}{r_{ij}} \tag{4.21}$$

which leads to the Dirac–Gaunt many–electron Hamiltonian H^{DG}:

$$H^{DG} = \sum_{i=1}^{n} h_D(i) + \sum_{i<j} V^G(ij) \tag{4.22}$$

The use of the Coulomb (4.18) Breit (4.19) or Gaunt (4.21) interaction operators in combination with Dirac Hamiltonians causes that the approximate relativistic Hamiltonians does not have any bound states, and thus, becomes useless in the calculations of relativistic interactions energies. This feature is known under the name of the Brown–Ravenhall disease and is a consequence of the spectrum of the Dirac Hamiltonian [38].

4.4   Two-Component Relativistic Theories

In solving the Dirac equation, it would be desirable to use as much as possible of the well–established techniques known from nonrelativistic theory. However, as it has been discussed above we quickly encounter a problem with the variational principle, since the Dirac equation describes both electronic and positronic states. The latter have much lower energies, and any attempt to minimize the energy without additional constraints is likely to result in a positronic–like solutions [12–14]. To avoid such a collapse, one must first take necessary precautions to ensure that the solutions are constrained to a space of proper electron–like solutions.

The easiest pragmatic solution is, by neglecting the so–called negative continuum of the Dirac spectrum. Once these negative energy (positronic) solutions are abandoned, the corresponding theory becomes what can be called the four–component Dirac theory 'for electrons only' with the one–electron solutions represented by four–component Dirac spinors. Such a theory would only neglect the QED effects and should be sufficient for most of interactions of chemically interesting systems. The neglect of positronic solutions of the Dirac equation is a rather artificial procedure which is carried out a posterior. However, this can be achieved in a formally correct way by passing to what is known as the two–component theory. If the relativistic interaction theory is considered as the theory 'for electrons' only, the four–component Dirac spinors can be replaced by two–component 'electronic' functions without any further approximation.

Historically, the first derivation of approximate relativistic operators of value in molecular science have become known as the Pauli (or Breit–Pauli) approximation [12, 13]. Still the easiest way to calculate the relativistic corrections originate from the elimination technique of the small component (ESC) which provided well–known operators like the spin–orbit or the mass velocity and Darwin operators. These operators are singular and variationally unstable and can only be employed within the framework of perturbation theory. Nowadays, these difficulties have been overcome by other relativistic Hamiltonians, so that operators like mass-velocity and Darwin terms are no longer needed in molecular calculations. However, these operators are still of great importance when one search for the limit of accuracy of the quantum electrodynamics of the light elements. Other schemes leading to the separation of electronic and positronic spectra have been proposed by Rutkowski et al. and Kutzelnigg. All of them are rooted in the particular form of the perturbation approach introduced by Rutkowski and shaped by Kutzelnigg into what is nowadays called the direct perturbation theory (DPT) [39–41]. Among the other methods based on the elimination of the small component procedure, the priority should be given to the so-called regular Hamiltonian approximation, which is nowadays, referred to as the ZORA, FORA etc. It has been originally proposed by Chang et al. [42] and later independently derived by Snijders and co-workers [43–45]. The major advantage of this approach is that the two–component Hamiltonian is non-singular. However, the energy eigenvalues of ZORA, FORA, or infinite-order RA (IORA) [46] are not invariant upon shifting the external potential by a constant.

Some of the other developments in the area have been derived by Dyall et al. [47, 48] and improved recently by Filatov et al. [49–51]. The method is known as the normalized elimination of the small component (NESC) method. One has to also mention the recent work of Liu et al. (see the review [52]) and Ilias [15].

Several excellent reviews of the relativistic two–component methods have recently appeared [12–15]. This review is not aimed at the completeness of the presentation of different two–component methods in relativistic quantum chemistry. In the article I will concentrate on the Infinite Order Two–Component Approach (IOTC) developed by Barysz et al. [17–19]. The method is exact and has been implemented in commonly used molecular codes (MOLCAS, GAMESS) and is ready for general applications.

4.4.1   Infinite Order Two–Component Method (IOTC)

The IOTC method presented in this review is obviously related to earlier attempts to reduce the four–component Dirac formalism to computationally much simpler two–component schemes. Among the different to some extent competitive methods, the priority should be given to the Douglas-Kroll-Hess method [13,53,54]. It was Bernd Hess and his work which was our inspiration to search for the better solutions to the two–component methodology.

The main relativistic contribution to the target two–component Hamiltonians comes from the one–electron terms, i.e., from the potential V in the Dirac Eq. 4.13 which can be understood as the usual electron-nuclear attraction operator. Also replacing V by the one-electron Dirac-Hartree-Fock potential would not make too much difference for the general theory. Hence it is sufficient to consider ordinary time–independent Dirac Eq. 4.12 for a single electron moving in the Coulomb field without any loss of generality, The idea to separate the so–called positive and negative parts of the Dirac eigenvalue spectrum goes back to Foldy and Wouthuysen (FW) [55]. The idea of Foldy and Wouthuysen was to completely separate the electronic and positronic solutions of the Dirac Eq. 4.12 by the van Vleck–type unitary transformation of the Dirac Hamiltonian:

$$H^U = U^\dagger h_D U, \tag{4.23}$$

with

$$H^U = \begin{pmatrix} h_+ & 0 \\ 0 & h_- \end{pmatrix}. \tag{4.24}$$

Such a transformation converts the 4×4 Dirac Eq. 4.12 into two separate eigenvalue problems. The one which corresponds to what is called the positive (electronic) part of the spectrum:

$$h_+ \Psi^L = \epsilon_+ \Psi^L, \tag{4.25}$$

where ψ^L is the upper 2–spinor of the transformed four–component function:

$$\Psi^U = U^\dagger \Psi_D = \begin{pmatrix} \Psi^L \\ 0 \end{pmatrix}, \tag{4.26}$$

should convert into the two-component nonrelativistic Schrödinger equation in the limit of $c \to \infty$.

The FW transformation U of Eq. 4.23 can be expressed as a product of a series of unitary transformations:

$$U = U_0 U_1 U_2 \dots \tag{4.27}$$

Once the final U is expressed as the product of the unitary operator matrices U_k ($k = 0, 1, \dots$) the transformation (4.23) can be achieved as a sequence of transformations:

$$H_{k+1}^{U} = H_{k+1} = U_k^{\dagger} H_k U_k, \quad k = 0, 1, 2, \dots, \tag{4.28}$$

initialized by the Dirac Hamiltonian $H_0 = h_D$ and some suitable operator U_0. The idea of using the sequence of unitary transformations defined in Eq. 4.27 has been a basis of the Douglas–Kroll–Hess method and its generalization formulated by Reiher et al. [13]. The block diagonal Hamiltonian (4.24) is constructed step by step. The philosophy of the IOTC method defined by Barysz et al. is different. The main idea was to find a unitary transformation which completely diagonalize the Dirac Hamiltonian (4.13) in one or two steps only. For a long time, it has been believed that the only way to avoid singular operators in the FW like transformation (Eq. 4.27), is the necessity to carry out the free–particle transformation U_0 (the transformation which diagonalize the free–particle Hamiltonian) at first, before any further unitary transformation is carried out:

$$U_0 = \begin{pmatrix} A & \alpha AB \\ \alpha AB & -A \end{pmatrix}, \quad U_0^{\dagger} U_0 = I, \tag{4.29}$$

where

$$A = \sqrt{\frac{e_p + 1}{2 e_p}}, \quad B = \frac{1}{e_p + 1} \sigma p, \tag{4.30}$$

and

$$e_p = \sqrt{1 + \alpha^2 p^2}. \tag{4.31}$$

Nowadays, it is known that this step is not really needed. In the time the IOTC method was defined the authors were not aware of this fact and that is why the idea to start the diagonalization of the Dirac Hamiltonian with the free–particle transformation is also the basis of the IOTC method. The resulting transformed hamiltonian $H_1 = U_0^{\dagger} h_D U_0$ is:

$$H_1 = \begin{pmatrix} T_p + A(V + \alpha^2 BVB)A & \alpha A[V, B]A \\ \alpha A[B, V]A & -2\alpha^{-2} - T_p + A(V + \alpha^2 BVB)A \end{pmatrix} \tag{4.32}$$

where

$$T_p = c^2(e_p - 1). \tag{4.33}$$

and $[X, Y] = XY - YX$ denotes the commutator of X and Y,

To continue the block–diagonalization of the Dirac Hamiltonian, one needs to find the form of the U_1 transformation. The form proposed by Heully et al. [56] was used in the IOTC approach. It is a general technique equally well applied to the initial Dirac Hamiltonian h_D or to the free-particle transformed operator (4.32). According to Heully et al., the unitary transformation U_1 can be written in the following form:

$$U_1 = \begin{pmatrix} \Omega_+ & R_-\Omega_- \\ R_+\Omega_+ & \Omega_- \end{pmatrix}, \tag{4.34}$$

where

$$\Omega_+ = (1 + R_+^\dagger R_+)^{-1/2}, \ \Omega_- = -(1 + R_-^\dagger R_-)^{-1/2}, \ R_- = -R_+^\dagger, \tag{4.35}$$

and R_+ is the root of the following operator equation:

$$R = [(H_1)_{22}]^{-1} [-(H_1)_{21} + R(H_1)_{11} + R(H_1)_{12} R], \tag{4.36}$$

which corresponds to the assumption that $R = R_+$ is a 'small' operator as compared to the other operators in the r.h.s. of Eq. 4.36.

After substituting the H_1 blocks into Eq. 4.36 one finds that the determination of R requires that the following operator equation:

$$2R + \alpha^2 T_p R + \alpha^2 R T_p = e_p R + R e_p = \alpha^3 A[B, V]A$$
$$+ \alpha^2 [AVA, R] + \alpha^4 [BAVAB, R] + \alpha^3 R[B, AVA]R, \tag{4.37}$$

is solved. However, no closed–form of the solution of this equation appeared to be available and that is why, the approximate theories have been defined [16, 17].

The problem of solving the Eq. 4.37 lies in the presence of linear terms in σp in this equation. The problem has been solved recently [18, 19] and it opened the possibility to solve the Eq. 4.37 in a purely numerical way. Let us multiply the Eq. 4.37 from the left by the operator $p^{-1}\sigma p$, where p^{-1} denote the inverse square root of p^2 and is a scalar operator. One obtains then:

$$e_p Y + Y e_p = \alpha^3 (pAbVA - p^{-1}A\sigma p V\sigma pbA)$$
$$+ \alpha^2 (p^{-1}A\sigma p V\sigma p p^{-1}AY - YAVA)$$
$$+ \alpha^4 (pAbVbApY - YAb\sigma p V\sigma pbA)$$
$$+ \alpha^3 Y(Ab\sigma p V\sigma pAp^{-1} - AVAbp)Y, \tag{4.38}$$

where

$$b = \frac{1}{e_p + 1} \tag{4.39}$$

and the operator Y is

$$Y = p^{-1} \boldsymbol{\sigma} p R. \tag{4.40}$$

No more linear terms are present in the operator Eq. 4.38 but it is still in the form which can not be used in the practical applications.

4.4.2 Matrix Formulation

The natural solution of the operator Eq. 4.38 would be in the momentum space due to the presence of momentum p which replaces the standard configuration space formulation by a Fourier transform. The success of the Douglas–Kroll–Hess and related approximations is mostly due to excellent idea of Bernd Hess [53, 54] to replace the explicit Fourier transformation by some basis set (discrete momentum representation) where momentum p^2 is diagonal. This is a crucial step since the unitary transformation U of the Dirac Hamiltonian can easily be accomplished within every quantum chemical basis set program, where the matrix representation of the nonrelativistic kinetic energy $T = p^2/2m$ is already available. Consequently, all DKH operator equations could be converted into their matrix formulation and they can be solved by standard algebraic techniques [13].

The first step is the generation of eigenvectors of the p^2; this is done by using a finite coordinate–dependent basis set $\{\zeta_\alpha(\mathbf{r})\}$ in which the eigenvectors $|k\rangle$ of p^2:

$$|k\rangle = \sum_\alpha c_{\alpha k} \zeta_\alpha(\mathbf{r}) \tag{4.41}$$

are determined, i.e.,

$$\langle k \mid p^2 \mid k' \rangle = w_k \delta_{kk'}. \tag{4.42}$$

In the basis set $\{| k \rangle\}$ the approximate identity resolution is built:

$$1 \approx \sum_k | k \rangle \langle k | \tag{4.43}$$

and inserted between operators defined in the coordinate and momentum spaces. Then, for any function of the p^2 operator, say $f(p^2)$ the following approximation is introduced:

$$\langle k \mid f(p^2) \mid k' \rangle = f(w_k)\delta_{kk'}, \tag{4.44}$$

whereas the matrix elements of V and $\sigma p V \sigma p$ can be evaluated directly in the **r**–space. What remains are the operators of the form:

$$\sigma p V f(p^2) V \sigma p. \tag{4.45}$$

According to Hess idea, this integrals can be evaluated by using the identity:

$$\frac{\sigma p \sigma p}{p^2} = 1, \tag{4.46}$$

whose insertion into (4.45) gives:

$$\sigma p V \sigma p p^{-2} f(p^2) \sigma p V \sigma p. \tag{4.47}$$

Converting the R–operator Eq. 4.38 into its matrix formulation gives the equation:

$$Y_{kk'} = \frac{1}{E_k + E_{k'}} \left[\alpha^3 \langle k \mid p A b V A - p^{-1} A \sigma p V \sigma p b A \mid k' \rangle \right.$$

$$+ \alpha^2 \sum_{k''} \left(\langle k \mid p^{-1} A \sigma p V \sigma p p^{-1} A \mid k'' \rangle Y_{k''k'} - Y_{kk''} \langle k'' \mid A V A \mid k' \rangle \right)$$

$$+ \alpha^4 \sum_{k''} \left(\langle k \mid p A b V b A p \mid k'' \rangle Y_{k''k} - Y_{kk''} \langle k'' \mid A b \sigma p V \sigma p b A \mid k' \rangle \right)$$

$$\left. + \alpha^3 \sum_{k''k'''} Y_{kk''} \langle k'' \mid A b \sigma p V \sigma p A p^{-1} - A V A b p \mid k''' \rangle Y_{k'''k'} \right] \tag{4.48}$$

which can be solved to arbitrarily high accuracy by using standard iterative techniques. In contrary to the DKHn approach [13], the numerical (algebraic) iterative solution for the matrix representation of Y means that order by order summation of the perturbative series could be done without the necessity to define the analytical form of the Hamiltonians in each step of the iteration. This step in the IOTC method is crucial and makes it possible to get the unitary transformation exact. Once the numerical solution for the matrix representation Y is found the corresponding matrix representation of h_+ can be calculated as well.

4.5   The Change of Picture Problem

The two–component methods, though much simpler than the approaches based on the 4–spinor representation, bring about some new problems in calculations of expectation values of other than energy operators. The unitary transformation U on the Dirac Hamiltonian h_D (Eq. 4.23 is accompanied by a corresponding reduction of the wave function to the two–component form (Eq. 4.26). The expectation value of any physical observable O in the Dirac theory is defined as:

$$\bar{O} = \langle \Psi_D | O | \Psi_D \rangle \tag{4.49}$$

and can easily be transformed within a two–component scheme employing the transformed two–spinor Ψ^L:

$$\bar{O} = \langle \Psi_D | UU^\dagger OUU^\dagger | \Psi_D \rangle = \langle \Psi_D^U | U^\dagger OU | \Psi_D^U \rangle =$$
$$= \langle \Psi^L | (U^\dagger OU)_{11} | \Psi^L \rangle \tag{4.50}$$

However, in two–component quantum chemical calculations of expectation values the unitary transformation of the operator O is often not taken into account. Instead, the quantity

$$\bar{o} = \langle \Psi^L | O_{11} | \Psi^L \rangle \tag{4.51}$$

is calculated. That is, the change of picture of the operator O is neglected completely, and the difference $\bar{O} - \bar{o}$ is referred to as the picture change effect on the expectation value. The picture change arising from the unitary transformations of the Dirac equation is well known among physicists but its full recognition in relativistic quantum chemistry has taken some time. The first study of this problem within the framework of the DKH approximation is to be most likely credited to Kellö et al. [57]. These authors have considered the picture change effect upon the calculation of dipole moment and dipole polarizabilities of the coinage metal atoms and their hydrides. The effect was found to be negligible and led to the incorrect conclusion that the difference $\bar{O} - \bar{o}$ is of the order of α^4 ($\alpha = 1/c$). Later, the picture change problem was carefully considered by Sadlej et al. [58] who found that the picture change effect is of the order of α^2 and may not be negligible for certain operator. The formal investigation of the picture change problem presented in Ref. [58] links the magnitude of the picture change effect to the form of the operator O. For operators which assume large values in the vicinity of nuclei, this effect is expected to be quite large. Indeed, it has been found [59] that for electric field gradient at heavy nuclei the picture change contribution is by no means negligible, see the Table 4.10. Since the calculated electric field gradients are used for the determination of nuclear quadrupole moments from molecular quadrupole coupling

Table 4.10 Change of picture effect on the electronic contribution to electric field gradients calculated in DKH2–SCF approximation [a.u.]. Reproduced from Ref. [15] with kind permission of Acta Phys. Slovaca

Molecule	q^a_{no-CP}	q^b_{CP}
HCl	3.540	3.511
HBr	7.806	7.520
HI	12.657	11.683
HAt	34.621	26.656

[a]Calculations with the change of picture error
[b]Calculations without the change of picture error

constants (see e.g., Ref. [60]), this finding leads to rather important consequences concerning the accuracy of the so–called molecular values of nuclear quadrupole moments [15, 60].

The picture change effect has been also found to be quite large for dipole moment derivatives with respect to nuclear coordinates. The same can be expected in the case of the dipole polarizabilities derivatives [15]. These findings show that two–component calculations of infrared and Raman intensities for heavy systems need to take into account the picture change of the relevant operators. It should be mentioned that for both derivatives the operator which is responsible for the large picture change effect is the intramolecular electric field at the heavy nucleus.

It is also appropriate to mention that the picture change problem is by no means restricted to one–electron operators, although the way of phrasing it in the case of the Coulomb repulsion terms is usually somewhat different from the one presented here. For instance, Sucher [61] considered the modification of the Coulomb repulsion operator caused by the no–pair approximation. The expression he derived is simply the interaction operator in the very approximate two–component formalism. This form of the Coulomb interaction operator was also investigated by Samzow et al. [62]. They found that for valence orbitals of heavy systems the corresponding modification of two–electron integrals is relatively small.

Finally let us add that the change of picture for different interaction operators which occurs upon transition from the four–component to two–component functions is quite relevant from the point of view of intermolecular interactions. The problem of the form of the operator which corresponds to the interaction between two "relativistic" subsystems which are described in the framework of the two–component formalism has been recently investigated [37] and the analytic form of the corresponding operator has been derived.

4.6 Quasi–Relativistic or Exact Two–Component Method

Since the infinite–order two–component theory is based on exact equations, it is obvious that it must reproduce all features of the positive–energy Dirac spectrum. However, the way this theory is used introduces the algebraic approximation.

The equivalence of the IOTC method to the four–component Dirac approach has been documented by calculations of spin orbital energies in several papers [18, 20, 63]. The unitary transformation does not affect the energy eigenspectrum, though it reduces the four–component bi–spinors to two–component spinor solutions. Due to this fact the two–component methods are frequently addressed as being quasi–relativistic and it is assumed that some information is lost. It can be demonstrated [22] that the two–component IOTC wave function which is the upper component of the unitarly transformed four–component Dirac spinor Ψ

$$\Phi = U_1^\dagger U_0^\dagger \Psi = \begin{pmatrix} \phi^{IOTC} \\ 0 \end{pmatrix}, \tag{4.52}$$

can be back transformed and one obtains the exact Dirac solution which corresponds to the given IOTC solution:

$$\Psi = U_0 U_1 \Phi \tag{4.53}$$

and with the unitary transformations U_0 (4.29) and U_1 (4.34) the form of the inverse transformation of the IOTC wave function becomes

$$|\Psi\rangle = \begin{pmatrix} A\Omega_+ + \alpha A \frac{1}{e_p+1} pY\Omega_+ \\ \sigma p \left(\alpha A \frac{1}{e_p+1}\Omega_+ - Ap^{-1}Y\Omega_+ \right) \end{pmatrix} |\phi^{IOTC}\rangle \tag{4.54}$$

Then the expectation value of the diagonal operator $\hat{O}$ for the Dirac state (4.54) will be given by:

$$\langle \hat{O}\rangle = \left\langle \Psi \left| U_1^\dagger U_0^\dagger \hat{O} U_0 U_1 \right| \Psi \right\rangle$$

$$= \left\langle \phi^{IOTC} \left| \left(\Omega_+ A + \alpha\Omega_+ Y_+^\dagger p \frac{1}{e_p+1} A \left(\alpha\Omega_+ A \frac{1}{e_p+1} - \Omega_+ Y_+^\dagger p^{-1}A \right) \sigma p \right) \right.\right.$$

$$\times \hat{O} \begin{pmatrix} A\Omega_+ + \alpha A \frac{1}{e_p+1} pY_+\Omega_+ \\ \sigma p \left(\alpha A \frac{1}{e_p+1}\Omega_+ - Ap^{-1}Y_+\Omega_+ \right) \end{pmatrix} |\phi^{IOTC}\rangle$$

$$= \left\langle \phi^{IOTC} \left| \left\{ \Omega_+ A + \alpha\Omega_+ Y_+^\dagger p \frac{1}{e_p+1}A \right\} \hat{O} \left\{ A\Omega_+ + \alpha A \frac{1}{e_p+1} pY_+\Omega_+ \right\} \right.\right.$$

$$+ \left\{ \alpha\Omega_+ A \frac{1}{e_p+1} - \Omega_+ Y_+^\dagger p^{-1}A \right\} \sigma p \hat{O} \sigma p \left\{ \alpha A \frac{1}{e_p+1}\Omega_+ \right.$$

$$\left. \left. -Ap^{-1}Y_+\Omega_+ \right\} |\phi^{IOTC}\rangle \right. \tag{4.55}$$

which can be calculated by using the approximate identity resolution in terms of eigenfunctions of the p^2 operator in the selected basis set. In order to check

Table 4.11 The calculated $\langle r^k \rangle$ expectation values and spinorbital energies for the $1s_{1/2}$ core shell of the fermium ion[a]. Atomic units have been used

Operator	Energy	BT IOTC	Dirac[b]
Exact energy	−0.5939195374		
$\langle r^{-2} \rangle$	−0.5939195374	7.9602724141	7.9604206663
$\langle r^{-1} \rangle$	−0.5939195374	1.4625665349	1.4625661529
$\langle r^1 \rangle$	−0.5939195374	1.1837296044	1.1837297568
$\langle r^2 \rangle$	−0.5939195374	1.9930806284	1.9930810155
$\langle r^3 \rangle$	−0.5939195374	4.3523493474	4.3523503212

[a]Barysz et al.[22]
[b]Andrae D. [64]

how exact is the back–transformed function Ψ, the transformed two–component solutions can be used to evaluate different moments $< r^k >$ of the electron distribution. Table 4.11 gives example of the calculations of different $< r^k >$ expectation values for the $1s_{1/2}$ core shell of the one electron fermium ion [22]. With sufficiently large basis of Gaussian functions, the Dirac values of different moments $< r^k >$ are fully recovered. Hence the IOTC approach is not quasi–relativistic. It is as relativistic as the four–component Dirac approach.

4.7 Summary

The life is relativistic and of the same kind should be the quantum chemistry. The four–component calculations involving large number of dynamic correlation are extremely time–consuming. The important thing, however, is that one is now able to formulate the fully equivalent two–component algorithms for those calculations. Since the final philosophy of the two–component calculations is similar to non–relativistic theory and most of mathematics are simply enough to comprehend the routine molecular relativistic calculations are possible. The standard nonrelativistic codes can be used with simple modification of the core Hamiltonian. The future is still in the development of the true two–component codes which will be able to deal with the spin–orbit interaction effect not in a posterior way as it is done nowadays.

Acknowledgements The author wishes to thank V. Kellö, M. Urban, M. Iliaš, and K. G. Dyall for their permission to use of their numerical data.

References

1. Styszyński J (2010) Challenges and advances in computational chemistry and physics. In: Barysz M, Ishikawa J (eds) Relativistic methods for chemists, vol 10. Springer, Dordrecht/Heidelberg/London/New York, pp 99–164

2. Pyykkö P (1988) Chem Rev 88:563
3. Sommerfeld A (1916) Ann Phys 51:1
4. Biedenharn LC (1983) Phys 13:13
5. Lide DR (ed) (2005) CRC handbook of chemistry and physics, CD-ROM. CRC, Boca Raton
6. Neogrady P, Kellö V, Urban M, Sadlej AJ (1997) Int J Quantum Chem 63:557
7. Iliaš M, Neogrády P (1999) Chem Phys Lett 309:441
8. Pyykkö P (1986) Relativistic theory of atoms and molecules I – A bibligraphy 1916–1985. Lecture notes in chemistry. Springer, Berlin
9. Pyykkö P (1993) Relativistic theory of atoms and molecules II – A bibligraphy 1986–1992. Lecture notes in chemistry, vol 60. Springer, Berlin
10. Pyykkö P (2000) Relativistic theory of atoms and molecules III – A bibligraphy 1993–1999. Lecture notes in chemistry, vol 76. Springer, Berlin
11. Pyykkö P (2009) Database 'RTAM' Version 15, Relativistic quantum chemistry database. http://rtam.csc.fi/
12. Dyall KG, Faegri K Jr (2007) Introduction to relativistic quantum chemistry. Oxford University Press, New York
13. Reiher M, Wolf A (2009) Relativistic quantum chemistry. The fundamental theory of molecular science. Wiley, Weinheim
14. Barysz M, Ishikawa Y (eds) (2010) Challenges and advances in computational chemistry and physics. Relativistic methods for chemists. Springer, London
15. Iliaš M, Kellö V, Urban M (2010) Acta Phys Slovaca 60:259
16. Barysz M (2000) J Chem Phys 113:4003
17. Barysz M (2001) J Chem Phys 114:9315
18. Barysz M, Sadlej AJ (2002) J Chem Phys 116:2696
19. Barysz M, Sadlej AJ (2002) J Mol Struct Theochem 573:181
20. Kędziera D, Barysz M (2004) Chem Phys Lett 393:521
21. Kędziera D, Barysz M, Sadlej AJ (2004) Struct Chem 15:369
22. Barysz M, Mentel L, Leszczyński J (2009) J Chem Phys 130:164114
23. Barysz M (2010) Two–component relativistic theories. In: Barysz M, Ishikawa Y (eds) Challenges and advances in computational chemistry and physics. Relativistic methods for chemists. Springer, London, p 165, the review article
24. Barysz M, Pyykkö P (1998) Chem Phys Lett 285:398
24a. Reprinted from publication, Barysz M, Pyykkö P (1998) Strong chemical bonds to gold. CASPT2 results for diatomic $AuBe^+$, AuC^+, $AuMg^+$, and $AuSi^+$, 285:398–403, with permission from Elsevier, Copyright 2011
25. Schwerdtfeger P (ed) (2004) Theoretical and computational chemistry. Relativistic electronic structures, part 2. Applications p 394
26. Frobe M, Schulze N, Kloss H (1983) Chem Phys Lett 99:500
27. Dyall KG (1992) J Chem Phys 96:1210
27a. Reprinted with permission from, Dyall KG (1992) J Chem Phys 96:1210. Copyright 2011, American Institute of Physics
28. Baes CF Jr, Messmer RE (1986) The hydrolysis of cations. Krieger, Malabar
29. Barysz M, Leszczyński J, Bilewicz A (2004) Phys Chem Chem Phys 6:4553
30. Siegbahn K, Nordling C, Johansson G, et al (1969) ESCA applied to free molecules. North-Holland, Amsterdam
31. Bakke AA, Chen AW, Jolly WL (1980) J Electron Spectrosc Relat Phenom 20:333
32. See the review papers in the Coord Chem Rev (2005) vol 249
33. Siegbahn K, Fahlman C, Nordberg R, Hamrin K, Hedman J, Johansson G, Bergamrk T, Karlsson SE, Lindgren L, Lindberg B (1967) ESCA, atomic, molecular and solid state structure studied by means of electron spectroscopy. Almqvist and Wiksell, Uppsala
34. Barysz M, Leszczyński J (2007) J Chem Phys 126:154106
34a. Reprinted with permission from, Barysz M, Leszczyński J (2007) J Chem Phys 126:154106. Copyright 2011, American Institute of Physics
35. Barysz M, Klobukowski M, Leszczynski J (2011) Struct Chem (to be published)

36. Moss RE (1973) Advanced molecular quantum mechanics. Chapman and Hall, London
37. Pestka G, Sadlej AJ (2002) J Mol Struct Theochem 592:7
38. Grant IP, Quiney HM (1988) Adv At Mol Phys 23:457, and references therein
39. Rutkowski A (1986) J Phys B 19:141
40. Kutzelnigg W (1989) Z Phys D 11:15
41. Kutzelnigg W (1990) Z Phys D 15:27
42. Chang Ch, Pélissier M, Durand P (1986) Phys Scr 34:394
43. van Lenthe E, Baerends EJ, Snijders JG (1993) J Chme Phys 99:4597
44. Faas S, Snijders JG, van Lenthe JH, van Lenthe E (1995) J Chem Phys Lett 246:632
45. van Lenthe E, van Leeuven, Baerends EJ, Snijders JG (1996) Int J Quantum Chem 57:281
46. Dyall KG, van Lenthe E (1999) J Chem Phys 111:1366
47. Dyall KG (1997) J Chem Phys 106:9618
48. Dyall KG (1998) J Chem Phys 109:4201
49. Filatov M, Cremer D (2002) Theor Chem Acc 108:168
50. Filatov M, Cremer D (2002) Chem Phys Lett 351:259
51. Filatov M, Cremer D (2002) Chem Phys Lett 370:647
52. Liu W (2010) Mol Phys 108:1679
53. Hess BA (1985) Phys Rev A 32:756
54. Hess BA (1986) Phys Rev A 33:3742
55. Foldy LL, Wouthuysen SA (1950) Phys Rev 78:29
56. Heully JL, Lindgren I, Lindroth E, Lundquist S, Mårtenson-Pendril AM (1986) J Phys B 19:2799–2815
57. Kellö V, Sadlej AJ, Hess BA (1996) J Chem Phys 105:1995
58. Barysz M, Sadlej AJ (1997) Theor Chem Acc 97:260
59. Kellö V, Sadlej AJ (1998) Int J Quantum Chem 68:159
60. Kellö V, Sadlej AJ (1999) Phys Rev A 60:3575, and references therein
61. Sucher J (1980) Phys Rev A 22:348
62. Samzow R, Hess BA, Jansen G (1992) J Chem Phys 96:1227
63. Kędziera D, Barysz M (2004) J Chem Phys 121:6719
64. Andrae D (1997) J Phys B At Mol Opt Phys 30:4435

Chapter 5
On the Electronic, Vibrational and Relativistic Contributions to the Linear and Nonlinear Optical Properties of Molecules

Aggelos Avramopoulos, Heribert Reis, and Manthos G. Papadopoulos

Abstract This article reviews some recent works, which deal with state-of-the-art results linear and nonlinear (L&NLO) properties of molecules (organic and inorganic). The electronic correlation, vibrational and relativistic contributions have been computed. Correlation has been taken into account by employing a series of methods e.g. MP2, CCSD, CCSD(T), CASPT2. The vibrational contributions have been computed by using the Bishop and Kirtman perturbation theory as well as the Numerov-Cooley integration. The relativistic correction has been computed by using the Douglas-Kroll method. It is noted that vibrational contribution has been calculated by taking into account both correlation and relativistic corrections. The properties are mainly static, but to a smaller extent, we have also considered frequency dependent properties. The main emphasis of the reviewed works was to use all the computational tools currently available in order to calculate property values as accurate as it is currently possible. The accuracy of the employed approximations is evaluated and discussed. Several interesting features and trends have been observed, for example the relativistic effects significantly reduce the magnitude of the vibrational corrections (coinage metal hydrides). An extensive study of fullerenes and endohedral fullerenes has been performed and reviewed. Of particular interest is the effect of the endohedral atom (e.g. Li) on the L&NLO properties of the system. The reviewed works have employed a large variety of systems (e.g. heavy metal -Group IIb- sulfides, coinage metal hydrides, ketones, thiones etc.).

Keywords Non-linear optical properties • Vibrational contributions • Relativistic corrections • Fullerene derivatives

A. Avramopoulos (✉) • H. Reis • M.G. Papadopoulos
Institute of Organic and Pharmaceutical Chemistry, National Hellenic Research Foundation,
48 Vas. Constantinou Ave, Athens 116 35, Greece
e-mail: aavram@eie.gr; hreis@eie.gr; mpapad@eie.gr

J. Leszczynski and M.K. Shukla (eds.), *Practical Aspects of Computational Chemistry I:* 129
An Overview of the Last Two Decades and Current Trends,
DOI 10.1007/978-94-007-0919-5_5, © Springer Science+Business Media B.V. 2012

5.1 Introduction

The linear and nonlinear optical (L&NLO) properties are of great importance, because they allow to understand a large number of physical effects (e.g. Kerr-effect, intermolecular forces, solvatochromism)[1, 2]. In addition they are key properties for the design and study of novel photonic materials, which have several significant applications (e.g. optical processing of information and optical computing etc.) [3, 4]. The static NLO properties have, in general, two important contributions: the electronic and the vibrational. Here we shall mainly review some of our recent work on these properties, but for completeness we shall also present some of the recently published articles by other teams.

The effect of the spatial confinement on the dipole moment and (hyper) polarizabilities of LiH has been studied by Bartkowiak and Strasburger [5]. It has been assumed that the confining model potential has the form of a penetrable spherical box. The first hyperpolarizability dispersion curve of an octupolar NLO molecule has been measured and modeled theoretically [6]. Quantum mechanical calculations have been used to model the relative intensities of the various resonances. Hrobarik et al. [7] synthesized and studied theoretically a series of push-pull chromophores which involve a dimethylamino or diphenylamino electron-donating group and a cationic benzothiazolium acceptor with an additional electron-withdrawing group (NO_2 or CN) at various positions of the heterocyclic benzene ring. Several experimental (e.g. UV-visible, spectroscopy, hyper-Rayleigh scattering) and theoretical (e.g. RI-CC2, MP2, CAM-B3LYP) techniques have been employed. They found that, in general, the studied systems have large static first hyperpolarizabilities.

Molina et al. [8] computed from first principles the dipole polarizabilities of a series of ions (e.g. Li^+, Na^+, Mg^{2+}, Ca^{2+}) in aqueous solutions. The technique they employed is based on the linear response of the maximally localized Wannier functions to an externally applied electric field. They found that proton transfer leads to instantaneous switch of the molecular polarizability. Sin and Yang [9] employed DFT to compute the first hyperpolarizability and other properties (e.g. excitation energies) of 20 silafluorenes and spirobisilafluorenes. They found that the nonlinearity increases with (increasing) number of branches. This effect has been attributed to a cooperative enhancement of the charge–transfer.

Haskopoulos and Maroulis [10] studied the interaction electric properties of $H2O\cdots Rg$ (Rg = He, Ne, Ar, Kr, Xe). Correlation effects have been taken into account by employing Møller-Plesset (MP2, MP4) and coupled-cluster theories (CCSD, CCSD(T)) in connection with flexible, carefully designed basis sets. Baranowska et al. [11] computed the interaction-induced axial static dipole moments, polarizabilities and first hyperpolarizabilities of $HCHO\cdots(HF)_n$ (n = 1,2). They employed a series of methods (e.g. MP2, CCSD(T)) in connection with various basis sets.

Rinkevicius et al. [12] have shown that when ultra-small metallic nanoparticles are combined with organic derivatives with large second- and third-order hyperpolarizabilitis, lead to enhancement of the overall third-order response of the system. Employing DFT they have demonstrated that a substantial increase in the second hyperpolarizability is produced by linking two para-nitro-aniline molecules with a tetrahedral gold cluster. Orlando et al. [13] extended to local-density, gradient-corrected and hybrid density functional, a computational technique, recently implemented in the CRYSTAL code, for the calculation of the second-order electric susceptibility in periodic systems. Karamanis and Pouchan [14] computed the static dipole (hyper)polarizabilities of gallium arsenic clusters composed of 72 atoms. They employed Hartree-Fock and a series of functionals. They found that the cluster shape has an important effect on the magnitude of the second hyperpolarizability and the performance of the DFT functionals.

5.2   The Correlation, Relativistic and Vibrational Contributions to the L&NLO Properties of ZnS, CdS and HgS

The correlation, vibrational and relativistic effects to L&NLO properties have been studied by selecting as model systems the Group IIb sulfides: ZnS, CdS and HgS [15]. These weakly bound systems are expected to have quite large vibrational (hyper) polarizabilities. To the best of our knowledge this was the first study which included the computation of all three contributions to the (hyper) polarizabilities.

The total value of a given property, P, partitioned into different contributions can be given by:

$$P_{tot}(M) = P_{tot,nr}(M) + P_{tot,rc}(M) \qquad (5.1)$$

where $P_{tot,nr}(M)$ is the nonrelativistic value and $P_{tot,nr}(M)$ is the total relativistic contribution $P_{tot,rc}(M)$; M is the method with which the computations have been performed.

The vibrational corrections can be accounted for at all levels of approximation. For example, in the case of the nonrelativistic SCF approximation the nonrelativistic term can be further partitioned in the following way:

$$P_{tot,nr}(SCF) = P_{el,nr}(SCF) + P_{vib,nr}(SCF) \qquad (5.2)$$

i.e., into pure electronic (el) and vibrational (vib) contributions. The results of all other methods employed in this study may be portioned in a similar way. The difference

$$P_{el,rc}(SCF) - P_{el,nr}(SCF) \qquad (5.3)$$

denotes the relativistic contribution to the pure electronic value of P. The correlation correction is given by:

$$P_{corr}(M) = P_{el}(M) - P_{el}(SCF) \qquad (5.4)$$

This decomposition is applicable to both nonrelativistic and relativistic results. The difference

$$P_{corr,rc}(M) - P_{corr,nr}(M) \qquad (5.5)$$

is the interference between the relativistic and the electron correlation effects. The property P may be further resolved in the following way:

$$P = <0|P^{el}|0> + P^{pv} = P^{el} + P^{ZPVA} + P^{pv}, \qquad (5.6)$$

where $-0>$ is the ground state vibrational wave function, and P^{el}, P^{ZPVA}, and P^{pv} are the electronic contributions, the zero-point vibrational averaging corrections and the pure vibrational property, respectively. These formulas will be helpful in the analysis of different computed results and in the discussion of their composition in terms of the electron correlation, relativistic, and vibrational contributions.

We shall briefly discuss our results of the polarizabilities and first hyperpolarizabilities of ZnS, CdS and HgS. In the original article [15] we have also reported the dipole moment and the second hyperpolarizability of the above derivatives.

Dipole polarizabilities. The electron correlation makes a negative contribution to α_{zz}. This is due to the shrinking of the diffuse distributions along the molecular axis. The importance of the factorizable higher order excitations is shown by the large difference between the MP2 and CCSD results. The correction due to triple excitations, accounted for at the CCSD(T) level, removes to a large extend this difference. The negative electron correlation contribution increases, in absolute value, with the nuclear charge of the metal atom. The negative sign indicates that the ionic character of the metal sulfides is overshoot by the SCF method. Similarly, the relativistic correction increases (in absolute value) with the nuclear charge of the metal atom. Both ZPVA and pure vibrational contributions are very small. At the CCSD(T) level, where all contributions are included, we observe the following sequence:

$$\alpha_{zz}(ZnS) < \alpha_{zz}(HgS) < \alpha_{zz}(CdS) \qquad (5.7)$$

Relativistic and nonrelativistic α_{zz} values for HgS, computed at CCSD(T), have a significant difference. These values are 86.934 a.u. (rel.) and 112.121 a.u. (nonrel.) (Table 5.1).

First hyperpolarizabilities. At the SCF level (nonrelativistic), the major change in β_{zzz} occurs between ZnS and CdS:

$$|\beta_{zzz}(ZnS)| - |\beta_{zzz}(CdS)| = 1136.2 \text{ a.u.} \qquad (5.8)$$

$$|\beta_{zzz}(HgS)| - |\beta_{zzz}(CdS)| = 222 \text{ a.u.} \qquad (5.9)$$

Table 5.1 Møller-Plesset and coupled cluster values of the polarizabilities of ZnS, CdS and HgS. DK relativistic and vibrational contributions are also presented

	α_{zz}		
	ZnS[a]	CdS[a]	HgS[a]
Electronic contribution			
SCF	87.325	116.123	124.972
Electron correlation corrections[b]			
MP2	−2.675	−9.918	−14.631
CCSD	−1.057	−5.734	−9.038
CCSD(T)	−2.602	−9.871	−13.688
DK relativistic corrections[c]			
SCF	−1.728	−6.582	−27.554
MP2	−1.805	−6.803	−5.312
CCSD	−1.073	−3.067	−0.747
CCSD(T)	−0.947	−1.799	3.051
ZPVA contribution			
SCF[d]	0.170	0.236	
CCSD(T)[e]	0.033	0.064	0.081
CCSD(T)[f]	0.037	0.065	0.064
Pure vibrational contribution			
SCF[d]	0.104	0.043	
CCSD(T)[e]	0.467	0.659	0.756
CCSD(T)[f]	0.458	0.554	0.089
Total			
CCSD(T)[e]	85.223	106.975	112.121
CCSD(T)[f]	82.543	98.490	86.934

All values are in a.u.

[a] The polarized basis set for S is [13 s.10p.4d.4f/7 s.5p.2d.2f]

[b] Property values calculated according to Eq. 5.4

[c] Values calculated according to Eqs. 5.3 and 5.5

[d] Non-relativistic results calculated at the SCF level of theory, using the SCF optimized geometry

[e] Non-relativistic results calculated at the CCSD(T) level of theory

[f] Relativistic results which include the relativistic correction and mixed relativistic-correlation contributions evaluated at the level of CCSD(T) approximation

The very large nonrelativistic electron correlation contribution at the MP2 level is rather disturbing. An interpretation of this pattern has been reported in ref. [15]. The CCSD and CCSD(T) methods lead to more correct energy denominators and diminish the electron correlation correction. However, the correlation correction to β_{zzz} remains large. The magnitude, in absolute value, of the Douglas –Kroll (DK) [16, 17] relativistic correction increases with the atomic number of the metal for HgS is quite significant, it is comparable with the correlation contribution (Table 5.2).

A. Avramopoulos et al.

Table 5.2 Møller-Plesset and coupled cluster values of the first hyperpolarizability parallel component of ZnS, CdS and HgS

	β_{zzz}		
	ZnS[a]	CdS[a]	HgS[a]
Electronic contribution			
SCF	−1095.8	−2232.0	−2454.0
Electron correlation corrections[b]			
MP2	1642.3	4171.0	4965.4
CCSD	685.6	1551.7	1788.9
CCSD(T)	906.5	1944.3	2130.8
DK relativistic corrections[c]			
SCF	130.5	697.4	1982.4
MP2	−126.8	−1076.6	−3794.2
CCSD	−58.1	−435.6	−1414.9
CCSD(T)	−112.9	−658.8	−1752.2
ZPVA contribution			
CCSD(T)[d]	3.5	5.3	4.0
CCSD(T)[e]	2.8	2.1	−1.4
Pure vibrational contribution			
SCF[f]	−99.8	−53.9	
CCSD(T)[d]	−98.7	−151.7	−176.6
CCSD(T)[e]	−96.1	−137.4	−44.8
Total			
CCSD(T)[d]	−284.5	−434.1	−495.8
CCSD(T)[e]	−265.0	−384.4	−139.2

The DK relativistic and vibrational corrections are also presented. All values are in a.u.
[a] The polarized basis set for S is [13s.10p.4d.4f/7s.5p.2d.2f]
[b] Property values calculated according to Eq. 5.4
[c] Values calculated according to Eqs. 5.3 and 5.5
[d] Non-relativistic results calculated at the CCSD(T) level of theory
[e] Relativistic results which include the relativistic correction and mixed relativistic-correlation contributions evaluated at the level of CCSD(T) approximation
[f] Non-relativistic results calculated at the SCF level of theory, using the SCF optimized geometry

The ZPVA contributions for the considered sulfides are very small, but the relativistic contribution has a significant effect on this small quantity. The pure vibrational contribution is small, but not negligible. The relativistic correction has a very significant effect on this contribution of HgS. One may compare the CCSD(T) values for the pure vibrational contribution to β_{zzz} of HgS, which are −176.6 a.u. (nonrelativistic) and −44.8 a.u. (relativistic). The following sequence is observed (CCSD(T) values with vibrational and relativistic corrections):

$$|\beta_{zzz}(\text{ZnS})| < |\beta_{zzz}(\text{HgS})| < |\beta_{zzz}(\text{CdS})|. \tag{5.10}$$

5.3 Vibrational Corrections by Numerov-Cooley Integration

We have investigated the possibility of using derivative Numerov-Cooley (NC) technique [18, 19] in order to present different contributions to the vibrational (hyper)polarizabilities for a given diatomic molecule [20]. For our pilot study we have selected the KLi dimmer. This weakly bound system is interesting, because it has very large electronic properties [21], due to the diffuse character of the valence shells. The rather weak bonding in connection with the high polarity, was expected to lead to large vibrational electric properties. The NC approach, which has been studied, is primarily applicable to the static electric properties. Its extension to frequency dependent properties may face some fundamental problems [20]. The proposed approach uses the property functions and avoids explicit numerical differentiation of field dependent vibrational energies. This method takes into account all orders of anharmonicities, at variance with the finite-order perturbation techniques. The DNC approach may be used with complicated potential and property curves. It may also be used to obtain benchmark results; these are of infinite order in mechanical and electric anharmonicities.

α_{zz}. The vibrational contribution is small, despite the large electronic property (Table 5.3). Both the pure vibrational contribution and the ZPVA correction are 1% of the total property. The present results allow to check the accuracy of the Bishop-Kirtman perturbation theory (BKPT) [22–27]. We observe that there is a satisfactory agreement between the CASSCF/NC [28] and CASSCF/BKPT results (Table 5.3). The effect of core correlation leads to reduction of the electronic and vibrational contributions (ZPVA correction and $|p^{vib}|$) at the CASPT2 [29, 30] level.

Table 5.3 Electronic and vibrational contributions to the parallel components of the polarizability α_{zz} and first hyperpolarizability β_{zzz} obtained using the Numerov-Cooley intergration scheme at the CASSCF and CASPT2 levels of theory

Method	K-Li r_{opt}	Total p^{total}	Average p^{ave}	Electronic p^e	Vibrational p^{ZPVA}	$p^{vib.}$
$p = \alpha_{zz}$						
CASSCF/NC	6.500	543.12	537.82	532.00	5.795	5.292
CASSCF/BKPT[a]	6.500	543.10	537.77	532.00	5.770	5.326
CASPT2/NC	6.268	520.08[b]	518.37	512.90	5.467	1.735
$p = \beta_{zzz}$						
CASSCF/NC	6.500	37,160	50,087	48,919	1,167	−12,926
CASSCF/BKPT[a]	6.500	37,112	50,089	48,919	1,170	−12,977
CASPT2/NC	6.268	23,353[b]	30,377	29,867	510	−6844

Corresponding CASSCF contributions calculated using the Bishop-Kirtman perturbation theory (BKPT) are also listed. All contributions were computed at the optimized bond length r_{opt} for the given method. All values are in a.u.

[a]The average contribution in the BKPT approach is the sum of the electronic and ZPVA contributions: $p^{ave} = p^{el.} + p^{ZPVA}$. The total contribution is the sum of the average and vibrational contributions: $p^{ave} = p^{ave} + p^{vib}$

[b]The total value is obtained as a sum of the average p^{ave} and vibrational p^{vib} contributions. The total is not evaluated directly using the VIBROT program as with CASSCF

β_{zzz}. The vibrational contribution is quite large. Specifically the pure vibrational correction is 35% of the total property. The ZPVA correction is about an order of magnitude smaller than the pure vibrational contribution. There is a perfect agreement between the CASSCF/NC and the CASSCF/BKPT results. Correlation (CASPT2) has a very significant effect on all contributions (Table 5.3) [20].

5.4   Relativistic Corrections of the L&NLO Properties of Coinage Metal Hydrides

Relativistic effects on the pure electronic contributions to (hyper)polarizabilities of coinage metal hydrides (CuH, AgH, AuH) have been studied extensively [31–34]. However, the vibrational contributions on the L&NLO properties have been mainly studied for systems involving light nuclei [35]. In systems composed of heavy nuclei one may expect significant differences between relativistic and nonrelativistic vibrational contributions. Thus the objective of this study was to find out the effect of the relativistic contribution on the vibrational contributions. The present study considered only the parallel components of the (hyper)polarizabilities.

Methodology. The NC technique [20] for the evaluation of vibrational contributions to the electric properties of MH (M = Cu, Ag, Au) requires computation of the Born-Oppenheimer potential energy and property curves [35]. The first step involves calculations of the potential energy and property curves at the CASSCF level [28]. The active space comprises the bonding and anti-bonding sigma orbitals for two valence electrons. This two-configuration CASSCF scheme is the smallest qualitatively correct active space. Subsequently we employed the CASPT2 method in order to improve upon the valence CASSCF results. It was considered that the field dependent CASPT2 [29, 30] potential energy curves are not accurate enough to use them in the numerical NC scheme [20]. Thus we used a hybrid method, according to which the electric property curves were computed over a narrow-range of internuclear distances using the CCSD(T) method [36–38]. This approach gives field dependent energies of very high accuracy which can be used to calculate satisfactory (hyper)polarizability values. The relativistic effects have been computed by employing the Douglas-Kroll (DK) approach [16, 17]. The non-relativistic (NR) property values have been computed by employing the first-order polarized basis sets [39]. For the relativistic calculations (DK) the counterparts of the above basis sets were employed and which were developed in the context of the DK method [33]. The vibrational contributions were computed by employing the Numerov Cooley (NC) integration of the vibrational equation [18, 19]. In addition the vibrational corrections have been calculated by employing the Bishop and Kirtman perturbation theory (BKPT) [22–27]. According to the BKPT approach the ZPVA correction is given by:

$$P^{ZPVA} = [P]^{0,1} + [P]^{1,0},$$

Table 5.4 Dipole moments of the coinage metal hydrides from nonrelativistic (NR) and relativistic (DK) calculations based on CASPT2 potential energy curves and CCSD(T) dipole moments functions

	CuH		AgH		AuH	
	NR[a]	DK[b]	NR[a]	DK[b]	NR[a]	DK[b]
μ^e	−1.1140	−1.0182	−1.3796	−1.1167	−1.3240	−0.5236
μ^{ZPVA}	−0.0076	−0.0111	−0.0113	−0.0062	−0.0107	−0.0006
μ	−1.1216	−1.0293	−1.3909	−1.1229	−1.3347	−0.5242

All values are in a.u.

[a]Results of the hybrid CASPT2/CCSD(T) nonrelativistic calculations with the ZPVA correction evaluated using the NC method

[b]Results of the hybrid CASPT2/CCSD(T) relativistic (DK) calculations with the ZPVA correction evaluated using the NC method

where P is μ_z, α_{zz} or β_{zzz}. The pure vibrational contributions to α_{zz} and β_{zzz} are given by:

$$\alpha^{pv} = \left[\mu^2\right]^{0,0} + \left[\mu^2\right]^{2,0} + \left[\mu^2\right]^{1,1} + \left[\mu^2\right]^{0,2} \tag{5.11}$$

$$\beta^{pv} = \left[\mu\alpha\right]^{0,0} + \left[\mu^3\right]^{1,0} + \left[\mu^3\right]^{0,1} + \left[\mu\alpha\right]^{2,0} + \left[\mu\alpha\right]^{1,1} + \left[\mu\alpha\right]^{0,2} \tag{5.12}$$

The superscripts n, m denote the level of mechanical and electric anharmonicity. This compact notation is described in [22–27]. Marti et al. [40–42] proposed another partition scheme:

$$P = P^e + P^{vib} = P^{el} + P^{pv} + P^{ZPVA} = P^{el} + P^{relax} + P^{curv} \tag{5.13}$$

where P^{relax} is the nuclear relaxation contribution. It represents the effect of the change in the equilibrium geometry in the presence of the electric field. P^{curv} is the curvature contribution and is associated with the change of the shape of the potential energy function. According to BKPT we have:

$$\alpha^{relax} = \left[\mu^2\right]^{0,0} \tag{5.14}$$

$$\beta^{relax} = \left[\mu\alpha\right]^{0,0} + \left[\mu^3\right]^{1,0} + \left[\mu^3\right]^{0,1} \tag{5.15}$$

$$\alpha^{curv} = \alpha^{ZPVA} + \left[\mu^2\right]^{2,0} + \left[\mu^2\right]^{1,1} + \left[\mu^2\right]^{0,2} \tag{5.16}$$

$$\beta^{curv} = [\beta]^{ZPVA} + \left[\mu\alpha\right]^{2,0} + \left[\mu\alpha\right]^{1,1} + \left[\mu\alpha\right]^{0,2}. \tag{5.17}$$

μ_z. The relativistic correction of the electronic contribution increases with the atomic number of the metal, as it should be expected, and for AuH is quite significant (Table 5.4). The NR and DK ZPVA values are very small and in particular the DK values for AgH and AuH are negligible.

Table 5.5 The paraller dipole polarizability (α) of the coinage metal hydrides from nonrelativistic (NR) and relativistic (DK) valence CASSCF calculations

Molecule		NR		DK
		BKPT[a]	NC[b]	NC[b]
CuH	α^e		40.42[c]	39.31
	α^{ZPVA}	1.43	1.48	1.36
	α^{pv}	0.90	0.88	0.85
	α^{relax}	0.87		
	α^{curv}	1.43		
AgH	α^e		49.84	46.37
	α^{ZPVA}	1.86	2.15	1.89
	α^{pv}	1.27	1.25	0.87
	α^{relax}	1.26		
	α^{curv}	1.87		
AuH	α^e		51.24	41.67
	α^{ZPVA}	1.88	2.33	0.95
	α^{pv}	1.09	1.09	0.29
	α^{relax}	1.07		
	α^{curv}	1.90		

All values are in a.u.
[a]The ZPVA and pure vibrational (pv) corrections from the Bishop-Kirtman perturbation theory
[b]The ZPVA and pure vibrational (pv) corrections from the Numerov-Cooley intergration
[c]Analytic calculation using the Dalton software

Table 5.6 Paraller dipole polarizability (α) of the coinage metal hydrides from nonrelativistic (NR) and relativistic (DK) calculations based on CASPT2 potential energy curves and CCSD(T) property functions

	CuH		AgH		AuH	
	NR[a]	DK[b]	NR[a]	DK[b]	NR[a]	DK[b]
α^e	37.33	36.42	47.52	44.48	49.84	39.51
α^{ZPVA}	0.97	1.08	1.66	1.13	1.63	0.52
α^{pv}	0.38	0.27	0.60	0.37	0.47	0.00
α	36.38	37.77	49.78	45.98	51.94	40.03

All values are in a.u.
[a]Results of the hybrid CASPT2/CCSD(T) nonrelativistic calculations with the ZPVA and pure vibrational corrections evaluated using the NC method
[b]Results of the hybrid CASPT2/CCSD(T) relativistic (DK) calculations with the ZPVA and pure vibrational corrections evaluated by using the NC method

α_{zz}. The NR electronic contribution increases with the atomic number of the metal. At the relativistic level, a maximum value is observed for AgH (Tables 5.5 and 5.6). The relativistic correction has a significant effect on the ZPVA and pure vibrational contributions of AgH, and in particular of AuH. The relaxation and curvature contributions are small (Table 5.5).

β_{zzz}. Both the NR and DK values of the electronic contribution get a maximum value for AgH. The relativistic effect is very significant for the electronic contri-

Table 5.7 The paraller first hyperpolarizability (β) of the coinage metal hydrides from nonrelativistic (NR) and relativistic (DK) valence CASSCF calculaions

Molecule		NR		DK
		BKPT[a]	NC[b]	NC[b]
CuH	$\beta^{el,c}$	826	837	745
	β^{ZPVA}	64	65	53
	β^{pv}	−210	−204	−174
	β	681	698	624
	β^{relax}	−183		
	β^{curv}	38		
AgH	$\beta^{el,c}$	1,085	1,113	804
	β^{ZPVA}	108	129	83
	β^{pv}	−358	−363	−236
	β	835	879	651
	β^{relax}	−318		
	β^{curv}	68		
AuH	$\beta^{el,c}$	970	994	351
	β^{ZPVA}	100	131	15
	β^{pv}	−315	−324	−59
	β	755	801	307
	β^{relax}	−275		
	β^{curv}	60		

All values are in a.u.
[a]The ZPVA and pure vibrational (pv) corrections from the Bishop-Kirtman perturbation theory [35]
[b]The ZPVA and pure vibrational (pv) corrections from the Numerov-Cooley intergration
[c]The electronic contributions (β^{el}) listed as BKPT follow from the analytic evaluation of the nonrelativistic CASSCF energy derivatives. The NC value correspond to the numerical differentiation of the field dependent CASSCF energy

Table 5.8 Paraller first hyperpolarizability (β) of the coinage metal hydrides from nonrelativistic (NR) and relativistic (DK) calculations based on CASPT2 potential energy curves and CCSD(T) property functions

	CuH		AgH		AuH	
	NR[a]	DK[b]	NR[a]	DK[b]	NR[a]	DK[b]
β^e	468	416	734	489	685	166
β^{ZPVA}	20	17	56	23	49	0
β^{pv}	−67	−52	−159	−90	−131	−35
β	421	381	632	422	603	131

All values are in a.u.
[a]Results of the hybrid CASPT2/CCSD(T) nonrelativistic calculations with the ZPVA and pure vibrational corrections evaluated using the NC method [35]
[b]Results of the hybrid CASPT2/CCSD(T) relativistic (DK) calculations with the ZPVA and pure vibrational corrections evaluated by using the NC method

bution in particular for AgH and AuH. The relativistic effects on both the ZPVA correction and the *pv* contribution are very significant (Tables 5.7 and 5.8). β^{relax} makes a large contribution. Less significant is the contribution of β^{curv} (Table 5.7).

5.5   Cyclopropenone and Cyclopropenethione

We shall discuss the electronic and vibrational contributions to the hype r polarizabilities of some ketones and thiones, that is cyclo prope none/cyclo propenthione,
cyclopropanone/cyclopropanthione, acetone/thioacetone [43]. Cyclo propenone and
cyclopropenthione have several structural features, which make the study of their
hyperpolarizabilities very appealing, that is there is a π-conjugation effect and the
three-membered carbon ring is highly strained. The effect of the strain is expected
to be significant on the vibrational contributions. Of course π-conjugation and strain
can not be separated from each other. However, their effect on the first hyperpolarizability can be estimated to some extend by comparing the results of some properly
selected model systems with those of cyclopropenone and cyclopropenthione. For
example, saturation of the carbon-carbon double bond removes the π-conjugation
effect, but two additional C-H bonds are introduced. The effect of strain may be
discussed by adopting as model compounds acetone and thioacetone (Fig. 5.1).

In this article we shall review the hyperpolarizability results. However, in the
original paper the dipole moment and the polarizability values were also reported
[43]. For completeness we note that the electronic contributions to the dipole moment and the (hyper) polarizabilities of acetone and thioacetone have been studied
by employing SCF, CASSCF, MP2, CCSD, CCSD(T) and CASPT2 methods [44].
The importance of the electron correlation was discussed and Eckart et al. [44] found
that cyclic conjugated thioketones have large NLO properties. The computations
presented here have been performed at the SCF-HF level by employing the polarized

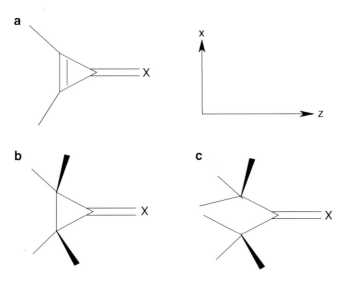

Fig. 5.1 Geometry and coordinate axes assumed in calculations for ketones (X═O) and thiones
(X═S): (**a**) cyclopropenone (cyclopropenethione), (**b**) cyclopropanone (cyclopropanthione), and
(**c**) acetone (thioacetone)

Table 5.9 Pure electronic values of the first dipole hyperpolarizability of cyclopropenone, cyclopropenethione, and related reference molecules

Property	Molecule		
	Cyclopropenone[a] (C_3H_2O)	Cyclopropanone (C_3H_4O)	Acetone (C_3H_4O)
β_{xxz}	25.0	19.5	28.2
β_{yyz}	68.1	22.4	−5.4
β_{zzz}	53.0	50.5	54.1
β^b	87.6	55.4	46.1
	Cyclopropenethione[a] (C_3H_2S)	Cyclopropanethione (C_3H_4S)	Thioacetone (C_3H_4S)
β_{xxz}	−65.6	−41.2	−31.3
β_{yyz}	−40.0	−85.0	−106.5
β_{zzz}	123.4	10.1	15.1
β^b	11.2	−69.7	−73.1

Results of the SCF-HF calculations with the PolX basis sets. All values in a.u.
[a] The reported values consistently correspond to the equilibrium geometry determined in SCF-HF calculations with the PolX basis sets
[b] Average values

basis sets [45, 46], proposed by Sadlej, which have been shown to perform well for the studied molecules [44]. These basis give also reasonable hyperpolarizabilities [47]. The coordinate system assumes that C = O (C = S) and the C-S bonds lie on the yz plane. The z-axis is the two-fold symmetry axis and its positive end points to the heteroatom.

Electronic contributions. Saturation of the ring reduces the electronic contribution to β of ketones and increases the above contribution to $|\beta|$ of thiones. A similar observation is made for the effect of strain. However, the resulting change is smaller. A more pronounced effect, related to strain, may be seen in some individual components (e.g. β_{yyz} of C_3H_4O/C_3H_6O) (Table 5.9).

Vibrational contributions. The pure vibrational term dominates the vibrational correction. The ZPVA correction is negligible. Saturation of the carbon ring brings a significant reduction of the *pv* contribution to β_{xxz} for the ketone series (the corresponding values are 36.5 a.u., −43.1 a.u.). In the thione series this effect is smaller. The *pv* contribution to β_{xxz} increases on passing from cyclopropanone (−43.1 a.u.) to acetone and becomes negligible (1.1 a.u.). An impressive difference is observed between the *pv* contribution to β_{zzz} for the ketone and thione series. In the first series this contribution is practically constant. In the thione series, saturation of the carbon ring leads to a change of −127 a.u. for β_{zzz}. The change on passing from cyclopropanone to acetone is −34.5 a.u. For these two molecules the vibrational contribution (*pv* + ZPVA) is considerably larger than the electronic one. For cyclopropenone $[\mu\alpha]^{0,0}$ and $[\mu^3]^{1,0}$ have approximately equal contribution. In cyclopropanone the dominant term, in absolute value, is $[\mu\alpha]^{0,0}$. In acetone the terms $[\mu\alpha]^{2,0} + [\mu\alpha]^{0,2}$, cancel $[\mu\alpha]^{1,1}$, approximately. In cyclopropenthione, the dominant term, in absolute value, is $[\mu^3]^{0,1}$, while in cyclopropanethione is $[\mu\alpha]^{0,0}$.

Table 5.10 Vibrational contributions to the first dipole hyperpolarizability of cyclopropenone, cyclopropenethione, and related reference ketones and thioketones

Property	Vibrational contribution[a]	Molecule		
		Cyclopropenone[a] (C_3H_2O)	Cyclopropanone (C_3H_4O)	Acetone (C_3H_4O)
β_{xxz}	ZPVA	0.5	0.3	−3.0
	PV	36.5	−43.1	1.1
β_{yyz}	ZPVA	0.8	0.3	2.3
	PV	33.8	38.0	17.0
β_{zzz}	ZPVA	−0.2	0.1	−2.8
	PV	−59.8	−57.2	−61.5
β^b	ZPVA	0.7	0.4	−2.1
	PV	6.3	−57.4	−26.0
		Cyclopropenethione[a] (C_3H_2S)	Cyclopropanethione (C_3H_4S)	Thioacetone (C_3H_4S)
β_{xxz}	ZPVA	0.3	0.2	−2.3
	PV	20.0	−8.4	−3.3
β_{yyz}	ZPVA	0.6	0.5	0.7
	PV	−2.5	17.6	−10.9
β_{zzz}	ZPVA	0.7	0.8	−3.1
	PV	−39.0	−165.7	−200.2
β^b	ZPVA	1.0	0.9	−2.8
	PV	−12.9	−93.9	−128.6

Results of the SCF-HF calculations with the PolX basis sets. All values in a.u.
[a]ZPVA and pure vibrational (PV) contribution calculated according with to Bishop-Kirtman PT expressions
[b]Average values

Table 5.11 Contributions to vibrational corrections to the parallel component (β_{zzz}) of the first dipole hyperpolarizability

Molecule	Contribution					
	$[\mu\alpha]^{0,0}$	$[\mu\alpha]^{2,0}$	$[\mu\alpha]^{1,1}$	$[\mu\alpha]^{0,2}$	$[\mu^3]^{1,0}$	$[\mu^3]^{0,1}$
Cyclopropenone	−33.3	1.2	1.3	−1.1	5.9	−33.8
Cyclopropanone	−60.9	−1.0	5.5	−2.6	13.7	−12.0
Acetone	−56.1	430.5	−871.7	443.5	0.4	−8.1
Cyclopropenethione	7.8	2.8	2.3	1.2	14.0	−67.0
Cyclopropanethione	−165.1	−2.4	2.8	15.8	−2.2	−14.5
Thioacetone	−179.9	88.2	−239.8	161.7	−13.0	−17.4

All values are in a.u.

In thioacetone the $[\mu\alpha]^{0,0}$, $[\mu\alpha]^{2,0}$ and $[\mu\alpha]^{1,1}$ have significant contribution (in absolute value). The above analysis corresponds to the static limit. We have also considered the vibrational contributions to $\beta(-\omega; \omega,0)$, at a frequency of 0.02 a.u. The frequency dependent vibrational correction for cyclopropenone is −1.2 a.u., while the static value is 7.0 a.u. For cyclopropenethione the respective property values are −7.7 a.u. and −12.0 a.u (Tables 5.10 and 5.11).

5.6   Hyperpolarizabilities of the Hydrides of Li, Na and K

We have performed an extensive study of the dipole moment and (hyper)polarizabilities of MeH, Me = Li, Na and K [48]. Here we shall briefly review the electronic and vibrational contributions to first and second hyperpolarizabilities, since these results show the more interesting trends. We will consider the average values [49]. The basis sets developed by Sadlej and Urban [45, 49] have been used for the computations. The electric properties of interest can be expressed by:

$$\gamma^{nr} = [\alpha^2]^{0,0} + [\mu\beta]^{0,0} + [\mu^2\alpha]^{1,0} + [\mu^2\alpha]^{0,1} + [\mu^4]^{2,0} + [\mu^4]^{1,1} + [\mu^4]^{0,2}$$
$$(5.18)$$

$$\gamma^{curv} = \gamma^{ZPVA} + [\alpha^2]^{2,0} + [\mu\beta]^{2,0} + [\alpha^2]^{1,1} + [\mu\beta]^{1,1} + [\alpha^2]^{0,2} + [\mu\beta]^{0,2} \quad (5.19)$$

$$\gamma^{pv} = [\alpha^2]^{0,0} + [\alpha^2]^{2,0} + [\alpha^2]^{1,1} + [\alpha^2]^{0,2} + [\mu\beta]^{0,0} + [\mu\beta]^{2,0} + [\mu\beta]^{1,1} + [\mu\beta]^{0,2}$$
$$+ [\mu^2\alpha]^{1,0} + [\mu^2\alpha]^{0,1} + [\mu^4]^{2,0} + [\mu^4]^{1,1} + [\mu^4]^{0,2}$$
$$(5.20)$$

Electronic hyperpolarizabilities. It is of interest to note that the correlation contribution, at the CCSD(T) level, to β^{el} is 50% (51%), 61% (64%) and 71% (71%) for LiH, NaH and KH, respectively. The corresponding percentages to γ^{el} are given in parenthesis. It is observed that the correlation contributions to both β^{el} and γ^{el}, in percentage terms, are very similar (Table 5.12).

Vibrational first hyperpolarizabilities. The double harmonic approximation ($[\mu\alpha]^{0,0}$) gives, in general satisfactory results for the *pv* contribution. The above approximation is less satisfactory for KH at the SCF level. The second largest contribution to *pv* (in absolute value) at both HF and CCSD(T) levels, comes from $[\mu^3]^{0,1}$. The reported results show that β^{nr} is a reasonable approximation to β^{pv}. The dominant contribution to β^{pv} comes from $[\mu\alpha]^{0,0}$. It is observed that $|[\mu^3]^{0,1}| > -[\mu^3]^{1,0}$ and $|[\mu\alpha]^{0,2}| > -[\mu\alpha]^{2,0}$. In general, β^{curv} has a small value. β^{ZPVA} is much smaller than $|\beta^{pv}|$. Correlation has a significant effect on the

Table 5.12 The electronic first and second hyperpolarizabilities of MeH computed using SCF and CCSD(T)[a] theories

Property	SCF			CCSD(T)		
	LiH	NaH	KH	LiH	NaH	KH
β	419.8	1146.8	1746.5	885.8	2944.8	6047.1 (6057.1)
$\gamma \times 10^{-3}$	41.9	80.7	201.2	86.1	224.35	702.2 (704.0)

The property values are in atomic units
[a]The computations have been performed by correlating 4 electrons for LiH, 10 for NaH and 20 KH. The property values for KH, which are given in parentheses have been computed by correlating 10 electrons

Table 5.13 Analysis of the first hyperpolarizabilities of LiH, NaH and KH, computed at the SCF and CCSD(T)[a] levels of theory

Contribution	LiH		NaH		KH	
	SCF	CCSD(T)	SCF	CCSD(T)	SCF	CCSD(T)
$[\mu\alpha]^{0,0}$	−231.8	−287.3	−390.6	−559.7	−618.9	−1113.6
$[\mu^3]^{0,1}$	−25.8	−20.1	−80.1	−41.0	−305.8	−202.8
$[\mu^3]^{1,0}$	−14.5	−5.3	−24.4	10.9	−52.4	50.7
$[\mu\alpha]^{1,1}$	−13.0	−18.5	−22.6	−31.1	−41.2	−97.5
$[\mu\alpha]^{2,0}$	0.7	11.3	−0.2	2.6	−1.2	3.0
$[\mu\alpha]^{0,2}$	−10.1	−14.9	−16.9	−20.5	−28.6	−60.9
β^{nr}	−272.1	−312.7	−495.1	−589.8	−977.1	−1265.7
β^{curv}	7.9	59.9	14.3	133.1	3.1	250.0
β^{pv}	−294.4	−335.0	−534.8	−638.8	−1048.1	−1421.2
β^{zpva}	30.3	82.0	54.0	182.1	74.2	405.5
β^{vib}	−264.1	−252.9	−480.8	−456.7	−973.9	−1015.7

The values are in atomic units

[a]Four, ten and twenty electrons are correlated for LiH, NaH and KH, respectively

Table 5.14 Analysis of the vibrational second hyperpolarizabilities of LiH, NaH and KH using SCF, and CCSD(T)[a] theories

Method Comp.	LiH		NaH		KH	
	SCF	CCSD(T)	SCF	CCSD(T)	SCF	CCSD(T)
$[\mu\beta]^{0,0}$	−4,338	−9,194	−12,018	−30,201	−26,800	−101,886
$[\alpha^2]^{0,0}$	3,795	7,131	4,979	14,440	6,004	23,423
$[\mu^2\alpha]^{1,0}$	1,331	1,380	3,164	2,593	9,045	13,480
$[\mu^2\alpha]^{0,1}$	1,924	2,321	5,039	5,531	13,175	21,685
$[\alpha^2]^{1,1}$	187	541	331	1,357	474	3,276
$[\mu\beta]^{1,1}$	−364	−922	−906	−2,126	−2,064	−9,977
$[\mu^4]^{1,1}$	240	86	629	−215	2,231	−1,974
$[\mu\beta]^{0,2}$	775	1,862	2,115	4,907	4,434	19,738
$[\mu\beta]^{2,0}$	22	375	11	85	−23	5
$[\alpha^2]^{0,2}$	−678	−1,444	−877	−2,346	−993	−4,538
$[\alpha^2]^{2,0}$	12	48	30	191	56	554
$[\mu^4]^{0,2}$	223	172	1,082	416	7,015	4,268
$[\mu^4]^{2,0}$	14	−149	35	−15	280	165
γ^{nr}	3,192	1,746	2,909	−7,451	10,951	−40,838
γ^{pv}	3,146	2,206	3,614	−5,383	12,834	−31,780
γ^{zpva}	2,283	6,965	3,430	14,329	6,138	51,295
γ^{curv}	2,237	7,426	4,134	16,396	8,022	60,354

The values are in atomic units

[a]The number of electrons, which have been correlated are given in parenthesis: LiH(4), NaH(10) and KH(20)

vibrational properties. A similar observation has been made for the electronic contributions. $|\beta^{pv}|$ is smaller than β^{el}, but it has a significant magnitude. $|\beta^{nr}|$, β^{curv}, $|\beta^{pv}|$ and β^{ZPVA} increase with the atomic number of Me (MeH), at the CCSD(T) level (Table 5.13).

Vibrational second hyperpolarizabilities. γ^{nr} and γ^{pv} of MeH decrease with correlation. The leading contribution comes from $[\mu\beta]^{0,0}$, which decreases with correlation. $\gamma^{ZPVA} > \gamma^{pv}$ for MeH, at the CCSD(T) level (Table 5.14). The reverse inequality is observed at the SCF level. We recall that β^{ZPVA} is small in comparison with β^{pv}. The double harmonic approximation $([\alpha^2]^{0,0} + [\mu\beta]^{0,0})$ does not satisfactorily approximate γ^{pv} (Table 5.14). In order to compare the effect of the various anharmonicities, collectively, we define:

$$1. \quad \Sigma[A]^0 = |\mu\beta|^{0,0} + [\alpha^2]^{0,0} \tag{5.21}$$

$$2. \quad \Sigma[A]^{1,1} = |\alpha^2|^{1,1} + [\mu\beta]^{1,1} + [\mu^4]^{1,1} \tag{5.22}$$

$$3. \quad \Sigma[A]^{0,2} = |\mu\beta|^{0,2} + [\alpha^2]^{0,2} + [\mu^4]^{0,2} \tag{5.23}$$

$$4. \quad \Sigma[A]^{2,0} = |\mu\beta|^{2,0} + [\alpha^2]^{2,0} + [\mu^4]^{2,0} \tag{5.24}$$

The presented results for MeH, at the CCSD(T) level, have shown the greater importance of the mechanical over the electrical anharmonicity terms and can be summarized by the following inequality:

$$|\Sigma[A]^0| > |\Sigma[A]^{0,2}| > |\Sigma[A]^{1,1}| > \Sigma[A]^{2,0} \tag{5.25}$$

The inequality $[\mu^2\alpha]^{0,1} > -\mu^2\alpha]^{1,0}$ confirms the greater importance of the mechanical over the electrical anharmonicity terms.

5.7   Electronic and Vibrational Contributions to Pyrrole

We shall review the electronic and the pure vibrational contributions to the hyperpolarizabilities of pyrrole [50]. The original article involved also dipole moment and polarizabilities. The molecule is placed on the yz plane (Fig. 5.2). The computations have been performed at the Hartree-Fock level, employing the Pol basis set [45].

Electronic contributions. In Table 5.15 the electronic contributions to hyperpolarizabilities. These have been computed with the HF, MP2, CCSD and CCSD(T) methods and the Pol basis set. Correlation at the CCSD(T) level increases β and γ by 50.2% and 17.4%, respectively [50] (Table 5.15).

Vibrational first hyperpolarizabilities. β_{xxz}^{pv} is the dominant component; the main contributions come from $[\mu^3]^{0,1}$ and $[\mu^3]^{1,0}$ [50]. The first of those is computed by using first-order dipole moment derivatives and cubic force constants. $[\mu^3]^{1,0}$ is calculated in terms of first- and second-order dipole moment derivatives. β_{yyz}^{pv} and β_{zzz}^{pv} are much smaller than $|\beta_{xxz}^{pv}|$. The large anharmonicity of β_{xxz}^{pv},

Fig. 5.2 The structure of
Pyrrole

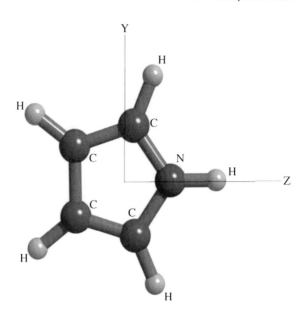

Table 5.15 Electronic static, first and second hyperpolarizability of pyrrole[a] at various level of theory, computed with the PolX basis set

Method Property[a]	HF	MP2	CCSD	CCSD(T)
β_{zxx}	35.4	38.2	35.7	37.1
β_{zyy}	−25.8	−19.9	−13.3	−15.2
β_{zzz}	33.6	34.1	41.0	43.0
β	25.9	31.4	38.0	38.9
γ_{xxxx}	16,200	18,600	15,800	16,900
γ_{yyyy}	14,200	17,000	15,400	16,400
γ_{zzzz}	6,760	9,490	8,690	9,430
γ_{zzxx}	5,010	6,960	5,910	6,420
γ_{zzyy}	2,970	4,590	3,880	4,300
γ_{xxyy}	9,250	10,900	9,160	9,960
γ	14,324	17,998	15,558	16,818

All the property values are in a.u.
[a]The HF/PolX optimized geometry was used to all computations. The molecule lies on the YZ plane with the dipole directed along z- axis

clearly shown by the large values of $[\mu^3]^{0,1}$ and $[\mu^3]^{1,0}$, is mainly, attributed to the NH group. For furan, $[\mu^3]^{1,0}$ and $[\mu^3]^{0,1}$, take the values −10.12 a.u. and 3.52 a.u., respectively. The corresponding values for thiophene are −12.39 a.u. 5.66 a.u., respectively. Comparison with the electronic contributions shows that $\beta_{zxx}^{pv}/\beta_{zxx}^{el} = -14.9$ (Table 5.16).

Vibrational second hyperpolarizabilities. γ_{xxxx}^{pv} and γ_{xxzz}^{pv} make the leading contributions to γ^{pv} [50]. The other components make a much smaller contribution. The larger contributions to γ_{xxxx}^{pv} and γ_{xxzz}^{pv} are made by $[\mu^4]^{1,1}$ and $[\mu^4]^{2,0}$

Table 5.16 Analysis of the pure vibrational contribution to the first hyperpolarizability components (a.u.) of pyrrole[a]

	β_{xxz}^{pv}	β_{yyz}^{pv}	β_{zzz}^{pv}
$[\mu\alpha]^{0,0}$	28.50	3.37	0.18
$[\mu^3]^{0,1}$	−213.06	−0.78	−2.41
$[\mu^3]^{1,0}$	−307.33	0.78	−5.56
$[\mu\alpha]^{1,1}$	−13.32	2.35	−2.00
$[\mu\alpha]^{2,0}$	−0.81	−1.94	−13.57
$[\mu\alpha]^{0,2}$	6.48	1.77	8.84
β_{iiz}^{pv}	−499.54	5.55	−14.52

[a]The computations have been performed using the Pol basis set

(Table 5.17). $[\mu^4]^{1,1}$ is computed in terms of the cubic force constant and dipole moment derivatives (first- and second-order), while $[\mu^4]^{2,0}$ is calculated in terms of the first-, second- and third-order dipole moment derivatives. For the analysis of the results we define:

$$[A]^0 = [\mu\beta]^{0,0} + [\alpha^2]^{0,0} \tag{5.26}$$

$$[A]^I = [\mu^2\alpha]^{1,0} + [\mu^2\alpha]^{0,1} \tag{5.27}$$

$$[A]^{II} = [\alpha^2]^{2,0} + [\alpha^2]^{1,1} + [\alpha^2]^{0,2} + [\mu\beta]^{2,0} + [\mu\beta]^{1,1} + [\mu\beta]^{0,2}$$
$$+ [\mu^4]^{2,0} + [\mu^4]^{1,1} + [\mu^4]^{0,2} \tag{5.28}$$

For the components γ_{yyyy}^{pv}, γ_{zzzz}^{pv}, γ_{xxyy}^{pv}, and γ_{yyzz}^{pv}, we have found that:

$$[A]^0 > |[A]^I| > [A]^{II} \tag{5.29}$$

For γ_{xxxx}^{pv} and γ_{xxzz}^{pv}, it has been found:

$$|[A]^{II}| > [A]^0 > [A]^I \tag{5.30}$$

In this case $[A]^{II}$ makes by far the larger contribution to γ_{xxzz}^{pv} (Table 5.17).

It is useful to understand the effect of the employed derivatives on the properties of interest [50]. We specifically consider, as a test case, the components β_{xxz}^{pv} and γ_{xxzz}^{pv}. The anharmonicities will be described by (mnop) [50], where m denotes the mechanical and nop defines the electrical anharmonicity [51]. m indicates the use of harmonic (m = 0), cubic (m = 1), and quartic (m = 2) approximation to the potential. Dipole moment derivatives: n = 0,1,2,3, that is no derivatives; first derivatives; first and second derivatives; first, second and third derivates are taken into account, respectively. Polarizability derivatives: o = 0,1,2 and first hyperpolarizability derivatives: p = 0,1; the notation is similar to that given for the dipole moment derivatives [52]. We will consider some specific examples. It

Table 5.17 Analysis of the pure vibrational contribution to the second hyperpolarizability components of pyrrole[a]

	γ_{xxxx}^{pv}	γ_{yyyy}^{pv}	γ_{zzzz}^{pv}	γ_{xxzz}^{pv}	γ_{yyzz}^{pv}	γ_{xxyy}^{pv}
$[\mu\beta]^{0,0}$	2366.3	447.0	33.3	1894.3	−178.5	2918.0
$[\alpha^2]^{0,0}$	686.3	4937.6	3059.7	499.8	1074.3	704.8
$[\mu^2\alpha]^{1,0}$	2576.5	393.2	94.9	7011.2	−86.8	1821.9
$[\mu^2\alpha]^{0,1}$	−5118.9	−4.9	208.1	−5397.1	2.4	−2941.4
$[\alpha^2]^{1,1}$	1.5	30.8	−98.6	−4.1	−3.1	2.3
$[\mu\beta]^{1,1}$	6.3	−6.2	−56.1	42.7	−13.6	38.9
$[\mu^4]^{1,1}$	15319.4	−6.8	24.5	27382.8	0.02	−509.2
$[\mu\beta]^{0,2}$	3200.1	−40.2	20.6	−1559.1	25.5	−848.4
$[\mu\beta]^{2,0}$	−61.6	−5.5	−3.3	−46.3	−8.1	−85.6
$[\alpha^2]^{0,2}$	−38.5	−289.3	−271.1	18.2	−80.9	−35.5
$[\alpha^2]^{2,0}$	15.7	271.6	517.5	29.1	82.5	27.9
$[\mu^4]^{0,2}$	−1268.5	−0.3	23.2	9191.1	6.6	7.5
$[\mu^4]^{2,0}$	−25034.7	36.9	53.1	21306.0	−9.7	679.3
γ_{iijj}^{pv}	−73501	5763.9	3606.8	60368.6	810.6	1780.5

[a]The computations have been performed using the Pol basis set. The property values are in a.u.

Table 5.18 Analysis of the effect of the various property deri vatives on β_{zxx}^{pv} and γ_{xxzz}^{pv} of pyrrole (all in a.u.)[a]

mnop	β_{xxz}^{pv}	γ_{xxzz}^{pv}
0100	0.0	0.0
0200	−307.33	21,068
0210	−278.83	21,041
0220	−279.11	28,608
0221	−279.11	30,503
0321	−279.64	30,694
1200	−520.39	57,224
1100	−213.06	8,773
1210	−468.08	51,579
1220	−482.76	59,323
1221	−482.76	60,338
1321	−482.29	60,529
2100	−213.06	21,068
2200	−520.39	57,642
2210	−484.32	52,236
2220	−499.00	59,799
2221	−499.00	60,177
2320	−499.54	60,037
2321	−499.54	60,368

[a]The computations have been obtained with the Pol basis set at the SCF level

is observed that cubic terms have a great effect ([(0200/1200)] on the considered properties (Table 5.18). The second-order dipole moment derivatives appear to have a significant effect (2100)/(2200). The effect of the polarizability derivatives may be seen, for example by (2200)/(2210) and (2210)/(2220), while the effect of the first hyperpolarizability derivatives is shown by (2320)/(2321) (Table 5.18).

Comparison with the electronic contributions shows that $\gamma_{xxxx}^{pv}/\gamma_{xxxx}^{el} = -4.5$ and $\gamma_{xxzz}^{pv}/\gamma_{xxzz}^{e} = 12.1$.

5.8   Linear and Nonlinear Optical Properties of Fullerene Derivatives and Endohedral Fullerenes

Since their discovery, fullerenes and their derivatives have been the subject of very extensive research. One of the topics investigated intensively are the linear and nonlinear optical (NLO) properties, owing to a variety of possible applications. Here we review some of the recent work of our group in this area, which is concerned with the *ab-initio* calculation of molecular NLO properties of two different kinds of fullerene derivatives, a) substituted 1,2-dihydro fullerenes and b) fullerenes endohedrally doped with atoms or small molecules. Apart from the purely electronic response, we also focus on the vibrational contributions to the NLO response, that is, to the response of the nuclei to the external electric fields.

The large size of the systems considered here places severe restrictions on the level of theory and the size of basis sets, which can be employed. Thus an important aspect of our investigations was to evaluate the accuracy and reliability of computationally 'cheaper' methods, like density functional theory (DFT) or semiempirical methods. Thus, whenever possible, values obtained with DFT or semiempirical methods were compared with results obtained with standard *ab-initio* methods as HF and MP2. With respect to DFT, the discovery of the so-called 'DFT catastrophe' in NLO properties of long linear chains about a decade ago [53–55] calls for special care when applying DFT methods to other large molecular systems of a more complicated shape, for which little is known about the behaviour of DFT with respect to NLO properties. Accordingly, we found a 'DFT catastrophe' in substituted fullerenes, whose magnitude correlates with the electron donating capability of the ligand, instead of the more common correlation with system size.

5.8.1   Substituted Dihydro-Fullerenes

In Ref. [56] the NLO properties of C_{60} covalently bound to benzothiadiazole, triphenylamine-benzothiazole and carbazole derivatives of different electron-donating power were investigated (Fig. 5.3). Geometries were optimized at the semiempirical PM3 level. To access the reliability of this method, the geometry of molecule 1 was additionally optimized at the B3LYP/6-31G* level, and found to be in good agreement with the PM3 optimized structure.

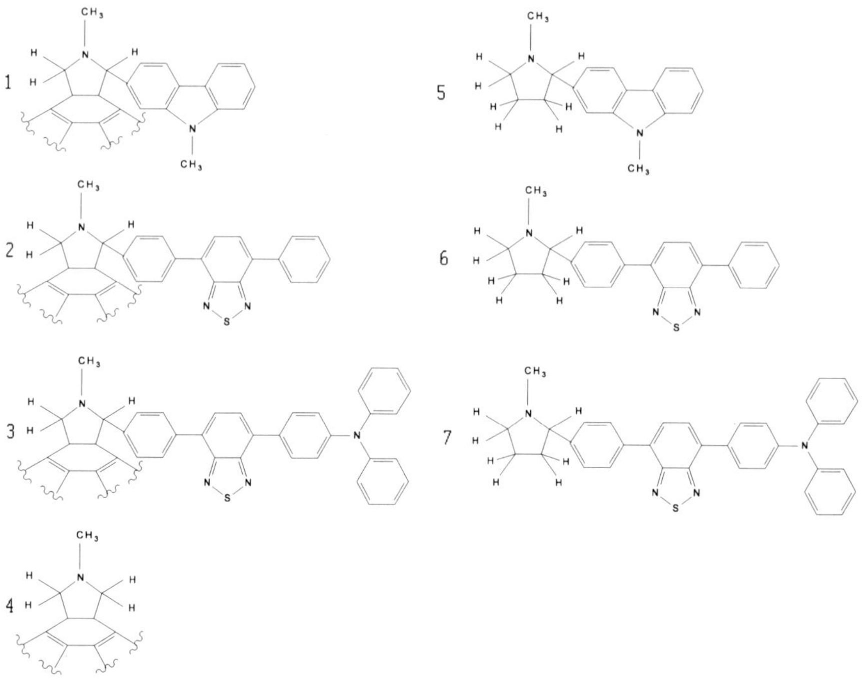

Fig. 5.3 Molecules investigated in Ref. [56]

The first aim of this study was a comparison of the performance of a large variety of DFT functionals among each other and with standard *ab-initio* methods (RHF and MP2) with respect to the accuracy of the (electronic) NLO properties. Thus the diagonal component in *x*-direction, which is roughly the direction of largest extension, of the dipole moment μ_x, polarizability α_{xx} and first hyperpolarizability β_{xxx} for compound **1** was computed at the HF and MP2 level and with DFT using different functionals. In addition, different basis sets were employed: 3-21G, DZP and 6-31G*, where DZP is a polarized double zeta basis of Slater-type, as employed in the ADF program [57], and is approximately of the same quality as 6-31G*. As the results reproduced in Table 5.19 show, standard DFT functionals like LDA, PW91, RevPBEx, GRAC and LB94 show the typical DFT overshoot as found previously for long linear chains [53–55]. On the other hand, the long-range corrected functionals CAM-B3LYP and LC-BLYP, which use Hartree-Fock exchange at long range and conventional functionals at short range and were devised specifically to address the overshoot problem, yield properties in reasonably good agreement with values computed at the MP2 level. More surprisingly, some non-standard functionals like the Krieger-Li-Iafrate functional (KLI) [58], or the current-DFT (cDFT) method [59], which were previously reported to alleviate the overshoot problem for some linear systems [60] also fail for most of the NLO

Table 5.19 Diagonal x-component of electronic dipole moment (μ_x), polarizability (α_{xx}), and first hyperpolarizability (β_{xxx}) of compound **1** (in a.u.)

Method	μ_x	α_{xx}	β_{xxx}
HF/3-21G	−1.79	681.9	−661
HF/6-31G*	−1.47	735.3	−558
MP2/3-21G	−1.88	719.0	−1,526
MP2/6-31G*	−1.67	783.0	−1,301
CAM-B3LYP/3-21G	−1.82	719.3	−1,672
LC-BLYP/3-21G	−1.77	702.8	−1,193
B3LYP/3-21G	−1.91	758.6	−4,485
cDFT/DZP	−2.09	672.1	−5,379
LB94/DZP	−2.09	901.3	−10,291
RevPBEx/DZP	−2.42	900.6	−15,196
GRAC/DZP	−2.64	926.2	−19,816
PW91/3-21G	−2.19	819.7	−20,126
LDA/3-21G	−2.32	830.1	−22,487
KLI/DZP	−3.65	975.1	−83,807

Table 5.20 Diagonal x-component of electronic dipole moment (μ_x), polarizability (α_{xx}), first- (β_{xxx}) and second hyperpolarizability (γ_{xxxx}) of compounds **1, 2** and **3**

Property	Method	1	2	3
μ_x	HF/3-21G	−1.79	1.09	−1.29
	MP2/3-21G	−1.88	1.00	−1.25
	LC-BLYP/3-21G	−1.77	0.94	−1.22
	PM3	−1.30	0.86	−1.02
α_{xx}	HF/3-21G	681.9	859.7	1051.7
	MP2/3-21G	719.1	867.3	1079.6
	LC-BLYP/3-21G	702.8	874.9	1091.9
	PM3	671.9	858.7	1061.9
β_{xxx}	HF/3-21G	−674	664	−4,552
	MP2/3-21G	−1,526	985	−8,267
	LC-BLYP/3-21G	−1,193	875	−7,457
	PM3	−1,028	1,374	−7,401
$\gamma_{xxxx} \times 10^{-3}$	HF/3-21G	171.7	596.5	1542.4
	MP2/3-21G	384.6	988.3	3103.6
	LC-BLYP/3-21G	291.5	826.3	2655.8
	PM3	261.7	1073.9	3051.1

properties of the fullerene system investigated in [56]; the overshoot of the KLI functional is indeed the largest for all the functionals investigated.

The electronic NLO properties computed for the substituted dihydro-fullerenes are reproduced in Table 5.20. The effect of electron correlation on β is significant, as well as the effect of extending the basis set from 3-21G to 6-31G*. It was also found that the NLO properties computed at the semiempirical PM3 level were in reasonably good agreement with MP2/3-21G. In general, the magnitude of the (hyper)polarizabilities correlated with the length of the substituent donor. The NLO properties of the dyads were compared with the corresponding properties of the

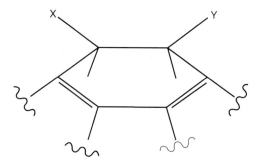

Fig. 5.4 General structure of the model compounds used for the investigation of the DFT overshoot. The employed addends X, Y are given in Table 5.21

Table 5.21 Dipole moment μ, average polarizability α and average first hyperpolarizability β for several model systems; the 6-31*G basis set was employed throughout

X,Y	μ	α	β	Method
H,H	1.00	449.23	−29.0	HF
	0.92	479.65	−58.2	BLYP
H,CH$_3$	1.07	461.00	−18.6	HF
	0.98	493.24	−36.0	BLYP
CH$_3$, CH$_3$	1.18	472.26	−3.4	HF
	1.10	506.32	12.45	BLYP
H,NH$_2$	1.09	457.12	−29.3	HF
	1.01	490.18	81.9	BLYP
NH$_2$, NH$_2$	1.21	465.04	−7.5	HF
	1.19	501.79	277.2	BLYP
H, N(CH$_3$)$_2$	1.11	481.76	−64.3	HF
	1.00	520.46	1019.1	BLYP
N(CH$_3$)$_2$, N(CH$_3$)$_2$	1.33	512.28	−74.8	HF
	1.28	562.06	3046.5	BLYP

addends **5–7** at the PM3 level, and found to be considerably larger. It should be noted, however, that the lack of diffused functions in all the basis sets employed probably makes the reported values only a qualitative estimate, as shown in our later work on endohedral fullerenes [61].

In order to investigate the role of the electron donating strength of the substituent on the DFT overshoot, a series of model compounds was devised based on 1,2-dihydro-fullerene (see Fig. 5.4) and the NLO properties computed with BLYP/3-21G and HF/3-21G. The substituents were selected according increasing Hammet σ_p constant, which may be used as a measure for the electron-donating strength of the donor group. A clear correlation between electron-donor strength and the DFT overshoot, measured as the ratio between the values at the DFT and RHF levels, was obtained for β (see Table 5.21).

The vibrational contributions to the NLO properties were computed at the RHF/6-31G level. Generally, the double harmonic (DH) approximation was applied. For one component of molecule **8** (see Fig. 5.4), the full nuclear relaxation (NR) term was computed, using the so-called 'field-induced coordinates' (FIC) approach [62, 63]. The curvature term, which added to the NR term would yield the total vibrational contribution, was assumed to be small and neglected. Comparison of the DH approximation $\beta_{xxx}{}^{nr}$ (128.7 au) with the full NR contribution (-608.7 au) shows that the DH approximation is insufficient for even a qualitative estimate, thus we will focus here only on the NR contributions to the diagonal x-components of **8**.

The static NR contribution to $\alpha_{xx}{}^{nr}$ was with 11.78 au considerably smaller than the corresponding electronic value (501.3 au). For β_{xxx}, however, the NR contribution (-608.7 au) was much larger than the electronic one (-35.8 au). The vibrational contribution to the Pockels effect, computed in the infinite optical frequency approximation was found to be 42.9 au, approximately equal in magnitude to the electronic static value. In the same approximation, the vibrational contribution to the EFISH property, $\gamma(-2\omega;\omega,\omega,0)$, was -40 au, thus much smaller than the electronic counterpart (40,201 au).

In addition to the nonresonant NLO properties, one- and two-photon spectra of several of the compounds in Fig. 5.3 were also computed in Ref. [56]. The averaged two-photon absorption cross section for isotropic media in the case of two parallel linearly polarized photons $<\delta^{0F}>$ is

$$< \delta^{0F} >= 1/15\Sigma_{ij}\left[S_{ii}{}^{0F}\left(S_{jj}{}^{0F}\right)^* + 2S_{ij}{}^{0F}\left(S_{ij}{}^{0F}\right)^*\right], \qquad (5.31)$$

where

$$S_{ij}^{0F}(\underline{\xi}) = 2\hbar^{-1}\sum_K \frac{< 0|\underline{\xi}\cdot\hat{\underline{\mu}}_i|K >< 0|\underline{\xi}\cdot\hat{\underline{\mu}}_j|K >}{\omega_K - \omega_F/2}, \qquad (5.32)$$

where ξ is the polarization vector, $<0|\xi.\mu|K>$ is the transition dipole between the ground state $-0>$ and the excited state $|K>$ and ω_F is the circular frequency of one absorbed photon.

The connection between the theoretical two-photon absorption cross section and the experimental spectrum is made via: $\sigma_{0F}^{(2)} = K\frac{\omega_F{}^2 g(\omega_F)}{\Gamma_F/2} <\delta^{0F}>$, where K is a constant, $g(\omega_F)$ is the spectral line profile (assumed to be a δ-function in Ref. [56]) and Γ_F is the lifetime broadening of the final state. The properties in Eq. 5.2 were computed in the single excitation configuration interaction (CIS) model, using the semiempirical GRINDOL [64] code, which is based on the INDO approximation. The computed one-photon spectra for molecules **1** and **4** were in good agreement with available experimental spectra in methylenchloride and cyclohexane, respectively. The GRINDOL/CIS data for compound **3** were compared with one one-photon spectra computed with time-dependent density functional theory at the B3LYP/6-31G* level. At this level, the most intense transition was found at 567 nm, thus substantially shifted from both the GRINDOL/CIS (418 nm) and the experimental (436 nm in CH_2Cl_2) values. This was explained by the fact that charge transfer excitations tend to be poorly described by the B3LYP functional [65]. The excited state dipole moments computed by GRINDOL/CIS were also reported. While in the ground state the fullerene cage acts as electron acceptor, in some of the excited states it functions as an electron donor. Finally, the computed two-photon absorption spectra, using a lifetime broadening of 6,000 cm^{-1}, were found to be in good agreement with available experimental data.

In a later paper [66], the NLO properties of N-methylfulleropyrrolidine **4** (Fig. 5.3), calculated at different *ab-initio*, DFT and semiempirical levels, was reported. The results were similar to those of Ref. [55], in that a large overshoot was found with the BLYP functional, while the properties computed with the long range corrected LC-BLYP functional were similar to those obtained with MP2. The semiempirical methods PM3 and PM6 were again in reasonable agreement with the MP2 values.

5.8.2   Endohedral Fullerenes

The electronic and vibrational linear and nonlinear polarizabilities of $Li@C_{60}$ and its cation $[Li@C_{60}]^+$ was studied in Ref. [61]. The interest in the vibrational properties of these fullerenes arises from earlier work of Whitehouse and Buckingham [67], who argued that large amplitude vibrational motions, due to weak interactions between the dopant atom and the cage may be responsible for large vibrational (hyper)polarizabilities. Indeed, they obtained large vibrational contributions for the cation, using a simplified potential in conjunction with a classical analysis of the field dependence of the (temperature-dependent) vibrationally averaged dipole moment. The electronic (hyper)polarizabilities of the two endohedral fullerenes had been computed previously only at very approximate levels.

The geometries were optimized at the UB3LYP/6-31G level for $Li@C_{60}$ and at RB3LYP/6-31G for $[Li@C_{60}]^+$. For the neutral, the minimum geometry was found at a structure with approximately C_{3v} symmetry, where the Li atom was located at about 1.5 Å away from the center of the cage. This structure was used for further analysis. Another minimum at slightly higher energy was found in a structure with approximately C_{5v} symmetry. It was always necessary to break the symmetry slightly, because the SCF procedure did not converge in the full symmetry. The actual symmetry of the lowest minimum was C_s. All the results were in good agreement with those of Ref. [68]. The geometry of $[Li@C_{60}]^+$ was optimized with RB3LYP/6-31G in approximately C_{3v} symmetry. In this case the Li atom was shifted about 1.4 Å away from the center of the cage.

Table 5.22 Electronic contribution to the diagonal polarizabilities along the dipole direction of Li@C_{60} and its cation, using different basis sets and levels of theory

Method	Basis set	Li@C_{60}			[Li@C_{60}]$^+$		
		α_{zz}^e	β_{zzz}^e	$\gamma_{zzzz}^e \times 10^{-3}$	α_{zz}^e	β_{zzz}^e	$\gamma_{zzzz}^e \times 10^{-3}$
B3LYP	6-31G	508.0	1,541	−39	468.9	−237	28
B3LYP	6-31+G	578.6	1,839	64	520.2	−53	99
B3LYP	6-31+G*	589.0	1,533	66	533.9	−118	99
HF	6-31G	479.0	5,540	900	445.8	−83	54
MP2	6-31G	527.1	1,363	−	463.6	−192	−

Table 5.23 Electronic (e) contribution to diagonal components of α, β and γ of Li@C_{60}, [Li@C_{60}]$^+$, C_{60}, C_{60}^-, calculated at the (U)B3LYP/6-31+G* level, and Li at the UB3LYP-aug-cc-pV5Z level

i=	Li@C_{60}			[Li@C_{60}]$^+$			Li
	x	y	z	x	y	z	x=y=z
α_{ii}^e	560.03	590.29	589.03	533.86	533.86	533.89	142.9[a]
β_{iii}^e	0.0	−290.0	1532.6	0	0	−118.1	0
$\gamma_{iiii}^e \times 10^{-3}$	293	−20	66	102	102	99	631[a]
		C_{60}^- [b]			C_{60}^c		Li
α_{ii}^e	575.9	617.7	620.5	550.5	550.5	551.3	
β_{iii}^e	0	−88	441	0	0	−11	
$\gamma_{iiii}^e \times 10^{-3}$	211	−46	−86	136	135	136	

[a] with aug-cc-pV5Z basis
[b] At the geometry of Li@C_{60}/UB3LYP/6-31G
[c] At the geometry of [Li@C_{60}]$^+$/B3LYP/6-31G

The electronic NLO properties were computed at the B3LYP/6-31+G* level (Table 5.22). Tests with smaller basis sets showed that this comparably large basis was the minimum required in order to arrive at a semiquantitative description of β and γ. The polarizabilities and first hyperpolarizabilities computed with the (U)B3LYP functional were in good agreement with those at the (U)MP2 level, suggesting that both methods account similarly for correlation effects, although only the smaller 6-31G basis set could be used. Calculations at the HF level revealed that the correlation effects on the polarizabilities are large, essentially showing that the HF approximation is inadequate for the hyperpolarizabilities of Li@C_{60} and its cation.

A general feature of fullerenes endohedrally doped with metal atoms is the charge transfer of electrons from the metal atom to the cage. To compare the NLO properties of the endohedral compounds with the corresponding noninteracting species, the static electronic properties of Li@C_{60}, [Li@C_{60}]$^+$, C_{60}^-, C_{60} and Li were computed in Ref. [60] (for an excerpt see Table 5.23). The properties of Li$^+$ were assumed to be negligible [69], and the magnitude of the properties of Li show that it is indeed unlikely to be present in either of the complexes. For the cation, the (hyper)polarizabilities are approximately comparable with those of the noninteracting system obtained by charge transfer from the Li atom to the cage, Li^+ + C_{60}, although the interaction between the two species leads to a slight reduction of the properties of the cage, probably due a contraction of the electron density by the charge of the cation. The same holds for the polarizability of Li@C_{60}, where the noninteracting system would now be Li^+ + C_{60}^-. For the hyperpolarizabilities, however, the noninteracting charge transfer model breaks down completely.

The obtained electronic (hyper)polarizabilities were compared with values published previously by Campbell et al. [70, 71] which were computed using an uncoupled approximation to coupled-perturbed HF theory, with the 6-31G* basis set. Their values were very different from those obtained in Ref. [60], e.g. for the first hyperpolarizabilities they reported $\beta_{yyy} \sim -7,000$ au, $\beta_{yyy} \sim 15,000$ au, about an order of magnitude larger than in Ref. [61]. For the second hyperpolarizability $\gamma_{xxxx} \sim 320 \times 10^3$ au, $\gamma_{yyyy} \sim 540 \times 10^3$ au, $\gamma_{zzzz} \sim -320 \times 10^3$ au. These large differences may be interpreted as an additional indication that HF theory is unreliable for the hyperpolarizabilities of Li@C_{60}.

The variation of the electric properties of Li@C_{60} with the position of the Li atom along the dipole axis was also investigated. The dependence of the diagonal components of the dipole moment, polarizability and first hyperpolarizability along the dipole moment direction (z) on the distance between the Li atom and the center of mass of the cage is small (μ,α) to modest (β). The second hyperpolarizability component γ_{zzzz}, on the other hand, is altered much more drastically, even undergoing a sign change at small distances from the cage center. This could possibly be used in a potential nonlinear "flip-flop" device, where an external electric field would shift the Li atom between different equilibrium positions. However, it has been shown [72] that the screening effect of the fullerene cage prevents large displacements of the Li atom due to an externally applied field. This is also corroborated by the field-optimizations in Ref. [60], in order to obtain the vibrational properties (see below), which show that a field of 0.0128 au results in a shift of only 0.03 au of the field-free position of the Li atom, much too small for any appreciable change of γ_{zzzz}.

5.8.2.1 Nuclear Relaxation Contribution to the Vibrational NLO Properties

The nuclear relaxation (NR) contributions were computed using a finite field approach [73, 74]. In this approach one first optimizes the geometry in the presence of a static electric field, maintaining the Eckart conditions. The difference in the static electric properties induced by the field can then be expanded as a power series in the field. Each coefficient in this series is the sum of a static electronic (hyper) polarizability at the equilibrium geometry and a nuclear relaxation term. The terms evaluated in Ref. [61] were the change of the dipole moment up to the third power of the field, and that of the linear polarizability up to the first power:

Table 5.24 Nuclear relaxation contributions to the diagonal components of the static electrical properties of Li@C_{60} and [Li@C_{60}]$^{+}$

	Li@C_{60}			[Li@C_{60}]$^{+}$		
i=	x	y	z	x	y	z
α_{ii}^{nr}	14.7[a]	10.3	10.2 [11.9]	10.4	9.4	4.5
β_{iii}^{nr}	–	−125.9	794.6 [(912–915)]	0	95	18
γ_{iiii}^{nr}	–	−90	(25–40) [(52–81)]	560	190	37

[a]Computed analytically

$$\Delta\mu_i(F, R_F) = a_{1,ij} F_j + \frac{1}{2} b_{1,ijk} F_j F_k + \frac{1}{6} g_{1,ijk} F_j F_k F_l$$

$$\Delta\alpha_{ij}(F, R_F) = b_{2,ijk} F_k \qquad (5.33)$$

where the argument R_F implies structure relaxation in the field F, and the coefficients are given by

$$a_{1,ij} = \alpha_{ij}^e(0;0) + \alpha_{ij}^{nr}(0;0)$$

$$b_{1,ijk} = \beta_{ijk}^e(0;0,0) + \beta_{ijk}^{nr}(0;0,0)$$

$$g_{1,ijkl} = \gamma_{ijkl}^e(0;0,0,0) + \gamma_{ijkl}^{nr}(0;0,0,0)$$

$$b_{2,ijk} = \beta_{ijk}^e(0;0,0) + \beta_{ijk}^{nr}(-\omega;\omega,0)_{\omega\to\infty} \qquad (5.34)$$

Here, P^{nr} denotes the nuclear relaxation part of P, and the subscript $\omega \to \infty$ invokes the "infinite optical frequency" approximation, which is generally considered to be good approximation to the frequency-dependent properties at the usual laser frequencies applied in experiments [75]. Due to the high computational cost of these calculations, in general the rather small 6-31G basis was used. For some control calculations, the 6-31+G basis set was used.

The finite field method works well as long as the field-dependent optimum structure corresponds to the same minimum as the field-free optimized structure. For the endohedral fullerenes considered in Ref. [60], especially for Li@C_{60}, it appears that there were several minima lying nearby on the potential energy surface, which are separated by low energy barriers. As a consequence it was not possible to determine the NR contribution perpendicular the symmetry plane of the C_s symmetry structure using the finite field method. However, α_{xx}^{nr} could be determined using alternative analytical formulae [24, 25, 27], which do not require field-optimized structures, but only dipole derivatives and the Hessian of the field-free equilibrium structure.

The obtained static values, reproduced in Table 5.24, show that the effect of diffused functions on the NR contributions is not negligible, although it is smaller than that on the electronic contributions. Considering that the changes brought about by the larger basis set were only quantitatively different than those obtained with

Table 5.25 NR contribution to the dc-Pockels effect $(\beta^{nr}(-\omega;\omega,0)_{\omega\to\infty})$ of Li@C$_{60}$ in the infinite optical frequency approximation

ij=	xx	yy	yz	zz
$\beta_{ijy}{}^{nr}$	−23	−33	51	−9
$\beta_{ijz}{}^{nr}$	−199	119	15	200

Table 5.26 Electronic and NR contributions to α and γ of Sc$_2$@C$_{72}$

	Electronic contrib.		NR contrib.
	Sc$_2$@C$_{72}$	C$_{72}$	Sc$_2$@C$_{72}$
$\alpha_{zz}(0;0)$	749.13	777.41	30.59
$\gamma_{zzzz}(0;0,0,0)$	193,100	297,000	
$\gamma_{zzzz}(-2\omega;\omega,\omega,0)_{\omega\to\infty}$			98,800

All values in a.u.

the 6-31G basis, the NR values with 6-31G may be considered to be qualitatively correct. The NR contributions to α are small in comparison with their electronic counterparts (see Table 5.23), in apparent contradiction to what Whitehouse and Buckingham, [66] (WB) have previously found. However, it is argued in Ref. [61] that the two calculations are not directly comparable, as WB calculated their values using a classical averaging procedure at a finite temperature (above 20 K) and in addition computed the total vibrational contribution, while in Ref. [61] the temperature is 0 K, and the curvature contribution to the vibrational NLO properties is missing. Although the latter is generally expected to be small, this may be different for the endohedral molecules investigated here. It is argued in Ref. [60] that at least one of the approximations in the WB treatment is questionable, i.e. the assumption that the field-free potential is spherical. This has been shown to be not in agreement with the low energy vibrational spectrum of the neutral [68].

In contrast to α, the NR contributions to β and γ are approximately of the same order of magnitude as the corresponding electronic properties.

In addition to the static vibrational properties, the NR contributions to the dc-Pockels effect of Li@C$_{60}$ in the infinite optical frequency approximation were also reported in Ref [61] (Table 5.25). Compared with their electronic counterparts the values were small, but not negligible.

In Ref. [76] the electronic and vibrational NR properties of the endohedral fullerene Sc$_2$@C$_{70}$ were reported at the HF level with the 6-31+G* basis set for C, while Sc was treated by an Stuttgardt-Dresden ECP [77] with 11 valence electrons. The optimized geometry was of D$_{2h}$ symmetry, and only the properties along the D$_2$ axis, on which the two Sc atoms are located, were computed. The results are shown in Table 5.26. The vibrational properties were calculated using FICs. Similarly as found for the Li@C$_{60}$ fullerenes, the vibrational NR contribution to the static α was very small compared to the electronic contribution. Due to heavy computational cost, for γ only the vibrational contribution to the EFISH signal in the IFA $\gamma(-2\omega;\omega,\omega,0)_{\omega\to\infty}$ was calculated. But already this value was substantial, compared with the static electronic contribution. Considering that generally frequency –dependent vibrational properties are considerably smaller than the corresponding static property, one can anticipate that this value may be at least as large or even larger than the static electronic γ. Comparison with C_{72} showed a quenching effect of the Sc_2 inclusion on both α and γ. It should be noted however, that the effect of correlation, which was seen to be large in the case of $Li@C_{60}$ and its cation, could not be explored for $Sc_2@C_{72}$, due to the heavy computational cost.

5.9   Nonlinear Optical Properties Due to Large Amplitude Vibrational Motions, with an Application to the Inversion Motion in NH_3

The treatment of the vibrational NLO properties in the previous sections employed either the Bishop-Kirtman perturbational theory (BKPT) or the finite field-nuclear relaxation (FF-NR) approach. These approaches may fail for molecules containing large amplitude anharmonic motions, as indeed was suspected to happen in $Li@C_{60}$. In such cases a more recently proposed variational method, based on analytical response theory [78, 79], would in principle be applicable, but is computationally extremely expensive, as it requires an accurate numerical description of the potential energy surface (PES), at least if the anharmonicity is so large that a power series expansion of the PES is inadequate [80].

In a large class of molecules with large amplitude motions, only a few internal coordinates are expected to contribute significantly to the vibrational NLO properties. This is the case where there is one or few motions governed by a potential with multiple minima separated by low barriers. Then one may explicitly compute the contribution of these motions to the NLO properties, while treating the rest in a simpler fashion. A first such treatment based on BKPT and using several approximations directed at the specific treatment of the torsional motion in H_2O_2 was presented by Santiago et al. [81]. A somewhat different approach was developed by Luis et al. [82], directed ultimately at the development of a feasible approach to the routinely calculation of vibrational NLO properties of both large and small molecules containing a few large amplitude vibrational modes. In this approach, the FF-NR method is coupled with a calculation of the PES of the specific large amplitude motion (LAM), for which the inversion motion of ammonia is taken as an example. The coupling between the LAM coordinate and the other coordinates is taken into account only in first order by nuclear relaxation, i.e. optimization at fixed value of the LAM coordinate. In this way the determination of the vibrational NLO properties reduces to an effectively one-dimensional problem.

As the approach utilizes internal coordinates, and requires the calculation of the effective 1D potential (double well potential in the case of the inversion motion of ammonia) two problems need to be solved beforehand: a) the selection of the LAM coordinate and b) the expression for the kinetic operator and its computation. The LAM coordinate needs to be able to describe the entire range of the motion in a continuous way, because derivatives with respect to this coordinate appear in the final equations. For the inversion of ammonia, two choices were tried out in Ref. [82]: the first one, θ, is defined as $\theta = 1/3(\theta_1 + \theta_2 + \theta_3) - \pi/2$, where the θ_i are the angles between a N-H bond and the trisector axis and has been used previously by Handy et al. [83]. The second choice, z, is defined as the distance between the plane containing the three hydrogen atoms and the N atom. The potential energy function $V(q_{inv})$ was computed as the electronic energy at a fixed value of q_{inv}, with the geometry optimized with respect to the other coordinates. Thus first-order couplings between q_{inv} and the remaining coordinates were eliminated.

The kinetic energy operator T as a function of q_{inv} was evaluated numerically, in order to be generally applicable. The general expression of $T(\mathbf{q})$ of a molecule with N atoms as a function of the $3N$ internal coordinates $q_i(i = 1, \ldots, 3N)$ is:

$$T(q) = \sum_{ij}^{3N} f_2^{ij}(q)\partial_{ij}^2 + \sum_{i}^{3N} f_1^{i}(q)\partial_i + v(q) \tag{5.35}$$

$$f_2^{ij}(q) = -\frac{\hbar^2}{2}G^{ij}(q)$$

$$f_1^{i}(q) = -\frac{\hbar^2}{2}\sum_{j}^{3N}\left[G^{ij}(q)\partial_j \ln J(q) + \partial_j G^{ij}(q)\right] \tag{5.36}$$

where

$$G^{ij}(q) = \left[g^{ij}(q)\right]^{-1},$$

$$g^{ij}(q) = \sum_{\lambda}^{3N}(\partial_i x^{\lambda}(q))(\partial_j x^{\lambda}(q))$$

$$J(q) = det\left[\partial_i x^{\lambda}(q)\right],$$

where the x^{λ} are the mass-weighted Cartesian coordinates in the molecule-fixed coordinate system. The third term in Eq. 5.5$v(\mathbf{q})$ is a pseudopotential which depends on the definition of the volume element used in the normalization of the wave function [84]. For the evaluation of the different terms $f_2^{ij}(\mathbf{q})$, $f_1^{i}(\mathbf{q})$, $v(\mathbf{q})$ for fixed values of $\mathbf{q}$ the program TNUM [85, 86] was employed. This program uses the same formulation for the kinetic energy operator for *reduced* dimensionality models, as required in the context of Ref. [82]. Then $G(\mathbf{q})$ is replaced by $G^{red}(\mathbf{q})$, which is calculated using a method proposed by Wilson et al. [87].

The different functions obtained, $f_1(q_{inv})$, $f_2(q_{inv})$, and $f_3(q_{inv}) = v(q_{inv}) + V(q_{inv})$ were fitted to a polynomial in q_{inv}. This procedure was repeated for several values of the electric field F_z, so that finally the field-dependent effective Hamiltonian for the inversion motion was obtained.

In order to obtain the final expressions for the evaluation of the vibrational (hyper)polarizabilities, the *quasi-degenerate* nature of the first two vibrational levels of the inversion coordinate needs to be taken into account. Without tunneling, the two levels are degenerate, but with tunneling the two levels split into a symmetric and an antisymmetric pair. The perturbing effect of a static electric field on the two quasi-degenerate levels was analyzed with the help of generalized Van Vleck perturbation theory [88, 89]. This method eliminates the coupling between the quasi-degenerate reference wave functions and the remaining vibrational states not only in first, but also in higher order perturbation theory. The symmetry adapted zero-order wave functions are the symmetric and antisymmetric combinations ϕ^+ and ϕ^-. Assuming that the matrix elements are expandable as a power series in the field and taking into account the inversion symmetry, the Hamiltonian matrix up to fourth order is then given by

$$H(F_z) = \begin{pmatrix} E^+(0) - \frac{1}{2}\alpha_{zz}^+ F_z^2 - \frac{1}{24}\gamma_{zzzz}^+ F_z^4 & -\mu_z^\pm F_z - \frac{1}{6}\beta_z^\pm F_z^3 \\ -\mu_z^\pm F_z - \frac{1}{6}\beta_z^\pm F_z^3 & E^-(0) - \frac{1}{2}\alpha_{zz}^- F_z^2 - \frac{1}{24}\gamma_{zzzz}^- F_z^4 \end{pmatrix},$$
(5.37)

where $P^+ = <\phi^+|P| \phi^+>$, etc., and $P^\pm = <\phi^+|P| \phi^->= <\phi^-|P| \phi^+>$. Due to the non-degeneracy of ϕ^+ and ϕ^-, the eigenvalues of this matrix can be expanded in the field only if either $(E^+(0) - E^-(0))^2 >> (2\mu_z^\pm F)^2$ or $(E^+(0) - E^-(0))^2 << (2\mu_z^\pm F)^2$. For ammonia, the latter case was found to be the case, for not too small field strengths. In this case, the energy eigenvalues of the Hamiltonian (5.6) are given by

$$E_n(F_z) = E(0) - (-1)^n \mu_z F_z - \frac{1}{2}\alpha_{zz} F_z^2 - (-1)^n \frac{1}{6}\beta_{zzz} F_z^3 - \frac{1}{2}\gamma_{zzzz} F_z^4, \quad (5.38)$$

with $n = 0,1$, $P = 1/2(P^+ + P^-)$, for $P = E(0)$, α_{zz}, γ_{zzzz} and $P = P^\pm$ for $P = \mu_z$, β_{zzz}. If the field-dependent energies are known for different field strengths, the dipole moment and diagonal (hyper)polarizability components can be obtained by numerical differentiation.

The field dependent energies $E_0(F_z)$, $E_1(F_z)$ are the solutions of the 1D vibrational Schrödinger equation

$$H(q_{inv}, F_z)\psi_i(q_{inv}, F_z) = [T(q_{inv}) + V(q_{inv}, F_z)] \psi_i(q_{inv}, F_z)$$
$$= E_i(F_z)\psi_i(q_{inv}, F_z); \text{ with } i = 0, 1 \quad (5.39)$$

for each fixed F_z value. In order to solve this equation with the numerical functions obtained from the fitting procedure mentioned above, a numerical solution method called the *shooting* method [90] was used in Ref. [82].

The electric properties in the above expressions include all electronic and vibrational contributions within the reduced dimensionality (1D) approximation. In terms of the usual partitioning of the total properties in the FF-NR approach they can be written as: $P = P^e + P^{nr} + P^{ZPVA} + P^{c-ZPVA}$. As the first three terms can be easily computed separately, the total P can be used to get the P^{c-ZPVA}. The P^{ZPVA} contributions were computed numerically as expectation values over the field-dependent vibrational wavefunctions, using property expansions as a function of q_{inv} obtained by fitting to a polynomial in q_{inv} in the same way as the PES (see above). *Dynamic* hyperpolarizabilities in the infinite optical frequency approximation can be obtained from numerical derivatives of the field-dependent static polarizabilities, as in the usual Bishop-Hasan-Kirtman (BHK) procedure [73].

This treatment was applied in Ref. [82] to the inversion motion of NH_3, which could be compared with previous work, where the variational treatment was applied in single well approximation, using series expansion of the potential energy surface and the electric properties [80]. In order to validate the chosen basis sets and methods, the vibrational energy levels and the inversion splittings were computed at the MP2/Pol, CCSD/aug-cc-pVTZ and CCSD(T)/aug-cc-pVTZ levels, for MP2/Pol with both coordinates (θ and z), for the other two levels only with the θ coordinate. Although in the series using the θ coordinate, the agreement with experimental values increased with increasing level of theory and size of basis set (mean absolute deviation (MAD) 27.60, 20.21, 12.39 cm^{-1} for the energies of the 7 lowest levels, and 25.85, 22.11, 14.04 cm^{-1} for the splittings computed at MP2/Pol, CCSD/aTZ, CCSD(T)/aTZ, respectively), the agreement was deemed to be satisfactory already at the lowest level applied, i.e. MP2/Pol/θ, which was chosen primarily for comparison purposes with the previous variational work [80]. The deviation of the energy levels obtained with the second coordinate z and MP2/Pol with an MAD of 94.15 cm^{-1} were substantially larger that for MP2/Pol/θ (MAD $= 27.60$ cm^{-1}), but smaller for the inversion splittings (MAD $= 10.22$ vs. 25.85 cm^{-1}).

The results for the vibrational electric properties are summarized in Table 5.27. Only the P^{ZPVA} and P^{c-ZPVA} values are shown, obtained as mentioned above, P^{ZPVA} by averaging over the vibrational wavefunction, and P^{c-ZPVA} from the total P by subtracting P^{ZPVA}, as well as the electronic and NR values, which were obtained independently. The computed values are compared with those of the variational results in the one-well approximation [80]. These were obtained by expanding the PES and the electronic electric properties in power series of the normal coordinates. While low-order expansions of the electric properties yielded converged vibrational properties at a give expansion of the PES, different orders of expansion of the PES yielded very different NLO properties. This is apparent in the values given in Table 5.27, where the results of fourth order and sixth order expansions of the PES are shown. The conclusion of Ref. [80], that the PES should not be treated in a series expansion, seems to be confirmed with the full 1D treatment, which shows much better agreement with the fourth order expansion than with the expansion to sixth order of the PES.

The sensitivity of the properties on the choice of the inversion coordinate turned out to be small, with maximum differences of about 7% in P^{c-ZPVA}, which was further analyzed using mixed calculations, where the potential for the θ coordinate

Table 5.27 ZPVA and c-ZPVA contributions to the diagonal components of electric properties in the direction of μ for NH_3; the full-valence CI values were obtained using fourth-order and sixth-order expansions of the PES and low-order expansions of the properties

	1D		(3N-6) modes	
	θ	z	4th-order PES	6th-order PES
μ^{ZPVA}	−0.0277	−0.0303	−0.0196	−0.0180
$\alpha_{zz}^{ZPVA}(0;0)$	0.13	0.15	0.46	0.46
$\alpha_{zz}^{c\text{-}ZPVA}(0;0)$	1.05	1.12	0.88	0.63
$\beta_{zzz}^{ZPVA}(0;0,0)$	0.8	0.8	−0.0	−0.0
$\beta_{zzz}^{c\text{-}ZPVA}(0;0,0)$	−111.3	−119.4	−114.6	−60.9
$\beta_{zzz}^{c\text{-}ZPVA}(-\omega;\omega,0)_{\omega\to\infty}$	−3.7	−3.6	−1.4	−0.5
$\gamma_{zzzz}^{ZPVA}(0;0,0,0)$	281	306	173	158
$\gamma_{zzzz}^{c\text{-}ZPVA}(0;0,0,0)$	25,234	26,430	25,041	9,504
$\gamma_{zzzz}^{c\text{-}ZPVA}(-\omega;\omega,0,0)_{\omega\to\infty}$	264	260	288	137
$\gamma_{zzzz}^{c\text{-}ZPVA}(-2\omega;\omega,\,\omega,0)_{\omega\to\infty}$	−46	−48	11	19

See Refs. [80, 82] for details

was combined with NLO calculations using the z coordinate. The results were very close to the pure z coordinate calculations, which may be interpreted as that the differences between the two coordinates is primarily determined by the kinetic energy rather than the potential energy.

5.10 Summary

We have reviewed several recent articles, which report state-of-the-art results of the L&NLO properties of a series of properly selected organic and inorganic molecules. The correlation, vibrational and relativistic contributions to the L&NLO properties have been computed. The correlation contribution has been calculated by a series of methods (e.g. CCSD(T), CASPT2). The vibrational contribution has been computed by employing the Bishop-Kirtman perturbation theory (BKPT). The Numerov-Cooley integration technique has also been employed for the calculation of the vibrational correction, since it is of infinite order and its results could be used to check the BKPT approach, which is finite order. The relativistic correction has been used by employing the Douglas-Kroll approximation. The reviewed selected pilot studies reveal several interesting features and trends.

References

1. Buckingham AD, Orr BJ (1967) Quart Rev 21:195
2. Bartkowiak W (2006) In: Papadopoulos MG, Sadlej AJ, Leszczynski J (eds) Nonlinear optical properties of matter: from molecules to condensed phase. Springer, Dordrecht, p 299

3. Ingamells VE, Raptis SG, Avramopoulos A, Papadopoulos MG (2003) In: Papadopoulos MG (ed) Nonlinear optical responses of molecules, solids and liquids: methods and applications. Research Signpost, Trivandrum
4. Prasad PN, Williams DJ (1991) Introduction to nonlinear optical effects. Wiley, New York
5. Bartkowiak W, Strasburger K (2010) J Mol Struct (THOECHEM) 960:93
6. Campo J, Painelli A, Terenziani F, van Regemorter T, Beljonne D, Goovaerts E, Wenseleers W (2010) J Am Chem Soc 132:16467
7. Hrobaric P, Sigmundova I, Zahradnik P, Kasak P, Arion V, Franz E, Clays K (2010) J Phys Chem C 114:22289
8. Molina JJ, Lectez S, Tazi S, Salanne M, Dufrêche JF, Roques J, Simoni E, Maden PA, Turque P (2011) J Chem Phys 134:014511
9. Si Y, Yang G (2011) Theor Chem Acc 128:249
10. Haskopoulos A, Maroulis G (2010) J Phys Chem A 114:8730
11. Baranowska A, Zawada A, Fernández B, Bartkowiak W, Kędziera D, Kaczmarek-Kędziera A (2010) Phys Chem Chem Phys 12:852
12. Rinkevicius Z, Autschbach J, Baev A, Swihart M, Ågren H, Prasad PN (2010) J Phys Chem A 114:7590
13. Orlando R, Lacivita V, Bast R, Ruud K (2010) J Chem Phys 132:244106
14. Karamanis P, Pouchan C (2011) Intl J Quant Chem Spec Issue 111:788
15. Raptis SG, Papadopoulos MG, Sadlej AJ (1999) J Chem Phys 111:7904
16. Douglas M, Kroll NM (1974) Ann Phys (NY) 82:89
17. Hess BA (1986) Phys Rev A 33:3742; Jansen G, Hess BA (1989) Phys Rev A 39:6016
18. Cooley JW (1961) Math Comput 15:363; Blatt JM (1967) J Comput Phys 1:382
19. Dykstra CE, Malik DJ (1987) J Chem Phys 87:2806
20. Ingamells VE, Papadopoulos MG, Sadlej AJ (2000) Chem Phys Lett 316:541
21. Urban M, Sadlej AJ (1995) J Chem Phys 103:9692
22. Bishop DM (1990) Rev Mod Phys 62:343, and references therein
23. Kirtman B, Bishop DM (1990) Chem Phys Lett 175:601
24. Bishop DM, Kirtman B (1991) J Chem Phys 95:2646
25. Bishop DM, Kirtman B (1992) J Chem Phys 97:5255
26. Bishop DM (1998) Adv Chem Phys 104:1, and references therein
27. Bishop DM, Luis JM, Kirtman B (1998) J Chem Phys 108:10013
28. Roos BO (1987) Adv Chem Phys 69:399, and references therein
29. Andersson K, Malmqvist PA, Roos BO, Sadlej AJ, Woliński K (1990) J Phys Chem 94:5483; Andersson K, Malmqvist P-A, Roos BO (1992) J Chem Phys 96:1218
30. Andersson K, Roos BO (1995) In: Yarkony DR (ed) Modern electronic structure theory, part I. World Scientific, Singapore, p 55
31. Kellö V, Sadlej AJ (1990) J Chem Phys 93:8122; Sadlej AJ, Urban M (1991) Chem Phys Lett 176:293; Sadlej AJ (1991) J Chem Phys 95:2614; Kellö V, Sadlej AJ (1991) J Chem Phys 95:8248
32. Kellö V, Sadlej AJ (1995) Theor Chim Acta 92:253
33. Kellö V, Sadlej AJ (1996) Theor Chim Acta 94:93
34. Kellö V, Sadlej AJ, Hess BA (1996) J Chem Phys 105:1995
35. Avramopoulos A, Ingamells VE, Papadopoulos MG, Sadlej AJ (2001) J Chem Phys 114:198
36. Urban M, Černušák I, Kellö V, Noga J (1987) Methods in computational chemistry, vol 1. Plenum, New York, p 117, and references therein
37. Bartlett RJ (1995) In: Yarkony DR (ed) Advanced series in physical chemistry, vol 2. Methods in computational chemistry. World Scientific, Singapore, p 1047
38. Lee TJ, Scuseria GE (1995) In: Langhoff SR (ed) Quantum mechanical electronic structure calculations with chemical accuracy. Kluwer, Dordrecht, p 47
39. Neogrady P, Kellö V, Urban M, Sadlej AJ (1996) Theor Chim Acta 93:101
40. Martí J, Andrés JL, Bertrán J, Duran M (1993) Mol Phys 80:625
41. Martí J, Bishop DM (1993) J Chem Phys 99:3860
42. Luis JM, Martí J, Duran M, Andrés JL (1997) Chem Phys 217:29

43. Eckart U, Ingamells VE, Papadopoulos MG, Sadlej AJ (2001) J Chem Phys 114:735
44. Eckart U, Fülscher M, Serrano-Andrés L, Sadlej AJ (2000) J Chem Phys 113:6235
45. Sadlej AJ (1988) Collect Czech Chem Commun 53:1995
46. Sadlej AJ (1991) Theor Chim Acta 79:123
47. Pluta T, Sadlej AJ (1998) Chem Phys Lett 297:391
48. Avramopoulos A, Papadopoulos MG (2002) Mol Phys 100:821
49. Sadlej AJ, Urban M (1991) J Mol Struct Theochem 234:147
50. Jug K, Chiodo S, Calaminici P, Avramopoulos A, Papadopoulos MG (2003) J Phys Chem A 107:4172
51. Ingamells VE, Papadopoulos MG, Handy NC, Willetts A (1998) J Chem Phys 109:1845
52. Cohen MJ, Willetts A, Amos RD, Handy NC (1994) J Chem Phys 100:4467
53. Champagne B, Perpete EA, van Gisbergen SJA, Baerends EJ, Snijders JG, Soubra-Ghaoui C, Robins KA, Kirtman B (1998) J Chem Phys 109:10489; erratum 110:11664 (1999)
54. Champagne B, Perpete EA, Jacquemin D, van Gisbergen SJA, Baerends EJ, Soubra-Ghaoui C, Robins KA, Kirtman B (2000) J Phys Chem A 104:4755
55. van Gisbergen SJA, Schipper PRT, Gritsenko OV, Baerends EJ, Snijders JG, Champagne B, Kirtman B (1999) Phys Rev Lett 83:694
56. Loboda O, Zalesný R, Avramopoulos A, Luis J-M, Kirtman B, Tagmatarchis N, Reis H, Papadopouls MG (2009) Linear and nonlinear optical properties of [60] fullerene derivatives. J Phys Chem A 113:1159–1170
57. ADF2007.01, SCM, Theoretical chemistry, Vrije Universiteit Amsterdam. http://www.scm.com
58. Krieger JB, Li Y, Iafrate GJ (1992) Phys Rev A 45:101
59. Vignale G, Kohn W (1996) Phys Rev Lett 77:2037
60. van Faassen M, de Boeij PL, van Leeuwen R, Berger JA, Snijders JG (2002) Phys Rev Lett 88:186401
61. Reis H, Loboda O, Avramopoulos A, Papadopoulos MG, Kirtman B, Luis J-M, Zalesný R (2011) Electronic and vibrational linear and nonlinear polarizabilities of Li@C_{60} and [Li@C_{60}]$^{+}$. J Comput Chem 32:908–914
62. Luis JM, Duran M, Champagne B, Kirtman B (2000) J Chem Phys 113:5203
63. Kirtman B, Champagne B, Luis JM (2000) J Comput Chem 21:1572
64. Lipiński J, Bartkowiak W (1991) J Phys Chem A 101:2159
65. Rudberg E, Salek P, Helgaker T, Ågren H (2005) J Chem Phys 123:184108
66. Oloboda O, Zalesný R, Avramopoulos A, Papadopoulos MG, Artacho E (2009) Linear-scaling calculations of linear and nonlinear optical properties of [60]fullerene derivatives. In: AIP conference proceedings, 1108:198–204
67. Whitehouse DB, Buckingham AD (1993) Chem Phys Lett 207:332
68. Zhang M, Harding LB, Gray SK, Rice SA (2008) J Phys Chem A 112:5478
69. Fowler PW, Madden PA (1984) Phys Rev B 30:6131
70. Campbell EEB, Fanti M, Hertel IV, Mitzner R, Zerbetto F (1998) Chem Phys Lett 288:131
71. Campbell EEB, Couris S, Fanti M, Koudoumas E, Kravez N, Zerbetto F (1999) Adv Mater 11:405
72. Delaney D, Greer JC (2004) Appl Phys Lett 63:431
73. Bishop DM, Hasan M, Kirtman B (1995) J Chem Phys 103:4157
74. Luis J-M, Duran M, Andres JL, Champagne B, Kirtman B (1999) J Chem Phys 111:875
75. Bishop DM, Dalskov EK (1996) J Chem Phys 104:1004
76. Skwara B, Loboda O, Avramopoulos A, Luis J-M, Reis H, Papadopoulos MG (2009) Electronic contributions to linear and nonlinear electric properties in fullerene-based molecular systems. In: ICCMSE 2009, to be published in AIP conference proceedings
77. http://www.theochem.uni-stuttgart.de/pseudopotentials/clickpse.en.html
78. Christiansen O (2005) J Chem Phys 122:194105
79. Christiansen O, Kongsted J, Paterson MJ, Luis J-M (2006) J Chem Phys 125:214309
80. Luis JM, Torrent-Sucarrat M, Christiansen O, Kirtman B (2007) J Chem Phys 127:084118
81. Santiago E, Castro MA, Fonseca TL, Mukherjee PK (2008) J Chem Phys 128:064310

82. Luis JM, Reis H, Papadopouls MG, Kirtman B (2009) J Chem Phys 131:034116
83. Handy NC, Carter S, Colwell SM (1999) Mol Phys 96:477
84. Chapuisat X, Nauts A, Brunet JP (1991) Mol Phys 72:1
85. Lauvergnat D, Nauts A (2002) J Chem Phys 116:8560
86. Lauvergnat D, Baloitcha E, Dive G, Desouter-Lecomte M (2006) Chem Phys 326:500
87. Wilson EB Jr, Decius JC, Cross PC (1955) Molecular vibrations. McGraw Hill, New York
88. Van Vleck JH (1929) Phys Rev 33:467
89. Kirtman B (1968) J Chem Phys 49:3890
90. Press WH, Teukolsky SA, Vetterling WT, Flannery BP (1992) Numerical recipes. Cambridge
    University Press, Cambridge

Chapter 6
Using Chebyshev-Filtered Subspace Iteration and Windowing Methods to Solve the Kohn-Sham Problem

Grady Schofield, James R. Chelikowsky, and Yousef Saad

Abstract Ground state electronic properties of a material can be obtained using density functional theory and obtaining a solution of the Kohn-Sham equation. The traditional method of solving the equation is to use eigensolver-based approaches. In general, eigensolvers constitute a bottleneck when handling systems with a large number of atoms. Here we discuss variations on an approach based on a nonlinear Chebyshev-filtered subspace iteration. This approach avoids computing explicit eigenvectors except to initiate the process. Our method centers on solving the original nonlinear Kohn-Sham equation by a nonlinear form of the subspace iteration technique, without emphasizing the intermediate linearized Kohn-Sham eigenvalue problems.

The method achieves self-consistency within a similar number of self-consistent field iterations as eigensolver-based approaches. However, the replacement of the standard diagonalization at each self-consistent iteration by a polynomial filtering step results in a significant speedup over methods based on standard diagonalization, often by more than an order of magnitude. Algorithmic details of a parallel

G. Schofield
Institute for Computational Engineering and Sciences, University of Texas, Austin, TX, 78712, USA
e-mail: grady@ices.utexas.edu

J.R. Chelikowsky (✉)
Institute for Computational Engineering and Sciences, University of Texas, Austin, TX, 78712, USA

Departments of Physics and Chemical Engineering, University of Texas, Austin, TX, 78712, USA
e-mail: jrc@ices.utexas.edu

Y. Saad
Department of Computer Science and Engineering, University of Minnesota, Minneapolis, MN, 55455, USA
e-mail: saad@cs.umn.edu

J. Leszczynski and M.K. Shukla (eds.), *Practical Aspects of Computational Chemistry I: An Overview of the Last Two Decades and Current Trends*, DOI 10.1007/978-94-007-0919-5_6, © Springer Science+Business Media B.V. 2012

implementation of this method are proposed in which only the eigenvalues within a specified energy window are extracted. Numerical results are presented to show that the method enables one to perform a class of challenging applications.

Keywords Kohn-Sham equation • Chebyshev-Fitered subspace interaction • Pseudopotential theory • PARSEC • Nonlinear eigenvalue problem

6.1 Introduction

Most properties of a material can be determined from knowledge of the electronic structure, i.e., the arrangement of electronic states, with the material. For example, the optical and dielectric properties can be predicted by knowing the energetic and spatial distribution of electrons in the material. Likewise, the structural properties such as phase stability and compressibilities can be known from the electron distributions. The electronic configurations in materials can often be obtained by a direct quantum mechanical approach using a combination of *density functional theory* (DFT) [1, 2] and *pseudopotential theory* [3–6]. DFT reduces the original multi-electron Schrödinger equation into an effective one-electron Kohn-Sham equation, where the non-classical electronic interactions are replaced by a functional of the charge density, i.e., the spatial distribution of electrons. Pseudopotential theory further simplifies the problem by replacing the "all electron" atomic potential with an effective "pseudopotential" that is smoother, but takes into account the effect of core electrons. Combining pseudopotentials with DFT greatly reduces the number of one-electron wave-functions to be computed, but more importantly the energy and length scales are set solely by the valence states. As such, species such as a carbon and tin can be treated on equal footing. However, even with these simplifications, solving the Kohn-Sham equation remains computationally challenging when the systems of interest contain a large number, e.g., more than a few hundred, atoms.

Several approaches have been advocated for solving the Kohn-Sham equations. They can be classified in two major groups: basis-free or basis-dependent approaches, according to whether they use an explicit basis set for electronic orbitals or not. Among the most popular basis-dependent approaches is the plane wave basis set, which is frequently used in applications of DFT to periodic systems where plane waves can easily accommodate the boundary conditions [7, 8]. Another popular approach uses localized basis sets such as Gaussian orbitals, which are commonly utilized in quantum-chemistry calculations[6, 9].

We focus on a different approach based on *real space* methods, which are "basis free." Real space methods have gained ground in recent years [10–13] owing in great part to their great simplicity and ease of implementation. In particular, these methods are readily implemented in parallel computing environments. A second advantage is that, in contrast with a plane wave approach, real space methods do not impose artificial periodicity in non-periodic systems. While plane wave basis techniques can be applied to clusters (or molecules), such applications often proceed by placing the cluster of interest in a large supercell. Provided the supercell is sufficiently large so that the cluster is removed from neighboring replicants, the electronic structure solution will correspond to that of the isolated cluster. However, because the potential from neighboring cells can be significant, in some cases this is not a practical solution, i.e., supercell solutions may converge slowly with the size of the cell [14]. A related, and perhaps more significant issue, is that supercells complicate the handling of systems that are not electronically neutral. Charged systems can be handled within plane wave methods by including a compensating uniform charge or a more complex procedure [14, 15]. Real space methods need not address such complications. A third advantage is that the application of the Hamiltonian to electron wave-functions is performed directly in real-space. Although the Hamiltonian matrix in real space methods is typically much larger than with plane waves, the Hamiltonians are *highly sparse* and never stored or computed explicitly. Only matrix-vector products that represent the application of the Hamiltonians on wave-functions need to be computed.

As in plane wave methods, the chief impediment to solving the Kohn-Sham problem is "diagonalizing" the Hamiltonian and obtaining a self-consistent field (SCF) solution. We present examples of a recently developed nonlinear Chebyshev-filtered subspace iteration (CheFSI) method, implemented in our own DFT solution package called PARSEC (Pseudopotential Algorithm for Real-Space Electronic Calculations) [10, 11]. Although described in the framework of real-space DFT, CheFSI can be employed to other SCF iterations. The subspace filtering method takes advantage of the fact that intermediate SCF iterations do not require accurate eigenvalues and eigenvectors of the Kohn-Sham equation.

The "standard" SCF iteration framework is used in CheFSI, and a self-consistent solution is obtained as with previous work, which means that CheFSI has the *same* accuracy as other standard DFT approaches. Unlike, some so-called "order-N" methods [16, 17] CheFSI is equally applicable to metals and insulators.

One can view CheFSI as a technique to tackle directly the original nonlinear Kohn-Sham eigenvalue problems by a form of nonlinear subspace iteration without emphasizing the intermediate linearized Kohn-Sham eigenvalue problems. In fact, within CheFSI, explicit eigenvectors are computed only at the first SCF iteration, in order to provide a suitable initial subspace. After the first SCF step, the explicit computation of eigenvectors at each SCF iteration is replaced by a single subspace filtering step. The method reaches self-consistency within a number of SCF iterations that is close to that of eigenvector-based approaches. However, since eigenvectors are not explicitly computed after the first step, a significant gain in execution time results when compared with methods based on explicit diagonalization.

When compared with calculations based on efficient eigenvalue packages such as ARPACK [18] and TRLan [19, 20] an order of magnitude speed-up is usually observed.

CheFSI enabled us to perform a class of challenging DFT calculations, including clusters with over 10,000 atoms, which were not feasible before without invoking additional approximations in the Kohn-Sham problem [21–23].

6.2   Eigenvalue Problems in Density Functional Calculations

The Kohn-Sham equation as defined in density functional theory is given by

$$\left[-\frac{\hbar^2}{2m} \nabla^2 + V_{total}(\rho(r), r) \right] \Psi_i(r) = E_i \Psi_i(r), \tag{6.1}$$

where $\Psi_i(r)$ is a wave function, E_i is a Kohn-Sham eigenvalue, $\hbar$ is the Planck constant, and m is the electron mass. (We will often use atomic units: $\hbar = m = e = 1$ in the following discussion.)

The *total potential* V_{total}, is the sum of three terms,

$$V_{total}(\rho(r), r) = V_{ion}(r) + V_H(\rho(r), r) + V_{xc}(\rho(r), r), \tag{6.2}$$

where V_{ion} is the ionic potential, V_H is the Hartree potential, and V_{xc} is the exchange-correlation potential. The Hartree and exchange-correlation potentials depend on the *charge density* $\rho(r)$, which is defined as

$$\rho(r) = 2 \sum_{i=1}^{n_{occ}} |\Psi_i(r)|^2. \tag{6.3}$$

Here n_{occ} is the number of occupied states, which is equal to half the number of valence electrons in the system. The factor of two comes from spin multiplicity, if the system is non-magnetic. Equation 6.3 can be easily generalized to situations where the highest occupied states have fractional occupancy or when there is an imbalance in the number of electrons for each spin component.

The most computationally expensive step of DFT is in solving the Kohn-Sham Eq. 6.1. Since V_{total} depends on the charge density $\rho(r)$, which in turn depends on the wave functions Ψ_i, Eq. 6.1, can be viewed as a *nonlinear eigenvalue problem*. The SCF iteration is a general technique used to solve this nonlinear eigenvalue problem. The iteration process begins with an initial guess of the charge density usually constructed from a superposition of free atomic charge densities, then obtains the initial V_{total} and solves Eq. 6.1 for $\Psi_i(r)$'s to update $\rho(r)$ and V_{total}. Then the Kohn-Sham (Eq. 6.1) is solved again for the new $\Psi_i(r)$'s and the process is iterated until V_{total} (and also the wave functions) becomes stationary. The standard SCF process is described in Algorithm 6.1 and illustrated in Fig. 6.1.

The number of eigenvectors needed in *Step 2* of Algorithm 6.1 is just the number of occupied states. In practice, a few more eigenvectors are usually computed.

Algorithm 6.1: Self-consistent-field iteration:

---

1. Provide initial guess for $\rho(r)$, get $V_{total}(\rho(r), r)$.
2. Solve for $\Psi_i(r)$, $i = 1, 2, ..., $ from

$$\left[-\frac{1}{2}\nabla^2 + V_{total}(\rho(r), r) \right] \Psi_i(r) = E_i \Psi_i(r). \tag{6.4}$$

3. Compute the new charge density $\rho(r) = 2 \sum_{i=1}^{n_{occ}} |\Psi_i(r)|^2$.
4. Obtain new Hartree potential V_H by solving: $\nabla^2 V_H(r) = -4\pi\rho(r)$.
5. Update V_{xc}; get new $\tilde{V}_{total}(\rho, r) = V_{ion}(r) + V_H(\rho, r) + V_{xc}(\rho, r)$ with a potential-mixing step.
6. If $\| \tilde{V}_{total} - V_{total} \| < tol$, stop; Else, $V_{total} \leftarrow \tilde{V}_{total}$, goto step 2.

Fig. 6.1 Flow diagram for obtaining a self-consistent solution of the Kohn-Sham equation

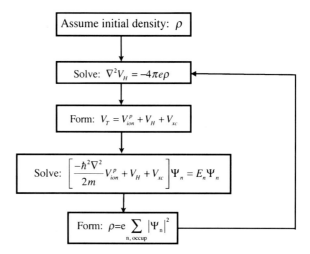

For complex systems, i.e., when the number of valence electrons is large, each of the linearized eigenvalue problems can be computationally demanding. This is compounded by the fact that Hamiltonian matrices can be of very large size.

For this reason, one hopes to lessen the burden of solving Eq. 6.4 in the SCF iteration. There are several options here. One could use some physical arguments to reduce the matrix size or zero some existing elements. Or, one could attempt to avoid diagonalization altogether, as is done in work represented by linear-scaling or order-N methods (see e.g. [16, 17]). This approach, however, has other limitations. In particular, the approximations involved rely heavily on some decay properties of the density matrix in certain function bases. In particular, they can be difficult to implement in real-space discretizations or for systems where the decay properties are not optimal, e.g., in metals. Another option is to use better (faster) diagonalization routines. However, this approach is limited as most diagonalization software is quite mature.

Our approach *avoids standard diagonalizations*, but otherwise makes no new approximations to the Hamiltonian. We take advantage of the fact that accurate eigenvectors are unnecessary at each SCF iteration, since Hamiltonians are only approximate in the intermediate SCF steps, and exploit the nonlinear nature of the problem. The main point of the new algorithm is that once we have a good starting point for the Hamiltonian, it suffices to *filter* each basis vector at each iteration. In the intermediate SCF steps, these vectors are no longer eigenvectors but together they represent a good basis of the desired invariant subspace.

6.3 Numerical Methods for Parallel Platforms

The motivation and original ideas behind our real space method (PARSEC) go back to the early 1990s, see [10, 11]. Within PARSEC, an uniform Cartesian grid in real-space is placed on the region of interest, and the Kohn-Sham equation is discretized by a high order finite-difference method [24] on this grid. Wave functions are expressed as values on grid positions. Outside a specified sphere boundary that encloses the physical system, wave functions are set to zero for non-periodic systems. In addition to the advantages mentioned in the introduction, another advantage of the real-space approach is that periodic boundary conditions are also reasonably simple to implement [25].

The latest version of PARSEC is written in Fortran 90/95. PARSEC has now evolved into a mature, massively parallel package, which includes most of the functionality of comparable DFT codes [26]. The reader is referred to [27, 28] for details and the rationale of the parallel implementation.

The following is a brief summary of the most important points. PARSEC allows for either parallel or sequential executions. When run in the parallel mode, PARSEC uses the standard Message Passing Interface (MPI) library for communication. Parallelization is achieved by partitioning the physical domain which can have various shapes depending on boundary conditions and symmetry operations. For a generic, confined system without symmetry, the physical domain is a sphere which contains all atoms plus some additional space (owing to delocalization of electron charge).

In recent years, PARSEC has been enhanced to take advantage of physical symmetry. If the system is invariant upon certain symmetry operations, the physical domain is replaced with an irreducible wedge constructed according to those operations. For example, if the system has mirror symmetry on the xy plane, the irreducible wedge covers only one hemisphere, either above or below the mirror plane. For periodic systems, the physical domain is the periodic cell, or an irreducible wedge of it if symmetry operations are present. In any circumstance, the physical domain is partitioned in compact regions, each assigned to one processor only. Good load balance is ensured by enforcing that the compact regions have approximately the same number of grid points.

Once the physical domain is partitioned, the physical problem is mapped onto the processors in a data-parallel way: each processor is in charge of a block of rows of the Hamiltonian corresponding to the block of grid points assigned to it.

The eigenvector and potential vector arrays are row-wise distributed in the same fashion. The program only requires an index function $indx(i, j, k)$ which returns the number of the processor in which the grid point (i, j, k) resides.

Because the Hamiltonian matrix is never stored, we need an explicit reordering scheme which renumbers rows consecutively from one processor to the next one. For this purpose we use a list of pointers that gives for each processor, the row with which it starts.

Since finite difference discretization is used, when performing an operation such as a matrix-vector product, communication will be required between nearest neighbor processors. For communication we use two index arrays, one to count how many and which rows are needed from neighbors, the other to count the number of local rows needed by neighbors. With this decomposition and mapping, the data required by the program is completely distributed. In other words, the code runs in the so-called "Single Program Multiple Data" approach. For large problems it is quite important to be able to distribute memory loads among processors on high performance computers. For example, certain large jobs can simply not be run on a small number of processors on good-size distributed memory machines.

Parallelizing subspace methods for the linearized eigenvalue problems (represented as Eq. 6.4) becomes quite straightforward with the above mentioned decomposition and mapping. Note that the subspace basis vectors contain approximations to eigenvectors, therefore the rows of the basis vectors are distributed in the same way as the rows of the Hamiltonian. In this way, all vector updates (e.g., linear combinations of vectors), can be executed locally (i.e., without communication). Matrix-vector products, and matrix-matrix products, can be easily executed in parallel but require some communication with a few neighbors. Reduction operations, e.g., computing inner products and making the result available in each processor, are efficiently handled by the MPI reduction function `MPI_ALLREDUCE()`.

6.4   The Nonlinear Chebyshev-Filtered Subspace Iteration

Since the Hamiltonians of the intermediate SCF steps are approximate, there is no need to compute eigenvectors of the intermediate Hamiltonians to a high accuracy. Moreover, as observed in Refs. [12, 17, 22, 29–31], the (discretized) charge density is the diagonal of the "functional" charge density matrix defined as $P = \Phi\Phi^T$, where the columns of the matrix Φ are discretized wave functions corresponding to occupied states. Notice that for any orthonormal matrix Q of a suitable dimension, $P = (\Phi Q)(\Phi Q)^T$. Therefore, explicit eigenvectors are not needed to calculate the charge density. Any orthonormal basis of the eigensubspace corresponding to occupied states can give the desired intermediate charge density.

The proposed method combines the outer SCF iteration and the inner iteration required for diagonalization at each SCF step into one nonlinear subspace iteration. In this approach an initial subspace is progressively refined by a low degree Chebyshev polynomials filtering. This means that each basis vector u_i is processed as follows:

$$u_{i,new} := p_m(H)u_i$$

where p_m is some shifted and scaled Chebyshev polynomial whose goal is to enhance eigencomponents of u_i associated with the occupied states. Throughout the article the integer m denotes the degree of the polynomial p_m which is used for filtering.

If it were not for the nonlinear nature of the SCF loop, i.e., if H were a fixed operator, this approach would be equivalent to the well-known Chebyshev accelerated subspace iteration proposed by Bauer [32], and later refined by Rutishauser [33, 34].[1]

Chebyshev polynomial filtering has long been utilized in electronic structure calculations (see e.g. [29, 35–39]), focussing primarily on approximating the Fermi-Dirac operator.

Chebyshev polynomials of rather high degree were necessary and additional techniques were required to suppress the Gibbs phenomena. In contrast, the polynomials used in our approach are of relatively low degree (say <20). They exploit the fast growth property of Chebyshev polynomials outside the interval $[-1, 1]$ to filter out undesired eigencomponents.

The main idea of CheFSI is to start with a good initial subspace V corresponding to occupied states of the initial Hamiltonian, this initial V is usually obtained by a diagonalization step. No diagonalizations are necessary after the first SCF step. Instead, the subspace from the previous iteration is filtered by a degree-m polynomial, $p_m(t)$, constructed for the current Hamiltonian H. The polynomial differs at each SCF step since H changes. Note that the goal of the filter is to make the subspace spanned by $p_m(H)V$ approximate the eigensubspace corresponding to the occupied states of the final H. At the intermediate SCF steps, the basis need not be an accurate eigenbasis since the intermediate Hamiltonians are not exact. The filtering is designed so that the resulting sequence of subspaces will progressively approximate the desired eigensubspace of the final Hamiltonian when self-consistency is reached. At each SCF step, only two parameters are required to construct an efficient Chebyshev filter, namely, a lower bound and an upper bound of the higher portion of the spectrum of the current Hamiltonian H in which we want $p_m(t)$ to be small. These bounds can be obtained with little additional cost, as will be seen in Sect. 6.4.2.

After self-consistency is reached, the Chebyshev filtered subspace includes the eigensubspace corresponding to occupied states. Explicit eigenvectors can be readily obtained by a *Rayleigh-Ritz refinement* [40] (also called *subspace rotation*) step.

---

[1] Rutishauser published an Algol routine called *ritzit* in the volume: "Handbook for automatic computations: linear algebra", see [34]. This volume was largely at the origin of the EISPACK package (which later became a part of LAPACK), but Rutishauser's *ritzit* Algol routine was not translated into EISPACK.

Algorithm 6.2: CheFSI for SCF calculation:

1. Start from an initial guess of $\rho(r)$, get $V_{total}(\rho(r), r)$.
2. Solve $\left[-\frac{1}{2}\nabla^2 + V_{total}(\rho(r), r)\right]\Psi_i(r) = E_i\Psi_i(r)$ for $\Psi_i(r)$, $i = 1, 2, ..., s$.
3. Compute new charge density $\rho(r) = 2\sum_{i=1}^{n_{occ}}|\Psi_i(r)|^2$.
4. Solve for new Hartree potential V_H from $\nabla^2 V_H(r) = -4\pi\rho(r)$.
5. Update V_{xc}; get new $\tilde{V}_{total}(\rho, r) = V_{ion}(r) + V_H(\rho, r) + V_{xc}(\rho, r)$ with a potential-mixing step.
6. If $\|\tilde{V}_{total} - V_{total}\| < tol$, stop; Else, $V_{total} \leftarrow \tilde{V}_{total}$ (update H implicitly), call the Chebyshev-filtered subspace method (Algorithm 6.3) to get s approximate wave functions; goto step 3.

6.4.1   Chebyshev-Filtered Subspace Iteration

The main structure of CheFSI, which is given in Algorithm 6.2, is quite similar to that of the standard SCF iteration (Algorithm 6.1). One major difference is that the inner iteration for diagonalization at *Step 2* is now performed only at the first SCF step. Thereafter, diagonalization is replaced by a single Chebyshev subspace filtering step, performed by calling Algorithm 6.3.

Although the charge density (Eq. 6.3) requires only the lowest n_{occ} states, the number of computed states, which is the integer s in Algorithm 6.2, is typically set to a value larger than n_{occ}, in order to avoid missing any occupied states. In practice we fix an integer n_{state} which is slightly larger than n_{occ}, and set $s = n_{state} + n_{add}$ with $n_{add} \leq 10$.

The parallel implementations of Algorithms 6.2 and 6.3 are quite straightforward with the parallel paradigm discussed in Sect. 6.3. We only mention that the matrix-vector products related to filtering, computing upper bounds, and Rayleigh-Ritz refinement, can easily execute in parallel. The re-orthogonalization at *Step 4* of Algorithm 6.3 uses a parallel version of the iterated Gram-Schmidt DGKS method [41], which scales better than the standard modified Gram-Schmidt algorithm. This process is illustrated in Fig. 6.2.

The estimated complexity of the algorithm is similar to that of the sequential CheFSI method in [22]. For parallel computation it suffices to estimate the complexity on a single processor. Assume that p processors are used, i.e., each processor shares N/p rows of the full Hamiltonian. The estimated cost of Algorithm 6.3 on each processor with respect to the dimension of the Hamiltonian denoted by N, and the number of computed states s, is as follows:

- The Chebyshev filtering in *Step 3* costs $O(s * N/p)$ flops. The discretized Hamiltonian is sparse and each matrix-vector product on one processor costs $O(N/p)$ flops. *Step 3* requires $m * s$ matrix-vector products, at a total cost of $O(s*m*N/p)$ where the degree m of the polynomial is small (typically between 8 and 20).
- The ortho-normalization in *Step 4* costs $O(s^2 * N/p)$ flops. There are additional communication costs because of the global reductions.

Algorithm 6.3: Chebyshev-filtered Subspace (CheFS) method:

---

1. Get the lower bounds b_{low} and γ from previous Ritz values (use the largest one and the smallest one, respectively).
2. Compute the upper bound b_{up} of the spectrum of the current discretized Hamiltonian H (call Algorithm 6.5 in Sect. 6.4.2).
3. Perform Chebyshev filtering (call Algorithm 6.4 in Sect. 6.4.2) on the previous basis Φ, where Φ contains the discretized wave functions of $\Psi_i(r)$, $i = 1, ..., s$:
   $\Phi = \text{Chebyshev\_filter}(\Phi, m, b_{low}, b_{up}, \gamma)$.
4. Ortho-normalize the basis Φ by iterated Gram-Schmidt.
5. Perform the Rayleigh-Ritz step:

   (a) Compute $\hat{H} = \Phi^T H \Phi$;
   (b) Compute the eigendecomposition of $\hat{H}$: $\hat{H} Q = QD$,
       where D contains non-increasingly ordered eigenvalues of $\hat{H}$, and Q contains the corresponding eigenvectors;
   (c) 'Rotate' the basis as $\Phi := \Phi Q$; return Φ and D.

- The eigen-decomposition at *Step 5* costs $O(s^3)$ flops.
- The final basis refinement step ($\Phi := \Phi Q$) costs $O(s^2 * N/p)$.

If a standard iterative diagonalization method is used to solve the linearized eigenproblem (Eq. 6.4) at each SCF step, then it also requires (1) the orthonormalization of a (typically larger) basis; (2) the eigen-decomposition of the projected Rayleigh-quotient matrix; and (3) the basis refinement (rotation). These operations need to be performed several times within this single diagonalization. But Algorithm 6.3 performs each of these operations only once per SCF step. Therefore, although Algorithm 6.3 scales in a similar way to standard diagonalization-based methods, the scaling constant is much smaller. For large problems, CheFS can achieve a tenfold or more speedup per SCF step, over using the well-know efficient eigenvalue packages such as ARPACK [18] and TRLan [19, 20].

In summary, a standard SCF method has an outer SCF loop – the usual nonlinear SCF loop, and an inner diagonalization loop, which iterates until eigenvectors are within specified accuracy. Algorithm 6.2 essentially bypasses the second loop, or rather it merges it into a single outer loop, which can be considered as a *nonlinear subspace iteration algorithm*. The inner diagonalization loop is replaced by a single Chebyshev subspace filtering step.

6.4.2 Chebyshev Filters and Estimation of Bounds

Chebyshev polynomials of the first kind are defined, for $k = 0, 1, \cdots$, by (see e.g., [40], p. 371 or [42], p. 142):

Fig. 6.2 Flow diagram for obtaining a self-consistent solution of the Kohn-Sham equation using damped Chebyshev subspace filtering

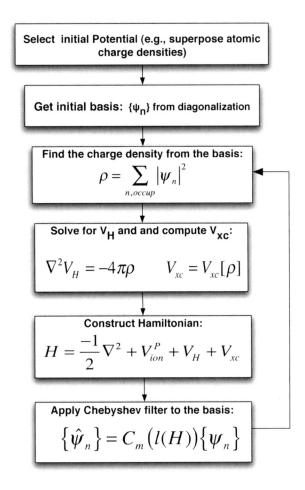

$$C_k(t) = \begin{cases} \cos(k\ \cos^{-1}(t)), & -1 \le t \le 1, \\ \cosh(k\ \cosh^{-1}(t)), & |t| > 1. \end{cases}$$

Note that $C_0(t) = 1, C_1(t) = t$. The following important 3-term recurrence is easy to derive from properties of the cosine function,

$$C_{k+1}(t) = 2t\ C_k(t) - C_{k-1}(t), \quad t \in \mathbb{R}. \tag{6.5}$$

By filtering we mean a process applied to a vector that has the effect of magnificent desired eigen-components of this vector relative to other, undesirable, components. If the process is repeated indefinitely, the resulting vector will have zero components in the undesirable part of the spectrum. In our context, we need to filter out all components associated with the non-occupied states, or, equivalently to enhance the components associated with occupied states, relative to other components.

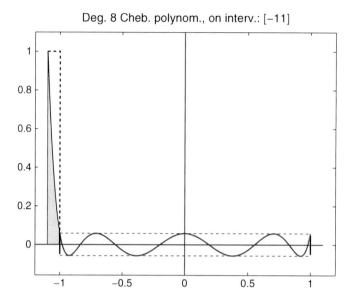

Fig. 6.3 Degree 8 Chebyshev polynomial on the interval $[-1, 1]$ scaled to one at $\gamma = -0.2$. The *shaded* area corresponds to eigen-components that will be amplified relative to the other eigencomponents, those corresponding to the interval $[-1, 1]$, which will be dampened

Filtering can be readily achieved by exploiting well-known properties of Chebyshev polynomials. It is known that among all polynomials of degree k, which have value one at a certain point $|\gamma| > 1$, the polynomial $C_k(t)/C_k(\gamma)$ is the one whose maximum absolute value in the interval $[-1, 1]$ is minimal. Thus, $C_k(t)/C_k(\gamma)$ can be viewed as an optimal polynomial if one wishes to dampen values of the polynomial in $[-1, 1]$ among all polynomials p of degree k, scaled so that $p(\gamma) = 1$. The 8th degree Chebyshev polynomial scaled at $\gamma = -0.2$ is shown in Fig. 6.3.

Assume that the full spectrum of H (denoted by $\Lambda(H)$) is contained in $[\gamma, b]$. Then, in order to approximate the eigensubspace associated with the lower end of the spectrum, say $[\gamma, a]$ with $\gamma < a < b$, it is necessary to map $[a, b]$ into $[-1, 1]$ before applying the Chebyshev polynomial. This can be easily realized by an affine mapping defined as

$$\mathcal{L}(t) := \frac{t - c}{e}; \quad c = \frac{a + b}{2}, \quad e = \frac{b - a}{2}$$

where c denotes the center and e the half-width of the interval $[a, b]$. The Chebyshev iteration utilizing the three-term recurrence (6.5) to dampen values on the interval $[a, b]$ is listed in Algorithm 6.4, see also [22]. The algorithm computes

$$Y = p_m(H)X \qquad \text{where} \qquad p_m(t) = C_m[\mathcal{L}(t)]. \tag{6.6}$$

Algorithm 6.4: $Y = \texttt{Chebyshev\_filte}\,(X, m, a, b, \gamma)$.

Purpose: Filter column vectors of X by an m degree Chebyshev polynomial in H that dampens on the interval $[a, b]$. Output the filtered vectors in Y.

1. $e = (b - a)/2;\quad c = (b + a)/2;$
2. $\sigma = e/(\gamma - c);\qquad \sigma_1 = \sigma;\qquad \gamma = 2/\sigma_1.$
3. $Y = \frac{\sigma_1}{e}(HX - cX);$
4. For $i = 2 : m$
5. $\qquad \sigma_2 = 1/(\gamma - \sigma);$
6. $\qquad Y_{new} = \frac{2\sigma_2}{e}(HY - cY) - \sigma\sigma_2 X;$
7. $\qquad X = Y;$
8. $\qquad Y = Y_{new};$
9. $\qquad \sigma = \sigma_2;$
10. End For

This yields the iteration

$$X_{j+1} = \frac{2}{e}(H - cI)X_j - X_{j-1}, \quad j = 1, 2, ..., m - 1.$$

with X_0 given and $X_1 = (H - cI)X_0$.

The above iteration is without any scaling. In the case of the interval $[-1, 1]$ we scaled the polynomial by $C_k(\gamma)$ in order to ensure that the value of the polynomial at γ equals one. For general intervals, this leads to the scaled sequence of polynomials [42]

$$\tilde{X}_j = \frac{C_j[\frac{2}{e}(H - cI)]}{C_j[\frac{2}{e}(\gamma - cI)]}X_0.$$

Thus, the scaling factor is $\rho_j = C_j[\frac{2}{e}(\gamma - cI)]$. Clearly this requires an estimate for γ which, in our case, is the smallest eigenvalue of the Hamiltonian. However, since this is used for scaling, for the purpose of avoiding overflow, only a rough value is needed. For the first SCF iteration, we can use the smallest Ritz value of T from the same Lanczos run (Algorithm 6.5 below) as used to obtain the upper bound b for γ. For the latter SCF steps, the smallest Ritz value from the previous SCF step can be used. Clearly, the vector sequence is not computed as shown above because ρ_j itself can be large and this would defeat the purpose of scaling. Instead, each $\tilde{X}_{j+1}$ is updated using the scaled vectors $\tilde{X}_j$ and $\tilde{X}_{j-1}$. The corresponding algorithm, discussed in [42] is shown in Algorithm 6.4 (the tildes and vector subscripts are omitted).

The eigen-components associated with eigenvalues in $[a, b]$ will be transformed to small values while those to the left of $[a, b]$ will be around unity owing to the properties of the Chebyshev polynomials. This is the desired filtering property when computing an approximation to the eigensubspace associated with the lower end of $\Lambda(H)$. As seen in Algorithm 6.4, a desired filter can be easily controlled by adjusting two endpoints that bound the higher portion of $\Lambda(H)$.

The wanted lower bound can be any value which is larger than the Fermi-level but smaller than the upper bound. It can also be a value slightly smaller than the Fermi-level; thanks to the monotonicity of the shifted and scaled Chebyshev polynomial on the spectrum of H, and the fact that we compute $s > n_{occ}$ number of Ritz values, the desired lowered end of the spectrum will still be magnified properly with this choice of lower bound.

Since the previous SCF iteration performs a Rayleigh-Ritz refinement step, it provides naturally an approximation for the lower bound a. Indeed, we can simply take the largest Rayleigh-quotient from the previous SCF iteration step as an approximation to the lower bound for the current Hamiltonian. In other words, a is taken to be the largest eigenvalue computed in step 5-(b) of Algorithm 6.3 from the previous SCF iteration, with no extra computation.

The upper bound for the spectrum (denoted by b) can be estimated by a k-step standard Lanczos method. As pointed out in [23], the higher endpoint b must be an upper bound for the full spectrum of H. This is because the Chebyshev polynomial also grows fast to the right of $[-1, 1]$. So if $[a, b]$ with $b < \lambda_{max}(H)$ is mapped into $[-1, 1]$, then the $[b, \lambda_{max}(H)]$ portion of the spectrum will also be magnified, which will cause the procedure to fail. Therefore, it is imperative that the bound b be larger than $\lambda_{max}(H)$. On the other hand it should not be too large as this would result in slow convergence. The simplest strategy which can be used for this is to use Gerschgorin's Circle Theorem. Bounds obtained this way can, however, overestimate $\lambda_{max}(H)$.

An inexpensive way to estimate an upper bound of $\Lambda(H)$ by the standard Lanczos [43] method is described in Algorithm 6.5, to which a safeguard step is added. The largest eigenvalue $\tilde{\lambda}$ of the tridiagonal matrix T is known to be below the largest eigenvalue λ of the Hamiltonian. If $\tilde{u}$ is the corresponding Ritz vector and $r = (H - \tilde{\lambda}I)\tilde{u}$ then there is an eigenvalue of H in the interval $[\tilde{\lambda} - \|r\|, \tilde{\lambda} + \|r\|]$ (see e.g. [40]). Algorithm 6.5 estimates λ_{max} by $max(\tilde{\lambda}) + \|f\|$, since it is known that $\|r\| \leq \|f\|$. This is not theoretically guaranteed to return an upper bound for λ_{max} – but it is generally observed to yield an effective upper bound. The algorithm for estimating b is presented in Algorithm 6.5 below. Note that the algorithm is easily parallelizable as it relies mostly on matrix-vector products. In practice, we found that $k = 4$ or 5 is sufficient to yield an effective upper bound of $\Lambda(H)$. Larger k values (e.g., $k > 10$) are not necessary in general.

We see that the extra work associated with computing bounds for constructing the Chebyshev polynomials is negligible. The major cost of filtering is in the three-term recurrences in Algorithm 6.4, which involve matrix-vector products. The polynomial degree m is left as a free parameter. Our experience indicates that an m between 8 and 20 is good enough to achieve overall fast convergence in the SCF loop.

Algorithm 6.5: Estimating an upper bound of $\Lambda(H)$ by k-step Lanczos:

---

1. Generate a random vector v, set $v \leftarrow v/\|v_2\|$;
2. Compute $f = Hv$; $\alpha = f^T v$; $f \leftarrow f - \alpha v$; $T(1,1) = \alpha$;
3. Do $j = 2$ to $min(k, 10)$
4. $\beta = \|f_2\|$;
5. $v_0 \leftarrow v$; $v \leftarrow f/\beta$;
6. $f = Hv$; $f \leftarrow f - \beta v_0$;
7. $\alpha = f^T v$; $f \leftarrow f - \alpha v$;
8. $T(j, j-1) = \beta$; $T(j-1, j) = \beta$; $T(j, j) = \alpha$;
9. End Do
10. Return $\|T_2\| + \|f_2\|$ as the upper bound.

6.5 Window Filtering

The Chebyshev Davidson method as outlined uses only one filter. This filter enhances the part of the spectrum associated with the occupied states and a limited number of virtual states or empty states. Any basis for the subspace of vectors generated using the filter has to be made orthonormal, and this comes at a cost of $O(N_{occ}^2 N_{dim})$ [23]. In addition to orthogonalization, the Rayleigh-Ritz procedure must be used to extract eigenvectors from an approximation subspace. This algorithm requires diagonalization of a dense $N_{occ} \times N_{occ}$ matrix, an procedure that scales as $O(N_{occ}^3)$. The idea behind a spectrum slicing method for the eigenproblem is that if the spectrum is partitioned into segments, and window filters are used in each segment, then the cost of algorithms that scale superlinearly in N_{occ} can be reduced. This is due to the fact that each slice of the spectrum contains a smaller number of eigenpairs than the whole. The price that must be paid for this savings is an increase in the cost of applying a filter operator. Filters used in the Chebyshev-Davidson method, such as that in Fig. 6.3, typically need not exceed a degree of 30. In general, a higher degree is need as the total width of the Hamiltonian's spectrum increases. The width will increase with increasing finite-difference order and decreasing grid spacing. In contrast, the window filters, as depicted in Fig. 6.4, will be around degree 100 for the same problem. Therefore a windowing method is most advantageous when N_{occ} is large. When spectrum slicing can be used, the work associated with each slice can be given to an independent group of processors. This adds another layer of parallelization on top of the scheme already implemented in PARSEC, see [21].

For our filter operators we employ the Chebyshev-Jackson approximations, used previously for electronic structure calculations [39], to the step function in (6.7). The step function

$$\phi(x) = \begin{cases} 1 & : \quad x \leq \lambda_{occ} \\ 0 & : \quad x > \lambda_{occ} \end{cases} \qquad (6.7)$$

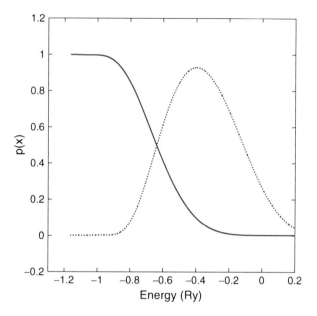

Fig. 6.4 Filters for slices at $[-1.16, -0.64]$ and $[-0.64, -0.11]$ used to compute the electronic structure of $Si_{275}H_{172}$. The left filter has degree 88, and the right filter has degree 121. The spectrum extends to 47.9 Ry

The Chebyshev-Jackson approximation smooths the ripples that occur in a standard Chebyshev approximation for any function with sharp features. This effect is obtained by damping the high order coefficients, c_i, in the sum (6.8) by the damping factors, $g_i^k \in [0, 1]$, given by (6.10).

$$f(x) \approx \sum_{i=0}^{k} c_i g_i^k T_i(x) \qquad (6.8)$$

$$c_i = \begin{cases} \dfrac{1}{\pi} \displaystyle\int_{-1}^{1} \dfrac{1}{\sqrt{1-x^2}} f(x) dx & : i = 0 \\[4mm] \dfrac{2}{\pi} \displaystyle\int_{-1}^{1} \dfrac{1}{\sqrt{1-x^2}} f(x) T_i(x) dx & : i > 0 \end{cases} \qquad (6.9)$$

$$g_i^k = \dfrac{\left(1 - \frac{i}{k+2}\right) \sin(\alpha_k) \cos(i\alpha_k)}{\sin(\alpha_k)} + \qquad (6.10)$$

Algorithm 6.6: Computing the eigenpairs in a given slice

---

input : f, filter polynomial over $[a, b]$
output: $W = \{(\lambda_i, \psi_i)\}_{i=1}^{N_{[a,b]}}$, eigenpairs contained in $[a, b]$
$V = \emptyset$;
$x_0 = \text{Random}()$;
while *outside < max_out* **do**
$\quad\left|\quad\begin{array}{l} t = f(H)x_i; \\[4pt] t^{\perp} = t - VV^H t; \\[4pt] x_{i+1} = t^{\perp}/\|t^{\perp}\| \, ; \\[4pt] V = V \cup \{x_{i+1}\}; \\[4pt] \textbf{if } x_{i+1}^H V x_{i+1} > b \textbf{ then} \\[4pt] \quad| \quad \text{outside} = \text{outside} + 1 \; ; \\[4pt] \textbf{end} \end{array}\right.$
end
$R = V^H H V$;
$Z^H \Lambda Z = R$;   // Compute diagonalization
$\Psi = ZV$;
$W = \emptyset$;
foreach $\psi_i \in \Psi$ **do**
$\quad\left|\quad\begin{array}{l} \lambda_i = \psi_i^H H \psi_i; \\[4pt] \textbf{if } \lambda_i \in [a, b] \textbf{ then} \\[4pt] \quad| \quad W = W \cup (\lambda_i, \psi_i) \; ; \\[4pt] \textbf{end} \end{array}\right.$
end

$$\frac{\frac{1}{k+2} \cos(\alpha_k) \sin(i\alpha_k)}{\sin(\alpha_k)} \tag{6.11}$$

Use of smoothed polynomials for the filter operators insures that states in the ripple local extremum do not converge prematurely before states closer to the segment of interest.

In order to partition the occupied states into slices, it is necessary to have a guess for the Fermi level. The guess does not need a high degree of accuracy. For this the Fermi level from the previous SCF iteration can be used. This means that the spectrum slicing method will need to be bootstrapped with a regular method similarly to Chebyshev-Davidson.

With a filter polynomial in hand, Algorithm 6.6 can compute the eigenpairs in a given slice

In line 1, Algorithm 6.6 exploits the fact that, by using a smooth filter, the Ritz values for the iterates, x_i, will stay within the interval $[a, b]$. Once the Ritz value has drifted outside of the the slice, the algorithm can safely proceed to extract eigenpair approximations from the subspace, W.

In a spectrum slicing method, there is a danger that adjacent slices will compute approximations to the same eigenpair. Given an approximation, (λ_i, ψ_i), to an

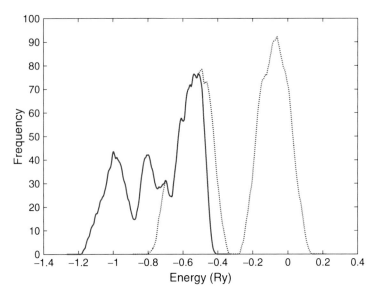

Fig. 6.5 Density of states of $Si_{275}H_{172}$ computed using two filters

eigenpair, the Ritz residual, $\sigma_i = \|H\psi - \lambda_i\psi_i\|_2$, can be used to ascertain the existence of an eigenpair in $[\lambda_i - \sigma_i, \lambda_i + \sigma_i]$. Near slice boundaries, this means that some eigenpairs with $\lambda_i \notin [a, b]$ may need to be included in the set of eigenpairs for $[a, b]$. So the possibility emerges that the algorithm run on an adjacent slice, $[b, c]$, might return an approximation to the same eigenpair in W. To rectify this situation eigenpairs near slice boundaries can be gathered into a subspace, and analysis of the principal angles, [44] will reveal the presence of any duplicate approximations.

As an example, we computed the spectrum of an unrelaxed $Si_{275}H_{172}$ cluster. This calculation was done with PARSEC employing a parallel implementation of the spectrum slicing algorithm. In the final iteration of the self-consistency cycle, the spectrum of the Hamiltonian for this system ranges from -1.16 to $47.9\,eV$ with the Fermi level at -0.37 and 636 occupied states. A histogram of the states is shown in Fig. 6.5. To apply Algorithm 6 we split the spectrum into slices at $[-1.16, -0.64]$ and $[-0.64, -0.11]$. The polynomial filters used in this calculation are those displayed in Fig. 6.4. The result agrees exactly with the standard PARSEC solver. The degree of the filters in Fig. 6.4 are 88 and 121, and for a problem of this size the majority of the time is spent in filtering. Given the result of this and similar calculations, we expect the spectrum slicing method to become competitive on systems with approximately 3,000 states.

6.6   Diagonalization in the First SCF Iteration

Within CheFSI, the most expensive SCF step is the first one, as it involves a diagonalization in order to compute a good subspace to initiate the nonlinear SCF loop. This section discusses options available for this task.

In principle, any effective eigenvalue algorithms can be used for the first SCF step. PARSEC originally had three diagonalization methods: Diagla, which is a preconditioned Davidson method [27, 28]; the symmetric eigensolver in ARPACK [18, 45]; and the Thick-Restart Lanczos algorithm called TRLan [19, 20]. For systems of moderate sizes, Diagla works well, and then becomes less competitive relative to ARPACK or TRLan for larger systems when a large number of eigenvalues are required. TRLan is about twice as fast as the symmetric eigensolver in ARPACK, because of its reduced need for re-orthogonalization. In [22], TRLan was used for the diagonalization at the first SCF step.

Another option suggested and tested in [31] but not implemented in PARSEC, is to resort to the Lanczos algorithm with partial reorthogonalization. Partial reorthogonalization Lanczos would run the Lanczos algorithm without restarting, reorthogonalizing the vectors only when needed, see [40]. This is a very effective procedure, some would even say optimal in some sense, except that it typically requires an enormous amount of memory. As illustrated in [31] the method can be five to seven times faster than ARPACK for moderate size problems. It is possible to address the memory problem by resorting to secondary storage, though parallel implementations would be tedious.

At the other extreme when considering memory usage, one can use the Chebyshev filtered subspace iteration *in its linear implementation*. This means that we will now add an outer loop to the procedure described by Algorithm 6.3 and test convergence for the same Hamiltonian (the initial one) without updating potential from one outer loop to the next. Practically, this is simply as a variant of Algorithm 6.2, whereby step 2 is replaced by as many filtering steps of Algorithm 6.3 as are required for the subspace to converge. This procedure is the most economical in terms of memory, so it is recommended if memory is an issue. However, it is well-known that subspace iteration methods (linear) are not as effective as the Lanczos algorithm, and other Krylov-based methods, see, e.g., [40], Chap. 14.

Even with standard restart methods such as ARPACK and TRLan, the memory demand can still remain too high in some cases. Hence, it is important to develop a diagonalization method that is less memory demanding but whose efficiency is comparable to ARPACK and TRLan. The Chebyshev-Davidson method [23] was developed with these two goals in mind. Details can be found in [23]. The principle of the method is to simply build a subspace by a procedure based on a form of Block-Davidson approach. The Block-Davidson approach builds a subspace by adding a 'window' of preconditioned vectors. In the Chebyshev-Davidson approach, these vectors are built by exploiting Chebyshev polynomials.

The first step diagonalization by the block Chebyshev-Davidson method, together with the Chebyshev-filtered subspace method (Algorithm 6.3), enabled us to perform SCF calculations for a class of large systems, including the silicon cluster $Si_{9041}H_{1860}$ for which over 19,000 eigenvectors of a Hamiltonian with dimension around three million were to be computed. These systems are practically infeasible with the other three eigensolvers (ARPACK, TRLan and Diagla) in PARSEC, using the current supercomputer resources available to us at the Minnesota Supercomputing Institute (MSI).

Though results obtained with the Chebyshev-Davidson method in the first step diagonalization are satisfactory, there is still much work to be done in this area. We do not know for example how accurate the subspace must be in order to be a good initial guess to ensure convergence. It may possible to further reduce execution times by changing the stopping criterion needed in the first SCF step. It may be also possible to exploit well-known "global convergence" strategies utilized for non-linear iterations (such as continuation, or damping) to avoid completely the first step diagonalization.

6.7 Numerical Results

PARSEC has been applied to study a wide range of material systems (e.g., [11, 25, 26]). Here we focus on hydrogenated Si nanocrystals containing at least 1,000 atoms, where relatively few numerical results exist because of the infeasibility of eigenvector-based methods. In particular, Zhao et al. [46] have examined clusters containing up to 1,100 silicon atoms, using the well-known package VASP [8, 47, 48]. Although VASP is a very powerful code, Zhao et al. found clusters with more than ~1,200 silicon atoms was "too computationally intensive." As a comparison, PARSEC using CheFSI, together with the currently developed symmetric operations of real-space pseudopotential methods can now routinely solve silicon clusters with several thousands of atoms. The largest examined with CheFSI exceeds 9,000 atoms.

The hardware used for the computations illustrated here was the SGI Altix cluster at MSI, it consists of 256 Intel Itanium processors at CPU rates of 1.6 GHz, sharing 512 GB of memory (but a single job is allowed to request at most 250 GB memory).

The goal of the computations is to use PARSEC to do SCF calculations for large systems which were not studied before. We did not use different processor numbers to solve the same problem. Scalability is studied in [28] for the preconditioned Davidson method, we mentioned that the scalability of CheFSI is better than eigenvector-based methods because of the reduced reorthogonalizations.

In the reported numerical results, the total energy per atom in electron-volts, this value can be used to assess accuracy of the final result; the #SCF is the iteration steps needed to reach self-consistency; and the #MVp counts the number of matrix-vector products. Clearly #MVp is not the only factor that determines CPU time, the orthogonalization cost can also be a significant component.

Table 6.1 $Si_{525}H_{276}$, using 16 processors. The Hamiltonian dimension is 292,584, where 1,194 states need to be computed at each SCF step. The first step diagonalization by Chebyshev-Davidson cost 79,755 #MVp and 221.05 CPU seconds; so the total #MVp spent on CheFS in CheFSI is 110,000. The polynomial degree used is $m = 17$ for Chebyshev-Davidson and $m = 8$ for CheFS. The fist step diagonalization by TRLan requires 14,909 #MVp and 265.75 CPU seconds

Method	#MVp	#SCF steps	Total_eV/atom	CPU(sec)
CheFSI	189,755	11	−77.316873	542.43
TRLan	149,418	10	−77.316873	2,755.49
Diagla	493,612	10	−77.316873	8,751.24

Table 6.2 Performance of the CheFSI methods for large hydrogenated silicon nanocrystals. All calculations were done on 16 processors

System	Dim. of H	n_{state}	#MVp	#SCF	Total_eV/ atom	1st CPU(h)	Total CPU (h)
$Si_{4001}H_{1012}$	1,472,440	8,511	1,652,243	12	−89.12, 338	18.63	38.17
$Si_{6047}H_{1308}$	2,144,432	12,751	2,682,749	14	−91.34809	45.11	101.02
$Si_{9041}H_{1012}$	2,992,832	19,015	4,804,488	18	−92.00412	102.12	294.36

The first example (Table 6.1) is a relatively small silicon cluster $Si_{525}H_{276}$, which is used to compare the performance of CheFSI with two eigenvector-based methods. All methods use the same symmetry operations in PARSEC.

We considered large nanocrystals of silicon, including $Si_{2713}H_{828}$, $Si_{4001}H_{1012}$ and $Si_{9041}H_{1860}$. For these clusters, Diagla became too slow to be practical. In the case of $Si_{2713}H_{828}$, we could still apply TRLan for the first step diagonalization for comparison, but we did not iterate until self-consistency was reached without excessive computational resources. Note that with the problem size increasing, Chebyshev-Davidson compares more favorably over TRLan. This is because we employed an additional trick in Chebyshev-Davidson, which corresponds to allowing the last few eigenvectors not to converge to the required accuracy. The number of the non fully converged eigenvectors is bounded above by act_{max}, which is the maximum dimension of the active subspace. Typically $30 \leq act_{max} \leq 300$ for Hamiltonian size over a million where several thousand eigenvectors are to be computed. The implementation of this trick is rather straightforward since it corresponds to applying the CheFS method to the subspace spanned by the last few vectors in the basis that have not converged to required accuracy.

For a larger cluster such as $Si_{6047}H_{1308}$, it became impractical to apply TRLan for the first step diagonalization because of too large memory requirements. For these large systems, using an eigenvector-based method for each SCF step is clearly not feasible. We note that the cost for the first step diagonalization by Chebyshev-Davidson is still rather high, it took close to 50% of the total CPU. In comparison, the CheFS method (Algorithm 6.3) saves a significant amount of CPU for SCF calculations over diagonalization-based methods, even if very efficient eigenvalue algorithms are used. Our results are summarized in Table 6.2.

6.8 Concluding Remarks

Here we illustrated a parallel CheFSI method for DFT SCF calculations, including a extension using a "multi-window" approach. Within CheFSI, only the first SCF step requires a true diagonalization, and we perform this step by the block Chebyshev-Davidson method. No diagonalization is required after the first step; instead, Chebyshev filters are adaptively constructed to filter the subspace from previous SCF steps so that the filtered subspace progressively approximates the eigensubspace corresponding to occupied states of the final Hamiltonian. The method can be viewed as a nonlinear subspace iteration method which combines the SCF iteration and diagonalization, with the diagonalization simplified into a single step Chebyshev subspace filtering.

Additional tests not reported here, have also shown that the subspace filtering method is robust with respect to the initial subspace. Besides self-consistency, it can be used together with molecular dynamics or structural optimization, provided that atoms move by a small amount. Even after atomic displacements of a fraction of the Bohr radius, the CheFSI method was able to bring the initial subspace to the subspace of self-consistent Kohn-Sham eigenvectors for the current position of atoms, with no substantial increase in the number of self-consistent cycles needed.

CheFSI significantly accelerates the SCF calculations, and this enabled us to perform a class of large DFT calculations that were not feasible before by eigenvector-based methods. As an example of physical applications, we discuss the energetics of silicon clusters containing up to several thousand atoms.

Acknowledgements We wish to acknowledge support from the Welch Foundation under grant No. F-1708 as well as from the U. S. Department of Energy, Office of Basic Energy Sciences and Office of Advanced Scientific Computing Research, under grant DE-SC0001878 and from the National Science Foundation under grant DMR 09-41645. We would also like to acknowledge computational support from the Minnesota Supercomputing Institute and the Texas Advanced Computing Center.

References

1. Hohenberg P, Kohn W (1964) Phys Rev 136:B864
2. Kohn W, Sham LJ (1965) Phys Rev 140:A1133
3. Phillips JC (1958) Phys Rev 112:685
4. Phillips JC, Kleinman L (1959) Phys Rev 116:287
5. Chelikowsky JR, Cohen ML (1992) Handbook on semiconductors, vol 1. Elsevier, Amsterdam, p 59
6. Martin RM (2004) Electronic structure: basic theory and practical methods. Cambridge University Press, Cambridge
7. Payne MC, Teter MP, Allan DC, Arias TA, Joannopoulos JD (1992) Rev Mod Phys 64:1045
8. Kresse G, Furthmüller J (1996) Phys Rev B 54:11169
9. Koch W, Holthausen MC (2000) A chemist's guide to density functional theory. Wiley, Weinheim

10. Chelikowsky JR, Troullier N, Saad Y (1994) Phys Rev Lett 72:1240
11. Chelikowsky JR, Troullier N, Wu K, Saad Y (1994) Phys Rev B 50:11355
12. Seitsonen AP, Puska MJ, Nieminen RM (1995) Phys Rev B 51:14057
13. Beck TL (2000) Rev Mod Phys 72:1041
14. Alemany MMG, Jain M, Chelikowsky JR, Kronik L (2004) Phys Rev B 69:075101
15. Makov G, Payne MC (1995) Phys Rev B 51:4014
16. Otsuka T, Miyazaki T, Ohno T, Bowler DR, Gillan MJ (2008) J Phys Condens Matter 20:294201
17. Goedecker S (1999) Rev Mod Phys 71:1085
18. Lehoucq RB, Sorensen DC, Yang C (1998) ARPACK users guide: solution of large scale eigenvalue problems by implicitly restarted Arnoldi methods. SIAM, Philadelphia
19. Wu K, Canning A, Simon HD, Wang L-W (1999) J Comput Phys 154:156
20. Wu K, Simon H (2000) SIAM J Matrix Anal Appl 22:602
21. Zhou Y, Saad Y, Tiago M, Chelikowsky JR (2006) Phys Rev E 74:066704
22. Zhou Y, Saad Y, Tiago ML, Chelikowsky JR (2006) J Comput Phys 243:1063
23. Zhou Y, Saad Y (2007) SIAM J Matrix Anal Appl 29:954
24. Fornberg B, Sloan DM (1994) In: Iserles A (ed) Acta numerica, vol 3. Cambridge University Press, Cambridge, p 203
25. Alemany MMG, Jain M, Kronik L, Chelikowsky JR (2004) Phys Rev B 69:075101
26. Kronik L, Makmal A, Tiago M, Alemany M, Jain M, Huang X, Saad Y, Chelikowsky JR (2006) Phys Status Solidi B 243:1063
27. Saad Y, Stathopoulos A, Chelikowsky J, Wu K, Öğüt S (1996) BIT 36:563
28. Stathopoulos A, Öğüt S, Saad Y, Chelikowsky JR, Kim H (2000) IEEE Comput Sci Eng 2:19
29. Stephan U, Drabold DA, Martin RM (1998) Phys Rev B 58:13472
30. Baroni S, Giannozzi P (1992) Europhys Lett 17:547
31. Bekas C, Saad Y, Tiago ML, Chelikowsky JR (2005) Comput Phys Commun 171:175
32. Bauer FL (1957) Z Angew Math Phys 8:214
33. Rutishauser H (1969) Numer Math 13:4
34. Rutishauser H (1971) In: Wilkinson JH, Reinsch C (eds) Handbook for automatic computation (Linear Algebra), vol II. Springer, New York, p 284
35. Sankey OF, Drabold DA, Gibson A (1994) Phys Rev B 50:1376
36. Goedecker S, Colombo L (1994) Phys Rev Lett 73:122
37. Baer R, Head-Gordon M (1997) J Chem Phys 107:10003
38. Baer R, Head-Gordon M (1998) J Chem Phys 109:10159
39. Jay LO, Kim H, Saad Y, Chelikowsky JR (1999) Comput Phys Commun 118:21
40. Parlett BN (1998) The symmetric eigenvalue problem. SIAM, Philadelphia
41. Daniel J, Gragg WB, Kaufman L, Stewart GW (1976) Math Comput 30:772
42. Saad Y (1992) Numerical methods for large eigenvalue problems. Wiley, New York
43. Lanczos C (1950) J Res Nat Bur Stand 45:255
44. Bjorck A, Golub GH (1973) Math Comput 27:123
45. Sorensen DC (1992) SIAM J Matrix Anal Appl 13:357
46. Zhao Y, Du M-H, Kim Y-H, Zhang SB (2004) Phys Rev Lett 93:015502
47. Kresse G, Hafner J (1994) J Phys Condens Matter 6:8245
48. Saad Y, Chelikowsky JR, Shontz S (2010) SIAM Rev 52:3

Chapter 7
Electronic Structure of Solids and Surfaces with WIEN2k

Karlheinz Schwarz and Peter Blaha

Abstract Density functional theory (DFT) in various modifications provides the basis for studying the electronic structure of solids and surfaces by means of our WIEN2k code, which is based on the augmented plane wave (APW) method. Several properties, which can be obtained with this code, are summarized and the application of the code is illustrated with four selected examples focusing on very different aspects from electron-structure relations, complex surfaces or disordered layer compounds to the dependence of the equilibrium lattice constants on the DFT functionals.

Keywords Quantum mechanics • Density functional theory • Augmented plane wave method • WIEN2k • Solids • Surfaces

7.1 Introduction

In many cases an understanding of materials on the atomic scale becomes an essential requirement. This is true for modern devices in the electronic industry or magnetic recording as well as for surface science and catalysis. When one comes to atomic dimensions measured in Å, all properties are determined (or critically influenced) by the electronic structure governed by quantum mechanics. This holds for solids, surfaces or molecules. One needs to consider a sequence of topics from chemical composition (including defects or vacancies), atomic structure (with the position of all atoms), the electronic structure (based on quantum mechanics) analyzed in terms of convergence and parameters all the way to properties which can

K. Schwarz (✉) • P. Blaha
Institute of Materials Chemistry, Vienna University of Technology, Getreidemarkt 9/165-TC, A-1060 Vienna, Austria
e-mail: kschwarz@theochem.tuwien.ac.at; pblaha@theochem.tuwien.ac.at

J. Leszczynski and M.K. Shukla (eds.), *Practical Aspects of Computational Chemistry I: An Overview of the Last Two Decades and Current Trends*, DOI 10.1007/978-94-007-0919-5_7, © Springer Science+Business Media B.V. 2012

be directly compared with experimental data (e.g. spectra). During the last decades a large variety of theoretical methods have been developed, which all have their advantages and disadvantages depending on the system in question.

We focus on the atomic scale, where one often starts with an ideal crystal that is studied at zero temperature. The unit cell contains several atoms (with their nuclei at specified positions) and is repeated with periodic boundary conditions. Quantum mechanics governs the electronic structure that is responsible for properties such as relative stability, chemical bonding, relaxation of the atoms, phase transitions, electrical, mechanical, optical or magnetic behavior, etc. Corresponding first principles calculations are mainly done within Density Functional Theory (DFT), according to which the many-body problem of interacting electrons and nuclei is mapped to a series of one-electron equations, the so-called Kohn-Sham (KS) equations. For the solution of the KS equations several methods have been developed, with the Linearized-Augmented-Plane-Wave (LAPW) method being among the most accurate. During the last 30 years we have developed a computer code – WIEN2k – that is now used worldwide to solve crystal properties on the atomic scale (see www.wien2k.at). The major steps in the development during the last four decades were described in detail in a recent review article [1].

Our presentation is oriented around that code. The paper is organized as follows: Sect. 7.2 describes the quantum mechanical aspect, Sect. 7.3 summarizes the major steps in the development of the augmented plane wave (APW) method and its implementation in WIEN2k, Sect. 7.4 discusses various properties that are derived from the electronic structure of a condensed matter system with illustrations using selected examples of published research; Sect. 7.5 summarizes the role of theory and gives a short conclusion.

7.2 Quantum Mechanics

The quantum mechanical treatment of systems on the atomic scale has been discussed in many papers and thus can be omitted here. However, a few general remarks are appropriate following [1]. Because electrons are indistinguishable Fermions, their wave functions must be antisymmetric when two electrons are interchanged leading to the phenomenon of exchange. In a variational wave-function description (with one Slater determinant) this can be treated exactly with the Hartree Fock (HF) approximation. The HF equations have the computational disadvantage that each electron moves in a different potential. Exchange is treated exactly but correlation effects, which occur because of the Coulomb interaction, are omitted by definition. The latter can be included by more sophisticated approaches such as configuration interaction (CI) or coupled cluster (CC) schemes [2] but such refinements progressively require more computer time with a scaling as bad as N^7, where the system size is proportional to N, the number of electrons. Therefore such highly accurate solutions can only be obtained for relatively small systems (atoms or small molecules), which are important test cases for finding a proper quantum mechanical treatment. When the system size is significantly bigger (as often in condensed matter applications), approximations are unavoidable.

The predominant scheme for calculating the electronic properties of solids (and often of large molecules) is based on density functional theory (DFT), a universal approach to the quantum mechanical many-body problem. It was shown by Hohenberg and Kohn [3], and Kohn and Sham [4] that the key quantity is the electron density ρ, which uniquely parameterizes the variational principle for the total energy E of the system. In DFT the system of interacting electrons is mapped uniquely onto an effective non-interacting system with the same total density. In practical DFT calculations exchange and correlation effects are included, but both approximately. Due to a compensation of errors DFT is better than HF (due to the inclusion of correlation) but worse since the exchange is only treated approximately (leading to the self interaction error). From a numerical point of view an important idea of Kohn-Sham [4] was to calculate the kinetic energy (a large quantity) of non-interacting electrons (quasi particles) by introducing orbitals, which allows computing this large number very accurately. The quantum mechanics is contained in the exchange-correlation energy E_{xc} and the corresponding potential V_{xc} that is defined as the functional derivative with respect to the density. The exact functional form of the exchange-correlation energy, and hence the potential V_{xc}, is not known, and thus one needs to make approximations. The results from quantum Monte Carlo calculations for the homogeneous electron gas, for which the problem of exchange and correlation can be solved exactly, led to the modern version of the local density approximation (LDA) [5]. LDA works reasonably well but has some shortcomings mostly due to its tendency to overbind, which often causes shortened lattice constants relative to experiment. Modern XC approximations, especially those using the generalized gradient approximation (GGA), often improve upon LDA by introducing an extra term that depends on the gradient of the electron density. For long time the Perdew-Burke-Ernzerhof (PBE) [6] version was believed to be the "best" GGA, but now new types of GGAs have been developed which perform better, at least for certain properties (as will be discussed in Sect. 7.4.4). There is an extensive literature about DFT, which we do not attempt to cover here. After LDA and GGA, meta-GGA functionals were proposed (for example in [7]), which depend not only on the density and its gradient, but also on the kinetic energy density τ. The main advantage of all these DFT schemes lies in the fact that they allow calculating the electronic structure of complex systems containing many atoms such as very large molecules or solids.

Besides wave-function based methods (HF, CI, CC) or DFT there is a third category that became important recently, namely many-body physics, which can handle correlation effects on a different level. Traditionally such schemes were often based on parameters but now they can be combined with DFT results. For example one can start with an LDA calculation and transform the basis set from a Bloch-picture to a Wannier description (see for example [8]). In the latter the correlated electrons can be described by the dynamical mean field theory (DMFT) which can account for the local correlation effects using a Hubbard U and hopping parameters that were extracted from LDA results. Such combinations are called LDA + DMFT as described in a recent review [9].

In general it can be said that theory has gained a lot by combining the expertise from the three (previously separated) fields, namely wave-function based methods, DFT, and many-body theory. All three have their strength and weaknesses but in a combined effort one can gain new insight.

7.3   The Augmented Plane Wave Based Method and WIEN2k

In the present paper we focus on crystals and surfaces. We choose DFT as the quantum mechanical treatment of exchange and correlation. This means that we must solve the Kohn-Sham (KS) equations by means of a proper basis set. For this purpose we use the augmented plane wave (APW) scheme, which originally was proposed by Slater [10]. The development of APW and its linearized version, which led to the WIEN code [11] and its present version WIEN2k [12], was described in detail in a recent review [1] and previous articles [13–15]. The main concepts are summarized below:

The unit cell is partitioned into (non-overlapping) atomic spheres that are centered at the atomic sites (region I) and an interstitial region (II), for which different basis functions are used. For the construction of these functions the *muffin tin* approximation (MTA) is used, i.e. the potential is assumed to be spherically symmetric within each atomic sphere but constant outside. Plane waves are used in region II. Each plane wave is augmented by corresponding atomic partial waves, i.e. atomic-like solutions inside each atomic sphere (region I) consisting of a radial function u_ℓ times spherical harmonics.

The energy dependence of the atomic-like radial functions can be treated in different ways. In the original APW this was done by choosing a fixed energy for each radial function, which led to a non-linear eigenvalue problem. In LAPW this energy dependence of each radial basis function $u_\ell(r,E)$ is linearized (that is, treated in linear order) according to Andersen's prescription [16] by taking a linear combination of a solution $u_\ell(r,E_\ell)$ at a fixed linearization energy E_ℓ (chosen at the center of the corresponding band) and its energy derivative $\dot{u}_\ell = \partial u_\ell / \partial \varepsilon$ computed at the same energy. Each plane wave is joined continuously (in value and slope) to the one-center solutions inside the atomic sphere, thereby defining the relative weights of the u_ℓ and $\dot{u}_\ell$ contributions. This LAPW basis set allows finding all needed eigenvalues with a single diagonalization, in contrast to APW. The more strict constraint (matching in value and slope) had the disadvantage that more PWs were needed to reach convergence.

The LAPW method made it computationally attractive to go beyond the MTA. It was important to treat the crystal potential (and charge density) without any shape approximation as pioneered by the Freeman group [17]. The potential and charge density are expanded inside each atomic sphere into a radial part times lattice harmonics (a symmetry-adapted linear combination of spherical harmonics) and as a Fourier series in the interstitial region. This scheme is termed a full-potential calculation.

WIEN2k is an all-electron scheme. Core states are low in energy and the corresponding KS orbitals (or densities) are (practically speaking) completely confined within the atomic spheres and can be obtained using the spherical part of the potential (but using a thawed core instead of a frozen core approximation). Valence states are high in energy with delocalized orbitals which are responsible for chemical bonding and form energy bands. However, between the core and valence states for some atoms there might be so called semi-core states, which reside mostly inside the spheres but have a "core-leakage" of a few per cent. For them Singh [18] proposed adding local orbitals (LO) to the LAPW basis set in order to accurately treat states with different principal quantum numbers (e.g. 3p and 4p states) while retaining orthogonality. For further details see review [1]. The concept of LOs fostered another idea, namely the APW plus local orbitals (APW + lo) method [19]. These local orbitals (lo) are denoted with lower case to distinguish them from the semi-core LOs just discussed. In APW+lo, one goes back to the APW basis but with the crucial difference that the radial wave functions are expanded at fixed energies. This new scheme is significantly faster (up to an order of magnitude) while keeping the convenience of LAPW [20]. The details of the three types of schemes (APW, LAPW, APW+lo) were described in [1, 15]. A combination of the latter two schemes provides the basis for the WIEN2k program [12].

In systems with heavier elements, relativistic effects must be included. In the medium range of atomic numbers (up to about 54) the so called scalar relativistic scheme is often used [21]. It describes the main contraction or expansion of various orbitals (due to the Darwin s-shift or the mass-velocity term), but omits spin-orbit interaction. The latter becomes important for the heavy elements or when orbital magnetism plays a significant role. In the present version of WIEN2k the core states always are treated fully relativistically by numerically solving the radial Dirac equation. For all other states, the scalar relativistic approximation is used by default, but spin-orbit interaction (computed in a second-variational treatment [22]) can be included if needed [23].

The computational aspects like parallelization (k-points or MPI), algorithms, accuracy and efficiency were discussed in the review [1]. WIEN2k can treat all atoms in the periodic table. The high accuracy of WIEN2k comes from a balanced mixed basis set of plane waves and atomic functions, whose radial functions are recalculated numerically in the new potential. This allows them (in each iteration) to expand or contract according to the potential and ionicity (charge state). The main control of basis size convergence is done via a single parameter, RK_{max}, the product of the smallest sphere radius R times the largest plane wave vector K_{max}. Therefore the convergence is easy to test. Integration in reciprocal space requires a proper k-point mesh in the irreducible Brillouin zone (BZ) which needs to be checked for convergence.

7.4 Properties and Applications

When the electronic structure of a condensed matter system is calculated with
WIEN2k several topics need to be considered. A short summary is given below but
the reader is referred to the review [1] for more references and details (especially
Sect. 6):

1. In a system with translational symmetry (a perfect infinite crystal) one makes
   use of periodic boundary conditions and thus can expand the wave functions in a
   plane wave basis set. The concept of a unit cell is appropriate for (nearly) perfect
   single crystals, but a real crystal has surfaces and may have imperfections such
   as impurities or vacancies. Such effects can approximately be treated with slabs
   or supercells.
2. KS eigenvalues with respect to the reciprocal k-vector can be represented as
   band structure. The corresponding wave functions contain the information how
   much various basis sets contribute to each state. In the APW framework this
   can be done by using the partial charges $q_{t\ell m}$, which define the fraction of the
   corresponding total charge density (normalized to unity in the unit cell) that
   resides in the atomic sphere t and comes from the orbital characterized by the
   quantum numbers ℓm. The fraction from the interstitial regions is contained in
   q_{out}. These numbers help to interpret each state in terms of chemical bonding.
   From all energy eigenvalues in the Brillouin zone the density of states (DOS)
   can be calculated and again decomposed into partial DOS.
3. The key quantity in DFT is the electron density. It contains the essential ingredient for understanding chemical bonding. By computing difference electron
   densities (with respect to superposed atomic densities) the bonding features
   become more apparent. Another possibility is to use the topological analysis by
   Bader [24] for example to define atomic charges within atomic basins, a relevant
   quantity for charge transfer estimates.
4. The electric field gradient (EFG) is a ground state property that is sensitive to
   the asymmetry of the charge distribution around a given nucleus. By measuring
   the nuclear quadrupole interaction (e.g. by NMR) the EFG can be determined
   experimentally. This local probe is often essential for distinguishing between
   different atomic arrangements.
5. The total energy of the system is a crucial quantity for any given atomic
   configuration. Often this can be a rather big number which nowadays can be
   calculated with high precision. Total energy differences, for example, tell which
   structure is more stable. The derivative with respect to nuclear coordinates yield
   the forces acting on an atom. These forces are needed to optimize the atomic
   positions towards an equilibrium geometry, which corresponds to a minimal total
   energy and vanishing forces. In addition they can be used to calculate phonons.
6. If a system is magnetic, the calculation must be carried out in a spin-polarized
   fashion. Often a collinear arrangement of the magnetic moments like in an
   (anti-) ferromagnet is assumed but there is the possibility to study non-collinear
   arrangements.

7. In connection with spectroscopy various properties can be calculated like x-ray emission or absorption spectra (XES, XAS), optical spectra or photoelectron spectra (UPS). The data for hyperfine interaction can be obtained from WIEN2k too.

Since the WIEN2k program package is used worldwide by more than 1,850 groups many papers have appeared that make use of this program. Here we illustrate from our own research how results obtained with WIEN2k can help to solve interesting problems in material sciences.

7.4.1   Verwey Transition in YBaFe$_2$O$_5$

A perovskite ABO$_3$ (like SrTiO$_3$) contains as the main building block the B atom that is octahedrally coordinated by oxygen. An oxygen-deficient double perovskite, however, has B with a pyramidal coordination, in which the sixth oxygen is missing. One member of this group is YBaFe$_2$O$_5$ whose crystal structure is well established [25]. It is particularly interesting, because it shows a temperature induced phase transition at about 309 K. At low temperatures it forms a charge ordered (CO) state (with Fe^{2+} and Fe^{3+} at the two crystallographic inequivalent sites) but above the transition temperature a valence-mixed (VM) state (sometimes called mixed valence) appears in which Fe has the formal oxidation state of Fe$^{2.5+}$ [26]. Such a change is called Verwey transition [27] as has originally been suggested for magnetite Fe$_3$O$_4$, a system that is still often discussed. In YBaFe$_2$O$_5$ the structure changes from a strongly distorted orthorhombic to a nearly tetragonal symmetry. With this structural change both the magnetic and conducting behaviour change significantly. In the CO phase (with space group Pmma) the Fe^{2+} and Fe^{3+} form chains along the a direction and have an antiferromagnetic (AFM) arrangement (Fig. 7.1 top).

A standard GGA calculation would lead to a metallic behavior and magnetic moments that are much smaller than the experimental values. Therefore one must go beyond GGA and include the local correlation effects (that are important for Fe oxides) by means of a Hubbard U. Although this introduces a parameter that is not strictly given on a first principles basis, a GGA+U calculation (with an effective U of around 7 eV) gives a proper description of the system as discussed in detail in [26]. A structural optimization of the atomic coordinates leads to different bond length around Fe^{2+} and Fe^{3+} (Fig. 7.1 bottom) in the CO phase, whereas they are similar in the VM phase. In addition a gap opens up making it a semiconductor and the magnetic moments obtained with GGA+U are in agreement with experiment. The AIM charges (according to Bader's atoms in molecules [24]) are +1.84 for Fe^{3+} and +1.36 for Fe^{2+} in the CO phase but +1.52 in the VM phase. The origin for this clear difference can be traced down to an orbital ordering which shows up in the partial DOS associated with various Fe-d-orbitals. The hyperfine fields and the electric field gradients are consistent with experimental data, provided a proper U

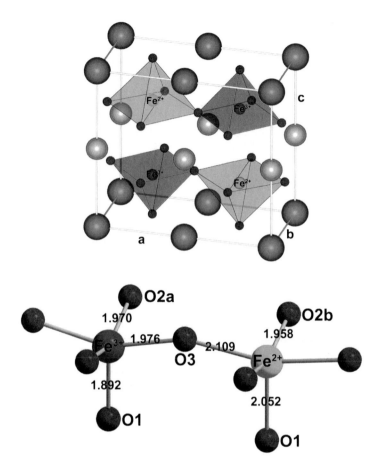

Fig. 7.1 The charged-ordered (CO) phase of the oxygen-deficient double perovskite YBaFe$_2$O$_5$: (*top*) orthorhombic unit cell with a chain of alternating Fe^{2+} and Fe^{3+} ions along the a direction; (*bottom*) the local coordination of the two iron sites giving the nearest neighbor distances as optimized by a GGA+U calculation

is used. In summary one can say that in the CO phase the Fe^{2+} is in a d^6 high-spin configuration in which a single spin-down electron of d-xz symmetry is occupied which triggers a cooperative Jahn-Teller distortion. The apparent strong electron-lattice coupling cause in the VM phase (with its similar bond lengths) that the Fe d-z^2 spin-down orbital (at the top of the valence band) become partly occupied.

Chemical bonding changes the total electron density only by small amounts and thus the difference between the final SCF density and the superposition of free atomic densities (the start of an SCF cycle) shows the main reorganization due to bonding. In such a difference electron density (see Figs. 8 and 9 of ref. [26]) the two phases, CO and VM, clearly differentiate between Fe^{2+} and Fe^{3+} in the former but not in the latter case.

One numerical detail shall be mentioned, namely the magneto-crystalline anisotropy, which is this difference in total energy when the magnetic moments point in different crystallographic directions [100], [010] or [001], a question for which spin-orbit interaction is essential. The lowest total energy is $-115{,}578.24065$ Ry, when the moments point in the y direction, while the other directions are about 0.4 mRy higher in energy; thus the difference is in the tenth decimal. This illustrates which numerical precision is needed for such a quantity. For more details see the original paper [26].

7.4.2   Nanomesh with h-BN on a Rh(111) Surface

When borazin is thermally decomposed on a Rh(111) surface a self-assembling structure is formed, which Corso et al. called a nanomesh [28]. It consists of a hexagonal boron nitride (h-BN) that binds to a Rh(111) surface. Originally these authors described the structure with a double layer of BN where the top layer has holes. However, DFT calculations [29] proposed a different atomic structure of this surface, consisting of a single but highly corrugated layer of h-BN. There is a lattice mismatch between h-BN and Rh(111) of about 8%, with the result that 13×13 unit cells of h-BN match 12×12 unit cells of the underlying Rh(111) with a periodicity of about 3.2 nm. Such a system is a real challenge for theory, since it is metallic and already a crude model of the surface contains many atoms. In order to simulate this complex structure, a slab was constructed containing three layers of Rh (corresponding to the three layers A, B, C of an fcc structure) and h-BN layers on both sides (top and bottom) of the metal layers. This is done for computational reasons to keep inversion symmetry, which makes the matrices real instead of complex. This supercell contains 1,108 atoms (and around 25.000 electrons), which makes the calculation rather demanding but feasible nowadays (see Fig. 4 in [1]).

The structure optimization started with a flat h-BN layer but allowed the atoms to relax, which led to a significant surface corrugation (Fig. 7.2). Due to the lattice mismatch and the relative rigidity of h-BN there are regions with different bonding situations. The preferred orientation for boron is a hollow site above three Rh atoms, whereas nitrogen likes to be on top of a Rh atom (Fig. 7.3). This situation is almost satisfied in the so called "low" region (Fig. 7.2 bluish region) where h-BN binds strongly to the metals leading to short distances to the Rh sublattice. Otherwise the more repulsive interaction between N and the surface cannot compensate the weaker B attraction and thus the h-BN is further away from the surface leading to the "high" region (Fig. 7.2 yellow region). An analysis has shown that the B atoms have predominantly attractive forces towards Rh (with bonding orbitals) whereas for N the repulsive forces dominate (due to partial occupation of antibonding orbitals). The lattice mismatch causes locally different lateral orientations of h-BN with respect to the Rh-subsurface. The combination of these two scenarios (with favorable and unfavorable bonding) caused a corrugation that can be seen

Fig. 7.2 The corrugation of the hexagonal boron nitride layer in the h-BN/Rh(111) nanomesh showing a 2 × 2 supercell. The "low" region (close to Rh) appears in *blue* but the "high" region (further away from the Rh sublayer) are in *yellow*. The B atoms (visible in the *front*) are shown with *small* but the N with *large spheres* (see also Figs. 4 and 5 in [1])

Fig. 7.3 The local atomic arrangement of h-BN on Rh(111) in the "low" region, where N (in *red*) is about on top of Rh (*grey*) and B (*blue*) is in the hollow position optimally binding to the three Rh underneath. Further details can be found in [29]

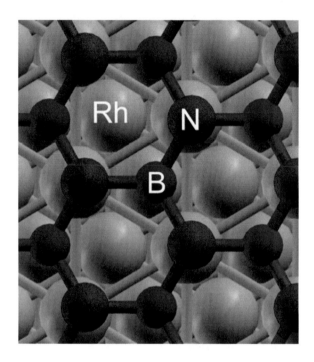

in experiment, e.g. by scanning tunneling microscopy (STM). Once the structure is unraveled, other experimental data can be explored, for example X-ray absorption spectroscopy, or N-1s core level shifts as discussed in [29, 30]. Some additional discussions can be found in [1].

7.4.3 The Misfit Layer Compounds

Hexagonal transition metal dichalcogenites such as TaS_2 are layered compounds, which easily can be intercalated for example with Li ions. When they are intercalated with a pseudocubic double layer such as PbS, they belong to the so called misfit layer compounds [31]. Because of the different crystal symmetries of the two subsystems, the lattice constants can match only in one direction forming a periodic structure (e.g. in the b direction). Perpendicular to this commensurate direction there is an incommensurate lattice mismatch due to the ratio between the two lattice constants, in our case between TaS_2 and PbS. This ratio can be approximated with 4/7 leading to periodic boundary conditions. This means that 4 lattice constants of PbS match to a good approximation 7 of TaS_2 corresponding to the misfit layer compound $(PbS)_{1.14}TaS_2$ as discussed in our paper and references therein [32]. These materials have interesting properties and are rather stable. The main question that remained open was to explain what causes the stability. In this context previous experimental studies proposed two possible binding mechanisms, namely non-stoichiometry or metal cross substitution that should be responsible for the stability.

This is an ideal starting point for theory, since one can try both schemes and find out which is more likely to be correct or consistent with experimental details. The idealized system (forced to be commensurate) contains 74 atoms per unit cell and consists of alternating perfect layers of TaS_2 and double layers of PbS. First DFT test calculations showed that the binding energy between the ideal layers of TaS_2 and the PbS-double layer is nearly zero. Therefore one of the proposed mechanisms may provide a clue for the stabilization and thus they need to be explored. Now disorder comes into play, either in form of defects or when one substitutes Pb into TaS_2 or Ta into PbS. With our WIEN2k we must enforce periodic boundary conditions and thus artificially introduce some order. In order to be more realistic and avoid artifacts (like rows of impurity atoms) an even larger supercell with 296 atoms was used in some of the calculations. In such large supercells one cannot explore all possible configurations of disorder. However, one can follow certain strategies. For example if one puts 2 Ta atoms replacing Pb in PbS, then they can be close together (clustering) or far apart (avoiding each other), or in the same PbS layer or in adjacent layers. For each of the explored configurations one optimizes the atomic positions (till the forces acting on all atoms vanish) and determines the corresponding total energy. Different configurations can be compared by their total energy, leading to an insight how the impurities prefer to distribute. For example, it turned out that it is energetically more favorable to have the Ta impurities in the same PbS layer than putting them in the adjacent layer. The computer experiments that were carried out made use of the rules that have been learned from previous results. Configurations which are likely to be unfavorable need not to be studied. This strategy reduces the effort and makes the investigation of such a complicated system (with so many atomic configurations) feasible. A representative configuration is shown in Fig. 7.4, in which one sees (upon substitution) relatively little changes in the TaS_2 layer but significant relaxations in the substituted PbS layer. This difference is caused by the size change (small Ta vs. larger Pb) and the structural details between the TaS_2 layer, in which only a small breathing of sulfur atoms around the large Pb is possible. However, the small Ta atom can move a lot towards the sulfur leading to large distortions of the pseudo-NaCl planes of PbS.

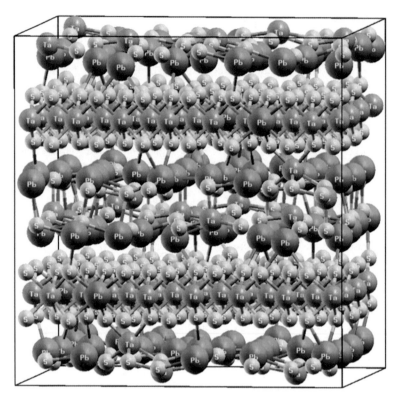

Fig. 7.4 The large supercell representing the $(PbS)_{1.14}TaS_2$ misfit layer compound containing 296 atoms. In the TaS_2 layer 4 Ta atoms are substituted by Pb and in the PbS double layers 12 Ta atoms substitute the corresponding Pb atoms leading to a strong relaxation. For further details see [32]

From the calculated total energies one can conclude [32] that the metal cross substitution alone cannot stabilize this compound whereas the nonstoichiometric model works, when Ta substitutes Pb mainly in one of the PbS double layers. This case shows large lattice relaxations (Fig. 7.4) which stabilize the misfit layer compounds but are accompanied by charge transfer effects. In addition the insulating behavior of PbS is lost in the stoichiometric compound, whereas Ta doping leads to a charge transfer that brings the PbS layer closer to be an insulator. For further details see [32]. A Ta impurity concentration of about $x = 0.13 - 0.19$ is energetically most favorable which is in excellent agreement with experimental findings. In this case DFT calculations could find an explanation for the stability and confirm one of the proposed mechanisms.

7.4.4 Performance of Various GGA Functionals

In recent years several attempts were made to improve the performance of various generalized gradient approximations. In this context one usually investigates small systems which can be well characterized as ideal crystals without any structural uncertainties such as defects, impurities or non-stoichiometry. It is well known that LDA gives too small lattice constants while the standard PBE version of GGA [6] always leads to larger values than LDA but often also with respect to experiment. By choosing crystals without structural uncertainties and a computational scheme which is highly accurate (as results with WIEN2k) one can test the quality of functionals when compared with experimental data. Otherwise one would have a combination of effects from structure over DFT to convergence of basis sets. In the latter case one would not be able to come to firm conclusions.

We were involved in one of such investigations of GGA functionals and want to summarize some results (see [33] and references therein). From the newly proposed GGA functionals (in addition to the standard PBE [6]) we mention the functional by Wu and Cohen (WC) [34], AM05 by Armiento and Mattssson [35] and PBEsol [36]. In GGA the exchange correlation energy can be expressed in terms of an enhancement factor F_{xc}

$$E_{xc}^{GGA}[\rho] = \int \varepsilon_x^{LDA}(r_s(r)) F_{xc}(r_s(r), s(r)) d^3r$$

which depends on the Wigner-Seitz radius r_s, a measure of the electron density $\rho(r)$,

$$r_s = [3/(4\pi\rho)]^{1/3}$$

and the reduced gradient density s

$$s = |\nabla\rho| / [2(3\pi^2)^{1/3}\rho^{4/3}]$$

The variation of the (exchange only) enhancement factor F_{xc} with s is shown in Fig. 7.5 for one example, namely $r_s = 0$, but more cases are depicted in [33]. For the chosen functionals there is a significant difference in the enhancement factors for larger s but they are all smaller than that of PBE. Now one can analyze various systems from metals, to insulators or covalently bonded systems and explore which s-values are relevant. The detailed analysis has shown that it is not only the value of the enhancement factor but also its derivative with respect to s and r_s which determine the equilibrium lattice constant. In addition it was found that in most solids (in contrast to molecules) vales of s larger than 1.5–2.0 or values of r_s larger than 4 bohr hardly occur. Let us illustrate this situation for the covalently bonded system of diamond (see Fig. 7.6) which shows how the s value changes in the unit cell. The surprising result is found that in the region of the covalent bond (between

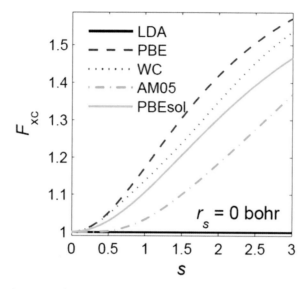

Fig. 7.5 The enhancement factors F_{xc} (for $r_s = 0$ bohr) with respect to the LDA exchange energy as a function of s, the reduced density gradient, is shown for the four functionals PBE, WC, AM05, and PBEsol. Additional plots for other r_s values are given in [33]

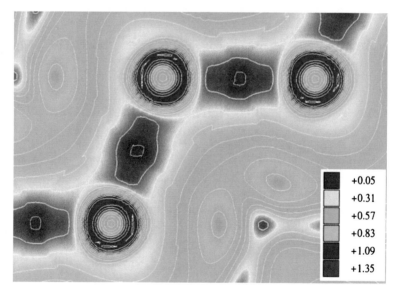

Fig. 7.6 A two-dimensional plot of the reduced density gradient s is shown for *diamond* in the (110) plane. The *color coding* indicating the value of s is specified in the insert

the carbon atoms) s is small and thus all enhancement factors remain small so that we are almost left with LDA. Large values of s appear in the range separating core from valence states around all carbon atoms, but also in the interstitial region. These are the important regions, in which the enhancement factors differ between the functionals and thus these – rather than the bonding regions – are responsible for the equilibrium lattice constant. A functional with a large enhancement factor in the important region usually leads to larger lattice constants.

7.5 Summary and Conclusion

In this paper we discussed the status of quantum mechanical calculations focusing on solids and surfaces. In the quantum mechanics section DFT was presented with respect to the alternative approaches such as wave function based methods or many-body physics. For the solution of the DFT Kohn Sham equations we use an adapted augmented plane wave method implemented in our WIEN2k code, which can be shortly summarized as a full-potential, all electron and relativistic code that is one of the most accurate for solids and is used worldwide by more than 1,850 groups in academia and industry.

In the spirit of Coulson ("Give me insight not numbers") many properties and detailed analyses are needed to solve complex material problems. For that purpose WIEN2k provides many tools to compute a large variety of properties and results, from energy bands, DOS, electron and spin densities, magnetic moments, optical spectra, total energies, forces, EFG, hyperfine interactions, spin-orbit coupling etc. Often several of these results are needed to explain the open questions.

In the four examples presented here different aspects were highlighted. In the first case (Verwey transition) the unit cell is small and well defined but the electron-lattice coupling (with a cooperative Jahn-Teller distortion) requires a high level description (GGA + U) to get the physics right. In the nanomesh example the size of the system (more than 1,100 atoms per unit cell) of a metallic system is a challenge. For the misfit layer compounds we demonstrate another area, namely performing computer experiments on disordered systems, which lead to strategies that allow us to focus on the likely structures that may be present in the real system. With such a scheme it became possible to describe the system and explain the stability between the layers of the misfit compound. In the last example we explore different GGA functionals in order to find out which system dependent parameters are essential for obtaining lattice constants in good agreement with experiment. In the latter case we restrict our efforts to very ideal crystals with small unit cells. The chosen examples are just a small selection to illustrate what can be obtained with the WIEN2k code. The used basis set (containing atomic orbitals) allows interpretations in chemical terms, which can be an important advantage over pseudo-potentials methods. The high accuracy is achieved partly by the numerical basis (for the radial wave functions) but also by the fact that the convergence (with the number of plane waves) can be controlled with one parameter (RK_{max}).

Therefore DFT calculations with approximate functionals can provide extremely useful information concerning the electronic structure of ordered crystal structures and surfaces irrespective whether they can be prepared or not. Nowadays relatively large systems can be simulated due to the increased computer power combined with improved algorithms and efficient parallelization.

Acknowledgements The authors express their thanks to the many people who have contributed to the development of the WIEN2k code. Special thanks go to Robert Laskowski for providing us Fig. 7.2 and Evgeniya Kabliman for Fig. 7.4.

References

1. Schwarz K, Blaha P, Trickey SB (2010) Mol Phys 108:3147
2. Bartlett RJ, Musial M (2007) Rev Mod Phys 79:291; Sode O, Keçeli M, Hirata S, Yagi K (2009) Int J Quantum Chem 109:1928
3. Hohenberg P, Kohn W (1964) Phys Rev 136B:864
4. Kohn W, Sham LS (1965) Phys Rev 140:A1133
5. Ceperley CM, Alder DJ (1980) Phys Rev Lett 45:566
6. Perdew JP, Burke K, Ernzerhof M (1996) Phys Rev Lett 77:3865
7. Perdew JP, Kurth S, Zupan A, Blaha P (1999) Phys Rev Lett 82:2544
8. Kunes J, Arita R, Wissgott P, Toschi A, Ikeda H, Held K (2010) Comput Phys Commun 181:1888
9. Held K (2007) Adv Phys 56:829
10. Slater JC (1937) Phys Rev 51:846
11. Blaha P, Schwarz K, Sorantin P, Trickey SB (1990) Comput Phys Commun 59:399
12. Blaha P, Schwarz K, Madsen GKH, Kvasnicka D, Luitz J (2001) An augmented plane wave plus local orbitals program for calculating crystal properties. Vienna University of Technology, Austria. ISBN 3-9501031-1-2
13. Singh DJ, Nordström L (2006) Plane waves, pseudopotentials and the LAPW method, 2nd edn. Springer, New York. ISBN 10:0-387-28780-9
14. Schwarz K, Blaha P (2003) Comput Mater Sci 28:259
15. Schwarz K, Blaha P, Madsen GKH (2002) Comput Phys Commun 147:71
16. Andersen OK (1975) Phys Rev B 12:3060
17. Wimmer E, Krakauer H, Weinert M, Freeman AJ (1982) Phys Rev B 24:4571
18. Singh DJ (1991) Phys Rev B 43:6388
19. Sjöstedt E, Nordström L, Singh DJ (2000) Solid State Commun 114:15
20. Madsen GHK, Blaha P, Schwarz K, Sjöstedt E, Nordström L (2001) Phys Rev B 64:195134
21. Koelling DD, Harmon BN (1977) Solid State Phys 10:3107
22. MacDonnald AH, Pickett WE, Koelling DD (1980) J Phys C Solid State Phys 13:2675
23. Laskowski R, Madsen GKH, Blaha P, Schwarz K (2004) Phys Rev B 69:140408
24. Bader RWF (1994) Atoms in molecules: a quantum theory. Oxford University Press, New York
25. Woodward P, Waren P (2003) Inorg Chem 42:1121
26. Spiel C, Blaha P, Schwarz K (2009) Phys Rev B 79:115123
27. Verwey E (1939) Nature 144:327
28. Corso M, Auwärter W, Muntwiler M, Tamai A, Greber T, Osterwalder J (2006) Science 303:217
29. Laskowski R, Blaha P, Gallauner Th, Schwarz K (2007) Phys Rev Lett 98:106802
30. Laskowski R, Blaha P (2010) Phys Rev B 81:075418
31. Wiegers GA (1996) Prog Solid State Chem 24:1

32. Kabliman E, Blaha P, Schwarz K (2010) Phys Rev B 82:125308
33. Haas P, Tran F, Blaha P, Schwarz K, Laskowski R (2009) Phys Rev B 80:195109
34. Wu Z, Cohen RE (2006) Phys Rev B 73:236116
35. Armiento R, Mattsson AE (2005) Phys Rev B 72:085108
36. Perdew JP et al (2008) Phys Rev Lett 100:136406

Chapter 8
Model Core Potentials in the First Decade of the XXI Century

Tao Zeng and Mariusz Klobukowski

Abstract During the past decade the method of model core potential has undergone a period of dynamic development and applications, which ranged from atomic to protein-scale studies. Incorporation of the relativistic effects became the centre of the model core potential development and the accuracy and applicability of this method were greatly increased. A breakthrough on this front of research was the development of the model core potential that can account for the spin-orbit coupling effect. In the present chapter we review the theoretical foundations of the pseudopotential approach to the molecular electronic structure. We then provide an overview of the model core potential method as well as its development and applications in the first decade of this century. A perspective on the future of this method is also given.

Keywords Model core potential • Pseudopotential • Scalar-relativistic effects • ECP spin-orbit calculation • Fragment molecular orbital

8.1 Introduction

The present section offers a brief introduction to the pseudopotential approach in general, followed by a review of the fundamentals of the effective core potentials and model core potentials, and starts from where the most recent review ended [1].

8.1.1 Separability of the Valence and Core Spaces

Mendeleev's Periodic Table of the chemical elements is one of the most important milestones in the history of chemistry. The fact that elements can be classified into

T. Zeng · M. Klobukowski (✉)
Department of Chemistry, University of Alberta, Edmonton, AL, Canada, T6G 2G2
e-mail: mariusz.klobukowski@ualberta.ca

J. Leszczynski and M.K. Shukla (eds.), *Practical Aspects of Computational Chemistry I:*
An Overview of the Last Two Decades and Current Trends,
DOI 10.1007/978-94-007-0919-5_8, © Springer Science+Business Media B.V. 2012

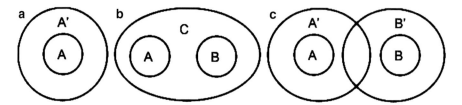

Fig. 8.1 Examples of valence and core separability adapted from Ref. [3]. In each case, the smaller regions are the "core" regions and the larger are the "valence" regions. (**a**) atoms; (**b**) covalent bondings; (**c**) weak interactions

different groups reveals that the chemical properties of an element are determined by its valence electronic structure. A straightforward application of this idea in chemistry results in the concept of valence electrons [2]. An imperative extension of this concept to computational chemistry is to develop algorithms with only few valence electrons. This idea is very tempting since, by doing so, all the computational resources consumed by considering the core electrons are saved and, for example, a calculation of a cesium cluster with one thousand atoms would ideally be equal to the calculation of one thousand hydrogens. Such pseudopotential methods are even more desired today as the calculations of DNA, hydrocarbons, proteins, and other biomolecules have become the main stream of the applied computational chemistry. Despite its simplicity, the concept of the valence electrons is intrinsically wrong in a quantum physicist's eyes. This is because electrons are indistinguishable fermions and it is illegal to tag some electrons as the valence and some as the core electrons.

Despite the indistinguishability of electrons, since the orbitals are the building blocks for the electronic wave function, one may switch to the separability of the core and valence parts of a wave function. The Auger electron spectroscopy indicates that probing an electron in the core region of an atom needs energy at the order of keV, which is by far greater than the order of chemical energetics (within 10 eV). Therefore, the energetic change caused by a chemical interaction is not strong enough to perturb the wave function close to the nuclei and one may assume that the core part of the wave function of an atom is frozen in all chemical processes. The separability of the valence and core wave functions is the theoretical foundation of the pseudopotential method and its three typical cases are schematically drawn in Fig. 8.1. For more details of the theoretical foundations for the separability of the valence and core spaces, one should refer to the comprehensive reviews written by Huzinaga and co-workers [3, 4].

In the pseudopotential method, the two-electron integrals between the core and valence orbitals are replaced by the one-electron integrals of the potential and this leads to a substantial saving of computational resources. The early history of using the pseudopotential method dates back to the pioneering works of Hellmann [5, 6] and Gombás [7] in mid-1930s. The strong orthogonal constraint between all the core and valence orbitals, i.e.

$$\langle \phi_i \mid \phi_j \rangle = 0 \qquad (8.1)$$

where ϕ_i is a core orbital and ϕ_j is a valence one, is the cornerstone for the valence-core separability. Two different ways to enforce the strong orthogonality constraint led to two families of the pseudopotential methods: effective core potential (ECP) and model core potential (MCP).

8.1.2 Effective Core Potential Method

Extensive introductions to the effective core potential method may be found in Ref. [8–19]. The theoretical foundation of ECP is the so-called Phillips-Kleinman transformation proposed in 1959 [20] and later generalized by Weeks and Rice [21]. In this method, for each valence orbital ϕ_v there is a pseudo-valence orbital χ_v that contains components from the core orbitals and the strong orthogonality constraint is realized by applying the projection operator on both the valence hamiltonian and pseudo-valence wave function (pseudo-valence orbitals). In the generalized Phillips-Kleinman formalism [21], the effect of the projection operator can be absorbed in the valence Fock operator and the core-valence interaction (Coulomb and exchange) plus the effect of the projection operator forms the core potential in ECP method.

The two main parameterization schemes extensively used in the last three decades are the shape consistent and energy consistent schemes. In the shape consistent scheme, the effective potentials are parameterized to reproduce the orbital energies and the shape of the radial functions of the valence orbitals. One should notice the difference between the valence energy and the valence orbital energies. The former is an observable under the frozen core approximation whereas the latter are only mathematical results of the variational treatment of the total or valence energy. Therefore, the reproduction of the valence orbital energies does not guarantee the reproduction of the valence energy and it is for this reason the radial shape of the valence orbitals is also needed in the parameterization. The reproduction of both the valence orbital energies and the valence orbitals shapes equals the reproduction of the valence energy under the Hartree-Fock approximation. The energy-consistent scheme parameterizes the effective potentials to reproduce the properties related to the valence directly, e.g., excitation energies to the low-lying states, ionization potentials and electron affinities to create the atomic ions with low charges, and the excitation energies to the low-lying states of the ions. Since the perturbation of an atom under a chemical environment could be viewed as the perturbation from the low-lying states of the neutral atom or low-charged ions, the effective potentials produced by this scheme contain most of the chemical properties of an atom.

The scalar-relativistic effects can be easily absorbed into the effective potential by taking the all-electron (AE) calculation results of the same order of relativistic approximation as the references to parametrize the potentials. Taking the two-component (or even four-component) form of the pseudo-valence orbitals, the spin-orbit coupling effect can also be absorbed into the ECP. Because the pseudo-valence orbitals are energetically the lowest-eigenvalue eigenvectors of the Fock

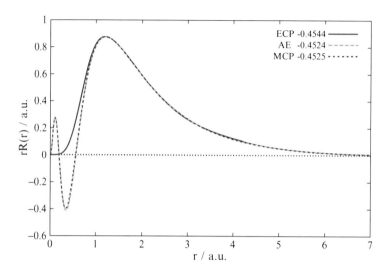

Fig. 8.2 The radial functions and orbital energies of the Au 5d orbital from the all-electron (AE), effective core potential (ECP) and model core potential (MCP) calculations. The AE calculation was at the DK3 relativistic level with uncontracted well-tempered basis functions [22]. The ECP calculation was performed using the ECP60MDF [23] potential and basis set. The MCP calculation was performed using our recently developed potentials (at the DK3 relativistic level) and basis set [24]. The orbital energies in the unit of eV are listed for comparison

operator with the effective potential, the radial nodal structure of the true valence orbitals is lost (the red curve in Fig. 8.2) and thus any properties determined by the wave function in the vicinity of the nuclei cannot be calculated by using the genuine operators and the nodeless wave function. Two such properties of chemical concern are the spin-orbit coupling and the nuclear magnetic resonance. Since the spin-orbit effect of the two-component ECP is approximated by a pseudopotential operator rather than the genuine dynamic operator, the change of this effect during a chemical process may not be well represented, i.e., it depends on how much of the atomic nature is preserved during the process and the less preserved it is, the less reliable are the results of the ECP spin-orbit calculation.

Another disadvantage rooted in the nodeless valence orbitals is the overestimation of the correlation energy. It has been reported that using the ECP pseudo-valence orbitals and pseudo-virtual orbitals overestimates the correlation energy by up to 23% [25–28]. Although this error can be reduced to 6% by using the large uncontracted basis set, the absolute error is still as large as 0.1 eV [29], larger than the chemical accuracy of 0.04 eV. (However, the differences in the correlation energies may be smaller [30].) The origin of this overestimation can be explained using a simple example of the $2s^2$ (^1S_g) ground state of the Be atom [26]. For this system, the most significant contribution to the electron correlation is the $2s^2$ to $2p^2$ excitation, i.e., the exchange type integral $\left[\phi_{2s}\phi_{2p} \mid \phi_{2s}\phi_{2p}\right]$ determines most of the magnitude of the correlation energy. (Here the chemist's notation [31] $[ij \mid kl]$

Fig. 8.3 Topology of the
2s2p overlap density of Be in
the (a) all-electron and (b)
effective core potential
calculations (Reproduced
from Ref. [26] with kind
permission of Springer)

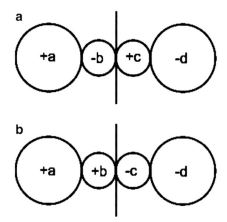

for the two-electron integral has been used.) This integral is actually the self-Coulomb interaction of the electron distribution $\phi_{2s}\phi_{2p}$. The 2s orbital should have one radial node and thus allow the distribution to have some negative contribution. However, the nodeless pseudo-2s orbital makes the distribution always positive and thus exaggerates both magnitude of the integral and the correlation energy. A schematic representation adapted from Ref. [26] of the distributions from AE and ECP calculations is given in Fig. 8.3. The reverse signs on the opposite sides of the vertical bar are the result of the p-type distribution of $\phi_{2s}\phi_{2p}$ and the different signs on the same side of the vertical bar but between panels (a) and (b) are the result of the different radial nodal structures of ϕ_{2s} and χ_{2s}. For the all-electron case, the exchange integral is

$$
\begin{aligned}
E_x^{AE} &= [a - b + c - d \mid a - b + c - d] \\
&= [a \mid a] - 2[a \mid b] + 2[a \mid c] - 2[a \mid d] + [b \mid b] \\
&\quad -2[b \mid c] + 2[b \mid d] + [c \mid c] - 2[c \mid d] + [d \mid d],
\end{aligned} \tag{8.2}
$$

and for the ECP case,

$$
\begin{aligned}
E_x^{ECP} &= [a + b - c - d \mid a + b - c - d] \\
&= [a \mid a] + 2[a \mid b] - 2[a \mid c] - 2[a \mid d] + [b \mid b] \\
&\quad -2[b \mid c] - 2[b \mid d] + [c \mid c] + 2[c \mid d] + [d \mid d].
\end{aligned} \tag{8.3}
$$

Subtracting Eq. 8.2 from Eq. 8.3 gives

$$
E_x^{ECP} - E_x^{AE} = 8[a \mid b] - 8[a \mid c], \tag{8.4}
$$

where the obvious equalities of $[a \mid b] = [c \mid d]$ and $[a \mid c] = [b \mid d]$ have been used. Because of the smaller distance between a and b than between a and c, Eq. 8.4

must be positive and ECP overestimates the exchange integral and the correlation energy. From Eqs. 8.2 to 8.4, $[a|b]$ means the integral of $\int \int a\left(\vec{r}_1\right) \frac{1}{r_{12}} b\left(\vec{r}_2\right) d\vec{r}_1 d\vec{r}_2$ and should not be confused with the chemist's notation.

Despite these disadvantages, there is one great advantage of using the nodeless ECP orbitals: the primitive functions describing the undulation of the valence orbitals in the vicinity of the nuclei are not needed and the computations are more economic. For more details of the recent developments and applications of the relativistic ECP method, one may refer to the references [14, 16, 23, 32–52].

8.1.3 Model Core Potential Method

Unlike ECP, the MCP method explicitly keeps the projection operator in the one-electron potential to maintain the strong orthogonality constraint between the core and valence wave functions. This method was proposed by Huzinaga and Cantu [3, 4, 53] in the early 1970s and the first applications for the elements lighter than and including the 3d block of the Periodic Table were reported in the mid-1970s [54–58]. In this section, the general formalism of the MCP method is briefly discussed, followed by the introduction of three versions of the method.

8.1.3.1 General Formalism of Model Core Potential Method

The derivation of the MCP formalism was detailed in the work of Höjer and Chung [59] and one of the present authors (TZ) gave a comprehensive discussion of their derivation in the second chapter of his Ph.D. thesis [60]. Interested readers should refer to those two works for a thorough exposition of this method while here, in the interest of brevity, we provide only the necessary information about MCP and focus on its level-shift operator, whose physical meaning needs more clarification. Following the same philosophy as the one behind the ECP, the MCP is naturally derived from all-electron atomic calculations at the Hartree-Fock level, and the final expression of the effective hamiltonian for the closed-shell valence electrons is

$$\hat{h}' = \hat{h} + \sum_g \left(2\hat{J}_g - \hat{K}_g\right) - \sum_g |g\rangle \, 2\epsilon_g \, \langle g| \,, \tag{8.5}$$

where g is the index for the core orbitals and ϵ_g is the eigenvalue of the atomic Fock operator. Different ways of approximating $\sum_g \left(2\hat{J}_g - \hat{K}_g\right)$ led to different versions of the potential which will be described in the next subsection. The operator $-\sum_g |g\rangle \, 2\epsilon_g \, \langle g|$ is the essence of the MCP and it maintains the nodal structure of the valence orbitals. A comparison of the Au 5d orbitals calculated from ECP, MCP, and AE calculations is shown in Fig. 8.2 and it may be seen that the overlap between the MCP and AE radial functions is nearly perfect. This feature of the MCP makes it a candidate for accurate spin-orbit coupling computation using the genuine operators, and the problem of correlation energy overestimation is also circumvented. The price of these improvements is that the undulation of the MCP valence orbitals requires more basis primitive functions than in the case of the ECPs, and one can expect smaller computer time reduction when using this method. However, calculations can be economized by contracting the primitives or fitting the inner portion of the radial functions by a small number of Gaussian functions.

The value of 2 in $-\sum_g |g\rangle 2\epsilon_g \langle g|$ is a result of maintaining the hermiticity of the Fock operator during the derivation and if we assume the core orbitals be the eigenfunctions of the Fock operator during the derivation, the value of 1 can be used [60]. Although the value of 1 is justified at the Hartree-Fock level, it will shift all the energies of the core orbitals to be 0 a.u., which is not high enough to prevent their contamination in the electron correlation treatment [61], and a larger value is needed. On the other hand, the use of $-\sum_g |g\rangle 2\epsilon_g \langle g|$ for the open-shell orbitals is only an approximation. Furthermore, a core orbital in the molecular environment may have the same or different orbital energy compared to the atomic calculation, depending on how much the core orbital is affected by the change of the chemical environment. Finally, the exact Fock equations are the foundations for the derivation to obtain Eq. 8.5. In reality, however, electron basis functions are used and the Fock equations are only approximately satisfied. Therefore, even for the closed-shell atomic valence orbitals, Eq. 8.5 is only an approximation. After all these considerations, we prefer to abandon the interpretation of $-\sum_g |g\rangle 2\epsilon_g \langle g|$ as an exact operator to maintain the valence-core orthogonality for the closed-shell system. Instead, it should be interpreted as the level shifter only, just as for the open-shell orbitals. If so, then the value B_g of $-\sum_g |g\rangle B_g \langle g|$ may be treated as a purely empirical parameter. It is obvious that the larger the magnitude of B_g, the stronger the orthogonality constraint, as demonstrated in calculations [54]. Experience accumulated in the last three decades shows that $B_g = 2\epsilon_g$ is a good choice. Sometimes, some "core" orbitals play a non-negligible role in a chemical process and less orthogonality is required to allow their participation in bonding or polarization. Under such circumstances, one may relax the requirement that B_g be $f\epsilon_g$ with $1 \leq f \leq 2$. Examples are the improved MCPs developed by Lovallo and Klobukowski [62]. Indeed, if f for a core orbital has too small a value, then this "core" orbital is energetically not deep enough to be frozen and should be included in the valence space.

The discussions of the ECP and MCP formalisms presented above are based on the wave function theory, as the approximations are made for the Fock operator. However, similar approximations can be directly generalized to the density functional theory (DFT) [63]. In the preparation of DFT pseudopotentials, one simply needs to replace the Fock equation by the Kohn-Sham equation [64] and then follow similar parameterization scheme to obtain the potential parameters [65–68].

8.1.3.2   Versions of the Model Core Potential Method

During the past three decades, three main versions of the MCP method have been developed [1, 53]. Version I is based on the local approximation. The core-valence Coulomb repulsion is a local interaction and can be satisfactorily approximated by a local potential function. For convenience of the integral evaluation, such a local potential function is chosen to be a linear combination of Gaussian type functions. The core-valence exchange operator is not a local operator. However, in Version I, this non-local interaction is also approximated by the local potential function of Gaussian type. This non-local to local approximation for the exchange operator shares the same concept with Slater's $X\alpha$ density functional model [69]. Under such an approximation, the one-electron hamiltonian for the valence space in an atom (Eq. 8.5) is rewritten as

$$
\hat{h}^{MCP}(i) = -\frac{1}{2}\nabla_i^2 - \frac{Z - N_c}{r_i}\left(1 + \sum_l A_l r_i^{n_l} \exp\left(-\alpha_l r_i^2\right)\right)
$$
$$
+ \sum_c B_c \left|\overline{\phi}_c(i)\right\rangle\left\langle\overline{\phi}_c(i)\right|, \tag{8.6}
$$

where the subscript "c" has been used to denote the core orbitals and a bar has been put on top of the orbitals to emphasize that they are fixed under all circumstances. $B_c = -f_c\epsilon_c$ and $1 \leq f_c \leq 2$. N_c denotes the sum of the occupation numbers of all the core orbitals. The number 1 in the parentheses represents the perfect screening of the nuclear charge by the electron density of the core orbitals and the $\sum_l$ term represents the correction to this screening, including the penetration effect of the valence orbitals and the core-valence exchange interaction. The exponents n_l are chosen to be 0 and 1 and the importance of selecting both values was analyzed by Höjer and Chung [59] who found that while the $n = 0$ terms represent the core-valence Coulomb potential with the correct core charge, they fail to describe the potential at the origin and the inclusion of the $n = 1$ terms is needed to solve this problem. The total number of terms in the summation $\sum_l$ is arbitrarily chosen and so is the number of terms with $n_l = 0$ or $n_l = 1$. Usually, the more terms are included the greater the flexibility of the potential, and better results would be obtained. However, this would complicate the parameterization of the potential as more parameters are involved. $\{A_l\}$ and $\{\alpha_l\}$ are the parameters of the potential and they are determined by the parameterization process based on atomic calculations, which is introduced in the next section. The core orbitals $\{\phi_c\}$ are naturally chosen to be the atomic core orbitals. Equation 8.6 can take a simpler form:

$$
\hat{h}^{MCP}(i) = -\frac{1}{2}\nabla_i^2 + \hat{V}^{core}(i) + \hat{\Omega}^{core}(i), \tag{8.7}
$$

where the order of operators matches that in Eq. 8.6.

In a molecular system, the potentials $\hat{V}^{core} + \hat{\Omega}^{core}$ of all cores are summed up to form the total core potential and the molecular MCP hamiltonian reads

$$\hat{H}^{MCP}(1,2,\cdots,N_v) = \sum_{i=1}^{N_v} \hat{h}^{MCP}(i) + \sum_{i<j}^{N_v} \frac{1}{r_{ij}} + \sum_{\alpha<\beta}^{N_n} \frac{(Z_\alpha - N_{\alpha,c})(Z_\beta - N_{\beta,c})}{R_{\alpha\beta}},$$

(8.8)

with

$$\hat{h}^{MCP}(i) = -\frac{1}{2}\nabla_i^2 + \sum_\alpha^{N_n} \left(\hat{V}_\alpha^{core}(i) + \hat{\Omega}_\alpha^{core}(i) \right),$$

(8.9)

where N_v is the number of valence electrons, $N_{\alpha,c}$ the core electrons of core α, and N_n the number of nuclei. The meaning of all the other symbols is clear enough and no specification is needed. The third term in Eq. 8.8 shows that the perfect screening has been assumed for the inter-nuclear repulsion. This is a good approximation as long as the core spaces are chosen correctly. Because of the simple form of the potential function (Eq. 8.6), Version I is quite popular and all the MCPs in this chapter refer to this version, unless further specified. Professor Miyoshi's research group at Kyushu University, Japan, and our research group are the main developers for this version of the MCPs. The earlier work of these groups on algorithm developments [70–73] laid the foundations for the recent progress. Applications of Version I during the past decade are described in Sect. 8.4.

Version II differs from Version I in the way of treating the core-valence exchange operator. Instead of approximating this non-local interaction by a local potential, Version II utilizes the atomic spectral resolution of identity operator to represent the exchange interaction. In this version, the core-valence exchange operator of an atom α

$$\hat{V}_{X,\alpha}^{core} = -\sum_{c\in\alpha} \hat{K}_c$$

(8.10)

is sandwiched by the identity operator formed by the basis functions from the same atom:

$$\hat{P}_\alpha = \sum_{p,q\in\alpha} |p\rangle \left(S^{-1}\right)_{pq} \langle q|,$$

(8.11)

to give the truncated non-local exchange operator:

$$\hat{\tilde{V}}_{X,\alpha}^{core} = \hat{P}_\alpha \hat{V}_{X,\alpha}^{core} \hat{P}_\alpha$$

$$= -\sum_{c,p,q,r,s\in\alpha} |p\rangle \left(S^{-1}\right)_{pq} [cq|cr] \left(S^{-1}\right)_{rs} \langle s|.$$

(8.12)

p,q,r, and s are the indices for the non-orthogonal basis functions of the atom α and the non-orthogonality is accounted for by the matrix $\underline{S}^{-1}$. c is the index for the orthonormal core orbitals of the atom α. The participial adjective "truncated" reflects the incompleteness of the atomic basis functions. This approximation employed the fact that $\hat{K}$ is mainly a short range operator [59] and the bulk of this interaction is well represented by the intra-atomic basis functions. One should have noticed that for an atomic system, $\hat{V}_{X,\alpha}^{core} = \hat{V}_{X,\alpha}^{core}$ exactly. For a molecular system, the inter-atomic exchange interaction is realized by the overlap integral of basis functions from different atoms. Because of the presence of $[cq|cr]$, more two-electron integrals are needed for this version. Nevertheless, because of the running indices $c, p, q, r, s \in \alpha$ in the summation in Eq. 8.12, only intra-atomic two-electron integrals are needed, and thus computational savings gained when using this version versus all-electron method are still substantial. Technically, any computer program containing Version II MCP may be employed for Version I calculations, because the formalism of Version II has just one more term ($\hat{V}_X^{core}$) than Version I and this term can be set zero by a condition statement in program.

Version II of MCP is also called the *ab initio model potential* (AIMP) and is mainly developed by the research group of Professors Seijo and Barandiarán of Universidad Autónoma de Madrid in Spain. For their recent work, please refer to the cited references [74–84]. Many researchers consider ECP, MCP (Version I) and AIMP to be three major branches of the pseudopotential method. Although in the present introduction the AIMP method is obviously within the category of the MCP, in the chapters following, we follow the well-accepted three schools categorization whenever the three methods are compared or cited together.

Version III is an extension of Version II in that the core-valence Coulomb interaction is also represented by the atomic spectral resolution of identity, i.e.,

$$
\hat{V}_\alpha^{core} = \hat{P}_\alpha \sum_{c \in \alpha} \left(\hat{J}_c - \hat{K}_c \right) \hat{P}_\alpha
$$

$$
= \sum_{c,p,q,r,s \in \alpha} |p\rangle \left(S^{-1} \right)_{pq} ([qr|cc] - [cq|cr]) \left(S^{-1} \right)_{rs} \langle s| . \quad (8.13)
$$

More of intra-atomic two-electron integrals are needed in this version than in Version II. This version was first proposed by Katsuki [85]. Because of the local nature of $\hat{J}_c$, its approximation by Gaussian type potential function is accurate enough and there is little need to employ the spectral technique. Thus, this version is much less popular than the other two.

8.1.4 Determination of the Model Core Potential Parameters

The MCPs are parameterized against results from atomic AE calculations in order to let the atomic orbital energies and radial structure in the all-electron calculation be perfectly reproduced. This is identical to the aforementioned shape-consistent scheme for ECP. As a consequence, the quality of the AE reference atomic calculation determines the quality of the model core potential, and the relativistic treatment in the reference calculation determines the relativistic level of the potential. Here we only introduce the parameterization procedure used during the recent years in our research group; other procedures are similar except for technical details.

In the parameterization procedure, the total error function Δ_{total} is defined as follows:

$$\Delta_{total} = \Delta_{sf} + \Delta_{SOC} \tag{8.14}$$

$$\Delta_{sf} = \sum_{vo}(w^{\epsilon}_{vo}\Delta^{\epsilon}_{vo} + w^{rad}_{vo}\Delta^{rad}_{vo}) \tag{8.15}$$

$$\Delta^{\epsilon}_{vo} = |\epsilon^{MCP}_{vo} - \epsilon^{AE}_{vo}|: \text{orbital energy deviation} \tag{8.16}$$

$$\Delta^{rad}_{vo} = \sqrt{\frac{\sum_{i=1}^{N_i}(r_i R^{MCP}_{vo}(r_i) - r_i R^{AE}_{vo}(r_i))^2}{N_i}} : \text{radial deviation} \tag{8.17}$$

$$r_k = 0.001\left(\frac{r_{max}}{0.001}\right)^{\frac{k-1}{N_i}} : \text{grid points for radial functions,} \quad k = 1, 2, \ldots, N_i \tag{8.18}$$

$$\Delta_{SOC} = \sum_{vo} w^{SOC}_{vo}\left|\frac{\langle\hat{K}\frac{1}{r^3}\hat{K}\rangle^{MCP}_{vo} - \langle\hat{K}\frac{1}{r^3}\hat{K}\rangle^{AE}_{vo}}{\langle\hat{K}\frac{1}{r^3}\hat{K}\rangle^{AE}_{vo}}\right| \tag{8.19}$$

Atomic units are used in the definitions. The subscript *sf* stands for *spin-free* and *vo* for *valence orbital*. $R(r)$ stands for the radial part of the valence orbital, r_{max} is the outer limit for the radial grid, and N_i is the number of radial grid points. The operator $\hat{K}$ (not the exchange operator) is a kinematic operator in the Douglas-Kroll first order (DK1) spin-orbit operator and will be defined later. By including this operator in the parameterization, the spin-orbit effect of the model core potentials is at the level of DK1. The remaining formulas are self-explanatory. By using Eqs. 8.17 and 8.18, more weight has been put in the small r region, where orbitals are more undulant. The spin-orbit coupling contribution to the total error function Δ_{SOC} can be omitted if the potential is not designed for the spin-orbit coupling calculations. The w are weights for each error quantity in the total error function. The values of the weights are subjective and are used to balance the various error contributions and facilitate the convergence of fitting. The total error function Δ_{total} is an implicit function of the potential parameters $\{A_l\}$ and $\{\alpha_l\}$ in Eq. 8.6. In our research, Δ_{total} was minimized by using the non-derivative numerical optimizer from Brent [86] to obtain the optimized parameters $\{A_l\}$ and $\{\alpha_l\}$ and MCP. A scheme for the determination of the model core potential parameters is illustrated in Fig. 8.4.

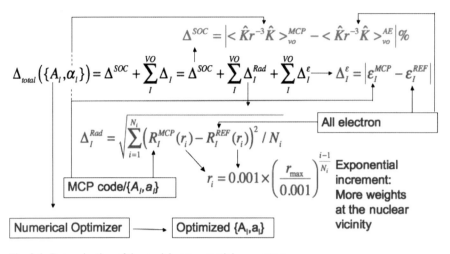

Fig. 8.4 Determination of the model core potential parameters

8.2   General Improvements in Parameterization and Basis Sets

Several approaches were used to generate new MCPs in the past decade and in the present section we summarize those approaches.

The first new family of the MCPs was developed by Mane who used the well-tempered basis sets [22, 87] to support extended basis sets for the MCPs: these so-called well-tempered MCPs (wtMCPs) were first developed for atoms from Groups 17 and 18 of the Periodic Table [88] and then for all the main-group elements from Li to Rn [89], leading to an excellent agreement between the results from wtMCP and AE calculations. Following the structure of the well-tempered basis sets, the basis sets for the wtMCPs shared exponents for the s and p spaces (the L-shell structure) that reduced integral evaluation time – a much welcome feature since the well-tempered basis sets contain a large number of primitive Gaussian functions. Modified wtMCPs were custom-made later for a few atoms in the work on interactions between rare gas and metal atoms [90–92].

The second new family of the MCPs, called the improved MCPs (iMCPs), was developed by Lovallo for the first-row transition metals [62] and then for the second- and third-row transition metals [93]. The purpose of the iMCPs was to provide fast and yet accurate MCPs for studies of transition metal compounds. In the process of atomic optimization, the standard procedure was followed (matching of orbital energies and radial functions obtained in all-electron calculations). Small basis sets were designed, with the exponents of basis set functions obtained by fitting to reference numerical atomic orbitals; the basis sets were contracted for easier use in production calculations. In contrast to basis sets used earlier in the MCP approach (but sharing this feature with the wtMCPs described above), the iMCP basis sets have the L-shell structure for the s and p spaces, leading to a significant reduction of the integral evaluation time. The iMCPs offered two new features: the basis sets were contracted and they were equipped with polarization and correlating functions, making them very easy to employ by using an internal library in GAMESS-US. The parameters of the core projector (B_g in the expression $-\sum_g |g\rangle B_g \langle g|$) were also optimized (see also Sect. 8.1.3.1) by requiring that the MCP results matched those for a small training set of molecules. Lovallo's work would not be possible if it were not for the availability of the analytical first and second derivatives of the MCP hamiltonian which were derived by Jayatilaka et al. [94] and implemented in the CADPAC suite of programs [95].

Both the iMCPs and wtMCPs enjoyed only limited applications. However, the flexibility of the wtMCPs ultimately led to the development of the most recent MCPs. The MCPs that became most popular are the ones designed in Miyoshi's laboratory. In the seminal reports [96, 97] Miyoshi and his co-workers reported MCPs that offered the user great flexibility and accuracy. They reparameterized the MCPs using small Gaussian basis sets, contracted them to make them ready for use, and, very importantly, equipped them with matching sets of polarization and correlating functions. These auxiliary functions, critically required for accurate results, had been prepared by Noro and co-workers [98–105] in their studies of polarization and correlating functions for relativistic calculations. Miyoshi's group subsequently re-developed the MCPs for the s-block elements [106]. Finally, in a series of three papers, Miyoshi's group reported new, compact, efficient, and very accurate MCPs for the first-row [107], second-row [108], and third-row [109] transition metal atoms covering, together with the earlier work on the MCPs for the lanthanides [110], nearly the entire Periodic Table. The basis sets were designed so as to provide various levels of accuracy (and associated computational cost): from small double-zeta valence plus polarization, through triple-zeta, to quadruple-zeta valence with polarization and they were named MCP-dzp, MCP-tzp, and MCP-qzp to parallel the correlation-consistent basis sets cc-pVDZ, cc-pVTZ, and cc-pVQZ [111–114], respectively; augmented with diffuse functions, they have been build into the libraries of GAMESS-US. These basis sets can be obtained on-line (http://setani.sci.hokudai.ac.jp/sapporo/Welcome.do), formatted for several major computational chemistry programs.

An important development related to the MCPs was the introduction of the model core potentials to the fragment molecular orbital (FMO) calculations [115]. The FMO/MCP method allows to carry out quantum mechanical calculations for large scale systems containing heavy metal atoms.

The latest family of the MCPs are the zfkMCPs which are designed for very accurate studies. Details of their development and performance in pilot studies are given in Sect. 8.3.

8.3   New Approach to Relativistic Effects

The pseudopotential methods are derived mainly for the systems with a large number of inert core electrons, i.e., heavy elements. The most pronounced characteristics of heavy elements are their relativistic effects [116–119] and, therefore, the development of pseudopotential method is closely related to the relativistic effects. The relativistic effects can be roughly classified into two categories, the scalar-relativistic and the spin-relativistic effects. The former effect stems from the mass-velocity and Darwin [120] effects and accounts mainly for the relativistic contraction of s and p orbitals and the expansion of d and f orbitals. No new symmetry property is introduced by the scalar-relativistic effects and the famous manifestations of this effect are the liquid state of mercury under ambient temperature and pressure [121], the golden color of gold [122–124], and the inert effect of the $6s^2$ electron pair [125]. The spin-relativistic effects, however, contain the electronic spin operator explicitly and introduce the double group symmetry [126, 127] related to the electron spin of $1/2$. The most important spin-relativistic effect is the spin-orbit coupling (SOC) [128–134], which couples the magnetic dipole moment induced by the electronic spin and the magnetic field induced by the electronic orbital motion around nuclei. SOC leads to three phenomena in physics and chemistry [133, 135]: fine structure splitting, intersystem crossings, and spin-forbidden radiative transitions, and their respective examples are: splitting of the visible yellow D-line of sodium atomic spectrum [136], the stereochemical control exerted by intersystem crossing on reactions in Paternò-Büchi radicals [137], and phosphorescence [138]. Depending on how the decoupling of the large and small components of the four-component hamiltonian in the Dirac equation [139, 140] is done, several versions of two-component relativistic hamiltonian have been derived and they are (chronologically arranged): Breit and Pauli's decoupling [141], Douglas-Kroll transformation (DK) [142], Cowan and Griffin's decoupling [143], Wood and Boring's decoupling [144], the zeroth order regular approximation (ZORA) [145–147] and its higher order extensions [148, 149], relativistic elimination of small components (RESC) [150], normalized elimination of small components (NESC) [151], and infinite-order two-component theory (IOTC) [152, 153]. The objective of the relativistic MCP is to approximate those two-component hamiltonians. Relativity in chemistry itself is an extensive and highly profound subject and we have no intention to present a detailed account of it in the present chapter. Interested readers may refer to several excellent textbooks [17, 154–156] for this subject. Any discussions of relativistic effects in this section are closely related to the MCP development. The improvement of the relativistic approximation formed the centre of the MCP development in the first 10 years of this century, and in this section we especially focus on the inclusion of the SOC effect into MCP.

8.3.1 Scalar-Relativistic Effect in Model Core Potential Method

As mentioned in Sect. 8.1, the implementation of the scalar-relativistic effect into any pseudopotential is easy and one just need to parameterize the potential against the all-electron reference calculations at the corresponding relativistic levels. The first relativistic MCP was published by one of the present authors (M.K.) in 1983 [157], when the MCP of Ag was parameterized against the Cowan and Griffin's [143] scalar-relativistic Hartree-Fock numerical AE calculation. The relativistic potential had been added [158] to a numerical AE program for atoms [159] to produce the reference quantities for this parameterization. This work demonstrated the feasibility of including the scalar-relativistic effect into MCP and immediately initiated a series of relativistic MCP studies of development and applications [160–167]. The success of these studies triggered the extension of the AIMP from non-relativistic [78, 168] to the version at the same level of relativistic effect [169]. Fourteen years later, a systematic study developed MCP at the same relativistic level for main group elements from Rb to Rn [72] and since then the development and applications of Cowan and Griffin's scalar-relativistic MCP has been one of the focuses of Miyoshi's research group [106–110]. As the result of their contributions, the Cowan and Griffin's relativistic effect has the greatest availability in all the relativistic MCP.

However, the Cowan and Griffin's decoupling scheme produces non-hermitian and energy-dependent operators [156] and this fact impairs the approximation of the Cowan and Griffin's relativistic MCP, as the MCP operator is hermitian and energy-independent. Therefore, implementing a higher order of relativistic approximation in the MCP is desired. The first attempt along this frontier was accomplished by Lovallo [92], who introduced the RESC scalar-relativistic effect into his MCP for rare gases (Ar, Kr, Xe), coinage metals (Cu, Ag, Au), and light halogens (F and Cl). After this initial work, he extended the RESC MCP to the second- and third-row transition metals [93]. However, almost immediately after the publication of the RESC method [150], Barysz performed a detailed study [170] of the method and demonstrated variational instability of this decoupling scheme. Furthermore, she pointed out that this scheme does not offer any particular advantages over the DK scheme, which is variational and convergent [171, 172] in its expansion. For this reason, we implemented the DK3 (Douglas-Kroll at the third order) scalar-relativistic effect in MCP.

Gold is considered to be the most relativistic element in the first six rows of the Periodic Table and this property is reflected by the term *Gold Maximum* [117, 122–124, 173]. On the other hand, because of its spatially-non-degenerate atomic ground electronic state (^2S_g), Au does not have significant SOC effect. Thus, we chose this element to develop the first DK3 scalar-relativistic MCP [24]. The real focus of that study was to select the right valence space, rather than parameterizing our MCP against the AE DK3 calculation. We developed DK3 MCP for valence spaces: 5p5d6s, 4f5p5d6s, 5s5p5d6s, and 5s4f5p5d6s and studied the necessity of including the 5s and 4f subshells in the valence space. The influence of these two subshells on Au chemistry is illustrated by Fig. 8.5. When the inter-electron dynamic

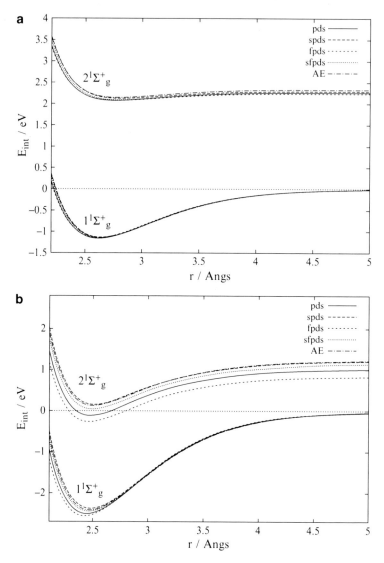

Fig. 8.5 Potential energy curves for the two lowest $^1\Sigma_g^+$ states of Au_2: (**a**) at the CASSCF level; (**b**) at the CASPT2 level (Reprinted from Ref. [24] with kind permission of The American Institute of Physics)

correlation is not considered (Fig. 8.5a, at the CASSCF level), all valence spaces produce similar potential energy curves for the two lowest $^1\Sigma_g^+$ states of Au_2 and all the MCP curves overlap satisfactorily with the AE reference curves. However, when the dynamic correlation is considered (Fig. 8.5b, at the CASPT2 level), only the valence spaces with the 5s subshell (5s5p5d6s, 5s4f5p5d6s) produce MCP curves close to the AE reference curves. This is even more obvious around the equilibrium

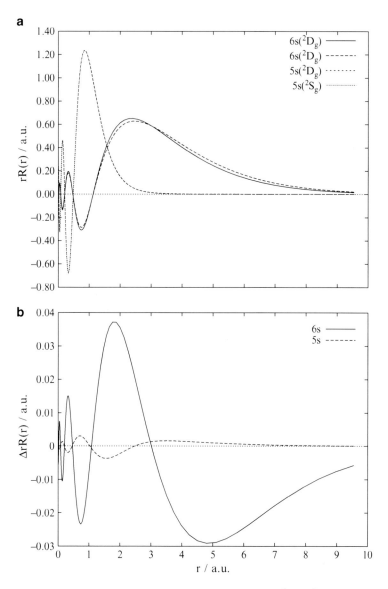

Fig. 8.6 (**a**) Radial functions of 6s and 5s before and after 5d→6s ($^2S_g \to ^2D_g$) excitation of Au atom. The orbitals came from all-electron MCSCF calculations; (**b**) The difference between the radial functions in ^2S_g and ^2D_g states, $\Delta rR(r) = rR^{^2D_g}(r) - rR^{^2S_g}(r)$ (Reprinted from Ref. [24] with kind permission of The American Institute of Physics)

internuclear distance (r_e) of the ground state ($1^1\Sigma_g^+$) and the whole r range of the excited state ($2^1\Sigma_g^+$). We found the ultimate reason for this phenomenon is that when one electron is excited from the 5d to 6s subshell to form the ^2D_g term of Au, the 6s orbital becomes more contracted (see Fig. 8.6), resulting in the change of (stronger) correlation between the 5s and 6s electrons. The ^2D_g character becomes more substantial at the r_e of the $^1\Sigma_g^+$ state and prevails over all the r range of the $^2\Sigma_g^+$ state. In order to account for this dynamic correlation, the 5s subshell must be included in the valence space. This work essentially solves the problem of the so-called 5s-4f puzzle: the 5s has lower energy than the 4f subshell but the 5s needs to be included in the valence space, not the 4f. The solution of this puzzle can be better explained by considering the competition between the electron correlation effect and the relativistic effect on the valence electrons. For more discussion on this subject, one should refer to Ref. [24].

The most important insight gained in developing the DK3 MCP for Au is that we need to consider the dynamic correlation when selecting an MCP valence space. Another remedy for this outer-core–valence correlation is to introduce a core-valence correlation function and this remedy has already been applied to ECP [174]. However, such approach would reduce the ab initio character of MCP and more importantly, so far we only found the importance of the $(n-1)$s subshell in the outer-core-valence dynamic correlation (vide infra) and including one more s subshell in the valence space would not substantially lower the computational efficiency. Therefore, we tend to enlarge the valence space to account for the dynamic correlation rather than introducing a new correlation function.

8.3.2 Spin-Orbit Coupling in Model Core Potential Method

Compared to the scalar-relativistic effect, the implementation of the SOC in MCP started much later. To the best of our knowledge, our recent publication in 2009 [175] was the first study to explicitly incorporate SOC into MCP. The most important reason for this delay, compared to implementation of SOC in ECP [9, 10, 32, 37, 38, 176] and AIMP [80, 82], was due to the assumption that such development was not needed because MCP always produces correct inner nodal structure of valence orbitals and, consequently, correct SOC matrix elements. However, as shown in our recent work [175] (and below), this assumption does not work for heavy elements.

The first application of the MCP in SOC calculations dates back to 1996 [61]. In this study, Krause and Klobukowski found the agreement between the MCP and AE SOC results to be within 1% for diatomic hydrides of P, As, and Sb. Later, a method to compute both one- and two-electron SOC using MCP was proposed [177] and applied to atoms, hydrides and homonuclear diatomic cations of P, As, and Sb. The agreement with results from AE calculations was within 3%. More recently, MCP was applied in SOC calculations of ions of S_2 [178] and the hydrides of C, Si, Ge, and Sn [179], and good agreement with experimental values was found for molecules containing light atoms, with the noticeable deviation for SnH. The error for SnH can be attributed to the truncation of the MCP basis set and the consequent distortion of the inner nodal structure of the valence orbitals [97]. All the

MCP-SOC calculations cited above employed the lowest-order Breit-Pauli (BP-) SOC operator and did not go beyond the fifth row of the Periodic Table. The BP-SOC operator works very well for light elements; however, because of its variational instability (discussed below), the operator tends to diverge for heavy elements and overestimates the SOC effect. Therefore, application of MCP with the higher order Douglas-Kroll (DK-) SOC operator, which exhibits good variational stability, is a better choice. Since there is no quantum chemistry program that combines MCP and the DK-SOC, developing a method for this combination was the first objective our work [175]. As mentioned above, the MCP-BP-SOC program [61, 177] had been developed before our study and employing such a program to perform MCP-DK-SOC calculations was highly desired. In order to do that, a profound understanding of the differences between the two types of SOC operators was needed.

8.3.2.1 Breit-Pauli and Douglas-Kroll Spin-Orbit-Coupling Operators

Two detailed discussions about the Breit-Pauli (BP-) and Douglas-Kroll (DK-) SOC operators are given in our recent publications [175, 180] and here we only give a short overview on this subject, as the background knowledge of our development of DK-SOC-adapted MCP. The one-electron BP-SOC and DK-SOC operators are related via

$$\hat{H}_{SO}^{BP1e} = \sum_i \hat{h}_{SO}^{BP1e}(i); \tag{8.20}$$

$$\hat{H}_{SO}^{DK1e} = \sum_i \hat{h}_{SO}^{DK1e}(i) = \sum_i \hat{K}_i \hat{h}_{SO}^{BP1e}(i) \hat{K}_i, \tag{8.21}$$

where

$$\hat{K}_i = \frac{2mc^2 \hat{A}_i}{\hat{E}_i + mc^2}; \tag{8.22}$$

$$\hat{A}_i = \sqrt{\frac{\hat{E}_i + mc^2}{2\hat{E}_i}}; \tag{8.23}$$

$$\hat{E}_i = \sqrt{\hat{p}_i^2 c^2 + m^2 c^4}. \tag{8.24}$$

$\hat{K}$ is a kinematic operator that damps the high momentum part of the wave function while keeps the low momentum part intact. Its appearance makes the DK-SOC operator bounded from below and variationally stable and a much better SOC operator than the BP-SOC. The effect of the $\hat{K}$ operator is illustrated by Fig. 8.7, where the 6p orbitals of Tl, At, and Rn are taken as examples. While the radial distribution function of the original Tl 6p orbital (Tl 6p) and the one after the operator of $\hat{K}$ (Tl K6p) are highly similar in Fig. 8.7a, the squares of these functions

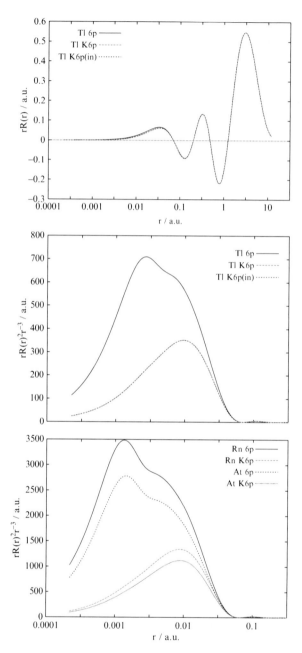

Fig. 8.7 (a) Radial functions and (b) integrands of the r^{-3} operator for the Tl 6p orbital (data labelled as 6p correspond to the SOC-BP operator (AE), while K6p and K6p(in) correspond to the relativstically contracted DK-SOC operator in AE and MCP, respectively). (c) Similar radial integrands for Rn and At. The logarithmic scale is used for the x axis; it is different in panel (a) and in panels (b) and (c) (Reprinted from Ref. [180] with kind permission of The American Institute of Physics)

Table 8.1 Comparison of intermediate coupling (within $6p^2$ configuration) SOC energy levels (cm^{-1}) for Pb atom relative to the ground level ($J = 0$) using different SOC treatments[a]

		DK-SOC[b]		BP-SOC[c]		BP-SOC ($\hat{K}6p$)[d]	
J	Exp. [266]	SO-CASCI	SO-MCQDPT	SO-CASCI	SO-MCQDPT	SO-CASCI	SO-MCQDPT
1	7819.26	6531.55	7401.75	11844.21	13201.30	6351.52	7401.57
2	10650.33	10161.41	11047.99	16010.76	17345.80	10161.38	11047.80
2	21457.80	20578.80	21722.55	31427.81	33624.88	20578.75	21722.18
0	29466.83	31962.11	29554.84	42587.44	41153.93	31962.06	29554.47

[a]Reprinted from Ref. [175] with kind permission of The American Institute of Physics

[b]DK3 spin-free calculation + DK SOC treatment

[c]DK3 spin-free calculation + BP SOC treatment

[d]DK3 spin-free calculation + BP SOC treatment with the $\hat{K}6p$ orbital

multiplied by by r^{-3} (the BP- and DK-SOC integrands) show significant difference in Fig. 8.7b. The comparison of Fig. 8.7a and b also indicates that even the smallest error in the radial function may result in a very large error in the SOC matrix element and therefore the inclusion of the SOC integral in the MCP error function (Eq. 8.19) is necessary. As the result of the $\hat{K}$-damping, the DK-SOC integrand has much smaller value than the BP-SOC one and, furthermore, it does not have the bump of the BP-SOC integrand that reflects the variational instability. Similar smaller values and bump-free integrands are also observed for the DK-SOC operator of At and Rn in Fig. 8.7c. Another important observation from Fig. 8.7b is that if we only apply $\hat{K}$ to the inner part of the Tl 6p radial function while keeping the outer two lobes untouched, we have similar DK-SOC integrand (Tl K6p(in)). This fact tells us that the effect of $\hat{K}$ can be stored in an additional set of contraction coefficients for the large exponent primitives and this forms the cornerstone to perform MCP-DK-SOC calculations with the MCP-BP-SOC program. The two-electron parts of the BP- and DK-SOC are more complicated and they are not written explicitly here. Compared to its one-electron analogue, the DK-SOC two-electron operator is damped by both $\hat{K}$ and $\hat{A}$ and this fact makes the one-electron DK-SOC operator more dominant. Usually, it is satisfactory to consider only the one-electron DK-SOC effect and we follow this convention in our research.

In order to numerically show the difference between the two SOC schemes, we present our recent results for the excitation energies of the low-lying excited states of Pb atom in Table 8.1, from which one sees that only the DK-SOC calculations at the correlation level of multi-configurational quasi-degenerate perturbation theory (MCQDPT) [181, 182] are comparable to the experimental data. It was this comparison that prompted us to develop the MCPs which are DK-SOC adapted. Also shown in Table 8.1 is the feasibility of using the contraction coefficients that store the effect of $\hat{K}$ and the BP-SOC program to perform the DK-SOC calculation, as the two columns under the heading of BP-SOC($\hat{K}6p$) are almost identical to the two columns under the heading of DK-SOC.

The idea of employing an additional set of contraction coefficients is equal to applying the $\hat{K}$ operator to damp the basis functions with large momentum at the start of calculations; this idea was inspired by Dyall's study on other relativistic operators [151]. Using the $\hat{K}$-damped basis functions and the BP-SOC program, we have developed DK-SOC adapted MCP for thirty p-block elements, from 2p to 6p [180, 183]. We chose to develop p-block MCP first because: (1) they have non-zero first-order SOC [133] and (2) their electronic structures are simpler than those of the transition metals. We stopped at 6p because there is evidence that the inter-electron Breit interaction becomes non-negligible for the heavier elements [41], and handling this interaction is still not standard in most of the quantum chemistry programs. In the following, we will switch from SOC back to the introduction of the development and performance of these MCPs.

8.3.2.2 Determination of the Model Core Potential Valence Space

As discussed for the Au MCP in Sect. 8.3.1, correct selection of the valence space is critically important for the performance of an MCP. For the np-block elements, ns and np subshells must be included in the valence space, since they are the primary valence subshells. The necessity of including $(n-1)$d, $(n-1)$p, or $(n-1)$s subshell was investigated by calculations of atomic term and level energies. In the present chapter, we follow the well-accepted convention to call an electronic state without (with) considering SOC a "term" ("level"). We may consider the electronic wave function of an atom under chemical environments as a mixture of the low-lying terms or levels, because from the view of perturbation theory, the large energetic gaps [184] in the denominators of the energy difference and the diffuse wave functions reduce the contributions from the high-lying states. Therefore, if a valence space can satisfactorily reproduce low-lying term and level energies, it should also work well for molecular calculations. By "low-lying", we mean the terms and levels coming from the ground state electronic configuration of an element, e.g., ^3P_g, ^1D_g, and ^1S_g terms and $^3P_{0,g}$, $^3P_{1,g}$, $^3P_{2,g}$, $^1D_{2,g}$, and $^1S_{0,g}$ levels of the $2p^2$ configuration of C atom. Our experience with the Au MCP also suggested to include dynamic correlation (at the MCQDPT level) in this investigation. AE MCQDPT and SO-MCQDPT calculations for Group 15 atoms and Group 13 cations were performed, employing multiple correlation spaces, and their term and level energies are compared with the AE calculations employing full correlation space. The calculated results are listed in Table 8.2.

The subshell $(n-1)$d is well-known to have substantial overlap with the $nsnp$ subshells and its inclusion is necessary. The results in Table 8.2 confirm that including $(n-1)$d in the correlation space reduces the error of term energies, e.g., from -165 to 20 cm^{-1} for the ^4S_u-^2P_u energy spacing in Sb. Therefore, $(n-1)$d is included in our MCP. On the other hand, Table 8.2 also indicates that the inclusion of $(n-1)$p reduces the error of level energies, e.g., from -191 to -12 cm^{-1} for the $^2P_{1/2,u}$-$^2P_{3/2,u}$ splitting in Sb. Therefore, $(n-1)$p is needed in the valence space. We presented a detailed discussion [175] about the importance of the $(n-1)$p subshell

Table 8.2 AE calculated atomic and ionic terms and levels (cm^{-1}) employing different correlation spaces for Groups 15 and 13 elements using the uncontracted well-tempered basis sets (WTBS) [22][a]

Correlation space[b]	All	A	B	C			All	C[c]	D[c]
P						Al$^+$			
^4S_u-^2D_u	14887	29	-3	-		^1S_g-^3P_u	37599	369	10
^4S_u-^2P_u	19446	17	-1	-		^1S_g-^1P_u	68016	464	14
$^2D_{3/2,u}$-$^2D_{5/2,u}$	23	-4	0	-		$^3P_{0,u}$-$^3P_{1,u}$	61	0	0
$^2P_{1/2,u}$-$^2P_{3/2,u}$	22	-5	0	-		$^3P_{1,u}$-$^3P_{2,u}$	123	0	0
As						Ga$^+$			
^4S_u-^2D_u	13102	-8	-31	7		^1S_g-^3P_u	45740	1001	2
^4S_u-^2P_u	19900	-92	-91	31		^1S_g-^1P_u	70494	783	23
$^2D_{3/2,u}$-$^2D_{5/2,u}$	307	-41	-3	-15		$^3P_{0,u}$-$^3P_{1,u}$	417	-10	-10
$^2P_{1/2,u}$-$^2P_{3/2,u}$	374	-52	-3	-18		$^3P_{1,u}$-$^3P_{2,u}$	862	-21	-21
Sb						In$^+$			
^4S_u-^2D_u	11856	-40	-63	3		^1S_g-^3P_u	41030	460	-11
^4S_u-^2P_u	18068	-165	-170	20		^1S_g-^1P_u	63015	490	5
$^2D_{3/2,u}$-$^2D_{5/2,u}$	1296	-132	-15	-71		$^3P_{0,u}$-$^3P_{1,u}$	1012	-29	-29
$^2P_{1/2,u}$-$^2P_{3/2,u}$	1711	-191	-23	-101		$^3P_{1,u}$-$^3P_{2,u}$	2214	-73	-72

[a]Reprinted from Ref. [183] with kind permission of The American Institute of Physics
[b]The column headings refer to different correlation spaces in the MCQDPT calculations: All = correlating all occupied orbitals; A = correlating the outermost s and p orbitals, $nsnp$; B = correlating $(n-1)pnsnp$ orbitals; C = correlating $(n-1)p(n-1)dnsnp$ orbitals; D = correlating $(n-1)s(n-1)p(n-1)dnsnp$ orbitals. The entries in the 'All' column are the calculated reference values, while the other entries are the differences from them
[c]For Al$^+$, Ga$^+$, and In$^+$, the correlation spaces under consideration are C = $(n-1)pnsnp$ and D = $(n-1)s(n-1)pnsnp$

for the np-block elements and this importance can be explained by the second order Rayleigh-Schrödinger perturbation theory formula [184]:

$$\frac{\langle (n-1)p|\hat{h}_{SO}|np\rangle}{\epsilon_{(n-1)p} - \epsilon_{np}}. \tag{8.25}$$

Since $\hat{h}_{SO}$ is a highly localized operator (see Fig. 8.7b and c), although the $(n-1)p$ and np do not have substantial overlap, their undulations at the nuclear vicinity still provide non-negligible numerator in Eq. 8.25. On the other hand, the energy gap between the two subshells in the denominator is not large enough to eliminate this coupling. Therefore, $(n-1)p$ subshell is needed in the valence space due to the SOC. Lastly, since Group 13 cations have an electron configuration of ns^2 and, as discussed above for the Au MCP, the $(n-1)s$ subshell is needed for correlating with the ns electrons. The importance of the $(n-1)s$ subshell is demonstrated by the Group 13 cation results in Table 8.2. Overall, the valence space for Group 14 to 18 elements was selected to be $(n-1)p(n-1)dnsnp$ and $(n-1)s$ is added for Group 13 elements. The subshell $(n-1)s$ is not included for Group 14 to 18 elements because of their high energy requirement ($>$ 15 electron volts (eV)) to form the ns^2 configuration. Exceptions for such valence space selections are for 2p elements, where only 2s2p is included, and for 3p elements, where there is no $(n-1)$d. Since the other core orbitals (replaced by potential) are not even dynamically correlated with the valence orbitals, they are not supposed to participate in any chemistry and the B_c in Eq. 8.6 was chosen to be $-2\epsilon_c$ for the new MCP development.

8.3.2.3 Basis Sets for the New Model Core Potentials

The completely uncontracted well-tempered basis sets (WTBS) [22] were used in our recent MCP parameterization [24, 175, 180, 183], including AE reference calculations and MCP calculations in the error function evaluations. Naturally, WTBS form the fundamental basis sets of our MCP. The large exponents of the WTBS primitives account for the relativistic effects, which are more prominent for the electronic motion with a large momentum. The WTBS basis set was augmented with the aug-cc-pVxZ (x =T, Q, 5) [111, 112, 114, 185, 186] primitives with $l = 0$ to 4 (s to g primitives) whose exponents are smaller than the smallest of the same l in WTBS in order to account for polarization and valence correlation. Depending on the x value, the new MCPs were named ZFKn-DK3, where $n = 3, 4, 5$ and ZFK is the acronym of the last names of the three developers of this series of MCPs. Together with the algorithm, the library of potentials and basis functions of ZFKn-DK3 has been implemented in GAMESS-US.

8.3.2.4 Performance of the DK-SOC Adapted Model Core Potential

Because of the novelty of the ZFKn-DK3 MCP series, they have not been extensively applied in quantum chemistry calculations. However, we have carried out calibration calculations to study their accuracy and efficiency. Since the MCPs were parameterized against the AE references, we took the agreement of the MCP results with their AE copunterparts as the ultimate indication of their accuracy. The comparison of MCP and AE calculations for Tl atomic terms and energies, TlH spectroscopic constants, and molecular term and level energies are listed in Tables 8.3, 8.4, and 8.5; the potential energy curves of TlH associated with Tables 8.4 and 8.5 are shown in Fig. 8.8.

For detailed discussion of those calculations, the reader should refer to the cited references and here we only highlight the most important conclusions of those comparisons. Although we only cited the results of Tl and TlH here, similar results were obtained for other systems and the conclusions are general. The typical errors are thousandths of an Å for molecular structure, within 40 cm^{-1} for vibrational frequencies, and within thousandths (hundredths) of an eV for energetics (atomic and molecular term and level energies; dissociation energies; ionization potentials; electron affinities) at the correlation levels of configuration interaction

Table 8.3 AE and MCP atomic calculation results for Tl[a]

Spin-orbit levels (in cm^{-1})[b]

J	AE CI	MCP CI	Δ^c_{CI}	AE PT	MCP PT	Δ_{PT}	Exp
Tl[d]							
3/2	6306	6295	−11	6516	6424	−92	7793
1/2	23765	23781	16	25302	25291	−11	26478
Tl[+ e]							
0	44097	44067	−30	49451	50343	892 (130)	49451
1	47060	47031	−29	52368	53217	849 (85)	52394
2	54766	54737	−29	60478	61213	735 (−74)	61728
1	73498	73495	−3	74177	74695	518 (18)	75663
Tl[−f]							
1	2400	2395	−5	2595	2555	−40	-
2	4009	4005	−4	3859	3811	−48	-
2	8184	8179	−5	7624	7536	−88	-
0	12664	12671	7	11270	11191	−79	-
Tl[g]							
3/2	-	-	-	-	7856	-	7793
1/2	-	-	-	-	27615	-	26478

Ionization potential (IP) and electron affinity (EA)[h]

-	AE CI	MCP CI	Δ_{CI}	AE PT	MCP PT	Δ_{PT}	Exp
IP / eV	5.722	5.724	0.002	6.167	6.126	−0.041	6.108[i]
EA / eV	0.089	0.089	0.001	0.125	0.152	0.027	0.2±0.2[j]

[a]Reprinted from Ref. [180] with kind permission of The American Institute of Physics

[b]All experimental levels are taken from Ref. [266]. CI and PT in this table denote SO-CASCI and SO-MCQDPT, respectively. The J values of the ground levels for Tl, Tl[+], and Tl[−] are 1/2, 0, and 0 respectively

[c]Δ denotes the difference between MCP and AE values: MCP−AE

[d]Intermediate coupling (IC) scheme of ^2P_u ($6p^1$) and ^2S_g ($7s^1$) was employed. The orbitals were obtained from a SA-MCSCF calculation including the same terms. The active space included 6s6p7s7p6d orbitals

[e]IC scheme of ^1S_g ($6s^2$), ^3P_u ($6s^1 6p^1$), and ^1P_u ($6s^1 6p^1$) was employed. Active orbitals from the same method[c] were used. The values in parentheses are from new MCP with 5s5p5d6s6p valence space [183]

[f]IC scheme of ^3P_g, ^1D_g, and ^1S_g terms (all of $6p^2$) was employed. Active orbitals from the same method[c] were used

[g]IC scheme of 1^2S_g, $2\,^2P_u$, $1\,^4S_u$, and $4\,^4P_u$ was employed. The same active orbitals[c] were used

[h]The ground state energies of Tl[+] and Tl[−] are from this work [d,e]. That of Tl was obtained from an IC-SOC calculation of ^2P_u only. A SA-MCSCF calculation of the term was used to generate orbitals

[i]Ref. [267]

[j]Ref. [268]

(CI) (perturbation theory (PT)). The PT energy errors are all within the chemical accuracy of 0.043 eV (1 kcal/mol). These error bounds are highly satisfactory and we also found that these error bounds do not increase with the number of MCP

Table 8.4 AE and MCP calculated spectroscopic constants of TlH[ab]

	AE CI[c]	MCP CI	Δ_{CI}	AE PT	MCP PT	Δ_{PT}	Exp[d]
$^1\Sigma^+$							
r_e / Å	1.970	1.968	−0.002	1.882	1.880	−0.002	-
D_e / eV	2.126	2.126	0.000	2.452	2.471	0.019	-
ω_e / cm^{-1}	1255	1255	0	1322	1339	17	-
$\omega_e x_e$ / cm^{-1}	18.51	18.71	0.20	21.50	24.01	2.51	-
0^+(I)							
r_e / Å	1.958	1.958	0.000	1.880	1.877	−0.003	1.872
D_e / eV	1.704	1.704	0.000	2.049	2.072	0.023	2.06
ω_e / cm^{-1}	1253	1251	−2	1310	1332	22	1390.7
$\omega_e x_e$ / cm^{-1}	20.64	19.35	−1.29	20.56	25.28	4.72	22.7
0^+(II)							
r_e / Å	-	-	-	1.838	1.840	0.002	1.91
D_e / eV	-	-	-	0.685	0.677	−0.008	0.74
ω_e / cm^{-1}	-	-	-	991	977	−14	760
$\omega_e x_e$ / cm^{-1}	-	-	-	158.14	197.01	38.87	-

[a]Only bound states with significant potential energy wells were considered

[b]Reprinted from Ref. [180] with kind permission of The American Institute of Physics

[c]For the spin-free molecular terms, CI and PT denote CASCI and MCQDPT, while for the spin-orbit molecular levels, CI and PT denote SO-CASCI and SO-MCQDPT

[d]The experimental r_e, D_e and ω_e of the 0^+(I) are summarized in Ref. [45] and were measured in Ref. [39, 40]. $\omega_e x_e$ is taken from Ref. [269]. The experimental r_e, D_e and ω_e of the 0^+(II) are summarized in Ref. [270]. An excellent MCP-Exp agreement was obtained by using a larger term space (see Ref. [180])

atoms in our study of Group 14 dimers [183]. Only when the dynamic correlation is considered at the PT level, the MCP results can be compared with the experimental data, and the PT results can be improved by adjusting active orbital space and multi-electron basis space [180].

We also compared the computational times of AE and MCP calculations [183] and the results are listed in Table 8.6. The calculations were performed with the augmented basis functions from the aug-cc-pV5Z basis sets. This means the MCP/AE basis ratios are the highest among the ZFKn series. However, the reduction of the CPU time by using MCP is very impressive (e.g., 87% and 71% for Te atom at the CI and PT levels), especially for the heavy elements, for which any pseudopotential method is mainly designed. These studies fully validated the accuracy and efficiency of the newly developed DK-SOC adapted MCP.

8.3.3 A Digression: From MCP to SOC

Since our recent research was about developing the DK-SOC adapted MCP, we performed a large number of SOC calculations. Therefore, improving the accuracy

Table 8.5 AE and MCP terms and levels of TlH (in cm^{-1})^a

Term or Level	AE CI^b	MCP CI	Δ_{CI}	AE PT	MCP PT	Δ_{PT}	Exp
$^1\Sigma^{+c}$	0	0	–	0	0	–	–
$^3\Pi$	17407	17404	−3	18448	18636	188	–
$^1\Pi$	21653	21652	−1	21585	21661	76	–
$^3\Sigma^+$	34161	34158	−3	37767	37862	95	–
$0^+(I)^d$	0	0	–	0	0	–	–
$0^-(I)$	15611	15606	−5	16640	16835	195	–
$0^+(II)$	16468	16461	−7	17423	17587	164	17723^e
$1(I)$	17130	17124	−6	17954	18114	160	–
$2(I)$	20462	20454	−8	21521	21648	127	–
$1(II)$	23169	23164	−5	23389	23438	49	24180^e
$0^-(II)$	35020	35015	−5	38765	38834	69	–
$1(III)$	35039	35034	−5	38767	38837	70	–

^{a}Reprinted from Ref. [180] with kind permission of The American Institute of Physics

^{b}For the spin-free molecular terms, CI and PT denote CASCI and MCQDPT, while for the spin-orbit molecular levels, CI and PT denote SO-CASCI and SO-MCQDPT

^{c}CI term energies were calculated at $r = 1.95$ Å, the grid point closest to the calculated r_e of the ground state. PT term energies were calculated at $r = 1.89$ Å, the grid point closest to the calculated r_e of the ground state

^{d}CI level energies were calculated at $r = 1.95$ Å, the grid point closest to the calculated r_e of the ground state. PT level energies were calculated at $r = 1.88$ Å, the grid point closest to the calculated r_e of the ground state

^{e}Ref. [269]

of SOC calculations and extracting information from an SOC wave function also became of interest to us. For the first subject, we carried out a study on the influence of different molecular orbital (MO) generation schemes on SOC calculations and for the latter, we proposed an algorithm to extract two-component natural orbitals (natural spinors) from an SOC wave function. Although these two studies are not directly related to MCP, they are briefly introduced in this section as their conclusions are very useful for future SOC investigations.

So far, all of our SOC calculations are based on the L-S coupling scheme (actually, most of the main-stream program packages only support L-S coupling); that is, a spin-free (non-SOC) calculation is carried out to generate a set of L-S states[1] and the SOC is added at a post-Hartree-Fock level taking the states as the multi-electron Ansatz [187, 188]. It should be noted that the L-S coupling calculations in this chapter are essentially of the *intermediate coupling* scheme as more than one set of L-S states are involved in the SOC calculations. The involvement of several states in the SOC calculations naturally raises the question about which L-S state(s) should be chosen for optimizing the molecular orbitals

---

[1]More generally, Γ-S, where Γ denotes the irreducible representation of the spatial wave function in the point symmetry group of the molecule; throught the chapter, we use L-S as more conventional although Γ-S is what is meant.

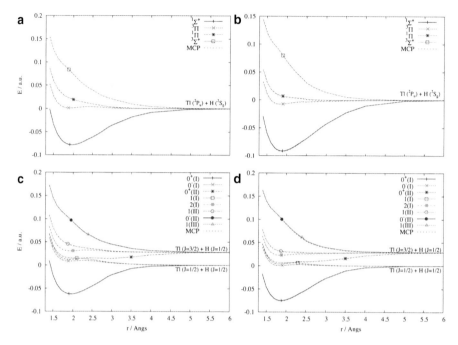

Fig. 8.8 Potential energy curves of TlH from (**a**) CASCI, (**b**) MCQDPT, (**c**) SO-CASCI, and (**d**) SO-MCQDPT calculations. The AE curves are labeled by term or level symbols whereas the MCP curves are shown as black dashed line for all terms, and they overlap nearly perfectly with the AE ones (Reprinted from Ref. [180] with kind permission of The American Institute of Physics)

Table 8.6 Timings (CPU time, in s), on one IBM p630 1.2 GHz processor[a]

	SO-CASCI			SO-MCQDPT			Basis ratio (%)[b]
	AE	MCP	MCP/AE (%)	AE[c]	MCP	MCP/AE (%)	
O	77	37	48	153	110	72	99
S	110	49	45	206	145	70	98
Se	393	86	22	568	202	36	95
Te	999	132	13	1525	438	29	90

[a]Reprinted from Ref. [183] with kind permission of The American Institute of Physics

[b]The number of MCP basis functions/the number of AE basis functions

[c]The core orbitals replaced by MCP were excluded from the AE correlation calculations of perturbation theory to keep the correlation spaces identical in AE and MCP calculations

(MO) for use in the subsequent SOC studies. Usually, either state-averaged (SA) MCSCF or state-specific (SS) MCSCF calculations are used to optimizes the orbitals. In SA-MCSCF calculations, several states are equally averaged [180, 189] while in SS-MCSCF calculations only the state that undergoes the first-order SOC splitting [178, 179] is used for orbital optimization. The former works better for the case of strong SOC, which usually occurs when several adiabatic energetic surfaces of the SO-coupled states are close to each other, while the latter is more reasonable for the weak case, when the adiabatic energy surfaces are far apart. However, in chemical processes the adiabatic energy surfaces can come close to (or get far from) each other as the nuclear configurations change (e.g., along a reaction path) and this requires an orbital optimization scheme to adjust according to the energy intervals between different states. Such an orbital optimization scheme was proposed by Deskevich et al. [190] in their studies of the reaction $F + H_2O \rightarrow HF + OH$, and this scheme is called dynamically-weighted (DW-) MCSCF. In the DW-MCSCF approach, the energy functional to be optimized is defined as

$$E^{DW-MCSCF} = \sum_i w_i (E_i - E_0) E_i, \qquad (8.26)$$

where E_0 is the energy of the target state, i indicates the other relevant states, with their energies being weighted by the respective weight w_i in the energetic functional. The w_i function is chosen to decrease as $|E_i - E_0|$ increase. The MO are obtained by an iterative calculation based on Eq. 8.26, i.e. the $\{w_i\}$ obtained from the previous MCSCF calculation are used in the next calculation to obtain $\{E_i\}$ and the next set of $\{w_i\}$, until the $\{E_i\}$ and $\{w_i\}$ values converge. In the SOC calculations, the target state is the state that has the most significant contribution to the L-S coupled wave function and the relevant states are those that have non-negligible but non-dominant contributions.

In our study [191] we applied all three orbital generation schemes to compute the SOC potential energy curves of Sn_2^+ and found that for the nuclear configurations where crossings or avoided crossings of electronic states occur, the SOC calculations using DW- or SA-MCSCF orbitals can produce smooth potential energy curves while discontinuities are likely to occur for the calculations using SS-MCSCF orbitals. The SO-MCQDPT calculations using DW- and SA-MCSCF orbitals brought about similar potential energy curves and predicted similar spectroscopic constants. We attribute this phenomenon to the smaller orbital dependence of the PT method as it is an approximation to full CI. The DW- and SA-MCSCF methods may be used as alternatives to each other whenever there is a convergence problem in one scheme but not in the other. The advantage (disadvantage) of SA-MCSCF (DW-MCSCF) is that the aforementioned weights-energies iteration is unnecessary (necessary) and the advantage (disadvantage) of DW-MCSCF (SA-MCSCF) is the partition of a state in orbital generation is (not) justified based on its energy.

Although the molecular orbital is a very useful tool to study a wave function, the L-S coupled wave function is constructed by multi-electron functions (L-S states) and the concept of MO cannot be employed for such a wave function. This fact limits the usage of L-S coupling in wave function studies compared to the j-j coupling scheme [146, 192–196], which includes the SOC in orbital optimization step and whose wave function is made of two-component orbitals, which can be analyzed based on the same method of analyzing MO. In our most recent studies [197], we proposed an algorithm to construct the L-S coupling density matrix and diagonalize the matrix to obtain the natural spinors. Using a properly chosen L-S coupling states in the construction of the density matrix, we can obtain natural spinors that share the same symmetry properties as the j-j two-component orbitals. Therefore, the L-S coupling wave function can be analyzed in the same way as for the j-j coupling through the natural spinors. By analyzing the natural spinors, we have studied the interplay between SOC and bonding and between SOC and the Jahn-Teller effect [198, 199] and we drew the conclusion [271] that the ultimate reason behind these interplays is the competition between the directionality of the covalent and Jahn-Teller interactions and the rotational nature of the SOC.

8.4   Model Core Potential Applications in the Last 10 Years

Several representative applications from the body of published reports are presented here to illustrate the scope of applications of the MCP methodology.

As mentioned earlier, the MCPs are available in only few programs: GAMESS-US [200, 201], CADPAC [95], MOLCAS (as a subset of the ab initio Model Potential modules) [202], ABINIT-MP [203], and DEMON2K [204].

The program DEMON2K has been primarily used in the studies of transition metal systems. Pereiro et al. [205] carried out detailed density functional studies of the dimer and trimer of cobalt taking into account several spin numbers and excited states. Russo and co-workers investigated neutral and charged rhodium clusters Rh_n (n = 2, 3, 4) of several spin multiplicities, focusing on their magnetic properties [206]. These studies were extended to complexes between molybdenum (both Mo and Mo_2) and ammonia, ethene, and propene [207] for which spin multiplicity of the ground electronic states was determined.

Using GAMESS-UK, Decker designed several sets of polarization/correlating functions for use with the old MCPs in comparative studies to assess the efficacy of the MCPs with respect to the ECPs in studies of the halogen complexes of Group 4 metals [208]. Tetrahalides MX_4 were studied, with M=Ti, Zr, and Hf and X=F, Cl, Br, I. Results showed that inclusion of electron correlation (at the MP2 level) was essential to improve agreement with experimental bond lengths.

Sakai et al. used several advanced post-Hartree-Fock methods (complete active space SCF, multi-reference singly- and doubly-excited conguration interaction, and multi-reference coupled pair approximation) in studies of low-lying electronic states of TiCl and ZrCl [209]. They found that the ground state for TiCl was $^4\Phi$ and the lowest excited states were $^4\Sigma^-$ and $^2\Delta$ (in that order). For ZrCl it was the doublet $^2\Delta$ that was the ground state, with the excited states $^4\Phi$ and $^4\Sigma^-$. The computed spectroscopic constants were reasonably close to the experimental ones.

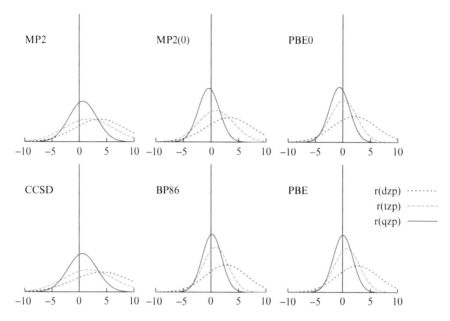

Fig. 8.9 Percent error distributions for equilibrium bond lengths of the total molecules sample (Reprinted from Ref. [210] with kind permission of John Wiley & Sons, Inc)

The latest MCPs for the main group elements [97] were calibrated against experimental structural data in careful studies to assess their reliability [210]. All the recent basis sets were used (MCP-dzp, MCP-tzp, and MCP-qzp) together with several methods of varying sophistication: post-Hartree-Fock (MP2 [211–216] and CCSD [217]) and density functional theory (PB86 [218, 219], PBE [220], and PBE0 [221]). It was found that the level of theory that is capable of providing reliable structural data is DFT/MCP-tzp. The MCP-dzp parameterization is most suited for qualitative studies. For the post-Hartree-Fock methods (MP2, CCSD) the basis set of MCP-qzp quality is required for quantitative studies. The results, summarized in Figs. 8.9 and 8.10 show that the MCP basis sets behave similarly to their all-electron counterparts [222] (see also Chap. 15 of Ref. [223]).

Miyoshi's group used the MCPs to pursue their interest in excited states of compounds containing transition metal atoms. They employed Tamm-Dancoff conguration interaction approximation to analyze the electronic excitation spectra of Mo and Re cluster complexes $[(Mo_6Cl_8)Cl_6]^{2-}$ and $[(Re_6S_8)Cl_6]^{4-}$, imbedded in a cage of 728 point charges (representing ionic crystal) [224]. The results roughly reproduced experimental absorption and emission spectra. The ALCHEMY II program [225–227] was used in this study.

The same program was used by Sakai to study low lying electronic states of GdO [228]: potential energy curves for the eight lowest states ($^9\Sigma^-$, $^9\Delta$, $^9\Pi$, $2\,^9\Sigma^-$, $^7\Sigma^-$, $^7\Delta$, $2\,^7\Sigma^-$) were obtained. Agreement between experimental and computed spectroscopic constants for the ground state $^9\Sigma^-$ was very good: calculated

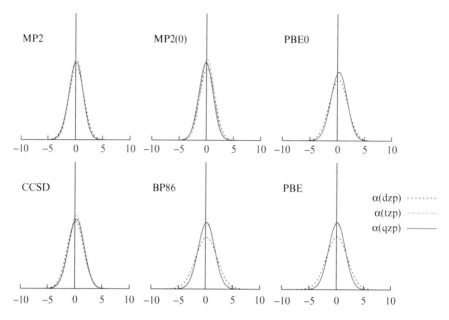

Fig. 8.10 Percent error distributions for equilibrium angles of the total molecules sample (Reprinted from Ref. [210] with kind permission of John Wiley & Sons, Inc)

$r_e = 1.818 \, \text{Å}$ and $\omega_e = 876 \, \text{cm}^{-1}$, experimental $r_e = 1.812 \, \text{Å}$ and $\omega_e = 824 \, \text{cm}^{-1}$. Good agreement was obtained also for the term energies T_e.

Miyoshi continued their earlier studies of excited states by analyzing electronic structure of the lower states of the $[Cr_2Cl_9]^{3-}$ and $[Mo_2Cl_9]^{3-}$ ions [229] using methodology similar to that used earlier [224] (multi-reference coupled-pair approximation and multi-reference configuration interaction with single and double excitations; the anions were surrounded by 11,326 point charges). Detailed analysis of wave functions indicated that while there is a single direct bond between the metal atoms in $[Mo_2Cl_9]^{3-}$, no such bond was found in $[Cr_2Cl_9]^{3-}$. Excitation energies could be improved if additional correlating functions were available (f- and g-type).

Ishikawa and Tanaka used the same approach to study electronic excitation spectra of $[(Re_6S_8)Cl_6]^{3-}$, finding good agreement between the calculated oscillator strength distribution and observed absorption curve [230].

The earlier studies of excited states in systems containing transition metal atoms were carried out using the older MCPs for the transition metal, with the valence subshells of only $(n - 1)p\,(n - 1)d\,ns$; the recent analysis of low-lying electronic states (the ground state $^1\Sigma_g^+$ and two excited states, $^{11}\Sigma_g^+$ and $^{11}\Pi_u$) of the weakly-bonded Mn_2 [231] used the more extensive spdsMCP with the additional subshell $(n-2)s$ electrons explicitly included. Presence of the $(n-2)s$ subshell was found to greatly improve reliability of the MCPs [107]. The manganese dimer was described using the multiconguration quasidegenerate perturbation theory (MCQDPT) [181,

182], with several MCP basis sets containing correlating functions up to g-type. The ground-state spectroscopic constants r_e and ω_e were in good agreement with previously reported results from all-electron calculations.

Lovallo studied thermochemistry of organoxenon complexes [232] using the MP2 method and two hybrid DFT functionals together with the iMCPs. The following reactions were studied:

$$(CH_3)_3SiL + XeF_2 \longrightarrow (CH_3)_3SiF + LXeF,$$

$$2(CH_3)_3SiL + XeF_2 \longrightarrow 2(CH_3)_3SiF + XeL_2,$$

where the ligand L = CN, CCH, CCF, C_6H_5, C_6F_5, CF_3, and CH_3. The first ligand substitution (leading to LXeF) was strongly exergonic, with ΔG_{rxn} between -22 and -44 kcal/mol; for the second ligand substitution ΔG_{rxn} was smaller (between -9 and -22 kcal/mol). The iMCPs augmented with polarization/correlating functions (two d and one f functions were used both on Xe and C; two p and one d on H) were very effective at predicting structures of the ligand molecules. Continuing his work on rare gases, Lovallo studied interactions between the coinage metal monohalides and a heavier rare gas atom RgMX (Rg = Ar, Kr, Xe; M = Cu, Ag, Au; X = F, Cl) [92]. Electron correlation was treated at the MP2 level, and the wtMCPs were used, with the RESC method [150] used to generate the reference data. Interaction energies were corrected for the basis set superposition error (BSSE) using the full counterpoise method (CP) [233]. The complexes were found to be linear and, for the lighter congeners Ar and Kr, the structural parameters agreed very well with the experimental results of Gerry [234–237]. One year later, one of the heavier congeners, XeAuF, has been identified and characterized using microwave rotational spectroscopy by Gerry [238], whose experimental structural data were in excellent agreement with Lovallo's theoretical prediction: r_e^{calc}(Xe-Au) = 2.545 Å, r_0^{exp}(Xe-Au) = 2.548 Å, r_e^{calc}(Au-F) = 2.922 Å, and r_e^{exp}(Au-F) = 2.918 Å.

Lovallo subsequently determined very accurate pair potentials of the interaction between alkaline earth atoms and helium: first for Be, Mg, Ca, Sr, and Ba [90], later improved for the heavier atoms Ca, Sr, and Ba [91]. The wtMCP basis sets were extended by multiple correlating functions; large set of bond functions was added as well. The MP2, CCSD, and CCSD(T) interaction energies were corrected for BSSE using full CP [233]. Using computed interaction parameters (equilibrium interatomic distance and well depth) together with a model proposed by Ancilotto [239], Lovallo predicted that Be and Mg would be fully solvated and reside inside a helium nanodroplet, Ba would be on the surface, while Ca and Sr were borderline cases that could occupy a dimple on the surface.

Miyoshi and co-workers carried out several studies related to adsorption. In an early study they modeled adsorption of SO_2 on Ni(111) and Cu(111) surfaces [240] using a hybrid DFT functional and a mixed basis set (ECP [241] and cc-pVDZ [112]), followed by single-point energy evaluations at the CASSCF level with MCPs. For structural studies, Ni_4SO_2 and Cu_4SO_2 were chosen as model clusters to represent SO_2 adsorbed on the metal M(111) surfaces. Single-point calculations were carried out for larger clusters, $M_{15}SO_2$. In a subsequent study the interactions between Si(111) surface and metal atoms [242] were modeled. Computations were carried out at the high level of theory (CASSCF and MRSDCI) using small MCP basis sets. Two cluster models were used to represent Si(111) surface: Si_4H_7 and Si_4H_{10} (with hydrogens used to cap the dangling Si bonds) and the metals studied were Na and Mg. Again, ALCHEMY II [225–227] was used (GAMESS-US was used to carry out population analysis). Dependence of results on cluster size was checked by using a larger cluster for Na, $NaSi_7H_{11}$; structural parameters and adsorption energies were essentially the same as for the smaller clusters.

One year later Miyoshi et al. published results of a molecular orbital study of Na, Mg, and Al adsorption on the Si (111) surface [243] using the same methodology as in their previous work [242], but extending the model cluster size: in addition to Si_4H_7 and Si_7H_{11}, $Si_{16}H_{18}$ was also employed. The most stable adsorption sites were determined and they were found to be different for the metals studied.

Gao and co-workers estimated spin-orbit splitting in the ground $X^2\Pi$ electronic state of the S_2^+ and S_2^- molecular ions [178] and studied spin-orbit interaction in the ground electronic states of XH molecules (X = C, Si, Ge, and Sn) [179]. Results from these papers were discussed earlier, in Sects. 8.3.2 and 8.3.3.

Zeng et al. carried out a detailed study of convergence of electron affinities of phosphoryl and thiophosphoryl halides APX_3 (A = O, S and X = Br, I) [244]. DFT, MP2, and CCSD(T) methods were used with three MCP basis sets: dzp, tzp, and qzp; convergence of results with basis set size was analyzed. Harmonic vibrational analysis was done leading to an assignment of vibrational modes. The chalcogen atom was found to play an insignificant role in the electron attachment process.

Gajewski studied the electronic structure and geometry of 18-crown-6 $(C_{12}H_{24}O_6)$, hexaaza[18]annulene $(C_{12}H_{12}N_6)$, and their complexes with cations of the heavier alkali (Rb^+ and Cs^+) and alkaline earth metals (Sr^{2+} and Ba^{2+}) [245]. She showed that the ions bind more strongly to hexaaza[18]annulene than to 18-crown-6, with affinity greater by 8-23 kcal/mol. Gajewski extended her work on host-guest chemistry to complexes between ethylenediamine tetraacetate and alkali (Na^+, K^+, and Rb^+) and alkaline earth cations (Mg^{2+}, Ca^{2+}, and Sr^{2+}) [246] and obtained binding affinity order $Mg^{2+} > Ca^{2+} > Sr^{2+} > Na^+ > K^+ > Rb^+$. Solvent effects on binding affinity were modeled by explicit water molecules.

Zeng used custom-made relativistic MCPs to determine the importance of inner subshells in the description of the Au^+Xe system [247]. Several high-level methods to describe electron correlation were used, with the most accurate being CR-CCSD(TQ)_B [217, 248–251]. At this level of theory, an excellent agreement for spectroscopic constants was obtained: $r_e^{calc} = 2.647$ Å, $r_e^{exp} = 2.61$ Å (in a crystal), $\omega_e^{calc} = 139\,cm^{-1}$, $\omega_e^{exp} = 138\,cm^{-1}$. The dissociation energy was predicted to be $D_e = 0.98$ eV, with half of the correct bonding energy due to the electron correlation. Zeng established that the Xe 4d orbital was not essential, while the 5p orbital of Au was necessary to obtain correct results.

Using the fspdsMCPs for lanthanides [110], Miyoshi and co-workers performed CASSCF and CASPT2 calculations for lanthanide trihalides LnX_3 [252], where

Ln = Pr, Nd, Pm, Sm, Dy, Ho, Er, and Tm; X = F, Cl, Br, and I. Ground electronic states for all trihalides were established, assuming that the molecular symmetry was planar (D_{3h}) rather than pyramidal (C_{3v}). Spin-orbit interaction was ignored. Comparison of calculated Ln-X bond lengths with experimental data showed that description of dynamic electron correlation was absolutely necessary for correct results. These studies on lanthanide systems were later extended to hydration models of trivalent rare-earth ions [253] for Y^{3+}, La^{3+}, Gd^{3+}, and Lu^{3+}; geometry optimization was carried out at the MP2 level for hydrates containing from one to ten water molecules. In addition, ab initio molecular dynamics simulations (by following the dynamical reaction coordinate) for the systems with more water molecules, $[M(H_2O)_{24}]^{3+}$ (M = Y, La) and $[La(H_2O)_{64}]^{3+}$, were done and both radial distribution function and coordination number were obtained.

A true *tour de force* in the studies of very large systems was presented in the form of the fragment molecular orbital method, coupled with MCPs [115]. The method was initially proposed by Kitaura et al. [254], implemented in the developer's version of ABINIT-MP [203, 255, 256], and, through the initiative of Fedorov [257], implemented in GAMESS-US. The method FMO/MCP was first applied to the studies of the Hg^{2+} ion surrounded by 256 water molecules at the MP2 level of theory [115]. In the same report the authors described application of the MP2/FMO/MCP method to the complex between the $[Pt(NH_3)_2]^{2+}$ moiety of cisplatin and DNA, hydrated with 980 water molecules (with 10,154 basis functions) [115]. They found that explicit hydration was needed in the modeling of the complex. The calculation on that system took slightly more than 8 days on 32 dual Opteron processors. In the following report [258] they carried out detailed study of polarizability of silicone-containing clusters in the context of potential applications in nanotechnology.

Barnett et al. examined the bond between Au^+ and ethene using both the ECP method and (in MCSCF calculations) the spdsMCPs for gold [259]. Subsequently, a series of five platinum(II) complexes of the form (N^N^N)PtCl were studied (where N^N^N represents the tridentate monoanionic ligands) using the time-dependent density functional theory [260].

Haiges et al. studied structure of the vanadium(IV) and vanadium(V) binary azides $V(N_3)_4$, $[V(N_3)_6]^{2-}$, and $[V(N_3)_6]^-$ [261], using both MP2 and DFT methods together with an extend MCP-tzp basis set. They obtained reasonable agreement between calculated and experimental structural parameters an vibrational frequencies.

Fitzsimmons et al. extended computational chemistry of the rare gases by studying the systems HRgF (Rg = Ar, Kr, Xe, Rn) [262]. In contrast to earlier predictions, they found that radon, being more polarizable than xenon, forms compounds that are more stable than those of xenon.

Shim et al. used the MCPs in the calibration of the ΔMP2 method for calculating core electron binding energies (CEBE) [263]. In that method, MCPs are used for all atoms except for the one from which the 1s electron is removed (that atom is treated using an all-electron basis set that included core polarization/correlating functions). The calibration, carried out for 114 core electron binding energies of C, N, O, and F in 55 molecules found the absolute average deviation from experimental CEBE of 0.16 eV.

We should add that there are numerous applications of the (related to the MCPs) core and embedding ab initio model potentials [79], with great possibilities, especially in the study of spectroscopy of metal ions in crystals; we refer the interested reader to http://www.uam.es/departamentos/ciencias/quimica/aimp/.

8.5 Summary and Outlook

Almost right after the dawn of quantum mechanics, the concept of pseudopotential was introduced in the 1930s in order to save computational effort and concentrate on the valence electronic structure of each element. This idea stems from the periodic principle for the elements, which states that the chemical properties of an element are determined primarily by its valence electrons only. However, congeners in the Periodic Table do not have the same chemical properties and this means the core electronic structure, despite its inertness, exerts a potential to influence the valence structure. The objective of any pseudopotential method is to approximate this core-valence potential accurately. Other than the core-valence potential, the anti-symmetry requirement for the electronic wave function imposes the strong orthogonality between the core and valence electronic structure, and the two different approaches to realize this orthogonality resulted in two main schools of pseudopotential methods: effective core potential (ECP) and model core potential (MCP). The ECP smoothens the inner nodal structure of the valence orbitals but forces the resultant pseudo-valence orbitals to have the same orbital energies by using the implicit Phillips-Kleinman potential. The smooth inner nodal structure of the valence orbitals requires fewer basis functions in practical calculations but fails to describe the wave function at the nuclear vicinity correctly. This means that any close-nuclear properties like spin-orbit coupling (SOC) or nuclear magnetic resonance (NMR) shielding factor cannot be calculated using the genuine operators. The MCP, on the other hand, enforces the core-valence orthogonality by using a level-shift operator, which makes any overlap with the core orbitals energetically unfavorable. By doing that, the inner nodal structure of the MCP valence orbitals is retained and the genuine operators can be used to calculate the close-nuclear properties. The correct inner-nodal structure is also important for the accurate evaluation of the valence correlation energy. Based on the treatment of the core-valence inter-electron exchange and Coulomb interaction, the MCPs can be classified into three versions. Version I approximates both the non-local exchange and local Coulomb interaction by a local potential; Version II approximates the exchange interaction by the spectral representation of the valence basis of each atom but still approximates the Coulomb interaction by a local potential; Version III approximates both the exchange and Coulomb interactions by the spectral representation. Version I is usually called "MCP", while Version II is known as the ab initio model potential (AIMP). Version III is much less popular than the other two and has disappeared from publications in the last 10 years.

The MCP basis set development, spearheaded by the research group under direction of Miyoshi, brought about three sets of compact MCP basis sets: dzp, tzp, and qzp, comparable to the correlation-consistent basis sets. The wtMCPs of Mane and Lovallo went in the opposite direction with very large basis sets both for the core functions and for the valence orbital expansions. In the same spirit (large basis sets, very accurate MCPs) Zeng led the effort to parameterize the zfkMCPs that created the foundation for accurate spin-orbit coupling calculations using the MCPs.

Another great leap forward in the MCP development is the inclusion of relativistic effects. In terms of the inclusion of the scalar-relativistic effects, our research group went beyond the Cowan and Griffin's relativistic level and successfully incorporated RESC and DK3 relativistic effects into the MCP. In our development of Au DK3-MCP, we established the importance of the 5s subshell due to its dynamic correlation with the 6s subshell. This was the first MCP development that considered dynamic correlation and the finding has been very useful for the valence space selection in subsequent MCP development. Another breakthrough was the inclusion of the spin-orbit coupling (SOC) in the MCP. Since the early days of the MCP history [54, 61, 177], researchers have been relying on the MCP to produce highly accurate inner nodal structure of the valence orbitals to produce accurate SOC matrix elements. However, we found that even the smallest error in the inner nodal structure may lead to incommensurately large error in the SOC calculations. For this reason, we decided to include the SOC matrix elements in the error function for MCP parameterization. On the other hand, since the conventional Breit-Pauli (BP-) SOC operator is non-variational and tends to overestimate the SOC interaction, we switched to the higher level approximation of Douglas-Kroll (DK-) SOC and implemented this level of SOC interaction into the new MCPs. In this procedure, we employed the highly local nature of the kinematic operator $\hat{K}$ to store its effect into an additional set of contraction coefficients and the BP-SOC program can be employed for the DK-SOC calculations. Using this algorithm we have developed DK-SOC adapted MCP for thirty p-block elements (from 2p to 6p). These new MCP were named ZFKn-DK3, where $n = 3, 4, 5$ indicates the basis set augmentation levels and DK3 indicates their scalar-relativistic level. The accuracy and efficiency of this series of MCPs have been validated in a series of atomic and molecular calculations. The new algorithm and the library of potentials and basis sets of the new MCPs have been implemented into the program package GAMESS-US for public use.

The applications of the MCP method in the last 10 years were made possible by significant coding improvements in the GAMESS-US program (such as the coding of analytical gradient for the MCP hamiltonian done by Professor Hirotoshi Mori) and by including libraries of the MCPs and corresponding basis sets (with a variety of polarization and correlating functions) in the standard GAMESS-US distribution. At present, using the library of MCPs in GAMESS-US is just as easy as using the library of ECPs.

Our future development of the MCPs can be divided into two frontiers. The first frontier is to extent the development of DK-SOC adapted MCP from p-block elements to the transition metals. The transition metals (d-block and f-block elements) have large angular momentum ($l = 2, 3$), their SOC is prominent, and therefore the SOC-adapted MCPs are needed for them. Also, because of their large angular momentum, they have larger orbital degeneracy ($2l + 1$ fold), which leads to a large number of closely-spaced atomic terms within the ground state electronic configurations. Incidentally, it is these closely-spaced atomic terms that are responsible for the rich chemistry of the transition metals. These closely-packed atomic terms, however, complicate the MCP parameterization procedure. In our studies we only carried out the shape-consistent parameterization for the ground atomic term, e.g., the ^3P_g term of C atom. This has to be changed for the transition metals due to the greater number of chemically active terms. We have conceived two ways to handle this problem. The first approach could be called the state-averaged shape-consistent scheme. In this scheme, the AE reference orbital energies and shapes come from a state-averaged MCSCF calculation including all chemically active atomic terms. Similar MCP MCSCF calculation is performed in the error function evaluation. With the correct state-averaged orbitals and orbital energies, MCP can reproduce the electronic spectrum of those terms from AE calculation. The electronic spectrum is important because the energetic gaps between the terms partly determine the SOC between them, and then a satisfactory MCP atomic electronic spectrum is necessary for accurate SOC calculation. The second approach could be called the energy-consistent scheme. Because of the importance of the atomic electronic spectrum, we may just follow the energy-consistent scheme in ECP parameterization and only include the deviation from the AE atomic electronic spectrum in the error function. A drawback of this method is that the orbitals and their SOC matrix elements are not included in the error function. This may result in the situation that the energetic gap between atomic terms are well reproduced but not the SOC matrix elements between them, which eventually may lead to erroneous results of the SOC calculations. Considering that, we prefer the first method over the second one. However, both are worth trying.

The second frontier is to improve the MCP operator. The model core potential hamiltonian employs the classical kinetic energy operator (see Eq. 8.6), however, a genuine DK hamiltonian employs the relativistic kinetic operator:

$$\sqrt{\hat{p}^2 c^2 + m^2 c^4} - mc^2. \tag{8.27}$$

Therefore, when parameterized against a DK relativistic atomic reference, the local potentials $\hat{V}^{core}$ not only approximates the core-valence interaction, but also all the differences between the two kinetic operators, which are non-local operators. The non-locality of the DK hamiltonian is hidden in the $\hat{K}$ (Eq. 8.22) and $\hat{A}$ (Eq. 8.23) operators and those operators show up in almost every term of the expansion of the DK hamiltonian [17, 156]. Thus, the non-local to local approximation of the Version I model core potential is more severe. Based on our experience, this error is negligible for the calculations at the Hartree-Fock level because all the occupied orbitals are reproducible by the potential. However, the calculations with the electron correlation involves virtual orbitals, which are energetically high-lying and have large magnitudes of kinetic energies. For those orbitals, the relativistic effects are not merely corrections, but determining factors. Therefore, the energetically high-lying virtual orbitals from MCP and AE calculations differ a lot in their shapes and energies. Although the Møller-Plesset perturbation theory [211] dictates that a virtual orbital with a higher energy will contribute less to the correlation energy, we suspect this non-local to local approximation is responsible for the typical energetic error of hundredths of eV at the perturbation theory level mentioned in Sect. 8.3.2.4. In the future, we will try to employ a new DK3-MCP hamiltonian

$$\hat{h}_{DK3}^{MCP}(i) = \hat{H}_{DK3}^{sr}\left(-\frac{Z-N_c}{r_i}\right) - \frac{Z-N_c}{r_i}\left(\sum_l A_l r_i^{n_l} \exp\left(-\alpha_l r_i^2\right)\right)$$
$$+ \sum_c B_c \left|\overline{\phi}_c(i)\right\rangle\left\langle\overline{\phi}_c(i)\right|, \tag{8.28}$$

where $\hat{H}_{DK3}^{sr}\left(\frac{Z-N_c}{r_i}\right)$ is the DK3 scalar-relativistic hamiltonian corresponding to the nuclear charge of $Z - N_c$. By doing this, most of the non-local nature of the DK3 kinematic operators is preserved in MCP and the potential is mainly for the core-valence electron interaction, as it was originally designed for. A formula similar to Eq. 8.28 has been first employed in the AIMP formalism [264, 265]. However, the discussion of its influence on dynamic correlation is still missing, and needed before the massive production of this type of MCP.

Acknowledgements This work would not be possible without the initial impetus provided by Professor Sigeru Huzinaga and his subsequent supervision and participation in the development of the consecutive versions of the model core potentials. We are grateful to Professors Yoshiko Sakai, Eisaku Miyoshi, and Hirotoshi Mori for many years of fruitful collaboration. We appreciate the support of Professor Mark Gordon and Dr. Mike Schmidt for the continuing development of the GAMESS-US program system which has been our principal computational tool. Recent developments in the spin-orbit adapted MCPs would be impossible without the inspiring and insightful participation of Dr. Dmitri Fedorov. Finally, a few words of gratitude to our financial supporters: TZ is grateful to Alberta Ingenuity Funds, Killam Trusts, and Alberta Scholarship Program for student scholarships during his PhD studies. MK thanks the Natural Sciences and Engineering Research Council of Canada for the support of the model core potential development under Research Grant No. G121210414.

References

1. Klobukowski M, Huzinaga S, Sakai Y (1999) Model core potentials: theory and applications. In: Leszczynski J (eds) Computational chemistry: reviews of current trends, vol 3. World Scientific, Singapore, p 49
2. Cotton FA, Wilkinson GFRS (1972) Advanced inorganic chemistry: a comprehensive text, 3rd edn. Interscience Publishers, New York

3. Huzinaga S, Cantu AA (1971) J Chem Phys 55:5543
4. Huzinaga S, McWilliams D, Cantu AA (1973) Adv Quantum Chem 7:187
5. Hellmann H (1934) Comput Rend Acd Sci URSS 3:444
6. Hellmann H (1935) J Chem Phys 3:61
7. Combas P (1935) Z Phys 94:473 8
8. Cao X, Dolg M (2006) Coordin Chem Rev 250:900
9. Lee YS, Ermler WC, Pitzer KS (1977) J Chem Phys 67:5861
10. Pitzer RM, Winter NW (1988) J Phys Chem 92:3061
11. Frenking G, Antes I, Böhme M, Dapprich S, Ehlers AW, Jonas V, Neuhaus A, Otto M, Stegmann R, Veldkamp A, Vyboishchikov SF (1996) Pseudopotential calculations of transition metal compounds – scope and limitations. In: Lipkowitz KB, Boyd DB (eds) Reviews in computational chemistry, vol 8, VCH Publishers, New York, p 63
12. Cundari TR, Benson MT, Lutz ML, Sommerer SO (1996) Effective core potential approaches to the chemistry of the heavier elements. In: Lipkowitz KB, Boyd DB (eds) Reviews in computational chemistry, vol 8. VCH Publishers, New York, p 145
13. Krauss M, Stevens WJ (1984) Annu Rev Phys Chem 35:357
14. Hülsen M, Weigand A, Dolg M (2009) Theor Chem Acc 122:23
15. Kahn LR, Hay PJ, Cowan RD (1978) J Chem Phys 68:2386
16. Figgen D, Peterson KA, Stoll H (2008) J Chem Phys 128:034110
17. Dyall KG, Faegri JK (2007) Introduction to relativistic quantum chemistry. Oxford University Press, New York
18. Stevens WJ, Krauss M, Basch H, Jasien PG (1991) Can J Chem 70:612
19. Dolg M, Cao X (2004) The relativistic energy-consistent ab initio pseudopotential approach and its application to lanthanide and actinide compounds. In: Hirao K, Ishikawa Y (eds) Recent advances in relativistic molecular theory. World Scientific, Singapore, p 1
20. Phillips JC, Kleinman L (1959) Phys Rev 116:287
21. Weeks JD, Rice SA (1968) J Chem Phys 49:2741
22. Huzinaga S, Klobukowski M (1993) Chem Phys Lett 212:260
23. Figgen D, Rauhut G, Dolg M, Stoll H (2005) Chem Phys 311:227
24. Zeng T, Klobukowski M (2009) J Chem Phys 130:204107
25. Teichteil C, Malrieux JP, Barthelat JC (1977) Mol Phys 33:181
26. Pittel B, Schwarz WHE (1977) Chem Phys Lett 46:121
27. Klobukowski M (1990) Chem Phys Lett 172:361
28. Seijo L, Barandiarán Z, Huzinaga S (1992) Chem Phys Lett 192:217
29. Dolg M (1996) Chem Phys Lett 250:75
30. Dolg M (1996) J Chem Phys 104:4061
31. Szabo A, Ostlund NS (1989) Modern quantum chemistry: introduction to advanced electronic structure theory. Dover Publications, Mineola
32. Hafner P, Schwarz WHE (1978) J Phys B 11:217
33. Andrae D, Haeussermann U, Dolg M, Stoll H, Preuss H (1990) Theor Chim Acta 77:123
34. Schwerdtfeger P, Dolg M, Schwarz W, Bowmaker G, Boyd P (1989) J Chem Phys 91:1762
35. Fuentealba P, Stoll H, Szentpaly LV, Schwerdtfeger P, Preuss H (1983) J Phys B 16:L323
36. Stoll H, Metz B, Dolg M (2002) J Comput Chem 23:767
37. Ermler WC, Lee YS, Christiansen PA, Pitzer KS (1981) Chem Phys Lett 81:70
38. Yabushita S, Zhang Z, Pitzer RM (1999) J Phys Chem A 103:5791
39. Grundström B, Valberg P (1938) Z Phys 108:326
40. Urban R-D, Bahnmaier AH, Magg U, Jones H (1989) Chem Phys Lett 158:443
41. Petrov AN, Mosyagin NS, Titov AV, Tupitsyn II (2004) J Phys B 37:4621
42. Isaev TA, Mosyagin NS, Titov AV, Alekseyev AB, Buenker RJ (2002) Int J Quantum Chem 88:687
43. Titov AV, Mosyagin NS (2000) Russ J Phys Chem 74:S376
44. Titov AV, Mosyagin NS (1999) Int J Quantum Chem 71:359
45. Titov AV, Mosyagin NS, Alekseyev AB, Buenker RJ (2001) Int J Quantum Chem 81:409
46. Mosyagin NS, Titov AV, Latajka Z (1997) Int J Quantum Chem 63:1107

47. Metz B, Stoll H, Dolg M (2000) J Chem Phys 113:2563
48. Han Y-K, Bae C, Lee YS (1999) J Chem Phys 110:9353
49. Lim IS, Stoll H, Schwerdtfeger P (2006) J Chem Phys 124:034107
50. Figgen D, Wedig A, Stoll H, Dolg M, Eliav E, Kaldor U (2008) J Chem Phys 128:024106
51. Peterson KA, Figgen D, Dolg M, Stoll H (2007) J Chem Phys 126:124101
52. Metz B, Schweizer M, Stoll H, Dolg M, Liu W (2000) Theor Chem Acc 104:22
53. Huzinaga S (1994) Can J Chem 73:619
54. Bonifacic V, Huzinaga S (1974) J Chem Phys 60:2779
55. Bonifacic V, Huzinaga S (1975) J Chem Phys 62:1507
56. Bonifacic V, Huzinaga S (1975) J Chem Phys 62:1509
57. Bonifacic V, Huzinaga S (1976) J Chem Phys 64:956
58. Bonifacic V, Huzinaga S (1976) J Chem Phys 65:2322
59. Höjer G, Chung J (1978) Int J Quantum Chem 14:1978
60. Zeng T (2010) Development and applications of model core potentials for the studies of spin-orbit effects in chemistry. Ph.D thesis, University of Alberta
61. Krause D, Klobukowski M (1996) Can J Chem 74:1248
62. Lovallo CC, Klobukowski M (2003) J Comput Chem 24:1009
63. Parr RG, Yang W (1994) Density-functional theory of atoms and molecules. Oxford University Press, Oxford
64. Kohn W, Sham LJ (1965) Phys Rev 140:A1133
65. Bachelet GB, Hamann DR, Schlüter M (1982) Phys Rev B 26:4199
66. Kleinman L, Bylander DM (1982) Phys Rev Lett 48:1425
67. Hamann DR (1989) Phys Rev B 40:2980
68. Troullier N, Martins JL (1991) Phys Rev B 43:1993
69. Slater JC (1951) Phys Rev 81:385
70. Sakai Y, Miyoshi E, Klobukowski M, Huzinaga S (1987) J Comput Chem 8:226
71. Sakai Y, Miyoshi E, Klobukowski M, Huzinaga S (1987) J Comput Chem 8:256
72. Sakai Y, Miyoshi E, Klobukowski M, Huzinaga S (1997) J Chem Phys 106:8084
73. Miyoshi E, Sakai Y, Tanaka K, Masamura M (1998) J Mol Struct (Theochem) 451:73
74. Ordejón B, Seijo L, Barandiarán Z (2007) J Chem Phys 126:194712
75. Diaz-Megias S, Seijo L (1999) Chem Phys Lett 299:613
76. Gracia J, Seijo L, Barandiarán Z, Curulla D, Niemansverdriet H, van Gennip W (2007) J Lumin 128:1248
77. Seijo L, Barandiarán Z, Harguindey E (2001) J Chem Phys 114:118
78. Huzinaga S, Seijo L, Barandiarán Z, Klobukowski M (1987) J Chem Phys 86:2132
79. Seijo L, Barandiarán Z (1999) The ab initio model potential method: a common strategy for effective core potential and embedded cluster calculations. In: Leszczynski J (ed) Computational chemistry: reviews of current trends, vol 4. World Scientific, Singapore, p 55
80. Seijo L (1995) J Chem Phys 102:8078
81. Tanner PA, Mak CSK, Edelstein NM, Murdoch KM, Liu G, Huang J, Seijo L, Barandiarán Z (2003) J Am Chem Soc 125:13225
82. Casarrubios M, Seijo L (1998) J Mol Struct (Theochem) 426:59
83. Ruipérez F, Roos BO, Barandiarán Z, Seijo L (2007) Chem Phys Lett 434:1
84. Ordejón B, Karbowiak M, Seijo L, Barandiarán Z (2006) J Chem Phys 125:074511
85. Katsuki S, Huzinaga S (1988) Chem Phys Lett 152:203
86. Brent P (1973) Algorithms for minimization without derivatives. Prentice-Hall, Englewood Cliffs
87. Huzinaga S, Klobukowski M (1988) J Mol Struct (Theochem) 44:1
88. Mane JY, Klobukowski M (2001) J Mol Struct (Theochem) 547:163
89. Mane JY, Klobukowski M (2004) Theor Chem Acc 112:33
90. Lovallo CC, Klobukowski M (2003) Chem Phys Lett 373:439
91. Lovallo CC, Klobukowski M (2004) J Chem Phys 120:246
92. Lovallo CC, Klobukowski M (2003) Chem Phys Lett 368:589
93. Lovallo CC, Klobukowski M (2004) J Comput Chem 25:1206

94. Jayatilaka D, Amos RD, Koga N (1989) Chem Phys Lett 163:151
95. CADPAC: The Cambridge Analytic Derivatives Package Issue 6, Cambridge, 1995. A suite of quantum chemistry programs developed by Amos RD with contributions from Alberts IL, Andrews JS, Colwell SM, Handy NC, Jayatilaka D, Knowles PJ, Kobayashi R, Laidig KE, Laming G, Lee AM, Maslen PE, Murray CW, Rice JE, Simandiras ED, Jones AJ, Su M-D, Tozer DJ.
96. Miyoshi E, Sakai Y, Osanai Y, Noro T (2004) Recent developments and relativistic model core potential method. In: Hirao K, Ishikawa Y (eds) Recent advances in relativistic molecular theory. World Scientific, Singapore, p 37
97. Miyoshi E, Hori H, Hirayama R, Osanai Y, Noro T, Honda H, Klobukowski M (2005) J Chem Phys 122:074104
98. Noro T, Sekiya M, Osanai Y, Miyoshi E, Koga T (2003) J Chem Phys 119:5142
99. Noro T, Sekiya M, Koga T (1997) Theor Chem Acc 98:25
100. Noro T, Sekiya M, Koga T, Matsuyama H (2000) Theor Chem Acc 104:146
101. Sekiya M, Noro T, Osanai Y, Koga T (2001) Theor Chim Acta 106:297
102. Osanai Y, Sekiya M, Noro T, Koga T (2003) Mol Phys 101:65
103. Noro T, Sekiya M, Koga T (2003) Theor Chim Acta 109:85
104. Osanai Y, Noro T, Miyoshi E, Sekiya M, Koga T (2004) J Chem Phys 120:6408
105. Osanai Y, Noro T, Miyoshi E (2002) J Chem Phys 117:9623
106. Anjima H, Tsukamoto S, Mori H, Mine M, Klobukowski M, Miyoshi E (2007) J Comput Chem 28:2424
107. Osanai Y, Mon MS, Noro T, Mori H, Nakashima H, Klobukowski M, Miyoshi E (2008) Chem Phys Lett 452:210
108. Osanai Y, Seijima E, Noro T, Mori H, Ma San M, Klobukowski M, Miyoshi E (2008) Chem Phys Lett 463:230
109. Mori H, Ueno-Noto K, Osanai Y, Noro T, Fujiwara T, Klobukowski M, Miyoshi E (2009) Chem Phys Lett 476:317
110. Sakai Y, Miyoshi E, Tatewaki H (1998) J Mol Struct (Theochem) 451:143
111. Dunning TH (1989) J Chem Phys 90:1007
112. Woon DE, Dunning TH (1993) J Chem Phys 98:1358
113. Woon DE, Dunning TH (1994) J Chem Phys 100:2975
114. Wilson AK, Woon DE, Peterson KA, Dunning TH (1999) J Chem Phys 110:7667
115. Ishikawa T, Mochizuki Y, Nakano T, Amari S, Mori H, Honda H, Fujita T, Tokiwa H, Tanaka S, Komeiji Y, Fukuzawa K, Tanaka K, Miyoshi E (2006) Chem Phys Lett 427:159
116. Desclaux JP, Pyykkö P (1980) Recherche 11:592
117. Pyykkö P (1988) Chem Rev 88:563
118. Pyykkö P, Desclaux JP (1979) Acc Chem Res 12:276
119. Pyykkö P, Desclaux JP (1981) Comput Rend Acd Sci 292:1513
120. Darwin CG (1928) Proc R Soc A 118:654
121. Norrby L (1991) J Chem Educ 68:110
122. Pyykkö P (2004) Angew Chem Int Ed 43:4412
123. Pyykkö P (2005) Inorg Chim Acta 358:4113
124. Pyykkö P (2008) Chem Soc Rev 37:1967
125. Schwerdtfeger P, Heath GA, Dolg M, Bennet MA (1992) J Am Chem Soc 114:7518
126. Visscher L (1996) Chem Phys Lett 253:20
127. Fedorov DG, Gordon MS (2002) Symmetry in spin-orbit coupling. In: Hoffmann MR, Dyall KG (eds) Low-lying potential energy surfaces. ACS symposium series, vol 828. American Chemical Society, Washington, pp 276–297
128. Blume M, Watson RE (1962) Proc R Soc A 270:127
129. Blume M, Watson RE (1963) Proc R Soc A 271:565
130. Blume M, Freeman AJ, Watson RE (1964) Phys Rev 134:A320
131. Richards WG, Trivedi HP, Cooper DL (1981) Spin-orbit coupling in molecules. Clarendon, Oxford

132. Hess BA, Marian CM, Peyerimhoff SD (1995) Ab initio calculation of spin-orbit effects in molecules including electron correlation. In: Yarkony DR (ed) Modern electronic structure theory, vol I. World Scientific, Singapore, p 152
133. Marian CM (2001) Spin-orbit coupling in molecules. In: Lipkowitz KB, Boyd DB (eds) Reviews in computational chemistry, vol 17. WILEY-VCH, New York, p 99
134. Fedorov DG, Koseki S, Schmidt MW, Gordon MS (2003) Int Rev Phys Chem 22:551
135. Marian CM (1997) Fine and hyperfine structure: spin properties of molecules. In: Wilson S, Diercksen GHF (eds) Problem solving in computational molecular science. Kluwer Academic Publishers, Dordrecht/Boston, p 291
136. Juncar P, Pinard J, Hamon J, Chartier A (1981) Metrolagia 17:77
137. Kutateladze AG (2001) J Am Chem Soc 123:9279
138. Bernath PF (1995) Spectra of atoms and molecules. Oxford University Press, New York
139. Dirac PAM (1928) Proc R Soc A 117:610
140. Dirac PAM (1928) Proc R Soc A 118:351
141. Breit G (1929) Phys Rev 34:553
142. Douglas M, Kroll NM (1974) Ann Phys 82:89
143. Cowan RD, Griffin DC (1976) J Opt Soc Am 66:1010
144. Wood JH, Boring AM (1978) Phys Rev B 18:2701
145. van Lenthe E, Baerends EJ, Snijders JG (1993) J Chem Phys 99:4597
146. van Lenthe E, Baerends EJ, Snijders JG (1994) J Chem Phys 101:9783
147. van Lenthe E, Ehlers AE, Baerends EJ (1999) J Chem Phys 110:8943
148. Dyall KG, van Lenthe E (1999) J Chem Phys 111:1366
149. Sadlej AJ (2005) Collect Czech Chem Commun 70:677
150. Nakajima T, Hirao K (1999) Chem Phys Lett 302:383
151. Dyall KG (2002) J Comput Chem 23:786
152. Barysz M, Sadlej AJ (2001) J Mol Struct (Theochem) 573:181
153. Barysz M, Sadlej AJ (2002) J Chem Phys 116:2696
154. Moss RE (1973) Advanced molecular quantum mechanics. Chapman and Hall, London
155. Balasubramanian K (1997) Relativistic effects in chemistry part A. Wiley, New York
156. Reiher M, Wolf A (2009) Relativistic quantum chemistry, the fundamental theory of molecular science. Wiley-VCH, Weinheim
157. Klobukowski M (1983) J Comput Chem 4:350
158. Sakai Y, Huzinaga S (1982) J Chem Phys 76:2537
159. Froese-Fischer C (1977) The Hartree-Fock method for atoms: a numerical approach. Wiley-VCH, New York
160. Huzinaga S, Klobukowski M, Sakai Y (1984) J Phys Chem 88:4880
161. Miyoshi E, Sakai Y, Mori S (1985) Chem Phys Lett 113:457
162. Miyoshi E, Sakai Y, Mori S (1985) Surf Sci 158:667
163. Andzelm J, Radzio E, Salahub DR (1985) J Chem Phys 83:4573
164. Andzelm J, Huzinaga S, Klobukowski M, Radzio E (1985) Chem Phys 100:1
165. Sakai Y, Miyoshi E (1987) J Chem Phys 87:2885
166. Miyoshi E, Sakai Y (1988) J Comput Chem 9:719
167. Musolino V, Toscano M, Russo N (1990) J Comput Chem 11:924
168. Seijo L, Barandiarán Z, Huzinaga S (1989) J Chem Phys 91:7011
169. Barandiarán Z, Seijo L, Huzinaga S (1990) J Chem Phys 93:5843
170. Barysz M (2000) J Chem Phys 113:4003
171. Wolf A, Reiher M, Hess BA (2002) J Chem Phys 117:9215
172. Brummelhuis R, Siedentop H, Stockmeyer E (2002) Doc Math 7:167
173. Autschbach J, Siekierski S, Seth M, Schwerdtfeger P, Schwarz WHE (2002) J Comput Chem 23:804
174. Schwerdtfeger P, Brown JR, Laerdahl JK (2000) J Chem Phys 113:7110
175. Zeng T, Fedorov DG, Klobukowski M (2009) J Chem Phys 131:124109
176. Ermler WC, Ross RB, Christiansen PA (1988) Adv Quantum Chem 19:139
177. Fedorov DG, Klobukowski M (2002) Chem Phys Lett 360:223
178. Zhang Y, Gao T, Zhang C (2007) Mol Phys 105:405

179. Song C, Gao T, Han H, Wan M, Yu Y (2008) J Mol Struct (Theochem) 870:65
180. Zeng T, Fedorov DG, Klobukowski M (2010) J Chem Phys 132:074102
181. Nakano H (1993) J Chem Phys 99:7983
182. Nakano H (1993) Chem Phys Lett 207:372
183. Zeng T, Fedorov DG, Klobukowski M (2010) J Chem Phys 133:114107
184. Shavitt I, Bartlett RJ (2009) Many-body methods in chemistry and physics: MBPT and coupled-cluster theory. Cambridge University Press, Cambridge, MA/New York
185. Peterson KA, Figgen D, Goll E, Stoll H, Dolg M (2003) J Chem Phys 119:11113
186. Peterson KA (2003) J Chem Phys 119:11099
187. Fedorov DG, Schmidt MW, Koseki S, Gordon MS (2004) Spin-orbit coupling methods and applications to chemistry. In: Hirao K, Ishikawa Y (eds) Recent advances in relativistic molecular theory. World Scientific, Singapore
188. Alekseyev AB, Liebermann H-P, Buenker RJ (2004) Spin-orbit multireference configuration interaction method and applications to systems containing heavy atoms. In: Hirao K, Ishikawa Y (eds) Recent advances in relativistic molecular theory. World Scientific, Singapore
189. Koseki S, Shimakura N, Fujimura Y, Asada T, Kono H (2009) J Chem Phys 131:044122
190. Deskevich MP, Nesbitt DJ, Werner H-J (2004) J Chem Phys 120:7281
191. Zeng T, Fedorov DG, Klobukowski M (2011) J Chem Phys 134:024108
192. Lee H-S, Han YK, Kim MC, Bae C, Lee YS (1998) Chem Phys Lett 293:97
193. Kim MC, Lee SY, Lee YS (1996) Chem Phys Lett 253:216
194. Rösch N, Krüger S, Mayer M, Nasluzov VA (1996) The Douglas-Kroll-Hess approach to relativistic density functional theory: methodological aspects and applications to metal complexes and clusters. In: Seminario JM (ed) Recent developments and applications of modern density functional theory. Elservier, Amsterdam
195. Nakajima T, Suzumura T, Hirao K (1999) Chem Phys Lett 304:271
196. Armbruster MK, Weigend F, van Wüllen C, Klopper W (2008) Phys Chem Chem Phys 10:1748
197. Zeng T, Fedorov DG, Schmidt MW, Klobukowski M (2011) J Chem Phys 134:214107
198. Bersuker IB (2006) The Jahn-Teller effect. Cambridge University Press, Cambridge, UK
199. Bersuker IB (2008) The Jahn-Teller effect and beyond. The Academy of Sciences of Moldova, Moldova; The University of Texas at Austin, Austin, TX
200. Schmidt MW, Baldridge KK, Boatz JA, Elbert ST, Gordon MS, Jensen JH, Koseki S, Matsunaga N, Nguyen KA, Su S, Windus TL, Dupuis M, Montgomery JA (1993) J Comput Chem 14:1347
201. Gordon MS, Schmidt MW (2005) Advances in electronic structure theory: GAMESS a decade later. In: Dykstra CE, Frenking G, Kim KS, Scuseria GE (eds) Theory and applications of computational chemistry: the first forty years. Elservier, Amsterdam/Boston
202. Karlström G, Lindh R, Malmqvist PA, Roos BO, Ryde U, Veryazov V, Widmark PO, Cossi M, Schmmelpfennig B, Neogrady P, Seijo L (2003) Comput Mat Sci 28:222
203. Nakano T, Kaminuma T, Sato T, Fukuzawa K, Akiyama Y, Uebayasi M, Kitaura K (2002) Chem Phys Lett 351:475
204. Koster AM, Calaminici P, Casida ME, Dominguez VD, Flores-Moreno R, Guedtner G, Goursort A, Heine T, Ipatov A, Janetzko F, del Campo JM, Reveles JU, Vela A, Zuniga B, Salahub DR (2006) deMon2k, Version 2, The deMon developers, Cinvestav, Mexico City
205. Pereiro M, Baldomir D, Iglesias M, Rosales C, Castro M (2001) Int J Quantum Chem 81:422
206. Lacaze-Dufour C, Mineva R, Russo N (2001) Int J Quantum Chem 85:162
207. Lacaze-Dufour C, Mineva R, Russo N (2001) J Comput Chem 22:1557
208. Decker SA, Klobukowski M (2001) J Chem Inf Comput Sci 41:1
209. Sakai Y, Mogi K, Miyoshi E (1999) J Chem Phys 111:3989
210. Zeng T, Mori H, Miyoshi E, Klobukowski M (2009) Int J Quantum Chem 109:3235
211. Møller C, Plesset MS (1934) Phys Rev A 46:618
212. Head-Gordon M, Pople JA, Frisch MJ (1988) Chem Phys Lett 153:503
213. Frisch MJ, Head-Gordon M, Pople JA (1990) Chem Phys Lett 166:275
214. Lee TJ, Jayatilaka D (1993) Chem Phys Lett 201:1

215. Head-Gordon M, Head-Gordon T (1994) Chem Phys Lett 220:122
216. Sæbø S, Almlöf J (1989) Chem Phys Lett 154:83
217. Piecuch P, Kucharski SA, Kowalski K, Musial M (2002) Comp Phys Commun 149:71
218. Becke AD (1988) Phys Rev A 38:3098
219. Perdew JP (1986) Phys Rev B 33:8822
220. Perdew JP, Burke K, Ernzerhof M (1996) Phys Rev Lett 77:3865
221. Adamo C, Barone V (1999) J Chem Phys 110:6158
222. Helgaker T, Gauss J, Jørgensen P, Olsen J (1997) J Chem Phys 106:6430
223. Helgaker T, Jørgensen P, Olsen J (2000) Molecular electronic-structure theory. Wiley, Chichester
224. Honda H, Noro T, Tanaka K, Miyoshi E (2001) J Chem Phys 114:10791
225. Lengsfield BH III (1980) J Chem Phys 73:382
226. Liu B, Yoshimine M (1981) J Chem Phys 74:612
227. Lengsfield BH III, Liu B (1981) J Chem Phys 75:478
228. Sakai Y, Nakai T, Mogi K, Miyoshi E (2003) Mol Phys 101:117
229. Tanaka K, Sekiya M, Tawada Y, Miyoshi E (2005) J Chem Phys 122:214315
230. Ishikawa T, Tanaka K (2004) Chem Phys Lett 395:166
231. Mon MS, Mori H, Miyoshi E (2008) Chem Phys Lett 462:23
232. Lovallo CC, Klobukowski M (2002) Int J Quantum Chem 90:1099
233. Boys SF, Bernardi F (1970) Mol Phys 19:553
234. Evans CJ, Lesarri A, Gerry MCL (2000) J Am Chem Soc 122:6100
235. Evans CJ, Gerry MCL (2000) J Chem Phys 112:9363
236. Evans CJ, Rubinoff DS, Gerry MCL (2000) Chem Phys Phys Chem 2:3943
237. Reynard LM, Evans CJ, Gerry MCL (2001) J Mol Spectrosc 206:33
238. Cooke SA, Gerry MCL (2004) J Am Chem Soc 123:17000
239. Ancilotto F, Lerner PB, Cole MW (1995) J Low Temp Phys 101:1123
240. Sakai Y, Koyanagi M, Mogi K, Miyoshi E (2002) Surf Sci 513:272
241. Hay PJ, Wadt WR (1985) J Chem Phys 82:299
242. Miyoshi E, Mori H, Tanaka S, Sakai Y (2002) Surf Sci 514:383
243. Miyoshi E, Iura T, Sakai Y, Touchihara H, Tanaka S, Mori H (2003) J Mol Struct (Theochem) 630:225
244. Zeng T, Jamshidi Z, Mori H, Miyoshi E, Klobukowski M (2007) J Comput Chem 28:2027
245. Gajewski M, Tuszynski J, Mori H, Miyoshi E, Klobukowski M (2008) Inorg Chim Acta 361:2166
246. Gajewski M, Klobukowski M (2009) Can J Chem 87:1492
247. Zeng T, Klobukowski M (2008) J Phys Chem A 112:5236
248. Piecuch P, Kucharski SA, Kowalski K, Musial M (2000) J Chem Phys 113:18
249. Piecuch P, Kucharski SA, Kowalski K, Musial M (2000) J Chem Phys 113:5644
250. Piecuch P, Wloch M (2005) J Chem Phys 123:224105
251. Wloch M, Gour JR, Piecuch P (2007) J Phys Chem A 111:11359
252. Tsukamoto S, Mori H, Tatewaki H, Miyoshi E (2009) Chem Phys Lett 474:28
253. Fujiwara T, Mori H, Mochizuki Y, Tatewaki H, Miyoshi E (2010) J Mol Struct (Theochem) 949:28
254. Kitaura K, Ikeo E, Asada T, Nakano T, Uebayasi M (1999) Chem Phys Lett 313:701
255. Mochizuki Y, Nakano T, Koikegami S, Tanimori S, Abe Y, Nagashima U, Kitaura K (2004) Theor Chem Acc 112:442
256. Mochizuki Y, Koikegami S, Nakano T, Amari S, Kitaura K (2004) Chem Phys Lett 396:473
257. Fedorov DG, Olson RM, Kitaura K, Gordon MS, Koseki S (2004) J Comput Chem 25:872
258. Ishikawa T, Mochizuki Y, Imamura K, Tokiwa H, Nakano T, Mori H, Tanaka K, Miyoshi E, Tanaka S (2006) Chem Phys Lett 430:361
259. Barnett NJ, Slipchenko LV, Gordon MS (2009) J Phys Chem A 113:7474
260. Hanson K, Roskop L, Djurovich PI, Zahariev F Gordon MS, Thompson ME (2010) J Am Chem Soc 132:16247
261. Haiges R, Boatz JA, Christe KO (2010) Angew Chem Int Ed 49:8008

262. Fitzsimmons A, Mori H, Miyoshi E, Klobukowski M (2010) J Phys Chem A 114:8786
263. Shim J, Klobukowski M, Barysz M, Leszczynski J (2011) Phys Chem Chem Phys 13:5703
264. Rakowitz F, Marian CM, Seijo L, Wahlgren U (1999) J Chem Phys 110:3678
265. Rakowitz F, Marian CM, Seijo L (1999) J Chem Phys 111:10436
266. Sansonetti JE, Martin WC (2005) Handbook of basic atomic spectroscopic data. J Phys Chem
     Ref Data 34(4):1559–2259
267. Baig MA, Connerade JP (1985) J Phys B 18:1101
268. Hotop H, Lineberger WC (1985) J Phys Chem Ref Data 14:731
269. Huber KP, Herzberg G (1979) Molecular spectra and molecular structure constants of
     diatomic molecules, vol 4. van Nostrand, New York
270. Pitzer KS (1984) Int J Quantum Chem 25:131
271. Zeng T, Fedorov DG, Schmidt MW, Klobukowski M (2011) J Chem Theory Comput 7:2864

Chapter 9
Practical Aspects of Quantum Monte Carlo for the Electronic Structure of Molecules

Dmitry Yu. Zubarev, Brian M. Austin, and William A. Lester Jr.

Abstract A family of quantum chemical methods under the common name "quantum Monte Carlo" are reviewed. QMC is one of the most powerful theoretical frameworks that can be applied to the problems of electronic structure theory. Theoretical and practical aspects of variational Monte Carlo and diffusion Monte Carlo approaches are discussed in detail due to their computational feasibility, robustness and quality of results. Several factors that contribute to the systematic improvement of the accuracy of QMC calculations are considered, including selection of accurate and flexible forms of the trial wave function and strategies for optimizing parameters within these trial functions. We also provide an analysis of the scaling properties that govern the growth of computational expense of QMC simulations for molecules of increasing size. Linear scaling algorithms for QMC are emphasized. Consideration of parallel efficiency is especially important in the view of proliferation of high-performance computing facilities, where the potential of QMC can be used to its fullest.

Keywords Electronic structure theory • ab initio quantum chemistry • Many-body methods • Quantum Monte Carlo • Fixed-node diffusion Monte Carlo • Variational Monte Carlo • Electron correlation • Massively parallel • Linear scaling

D.Y. Zubarev • B.M. Austin • W.A. Lester Jr.
Kenneth S. Pitzer Center for Theoretical Chemistry, Department of Chemistry,
University of California, Berkeley, CA 94720-1460, USA
e-mail: dmitry.zubarev@berkeley.edu

B.M. Austin
National Energy Research Scientific Computing, Lawrence Berkeley National Laboratory,
Berkeley, CA 94720, USA
e-mail: baustin@lbl.gov

W.A. Lester Jr. (✉)
Chemical Sciences Division, Lawrence Berkeley National Laboratory, Berkeley, CA 94720, USA
e-mail: walester@lbl.gov

J. Leszczynski and M.K. Shukla (eds.), *Practical Aspects of Computational Chemistry I:*
An Overview of the Last Two Decades and Current Trends,
DOI 10.1007/978-94-007-0919-5_9, © Springer Science+Business Media B.V. 2012

9.1   Introduction

Quantum Monte Carlo (QMC) is a collective name of a group of stochastic approaches to the treatment of problems in quantum mechanics [1]. MC methods are used for multiple purposes in QMC, most significantly: integration, simulation, and solution of an integral equation. QMC methods are different from expansion and perturbative approaches such as Hartree-Fock (HF), configuration interaction (CI), many-body perturbation theory (MBPT), coupled cluster (CC) theory, and from density functional theory (DFT) that dominate the field of ab initio quantum chemistry. QMC can, in principle, provide exact solutions to the Schrodinger equation, including complete treatment of electron correlation (both static and dynamic correlation effects). Basis set truncation and basis set superposition errors are absent from the diffusion MC (DMC) variant of QMC. The method is intrinsically parallel enabling QMC codes to take advantage of the largest supercomputers with little modification. The compute time for each step of a QMC calculation increases roughly as N^3, where N is a measure of system size (i.e., the number of particles of the system). Recent developments have achieved linear scaling in certain cases. Computer memory requirements are small and grow modestly with system size. Computer codes for QMC are significantly smaller than those of basis set molecular quantum mechanics computer programs. Any MC calculation can be improved to achieve whatever degree of precision is needed. The random error of QMC calculations decreases in proportion to the square root of the computational time. This enables one to estimate the compute time needed for performing a calculation with a prescribed error bar.

The first description of QMC is attributed to E. Fermi in a classic paper by Metropolis and Ulam [2]. Some years later, Kalos [3] proposed Green's Function QMC which was applied to the calculation of the ground state of three- and four-body nuclei. In retrospect this paper was indicative of the versatility of QMC for studies of nuclear, condensed matter as well as atomic and molecular systems. It remained for Anderson [4] to make the initial significant foray into atomic and molecular systems in the mid-seventies. These classic papers introduced the fixed-node approximation and served as a bellwether for the DMC described below.

The focus of this review is the rigorous solution of the electronic Schrödinger equation using QMC for atoms and molecules. The QMC methods described here are readily and regularly applied to vibrational and solid state electronic structure calculations as well. Earlier reviews [1, 5–15] complement the coverage provided in this contribution. Several computational physics texts also contain chapters about QMC [16–18]. Here, central ideas of QMC theory pertinent to zero-temperature methods including variational MC (VMC), DMC and some other are described. The related energy path integral MC (PIMC) approach will not be discussed here; see, for example [19, 20], The most common strategies of construction and improvement of trial wave functions are presented in detail, as these aspects of QMC are critical to systematically improving QMC results. Scaling properties of QMC are also discussed with emphasis on methods of reduction of computational cost.

9.2   Quantum Monte Carlo Approaches

In this chapter we provide the theoretical background for several established and emerging QMC methods. The applicability of general MC methods is diverse and spans classical statistical mechanics simulations, operations research, and applied statistical analysis. Details of common MC techniques such as importance sampling, correlated sampling, and MC optimization are available from multiple sources [21–31]. Here, the technical description of variational and diffusion MC approaches is presented in more detail due to their high popularity. Our discussion of alternative QMC techniques focuses on conceptual foundations and related advantages and challenges compared to DMC.

9.2.1   Variational Monte Carlo (VMC)

The VMC method uses Metropolis MC [32] to compute molecular properties by MC integration using a known trial wave function Ψ_T. One can estimate the energy of a trial wave function by MC integration and obtain a more accurate wave function by varying the parameters of the trial function to minimize the energy estimate. The following discussion shows how the MC integration techniques may be adapted to the evaluation of expectation values and matrix elements, and describes methods and criteria for optimizing wave functions.

The variational energy of an arbitrary wave function is given by the expectation value of the Hamiltonian $E = \langle \Psi | \hat{H} | \Psi \rangle / \langle \Psi / \Psi \rangle$. To evaluate the energy by VMC, the Hamiltonian and overlap integrals that appear in the expectation value must be rewritten in the form of Eq. 9.1.

$$\int f(X)\rho(X)dX = \lim_{K \to \infty} \frac{1}{K} \sum_{k:X_k \in \rho}^{K} f(X_k) \qquad (9.1)$$

where k enumerates points sampled from the probability density function (PDF) $\rho(X)$ of a D-dimensional random variable X. A natural choice for the PDF corresponds to the electron density $\rho(X) = \Psi^2(X)$. This density may be sampled using the Metropolis method. The energy is then computed using

$$E = \int \Psi^2(X) \frac{\hat{H} \Psi(X)}{\Psi(X)} dX = \frac{1}{K} \sum_{k:X_k \in \Psi^2}^{K} E_L(X_k) \qquad (9.2)$$

where the 'local energy' of Ψ at X is defined by $E_L(X) \equiv H \Psi(X)/\Psi(X)$.

The simple form of Eq. 9.2 is due to the selection of Ψ^2 as the PDF. The MC estimate of the denominator in the Hamiltonian expectation value is exactly one and has zero variance. Because Ψ_T is an approximate eigenfunction of the Hamiltonian,

$H\Psi_T \approx E\Psi_T$, the variance of the local energy approaches zero as Ψ_T approaches the exact eigenfunction. The relatively small variance of the local energy allows VMC to estimate energies much more efficiently than other properties whose operators do not commute with the Hamiltonian. VMC can also be used to evaluate more general matrix elements of the form $A_{ij} = \langle \Psi_i | \hat{A} | \Psi_j \rangle$. This integral may be rewritten in the following form.

$$N[\rho]\langle \Psi_i | \hat{A} | \Psi_j \rangle = N[\rho] \int \rho(X) \frac{\Psi_i(X) \hat{A}\Psi_j(X)}{\rho(X)}$$

$$d X = \frac{1}{K} \sum_{k:X_k \in \rho}^{k} \frac{\Psi_i(X_k)\Psi_j(X_k)}{\rho(X_k)} \frac{\hat{A}\Psi_j(X_k)}{\Psi_j(X_k)} \tag{9.3}$$

Walkers are sampled from ρ and the matrix element is evaluated using the average indicated in Eq. 9.3. The factor $N[\rho] \equiv 1/\int \rho$ accounts for the unknown normalization of ρ. Several new complications arise when computing off-diagonal matrix elements. First, there is no obvious choice for ρ when computing off-diagonal matrix elements. Averages of Ψ_i^2 and Ψ_j^2 are sometimes used [33] so that the densities of both wave functions are sampled, but this choice is somewhat arbitrary. Second, there is little reason to expect the summand of Eq. 9.3 to have a particularly small variance, so accurate evaluations of matrix elements are more computationally demanding than energies.

9.2.2 Diffusion Monte Carlo (DMC)

Although considerable research focuses on the development of high quality trial functions, the accuracy of the VMC method will always be limited by the flexibility and form of the approximate trial wave function. Highly accurate solutions to the Schrodinger equation can be computed by DMC which is rooted in the time-dependent Schrodinger equation and its solutions.

$$\frac{d}{dt}\Phi(X,t) = -i H \Phi(X,t) \tag{9.4}$$

$$\Phi(X,t) = e^{-itH}\Phi(X)$$

By expanding the wave function $\Phi(X, t)$ in eigenfunctions of H and shifting the potential by the ground state energy E_0, one obtains solutions that converge to the ground state when the 'imaginary time', $\tau = it$, becomes large:

$$\Phi(X,\tau) = \sum_j c_j e^{-\tau(E_j - E_0)} \Psi_j(X) \tag{9.5}$$

$$\lim_{\tau \to \infty} \Phi(X, \tau) = c_0 \Psi_0(X)$$

The DMC method extracts the ground state wave function by propagating the Schrodinger equation in imaginary time. This is accomplished by exploiting an isomorphism between the kinetic energy term in the Hamiltonian and a classical diffusion equation; the latter can be simulated by a random walk. There is the second isomorphism between the potential energy and a spatially inhomogeneous first-order rate equation that can be simulated by a branching process.

$$\frac{d}{d\tau}\Phi(X, \tau) = ((1/2\nabla^2)_{Diffusion} - (V(X) - E_0)_{Branching})\Phi(X, \tau) \qquad (9.6)$$

The connection between diffusion and random walks was given by Einstein in a study of Brownian motion [34]. Suppose a collection of points is initially sampled from $\rho(X, t=0)$ and that after a short time, each point takes an independent random step sampled from a three dimensional Gaussian distribution:

$$g(\eta, t) = [4\pi Dt]^{-3/2} e^{-\eta^2/4Dt} \qquad (9.7)$$

which is symmetric $g(\eta, t) = g(-\eta, t)$ and normalized $\int g(\eta) d\eta = 1$. The updated density function will then be

$$\rho(X, t) = \int \rho(X - \eta, t) g(\eta, t) d\eta \qquad (9.8)$$

Expanding the left hand side in terms of t and the right hand side in terms of η, the preceding equation becomes

$$\rho(X, t) = \rho(X, 0) + \frac{\partial \rho}{\partial t}|_{t=0} t + \dots = \int \left[\rho(X, t) + \sum_i \eta_i \frac{\partial \rho}{\partial x_i} \right.$$
$$\left. + \sum_{i,j} \frac{1}{2} \eta_i \eta_j \frac{\partial^2 \rho}{\partial x_i \partial x_j} + \dots \right] g(\eta, t) d\eta \qquad (9.9)$$

Due to the symmetry of $g(\eta)$, only even derivatives on the right hand side are nonzero. For small t, the dominant terms in this equation are

$$\frac{\partial \rho}{\partial t} = \sum_i \frac{\partial^2 \rho}{\partial x_i^2} \frac{1}{t} \int \frac{1}{2} \eta^2 g(\eta, t) d\eta = D \nabla^2 \rho \qquad (9.10)$$

which can be recognized as a classical diffusion equation.

Neglecting the diffusion term in Eq. 9.6 leads to an equation that is easily solved,

$$\frac{d}{d}\tau \Phi(X, \tau) = -(V(X) - E_0)\Phi(X, \tau) \qquad (9.11)$$

$$\Phi(X,\tau) = e^{-(V(X)-E_0)\tau} \, \Phi(X,\tau)$$

Branching can be incorporated into the MC algorithm by evaluating the potential at each point visited and associating a weight, w_k,

$$w_k = e^{[-(V(X_k)-E_0)\Delta\tau]} \tag{9.12}$$

with each walker k. The weights of a collection of walkers tend to diverge, so only a small number of walkers contribute to any MC average [35]. For this reason, it is computationally advantageous to replace weights by a stochastic birth/death process in which walkers with weights greater than 1 create offspring at the same position with a probability of $w_k - 1$ and walkers with weights less than 1 will die with probability $1 - w_k$.

The random walk and branching processes provide exact solutions to their respective differential equations, so it is tempting to combine the two by advancing one step according to the diffusion equation, then another by the branching equation which DMC achieves. The Trotter-Suzuki formula (Eq. 9.13) shows that even if the alternating steps are performed symmetrically, separating the diffusion and branching process creates an $O(\tau^3)$ error.

$$e^{-(T+V)\tau} = e^{[-1/2V\tau]} \, e^{[-T\tau]} \, e^{[-1/2V\tau]} + O(\tau^3) \tag{9.13}$$

The large-τ limit is instead reached by iterative application of short time solutions:

$$\Phi(X,\tau) = e^{-(T+V)\tau}$$

$$\Phi(X,\tau) = \lim_{n\to\infty} \prod_n e^{-(T+V)\frac{t}{n}}$$

$$\Phi(X,\tau) = \lim_{n\to\infty} \prod_n e^{-\frac{V}{2}\frac{t}{n}} \, e^{-T\frac{t}{n}} \, e^{-\frac{V}{2}\frac{t}{n}} \tag{9.14}$$

A series of DMC calculations with progressively smaller time-steps may then be used to extrapolate to the zero-time-step result that removes error associated with the use of a finite time-step which is referred to as time-step bias. The requirement of a short time-step makes DMC calculations significantly more costly than VMC and several modifications to the random walk algorithm have been introduced to minimize time-step errors [36, 37].

9.2.3 Fixed-Node DMC (FN-DMC)

Typically, DMC calculations use a trial function and sample the mixed density $\rho = \Phi\Psi_T$ instead of $|\Phi|$ or Φ^2. The mixed density distribution evolves according to a transformation of the time dependent Schrodinger equation that can be derived by multiplying Eq. 9.6 by Ψ_T:

$$\frac{d}{d\tau}\rho(X,\tau) = \left(\frac{1}{2}\nabla^2\rho\right)_{Diffusion} - \frac{1}{2}\nabla\left(\frac{2\nabla\Psi_T(X)}{\Psi_T(X)}\rho\right)_{Drift}$$
$$- ((E_L(X) - E_0)\rho)_{Branching} \tag{9.15}$$

The introduction of the trial function also modifies the branching term so that the branching rate depends on the local energy $E_L(X)$ instead of the potential $V(X)$. This leads to a suppression of the branching process because the local energy of a good trial function is nearly constant and is generally a good approximation to the exact energy E_0. In particular, uncontrolled branching at points where the Coulomb potential is singular can be curtailed by choosing a trial function that satisfies cusp conditions. As the variance of the local energy decreases, the branching factor $e^{-(E_L(X)-E_0)\Delta\tau}$ approaches a constant value. In this limit, the diffusion and drift operators commute. Time step errors can therefore be reduced by improved trial functions and, of course, eliminated by the exact one.

The trial function can also be used to define the fixed-node approximation (FNA), which constrains the DMC solution to have the nodes of Ψ_T. The FNA is enforced during the random walk by forbidding a walker from taking a step that generates a sign change. Without the fixed-node constraint, the DMC solution would converge to the ground state without symmetry restriction which is a Bosonic state. In the FNA, the stationary states of Eq. 9.6 are exact solutions of the Schrodinger equation subject to the fixed-node boundary condition. The nodes of the trial function exclusively determine the accuracy of FN-DMC. Away from the nodes, where the trial function is nonzero, the variance of the local energy determines the efficiency of the calculation. More specifically, when the trial function has the Slater-Jastrow form: $\Psi_T = \Psi_A e^F$, the nodes are given solely by the antisymmetric Slater part Ψ_A because the Jastrow function F is exponentiated. For a given Ψ_A, the FN-DMC energy will be the same for all forms of F. The correlation function e^F is important for reducing the variance of the local energy and the number of points needed to reduce the statistical error to an acceptable level.

The nodes of the exact wave function are known for only a few simple systems, so the fixed-node constraint is an approximation. Unlike time step error, it is not possible to correct for fixed-node error by extrapolation. The error introduced by FNA is the only uncontrolled factor in the FN-DMC method. Fortunately, the FNA is found to perform well, even when modest trial functions are used. Errors due to the FNA are typically less than 3 kcal/mol, even when simple single-determinant trial functions are used [38].

The centrality of the FNA has spawned considerable research into improvement of the approach. The strategies for obtaining better nodes are numerous. Canonical HF orbitals, Kohn-Sham orbitals from density functional theory (DFT), and natural orbitals from post-HF methods have been used. The latter do not necessarily yield better nodes than single configuration wave functions [39–41]. More success has been found with alternative wave function forms that include correlation more directly than sums of Slater determinants. These include antisymmetrized geminal power functions [42, 43], valence-bond [44, 45] and Pfaffian [46] forms as well as backflow transformed determinants [47]. Improvements to nodes have been made by optimization of Ψ_A in the presence of a correlation function [48–50].

A different approach to the node problem is to develop a mathematical or physical understanding of the differences between good and bad nodal surfaces so that trial functions with better nodal surfaces can be constructed. Relatively little is known about wave function nodes [40, 51, 52]. One often mentioned exception is the tiling theorem [53]: the nodal pockets of the Fermi ground state are the same within permutational symmetry. Plotting and comparing nodes may provide physical insight into some features of the exact nodes [54, 55]. Plots of a Pfaffian wave function have revealed tunnels between nodal pockets that are not present in HF wave functions [46], but the high dimensionality and complicated structure of the nodes make it difficult, in general, to translate these observations into an improved ansatz.

A few alternatives to the FN constraint are available within QMC. The released node method starts from a FN-DMC calculation [56, 57]. When the FN constraint is relaxed (released node QMC) an estimate of the exact ground state energy may be obtained by incorporating a factor of -1 for each walker that crosses the nodal surface:

$$E = \frac{\sum_k sgn\left(\Psi_T\left(X_k(\tau)\right)\Psi_T\left(X_k(0)\right)\right) w_k E_L(X_k)}{\sum_k sgn\left(\Psi_T\left(X_k(\tau)\right)\Psi_T\left(X_k(0)\right)\right)} \qquad (9.16)$$

As the number of walkers that cross nodes increases, so does the cancellation between positive and negative contributions to the averages, leading to rapid growth of the variance as the denominator approaches zero. Furthermore, the rate of this degeneration increases with the difference in energy between the Fermi and Bose ground states, which may be unsuitably fast for molecules of substantial size.

Other alternatives to the FNA include the A-function method, [58–60]. The method builds nodal surfaces from a sum of Gaussian functions centered on the DMC walkers. There is also lattice-regularized DMC (LR-DMC) in which the kinetic energy is discretized by a finite difference Laplacian with two meshes. The use of a regularized Hamiltonian also permits a consistent variational treatment of non-local potentials [61].

Green's Function Monte Carlo (GFMC) [3, 26, 62] relies on the standard resolvent operator of Schroedinger equation

$$\hat{P}(\hat{H}) \equiv \frac{1}{1 + \delta\tau(\hat{H} - E_0)} \qquad (9.17)$$

instead of the imaginary-time evolution operator

$$\hat{P}(\hat{H}) \equiv e^{-\delta t(\hat{H} - E_0)} \qquad (9.18)$$

GFMC does not have time-step error but its requirements for computational time are higher than DMC. This makes GFMC a less common choice for treatment of molecular or atomic electronic structure problem.

Another method is Fermion Monte Carlo (FMC) [63] for which nodes of the trial function play no special role. In the approach, antisymmetry is maintained by careful cancellation of positive and negative walkers. The FMC methods are under development.

9.2.4   Self-Healing DMC (SH-DMC)

As was mentioned, improvement of DMC energies is possible only by improvement of the nodal surface. The quality of VMC optimization is limited by the feature that VMC sampling occurs far from the nodes. SH-DMC is a route to direct optimization of nodal surfaces [64–66] as by locally removing kinks in the FN ground-state wave function. These kinks are manifestations of discontinuities in the gradient at the nodal surface of the trial wave function. Node improvement in SH-DMC relies on the ability to obtain an antisymmetric FN ground-state wave function from a DMC ensemble of configurations after convolution with a smoothing function. SH-DMC is applicable to both ground [65] and excited [66] states. It was shown to converge systematically to a high-accuracy CI solution for the ground state of oxygen atom and yielded a binding energy of N_2 to near chemical accuracy. The computational cost of the SH-DMC approach scales linearly with the number of degrees of freedom of the nodes and molecules as large as C_{20} have been studied [64]. Its accuracy depends only on the size of the available statistics and the flexibility of the form of trial wave function.

9.2.5   Auxiliary Field QMC (AF-QMC)

The FN constraint is avoided entirely in the AF-QMC method [67–69] where random walks are performed in the space of Slater determinants. The sign problem manifests differently in AF-QMC and analogous constraints must be used. Similarly to other QMC methods, AF-QMC uses an imaginary-time propagator to project out contributions of excited states in the trial wave function to obtain the ground state solution. Unlike FN-DMC, the random walk occurs in a space of non-orthogonal Slater determinants rather than in the position space of n-particle configurations. The many-body problem is mapped onto a linear combination of one-body problems via the Hubbard-Stratonovich transformation leading to many-body effects being accounted for through interactions among one-body operators and external auxiliary fields [70]. This can be done using any single-particle basis set, including plane waves and Gaussian-type functions [70, 71]. Because each walker in a QMC ensemble is a Slater determinant whose orbitals evolve during the random walk, antisymmetry of each walker is achieved without special treatment. A severe problem emerges, however, in the treatment of two-body interactions because, in general, they involve complex one-body operators and for large projection times the phase of each orbital becomes random leading to the AF-QMC wave function being dominated by noise. This phase problem is analogous to the fermion sign problem. The phaseless AF-QMC approach [72, 73] addresses this issue by confining the random walk based on overlap with a trial wave function. The ground-state energy in phaseless AF-QMC is approximate and non-variational. In addition, there is pronounced sensitivity of the method to basis set truncation. The AF-QMC method has been successful in applications to the energetics of selected atomic, molecular, and hydrogen-bounded systems [70–73].

9.2.6   Reptation QMC (RQMC)

The RQMC approach is directed at the calculation of observables that do not commute with Hamiltonian. In the framework of DMC this task is challenging due to bias introduced by trial wave function when mixed estimators are used. Elimination of such bias, along with effects of walker population control, is a source of decreased variance [74]. The RQMC method maps dynamical properties of a classical diffusion process onto the imaginary-time dynamical properties of the corresponding quantum system. It can be seen as a modification of a pure diffusion QMC method [75]. It samples the product of a joint probability distribution for a random-walk path and a Boltzman factor involving the discretized integral of the local energy along the path. A random-walk path, or a "reptile", is a basic variable of RQMC. Modification of "reptiles", or "reptation", is performed based on the time-discretized Langevin equation. The transition probability of a reptation step is used in the Metropolis acceptance test. The averaging is performed over reptile random walks [67]. In the FN variant of RQMC approach, estimates of the exact energy are obtained from the first and last configurations of the reptiles. The middle configurations are used to estimate expectation values of operators that do not commute with the Hamiltonian. The modifications of the original RQMC method include Metropolis-Hastings (RQMC-MH), "no-compromise" (RQMC-NC), and "head-tail adjusted" (RQMC-HT) algorithms [76, 77]. The RQMC-MH method corrects for irreversibility not properly addressed in the original RQMC, but fails to meet the assumed criterion of microreversibility. The resulting accumulation of the time-step bias is resolved in RQMC-NC which relaxes the microreversibility requirement and leads to stabilization of the middle configurations of reptiles. Further, RQMC-HT addresses the issue of low practical efficiency of RQMC-NC due to error accumulation and the related high ratio of reptation rejections.

9.2.7   Full CI QMC (FCI-QMC)

QMC methodology can be efficiently used within a framework of the FCI formalism, as demonstrated recently [78–80]. In FCI-QMC the random walk occurs in the space of the Slater determinants that form the FCI expansion. Unlike DMC, there is no diffusion associated with the random walk. The simulation ensemble consists of signed walkers, each associated with a Slater determinant. The ensemble of walkers evolves according to a set of population dynamics rules that resemble those of a cellular automaton. A walker can die or be cloned with a certain probability and it can spawn a new walker at a connected Slater determinant. Pairs of walkers of opposite sign on the same determinant are annihilated. The CI amplitude on a determinant is then defined to be proportional to the signed sum of walkers on the determinant. The stochastic description of a FCI wave function eliminates the requirement for concurrent storage of all amplitudes. The entire space of symmetry-allowed determinants is accessible for the random walk, so that the simulation ultimately converges to the FCI wave function. Evolution in the space of Slater determinants prevents convergence to a Bosonic solution, but does not eliminate sign problem. The latter recurs through the inability to predict *a priori* signs of CI coefficients. Scaling of FCI-QMC with the number of spanned determinants is similar to conventional FCI, but features a smaller prefactor making larger FCI spaces computationally accessible. The FCI-QMC approach has been successfully used to obtain total energies of small molecules [80], including some systems too challenging for conventional FCI, and ionization potentials of atoms from Li to Mg [78].

9.2.8   Time-Dependent QMC (TD-QMC)

The previously described QMC methods rely on random walks in the space of electron configurations. Those methods that utilize the time-dependent Schroedinger equation in imaginary time do not yield real-time dynamics of electrons in physical space. In TD-QMC each physical particle is a walker guided by an individual de Broglie-Bohm pilot-wave [81–85] TD-QMC can be seen as a set of coupled time-dependent Schroedinger equations for the guiding waves in physical space, and a de Broglie-Bohm guiding equation without quantum potentials for the walkers in physical space. The density of walkers represents density of corresponding physical particles, and the intrinsic statistical nature of quantum objects is due to guiding waves. Interactions between electrons are accounted for using explicit Coulomb potentials and pseudo-potentials in order describe non-local correlation effects. One of the advantages of TD-QMC is its insensitivity to the sign-problem which is due to correspondence between walker distribution and quantum probability density. TD-QMC is capable of describing real-time evolution of quantum systems including their interaction with external fields at a fully correlated level.

9.2.9 *Applications*

Ground-state QMC calculations are the most abundant. Preparation of trial wave functions of excited states in VMC faces the same difficulties as in standard orbital-based methods. In DMC, there are no bounds for excited states analogous to the variational principal for the ground state. No tiling theorem exists in FN-DMC treatment of excited states. Availability of accurate experimental data and, especially in the case of small systems, computational results from alternative ab initio approaches makes atoms and molecules in their ground states ideal objects for testing and benchmarking QMC techniques. The aspects of QMC currently being tested vary – from the application of different types of basis sets to new importance sampling strategies [86].

The main challenge in the case of FN-DMC calculations is to improve upon the quality of nodes of the trial wave function. This objective motivates investigations of novel optimization methods [87, 88] and tests of more flexible trial wave functions. For example, a systematic study of the effect of analytic structure of the trial wave function on the quality of nodes [40] demonstrated, that sophisticated single-particle based wave functions beyond restricted Hartree-Fock do not necessarily improve the nodes and FN-DMC accuracy, while recovering more energy at the VMC level. Considerable effort has been expended in the pursuit of the fundamental aspects of nodal surface structure [53, 55, 58, 89, 90]. General consideration of increased efficiency has stimulated development of the effective core potentials (ECP) and corresponding basis sets. Relativistic ECPs and relativistic approximations within QMC become critically important for understanding the chemistry of heavy elements, including transition metals [91]. FN-DMC samples a mixed distribution, which introduces a bias for computing properties that do not commute with the Hamiltonian. This issue has been addressed in studies of expectation values of moments [92, 93], differential operators [94, 95], relativistic corrections [91], and Born-Oppenheimer forces on nuclei [96]. Methods for sampling the "pure" density must overcome the increased variance and higher computational demands associated with computing weighting factors that correct for the mixed distribution. Alternatives to weighting factors has been proposed [97], in which derivatives of the DMC energy are computed from a mixed density by analyzing the serial correlation of walker weights. The Hellmann-Feynman sampling method can be used to sample a large class of operators diagonal in real space, including densities and interaction energies, exactly within FN-DMC [98].

Estimators of operators other than energy suffer from substantial fluctuations and the zero-variance principle does not hold for them. Variance can be reduced by construction of new estimators that decrease fluctuations without biasing the resulting estimate [99, 100]. Calculations of forces are particularly sensitive to the magnitude of the variance because the estimators are related to derivatives of the total energy. Various aspects of force calculations have been investigated, including sensitivity to electron-nucleus cusp quality [101], effect of Pulay's correction on calculations based on the Hellmann-Feynman theorem [102], and applicability of adjoint algorithmic differentiation [103]. Stochastic line minimization algorithm relying on Bayesian inference was devised to find precise structural minima [104].

Mixed estimators are avoided in the RQMC approach discussed earlier. Recent developments increase RQMC efficiency in studies of large systems [76]. Direct assessment of the quality of electron distributions in atoms and molecules showed that the highest accuracy among a variety of QMC methods is achieved by NC-RQMC algorithm [105]. The AFMC method has demonstrated excellent performance in traditionally challenging description of stretched bonds and a related issue of spin-contamination [106].

QMC calculations of excited states that are the lowest of a given symmetry are straightforward as long as a trial wave function of the appropriate symmetry is available. It is more challenging to obtain excited states of the same symmetry as a lower state. No theorem exists that insures that such a calculation evolves to the appropriate state. An early calculation of the $(E^1\Sigma_g{}^+)$ state of H_2 using a MCSCF trial function confirmed ability of FN-DMC to successfully resolve this problem, because the target excited state has the same symmetry as the ground state [107]. Challenging calculations of excited states for small hydrocarbon system have been carried out by Schautz and Filippi [108, 109]. The difficulty arises from close energy level spacing of a number of excited states that are sensitive to the treatment of electron correlation. Careful optimization of the variational trial functions used as importance functions for both ground and excited states enabled the authors to identify the correct excitations within the given symmetry types and enabling the accurate computation of relative energy differences.

9.3 Trial Wave Functions

Although QMC methods are capable of computing accurate molecular properties, they do not yield a wave function per se, but they can sample the many-electron density. Trial wave functions that are used to guide these calculations have strong influence on computational efficiency and subtle influence on accuracy. It is therefore desirable to have an accurate analytical expression for the wave function. Although the QMC energy is determined by nodes of the wave function, neither the Metropolis algorithm nor MC integration depends on the form of Ψ_T. Even the most elaborate trial wave functions can be explored with VMC without modifying the VMC algorithm at low cost. For example, feasibility of VMC computations with matrix-product states, a simple example of tensor networks, was demonstrated [110]. The accuracy of a flexible trial function section cannot be realized without optimizing its parameters. This section presents several common forms of the electronic wave function and describes a few of the approaches developed to optimize the trial function.

Typically QMC calculations are carried out in the FNA using a Slater-Jastrow wave function which is written as a product of an antisymmetric function, Ψ^A and an exponentiated Jastrow function, F: $\Psi_{SJ} = \Psi_A e^F$. The antisymmetric function describes the fundamental properties of the wave function including permutational or spin symmetry. The symmetric Jastrow function depends explicitly on inter-electronic distance and makes possible a more compact descriptions of short range, dynamically correlated electronic motion.

Although the product form of a SJ wave function has typically been used with a single global Jastrow factor, recently [111] the benefit of separate Jastrow terms for each molecular orbital of the antisymmetric function was demonstrated. The latter approach improves treatment of local electron correlation by facilitating adjustment to the local molecular environment. Also, the nodal structure of the trial wave function can better reflect the parameters of multiple Jastrow functions than the single global Slater-Jastrow wave function. This modification notably improves the nodal structure of trial wave functions and FN-DMC energies.

9.3.1 Antisymmetric Wave Functions

Most basis set ab initio methods build many-electron wave functions from atomic orbitals (AOs) or molecular orbitals (MOs). The simplest many-electron wave function constructed from MOs is obtained from HF theory. The multi-configuration wave functions are more general than HF wave functions and may approach the exact solution in certain limits.

Historically, MO methods have dominated trial-function construction because these functions are readily obtained from widely distributed computer codes. Recently, however, some QMC practitioners have renewed interest in a broader variety of wave functions including valence bond (VB) functions [44, 45], pairing wave functions, such as the antisymmetrized geminal power (AGP) [42, 43], Pfaffian, and perfect pairing forms [46].

A generic spin-orbital is a function of electron spatial coordinates, $r = (x,y,z)$, and spin coordinates $x = (r, \omega)$

$$\psi(x) = \psi^{\alpha}(r)\alpha(w) + \psi^{\beta}(r)\beta(w) \tag{9.19}$$

Spatial orbitals are typically (but not necessarily) expanded in a basis set. The choice of the latter expansion is somewhat arbitrary, but the quality of the possible choices can be judged by considering completeness of the basis set and how quickly the basis converges to eigenfunctions of the Hamiltonian. Alternatives include plane-wave basis sets, Slater-type orbitals (STO), Gaussian-type orbitals (GTO), and numerical orbitals.

The plane wave basis is not as well suited for the molecular Hamiltonian as GTOs and STOs. The electron-nucleus cusp conditions cannot be satisfied for any finite size of a plane wave basis, although this problem can be avoided by using effective core potentials. The periodicity of a plane wave basis makes it preferable for solids compared to isolated molecules [112].

Several desirable wave function properties are obtained with the use of STOs [113]. The long-range tail of each function has exponential dependence, making it easier to match the correct asymptotic behavior of the electron density. The electron-nucleus cusp condition is readily satisfied by STO functions.

Many of the integrals needed to evaluate molecular energies cannot be evaluated analytically with STOs which led to the use of GTOs owing to ease of integration [114]. Several advantages of STOs are not retained with GTOs. The radial component of the GTO has a zero derivative at the origin so a single GTO cannot satisfy the electron-nucleus cusp condition [115]. In addition, GTO basis sets are not complete and there is no simple prescription for their systematic improvement. Nevertheless, carefully tuned linear combinations of GTO's with assorted exponents (contracted GTO's) have been developed to give adequate descriptions of the molecular Hamiltonian [114] and advantageous trade-offs between expedience and accuracy make contracted GTOs the dominant basis sets used in quantum chemistry. Recently, mixed Gauss-Slater basis sets have been proposed for calculations with pseudopotentials [116].

In addition to basis set expansions, there are various numerical methods for parameterizing orbitals including numerical basis sets of the form $\varphi(r) = Y_{lm}(r) f(r)$, in which the radial function, $f(r)$ does not have an analytical form, but is evaluated by a spline procedure [117]. Numerical orbitals may be more flexible than STO or GTO basis sets, but their use is more computationally demanding. Wavelet representations of orbitals [118] are exceptionally flexible as well and have an intriguing multi-resolution property: wavelet algorithms adaptively increase the flexibility of the orbital in regions where the molecular energy depends sensitively on the precision of the orbital and use coarser descriptions where precision is less essential.

With the various orbital-types described, one can construct a simple many-electron wave function by forming the antisymmetrized product of orbitals. This is conveniently evaluated using the determinant of the Slater matrix, whose elements are the values of each orbital (in rows) evaluated at the coordinates of each electron (in columns).

$$\Psi(x_1,\ldots,x_N) = \frac{1}{\sqrt{N!}} \begin{vmatrix} \psi_1(x_1) & \psi_1(x_2) & \ldots & \psi_1(x_N) \\ \psi_2(x_1) & \psi_2(x_2) & \ldots & \psi_2(x_N) \\ \vdots & \vdots & \ddots & \vdots \\ \psi_N(x_1) & \psi_N(x_2) & \ldots & \psi_N(x_N) \end{vmatrix} = |\psi_1(x_1)\psi_2(x_2)\ldots\psi_N(x_N)\rangle$$

(9.20)

Linear combinations of Slater determinants can be used to construct configuration state functions (CSFs) that satisfy the total spin constraints (plus any additional symmetry-related requirements).

HF wave functions are obtained by minimizing the energy of a single CSF with respect to variations of the spatial orbitals. The HF energy is not exact because the HF wave function does not account for electron correlation in singlet wave functions and for triplet wave functions only through the antisymmetrizer. The difference between the HF energy and the exact energy is known as the correlation energy, $E_{corr} = E_{HF} - E_{exact}$. The correlation energy is typically a small percentage of the total energy [119, 120], but this relatively small error is often larger than the energy that accounts for a large number of chemical phenomenon. The need for chemical accuracy emphasizes the importance of exploring more elaborate wave functions and the HF wave function provides a starting point for many of these methods.

The simplest approach to evaluating the correlation energy is to increase the number of configurations that contribute to the wave function. Once a reference configuration has been obtained from a HF calculation, a set of related substituted determinants may be identified. The substituted determinants are related to the reference by replacing the orbitals from the occupied space with orbitals from the unoccupied space. The CI method then minimizes the variational energy of a wave function formed from a linear combination of determinants (or, equivalently, CSFs) with respect to the CI coefficients.

$$|\Psi_{CI}\rangle = \sum_{ia} c_i^a |\Psi_i^a\rangle + \sum_{ijab} c_{ij}^{ab} |\Psi_{ij}^{ab}\rangle + \dots \qquad (9.21)$$

where subscript and superscript indexes specify occupied and virtual orbitals involved in substitutions. In principle, CI calculations can provide exact results if the single-particle basis is complete and the full set of CSFs is used in the CI expansion. The second condition is exceptionally demanding. The number of determinants included in a calculation with N electrons and K spatial basis functions is $(2\ K)!/(N!(2\ K - N)!)$, making FCI calculations prohibitive for all but the smallest molecules. Truncated CI calculations in which a small number of orbitals are substituted into the reference configuration are feasible and a hierarchy of CI methods is formed by adding increasingly substituted configurations to the CI expansion. CI with single substitutions (CIS) is commonly used to estimate excitation energies, but truncated CI does not provide a size consistent treatment of the correlation energy. In practice, however, CI calculations that include up to four substitutions (CISDTQ) are nearly size consistent [121]. The MCSCF (multi-configuration self-consistent field) method is akin to truncated CI in the sense that the MCSCF also expresses the wave function as a linear combination of CSFs, but differs from CI in that the orbitals are optimized simultaneously with the CI coefficients [122].

Among MO-based theories, dynamic correlation is better described by perturbation theory or coupled cluster approaches [123] which are both accurate and computationally affordable, at least for small to medium-size molecules. Despite their prevalence in quantum chemistry, these methods will not be described here because the number of determinants required to evaluate their wave functions is presently too large for use as QMC trial functions.

Valence bond theory provides another class of wave functions [124]. General VB wave functions are linear combinations of 'structures' of the form

$$\Omega_{S,M,k}(X) = \hat{A}\left([\varphi_{k,1}(r_1) \dots \varphi_{k,N}(r_N)] \Theta_{S,M,k}(w_1, \dots, w_N)\right) \qquad (9.22)$$

Here, A is the antisymmetrizer and $\varphi_{k,v}$ are the atomic basis functions that participate in the k-th structure. The spin function $\Theta_{S,M}$ is constructed to make each structure an eigenfunction of S^2 and S_z operators with quantum numbers S and M. VB theory has many variants that can be classified by their method of selection of structures for the calculation. A more complete enumeration of VB wave functions can be found in references [124] and [125].

An alternative approach to improving upon HF wave functions is to include correlations between pairs of electrons more directly by means of two particle 'geminal' functions, $G(x_i, x_j)$. The antisymmetrized geminal power (AGP), Pfaffian and perfect pairing wave functions are all examples of pairing wave functions; each can be written as an antisymmetrized product of geminals,

$$\psi = \hat{A} \prod_i G_i(x_{2i}, x_{2i+1}) \tag{9.23}$$

that differ in the form of G.

The singlet AGP wave function [42] for $2N$ electrons shares the same singlet coupled spin geminal for each pair:

$$^1\Psi_{AGP} = \hat{A}\left[G(x_1, x_2) \ldots G(x_{2N-1}, x_{2N})\right]$$

$$G_{AGP}(x_i, x_j) = g(r_i, r_j)\left[\alpha(w_1)\beta(w_2) - \alpha(w_2)\beta(w_1)\right] \tag{9.24}$$

The spatial geminals are symmetric with respect to the exchange of particle coordinates and may be expressed as a linear combination of basis functions, φ_μ, in which the matrix of coefficients is symmetric (i.e., $c_{\mu v} = c_{v\mu}$). An equivalent expression for g, in terms of its natural orbitals, ψ, may be found from the unitary transformation that diagonalizes $[c_{\mu v}]$.

$$g(x_i, x_j) = \sum_{\mu, v} c_{\mu,v}\, \varphi_\mu(x_i)\, \varphi_v(x_j) = \sum_k \lambda_k\, \psi_k(x_i)\psi_k(x_j) \tag{9.25}$$

If only N of these eigenvalues, λk, are nonzero, the AGP wave function can be simplified to the closed shell HF form.

An additional K unpaired electrons may be included in an extended AGP wave function by appending the unpaired orbitals to the geminal product:

$$^{K+1}\Psi_{AGP} = \hat{A}\left[G(x_1, x_2) \ldots G(x_{2N-1}, x_{2N})\psi_1(x_{2N+1}) \ldots \psi_K(x_{2N+K})\right] \tag{9.26}$$

The AGP wave function can be efficiently evaluated as the determinant of an $(N + K) \times (N + K)$ determinant [43, 126].

Pfaffian wave functions [46] are distinguished from AGP wave functions by the addition of triplet coupled terms to the spin geminal:

$$\Psi_{Pf} = \hat{A}\left[G_{Pf}(x_1, x_2) \ldots G_{Pf}(x_{2N-1}, x_{2N})\right] \tag{9.27}$$

$$G_{Pf}(x_1, x_2) = g^{\alpha\beta}(r_1, \ r_2)(\alpha(\omega_1)\beta(\omega_2) - \alpha(\omega_2)\beta(\omega_1))/\bar{2}$$
$$+ h^{\alpha\alpha}(r_1, \ r_2)\,(\alpha(\omega_1)\alpha(w_2)) + h^{\beta\beta}(r_1, r_2)(\beta(\omega_1)\beta(w_2))$$

Spatial geminals for the triplet pairs, $h^{\sigma\sigma'}$, can also be expressed as linear combinations of basis functions, but their coefficient matrices must be antisymmetric. Clearly, the Pfaffian form reduces to the AGP wave function when $h^{\alpha\alpha} = h^{\beta\beta} = 0$.

The Pfaffian wave function can be computed from the Pfaffian of a matrix composed of blocks of geminals:

$$\Psi_{Pf} = Pf \begin{bmatrix} H^{\alpha\alpha} & G^{\alpha\beta} & \psi^{\alpha} \\ -G^{\alpha\beta T} & H^{\beta\beta} & \psi^{\beta} \\ -\psi^{\alpha T} & -\psi^{\beta T} & 0 \end{bmatrix} \tag{9.28}$$

$$H^{\alpha\alpha}_{ij} = h^{\alpha\alpha}(r_i, r_j) \ H^{\beta\beta}_{ij} = h^{\beta\beta}(r_i, \ r_j) \ G^{\alpha\beta} = g^{\alpha\beta}(r_i, r_j) \ \psi^{\sigma}_i = \psi^{\sigma}(r_i)$$

The ψ blocks are composed of the unpaired orbitals and are used only when the number of electrons is odd. The Pfaffian of this $N \times N$ skew symmetric matrix is defined by an antisymmetric product of its elements:

$$Pf[A] = \hat{A}[a_{1,2}\,a_{3,4}\ldots a_{N-1,N}] = \sum_p sgn\,(p)a_{i_1 j_1}\,a_{i_2 j_2}\ldots a_{i_N j_N} \tag{9.29}$$

The summation includes all possible products in which each row or column index appears only once (i.e. $i_k \neq i_m$, $i_k \neq j_m$ and $j_k \neq j_m$) and $i_k < j_k$. The sign of p is the sign of the permutation of i's and j's in the p-th term of the sum. Like determinants, Pfaffians may be evaluated by cofactor expansion:

$$Pf[A] = \sum_{j=2}^{N} a_{ij}\,P_c(a_{1,j}) \tag{9.30}$$

$$P_c(a_{j,k}) = (-1)^{j+k+1} Pf\,[A\,(j,k;j,k)]$$

The Pfaffian cofactor, $P_c(a_{j,k})$, is (within a sign) the Pfaffian of $A(j, k; j, k)$, the submatrix obtained by striking out rows j and k and columns j and k from A. To compute the Pfaffian of a large matrix by the cofactor procedure would be unduly slow due to the large number of terms in the expansion. Reference [46] provides a more efficient algorithm for evaluating the Pfaffian of a matrix by first bringing the matrix into a block diagonal form.

In the perfect pairing (PP) model, each pair of electrons is described by its own geminal. In contrast, the AGP and Pfaffian functions share the same geminal for all electron pairs. The perfect pairing geminal has a more constrained form than the previously described wave functions. In each geminal, a pair of 'active occupied' orbitals, $\psi_i \psi_{\bar{i}}$, has a corresponding pair of 'active virtual' orbitals, $\psi_{i*}\,\psi_{\bar{i}*}$:

$$G_{PP_i} = \psi_i \, \psi_{\bar{i}} + c_i \, \psi_{i*} \, \psi_{\bar{i}*} \qquad (9.31)$$

The significance of this pairing function is that it is capable of describing the static correlations that are most important when chemical bonds are broken. In most uses, the core orbitals are uncorrelated ($c_i = 0$), resulting in the closed shell generalized valence bond (GVB) wave function of Goddard et al. [127].

$$\Psi_{PP} = \hat{A} \Big[\underbrace{\psi_1 \, \psi_{\bar{1}} \ldots \psi_p \, \psi_{\bar{p}}}_{core} \, \underbrace{(\psi_i \, \psi_{\bar{i}} + c_i \, \psi_{i*} \psi_{\bar{i}*}) \ldots (\psi_N \, \psi_{\bar{N}} + c_N \, \psi_{N*} \, \psi_{\bar{N}*})}_{valence} \Big]$$

$$(9.32)$$

The relative simplicity of the PP wave function also allows it to be written in the coupled cluster form [128, 129]. The coupled cluster approach allows the PP wave function to be determined with relative ease. Presently, there is no method for evaluating the PP wave function apart from its determinant expansion. This expansion includes 2^N determinants, which is substantial, but far less than the factorial number generated in a CASSCF calculation that includes the same set of active orbitals. This limits the use of PP wave functions in QMC to wave function involving small numbers of active pairs.

9.3.2 Backflow Transformed Wave Functions

Yet another approach to incorporating correlation effects into an antisymmetric function is to allow the orbitals (or geminals) to depend on the coordinates of the other electrons. This is accomplished by the backflow transformation in which the coordinates of the electrons are modified by a backflow displacement:

$$\Psi_{BF}(X) = \Psi_A(X + \xi) \qquad (9.33)$$

The displacement, ξ, is analogous to an eddy of a classical fluid moving around a large impurity. Feynman and Cohen [130] used the variational principle to show that the energy of a quantum fluid is minimized by the transformation that conserves the current around the impurity.

The displacement function must be a symmetric function in order to preserve the overall antisymmetry of Ψ_A, but there remains a great deal of flexibility in the form of ξ. For homogeneous systems, it is common for the displacement of electron i to be determined by the sum of pairwise displacements in the directions of the other electrons, j:

$$\xi_i = \sum_j r_j + \eta(r_{ij})(r_i - r_j) \qquad (9.34)$$

where $\eta(r_{ij})$ is a function of interparticle distance. Various forms of $\eta(r_{ij})$, including rational [131], Gaussian [132], and polynomial [47], functions have been suggested. For molecular systems, the inhomogeneous backflow function introduced by Lopez Rios et al. [47] also includes displacements in directions of the nuclei.

9.3.3 Effective Core Potentials (ECP)

The purpose of the ECP approximation is to remove the core electrons from the calculation so that computational effort is focused on the valence electrons. Obviously, the ECP must account for changes to the Coulomb potential arising from removal of the core electrons. There must also be a commensurate reduction of nuclear charge. Additional terms are needed to account for the exchange energy and to maintain orthogonality between the valence orbitals and the core orbitals that have been removed. The non-locality of the latter terms underly the unusual form of the ECP:

$$V^{ECP}(r) = w_{L+1}(r) + \sum_{l=0}^{L} \sum_{m=-l}^{l} |y_{lm}\rangle \left(w_{L+1}(r) - w_l(r)\right) \langle Y_{lm}| \qquad (9.35)$$

The value of the ECP depends not only on an electron's coordinates, but also on the projection of the wave function of one electron (holding the other electron coordinates constant) onto the spherical harmonics, Y_{lm}. The local term, $w_{L+1}(r)$, depends only on the distance of the electron from the nucleus. The angular potentials $w_l(r)$ are determined so that, beyond some cutoff distance, the 'pseudo-orbitals' obtained from an ECP calculation match those of an all-electron calculation, but are nodeless and smoothly go to zero within the cutoff radius [133]. The w_l are then fit to a Gaussian expansion [134] so that the potential can be rapidly integrated over Gaussian basis functions.

$$w_l(r) = \sum_k A_{nlk} \, r_{lk}^n \, e^{-B_{lk} r^2} \qquad (9.36)$$

Some flexibility remains in the selection of terms to be included in Eq. 9.36. An assortment of ECPs are possible within this framework because the form of the pseudo-orbital within the cutoff radius is not completely defined. So-called soft ECPs have been designed so that w_l cancels the Coulomb singularity at the nucleus [133, 135]. This is valuable for QMC calculations because their efficiency is sensitive to rapid changes of the potential. Several sets of soft ECPs have been designed specifically for QMC so that Gaussian basis function can be used in QMC calculations without special consideration of the electron-nucleus cusp conditions [136, 137].

An alternative to the locality approximation has been proposed [138]. In the "T-moves" approach an additional displacement of walkers is performed using a heat-bath algorithm according to the negative matrix elements of the non-local potential. The positive matrix elements contribute to the local potential. The locality approximation is known to lead to numerical instabilities associated with divergences of the fully localized potential in the nodal regions of the trial wave function. T-moves effectively push walkers away from such regions increasing stability of simulations. Also, the ground-state energy of the effective Hamiltonian of the T-moves algorithm is an upper-bound to the exact ground-state energy, which makes the algorithm variational.

9.3.4 Jastrow Wave Functions

The antisymmetric wave functions in the previous section account for electron correlation indirectly through correlation among the coefficients of the geminal or CI expansions. More compact descriptions of electron correlation are achieved by Jastrow correlation functions that depend explicitly on interelectronic distance. A Jastrow correlation function F can be parameterized in an infinite number of ways. F can be partitioned into a hierarchy of terms $F_1 \ldots F_N$ in which F_n describes correlations among n electrons.

$$F = \sum_n F_n; \quad F_1 = \sum_i f_1(\mathbf{r}_i); \quad F_2 = \sum_{i<j} f_2(\mathbf{r}_i, \mathbf{r}_j) \qquad (9.37)$$

The electron-nucleus (e-n) correlation function does not describe electron correlation *per se* because it is redundant with the orbital expansion of the antisymmetric function. If the correlation function expansion is truncated at F_1 and the antisymmetric wave function is optimized with respect to all possible variations of the orbitals, then F_1 would be zero everywhere. There remain two strong reasons for including F_1 in the correlation function expansion. First, the molecular orbitals are typically expanded in Gaussian basis sets that do not satisfy the e-n cusp conditions. The e-n correlation function can satisfy the cusp conditions, but F_1 influences the electron density in regions beyond the immediate vicinity of the nucleus, so simple methods for determining F_1 solely from the cusp conditions may have a detrimental effect on the overall wave function. Careful optimization of a flexible form of F_1 is required if the e-n cusp is to be satisfied by the one-body correlation function [115].

Second, F_2 depletes more of electron density near the nucleus (where the density is high) than in the tail of the wave function (where the density is low). This leads to a net shift of electron density away from the nucleus, which reduces the energetically favorable e-n interactions. The e-n correlation function can be used to readjust the electron density without reoptimizing molecular orbitals [139, 140].

The basic form of f_1 is a function of the e-n distances,

$$f_1(\mathbf{r}_i) = \sum_A \bar{r}_A(|\mathbf{r}_i - \mathbf{r}_A|) \qquad (9.38)$$

$$\bar{r}(r)^{Pade} = (ar)/(1 + br)$$

$$\bar{r}(r)^{Sun} = -be^{(-ar)}$$

$$\bar{r}(r)^{Exponential} = b(1 - e^{(-ar)})$$

$$\bar{r}(r)^{PolynomialPade} = (1 - z(r/r_{cut}))/(1 + z(r/r_{cut})); \quad z(x) = x^2(6 - 8x + 3x^2),$$

where A enumerates nuclei.

The Pade function has a cusp at $r = 0$ that can be adjusted to match the Coulomb cusp conditions by adjusting the a parameter. The Sun form also has a cusp, but approaches its asymptotic value far more quickly than the Pade function, which is useful for the linear scaling methods. An exponential form proposed by Manten and Luchow is similar to the Sun form, but shifted by a constant. By itself, the shift affects only the normalization of the Slater-Jastrow function, but has other consequences when the function is used to construct more elaborate correlation functions. The polynomial Pade function does not have a cusp, but its value goes to zero at a finite distance.

Two electron correlations provide the largest contributions to the correlation energy [141]. The simplest e-e correlation functions are spatially homogeneous and depend only on the distances between the electron pairs. Scaled distance functions are of the same form as the e-n correlation functions, but because the e-e correlation causes electrons to avoid each other while e-n correlation causes electrons to approach nuclei, the multiplicative factors used to scale r will have opposite signs for e-e and e-n correlation.

Inhomogeneous contributions to F_2 can be described by an expansion in powers of the e-e and e-n scaled distance functions.

$$F_2(\mathbf{r}_i, \mathbf{r}_j) = \sum_{mno} \sum_{i<j}^{electrons} c_{mno} \left(\bar{r}_{iA}^m \bar{r}_{jA}^n + \bar{r}_{iA}^n \bar{r}_{jA}^m \right) \bar{r}_{ij}^o \qquad (9.39)$$

This ansatz was first proposed by Boys and Handy (BH) [142] and Schmidt and Moskowitz (SM) later arrived at the same form by considering averaged backflow effects [143]. The SMBH form includes both the e-n function (via the m, n, $o = m$, 0, 0 terms) and the homogeneous e-e Jastrow function (via the m, n, $o = 0$, 0, o terms). The remaining e-e-n terms modulate the e-e correlation function according to the e-n distances.

In SM's original work, $\bar{r}$ took the Pade form. Any of the forms of the distance functions (Eq. 9.38) are acceptable, but there are qualitative differences between the powers of these functions. If $\bar{r}$ is zero at the origin then $\bar{r}^n$ cannot have a cusp for $n > 1$ and higher powers of $\bar{r}$ can be used to fit longer range behavior. If $\bar{r}$ is nonzero at the origin, then all powers of $\bar{r}$ can have cusps (if $\bar{r}$ has a cusp) and higher powers of $\bar{r}$ have decreasing range. Schmidt and Moskowitz used this correlation function to recover $\approx 75\%$ of the correlation energy for the first row atoms, 25% more than the homogeneous e-e Jastrow [143].

By increasing the number of terms in the BH expansion (SM used physical arguments to select only a subset of the possible terms) and adding terms designed to satisfy the cusp conditions due to the simultaneous approach of two electrons to the nucleus, Filippi and Umrigar obtained between 79% and 94% of the correlation energy for the first-row homonuclear diatomic molecules [144]. The two-electron terms in their correlation function are

$$F_2 = \frac{b\bar{r}_{ij}}{1 + b'\bar{r}_{ij}} + P(U, S, T) + \bar{F}(U, S, T) + \bar{F}'(U, S, T)$$

$$U = \bar{r}_{ij} \quad S = \bar{r}_{iA} + \bar{r}_{jA} \quad T = \bar{r}_{iA} - \bar{r}_{jA} \tag{9.40}$$

The function P is a complete fifth order polynomial of U, S and T. The terms in the BH expansion are equivalent to those in P, although the polynomial coefficients will differ due to the transformation from $\bar{r}_{iA}$ and $\bar{r}_{jA}$ to S and T. The $\bar{F}$ and $\bar{F}'$ functions are composed of the leading terms in Fock's expansion for the helium atom [144].

The use of three-electron correlation functions was initially explored by Huang et al. [141] Their study used a complete fifth order polynomial of the interparticle distances and so describe F_3, but the energetic improvements over Filippi and Umrigar's F_2 function were less than 0.5% for the Li, Be and Ne atoms studied. The physical interpretation for the limited improvement due to F_3 is that correlations among three or more particles must include at least two particles of the same spin. The antisymmetry of Ψ_A ensures that the wave function approaches zero where two electrons of the same spin meet, so the many electron density will be very small wherever high-order correlations are most significant. Furthermore, the largest three body effects are accounted for by products of two-body functions. Most correlation functions are therefore truncated at F_2. Many-body terms such as F_3 are considerably more expensive to compute than F_2 but yield only slight improvement in the accuracy of the calculation, particularly if the results will be refined at the DMC level.

9.3.5 Trial Wave Function Optimization

Previous section described how the energy of a trial function can be computed by sampling walkers from the electron density, Ψ^2_T, and averaging the local energies of these walkers. The fixed sample optimization method [145] takes walkers sampled from $\Psi^2_T(\Lambda_0)$, the electron density at the initial set of parameter values, to compute the energy (or another optimization criterion) at different values of Λ.

For example, the VMC energy may be rewritten in terms of the fixed sample density by incorporating weights, $w_k = \Psi^2(X_k, \Lambda)/\Psi^2(X_k, \Lambda_0)$ to account for changes of the density due to parameter changes.

$$E(\Lambda) = \frac{\int \psi_T(\Lambda)\hat{H}\psi_T(\Lambda)d\mathbf{X}}{\int \psi^2_T(\Lambda)d\mathbf{X}} = \frac{1/K \sum\limits_k w_k(\Lambda)E_{Lk}(\Lambda)}{1/K \sum\limits_k w_k(\Lambda)} \tag{9.41}$$

Minimization of the Monte Carlo energy estimate minimizes the sum of the true value and the error due to the finite sample. Although the variational principle provides a lower bound for the energy, there is no lower bound for the error of an energy estimate. Fixed sample energy minimization is therefore notoriously unstable [140, 146]. Optimization algorithms based on Newton's, linear and perturbative methods have been proposed [43, 48, 140, 147–151].

More suitable optimization functions can be found by using the property that the exact wave function is an eigenfunction of the Hamiltonian. The variance of the local energies of an exact eigenfunction is zero because the local energy is a constant function of X. Accounting for the weights introduced in Eq. 9.41, the variance of the local energy as a function of Λ is

$$\sigma^2(\Lambda) = \frac{\Sigma w_k}{(\Sigma w_k)^2 - (\Sigma w_i^2)} \sum_k w_k (E_L(X_k, \Lambda) - E(\Lambda))^2 \qquad (9.42)$$

The summands for Eq. 9.42 must be positive (or zero) everywhere, so the variance of an approximate wave function is bounded from below by zero, even for a finite sample. The coincidence between the lower bound of the variance and the variance of the exact wave function enables the trial wave function to be optimized by variance minimization. The existence of an absolute lower bound makes variance minimization more robust than energy minimization [49].

Similar arguments provide grounds for minimizing the absolute deviation of the local energy [152–154]:

$$M(\Lambda) = 1/K \sum_k^K w_k(\Lambda)|E_L(X_k, \Lambda) - E(\Lambda)| \qquad (9.43)$$

Like the variance, the mean absolute deviation also has a lower bound of zero, but two arguments suggest that it may be a more effective optimization criterion than the variance. The contribution of each walker to the variance is a quadratic function of the local energy, but its contribution to the absolute deviation has a linear dependence on the local energy. It has been suggested that this difference makes the mean absolute deviation easier to optimize than the variance because the derivative of the absolute value is always ± 1, but the derivative of a quadratic function approaches zero near its minimum. Also, the quadratic contribution of each walker to the variance may allow a few particularly errant walkers to dominate the variance, but the absolute deviation is not as strongly influenced by outlying local energies.

Several authors, including Schmidt and Moskowitz and Greeff and Lester, have explored minimization of the unreweighted variance σ^2_u of the local energies without accounting for the changes to the density due to changes in Λ [133, 143].

$$\sigma^2_u(\Lambda) = 1/K \sum_k^K (E_L(X_k) - E(\Lambda))^2 \qquad (9.44)$$

The unreweighted variance does not minimize the variance of $\Psi(\Lambda)$, but it is nevertheless a valid minimization criterion because the variance of the exact wave function will be zero regardless of the distribution of walker coordinates. The renewed interest of Drummond and Needs [154] stems from their observation that the unreweighted variance is sometimes especially easy to minimize. When the only parameters being optimized are linear parameters of the Jastrow function the local energy is

$$E_L(\Lambda) = \sum_i \Lambda_i g_{ij}^{(2)} \Lambda_j + \sum_i \Lambda_i g_i^{(1)} + g^{(0)} \qquad (9.45)$$

$$g^{(0)}(R) = \frac{\nabla^2 \psi_A}{\psi_A}$$

$$g_i^{(1)}(R) = \nabla^2 f_i + 2\frac{\nabla \psi_A}{\psi_A} \nabla f_i$$

$$g_{ij}^{(2)}(R) = \nabla f_i \nabla f_j$$

For lines in parameter space $\Lambda = \Lambda_0 + Bt$, the local energy is a quadratic function a single variable, t, and the unreweighted variance is a quartic polynomial of t.

$$E_L(t; \ R, \ \Lambda_0, B) = p_2 t^2 + p_1 t + p_0$$

$$\sigma_u^2(t) = (\langle p_2^2 \rangle - \langle p_2 \rangle \langle p_2 \rangle)t^4 + 2(\langle p_2 p_1 \rangle - \langle p_2 \rangle \langle p_1 \rangle)t^3$$
$$+ (2\langle p_2 p_0 \rangle - 2\langle p_2 \rangle \langle p_0 \rangle + \langle p_1^2 \rangle - \langle p_1 \rangle \langle p_1 \rangle)t^2$$
$$+ 2(\langle p_1 p_0 \rangle - \langle p_1 \rangle \langle p_0 \rangle)t + (\langle p_0^2 \rangle - \langle p_0 \rangle \langle p_0 \rangle) \qquad (9.46)$$

$$p_0(R; \Lambda_0, B) = \frac{1}{2} A^i g_{ij}^{(2)}(R) A^j - \frac{1}{2} g_j^{(1)}(R) A^j - \frac{1}{2} g^{(0)}(R) + V(R)$$

$$p_1(R; \Lambda_0, B) = -\frac{1}{2} g_j^{(1)}(R) B^j - \Lambda^i g_{ij}^{(2)}(R) B^j$$

$$p_2(R; \Lambda_0, B) = -\frac{1}{2} B^i g_{ij}^{(2)}(R) B^j$$

The identification of Eq. 9.46 as a fourth order polynomial simplifies the minimization of the unreweighted variance. The line-minimization step of conjugate gradient-type algorithms becomes trivial because the minima of quartic polynomials can be found algebraically.

Snajdr and Rothstein compared a number of properties including the average interelectronic distances and multipole moments of wave functions optimized by variance minimization to those optimized by energy minimization [155]. They found that energy minimized wave functions provided more accurate results for non-energetic properties. Filippi and Umrigar also found that energy-optimized wave functions have lower energies and higher variances than variance minimized wave functions [156]. Considering that chemical reactivity is determined by energetic differences and not variance differences, these results suggest that energy minimization should be revisited.

Several new algorithms avoid the pitfalls of fixed sample energy minimization by devising alternative estimates of energy changes due to parameter modifications. These model energies are constructed from derivatives of the energy of the wave function and do not correspond to the fixed sample energy of Eq. 9.41 except when $\Lambda = \Lambda_0$. Unlike the fixed sample energy, the model energies have lower bounds, making them more stable metrics for optimization. This description focuses on the 'linear' method of Toulouse and Umrigar [140], which avoids the computationally demanding step of evaluating second derivatives of the local energy.

The construction of the linear energy model starts from an explicitly normalized expression for the trial wave function.

$$\tilde{\Psi}(\Lambda) = \frac{\Psi(\Lambda)}{\langle \Psi(\Lambda) | \Psi(\Lambda) \rangle^{1/2}} \tag{9.47}$$

a linearized wave function, $\Psi(\Lambda)$, is obtained by truncating its Taylor series expansion at first order:

$$\tilde{\Psi}(\Lambda_0 + \Delta) \approx \bar{\Psi}(\Lambda_0 + \Delta) = \tilde{\Psi}(\Lambda_0) + \sum_i \frac{\partial \Psi}{\partial \Lambda_i} \Delta_i + O(\Delta^2) \tag{9.48}$$

The initial wave function and its derivatives form a basis set for describing $\bar{\Psi}(\Lambda_0 + \Delta)$:

$$\bar{\Psi}_0 = \tilde{\Psi}(\Lambda_0) \quad \bar{\Psi}_i = \frac{\partial \tilde{\Psi}}{\partial \Lambda_i} = \frac{\partial \Psi}{\partial \Lambda_i} - \left\langle \frac{\partial \Psi}{\partial \Lambda_i} \middle| \Psi(\Lambda_0) \right\rangle \Psi(\Lambda_0) \tag{9.49}$$

When the Hamiltonian and overlap matrices are expanded in the Ψ_i basis, the energy of the linearized wave function becomes

$$\bar{E}(\Delta) = \frac{\langle \bar{\Psi} | \hat{H} | \bar{\Psi} \rangle}{\langle \bar{\Psi} | \bar{\Psi} \rangle} = \frac{\sum\limits_{i,j} \Lambda_i \hat{H}_{ij} \Lambda_j}{\sum\limits_{i,j} \Lambda_i \hat{S}_{ij} \Lambda_j} \tag{9.50}$$

Matrix elements can be computed using fixed sample averages at points sampled from $\Psi_T^2(\Lambda_0)$

$$\bar{H}_{00} = \langle E_L \rangle \tag{9.51}$$

$$\bar{H}_{i0} = \left\langle \frac{\Psi_i}{\Psi_0} E_L \right\rangle - \left\langle \frac{\Psi_i}{\Psi_0} \right\rangle \langle E_L \rangle$$

$$\bar{H}_{0j} = \left\langle \frac{\Psi_j}{\Psi_0} E_L \right\rangle - \left\langle \frac{\Psi_i}{\Psi_0} \right\rangle \langle E_L \rangle + \langle E_{L,j} \rangle$$

$$\bar{H}_{ij} = \left\langle \frac{\Psi_i}{\Psi_0} \frac{\Psi_j}{\Psi_0} E_L \right\rangle - \left\langle \frac{\Psi_i}{\Psi_0} \right\rangle \left\langle \frac{\Psi_j}{\Psi_0} E_L \right\rangle - \left\langle \frac{\Psi_j}{\Psi_0} \right\rangle \left\langle \frac{\Psi_i}{\Psi_0} E_L \right\rangle$$

$$+ \left\langle \frac{\Psi_i}{\Psi_0} \right\rangle \left\langle \frac{\Psi_j}{\Psi_0} \right\rangle \langle E_L \rangle + \left\langle \frac{\Psi_i}{\Psi_0} E_{L,j} \right\rangle - \left\langle \frac{\Psi_i}{\Psi_0} \right\rangle \langle E_{L,j} \rangle$$

$$\bar{S}_{00} = 1$$

$$\bar{S}_{i0} = \bar{S}_{0j} = 0$$

$$\bar{S}_{ij} = \left\langle \frac{\Psi_i}{\Psi_0} \frac{\Psi_j}{\Psi_0} \right\rangle - \left\langle \frac{\Psi_i}{\Psi_0} \right\rangle \left\langle \frac{\Psi_j}{\Psi_0} \right\rangle$$

where subscripts on Ψ_j and $E_{L,j}$ indicate differentiation with respect to the j-th parameter.

The model energy due to the approximate linearized wave function is bounded below by the minimum eigenvalue of H. The minimum of $E(\Delta)$ can be found by requiring that its derivatives with respect to Δ are zero. This condition leads to the generalized eigenvalue problem

$$\sum_j \bar{H}_{ij} \Delta_j = \bar{E} \sum_j S_{ij} \Delta_j \tag{9.52}$$

which is readily solved by numerical linear algebra software methods [157]. The solution to Eq. 9.52 is then used to update the trial function according to $\Psi(\Lambda) \rightarrow \Psi(\Lambda + \Delta)$.

9.4 Computational Considerations

There is an intimate connection between successes of ab initio quantum chemistry and development of computational resources. The tremendous growth of power and the availability of the high-performance computers enables computational studies of increasingly challenging problems. This Section provides an overview of QMC scaling properties which are pertinent to parallel computations with any degree of concurrency, up to 10^5 threads. Studies of various alternatives to traditional super-computers and clusters are discussed along with relevant algorithmic improvements and code development efforts.

9.4.1 Scaling Analysis

While QMC may appeal to quantum chemists primarily for its high accuracy, these methods also have computational advantages. The slowest step of a QMC calculation requires $O(M^3)$ floating point operations (flops) for a molecules of size M. In contrast, the cost of coupled-cluster methods that are sometimes referred to as the "gold-standard" for quantum chemistry can increase as fast as $O(M^7)$, which becomes prohibitively expensive for molecules with more than about 12 first row atoms [120, 158]. Due to their reduced scaling and natural affinity for parallel computers, QMC methods hold special promise for evaluating electronic properties of large molecules. QMC calculations are nevertheless a significant undertaking. The requirement of small statistical errors adds a large prefactor to their cost, even for small molecules. A variety of methods can help to reduce these costs so that QMC can be applied to a broader range of molecules. Several of these factors are determined by the statistical nature of the QMC method.

The vast majority of computing time used by a VMC or DMC calculations is spent evaluating the local energy of the specified trial wave function. The total compute time is roughly the product of T_{Elocal}, the time required for each evaluation of the local energy, and N_{MC}, the number of Monte Carlo points (walkers positions) where the local energy is evaluated. The walkers and the cost of evaluating their local energies are easily distributed among N_{proc} nodes of a parallel computer. There is also a penalty, T_{comm} incurred for communication between the nodes. The total time required to complete a QMC calculation is roughly

$$T_{wall} = \frac{T_{Elocal} N_{MC}}{N_{proc}} + T_{comm} \qquad (9.53)$$

The T_{Elocal} and T_{comm} terms can be directly reduced by algorithmic improvements such as those presented in later sections of this chapter. On the other hand, N_{MC} is, to a very large extent, an intrinsic property of the underlying QMC theory.

The factors that contribute to N_{MC} have been examined previously by Ceperley [159]. First, the variance of the local energy increases with molecular size, so the number of independent MC points must increase to reduce the error to the same tolerance. Second, the MC points will not be independent of each other until the random walk has been given sufficient time for the walker positions to decorrelate. Third, the time step bias of a DMC calculation increases with system size, so a larger number of steps must be taken before the decorrelation time is achieved.

Suppose the local energies for a monatomic system have a variance σ_1^2. If the system is extended to include M identical, non-interacting atoms, standard error analysis indicates that the variance increases to $\sigma_M^2 = M\sigma_1^2$. The MC error decreases as $K^{-1/2}$, so if K_1 points are needed to reduce the error bar to an acceptable level for the one-atom system, then $K_M = MK_1$ points are needed to achieve the same error for the M-atom system.

$$\epsilon_1 = \sqrt{\frac{\sigma_1^2}{K_1}} \quad \epsilon_M = \sqrt{\frac{\sigma_M^2}{K_M}} = \sqrt{M \frac{\sigma_1^2}{K_M}} \tag{9.54}$$

$$\epsilon_1 = \epsilon_M \quad K_M = MK_1$$

For small and moderately sized molecules, K_M may grow faster than $O(M)$ because interactions between the atoms cause σ_M to increase.

The number of independent points is less than the total number of points because points sampled by the same walker are serially correlated. A decorrelation time of κ_1 must pass before the local energies of a one-atom system are independent from their earlier values. The local energy of the non-interacting M-atom system is just the sum of the local energies of the independent atoms, so the total energy will decorrelate at the same rate as the component energies: $\kappa_M \approx \kappa_1$.

Although the decorrelation time is independent of system size, the degree of serial correlation between steps is determined largely by the time step size, τ. DMC calculations require small time steps to maintain reasonable branching factors. Assuming that the local energies are normally distributed with variance of $\sigma_M{}^2$, the average branching factor for an M-atom system is

$$\langle G_B \rangle = \int \frac{1}{\sigma_M \sqrt{2\pi}} e^{\frac{-(E_L - E_0)^2}{2\sigma_M^2}} e^{-(E_L - E_0)\tau M} \, dE_L \approx 1 + \frac{1}{2} M \, \sigma_1^2 \, \tau_M^2 \tag{9.55}$$

The last line relates the time step to system size by assuming the linear growth of the variance for increasing M. A time step of $O(M^{-1/2})$ is therefore needed to keep $\langle G_B \rangle$ constant.

The number of independent points required for DMC computations is found by combining the factors just described.

$$N_{MC} = K_M \frac{k_M}{\tau_M} = O(M) \frac{O(1)}{O(M^{-1/2})} = O(M^{3/2}) \tag{9.56}$$

Assuming a linear scaling algorithm for the local energy, the compute time for molecular DMC calculations increase at least as fast as $O(M^{5/2})$. For bulk properties that are reported in energies per atom or per unit cell, the computed energies and errors may both be reduced by $1/M$, so those calculations scale as $O(M^{3/2})$.

Several measures can be used to reduce the magnitude of N_{MC} (but not its scaling). Optimization of the trial wave function decreases N_{MC} in two ways. If the variance of the local energies is reduced then a smaller number of points will be needed to achieve the desired error bar. A reduced variance also permits a larger time step to be used c.f. Eq. 9.55. Effective core potentials decrease the variance of local energies by smoothing the Coulomb potential near the atomic core, thereby reducing N_{MC} via mechanisms similar to those of wave function optimization. Improved random walk algorithms with small time step errors may also be used to decrease serial correlation [36, 37].

9.4.2 Molecular Orbital Evaluation

The correlated molecular orbital (CMO) ansatz is the most common type of trial wave function used in QMC. Evaluating the MOs and their derivatives is usually the most time consuming step in evaluating the local energy for CMO wave functions, requiring up to $O(N^3)$ flops for an N-electron wave function. An assortment of methods for reducing the cost of MO evaluation to $O(N)$ has been developed by several groups [160–162]. The algorithm presented in this section provides a 50–75% speedup compared to an earlier O(N) MO evaluation procedure [163].

The Slater matrix for an N-electron determinant will contain N^2 elements consisting of the N occupied orbitals evaluated at each electron's coordinates. If the molecular orbitals are expanded using a linear combination of basis functions (i.e. $\varphi_i(x) = \Sigma_\mu C_{\mu i}\chi_\mu(x)$) and the number of basis functions is $O(N)$ then $O(N^3)$ flops are required to evaluate all of the entries in the Slater matrix.

Linear scaling can be achieved when localized molecular orbitals (LMOs) are used to create sparsity within the Slater matrix. The density of a LMO is confined to limited region of space around its centroid, so only a few LMOs need to be evaluated for each electron. The second consideration on the way to linear scaling is to accelerate the transformation from basis function to LMO's. Linear scaling QMC methods have been a popular field of research and several algorithms have already been published.

Williamson et al. used maximally localized Wannier functions to express the LMO's [160]. The LMOs were truncated by setting the value of the orbital to zero outside the sphere containing 99.9% of the orbital's density. The transformation from basis functions to MOs was sidestepped by tabulating the orbitals on a 3-D grid and using a spline procedure for orbital evaluation.

The nonorthogonal LMOs (NOLMOs) used by Alfe and Gillan were found by dividing the volume of the molecule into a set of overlapping localization regions and maximizing the self-overlap of the orbitals within the localization region [161]. Any part of the orbital outside its localization region was truncated. The transformation step was accelerated by re-writing the LMOs in a basis of 'blip' functions. Each blip is nonzero over a small domain so only 64 blips need to be evaluated and transformed.

The first linear scaling QMC method to use Gaussian basis functions is due to Manten and Luchow [162]. Their method truncated Boys-localized orbitals [164] by neglecting basis functions centered on atoms more than three bond lengths away from the centroid of the LMO. The deletion of basis functions simultaneously reduces both the number of LMOs that must be evaluated and the number of basis functions that must be evaluated and transformed.

The MO evaluation algorithm devised by Aspuru-Guzik et al. [163] truncated the orbital transformation using a numerical cutoff rather than a spatial one. Their algorithm uses a 3-D grid to identify sparsity in the Slater matrix. For each point on the grid, the threshold $C_{\mu i}\chi_\mu(x) > 10^{-12}$ is applied to create a list of relevant coefficients and basis functions. In contrast to the truncation schemes used by other groups, this method neglects the smallest contributions to the MOs instead of the most distant. For large molecules, the grid algorithm is substantially faster than a dense linear transform. However, this grid algorithm performs poorly for small molecules due to irregular memory access patterns.

The NOLMOs used in earlier linear scaling QMC calculations have used finite localization regions to truncate each orbital [161, 165]. The 'spread' functional used by Liu, Perez-Jorda and Yang [166] is equivalent to the Boys method for orthogonal orbitals, but in the nonorthogonal case, it minimizes the spatial extent of each orbital rather than maximizing the distance between orbital centers. Liu was able to reduce the spread functional by 83%, on average, by relaxing the orthogonality constraint, but their minimization algorithm required elaborate machinations to avoid singular transformations [166].

9.4.3 Correlation Function Evaluation

The SMBH function described previously is frequently used for its compact and accurate description of inhomogeneous electron correlation.

$$F^{SMBH} = \sum_A \sum_{mno} \sum_{i<j} c_{mno}^A \left(\bar{r}_{iA}^m \bar{r}_{jA}^n + \bar{r}_{iA}^n \bar{r}_{ja}^m \right) \bar{r}_{ij}^o \tag{9.57}$$

Evaluating the SMBH correlation function requires $O(M^3)$ flops due to the summation over three particles. It can be the rate limiting step in the evaluation of the local energy even if the BH expansion includes only a small number of terms.

Manten and Luchow (ML) created a linear scaling algorithm for evaluating two-body terms in the BH expansion. [162] The short range scaled distance function used by ML, $\bar{r}^{ML} = 1 - e^{-\alpha r_{ij}}$, quickly approaches a constant value so that only linear number of terms in Eq. 9.57 need to be evaluated explicitly.

A linear scaling algorithm for evaluation of three-body terms in the BH expansion has been described by Austin et al. [167, 168] Rewriting each term as a trace over a matrix product

$$F_{mno}^A = \sum_{i \neq j} \bar{r}_{Ai}^m \bar{r}_{ij}^0 \bar{r}_{jA}^n \tag{9.58}$$

and taking advantage of the sparsity in the $\bar{r}$ matrices. There is no such sparsity if $\bar{r}^{ML}$ is used because the asymptotic value of this function is one. To create sparsity, $\bar{r}$ is shifted so that its asymptotic value is zero. The result is identical to the correlation function used by Sun and Lester [169], $\bar{r}^{SL} = -e^{-\alpha r_{ij}}$.

9.4.4   Load Balancing

The independent motion of random walkers makes it exceptionally simple to use parallel computers for QMC calculations. The ensemble of walkers can be distributed among many processors and inter-processor communication is needed only for occasional averaging. A complication arises in DMC because the branching process can cause unpredictable imbalances in the number of walkers on each processor. This can severely degrade parallel efficiency if processors with fewer walkers must wait for others to complete before continuing past the parallel barrier associated with averaging. The cost of idle processor increases in significance because petascale computers have tens of thousands of processor cores. In order to maintain high parallel efficiency, a load balancing procedure should redistribute walkers after branching. For example, it is possible to send walkers from the processor with the greatest number of walkers to the one with the least until they have the same number of walkers. The process is repeated until all nodes have equal numbers of walkers. This procedure is less than ideal because iterative communication creates multiple communication barriers. The high cost of communication relative to flops magnifies the need for an alternative strategy. Early versions of the Zori program [170] used the same procedure, but the result of the iterative procedure was computed before any walkers were exchanged; this reduced the amount of communication, but the $O(N^2_{proc})$ memory required to store the 'transfer-matrix' is impossibly large when several thousand processors are used. A recently developed load balancing algorithm requires only one global communication step and two $O(N_{proc})$ arrays [168].

9.4.5   Parallelization and Hardware Acceleration

The advances in the design of high performance computing (HPC) systems and evolution of hardware solutions are of immediate relevance especially in view of the intrinsic parallelism of QMC. Traditional HPC is represented by supercomputers and computer clusters. Graphics processing units (GPU) are gaining momentum as a major component of such systems [171–173]. The major factors here are their high floating point throughput, high memory bandwidth, and low price. The tradeoffs are the accuracy due to single precision arithmetic and lack of code transferability between platforms due to dominance of platform-specific languages. The overall speed-up of a QMC application on a GPU platform can be up to 6x, with up to 30x speed-up achieved on individual kernels [172, 173]. Using QMC as a benchmark scientific application, a model for prediction of any GPU performance has been developed [171]. Programming and debugging efforts required to achieve parallelism in different GPU-based platforms using various languages were assessed [171]. Importantly, it was demonstrated [172] that single-precision GPU results are comparable to single-precision CPU results and any remaining inaccuracy was negligible.

The potential of reconfigurable systems exploiting polygranular parallelism of platforms based on field-programmable gate arrays (FPGA) and microprocessors is being explored [171, 174]. The most computationally intensive parts of QMC algorithm are mapped onto FPGA hardware so that it is coupled to the remaining software implementation. With hardware acceleration of the wave function and potential energy calculations, a speed up of a factor of 25 was observed in comparison to a serial reference code [174]. It was attributed to pipelining, use of fixed-point arithmetic, and fine-grained parallelism of FPGA-based systems. Significant modification of a QMC algorithm are required to make it work with fixed-point arithmetic. This increases development and debugging costs compared to GPU-based systems [174].

Public resource computing (PRC) can be seen as an alternative to HPC centers. It relies on the donation of the spare time by personal computer owners worldwide. In the first large scale distributed quantum chemical project, QMC@HOME, FN-DMC was used to obtain interaction energies in the stacked and Watson-Crick adenine/thymine and cytosine/guanine DNA base pair complexes [175]. By 2008 the project had access to 15 Tflop/s sustained computing power facilitating QMC studies of medium to large systems. This computing power was in the range of a contemporary Top500 supercomputer for the price of a mid-size server-system. A specially adopted version of the Amolqc [176] software was used for the QMC calculations. Work-scheduling, data-handling, accounting and community features for this work were provided by Berkeley Open Infrastructure for Network Computing (BOINC) which is based on standard web-server components [177].

The design of very low cost QMC specific parallel systems can be enabled by an efficient "manager-worker" parallelization algorithm [178]. Its performance on a homogeneous platform is essentially equal to the standard pure iterative parallelization algorithm, but is more favorable in the case of loosely coupled heterogeneous systems built with commodity hardware, including networking.

9.5 Conclusions

In this review we attempted to present a variety of QMC methods for electronic structure including basic theoretical aspects, considerations of computational efficiency and applicability to a wide range of chemical problems and systems. The latter include ground and excited state properties of atoms, molecules, and molecular assemblies. Computations of various energetic characteristics with FN-DMC are emphasized owing to the exceptional accuracy, reliability and robustness of this method. Until recently, the applications of the QMC method were limited to small systems of light atoms. Recent developments are presented that clearly indicate feasibility of QMC computations of large systems that often are beyond the rich of other ab initio methods.

Recent progress in algorithms and growth of available computational power in combination with inherent parallelism of QMC is pivotal in overcoming limitations imposed by the statistical nature of simulations. Even large systems in a very challenging chemical context can be simulated with statistical errors reduced to the level of chemical accuracy. Methodological advance have enabled application of QMC in its different forms to other electronic properties besides the energy. The ability of QMC to sample numerically quantum operators and wave functions that cannot be integrated or represented analytically remains the greatest strength of the approach. This capability has facilitated significant contributions of the QMC method to fundamental understanding of the electronic wave function and its properties. The rigorous nature of the concepts behind QMC amounts to reliability of it results that are considered as reference for other methods to be compared against.

The following considerations are pertinent to the future of QMC methods. The fermion sign problem is yet to be resolved. As far as a related fixed-node approximation to DMC, it recovers from 90% to 95% of the correlation energy and therefore captures enough many-body effects to predict significant quantities such as cohesive energies, barrier heights, optical gaps, and similar quantities to within a few percent of experiments. The missing correlation is nevertheless crucial for many subtle effects such as magnetic phenomena, differences between low-lying near degenerate states and macroscopic quantum phenomena such as superconductivity observed in, for example, certain transition metal oxides.

Another issue is development of methods suitable for very large systems. The challenges here are associated with necessity for more efficient and accurate representation of orbitals, more robust elimination of finite size errors for solids and surface calculations with periodic boundary conditions. Advancement in this area also imply certain progress in QMC-based dynamical methods. The issue of QMC approaches that treat nuclear and electronic problem on the same footing remains open.

Acknowledgements WAL was supported by the Director, Office of Energy Research, Office of Basic Energy Sciences, Chemical Sciences, Geosciences and Biosciences Division of the US Department of Energy, under Contract No. DE-AC03-76 F00098. DYZ was supported by the National Science Foundation under grant NSF CHE-0809969.

References

1. Hammond BL, Lester WA Jr, Reynolds PJ (1994) Monte Carlo methods in ab initio quantum chemistry. World Scientific, Singapore
2. Metropolis N, Ulam S (1949) J Am Stat Assoc 44:335
3. Kalos MH (1962) Phys Rev 128:1791
4. Anderson JB (1975) J Chem Phys 63:1499
5. Schmidt KE (1986) Variational and green's function Monte Carlo calculations of few body systems. Conference on models and methods in few body physics, Lisbon
6. Lester WA Jr, Hammond BL (1990) Annu Rev Phys Chem 41:283
7. Aspuru-Guzik A, Kollias AC, Salomon-Ferrer R, Lester WA Jr (2005) Quantum Monte Carlo: theory and application to atomic, molecular and nano systems. In: Rieth M, Schommers W

(eds) Handbook of theoretical and computational nanotechnology. American Scientific Publishers, Stevenson Ranch

8. Ceperley DM, Mitas L (1996) Quantum Monte Carlo methods in chemistry. In: Prigogine I, Rice SA (eds) New methods in computational quantum mechanics, vol XCIII, Advances in chemical physics. Wiley, New York

9. Acioli PH (1997) J Mol Struct (Theochem) 394:75

10. Bressanini D, Reynolds PJ (1998) Adv Chem Phys 105:37

11. Mitas L (1998) Diffusion Monte Carlo. In: Nightingale MP, Umrigar CJ (eds) Quantum Monte Carlo methods in physics and chemistry. Kluwer Academic Publishers, Dordrecht

12. Anderson JB (1999) Quantum Monte Carlo: atoms, molecules, clusters, liquids and solids. In: Lipkowitz KB, Boyd DB (eds) Reviews in computational chemistry. Wiley, New York

13. Luchow A, Anderson JB (2000) Annu Rev Phys Chem 51:501

14. Foulkes M, Mitas L, Needs R, Rajagopal G (2001) Rev Mod Phys 73:33

15. Aspuru-Guzik A, Lester WA Jr (2003) Quantum Monte Carlo methods for the solution of the Schroedinger equation for molecular systems. In: Le Bris C (ed) Computational chemistry, vol X, Handbook of numerical analysis. Elsevier, Amsterdam

16. Koonin SE, Meredith DC (1995) Computational physics, FORTRAN version. Addison Wesley, Reading

17. Gould H, Tobochnik J (1996) An introduction to computer simulation methods: applications to physical systems. Addison Wesley, Reading

18. Thijssen JM (1999) Computational physics. Cambridge University Press, Press

19. Ceperley DM (1995) Rev Mod Phys 67:279

20. Sarsa A, Schmidt KE, Magro WR (2000) J Chem Phys 113:1366

21. Bauer WF (1958) J Soc Ind Appl Math 6:438

22. Hammersley JM, Handscomb DC (1964) Monte Carlo methods. Methuen, London

23. Halton JH (1970) SIAM Rev 12:1

24. Wood WW, Erpenbeck JJ (1976) Annu Rev Phys Chem 27:319

25. McDowell K (1981) Int J Quant Chem: Quant Chem Symp 15:177

26. Kalos MH, Whitlock PA (1986) Monte Carlo methods volume 1: basics. Wiley, New York

27. Sobol IM (1994) A primer for Monte Carlo method. CRC Press, Boca Raton

28. Fishman GS (1996) Monte Carlo: concepts, algorithms and applications. Springer, New York

29. Manno I (1999) Introduction to the Monte Carlo method. Akademiai Kiado, Budapest

30. Doucet A, de Freitas N, Gordon N, Smith A (eds) (2001) Sequential Monte Carlo methods in practice. Springer, New York

31. Liu JS (2001) Monte Carlo strategies in scientific computing. Springer, New York

32. Metropolis N, Rosenbluth AW, Rosenbluth MN, Teller NM, Teller E (1953) J Chem Phys 21:1087

33. Ceperley DM, Bernu B (1988) J Chem Phys 89:6316

34. Einstein A (1987) The collected papers of Albert Einstein, investigations on the theory of Brownian Movement. Princeton University Press, Princeton

35. Assaraf R, Caffarel M, Khelif A (2000) Phys Rev E 61:4566

36. Umrigar CJ, Nightingale MP, Runge KJ (1993) J Chem Phys 99:2865

37. DePasquale MF, Rothstein SM, Vrbik J (1988) J Chem Phys 89:3629

38. Grossman JC (2002) J Chem Phys 117:1434

39. Flad HJ, Caffarel M, Savin A (1997) Recent advances in quantum Monte Carlo methods. World Scientific, Singapore

40. Bressanini D, Morosi G (2008) J Chem Phys 129:054103

41. Bressanini D, Morosi G, Tarasco S (2005) J Chem Phys 123:204109

42. Ortiz JV, Weiner B, Ohrn Y (1981) Int J Quant Chem S15:113

43. Casula M, Sorella S (2003) J Chem Phys 119:6500

44. Domin D, Braida B, Lester WA Jr (2008) J Phys Chem A 112:8964

45. Anderson AG, Goddard WA III (2010) J Chem Phys 132:164110

46. Bajdich M, Mitas L, Drobny G, Wagner LK, Schmidt KE (2006) Phys Rev Lett 96:130201

47. Lopez Rios P, Ma A, Drummond ND, Towler MD, Needs RJ (2006) Phys Rev E 74:066701

48. Reynolds PJ, Dupuis M, Lester WA Jr (1985) J Chem Phys 82:1983
49. Umrigar CJ, Toulouse J, Filippi C, Sorella S, Hennig RG (2007) Phys Rev Lett 98:110201
50. Umrigar CJ, Wilson KG, Wilkins JW (1988) Phys Rev Lett 60:1719
51. Klein DJ, Pickett HM (1976) J Chem Phys 64:4811
52. Korsch HJ (1983) Phys Lett A 97:77
53. Ceperley DM (1991) J Stat Phys 63:1237
54. Glauser WA, Brown WR, Lester WA Jr, Bressanini D, Hammond BL, Koszykowski ML (1992) J Chem Phys 97:9200
55. Bressanini D, Reynolds P (2005) J Phys Rev Lett 95:110201
56. Ceperley DM, Alder B (1980) J Phys Rev Lett 45:566
57. Ceperley DM, Alder BJ (1984) J Chem Phys 81:5833
58. Bianchi R, Bressanini D, Cremaschi P, Morosi G (1993) Comput Phys Commun 74:153
59. Bianchi R, Bressanini D, Cremaschi P, Morosi G (1991) Chem Phys Lett 184:343
60. Bianchi R, Bressanini D, Chremaschi P, Morosi G (1993) J Chem Phys 98:7204
61. Koseki J, Maezono R, Tachikawa M, Towler MD, Needs RJ (2008) J Chem Phys 129:085103
62. Ceperley DM, Kalos MH (1986) Quantum Many-Body problems. In: Binder K (ed) Monte Carlo methods in statistical physics. Springer, New York
63. Kalos MH, Pederiva F (1999) Fermion Monte Carlo. In: Nightingale MP, Umrigar CJ (eds) Quantum Monte Carlo methods in physics and chemistry. NATO advanced study institute on quantum Monte Carlo methods in physics and chemistry. Springer, Dordrecht
64. Bajdich M, Tiago ML, Hood RQ, Kent PRC, Reboredo FA (2010) Phys Rev Lett 104:193001
65. Reboredo FA, Hood RQ, Kent PRC (2009) Phys Rev B 79:195117
66. Reboredo FA (2009) Phys Rev B 80:125110
67. Baer R, Head-Gordon M, Neuhauser D (1998) J Chem Phys 109:6219
68. Baer R (2000) Chem Phys Lett 324:101
69. Zhang SW, Krakauer H (2003) Phys Rev Lett 90:2003
70. Zhang SW, Krakauer H (2003) Phys Rev Lett 90:136401
71. Al-Saidi WA, Zhang SW, Krakauer H (2006) J Chem Phys 124:224101
72. Al-Saidi WA, Krakauer H, Zhang SW (2006) J Chem Phys 125:154110
73. Al-Saidi WA, Krakauer H, Zhang SW (2007) J Chem Phys 126:194105
74. Baroni S, Moroni S (1999) Phys Rev Lett 82:4745
75. Caffarel M, Claverie P (1988) J Chem Phys 88:1088, 88:1100
76. Yuen WK, Farrar TJ, Rothstein SM (2007) J Phys A: Math Theor 40:F639
77. Yuen WK, Oblinsky DG, Giacometti RD, Rothstein SM (2009) Int J Quant Chem 109:3229
78. Booth GH, Alavi A (2010) J Chem Phys 132:174104
79. Cleland D, Booth GH, Alavi A (2010) J Chem Phys 132:041103
80. Booth GH, Thom AJW, Alavi A (2009) J Chem Phys 131:054106
81. Christov IP (2006) Opt Express 14:6906
82. Christov IP (2007) J Chem Phys 127:134110
83. Christov IP (2008) J Chem Phys 129:214107
84. Christov IP (2008) J Chem Phys 128:244106
85. Christov IP (2009) J Phys Chem A 113:6016
86. Luan T, Curotto E, Mella M (2008) J Chem Phys 128:164102
87. Ramilowski JA, Farrelly D (2010) Phys Chem Chem Phys 12:12450
88. Toulouse J, Umrigar CJ (2008) J Chem Phys 128:174101
89. Luchow A, Petz R, Scott TC (2007) J Chem Phys 126:144110
90. Scott TC, Luchow A, Bressanini D, Morgan JD III (2007) Phys Rev A 75:060101(R)
91. Vrbik J, DePasquale MF, Rothstein SM (1988) J Chem Phys 88:3784
92. Barnett RN, Reynolds PJ, Lester WA Jr (1991) J Comput Phys 96:258
93. Barnett RN, Reynolds PJ, Lester WA Jr (1992) J Chem Phys 96:2141
94. Vrbik J, Legare DA, Rothstein SM (1990) J Chem Phys 92:1221
95. Vrbik J, Rothstein SM (1992) J Chem Phys 96:2071
96. Assaraf R, Caffarel M (2000) J Chem Phys 113:4028
97. Vrbik J (2008) Int J Quant Chem 108:493

98. Gaudoin R, Pitarke JM (2007) Phys Rev Lett 99:126406
99. Assaraf R, Caffarel M (2003) J Chem Phys 119:10536
100. Toulouse J, Assaraf R, Umrigar CJ (2007) J Chem Phys 126:244112
101. Per MC, Russo SP, Snook IK (2008) J Chem Phys 128:114106
102. Lee MW, Levchenko SV, Rappe AM (2007) Mol Phys 105:2493
103. Sorella S, Capriotti L (2010) J Chem Phys 133:234111
104. Wagner LK, Grossman JC (2010) Phys Rev Lett 104:210201
105. Coles B, Vrbik P, Giacometti RD, Rothstein SM (2008) J Phys Chem A 112:2012
106. Purwanto W, Al-Saidi WA, Krakauer H, Zhang S (2008) J Chem Phys 128:114309
107. Grimes RM, Hammond BL, Reynolds PJ, Lester WA Jr (1986) J Chem Phys 84:4749
108. Schautz F, Filippi C (2004) J Chem Phys 120:10931
109. Schautz F, Buda F, Filippi C (2004) J Chem Phys 121:5836
110. Sandvik AW, Vidal G (2007) Phys Rev Lett 99:220602
111. Bouabca T, Braida B, Caffarel M (2010) J Chem Phys 133:044111
112. Andrews SB, Burton NA, Hillier IH, Holender JM, Gillan MJ (1996) Chem Phys Lett 261:521
113. Slater JC (1930) Phys Rev 36:57
114. Davidson ER, Feller D (1986) Chem Rev 86:681
115. Ma A, Towler MD, Drummond ND, Needs RJ (2005) J Chem Phys 122:224322
116. Petruzielo FR, Toulouse J, Umrigar CJ (2010) J Chem Phys 132:094109
117. Junquera J, Paz O, Sanchez-Portal D, Artacho E (2001) Phys Rev B 64:235111
118. Harrison RJ, Fann GI, Yanai T, Gan Z, Beylkin G (2004) J Chem Phys 121:11587
119. Lowdin PO (1959) Adv Chem Phys 2:59
120. Head-Gordon M (1996) J Phys Chem 100:13213
121. Olsen J, Helgaker T, Jorgensen P (2000) Molecular electronic-structure theory. Wiley, New York
122. Szabo A, Ostlund NS (1996) Modern quantum chemistry. Courier Dover Publications, New York
123. Bartlett RJ (1981) Annu Rev Phys Chem 32:359
124. Shaik SS, Hiberty PC (2008) A chemist's guide to valence bond theory. Wiley, Hoboken
125. Cooper DL (2002) Valence bond theory. Elsevier, Amsterdam/Boston
126. Bouchaud JP, Georges S, Lhuillier C (1988) J De Phys 49:553
127. Goddard WA III, Dunning TH Jr, Hunt WJ, Hay PJ (1973) Acc Chem Res 6:368
128. Cullen J (1996) Chem Phys 202:217
129. Van Voorhis T, Head-Gordon M (2001) J Chem Phys 115:7814
130. Feynman RP, Cohen M (1956) Phys Rev 102:1189
131. Kwon Y, Ceperley DM, Martin RM (1993) Phys Rev B 48:12037
132. Holzmann M, Ceperley DM, Pierleoni C, Esler K (2003) Phys Rev E 68:046707
133. Greeff CW, Lester WA Jr (1998) J Chem Phys 109:1607
134. Stevens WJ, Basch H, Krauss M (1984) J Chem Phys 81:6026
135. Bachelet GB, Hamann DR, Schluter M (1982) Phys Rev B 26:4199
136. Burkatzki M, Filippi C, Dolg M (2008) J Chem Phys 129:164115
137. Burkatzki M, Filippi C, Dolg M (2007) J Chem Phys 126:234105
138. Casula M (2006) Phys Rev B 74:161102(R)
139. Umezawa N, Tsuneyuki S (2004) J Chem Phys 121:7070
140. Toulouse J, Umrigar CJ (2007) Chem Phys 126:084102
141. Huang CJ, Umrigar CJ, Nightingale MP (1997) J Chem Phys 107:3007
142. Boys SF, Handy NC (1969) Proc R Soc Lon Ser A 310:43
143. Moskowitz JW, Schmidt KE (1990) J Chem Phys 93:4172
144. Filippi C, Umrigar CJ (1996) J Chem Phys 105:213
145. Sun ZW, Huang SY, Barnett RN, Lester WA Jr (1990) J Chem Phys 93:3326
146. Nightingale MP, Melik-Alaverdian V (2001) Phys Rev Lett 87:043401
147. Lin X, Zhang H, Rappe AM (2000) J Chem Phys 112:2650
148. Press WH, Teukolsky SA, Vetterling WT, Flannery BP (1992) Numerical recipes. Cambridge University Press, Cambridge

149. Casula M, Attaccalite C, Sorella S (2004) J Chem Phys 121:7110
150. Sorella S (2001) Phys Rev B 64:024512
151. Luo H, Hackbusch W, Flad HJ (2009) J Chem Phys 131:104106
152. Aspuru-Guzik A (2004) Solving Schrodinger's Equation Using Random Walks. Ph.D. thesis, UC Berkeley
153. Bressanini D, Morosi G, Mella M (2002) J Chem Phys 116:5345
154. Drummond ND, Needs RJ (2005) Phys Rev B 72:085124
155. Snajdr M, Rothstein SM (2000) J Chem Phys 112:4935
156. Umrigar CJ, Filippi C (2005) Phys Rev Lett 94:150201
157. Angerson E, Bai Z, Dongarra J, Greenbaum A, Mckenney A, Du Croz J, Hammarling S, Demmel J, Bischof C, Sorensen D (1990) Lapack: a portable linear algebra library for high-performance computers. In: Proceedings of supercomputing, New York
158. Pollack L, Windus TL, de Jong WA, Dixon DA (2005) J Phys Chem A 109:6934
159. Ceperley DM (1986) J Stat Phys 43:815
160. Williamson AJ, Hood RQ, Grossman JC (2001) Phys Rev Lett 87:246406
161. Alfe D, Gillan MJ (2004) J Phys Condens Matter 16:L305
162. Manten S, Luchow A (2003) J Chem Phys 119:1307
163. Aspuru-Guzik A, Salomon-Ferrer R, Austin B, Lester WA Jr (2005) J Comput Chem 26:708
164. Foster JM, Boys SF (1960) Rev Mod Phys 32:305
165. Reboredo FA, Williamson AJ (2005) Phys Rev B 71:121105
166. Liu SB, Perez-Jorda JM, Yang W (2000) J Chem Phys 112:1634
167. Austin B, Aspuru-Guzik A, Salomon-Ferrer R, Lester WA Jr (2006) In: Anderson JB, Rothstein SM (eds) Advances in quantum Monte Carlo, ACS symposium series 953. American Chemical Society, Washington, DC
168. Austin BM (2009) Enhancing the quantum Monte Carlo method for electronic properties of large molecules and excited tates. PhD thesis, UC Berkeley
169. Sun ZW, Reynolds PJ, Owen RK, Lester WA Jr (1989) Theor Chim Acta 75:353
170. Aspuru-Guzik A, Salomon-Ferrer R, Austin B, Perusquia-Flores R, Griffin MA, Oliva RA, Skinner D, Domin D, Lester WA Jr (2005) J Comput Chem 26:856
171. Weber R, Gothandaraman A, Hinde RJ, Peterson GD (2011) IEEE Trans Parallel Distributed Syst 22:58
172. Meredith JS, Alvarez G, Maier TA, Schulthess TC, Vetter JS (2009) Parallel Comput 35:151
173. Anderson AG, Goddard WA III, Schröder P (2007) Comput Phys Commun 177:298
174. Gothandaraman A, Peterson GD, Warren GL, Hinde RJ, Harrison RJ (2008) Parallel Comput 34:278
175. Korth M, Luchow A, Grimme S (2008) J Phys Chem A 112:2104
176. Luchow A (2002) Amolqc. Universitat Dusseldorf, Dusseldorf
177. Anderson D BOINC; see http://boinc.berkeley.edu
178. Feldmann MT, Cummings JC, Kent DR IV, Muller RP, Goddard WA III (2008) J Comput Chem 29:8

Chapter 10
Relativistic Quantum Monte Carlo Method

Takahito Nakajima and Yutaka Nakatsuka

Abstract We have recently proposed a new relativistic treatment in the quantum Monte Carlo (QMC) technique using the zeroth-order regular approximation (ZORA) Hamiltonian. We derived a novel ZORA local energy expression and examined its effectiveness in variational Monte Carlo calculations. A cusp correction scheme for the relativistic ZORA-QMC method has also been proposed by extending the non-relativistic cusp correction scheme. In this scheme, molecular orbitals that appear in Jastrow–Slater wave functions are replaced with the exponential-type correction functions within a correction radius. This method is shown to be useful for improving the numerical stability of the ZORA-QMC calculations using both Gaussian-type and Slater-type basis functions. By using our ZORA-QMC method with Jastrow–Slater wave functions, we optimize the wave functions variationally and evaluate the relativistic and correlation effects simultaneously so that our approach can recover not only relativistic effects but also electron correlations.

Keywords Cusp correction • Electron correlation • Quantum Monte Carlo • Relativistic effect • Zeroth-order regular approximation

10.1 Introduction

An efficient inclusion of electron correlation is one of the most important problems in quantum chemistry. The quantum Monte Carlo (QMC) method [1–3] is a promising candidate for solving this problem. The QMC method provides an efficient numerical integration of high-dimensional wave functions with a statistical

T. Nakajima (✉) • Y. Nakatsuka
Computational Molecular Science Research Team, Advanced Institute for Computational Science, RIKEN, 7-1-26, Minatojima-minami, Cyuo, Kobe, Hyogo 650-0047, Japan
e-mail: nakajima@riken.jp; yutakana@riken.jp

J. Leszczynski and M.K. Shukla (eds.), *Practical Aspects of Computational Chemistry I: An Overview of the Last Two Decades and Current Trends*,
DOI 10.1007/978-94-007-0919-5_10, © Springer Science+Business Media B.V. 2012

treatment using random numbers; moreover, it can treat any arbitrary form of the wave function, even a form that is difficult to integrate, e.g., the explicitly correlated wave function containing electron–electron distances. The QMC method can easily and efficiently be parallelized, and its scaling behavior is moderate [estimated as $O(N^3)$] [2], because of its statistical character. This moderate scaling is an advantage over conventional correlation methods such as the coupled-cluster method. Among various QMC methods, two variants are often used for solving chemical problems. One is the variational Monte Carlo (VMC) method that directly integrates a given wave function, and the other is the diffusion Monte Carlo (DMC) method that propagates a trial function in imaginary time using the similarity between the Schrödinger equation and the diffusion equation. In both methods, multi-determinant and explicitly correlated wave functions can be used for expressing the static and dynamical electron correlation, and highly accurate total energies can be obtained with error bars originating from the statistical character of the methods.

The QMC method was originally developed in physics [1, 4] and was later used for solving chemical problems [5]. Although there are many studies on QMC development and its application to chemical problems [6, 7], few attempts have been made to deal with relativistic effects in the QMC approach. One of the studies on the application of the QMC method to chemical problems was carried out by Vrbik et al. in 1988 [8], where the relativistic correction terms in the Breit–Pauli (BP) approx-imation [9], such as the Darwin term and the mass-velocity term, were estimated for a LiH molecule using the DMC technique. In 1995, Kenny et al. applied the same technique to estimate the relativistic correction for light atoms up to Ne [10]. Although satisfactory results were obtained in both studies, the BP-QMC approach is only applicable to light atoms and cannot be applied to systems containing heavy elements where the relativistic effects are really important. One familiar relativistic treatment is the use of relativistic effective core potentials (RECPs) [11]. The QMC approach with RECPs is advantageous because RECPs reduce the number of electrons to be treated explicitly in the QMC calculation. The reduction in the number of electrons leads to a shorter computational time and smaller statistical errors. However, such an approach completely approximates core electrons and treats valence electrons with the non-relativistic (NR) local energy. Other treatment was adopted in the recent work by Caffarel et al. on the all-electron calculation of the Cu atom and its cation [12]. In their work, the relativistic effect on the total energy was evaluated using the molecular orbital (MO) theory as the difference between Dirac–Hartree–Fock (DHF) and Hartree–Fock (HF) total energies, and the electron correlation effect was evaluated separately from the NR DMC calculation. Their approach is not consistent from the theoretical point of view. A simultaneous treatment of the relativistic and electron correlation effects is desirable. In all studies except the work of Aissing in the one-electron system [13], the NR local energy is used to evaluate wave functions in the relativistic treatments. To treat the relativistic effect appropriately, we should begin with a relativistic Hamiltonian and a corresponding relativistic local energy. However, to our knowledge, there has been no attempt to derive the relativistic local energy in the quantum chemical application of the QMC method. Thus, we have recently proposed a new all-electron relativistic

VMC method [14, 15] by deriving the relativistic local energy of the scalar version of the zeroth-order regular approximation (ZORA) Hamiltonian [16–19] as the first attempt to develop the relativistic QMC method.

This review article is composed as follows. In Sect. 10.2, the QMC method is briefly reviewed. In Sect. 10.3, our relativistic variant of the QMC method is introduced. The R4QMC code, which is our current workhorse for the QMC calculation, is briefly described in Sect. 10.4. Several applications of our ZORA-VMC method are presented in Sect. 10.5.

10.2   Qumatum Monte Carlo Method

10.2.1   Quantum Monte Carlo Foundations

First, we briefly review statistical foundations of the Monte Carlo (MC) integration. The heart of the MC integration is the central limit theorem, which states the characterization for the average of random samples. Suppose d-dimensional space, the corresponding position vector $\mathbf{X} = (x_1, x_2, \ldots, x_d)$, and a set of mutually independent M position vectors $\{\mathbf{X}_i; i = 1, 2, \ldots, M\}$ which are distributed according to the probability density $P(\mathbf{X})$. The probability density $P(\mathbf{X})$ should satisfy the following relations,

$$P(\mathbf{X}) \geq 0, \tag{10.1}$$

$$\int d\mathbf{X} P(\mathbf{X}) = 1. \tag{10.2}$$

For any function $f(\mathbf{X})$ defined in this space, we can find the (ideal) average,

$$\mu_f = \int d\mathbf{X} f(\mathbf{X}) P(\mathbf{X}), \tag{10.3}$$

the variance,

$$\sigma_f^2 = \int d\mathbf{X} \left(f(\mathbf{X}) - \mu_f \right)^2 P(\mathbf{X}), \tag{10.4}$$

and the average on the random position vectors $\{\mathbf{X}_i; i = 1, 2, \ldots, M\}$,

$$Z_f = \frac{1}{M} \sum_{i=1}^{M} f(\mathbf{X}_i). \tag{10.5}$$

The central limit theorem indicates that in a large enough sample size M the average Z_f is normally distributed with the mean μ_f and the variance σ_f^2 / M

regardless of the form of the probability density P(**X**). As a result, by using the random sampling we can estimate the integration with the error which is estimated by the square-root of the variance decreasing as $1/\sqrt{M}$ irrespective of the dimension d.

The practical problem is how to generate a set $\{\mathbf{X}_i; i = 1, 2, \ldots, M\}$ according to the desired probability density P(**X**). In general, the complicated probability density is not completely known in advance. Especially the normalization is often unknown. The Metropolis algorithm offers an efficient way of generating such a set as a sequence of weakly correlated position vectors (the Markov chain) without any knowledge of normalization. The Metropolis method consists of the following steps:

1. Generate a random position vector $\mathbf{X} = \mathbf{X}_0$.
2. Choose a "trial" next position vector $\mathbf{X}'$ with some probability density function $T(\mathbf{X}'; \mathbf{X})$.
3. Accept the next position vector $\mathbf{X}'$ with probability,

$$A(\mathbf{X}' \leftarrow \mathbf{X}) = \min\left(1, \frac{T(\mathbf{X}; \mathbf{X}')P(\mathbf{X}')}{T(\mathbf{X}'; \mathbf{X})P(\mathbf{X})}\right). \tag{10.6}$$

If the trial next vector is accepted, $\mathbf{X}'$ is stored in the set $\{\mathbf{X}_i\}$ and the position is updated as $\mathbf{X} = \mathbf{X}'$; if rejected, $\mathbf{X}$ is unchanged and stored in the set $\{\mathbf{X}_i\}$.
4. Return to Step (2) and repeat until a sufficient number of vectors are stored in the set $\{\mathbf{X}_i\}$.

The generated vector set is distributed according to the probability density P(**X**) after sufficiently long steps. The initial points generated in the early stage should be discarded. If the equilibrium is established, the detailed balance should hold with the number of vectors in the volume element $d\mathbf{X}$ at $\mathbf{X}$ (written as $n(\mathbf{X})d\mathbf{X}$),

$$A(\mathbf{X} \leftarrow \mathbf{X}')T(\mathbf{X}; \mathbf{X}')n(\mathbf{X}')d\mathbf{X}d\mathbf{X}' = A(\mathbf{X}' \leftarrow \mathbf{X})T(\mathbf{X}'; \mathbf{X})n(\mathbf{X})d\mathbf{X}d\mathbf{X}', \tag{10.7}$$

where the left hand side corresponds to the number moving from the volume element $d\mathbf{X}'$ at $\mathbf{X}'$ to the volume element $d\mathbf{X}$ at $\mathbf{X}$ in one step. From Eq. 10.7 with Eq. 10.6, the relation $n(\mathbf{X})/n(\mathbf{X}') = P(\mathbf{X})/P(\mathbf{X}')$ is easily shown which means that the equilibrium distribution $n(\mathbf{X})$ is proportional to the desired density function P(**X**).

The probability density function $T(\mathbf{X}'; \mathbf{X})$ can be any function which satisfies the condition that if $T(\mathbf{X}'; \mathbf{X})$ is not zero $T(\mathbf{X}; \mathbf{X}')$ is also not zero. Computationally feasible functions are chosen as the probability density function $T(\mathbf{X}'; \mathbf{X})$ for the computational convenience. The vector set $\{\mathbf{X}_i\}$ generated with this algorithm has the serial correlation, which means that the sequential position vectors such as $\mathbf{X}_i$ and $\mathbf{X}_{i+1}$ are correlated. Thus, the error estimation may be modified with the auto-correlation time [20, 21] in the practical QMC calculations. In addition, the modification to the central limit theorem is also proposed in recent QMC study [22].

10.2.2   *Variational Monte Carlo Method*

The variational Monte Carlo (VMC) method is the simplest QMC method. Suppose the time-independent Schrödinger equation,

$$\hat{H}|\Phi\rangle = E|\Phi\rangle, \tag{10.8}$$

with eigenfunctions $\{\Phi_i; i = 0, 1, 2, \cdots\}$ and corresponding eigenvalues $\{E_i; i = 0, 1, 2, \cdots\}$. The energy expectation value of an arbitral given wave function Ψ_T gives an upper bound of the exact ground state energy E_0:

$$E = \frac{\langle \Psi_T|\hat{H}|\Psi_T\rangle}{\langle \Psi_T|\Psi_T\rangle} = \frac{\int d\mathbf{R}\Psi_T^*(\mathbf{R})\hat{H}\Psi_T(\mathbf{R})}{\int d\mathbf{R}\Psi_T^*(\mathbf{R})\Psi_T(\mathbf{R})} \geq E_0. \tag{10.9}$$

This condition is the variational principle with respect to the energy. In the VMC calculation, the energy expectation value is evaluated with the MC integration technique,

$$E = \frac{\int d\mathbf{R}|\Psi_T(\mathbf{R})|^2\Psi_T^{-1}(\mathbf{R})\hat{H}\Psi_T(\mathbf{R})}{\int d\mathbf{R}|\Psi_T(\mathbf{R})|^2}. \tag{10.10}$$

The term $|\Psi_T(\mathbf{R})|^2/\int d\mathbf{R}|\Psi_T(\mathbf{R})|^2$ and the local energy $E_L(\mathbf{R}) = \Psi_T^{-1}(\mathbf{R})\hat{H}\Psi_T(\mathbf{R})$ in this equation correspond to the probability density function $P(\mathbf{X})$ and $f(\mathbf{X})$ in Eqs. 10.1–10.5, respectively. The position vector $\mathbf{R} = \{\mathbf{r}_1, \mathbf{r}_2, \ldots, \mathbf{r}_N\}$ evolving in the Markov chain is treated as an imaginary particle called a "walker". The walkers are the discrete representation of $|\Psi_T(\mathbf{R})|^2$. The energy expectation value is evaluated by using the expression,

$$E \cong \frac{1}{M}\sum_{m=1}^{M} E_L(\mathbf{R}_m), \tag{10.11}$$

where $E_L(\mathbf{R}) = H\Psi(\mathbf{R})/\Psi(\mathbf{R})$ is called the local energy and M is the number of sample points. The sample points $\{\mathbf{R}_m\}$ are distributed according to the probability $\rho(\mathbf{R}) = |\Psi(\mathbf{R})|^2$. The NR local energy is written as the sum of local kinetic energy, electron–nuclear Coulomb, and electron–electron Coulomb terms,

$$E_L^{NR}(\mathbf{R}) = -\frac{1}{2}\sum_i \frac{\nabla_i^2\Psi(\mathbf{R})}{\Psi(\mathbf{R})} - \sum_i\sum_A \frac{Z_A}{R_{iA}} + \sum_{i<j}\frac{1}{r_{ij}}. \tag{10.12}$$

An error in the VMC result can be evaluated as

$$\sqrt{\sigma_{\text{est}}^2} = \sqrt{\sigma^2/N^{\text{eff}}}, \tag{10.13}$$

where $\sigma^2 = \langle E_L^2 \rangle - \langle E_L \rangle^2$ is the variance of the local energy and N^{eff} is the effective number of samples.

Another expression of the variational principle may be derived. From Eq. 10.8 eigenfunctions must satisfy the following condition for any position vector $\mathbf{R}$:

$$\langle \mathbf{R}|\hat{H}|\Phi \rangle = E \langle \mathbf{R}|\Phi \rangle \Rightarrow \frac{\langle \mathbf{R}|\hat{H}|\Phi \rangle}{\langle \mathbf{R}|\Phi \rangle} = E(\text{const.}). \tag{10.14}$$

This condition indicates that the variance of the local energy $E_L(\mathbf{R})$ is zero if and only if the function Ψ_T is a true eigenfunction of the Hamiltonian $\hat{H}$:

$$\sigma_{E_L}^2 = \frac{\int d\mathbf{R} |\Psi_T(\mathbf{R})|^2 [E_L(\mathbf{R}) - \langle E_L \rangle]^2}{\int d\mathbf{R} |\Psi_T(\mathbf{R})|^2} = \langle E_L^2 \rangle - \langle E_L \rangle^2 \geq 0. \tag{10.15}$$

This is the variational principle for the variance of the local energy. The evaluation of the variance is easy with the MC method. Thus, the variance is often adopted as the objective function in the optimization of the wave function within the QMC method. The efficient optimization of the wave function is an important problem in the QMC method and several works can be found in this line [23–30].

10.2.3   Diffusion Monte Carlo Method

The diffusion Monte Carlo (DMC) method is based on the simulation of the diffusion equation. Let us start with the primitive DMC method. By replacing the real time t with the imaginary time $\tau = it$, the time-dependent Schrödinger equation is rewritten as

$$\frac{\partial}{\partial \tau} \Psi(\mathbf{R}, \tau) = -\hat{H} \Psi(\mathbf{R}, \tau), \tag{10.16}$$

where $\hbar = 1$ is omitted for the ease of notation and the operator $\hat{H}$ is the time-independent electronic Hamiltonian. Since the non-relativistic Hamiltonian contains at most the second derivative, this equation can be considered as a diffusion equation. The formal solution of Eq. 10.16 is written as

$$\Psi(\mathbf{R}, \tau) = \exp(-\tau \hat{H}) \Psi(\mathbf{R}, 0). \tag{10.17}$$

With the spectral decomposition, the initial trial wave function is written as

$$\Psi(\mathbf{R}, 0) = \sum_{i=0} c_i(0)\Phi_i(\mathbf{R}),$$
(10.18)

where the functions $\{\Phi_i; i = 0, 1, \cdots\}$ are the time-independent eigenfunctions of $\hat{H}$ with the corresponding eigenvalues $\{E_i; i = 0, 1, \cdots\}$. Thus, Eq. 10.18 becomes

$$\Psi(\mathbf{R}, \tau) = \sum_{i=0} \exp(-\tau E_i) c_i(0)\Phi_i(\mathbf{R}).$$
(10.19)

The exponential factor grows for eigenfunctions with negative eigenvalues and decays for eigenfunctions with positive eigenvalues. In the long τ limit, the normalized wave function reaches the eigenfunction of the lowest eigenvalue,

$$\lim_{\tau \to \infty} \Psi^{\text{normalized}}(\mathbf{R}, \tau) = \Phi_0(\mathbf{R}).$$
(10.20)

Note that there are no constraints on $\{c_i(0)\}$ except for $c_0(0) \neq 0$, which guarantees the arbitrariness of the initial function. The $\exp(-\tau E_0)$ oscillation of $\Psi(\mathbf{R}, \tau)$ can be removed with the aid of the energy offset $E_T \cong E_0$, i.e.,

$$\frac{\partial}{\partial \tau}\Psi(\mathbf{R}, \tau) = -\left(\hat{H} - E_T\right)\Psi(\mathbf{R}, \tau),$$
(10.21)

is solved instead of Eq. 10.16.

The DMC method simulates this (imaginary) time evolution by repeating the short time operation:

$$\Psi(\mathbf{R}, \tau) = \exp\left[-\Delta\tau\left(\hat{H} - E_T\right)\right]\cdots\exp\left[-\Delta\tau\left(\hat{H} - E_T\right)\right]\Psi(\mathbf{R}, 0).$$
(10.22)

In the integral form, the evolved function $\Psi(\mathbf{R}, \tau + \Delta\tau)$ is written as

$$\Psi(\mathbf{R}, \tau + \Delta\tau) = \int d\mathbf{R}' G(\mathbf{R} \leftarrow \mathbf{R}'; \Delta\tau)\Psi(\mathbf{R}', \tau),$$
(10.23)

where $G(\mathbf{R} \leftarrow \mathbf{R}'; \Delta\tau) = \langle\mathbf{R}|\exp[-\Delta\tau(\hat{H} - E)]|\mathbf{R}'\rangle$ is the Green's function of the short-time operator $\exp[-\Delta\tau(\hat{H} - E_T)]$.

In the Boson case where $\Psi(\mathbf{R}, \tau)$ is positive (or negative) for all $\mathbf{R}$ and τ, the function $\Psi(\mathbf{R}, \tau)$ can be considered as the non-normalized probability density. Suppose that sufficient number of walkers are distributed according to the probability $\Psi(\mathbf{R}', \tau)$ at the time τ. If these walkers are moved according to the Green's function $G(\mathbf{R} \leftarrow \mathbf{R}'; \Delta\tau)$, they will be distributed according to the probability $\Psi(\mathbf{R}, \tau + \Delta\tau)$ in the next step, and finally distributed according to the true ground state $\Phi_0(\mathbf{R})$ after long steps. Now the problem is the practical operation corresponding to

$G(\mathbf{R} \leftarrow \mathbf{R}'; \tau)$. If we use the non-relativistic Hamiltonian $\hat{H} = \hat{T} + \hat{V}$, where $\hat{T}$ is the kinetic operator and $\hat{V}$ is the total potential energy operator, $\exp[-\tau(\hat{H} - E_T)]$ can be divided as

$$
\exp\left[-\tau\left(\hat{T} + \hat{V} - E_T\right)\right]
$$
$$
\cong \exp\left[-\frac{\tau\left(\hat{V} - E_T\right)}{2}\right] \exp\left(-\tau\hat{T}\right) \exp\left[-\frac{\tau\left(\hat{V} - E_T\right)}{2}\right] + O\left[\tau^3\right] \quad (10.24)
$$

by using Suzuki–Trotter formula. Thus, the approximate Green's function for small τ is given by

$$
G(\mathbf{R} \leftarrow \mathbf{R}'; \tau) = \left\langle \mathbf{R} \left| \exp\left[-\tau\left(\hat{T} + \hat{V} - E_T\right)\right] \right| \mathbf{R}' \right\rangle
$$
$$
\cong (2\pi\tau)^{-3N/2} \exp\left[-(\mathbf{R} - \mathbf{R}')^2 /2\tau\right] \exp\left[-\tau\left(V(\mathbf{R}) + V(\mathbf{R}') - 2E_T\right)/2\right].
$$
$$
(10.25)
$$

The former term $(2\pi\tau)c^{-3N/2} \exp[-(\mathbf{R} - \mathbf{R}')^2 /2\tau]$ is a random walk with the Gaussian distribution, and the latter term $\exp[-\tau\,[V(\mathbf{R}) + V(\mathbf{R}') - 2E_T]/2]$ is a weighting factor. The weighting factor is incorporated by assigning weights to walkers or by using the branching algorithm.

The above mentioned DMC method is applicable to bosons only, since it requires that the function $\Psi(\mathbf{R}; \tau)$ can be considered as a (non-normalized) probability density. For the fermionic system, the fixed-node (FN) approximation with a trial function is often used. By using a time-independent trial function $\Psi_T(\mathbf{R})$, the time-independent Schrödinger equation is rewritten as

$$
\hat{H}\,\Phi(\mathbf{R}) = E\Phi(\mathbf{R})
$$
$$
\Rightarrow \Psi_T^*(\mathbf{R})\hat{H}\,\Phi(\mathbf{R}) = E\Psi_T^*(\mathbf{R})\Phi(\mathbf{R})
$$
$$
\Rightarrow \hat{H}'\rho(\mathbf{R}) = E\rho(\mathbf{R}), \quad (10.26)
$$

where $\rho(\mathbf{R}) = \Psi^*(\mathbf{R})\phi(\mathbf{R})$ and $\hat{H}'$ is the effective Hamiltonian. The FN-DMC method is a modified version of the above mentioned DMC method: The basic equation $H'\rho_i(\mathbf{R}) = E_i\rho_i(\mathbf{R})$ and corresponding functions $\{\rho_i; i = 0, 1, \cdots\}$ are used and the Green's function is derived from the effective Hamiltonian H'. For the $\rho(\mathbf{R}) = \Psi_T^*(\mathbf{R})\Phi(\mathbf{R})$ to be positive-definite, the node structures of permitted functions $\{\tilde{\Phi}_i; i = 0, 1, \cdots\}$ which satisfy $H\tilde{\Phi}_i(\mathbf{R}) = \tilde{E}_i\tilde{\Phi}_i(\mathbf{R})$ and $\Psi_T^*(\mathbf{R})\tilde{\Phi}_i(\mathbf{R}) = \rho_i(\mathbf{R})$ should be the same as that of the trial function $\Psi_T(\mathbf{R})$. The energy estimation in the FN-DMC method is usually performed with the mixed estimator method [31]. After the equilibrium step, the probability density

$\rho_0(\mathbf{R}) = \Psi_T^*(\mathbf{R})\tilde{\Phi}_0(\mathbf{R})$ is obtained as the walker distribution. If the local energy of the trial function $E_L(\mathbf{R}; \Psi_T)$ is averaged over the probability density $\rho_0(\mathbf{R})$, the ground state energy $\tilde{E}_0$ is obtained by

$$
\begin{aligned}
\langle E_L(\mathbf{R}; \Psi_T)\rangle_{\rho_0} &= \int d\mathbf{R}\Psi_T^*(\mathbf{R})\tilde{\Phi}(\mathbf{R}) E_L(\mathbf{R}; \Psi_T)\big/\langle\Psi_T \mid \tilde{\Phi}\rangle \\
&= \int d\mathbf{R}\tilde{\Phi}^*(\mathbf{R})\Psi_T(\mathbf{R})\left[\hat{H}\,\Psi_T(\mathbf{R})\big/\Psi_T(\mathbf{R})\right]\big/\langle\tilde{\Phi} \mid \Psi_T\rangle \\
&= \int d\mathbf{R}\tilde{\Phi}^*(\mathbf{R})\hat{H}\,\Psi_T(\mathbf{R})\big/\langle\tilde{\Phi} \mid \Psi_T\rangle \\
&= \langle\tilde{\Phi}\big|\hat{H}\big|\Psi_T\rangle\big/\langle\tilde{\Phi} \mid \Psi_T\rangle = \tilde{E}_0.
\end{aligned} \tag{10.27}
$$

10.2.4   Wave Functions and Selective Sampling in Optimization

The QMC method is thought to be promising for the treatment of dynamical and static electron correlation effects with the compact functional form of wave functions. One standard form of the wave function is the Jastrow–Slater wave function. The Jastrow–Slater wave function is defined by

$$
\Psi = J \cdot D^{\uparrow} D^{\downarrow}, \tag{10.28}
$$

where $J = \exp(S)$ is the symmetric Jastrow correlation factor [32] and $D^{\uparrow(\downarrow)}$ is the anti-symmetric Slater determinant for the up (down) spin electrons generated by the restricted HF/KS (RHF/RKS) or unrestricted HF/KS (UHF/UKS) calculation. Among several types of Jastrow factors developed for use in the QMC calculations, we adopted the rather simple Jastrow factor of Schmidt and Moskowitz [33, 34]. It is composed of several terms;

$$
S = \sum_{I, i<j} U_{Iij}, \tag{10.29}
$$

$$
U_{Iij} = \sum_{k}^{N(I)} \left(1 - \frac{1}{2}\delta_{l_{kI}, m_{kI}}\right) c_{kI} \left(\bar{r}_{iI}^{l_{kI}} \bar{r}_{jI}^{m_{kI}} + \bar{r}_{iI}^{m_{kI}} \bar{r}_{jI}^{l_{kI}}\right) \bar{r}_{ij}^{n_{kI}}, \tag{10.30}
$$

where the index I runs over the nuclei, ij runs over electron pairs, and k runs over the $N(I)$ terms centered on the I-th nucleus. $\bar{r}$ is the scaled distance $\bar{r} = r/(1 + br)$ whose parameter b is fixed to one in this work. c_{kI} is an adjustable parameter. Each term can be classified as electron–electron (e–e), electron–nucleus (e–n), and electron–electron–nucleus (e–e–n) terms according to three integers (l,m,n). For example, the $(0,0,1)$ term is a pure e–e term; the $(2,0,0)$ term is a pure e–n term; and the $(2,0,2)$ term is an e–e–n term.

In the optimization of the wave function, we use the modification of the variance minimization technique [35]. In the original scheme, the sample points are generated according to the initial wave function and are fixed during optimization. In our study, we select the sample points used in the optimization as follows: First, we generate sample points $\{\mathbf{R}_i; i = 1, 2, \ldots, M\}$ in the real space according to the initial wave function Ψ_0 by the random walk of a walker. If the local energy of Ψ_0 on a sample point $\mathbf{R}_i$ satisfies the condition $E_L(\mathbf{R}_i) < 2\bar{E}_L$ or $\bar{E}_L/2 < E_L(\mathbf{R}_i)$, we consider $\mathbf{R}_i$ as a bad sample point and discard it from $\{\mathbf{R}_i\}$, where $\bar{E}_L$ is the average of the local energies obtained from the preceding calculation. Our scheme is useful when the initial wave function that we use in the optimization is not sufficiently adequate and causes the instability of the optimization.

10.2.5   Electron–Nucleus Coalescence Condition

The electron–nucleus cusp condition depends on the cusp behavior of the wave function. In QMC calculations, the variance of the local energy is known to be seriously affected by the behavior of wave functions in two-particle collisions. When an electron approaches a nucleus of charge Z, the electron–nuclear Coulomb potential term in the local energy diverges as $-Z/r$. Because E_L is a constant for an exact wave function, the kinetic energy term of the local energy must show an equal divergence with the opposite sign. Kato's cusp condition [36] for the NR case is derived from a Taylor expansion of the NR local energy with respect to the electron–nucleus distance r. The requirement that the expansion coefficient of r^{-1}, $C_{-1} = -Z - \psi'(r)/\psi(r)|_{r \to 0}$, is zero in the $r \to 0$ limit ensures non-divergent local energy. Here, ψ and ψ' are the spherical average of the wave function at the nucleus and its radial derivative, respectively. This requirement leads to the condition,

$$\left.\frac{\partial \psi(r)}{\partial r}\right|_{r \to 0} = -Z\psi(r). \tag{10.31}$$

In the atomic case, the condition can be satisfied by using the Slater-type function $\exp(-\zeta r)$ with the exponent $\zeta = Z$. A similar cusp condition exists for the collision of two electrons. In the QMC method, the electron–electron cusp condition can easily be satisfied by using the Jastrow correlation factor where the inter-electron distance is explicitly included. On the other hand, some considerations are required for the electron–nucleus cusp condition. When using the Gaussian-type orbital (GTO) whose gradient is zero at the origin, the cusp condition cannot be satisfied and the local energy diverges at the nucleus. Even worse in the practical sense, the GTO wave function often gives a badly oscillating local energy surface around the nucleus, and this oscillation causes large variance. To solve this problem, several cusp correction schemes have been proposed. Manten and Lüchow [37] developed a scheme to correct the near-nucleus behavior of individual basis functions. Ma et al. [38] and Per et al. [39] developed the schemes to correct the MOs using analytical functions and quintic splines, respectively. Following Ma et al., Kussmann and Ochsenfeld [40] developed a correction scheme for the basis functions in density-matrix-based QMC method [41]. These correction schemes work well and considerably reduce the variance of the local energy of the GTO wave functions. These correction schemes were applied to the conventional NR QMC methods, and the exact cusp condition can be satisfied with the Slater-type orbitals (STOs) in atomic systems. In fact, Per et al. found that the correction did not improve the STO wave functions as far as the variance of the local energy was concerned [39].

10.3   Relativistic Quantum Monte Carlo Method

10.3.1   Dirac Hamiltonian

The relativistic quantum mechanics is based on the four-component Dirac equation in the external potential V,

$$H^{\text{Dirac}} \begin{pmatrix} \Psi^L \\ \Psi^S \end{pmatrix} = \begin{pmatrix} V & c\boldsymbol{\sigma} \cdot \mathbf{p} \\ c\boldsymbol{\sigma} \cdot \mathbf{p} & V - 2mc^2 \end{pmatrix} \begin{pmatrix} \Psi^L \\ \Psi^S \end{pmatrix} = E \begin{pmatrix} \Psi^L \\ \Psi^S \end{pmatrix}, \tag{10.32}$$

where Ψ^L and Ψ^S are the large and small components, respectively; $\boldsymbol{\sigma}$ is the 2×2 Pauli spin matrix vector;

$$\sigma_x \equiv \begin{pmatrix} 0 & \sigma_t \\ \sigma_t & 0 \end{pmatrix}, \quad \sigma_y \equiv \begin{pmatrix} 0 & -i \\ i & 0 \end{pmatrix}, \quad \sigma_z \equiv \begin{pmatrix} 1 & 0 \\ 0 & -1 \end{pmatrix}, \tag{10.33}$$

c is the speed of light; and E is the energy of a particle. Since the Dirac equation is valid only for the one-electron system, the one-electron Dirac Hamiltonian has to be extended to the many-electron Hamiltonian in order to treat the chemically interesting many-electron systems. The straightforward way to construct the relativistic many-electron Hamiltonian is to augment the one-electron Dirac operator with the Coulomb or Breit (or its approximate Gaunt) operator as a two-electron term. This procedure yields the Dirac–Coulomb (DC) or Dirac–Coulomb–Breit (DCB) Hamiltonian derived from quantum electrodynamics (QED). Recently, the Hartree–Fock (HF) and the Kohn–Sham (KS) methods with the DC(B) Hamiltonian have become familiar and powerful relativistic approaches with the continuous development of efficient computational algorithms [42–48].

10.3.2 Breit–Pauli Hamiltonian

Although the full four-component treatment with the Dirac Hamiltonian is ideal, the computation of four-component wave functions is expensive. Thus, since small components have little importance in most chemically interesting problems, various two- or one-component approximations to the Dirac Hamiltonian have been proposed. From Eq. 10.32, the Schrödinger–Pauli equation composed of only the large component is obtained as

$$\left[V + (\boldsymbol{\sigma} \cdot \mathbf{p}) \frac{c^2}{2c^2 - (V - E)} (\boldsymbol{\sigma} \cdot \mathbf{p}) \right] \Psi^{\mathrm{L}} = E \Psi^{\mathrm{L}}, \tag{10.34}$$

with the normalization condition,

$$\left\langle \Psi^{\mathrm{L}} \middle| 1 + X^\dagger X \middle| \Psi^{\mathrm{L}} \right\rangle = 1, \tag{10.35}$$

where the X operator is defined by

$$X = \frac{c \boldsymbol{\sigma} \cdot \mathbf{p}}{2c^2 - (V - E)}. \tag{10.36}$$

Note that no approximation has been made so far. The Breit–Pauli (BP) approximation [49] is introduced by expanding the inverse operators in the Schrödinger–Pauli equation in powers of $(V - E)/2c^2$ and ignoring the higher-order terms. Instead, the BP approximation can be obtained truncating the Taylor expansion of the FW transformed Dirac Hamiltonian up to the $(p/c)^2$ term. The one-electron BP Hamiltonian for the Coulomb potential $V = Z\mathbf{r}/r^3$ is represented by

$$H_{\mathrm{BP}} = T + V - \frac{\mathbf{p}^4}{8c^2} + \frac{Z\delta(r)}{8c^2} + \frac{Z\mathbf{s} \cdot \mathbf{l}}{2c^2 r^3}, \tag{10.37}$$

where $\mathbf{r} \times \mathbf{p} = \mathbf{l}$ and $\boldsymbol{\sigma} = 2\mathbf{s}$. The T operator is the non-relativistic kinetic energy. The BP equation has the well-known mass-velocity, Darwin, and spin–orbit operators. Although the BP equation gives reasonable results in the first-order perturbation calculation, it cannot be used in the variational treatment so that the BP-QMC approach would break down for the systems containing heavy elements.

10.3.3 ZORA Hamiltonian

One of promising relativistic approximations to the Dirac Hamiltonian is the regular approximation (RA). One of the shortcomings of the Breit–Pauli (BP) approach, which is a traditional relativistic approximation, is that the expansion in $(p/c)^2$ is not justified in the case where the electronic momentum is too large, e.g., for a

Coulomb-like potential. The zeroth-order regular approximation (ZORA) avoids this disadvantage by expanding in $E/(2c^2 - V)$ up to the first order so that the ZORA Hamiltonian is variationally stable. The ZORA Hamiltonian was first derived by Chang et al. in 1986 [16], and later rediscovered as an approximation to the FW transformation by van Lenthe et al. [17–19]. The ZORA Hamiltonian of one electron in the external potential V is given by

$$H^{\mathrm{ZORA}} = (\boldsymbol{\sigma}\mathbf{p})\frac{c^2}{2c^2 - V}(\boldsymbol{\sigma}\mathbf{p}) + V. \tag{10.38}$$

The derivation of the ZORA approach is valid only for the one-electron Dirac equation with an external potential. Thus, the theory must be extended in order to obtain the relativistic many-electron ZORA Hamiltonian with the electron–electron Coulomb or Breit interaction. The many-electron ZORA Hamiltonian may be defined in several ways. In the present study, we neglect the relativistic kinematics correction to the electron–electron interaction, which yields the simplest many-electron ZORA Hamiltonian, that is, the one-electron ZORA Hamiltonian with the electron–electron Coulomb operator in the non-relativistic form,

$$H^{\mathrm{ZORA}} = \sum_i^N (\boldsymbol{\sigma}\mathbf{p}_i)\frac{c^2}{2c^2 - V_i}(\boldsymbol{\sigma}\mathbf{p}_i) - \sum_i^N \sum_A \frac{Z_A}{R_{iA}} + \sum_{i<j}^N \frac{1}{r_{ij}}, \tag{10.39}$$

where V_i is the nuclear attraction potential with the i-th electron. The relativistic modification of the two-electron Coulomb term will be discussed in a future study. This kind of the Hamiltonian is the two-component Hamiltonian and can be separated into the spin-free and spin-dependent parts. The spin-dependent part includes the spin–orbit (SO) interaction. In the present study, furthermore, we neglect the spin-dependent effect;

$$
\begin{aligned}
H^{\mathrm{ZORA}} &= \sum_i^N \mathbf{p}_i \frac{c^2}{2c^2 - V_i}\mathbf{p}_i - \sum_i^N \sum_A \frac{Z_A}{R_{iA}} + \sum_{i<j}^N \frac{1}{r_{ij}} \\
&= \sum_i^N T_i^{\mathrm{ZORA}}(V_i) - \sum_i^N \sum_A \frac{Z_A}{R_{iA}} + \sum_{i<j}^N \frac{1}{r_{ij}},
\end{aligned} \tag{10.40}
$$

where T^{ZORA} is the spin-free ZORA kinetic energy term.

The higher-order RA Hamiltonians such as the first-order RA (FORA) have been also derived, but they cannot be used variationally because of the higher order derivatives. Instead, starting from an un-normalized FW transformation, Dyall and van Lenthe [50] introduced the infinite-order RA (IORA) equation,

$$H^{\mathrm{ZORA}}\Phi^{\mathrm{IORA}} = E^{\mathrm{IORA}}\left(1 + X_0^\dagger X_0\right)\Phi^{\mathrm{IORA}} \tag{10.41}$$

with

$$X_0 = \frac{c\boldsymbol{\sigma} \cdot \mathbf{p}}{2c^2 - V}. \tag{10.42}$$

The IORA equation corresponds to the ZORA equation with a modified metric operator. The numerical results with the IORA method show a considerable improvement over ZORA and for a many-electron system superior performance to FORA. While one of disadvantages of the ZORA and IORA methods is an incorrect dependence of energy eigenvalues on the choice of gauge in the electrostatic potential, the RA approach has an advantage of easier implementation in the MO calculation than other relativistic approximate approaches as will introduce in the following session.

10.3.4 Implementation of the ZORA Method into the MO Program

In usual MO calculations with the ZORA Hamiltonian, the atomic orbital integrals derived from the ZORA Hamiltonian are simple and are evaluated numerically in direct space. In our study, however, we use the resolution of identity (RI) approximation with finite basis functions to evaluate them. To this end we use the relation,

$$
\begin{aligned}
(\boldsymbol{\sigma}\mathbf{p})\frac{c^2}{2c^2 - V}(\boldsymbol{\sigma}\mathbf{p}) &= (\boldsymbol{\sigma}\mathbf{p})(\boldsymbol{\sigma}\mathbf{p})(\boldsymbol{\sigma}\mathbf{p})^{-1}\frac{c^2}{2c^2 - V}(\boldsymbol{\sigma}\mathbf{p})^{-1}(\boldsymbol{\sigma}\mathbf{p})(\boldsymbol{\sigma}\mathbf{p}) \\
&= \frac{1}{2}p^2\left[(\boldsymbol{\sigma}\mathbf{p})\left(1 - \frac{V}{2c^2}\right)(\boldsymbol{\sigma}\mathbf{p})\right]^{-1}p^2 \\
&= \frac{1}{2}p^2\left[p^2 - \frac{1}{2c^2}(\boldsymbol{\sigma}\mathbf{p})V(\boldsymbol{\sigma}\mathbf{p})\right]^{-1}p^2,
\end{aligned}
\tag{10.43}
$$

for the kinetic energy part of the ZORA Hamiltonian. For a practical calculation of Eq. 10.43, as in the implementation of the Douglas–Kroll Hamiltonian [51, 52] and the RESC Hamiltonian [53], the matrix elements with the ZORA Hamiltonian were evaluated using the RI in the space spanned by the eigenfunctions of the square momentum p^2 following Buenker et al. [54]. To evaluate the Eq. 10.43, the matrix elements of kinetic energy and $(\boldsymbol{\sigma}\mathbf{p})V(\boldsymbol{\sigma}\mathbf{p})$ operators are necessary.

10.3.5 Local Energy for ZORA Hamiltonian

In most of the approximate relativistic treatments, the first-order relativistic term can generally be written as $P^\dagger Q P$, where P is a functional of the momentum operator and Q is a functional of the position operator. In the QMC method, the real space is explored by the walkers so that the relativistic Hamiltonian that has a simpler form of P is adequate for application to the QMC approach. In our study, thus, the ZORA Hamiltonian is adopted because the momentum part is simple. Let us deal with an extension of the VMC approach to the relativistic treatment. With the Hamiltonian of Eq. 10.40 and the relation $\mathbf{p} = i\hbar\nabla$, the corresponding ZORA local energy is defined as

$$E_{\mathrm{L}}^{\mathrm{ZORA}}(\mathbf{R}) = \frac{\langle \mathbf{R}| H^{\mathrm{ZORA}} |\Psi\rangle}{\langle \mathbf{R}|\Psi\rangle} = \sum_i T_{\mathrm{L},i}^{\mathrm{ZORA}}(\mathbf{R}) - \sum_A \sum_i \frac{Z_A}{R_{iA}} + \sum_{i<j} \frac{1}{r_{ij}},$$
(10.44)

where

$$T_{\mathrm{L},i}^{\mathrm{ZORA}}(\mathbf{R}) = -\frac{c^2}{2c^2 - V_i(\mathbf{R})} \left[\frac{\nabla_i V_i(\mathbf{R})}{2c^2 - V_i(\mathbf{R})} \cdot \frac{\nabla_i \Psi(\mathbf{R})}{\Psi(\mathbf{R})} + \frac{\nabla_i^2 \Psi(\mathbf{R})}{\Psi(\mathbf{R})} \right] \quad (10.45)$$

is the ZORA local kinetic energy term. In the present approximations, that is, the neglect of the SO interaction and the kinematics correction to the two-electron interaction, the difference between ZORA and NR local energies is the local kinetic energy term. The additional operation is an evaluation of the first derivatives of the potentials. The ZORA local energy is constant for the exact wave function as in the NR case; therefore, variance minimization can be used to optimize the wave functions. Another local quantity may be derived by using the resolution of identity. It is referred to as the ZORA pseudo local energy and is written by

$$E_{\mathrm{pL}}^{\mathrm{ZORA}}(\mathbf{R}) = \sum_i \left[\frac{c^2}{2c^2 - V_i(\mathbf{R})} \left| \frac{\nabla_i \Psi(\mathbf{R})}{\Psi(\mathbf{R})} \right|^2 \right] + V_{\mathrm{e-n}}(\mathbf{R}) + V_{\mathrm{e-e}}(\mathbf{R}). \quad (10.46)$$

This quantity is not constant even for the exact wave function and gives larger error bars in the practical calculations. However, it can be useful for checking the quality of MC integration points; if the entire space is well explored in an MC walk according to the probability density $\rho(\mathbf{R}) = |\Psi(\mathbf{R})|^2$, the average of $E_{\mathrm{pL}}^{\mathrm{ZORA}}$ should be equal to that of $E_{\mathrm{L}}^{\mathrm{ZORA}}$. In our study, we discard the set of sample points in which the difference in the average values between $E_{\mathrm{L}}^{\mathrm{ZORA}}$ and $E_{\mathrm{pL}}^{\mathrm{ZORA}}$ is larger than a criterion.

10.3.6 Electron–Nucleus Cusp Condition in ZORA-QMC Method

As mentioned in Sect. 10.2.5, the exact cusp condition can be satisfied with the STOs in the NR QMC method. Unlike the NR QMC method, however, the divergence of the ZORA local energy at the nucleus cannot be avoided even with the STOs. Let us derive a condition that should be satisfied by the ZORA wave function. Following the derivation of the NR cusp condition, we simplify the problem as a two-body problem of one electron and one nucleus of charge Z. The ZORA local energy is written as

$$E_L^{ZORA}(\mathbf{r}) = -\frac{c^2 r}{2c^2 r + Z}\left[\frac{Z}{2c^2 r + Z}\frac{\mathbf{r}}{r^2}\cdot\frac{\nabla\phi(r)}{\phi(r)} + \frac{\nabla^2\phi}{\phi(r)}\right] - \frac{Z}{r}. \qquad (10.47)$$

Assuming ϕ to be a spherical function, $\nabla\phi = \mathbf{r}\phi'/r$ and $\nabla^2\phi = 2\phi'/r + \phi''$, where $\phi' = \partial\phi/\partial r$ and $\phi'' = \partial^2\phi/\partial r^2$, respectively. Equation 10.47 then becomes

$$E_L^{ZORA}(r) = -\frac{c^2 r}{2c^2 r + Z}\left[\frac{4c^2 r + 3Z}{r(2c^2 r + Z)}\frac{\phi'}{\phi} + \frac{\phi''}{\phi}\right] - \frac{Z}{r}. \qquad (10.48)$$

If the terms ϕ'/ϕ and ϕ''/ϕ do not diverge at $r = 0$, the kinetic energy term cannot cancel the divergence of the potential term. Therefore, to ensure the divergence of ϕ'/ϕ, we assume a 1s-like orbital $\phi = r^\alpha \exp(-\zeta r)$ with $\alpha < 0$. This form of function is similar to the exact solution to the Dirac equation for a hydrogen atom [55]. Thus the ϕ'/ϕ and ϕ''/ϕ become

$$\frac{\phi'}{\phi} = \frac{\alpha}{r} - \zeta, \qquad (10.49)$$

$$\frac{\phi''}{\phi} = \frac{\alpha(\alpha-1)}{r^2} - 2\alpha\zeta + \zeta^2. \qquad (10.50)$$

With this orbital, Eq. 10.48 is written as

$$E_L^{ZORA}(r) = -\frac{c^2 r}{2c^2 r + Z}\left[\frac{4c^2 r + 3Z}{r(2c^2 r + Z)}\left(\frac{\alpha}{r} - \zeta\right) + \frac{\alpha(\alpha-1)}{r^2} - 2\alpha\zeta + \zeta^2\right] - \frac{Z}{r}. \qquad (10.51)$$

By the Taylor expansion of $(2c^2 r + Z)^{-1}$ and $(2c^2 r + Z)^{-2}$ in the $r \to 0$ limit, we obtain

$$E_L^{ZORA}(r) = C_{-1}r^{-1} + C_0 + C_1 r + \dots, \qquad (10.52)$$

where $C_{-1} = -\alpha(\alpha + 2)c^2/Z - Z$. To cancel the divergence, we set $C_{-1} = 0$, which leads to

$$\alpha = -1 \pm \sqrt{1 - (Z/c)^2}. \qquad (10.53)$$

When considering the NR limit $c \to \infty$, α should be zero because it corresponds to the 1s orbital. Therefore a positive sign is adopted in Eq. 10.53,

$$\alpha = -1 + \sqrt{1 - (Z/c)^2}. \tag{10.54}$$

It should be noted that unlike the NR case the condition in the $r \to 0$ limit is independent of the exponent ζ.

10.3.7 Cusp Correction Algorithm

The behavior of the local energy around the nucleus is more severe for the ZORA-QMC results, because relativistic effects cause the contraction of s and p orbitals and increase the electron density near the nucleus. Thus, an adequate relativistic cusp correction scheme is necessary to treat heavier elements where the relativistic effects are important. Recently, we proposed a cusp correction scheme for the ZORA-VMC formalism [15]. Strictly speaking, the term "cusp correction" may not be adequate for our orbital correction scheme because relativistic wave function does not have a cusp but divergences at electron–nucleus collision. However, we call our relativistic orbital correction scheme as a "relativistic cusp correction scheme" because the term "cusp correction" makes clearer that our scheme is a relativistic extension of the NR cusp correction scheme.

Let us consider the cusp correction scheme for the Jastrow–Slater wave function. This type of wave function allows us to use a MO correction scheme. The s type component of the 1s MO inside a given radius r_c is replaced with the correction function,

$$\psi = \phi + \eta \to \tilde{\psi} = \tilde{\phi} + \eta, \tag{10.55}$$

where ψ is the whole MO, ϕ is the spherically symmetric component, η is the remaining component, and corrected functions are indicated by a tilde. In our ZORA-QMC study, the exponential-type function of Ma et al. [38] was adopted as the functional form of the correction, and a logarithmic term was newly added to satisfy the condition of Eq. 10.52. The functional form is written as

$$\tilde{\phi} = C + \text{sgn}\,[\phi(0)]\exp\,[p(r)], \tag{10.56}$$

where

$$p(r) = \alpha_{\ln}\ln r + p_0(r) = \alpha_{\ln}\ln r + \alpha_0 + \alpha_1 r + \alpha_2 r^2 + \alpha_3 r^3 + \alpha_4 r^4, \tag{10.57}$$

and C is a constant shift to ensure that $\tilde{\phi} - C$ is of one sign within r_c. The $p_0(r)$ term is the NR term of Ma et al. and the following terms are used for convenience of notation,

$$R(r) = \exp[p(r)], \tag{10.58}$$

$$R_0(r) = \exp[p_0(r)]. \tag{10.59}$$

One constraint in determining the coefficients is the continuity at r_c;

$$\ln|\phi(r_c) - C| = p(r_c) = X_1, \tag{10.60}$$

$$\frac{1}{R(r_c)}\frac{d\phi}{dr}\bigg|_{r_c} = p'(r_c) = X_2, \tag{10.61}$$

$$\frac{1}{R(r_c)}\frac{d^2\phi}{dr^2}\bigg|_{r_c} = p''(r_c) + p'^2(r_c) = X_3. \tag{10.62}$$

A further constraint is the condition given by Eq. 10.54,

$$\alpha_{ln} = -1 + \sqrt{1 - (Z/c)^2}. \tag{10.63}$$

The following two constraints, which require that the coefficients reach the NR coefficients in the $c \rightarrow \infty$ limit are also used,

$$p_0'(0) = -Z\left[\frac{C + R_0(0) + \eta(0)}{R_0(0)}\right] = X_4, \tag{10.64}$$

$$p_0(0) = X_5. \tag{10.65}$$

As in the NR case, the nonlinear constraints can be solved analytically, giving

$$\alpha_{ln} = -1 + \sqrt{1 - (Z/c)^2},$$

$$\alpha_0 = X_5,$$

$$\alpha_1 = X_4,$$

$$\alpha_2 = 6\frac{X_1'}{r_c^2} - 3\frac{X_2'}{r_c} + \frac{X_3'}{2} - 3\frac{X_4}{r_c} - 6\frac{X_5}{r_c^2} - \frac{X_2'^2}{2},$$

$$\alpha_3 = -8\frac{X_1'}{r_c^3} + 5\frac{X_2'}{r_c^2} - \frac{X_3'}{r_c} + 3\frac{X_4}{r_c^2} + 8\frac{X_5}{r_c^3} + \frac{X_2'^2}{r_c},$$

$$\alpha_4 = 3\frac{X_1'}{r_c^4} - 2\frac{X_2'}{r_c^3} + \frac{X_3'}{2r_c^2} - \frac{X_4}{r_c^3} - 3\frac{X_5}{r_c^4} - \frac{X_2'^2}{2r_c^2}. \tag{10.66}$$

where

$$X_1' = X_1 - \alpha_{\ln} \ln(r_c),$$

$$X_2' = X_2 - \alpha_{\ln} r_c^{-1},$$

$$X_3' = X_3 + \alpha_{\ln} r_c^{-2}. \tag{10.67}$$

Because X_5 cannot be determined by the constraints, it is optimized to make the following effective one-electron local energy flat;

$$E_L^{s,ZORA}(r) = \tilde{\phi}^{-1} \left[-\frac{c^2}{2c^2 - V_i} \left(\frac{\nabla V_i}{2c^2 - V_i} \cdot \nabla + \nabla^2 \right) - \frac{Z_{eff}}{r} \right] \tilde{\phi}. \tag{10.68}$$

The effective nuclear charge Z_{eff} is given by

$$Z_{eff} = Z \left[1 + \frac{\eta(0)}{C + R_0(0)} \right]. \tag{10.69}$$

Unlike the NR case, this definition is only an approximation because $R_0(0)$ is used instead of $R(0)$. Beginning with $X_5 = \phi(0)$, X_5 was optimized by minimizing the variance of $E_L^{s,ZORA}$ for $r_{min} < r < r_c$. Here $r_{min} \neq 0$ is determined by examining of the local energy surface, because the correction is more important near the correction radius.

10.4   R4QMC Program

We are currently developing the relativistic and non-relativistic QMC program for atomic and molecular QMC calculations called "R4QMC". The R4QMC code is written in Fortran90. The current capability of the code is as follows:

1. Non-relativistic QMC: Variational Monte Carlo and fixed–node Diffusion Monte Carlo.
2. Relativistic QMC: ZORA-VMC and fixed-node ZORA-DMC. The IORA version of the QMC method is under development.
3. Wave function: Single or Multi-determinant Jastrow–Slater wave function and their linear combinations are available. Jastrow factors of Schmidt–Moskowitz type are available.
4. Basis set: Slater type and Gaussian type basis functions.
5. Wave function optimization: Variance minimization, energy minimization, and their combination with fixed sample points.
6. Local and non–local pseudo potential.

10.5   Illustrative Results

10.5.1   Cusp Correction Effects

In Fig. 10.1, the 1s orbitals of the Ne atom are plotted with and without the cusp correction. GTO and STO orbitals are used in this calculation. The condition $(0.08|30)$, where $r_c = 0.08$a.u. and r_{min} is 30% of r_c, is used. Note that the corrected orbitals are not normalized. The original and corrected orbitals have similar shapes except in the region of $r \cong 0$. This observation in the relativistic case is the same as that seen for the case of NR cusp correction scheme of Ma et al. in the GTO case.

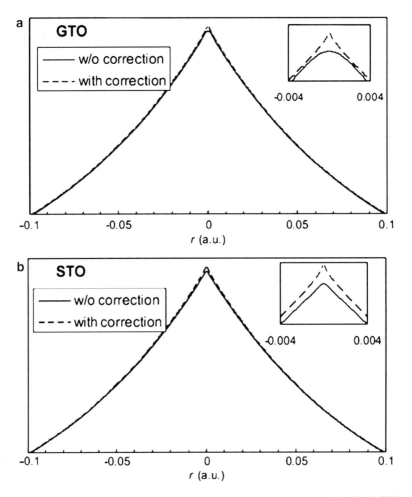

Fig. 10.1 Relativistic 1s orbital of Ne atom with and without the cusp correction. Both GTO and STO orbitals are obtained from the ZORA-HF calculation

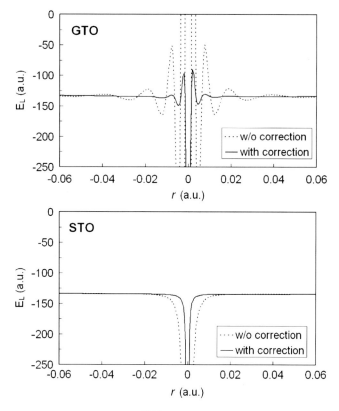

Fig. 10.2 Variation in the local energy E_L^{ZORA} as an electron moves through nucleus of Ne atom at the origin

Similar shapes are also observed in the STO case. Near the nucleus, on the other hand, our corrected orbitals have different shapes from the original orbitals in both STO and GTO cases. The local energies of Ne and Ar atoms are shown as a function of the separation of an electron from the nucleus in Figs. 10.2 and 10.3, respectively. The correction parameters are $(0.08|30)$ for the Ne atom and $(0.05|50)$ for the Ar atom. The plot is generated by taking an electron configuration from a VMC run and then calculating the local energy as the electron closest to the nucleus moves in a straight line through the nucleus. The figures show that the cusp correction scheme clearly reduces the local energy fluctuations in the near-nucleus region for both the GTO and STO cases. In the case of Ar, the uncorrected STO HF wave function shows an oscillation similar to the GTO wave function. The oscillation is produced because of the addition of the 1s basis functions. When calculated without the additional 1s orbitals, a smoothly divergent local energy surface is obtained, as in the case of STO-Ne. This suggests that a linear combination of non-adequate functions can hardly improve the smoothness of the local energy. On the other hand, the corrected STO wave function shows a local energy surface without oscillations. This indicates the necessity of correcting to the conventional STO basis functions.

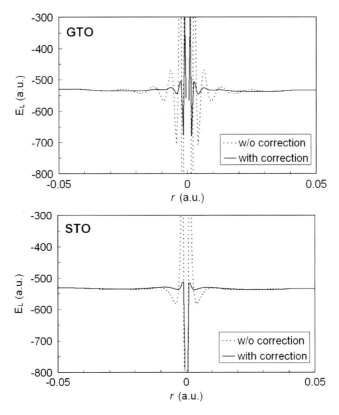

Fig. 10.3 Variation in the local energy E_L^{ZORA} as an electron moves through nucleus of Ar atom at the origin

10.5.2 Cu Systems

With the relativistic ZORA-QMC method, total energies of Cu systems containing Cu (^{2}S) atom, Cu$^+$ (^{1}S) cation, and CuH molecule with an experimental bond length of 1.463 Å [56] were calculated. The modified version of the GAMESS program package has been used for generating UHF and RHF wave functions for Cu atom and other systems, respectively. In these calculations, relativistic cc-pVTZ-DK basis sets were used. The Jastrow–Slater type wave functions were used in the QMC calculation. All QMC calculations including the optimization and the VMC calculation have been performed with our R4QMC program and the relativistic cusp correction of (0.03|60) for Cu 1s orbital has been employed for all systems. Schmidt–Moskowitz type functions with two- and three-body terms, (0,0,1), (0,0,2), (0,0,3), (0,0,4), (2,0,0), (3,0,0), (4,0,0), (2,2,0), and (2,0,2), were used as the Jastrow correlation function. Cusp conditions for electron–electron collision were approximated with non-relativistic parameters $c = 0.25$ for parallel

Table 10.1 Total energy values calculated by the ZORA-VMC method (in Hartree)

Cu$^+$	Cu	CuH
$-1665.582(23)$	$-1665.843(18)$	$-1666.530(20)$

Table 10.2 Ionization potentials (IPs) of Cu

Method	IP(Hartree)
ZORA-HF	0.242
ZORA-VMC	0.261(41)
NR DMC (Rel. corr.)[a]	0.279(11)
Exptl.	0.284

[a]Ref. [12]

and $c = 0.5$ for anti-parallel spins for the (0,0,1) electron–electron cusp terms. The distance parameter b was fixed to 1.0 for all terms. Other coefficients were optimized using both the correlated method of Umrigar and the selective sampling method with 2.0×10^5 sample points. With optimized wave functions, VMC runs were performed with approximately 4.0×10^8 sample points. In the VMC runs, one electron has been moved in one MC step by the Langevin-type random walk based on the Fokker–Planck equation. The small time step of 0.0005 has been used.

Table 10.1 lists the total energy values for the Cu atom, Cu cation, and CuH molecule. The experimental and calculated ionization potentials (IPs) of Cu are summarized in Table 10.2. The ionization potential calculated with corrected wave functions are 0.261(41) a.u. The ZORA-VMC calculation improves the analytical ZORA-HF result (0.242 a.u.) because of the inclusion of the electron correlation, but it only recovers about half of the correlation energy. This insufficient inclusion of the correlation energy will be improved by the ZORA-DMC method. We will investigate the improvements obtained by the relativistic DMC treatment in a future work.

10.6   Conclusions

We have developed a new relativistic treatment in the QMC technique using the ZORA Hamiltonian. We derived a novel relativistic local energy using the ZORA Hamiltonian and tested its availability in the VMC calculation. In addition, we proposed a relativistic electron–nucleus cusp correction scheme for the relativistic ZORA-QMC method. The correction scheme was a relativistic extension of the MO correction method where the 1s MO was replaced by a correction function satisfying the cusp condition. The cusp condition for the ZORA wave function in electron–nucleus collisions was derived by the expansion of the ZORA local energy and the condition required the weak divergence of the orbital itself. The proposed relativistic correction function is the same as the NR correction function of Ma et al. in the NR limit, but it is different when the parameter of the speed of light is finite. With our present approach, we can simultaneously deal with the electron correlation and relativistic effects in the QMC framework.

Acknowledgements This research was supported in part by a Grant-in-Aid for Scientific Research (B) 19350007 from Ministry of Education, Culture, Sports, Science and Technology-Japan (MEXT).

References

1. Ceperley D, Chester GV, Kalos MH (1977) Phys Rev B 16:3081
2. Foulkes WMC, Mitas L, Needs RJ, Rajagopal G (2001) Rev Mod Phys 73:33
3. Anderson JB (2007) Quantum Monte Carlo: origins, development, applications. Oxford University Press, Oxford
4. Ceperley DM, Alder BJ (1980) Phys Rev Lett 45:566
5. Conroy H (1964) J Chem Phys 41:1331
6. Hammond BL, Lester WA Jr, Reynolds PJ (1994) Monte Carlo methods in Ab initio quantum chemistry. World Scientific, Singapore
7. Lester WA Jr (1997) Recent advances in quantum Monte Carlo methods. World Scientific, Singapore
8. Vrbik J, DePasquale MF, Rothstein SM (1988) J Chem Phys 88:3784
9. Das TP (1973) Relativistic quantum mechanics of electrons. Harper and Row, New York
10. Kenny SD, Rajagopal G, Needs RJ (1995) Phys Rev A 51:1898
11. Trail JR, Needs RJ (2005) J Chem Phys 122:174109
12. Caffarel M, Daudey J-P, Heully J-L, Ramírez-Solís A (2005) J Chem Phys 123:094102
13. Aisssing G (1991) Phys Rev A 44:R2765
14. Nakatsuka Y, Nakajima T, Nakata M, Hirao K (2010) J Chem Phys 132:054102
15. Nakatsuka Y, Nakajima T, Hirao K (2010) J Chem Phys 132:174108
16. Chang Ch, Pelissier M, Durand Ph (1986) Phys Scr 34:394
17. van Lenthe E, Baerends EJ, Snijders JG (1993) J Chem Phys 99:4597
18. Faas S, Snijders JG, van Lenthe JH, van Lenthe E, Baerends EJ (1995) Chem Phys Lett 246:632
19. van Lenthe E, van Leeuven R, Baerends EJ, Snijders JG (1996) Int J Quantum Chem 57:281
20. Morales JJ, Nuevo MJ, Rull LF (1990) J Comput Phys 89:432
21. Umrigar CJ (1993) Phys Rev Lett 71:408
22. Trail JR (2008) Phys Rev E 77:016703
23. Filippi C, Fahy S (2000) J Chem Phys 112:3523
24. Schautz F, Fahy S (2002) J Chem Phys 116:3533
25. Prendergast D, Bevan D, Fahy S (2002) Phys Rev B 66:155104
26. Schautz F, Filippi C (2004) J Chem Phys 120:10931
27. Umrigar CJ, Filippi C (2005) Phys Rev Lett 94:150201
28. Sorella S (2005) Phys Rev B 71:241103(R)
29. Toulouse J, Umrigar CJ (2007) J Chem Phys 126:084102
30. Toulouse J, Umrigar CJ (2008) J Chem Phys 128:174101
31. Umrigar CJ, Nightingale MP, Runge KJ (1993) J Chem Phys 99:2865
32. Jastrow R (1955) Phys Rev 98:1479
33. Schmidt KE, Moskowitz JW (1990) J Chem Phys 93:4172
34. Moskowitz JW, Schmidt KE (1992) J Chem Phys 97:3382
35. Umrigar CJ, Wilson KG, Wilkins JW (1988) Phys Rev Lett 60:1719
36. Kato T (1957) Commun Pure Appl Math 10:151
37. Manten S, Lüchow A (2001) J Chem Phys 115:5362

38. Ma A, Towler MD, Drummond ND, Needs RJ (2005) J Chem Phys 122:224322
39. Per MC, Russo SP, Snook IK (2008) J Chem Phys 128:114106
40. Kussmann J, Ochsenfeld C (2007) Phys Rev B 76:115115
41. Kussmann J, Riede H, Ochsenfeld C (2007) Phys Rev B 75:165107
42. Visscher L, Visser O, Aerts H, Merenga H, Nieuwpoort WC (1994) Comput Phys Commun 81:120
43. Saue T, Fægri K, Helgaker T, Gropen O (1997) Mol Phys 91:937
44. Yanai T, Nakajima T, Ishikawa Y, Hirao K (2001) J Chem Phys 114:6526
45. Yanai T, Iikura H, Nakajima T, Ishikawa Y, Hirao K (2001) J Chem Phys 115:8267
46. Yanai T, Nakajima T, Ishikawa Y, Hirao K (2002) J Chem Phys 116:10122
47. Quiney HM, Skaane H, Grant IP (1999) Adv Quantum Chem 32:1
48. Grant IP, Quiney HM (2000) Int J Quantum Chem 80:283
49. Bethe HA, Salpeter EE (1957) Quantum mechanics of one- and two-electron atoms. Springer, Berlin/Heidelberg/New York
50. Dyall KG, van Lenthe E (1999) J Chem Phys 111:1366
51. Hess BA (1986) Phys Rev A 33:3742
52. Nakajima T, Hirao K (2000) J Chem Phys 113:7786
53. Nakajima T, Hirao K (1999) Chem Phys Lett 302:383
54. Buenker RJ, Chandra P, Hess BA (1984) Chem Phys 84:1
55. Kutzelnigg W (1989) Z Phys D 11:15
56. Ram RS, Bernath PF, Brault JW (1985) J Mol Spectrosc 113:269

Chapter 11
Computer Aided Nanomaterials Design – Self-assembly, Nanooptics, Molecular Electronics/Spintronics, and Fast DNA Sequencing

Yeonchoo Cho, Seung Kyu Min, Ju Young Lee, Woo Youn Kim, and Kwang S. Kim

Abstract Using diverse computer-aided molecular/material/device design approaches, we have successfully designed, synthesized, and characterized diverse new functional molecular systems, and measured their device characteristics. Here, we discuss why and how such self-assembled nanostructures are designed, and elucidate their unusual electronic and optical properties and device performance. In particular, nanorecognition phenomena of ionophores/receptors and nano-optical phenomena of self-assembled organic nanolenses are discussed. Given that dynamic and transport properties beyond the common static properties are very important for dynamic control of molecular systems toward the mechanical and electrical devices, we have investigated functionalized graphene which shows intriguing transport properties. We discuss the super-magnetoresitance and ultrafast DNA sequencing utilizing the unusual transport properties of graphene.

Keywords Computer aided molecular design • Nanooptics • Transport • Molecular electronics • DNA sequencing

11.1 Introduction

Experimental techniques, such as scanning tunnelling microscopes and atomic force microscopes, which can resolve individual molecules, have opened a new era towards single molecule devices. Despite sophisticated instruments capable of

Y. Cho • S.K. Min • J.Y. Lee • K.S. Kim (✉)
Center for Superfunctional Materials, Department of Chemistry and Department of Physics,
Pohang University of Science and Technology, Hyojadong, Namgu, Pohang 790-784,
South Korea
e-mail: kim@postech.ac.kr

W.Y. Kim
Department of Chemistry, KAIST, Daejeon 305-701, South Korea

J. Leszczynski and M.K. Shukla (eds.), *Practical Aspects of Computational Chemistry I: An Overview of the Last Two Decades and Current Trends*, DOI 10.1007/978-94-007-0919-5_11, © Springer Science+Business Media B.V. 2012

observing phenomena at the nanometer scale, the limitations of such practice arise from the requirement of incorporation of quantum effects in the intuition-based design approach and the difficulty in analyzing the observed phenomena. This is the reason the calculation methods based on quantum mechanics emerge to play an important role in investigating functional nanomaterials and become an inevitable tool for the material and device design approach [1]. The synthesis of nanostructures, for example, the bottom-up approaches, based on the self-assembly, can be understood by the quantum mechanical description of intermolecular interactions.

Based on understanding nano-scale phenomena, it is possible to design functional nanomaterials. Individual interaction forces can be exploited to build novel molecular architectures toward functional materials. As manifestations of cooperative effects and competitive effects in intermolecular interactions, we synthesized interesting nanostructures. Some intriguing structures of organic nanotubes, nanospheres, and nanolenses were utilized to investigate their unique properties [2–4]. We investigated new optical phenomena of nanolenses, which show near field focusing and magnification beyond the diffraction [4, 5]. Based on nanorecognition, we have designed diverse ionophores/receptors, nanomechanical devices, and non-linear optical switches [6–8]. Theory-based molecular sensors and nano-mechanical materials exploiting various types of aromatic interactions show selective detection of a target species and specific motion induced by external/internal means [9], which can be used to bio-medical applications. The quantum transport calculation [10–13] for graphene nanoribbon (GNR) shows extraordinary magnetic properties promising a future spintronic device [14], and the theoretical description of DNA sequencing exploiting GNR-based nanoscale device [15] suggests an ultrafast DNA sequencing in the near future.

In this chapter, we will discuss selected examples of the materials and devices we have designed. Many of them are experimentally realized, showing the functional properties predicted from our calculations. This chapter consists of three categories. First, we look into self-assembly phenomena for organic nanostructures (nanotubes, nanospheres, and nanolenses). Then, we discuss the real-time propagation of electromagnetic fields in the presence of a nanolens to explain 'super-refraction' phenomena in the near-field region. Second, we introduce the electron/spin transport phenomena through nanostructures. Nanoscale transport phenomena are based on quantum transport theory which we describe using non-equilibrium Green's function (NEGF) theory based on density functional theory (DFT) [10]. A few applications including 'super-magnetoresistance' phenomena in GNR are studied. Finally, diverse chemical nanosensors (ionophores, receptors and sensors) based on nanorecognition are investigated, and in particular, ultrafast DNA sequencing device based on GNR is discussed since this sequencing would give immense impacts on future human life.

11.2 Self-assembled Materials

11.2.1 Organic Nanotubes

A key issue in nanotechnology is fabrication of nanoscale architectures with the predicted size, shape, structure, and chemical composition. However, the top-down approach has often reached the limit of size. A new strategy of miniaturization is a bottom-up approach which forms organized structures from building blocks of atoms and molecules. Self-assembly, which is common throughout nature, is the autonomous association of components into well-defined aggregates by non-covalent interactions such as hydrogen bonds and van der Waals interactions [16]. In particular, we studied the self-assembly of calix-4-hydroquinone (CHQ), a calixarene-based molecule, in aqueous or non-aqueous environment. It turns out that CHQ is an excellent building block for supramolecular self-assembly. Composed of four hydroquinone subunits, a CHQ molecule has eight hydrogen bond donors, eight receptors, and four $\pi-\pi$ stacking pairs. The cone shape is stabilized by the four inner $-OH$ groups though the circular proton tunnelling resonance. The other $-OH$ groups contribute to the growth in the longitudinal directions by producing infinitely long arrays of hydrogen bonds. The growth in lateral directions, leading to the formation of tubular bundles, is led by the $\pi-\pi$ stacking interactions [17].

We investigated the assembling energetic of CHQs using DFT calculations for many possible assembled structures (Fig. 11.1) [18]. We found that a hexamer structure is stable in the absence of water. In the presence of water, on the other hand, a linear tubular polymeric chain is stable with H-bonded bridges between repeating tubular octamer units. Indeed, it was demonstrated by experiments that thin needle-like nanotube bundles are stabilized in the presence of water (Fig. 11.2) [2, 18]. Four dangling H atoms of CHQs are a key factor in forming a HQ-(water-HQ-HQ-)$_n$-water chain in the presence of water. Because one-dimensional short H-bonding interaction ($\sim$10 kcal/mol) [19] is stronger than the $\pi-\pi$ stacking interaction, the 1-dimensional short H-bond relay structure is formed.

11.2.2 Organic Nano-Scale Lens and Optical Properties

Self-assembly of CHQ molecules can be used to fabricate nanoscale lenses [4]. The fabrication method depends on the size of the lens. Smaller ones are intermediates in the self-assembling process forming nanospheres. The nanospheres spring anisotropically from the surfaces of the CHQ nanostructures including nanotubes. Small sized nanolenses can be collected in the middle of the process. On the other hand, larger CHQ nanolenses are formed from the evaporation process of the CHQ solution. During evaporation, film-like structures of CHQ cover the surface of the crystals. Evaporation causes the CHQ molecules to be released from the surface, and they are accumulated in a small volume under the film, leading to the nucleation

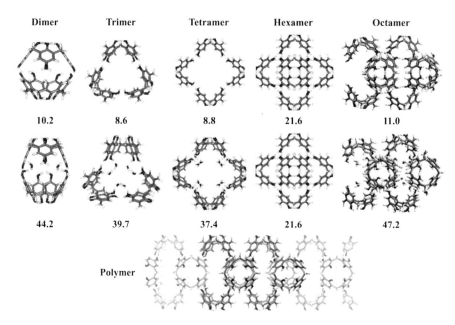

Dimer	Trimer	Tetramer	Hexamer	Octamer
10.2	8.6	8.8	21.6	11.0
44.2	39.7	37.4	21.6	47.2

Polymer

Fig. 11.1 Complexation energies of self-assembled CHQ *N*-mers without (*top*) and with (*below*) water molecules: dimer, trimer, tetramer, octahedral hexamer, and tubular octamer as a repeating unit of the tubular polymer of CHQs. The value below each structure is the calculated binding energy (kcal/mol) at the B3LYP/6-31G* level of theory. In the absence of water, the hexamer structure is favoured, while in the presence of water, the polymer chain is favoured [18]

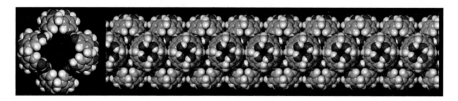

Fig. 11.2 Calix [4] hydroquinone nanotubes showing the longitudinal one-dimensional H-bond arrays. Each tube has four thin threads of infinitely long one dimensional hydogen bond relays, and the diameter of each is 8 Å (Reproduced from Ref. [2] with kind permission of The American Chemical Society)

and growth of two dimensional disk-shaped structures. As more of the released CHQ molecules re-assemble in three dimensions, spherical curvatures are then gradually formed. These lenses are 'plano-spherical convex' (PSC) structures with a spherical face on one side and a flat face on the other side (Fig. 11.3). Typically, PSC lenses with nanoscale thickness H <800 nm and diameter $D = 0.05$–3 μm can be synthesized and separated from the aqueous suspension for further experiments. Figure 11.3 shows the magnification effects obtained through the nanolens, with the images taken from the different distances from the top of face-up CHQ lens, corresponding to the SEM image.

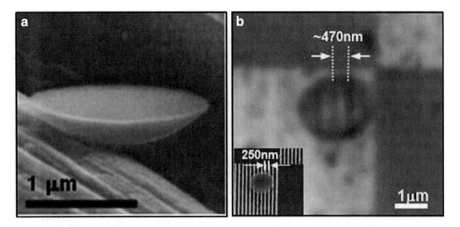

Fig. 11.3 CHQ plano-spherical convex lenses. (**a**) Scanning electron microscope (SEM) image of a CHQ nanolens. (**b**) Optical microscope (OM) image of a CHQ nanolens placed on a glass substrate with Palladium stripe patterns. It shows that the high resolution of CHQ nanolens to resolve sub-diffraction limit patterns. Inset, SEM image corresponding to (**b**) (Reproduced from Ref. [4] with kind permission of Nature Publishing Group)

Enhanced magnification is achieved by the reduction of the focal length in the CHQ nanolens. The magnifying effect allows the lens to resolve the stripe patterns of 250/220 nm spacing, which is beyond the diffraction limit. The magnifying effect of CHQ lens is as much as 2.5. The diffraction limit is a fundamental restriction on the optical resolution in geometrical optics. It means that two object located at a distance below half a wavelength of light cannot be resolved in a far-field optical system. To overcome the limitation, there have been several studies such as superlens/hyperlens systems driven by surface-plasmon excitation [20–22] and fluorescence microscopy driven by molecular excitation [23]. However, they are not geometrical lens-based systems. So far, any lens-based optical system has not been free from the diffraction limit even for the immersion technique [24] which enhances the resolution as much as the refractive index. We here discuss the super-resolution by nanoscale lens, using both experimental measurements and electromagnetic (EM) wave simulations (finite-difference time-domain (FDTD) simulations) [5, 25]. Based on the accurate solution of Maxwell's equation, we found the substantial difference in focal length between the far-field geometrical optics and the near-field wave optics. The focal length F obtained from EM wave simulations was much shorter than that obtained from the ray tracing method (Fig. 11.4).

The focusing phenomenon of a nanolens is due to the near-field focusing whose origin arises from the stringent phase matching phenomenon at the interface of a nanolens with extremely high curvature. To investigate the size-dependent focusing formation in the near-field regime, we performed FDTD simulations for nanolenses and microlenses of different diameters ($D = 0.8$–4 μm) with fixed ratio of lens thickness/heigh (H) to lens diameter (D) ($H/D = 0.35$). The incident plane waves are polarized along the x-direction, and Fig. 11.5 shows the spatial distribution of

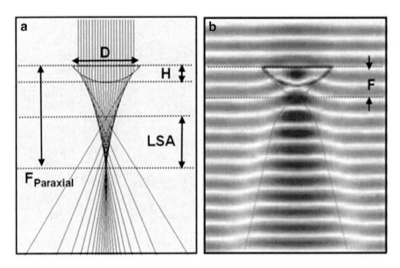

Fig. 11.4 (**a**) A ray-tracing simulation result of the lens calculated by OSLO program packages (Sinclair Optics, Inc). (**b**) A finit-difference time-domain (FDTD) simulation result (Ex) obtained by FullWAVE 4.0 program (RSoft Design Group). $D = 800$ nm, $H = 280$ nm and $\lambda = 365$ nm (Reproduced from Ref. [4] with kind permission of Nature Publishing Group)

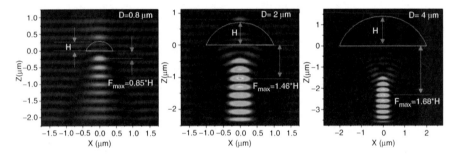

Fig. 11.5 Focal length changes for various sizes of CHQ lenses by using the FDTD simulation results. The thickness/diameter ratios are constant (H/D $= 0.35$). ($\lambda = 472$ nm) (Reproduced from Ref. [4] with kind permission of Nature Publishing Group)

$|E_x|^2$ where E_x is the x-component of the electric field. The results show that the ratio of the focal length (F_{max}) showing the maximum light intensity to the lens height H decreases remarkably as the lens size approaches the wavelength ($\lambda = 472$ nm). The near-field focusing phenomenon is due to superposition of the diffraction on the flat aperture and the interference of secondary Fresnel waves on the spherical surface of the PSC lens when the size of the lens is comparable to the wavelength.

To confirm the short focal length of nanolens through an experiment, we investigate the optical experiment to take transmitted images through the CHQ nanolens ($D = 1.7$ μm, $H = 0.48$ μm), which were recorded by an optical microscope with CCD camera focused on different image planes at a distance Z_{ip} from the lens/disk

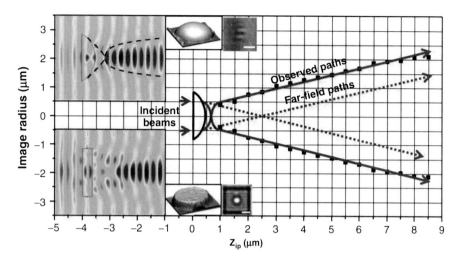

Fig. 11.6 Beam trajectories with the reduced focal length in the near field PSC lens. The small upper and lower insets on the left are the AFM images of the CHQ lens and the PMMA disk, respectively. The solid lines are guided to the eye, following the square dots which are the measured optical beam trajectories obtained from the magnified images along different image plane Z_{ip}. The large top and bottom insets on the left are the FDTD simulation results of radial component of the electric field (E_x) of the PSC lens and PMMA disk ($\lambda = 472$ nm) obtained by FullWAVE 4.0 program (RSoft Design Group). The other insets are the optical images created through the PSC lens and PMMA disk taken at the $Z_{ip} \sim 3$ μm. Scale bars, 2 μm (Reproduced by permission of NPG [4])

bottom (Fig. 11.6). The result shows the nanolens has very short focal length according to the curvilinear trajectories, which are in agreement with accurate FDTD simulations (top left inset of Fig. 11.6). In addition, the shape-dependent focusing and imaging effect are performed by comparing with a flat patterned poly(methylmethacrylate) (PMMA) disk. The light propagating through the disk shows no image while CHQ nanolens creates the clear an alphabetical character 'E' image. The difference between the images created through the PSC lens and the flat disk is clearly confirmed by electromagnetic simulations, and the image formation depends on the shape (surface curvature) of the optical elements.

The focal point of a nanolens originates mainly from surface waves at the spherical interface of the lens, and the near-field focal length is shortened by the interference of the propagating waves from the lens edges due to the scale of nanolens comparable to the wavelength. Due to high curvature of convex nanolens, the wave path must have shorter focal length than that of a common lens. The evanescent waves with optical information of fine features in a near-field region can be conveyed to a far-field region through the near-field focusing and magnification of the CHQ nanolens, so that it would be possible to achieve high resolution of the CHQ nanolens beyond diffraction limit. Our further accurate FDTD simulations predicted that the super-resolution is possible beyond the diffraction limit by ~25% [5].

11.2.3   Nano-Mechanical Devices

The design for nano-mechanical devices implies that one has to induce dynamic motions in a system using external or internal means. These include changes in pH, voltage application, laser excitation, irradiation, etc. Here, we discuss a molecular flipper [9]. We exploited the conformational changes from the edge-to-face (T-shaped) conformation to the face-to-face (stacked) one or vice versa to mimic the flipping/flapping motion. It is interesting to note that this conformational change can be electrochemically controlled by reduction/oxidation of the quinone moiety in the molecular system. This system was designed based on a theoretical investigation of the conformational characteristics of p-benzoquinon-benzene complexes. We found that the energy difference between the stacked and edge-to-face conformations of cyclophane molecules is substantial. The stacked conformer is 7 kcal/mol more stable than the edge-to-face conformer in the oxidized state, whereas the edge-to-face conformer is 9 kcal/mol more stable than the stacked conformer in reduced state. Thus, the subtle control of the conformational characteristics of 2,11-dithio[4, 4] metametaquinocyclophane (MQC) and 2,11-dithio[4, 4]metametahydroquino-cyclophane (MHQC) by electrochemical and/or photochemical means leads to a very interesting model of a potential molecular device. The cyclic voltammograms of MQC showed two reversible redox reactions. The electronic states of MQC and MHQC can be easily transformed into each other by simple electrochemical control of the redox reaction, which results in large conformational flapping motions due to a preference for the stable conformation caused by the change in the electronic state of the quinone moiety. Thus, a cyclophane system composed of quinone and benzene rings exhibits a flapping motion involving squeezing and thrusting motions in the presence of solvent molecules by electrochemical redox processes. This case illustrates a promising pathway for harnessing the differences in the relative magnitudes of different kinds of intermolecular interactions to design a mobile nanomechanical device for drug delivery and nanosurgery.

11.3   Nano-Scale Electronic Materials

11.3.1   Theoretical Description of Nano-Scale Electronic Transport Phenomena

Controlling the nano-scale transport phenomena of electrons or spins has been one of the most challenging subjects for the ultimate miniaturization of future electronic devices. Diverse measurements on currents through a single or a few molecules have been investigated based on molecular electronics which is considered as a promising candidate for the replacement of the current lithography-based electronics. As the size of devices is reduced, quantum effects in electron/spin transport are of paramount importance. The atomistic change in molecular junctions shows significant changes in transport phenomena. This can be utilized to tune electronic properties by modifying atomic structures or applying external potentials [11, 12]. In this regard, an accurate theoretical understanding of the transport phenomena is essential.

In the case of the ballistic transport regime, an incoming electron is elastically scattered in the metal-molecule contact regions, which shows fractional transmission. Non-homogeneous small molecular devices show elastic/coherent transport phenomena. As the molecular device gets larger, the inelastic effects, such as electron-phonon couplings, are important since the traversal time is long enough to interact with molecular vibrations.

For the theoretical treatment of the nano-scale transport, we use open boundary systems (stemming from the source and drain leads) under non-equilibrium condition. The most popular method is based on the Keldysh NEGF coupled to the DFT (NEGF-DFT) method. While the Landauer-Büttiker formalism enables us to describe non-interacting quantum transport corresponding to the ballistic or elastic transport regimes, the NEGF method provides a quantum mechanical description for the electron transport in all three transport regimes. For the non-interacting electron systems, both methods are equivalent. In the following paragraphs, we discuss the detailed implementation schemes based on the NEGF method for the elastic and inelastic transport regime subsequently [10–12]. The ballistic transport regime is nothing more than the extreme case of the elastic and coherent transport regime without scattering.

The system that we are interested in is divided into three partitions: a molecular part (M) and left/right electrodes (L/R). The total Hamiltonian (H) and overlap (S) matrix within the localized atomic orbital basis representation are written as follows:

$$H = \begin{bmatrix} H_L & H_{LM} & 0 \\ H_{ML} & H_M & H_{MR} \\ 0 & H_{RM} & H_R \end{bmatrix}$$

and

$$S = \begin{bmatrix} S_L & S_{LM} & 0 \\ S_{ML} & S_M & S_{MR} \\ 0 & S_{RM} & S_R \end{bmatrix}.$$

Since each electrode is a semi-infinite system along the direction of current, it cannot be treated as a simple periodic bulk system. We exploit the renormalization technique to deal with the semi-infinite electrodes so that the semi-infinite properties of electrodes are projected onto the molecular Hamiltonian, which is called "self-energy". The self-energy is a main ingredient to determine a renormalization of molecular energy levels. The self-energy for the left/right metal-molecule contact ($\Sigma_{L/R}$) is given by the following equation:

$$\Sigma_{L/R}(E) = v_{L/R}(E) g^r_{L/R}(E) v^\dagger_{L/R}(E)$$

where $v_L(E) = [H_{ML} - ES_{ML}]$ and $v_R(E) = [H_{RM} - ES_{RM}]$ are the interaction term between the left and right electrode and molecule, respectively. $g^r_{L/R}$ is the surface Green's function for the left/right electrode which can be calculated by various ways such as the transfer matrices method. Once we obtain the self-energy matrix, the effective Hamiltonian will be given by the following equation:

$$H_{eff} = H_M + \Sigma_L + \Sigma_R.$$

Then, we can calculate the retarded Green's function (G^r) matrix using the following equation:

$$G^r(E) = [ES - H_{eff}]^{-1}.$$

The electron density matrix (D) of the molecular system can be directly obtained from the retarded Green's function.

$$D = -\frac{1}{\pi}\text{Im} \int_{-\infty}^{\infty} [G^r(E) f(E - \mu)]dE$$

where $f(E-\mu)$ is the Fermi-Dirac distribution function and μ is a chemical potential.

Then, the conventional density matrix calculation procedure can be replaced by the above G^r-based method (Fig. 11.7). Using the converged electron density, we can calculate transmission coefficients as a function of energy and current (I) at the given bias voltage (V_b) through the Landauer-Büttiker formula as follows:

$$T(E, V_b) = \text{Tr}[\Gamma_L(E, V_b) G^a(E, V_b) \Gamma_R(E, V_b) G^r(E, V_b)]$$

and

$$I = \frac{2e}{h} \int [T(E, V_b)\{f_L(E, V_b) - f_R(E, V_b)\}]dE$$

where $\Gamma_{L/R}$ is the imaginary value of the self energy of the left/right junctions, $f_{L/R}$ is the Fermi-Dirac distribution function for the left/right electrodes and G^a is the advanced Green's function. In the presence of a finite bias, a system becomes non-equilibrium because of the difference in chemical potential between the left and right electrodes. The present NEGF-DFT method gives a steady-state current under the non-equilibrium condition.

In the case of the inelastic/incoherent transport regime, the phonon and the electron-phonon (e-ph) coupling term (H_{ph} and H_{e-ph}, respectively) should be added to the molecular Hamiltonian:

$$H = H_e + H_{ph} + H_{e-ph}.$$

Fig. 11.7 Flowchart of
the NEGF-DFT
self-consistent-field scheme
(Reproduced from Ref. [10]
with kind permission of
Wiley VCH)

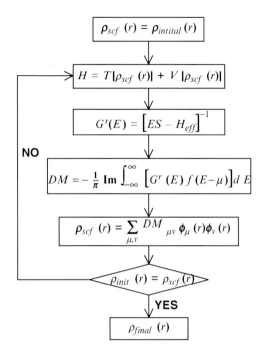

In the following several sections, we introduce a few examples of the nano-scale transport phenomena within the above theoretical descriptions using the POSTRANS program package [10, 11, 14].

11.3.2   Electron Transport in 1-Dimensional Nanowire

As the dimensionality of metals is reduced from the 3-dimensional bulk system to the 1-dimensional wire, the electronic or magnetic properties are changed drastically [26]. With the advancement in the experimental techniques to fabricate the metal nanowires, we anticipate the possibilities of incorporating them in futuristic electronic/optoelectronic devices such as quantum devices, magnetic storage, nanoprobes, and spintronics.

The conducting property of metallic nanowires can be measured by mechanically controllable break-junction experiments (MCBJE). Since the nanowires adopted in the MCBJE have been found to undergo thinning on a time scale, the real time changes in conductance show favourable structures, so called "magic structures", during the thinning process. To find the pathway of the thinning process for Ag nanowires, we carried out the DFT calculations of Ag nanowires for various circumstances. From the calculation of binding energies and the wire tensions, we found that the "magic" nanowires are 11/8, (9/8), 8/6, 5/4, (4/3), 2/2, (1/1), and 1/0 structures (Fig. 11.8) [27].

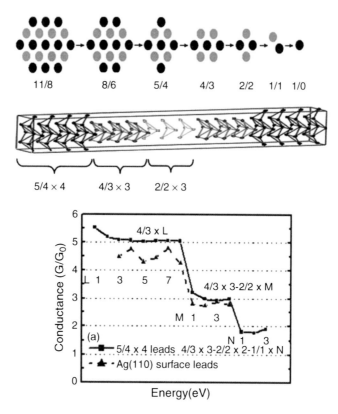

Fig. 11.8 Magic numbers for Ag nanowires for the thinning process (changes in cross section) and the fractional quantum conductance (Reproduced from Ref. [27] with kind permission of The American Physical Society)

Additionally, we calculated the conductance for the thin wires using the NEGF-DFT method to understand the fractionally quantized conductance value at 2.4 G_0. Since the pure Ag nanowires of uniform cross section shows quantized conductance values (6, 5, 4, 3, 3, 2, and 1 G_0 for 4/3, 4/2, 4/1, 2/2, 2/1, 1/1, and 1/0, respectively), the fractional conductance does not correspond to any pure structures. In this sense, we suggested the existence of mixed structures. Since the thinning process is a transient process, a new structure might evolve before the old structure has actually fully faded away. Hence, we calculated the transport property of mixed structures composed of several pure structures. Figure 11.8 shows one of the mixed structures which corresponds to $5/4 \times 4\text{-}4/3 \times 3\text{-}2/2 \times 3$ and the conductance according to our scenario of the thinning process. Based on our results, we conclude that the mixed structures show small conductance values. Thus, the mixed structures of 2/2 and 2/1 (with 3 G_0 for the pure) could be 2.4 G_0 in experiments.

11.3.3   Role of Electrodes in Molecular Electronics

Even though molecular electronics is a promising technology to replace the current silicon-based one, it seems that the endeavors for molecular electronics have partially declined, mainly due to contact problems. In a fabrication process, it is extremely difficult to control the local structure of molecule-electrode contact, and randomness in the structure results in random electronic behavior of devices. To understand contact problems, it is important to clarify what role each part plays in a molecular electronic device. We can divide the contact in two parts: an electrode and a linker. A linker denotes a functional group connecting between a molecule and electrodes at both sides. Its effects have been intensively studied because it, being more than a simple glue, significantly modifies the electronic structure of a device-molecule and thereby the transport properties. In contrast to linkers, few studies have been devoted to effects of electrodes. Most of them regarded an electrode just as a part of contact. Namely, electrode effects are reduced to those of a few metal atoms on contact, limiting the electrode effects to determining contact geometry and chemical bonding between individual atoms. However, we can obviously expect that as a linker is more than just a glue, an electrode plays a more profound role than a mere constituent of contact, as evident from its semi-infinite nature.

In this regard, we compared Au, Ru, and carbon nanotube (CNT) electrodes on molecular electronic devices. (This sub-chapter is mainly based on Ref. [28].) These are a few materials among the employed as an electrode in experiments. Au is the most popular in molecular electronics [29]. Tulevski et al. studied the Ru surface to which a carbon atom was bound by forming a multiple covalent bond [30]. Guo et al. studied CNT electrodes with an amide linkage [31]. Au, CNT, and Ru also represent a material mainly having s, p, and d band characters in the vicinity of their Fermi energies, respectively.

We studied the three electrodes using the DFT coupled with the NEGF scheme. To compare different electrode materials, we used a simple conjugated molecule, 1,3,5-hexatriyne (which we will call "alkyne" for convenience's sake). Transmission calculations were made to extract influences solely from electrodes by modification of a molecular core and linkers. The band structure and surface density of states (SDOS) of each electrode material were compared to the transmission function of each system under a zero bias.

Figure 11.9 shows the selected optimized structures of the whole device. The scattering region of the system includes a few layers of the leads and a sandwiched device molecule. Au(111) and Ru(111) surfaces are employed. For CNT, the molecule is connected to the metallic CNT(5,5) electrodes. Detailed options in calculation are described in the reference [28].

We briefly note from our computational approach the part relevant to electrode contributions. The contact effects due to the electrodes are to renormalize the electronic structure of the device molecule. The self-energy explicitly involves the contact effects on the molecule region due to the electrodes, so it plays a critical role in determining the transport characteristics; its real/imaginary parts give rise to shift and broadening of the molecular energy levels.

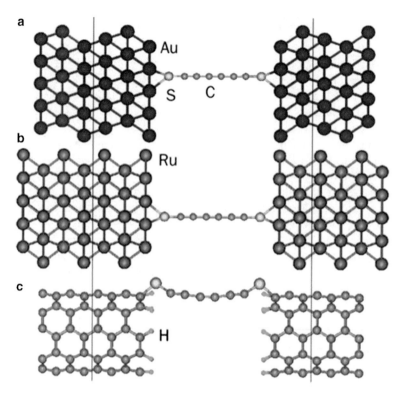

Fig. 11.9 Selected optimized geometries of (**a**) the Au-S-alkynes, (**b**) Ru-S-alkynes, and (**c**) CNT-S-alkynes systems. Size scales are arbitrary. Lines divide the scattering region from the electrode region (Reproduced from Ref. [28] with kind permission of The American Chemical Society)

While the electrodes provide the surface Green's function (g^s) in the self-energy term, the linkers determine the interaction term. It is also used when calculating SDOS, i.e. $SDOS = -\pi^{-1}\mathrm{Im}[Tr(g^s S)]$.

Since this study attempts to reveal the electrode effect, a realistic modeling of electrodes is of utmost importance. First, the electrode part is treated on an equal footing with the scattering region at the DFT level. Second, the surface Green function from which the self-energy matrix is computed is calculated from the realistic atomic crystal structure without resorting to the cluster approximation and the tight-binding approximation. Finally, the sufficient number of k-points is sampled to accurately take into account the semi-infinite nature of electrodes. In the calculations, we first increase the number of k-points sampled along the surface normal directions till the transmission curve converges. Given that the bond length between a molecule and an electrode is short, the systems studied here are close to the strong coupling limit, where the self-interaction errors give small contributions to the conductance.

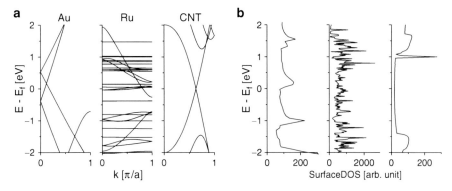

Fig. 11.10 (**a**) Band structures of the Au, Ru, and CNT unit cells presented as electrode parts in Fig. 11.9. The direction of k is parallel to the direction of electron flow. (**b**) SDOS of Au, Ru, and CNT. The "surface" means the interface between the semi-infinite electrodes and the scattering region including a few layers of electrodes at both ends (refer to Fig. 11.9) (Reproduced from Ref. [28] with kind permission of The American Chemical Society)

We first discuss the properties of electrode materials relevant to device's characteristics: (i) the band structures which show the characteristics of electronic states involved in electron transport and (ii) SDOS which gives the information how many states can accept or supply electrons that constitute current flow (Fig. 11.10). The 6s orbitals of Au form broad bands over a wide energy range. The SDOS curve of Au is smooth even except for small bumps from the surface. The d orbitals of Ru appear as flat lines in band structure and mainly contribute to SDOS. In the case of CNT, the p bands solely contribute to the band structure and SDOS around the Fermi energy. Other bands begin to play when the energy is different from the Fermi energy by more than 1 eV. The SDOS of the metallic CNT(5,5) has relatively large and broad peaks above 1 eV and below −1 eV, whereas such peaks are absent between −1 and 1 eV. This is the unique feature of CNT in contrast with the other metals.

We find that the position and broadening of peaks in the transmission curve depend on electrodes, though the same device molecule and linkages are employed. Figure 11.11a shows transmission curves where sulfur linkages are employed in every case. Since the work function of Au is large compared to the others, the HOMO peak is closer and the LUMO peak is farther in Au than in others. Regarding the extent of broadening, the transmission curve of the Au-contacted system has conspicuous peaks, but these peaks are too broad to be completely separated. In the Ru case, broad peaks are observed, but they seem to be made up of many sharp peaks. The HOMO and LUMO peaks can be distinguished, though the exact positions are not well defined. Since metals have many states nearby the Fermi energy, it is probable to have appreciable transmission values through this range. However, in CNT electrodes, discrete and narrow peaks appear, and transmission peaks are in the region of small SDOS between the peaks, i.e., between −1 and 1 eV.

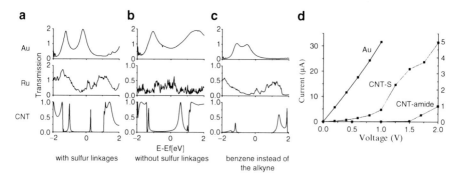

Fig. 11.11 Transmission curves of Au, Ru, and CNT electrodes for an alkynes molecule (**a**) with the sulfur linkage and (**b**) without the sulfur linkage. The transmission curves in (**c**) are calculated after the alkynes molecule is replaced by a benzene molecule. I–V curves of the alkynes with the gold electrode and two types of CNT electrodes in (**d**). "CNT-S" and "CNT-amide" denote the cases with the sulfur linkages and the amide linkages, respectively. The scale of the current on the left is for Au, while that on the right is for CNTs (Reproduced from Ref. [28] with kind permission of The American Chemical Society)

The features we discussed may be originated from the linkage and device molecule rather than the electrodes. Thus, we performed the same calculations without sulfur linkers (Fig. 11.11) and we also employed a different device molecule to confirm the trend of peak position and broadening. In Fig. 11.11 *p*-diethyl benzene molecule is sandwiched by Au, Ru, and CNT electrodes, respectively. Relative HOMO and LUMO positions in the transmission curves according to the type of electrodes do not change; the HOMO peak is much closer to the Fermi energy in Au than the LUMO peak, and both HOMO and LUMO peaks are almost equidistant from the Fermi energy in systems with Ru or CNT electrodes. The broadening tendency of peaks remains more or less the same as before, suggesting that they are strongly influenced from the nature of electrode materials.

We note that these characteristics given by electrodes are most evident when it comes to experimentally realized linkage-electrode combinations. As stated earlier, gold electrodes are mostly used with sulfur linkages, Ru with direct covalent bonds to the end carbons, and CNT with amide linkages. The Au, Ru, and CNT cases correspond to Fig. 11.11a–c, respectively. Figure 11.11d clearly demonstrates how such differences in transmissions are manifested in I–V characteristics. Indeed, the gold electrode shows an almost linear I–V curve, whereas the CNT electrodes lead to non-linear curves. Note also that Au gives about an order of magnitude larger current than CNT. It results from large broadening of the transmission curve. The "CNT-amide" case clearly shows the signature of the zero transmission gap; the current is almost zero till 1.5 V, but it surges thereafter. It is the LUMO of the alkynes which contributes to the surge since the peak corresponding to the HOMO is small compared to that of the LUMO. The "CNT-S" case also displays a similar behavior. This is because CNT, due to its bulk properties, makes discrete molecular peaks in the transmission. In summary, the alkynes molecule is merely a resistor when it is connected to the Au or Ru electrode, whereas it shows the highly non-linear behavior exhibiting unique characteristics of the molecule when it is attached to the CNT electrode; the characteristics of a nano device can be attributed strongly to the choice of an electrode beside the choice of linkage.

Our study shows that the CNT electrodes are more suitable to derive non-linear I–V curves from a particular molecule, as compared with Au or Ru electrodes showing almost linear I–V characteristics. On the other hand, many experimental observations have shown that electric currents strongly depend on a device molecule even for the same electrode and linkers. It is because electrical currents through a molecule are a consequence of complicated interplay between molecular energy level spacing, characteristics of each molecular orbital, their interactions with linkages and electrodes, change of a molecular structure under bias, and so on. Thus, the electrode effect should be considered not as a deterministic factor but as a useful factor characterizing molecular transport properties. An electrode in molecular electronics should show molecular features and highly reproducible current-voltage characteristics. Looking for such an electrode is still on the way and will also be a way to make molecular electronics in use at real life.

11.3.4   Graphene Nanoribbon as a Spintronic Memory Device

Since the first isolation of graphene [32], mono and a few layers of graphene can nowadays be synthesized in large quantity [33, 34], and would be utilized in diverse research fields [35–43]. Here, we discuss a spin valve device using a GNR [14]. Especially, GNR with zigzag shape edges (ZGNR) has received a lot of attention due to its unusual magnetic structure. It has been known that the ZGNR has a ferromagnetic spin ordering along both edges, whose relative spin orientation can be either parallel or anti parallel. The anti parallel spin configuration (ferromagnetic anti parallel; FMA) is the lowest energy structure. The small energy difference between FMA and the parallel spin configuration (ferromagnetic parallel; FMP) shows a power law as a function of the width of the ribbon [14]. In the presence of an external magnetic field, however, the FMP state is the most stable.

It turns out that a magnetic field can also control molecular orbital's (MOs) of GNR. MOs are a key to determine transport properties of a molecular device, since they act as conducting channels when electrons transport. Therefore, modulation of the MOs enables us to tune electron transport through the device. Here, we show the importance of such modulations in ZGNR. Their spin states have unique orbital symmetries. Normal spin-valve devices filter spins by spin matching/mismatching between both electrodes. Our device filters them by the orbital symmetry matching/mismatching in addition to spin. This is distinguished from the conventional tunneling or giant magnetoresistance. Therefore, we call this new type of magnetoresitance (MR) as super-magnetoresistance (SMR) showing an extremely large value.

To facilitate our discussion, we focus our attention on ZGNR spin-valve devices. The spin-dependent transmission is given by the Fisher-Lee relation [44]:

$$T_\sigma(\varepsilon) \equiv \text{Tr}\left[\text{Im}\left(\Sigma^r_{L\sigma}(\varepsilon)\right) G^r_\sigma(\varepsilon)\text{Im}\left(\Sigma^r_{R\sigma}(\varepsilon)\right) G^a_\sigma(\varepsilon)\right] \qquad (11.1)$$

where $G^{a/r}(\varepsilon)$ is the advanced/retarded Green's function and $\Sigma^r_{L/R}(\varepsilon)$ is the retarded self-energy for the left/right contact. To describe a magnetic domain wall, however, we need to deal with a spin vector to represent an arbitrary direction [11, 14]. In this non-collinear spin-polarized case, the Green's function matrix is given by the four spin-component block matrices as follows:

$$\widetilde{G}^r(\varepsilon) = \begin{bmatrix} G^r_{\alpha\alpha}(\varepsilon) & G^r_{\alpha\beta}(\varepsilon) \\ G^r_{\beta\alpha}(\varepsilon) & G^r_{\beta\beta}(\varepsilon) \end{bmatrix}$$

$$\widetilde{G}^r(\varepsilon) = \frac{1}{\varepsilon\widetilde{S}^r - \widetilde{H}^r - \widetilde{\Sigma}^r(\varepsilon)} \qquad (11.2)$$

where $\widetilde{S} = \begin{bmatrix} S & 0 \\ 0 & S \end{bmatrix}$, $\widetilde{H} = \begin{bmatrix} H_{\alpha\alpha} & H_{\alpha\beta} \\ H_{\beta\alpha} & H_{\beta\beta} \end{bmatrix}$, and $\widetilde{\Sigma}^r(\varepsilon) = \begin{bmatrix} \Sigma^r_{\alpha\alpha}(\varepsilon) & \Sigma^r_{\alpha\beta}(\varepsilon) \\ \Sigma^r_{\beta\alpha}(\varepsilon) & \Sigma^r_{\beta\beta}(\varepsilon) \end{bmatrix}$ are the overlap, Hamiltonian, and retarded self-energy matrices, respectively. The block Hamiltonian matrix $H_{\alpha\beta}$ is obtained from density functional theory with the four spin component density matrix given by

$$\rho_{\alpha\beta} = -\frac{1}{\pi}\text{Im}\int_{-\infty}^{\infty} G^r_{\alpha\beta}(\varepsilon)f(\varepsilon)d\varepsilon. \qquad (11.3)$$

The Green's function in the integrand of Eq. 11.3 is calculated from the Hamiltonian in Eq. 11.2. Thus, fully unconstrained non-collinear spin calculation is achieved in a self-consistent manner within DFT.

To discuss the case of the FMP state i.e. the parallel spin configuration between both edges, we show in Fig. 11.12 the band structure of the 8-ZGNR (N-ZGNR denotes a ZGNR with N zigzag chains) and its corresponding orbital symmetry. Occupied and unoccupied bands have orbital symmetries orthogonal to each other. For instance, the occupied bands below the Fermi energy (E_F) have 'C_2' symmetry regardless of their spins, while the unoccupied bands have 'σ' symmetry. More interestingly, one can change the spin polarization of the occupied bands by applying magnetic fields. As a result, for the same spins, their orbital symmetries are orthogonal. This is quite unusual, as compared with the magnetic control of the spin polarization in usual ferromagnetic materials. Figure 11.12 clearly shows their difference. Figure 11.12b presents typical switching behavior of the spin polarization due to the magnetic field in the density of states (DOS) of a usual ferromagnetic metal, while Fig. 11.13c exhibits switching behavior of both spin polarization and orbital symmetries in the DOS of a ZGNR. Consequently, for ZGNRs, their orbital symmetries as well as the spin symmetries can be manipulated by magnetic control.

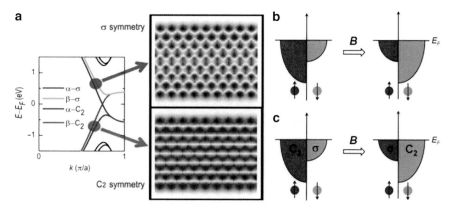

Fig. 11.12 (**a**) Orbital symmetries of the band structure of the 8-ZGNR. The *upper* and *lower panels* on the right exhibit the orbitals (wavefunctions) corresponding to the *blue/cyan* and *red/pink bands* on the *left panel*, respectively. The upper panel shows "σ" symmetry with respect to the middle horizontal line, while the lower panel shows "C_2" symmetry. Schematic representations of density of states (DOS) for a usual ferromagnetic metal (**b**) and the ferromagnetic state of a ZGNR (**c**). For the usual ferromagnetic metal, the spin symmetries in the DOS are changed by the applied magnetic field, while for the ZGNR, spin symmetries as well as their orbital symmetries in the DOS are simultaneously switched (Reproduced from Ref. [11, 14] with kind permission of respectively The American Chemical Society and Nature Publishing Group)

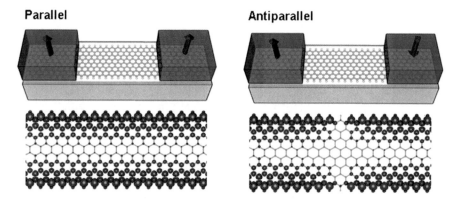

Fig. 11.13 Schematic ZGNR-based spin-valve device with parallel and anti-parallel spin configurations and the corresponding spin-magnetization density isosurfaces. The blue boxes represent ferromagnetic electrodes to control spin polarization of the ZGNR device and the red arrows indicate the directions of applied magnetic fields. In the isosurfaces, red/blue color denotes up/down spin and a ZGNR skeleton is drawn in green color (Reproduced from Ref. [14] with kind permission of Nature Publishing Group)

For a spin-valve device based on a ZGNR (Fig. 11.13) which is placed between two ferromagnetic (FM) electrodes, we can control the spin orientation of the ZGNR with magnetic fields on FM electrodes. If the directions of magnetic fields at both

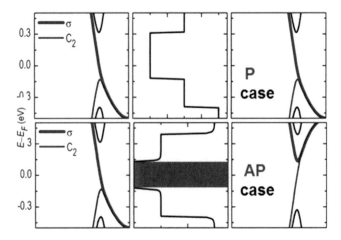

Fig. 11.14 Band structure for the left lead (*left*), transmission curve (*middle*), and band structure for the right lead (*right*) for the spin in the parallel (*P*: upper panel) and antiparallel (*AP*: lower panel) configurations of the 32-ZGNR at zero bias (Reproduced from Ref. [14] with kind permission of Nature Publishing Group)

FM electrodes are parallel, the spin polarizations of a ZGNR will be uniform, whereas if the directions are antiparallel, the spin polarizations at both sides of the ZGNR will be antiparallel with the formation of a magnetic domain wall between them. This causes spin-dependent conductance through the ZGNR device. For a ZGNR spin-valve device, the DOS in each spin configuration has additional labels denoting the orbital symmetry of the spin state. Therefore, the MR is determined by both spin matching/mismatching and orbital symmetry matching/mismatching.

Quantitative investigation is made by calculating transmission values and currents of the ZGNR spin-valve device at the first-principles level. Figure 11.14 exhibits the transmission curves sandwiched by band structures at the left and right leads for the ZGNR. The blue and red lines represent the corresponding orbitals that have the C_2 and σ symmetries, respectively. For the parallel configuration, bands having the same symmetry are aligned for all energy ranges, showing perfect transmission. On the other hand, for the anti-parallel configuration, the transmission curve shows perfect reflection, with zero transmission values within a particular energy range around the E_F where the orbital symmetries are mis-matching. MR can be calculated from the I–V characteristics for the 32-ZGNR spin-valve device. The MR value of the ZGNR device at low temperature is over a million %, which approaches to the infinite as the temperature decreases to zero. Indeed, both spin and symmetry filtering effect leads to an ideal spin-valve device, so called SMR.

For the antiparallel spin configuration, we need to address the formation of a magnetic domain wall between two ferromagnetic leads in the ZGNR devices in Fig. 11.15a. To this end, we have calculated transmission function with various domain wall sizes in the non-collinear spin-polarized mode. In Fig. 11.15b, transmissions for the non-collinear spin states result in reduced sharpness of corners in

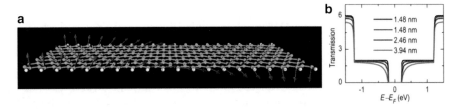

Fig. 11.15 (**a**) The non-collinear spin orientations in a domain wall of the 8-ZGNR for the antiparallel case. (**b**) Transmission curves depending on the domain wall size. The black line is for the collinear spin configuration, while the others are for the non-collinear spin configurations (Reproduced from Ref. [14] with kind permission of Nature Publishing Group)

the curves depending on the domain wall size; the thicker the domain wall, the more rounded the curve is. However, the zero transmission regions leading to the infinite resistance at a low bias are not affected by the domain wall size. Consequently, the conclusion based on the collinear spin state should be the same with that based on the non-collinear spin state.

11.4 Nano-Scale Molecular Sensors and DNA Sequencing

11.4.1 Ionophores/Receptors and Chemical Sensors

The design of novel ionophores and receptors is of great interest for environmental, biological, and medical applications [6, 7]. Though various receptors showing either cation or anion recognition have been synthesized in the past decades, the ultimate goal, the selective recognition of a target species, is still in progress. To this end, theoretical understanding and analysing the conformational changes of receptors upon complexing with a cation/anion are highly required. Since capturing/binding a target species is related to the total free energy of binding, the high level *ab initio* characterized interaction energies for an ion interacting with diverse receptors are useful. In this section, we introduce the *de novo* design for cation and anion ionophores/receptors with interaction energies of various ions with diverse synthetic organic receptors. The design strategy is complicated because the affinity and selectivity have to be displayed in the presence of a number of competing factors like counter ions, solvent molecules, etc.

First, we focussed on the selective recognition of the ammonium cation (NH_4^+) relative to the potassium cation (K^+), which is of nearly equivalent size. The selectivity is attributed to the different favourable coordination numbers between NH_4^+ and K^+. K^+ favors a coordination number of six, while NH_4^+ favors only four. Exploiting π-cation interactions, we designed various benzene based tripo-dal systems with various binding moieties (pyrazole, dihydro-pyrazole, oxazole, dihydro-oxazole, imidazole, dihydro-imidazole) were investigated. A system with dihydro-imidazole moiety and N-methyl substituted dihydro-imidazole had the best selectivity for NH_4^+ over K^+. The π-electron density of the receptor needed to be maximized for improving the binding affinity with cations, though its contribution to the selectivity was minimal.

The above concept can be extended to the receptor design for acethylcholine (Ach). Since Ach is a large molecule, dispersion interactions and diminished ionic interactions are utilized. Replacing the imidazole/pyrazole arms of the NH_4^+ receptors by pyrrole shows the high selectivity of Ach over NH_4^+. To enhance dispersion interactions between the receptor and Ach, we also tried to replace the pyrrole moiety by a bigger indole ring. When the indole rings were skewed while facing toward the center of the benzene ring for the interaction with NMe_4^+ (binding site of Ach), the cation-π interaction was diminished. The 1,3,5-tris(pyrrolyl)-benzene showed strong binding affinities for acetylcholine and NMe_4^+, with much better selectivity over NH_4^+ in aqueous solution. These were confirmed by experiments using ion-selective electrode (ISE). The system prefers to bind Ach selectively over NH_4^+ (24 times) at pH 8.0 in buffered solution.

For anion receptors, we should note that the properties of anions are highly different from those of cations: anions are more polarizable and hence more susceptible to polar solvents than cations. Thus, the solvent effect is important. We designed highly selective anionophores based on imidazolium moieties. The $(C–H)^+–X–$ ionic hydrogen bonding involves the dominating electrostatic interaction, and thus stabilizes the anion better than the normal one. An imidazolium-based receptor, calix-[4]-imidazolium-[2]-pyridine, showed high selectivity for fluoride anion due to the unique 1-to-1 binding mode of interaction compared to 1-to-2 binding profile with other halide anions. From *ab initio* study, we found that the most stable structure of ten with F^- was in good agreement with that of the crystal structure where F^- was captured at the center of the macrocycle by the strong ionic hydrogen bonding [45].

We also designed biological chemosensors able to capture the important phosphates GTP and ATP. Trimethyl-[4(3-methyl-imidazol-1-ium)-butyl]-ammonium substituted at 1,8 anthracene position showed a chelation-enhanced fluorescence quenching effect for GTP whereas it displayed a chelation-enhanced fluorescence effect for ATP, ADP and AMP. Therefore, it not only distinguished the structurally similar compounds GTP and ATP but also acted as a potential fluorescent chemosensor for GTP in 100% aqueous solution (pH = 7.4, 10 mM HEPES). The different strength of T-shape interaction via strong π-H interaction (π-HN for GTP vs. π-HC for ATP) with the central ring of the anthracene moiety of the nucleic base makes the selectivity of GTP six times higher than that of ATP [46]. Further, using a unique sandwich stacking of pyrene-adenine-pyrene, we were able to do selective and ratiometric fluorescent sensing for ATP at physiological pH [47].

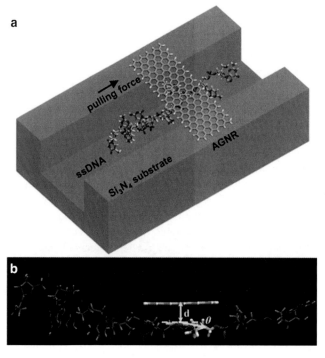

Fig. 11.16 DNA base stacking on a graphene nanodevice during the passage through a fluid nanochannel. (**a**) Schematic of a nanochannel device with GNR (AGNR) through which a ssDNA passes. The water molecules and counterions in the nanochannel are not depicted. (**b**) Instantaneous snapshot in a simulation (*d* stacking distance, *θ* tilt angle) (Reproduced from Ref. [15] with kind permission of Nature Publishing Group)

11.4.2   Graphene Nanoribbon as a Future DNA Sequencing Device

After the success of the Human Genome Project, fast and inexpensive DNA sequencing is desirable for the future biotechnology and bio-industry such as a personal medicine, gene therapy, and ubiquitous diagnosis. At the current stage, DNA sequencing based on cyclic-array method, so called the second generation DNA sequencing, requires over US$ one million and several weeks to decode the overall human genome. Even though the cyclic-array method reduces much cost and time compared to the original Sanger method, the faster and cheaper method is still in demand. In 2004, the National Institute of Health announced that US$ 1,000 DNA sequencing within 24 h is a goal of the new method. With a great advance of nanotechnology, several approaches have been proposed to achieve the goal [48–51].

We proposed a novel DNA sequencing method of a nanopore type. The novelty lies at the use of GNR in sensing (Fig. 11.16a) [15]. While a single stranded DNA (ssDNA) passes beneath the GNR, a single base interacts with the GNR via π-π interaction (Fig. 11.16b), signalling characteristic conductance. With a careful analysis of the conductance data, the clear sequence of the target ssDNA bases is decoded. We studied the system in four steps: (1) the interaction energies between each DNA base and GNR were studied by using *ab initio* quantum chemical calculations, (2) the molecular dynamics simulations were carried out in consideration of temperature, solvent molecules, and counterions, (3) the electron transport properties of DNA base-GNR complex by using NEGF-DFT method were performed to find the conductance differences between different types of nucleobases, and (4) the sequence of the target ssDNA was analysed by using our new data-mining technique and newly defined two-dimensional transient autocorrelation functions (2D-TACF).

More realistic molecular dynamics (MD) simulations were made to confirm whether the well stacked structures were still kept in the presence of other environmental perturbations. The ssDNA with eight bases (5'-GCATCGCT-3') was used as a test DNA. This MD simulation has been done at room temperature with 1 fs timestep. During the simulation, ssDNA was dragged with a force 35 kcal/mol/Å (~2,400 pN) to mimic solvent flows or electric field effects. As ssDNA passes along the silicon nitride nanogap and passes through AGNR, each bases of ssDNA forms stable stacking structures with AGNR devices. It took 1.6 ns for eight bases to pass through the AGNR device.

Based on the molecular dynamics simulation trajectory, we investigated the conductance of GNR upon the binding of nucleobases. The band gap of a GNR depends strongly on the width of the nanoribbon [15]. We uses an armchair GNR (AGNR) with ~1 nm width (1.16 eV band gap). For all the trajectories, we performed first principles calculations based on the Keldysh NEGF method coupled to DFT. Local density approximation (LDA) and generalized gradient approximation (GGA) for the exchange and correlation interaction of electrons are used in the calculations with double-ζ plus polarization basis sets. Both functionals showed almost same results so that only the LDA results are displayed here.

Using the NEGF method without bias voltage (V_b), the pristine GNR shows stepwise integer transmission coefficients on the whole energy range. For the GNR adsorbing a DNA base, the transmission curve shows a sharp drop at a characteristic energy related to each DNA base. This enabled us to obtain the real-time conductance map from the NEGF-DFT calculations of MD geometries. Then, the real-time histogram showed distributions whose values peak at certain energy depending on nucleobases (Ade, Cyt, Gua, and Thy), which helps to distinguish different types of bases (Fig. 11.17). This sequencing analysis is further refined with clear recognition of the interaction interval of each individual base with the GNR. We use the normalized 2D-TACF for the given time span τ:

$$C(t, t_0; \tau) = \int_{t_0}^{t_0+\tau} dt' \int_{-\infty}^{\infty} dE \, J(E, t') J(E, t+t') \Big/ \int_{t_0}^{t_0+\tau} dt' \int_{-\infty}^{\infty} dE \, [J(E, t')]^2,$$

which gives the autocorrelation of a given event function J(E,t) between two time intervals of duration separated by time t. Here, τ is smaller than the passage time of

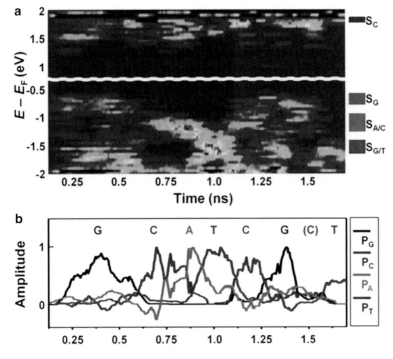

Fig. 11.17 Simulation results of the transport property for 5′-GCATCGCT-3′. **(a)** Time-dependent histogram of the transmission peak positions (*red* (*blue*) denoting maximal (minimal) values). The histogram shows features of Cyt (S_C), Gua (S_G), Ade/Cyt ($S_{A/C}$), and Gua/Thy ($S_{G/T}$) corresponding to the band centred around $E-E_F = E_s = 1.8$, -0.65, -1.2, and -1.65 eV, respectively. The height of each right box represents the energy range of integration for S_{base}. **(b)** The resolved sequence of GCATCGCT which was inferred from the probability curves of four types of nucleobase. For further refinement, see the 2D-TACF in the text (Reproduced from Ref. [15] with kind permission of Nature Publishing Group)

any single base. In the case where the transient correlation between different bases is zero, the 2D-TACF helps us determine the passage time of each base. Then, the base sequencing becomes very reliable.

In summary, the AGNR DNA sequencing device has the following merits:

1. The DNA bases-AGNR complexes show π-π stacking interactions between a nucleobase and a GNR. The interaction energy between a nucleobase and graphene is ∼20 kcal/mol. It should be noted that the dispersion energy does not seriously affected even by the solvent effects. This is the reason the dispersion interaction is often utilized for the self-assembly of macromolecules.
2. A well washed DNA single strand would be negatively charged due to the phosphate groups as the counteranions are washed away. Then, in the presence of applied electric field, a negatively charged DNA strand moves against the applied electric field, while the counterions move along the electric field.

3. Since each base of ssDNA is highly flexible, the strand allows each nucleobase, one by one, to interact with a GNR in the π-π stacked form. The electric field, though weak, is strong enough to release the stacked base from the GNR.
4. The conductance measurements of the GNR for the DNA base-GNR binding/release is highly reproducible for the sequence analysis.
5. The conductance of the GNR can be measured continuously as the ssDNA passes through a nanochannel.
6. The transmission curves show sharp drops at characteristic bias voltages depending on nucleobases. The 2D conductance measured with bias or gate voltage control differentiates different types of DNA bases.
7. The sequencing process is reversible. Since we measure the current through AGNR instead of DNA bases, the physically adsorbed DNA bases can be easily detached after sequencing analysis.
8. The sequencing is very fast because the binding/releasing of the DNA base to/from GNR is very fast (in μs scale), and so the total human genome sequencing can be carried out in an hour, i.e., on the spot.
9. This method could be applied to other types of molecular sensing and analysis.

11.5 Concluding Remarks

We have illustrated our efforts in nanomaterial design using a wide variety of examples. We have discussed the design approach for self-assembled static nanostructures with molecular building blocks, dynamically controllable molecular mechanical devices, and transport properties of metal nanostructures and carbon-based materials. In particular, here we detailed intriguing molecular architecture assembled with novel molecular building blocks, the intriguing nano-optical phenomena of the self-assembled nanoscale lenses, extremely large magnetoresitance of graphene nanoribbon spin valves, and ultrafast DNA sequencing based on graphene nanoribbon and nanochannel. Given the success of our approach in designing experimentally viable nanomaterials, we believe that the coming years would see the use of this approach in the development of novel nanosystems with potential applications in optics, opto-electronics, information storage, sensors, biotechnology, and nanoelectronics. Furthermore, we need a better understanding of the science behind most nanoscale processes so that we can harness these nanosystems as useful machines and electronic devices.

Acknowledgements This work was supported by NRF (National Honor Scientist Program: 2010-0020414) and KISTI (KSC-2008-K08-0002).

References

1. Dykstra CE, Frenking G, Kim KS, Scuseria GE (2005) Theory and applications of computational chemistry: the first 40 years. Elsevier, Amsterdam
2. Hong BH, Lee JY, Lee C-W, Kim JC, Bae SC, Kim KS (2001) J Am Chem Soc 123:10748
3. Hong BH, Bae SC, Lee C-W, Jeong S, Kim KS (2001) Science 294:348
4. Lee JY, Hong BH, Kim WY, Min SK, Kim Y, Jouravlev MV, Bose R, Kim KS, Hwang I-C, Kaufman LJ, Wong CW, Kim P, Kim KS (2009) Nature 460:498
5. Mason DR, Jouravlev MV, Kim KS (2010) Opt Lett 35:2007
6. Singh NJ, Lee HM, Hwang I-C, Kim KS (2007) Supramol Chem 19:321
7. Singh NJ, Lee HM, Suh SB, Kim KS (2007) Pure Appl Chem 79:1057
8. Singh NJ, Lee EC, Choi YC, Lee HM, Kim KS (2007) Bull Chem Soc Japan 80:1437
9. Kim HG, Lee C-W, Yun S, Hong BH, Kim Y-O, Kim D, Ihm H, Lee JW, Lee EC, Tarakeshwar P, Park S-M, Kim KS (2002) Org Lett 4:3971
10. Kim WY, Kim KS (2008) J Comput Chem 29:1073
11. Kim WY, Kim KS (2010) Acc Chem Res 43:111
12. Kim WY, Choi YC, Kim KS (2008) J Mater Chem 18:4510
13. Kim WY, Choi YC, Min SK, Cho Y, Kim KS (2009) Chem Soc Rev 38:2319
14. Kim WY, Kim KS (2008) Nat Nanotechnol 3:408
15. Min SK, Kim WY, Cho Y, Kim KS (2011) Nat Nanotechnol 6:162
16. Kim KS, Tarakeshwar P, Lee JY (2000) Chem Rev 100:4145
17. Lee EC, Kim D, Jurečka P, Tarakeshwar P, Hobza P, Kim KS (2007) J Phys Chem A 111:3446
18. Kim KS, Suh SB, Kim JC, Hong BH, Lee EC, Yun S, Tarakeshwar P, Lee JY, Kim Y, Ihm H, Kim HG, Lee JW, Kim JK, Lee HM, Kim D, Cui C, Youn SJ, Chung HY, Choi HS, Lee C-W, Cho SJ, Jeong S, Cho JH (2002) J Am Chem Soc 124:14268
19. Suh SB, Kim JC, Choi YC, Kim KS (2004) J Am Chem Soc 126:2186
20. Smolyaninov II, Hung Y-J, Davis CC (2007) Science 315:1699
21. Liu Z, Lee H, Xiong Y, Sun C, Zhang X (2007) Science 315:1686
22. Fang N, Lee H, Sun C, Zhang X (2005) Science 308:534
23. Hell SW (2007) Science 316:1153
24. Yano T, Shibata S, Kishi T (2006) Appl Phys B 83:167
25. Taflove A, Hagness SC (2000) Computational electrodynamics: the finite-difference time-domain method. Artech House, Boston
26. Nautiyal T, Rho TH, Kim KS (2004) Phys Rev B 69:193404
27. Cheng D, Kim WY, Min SK, Nautiyal T, Kim KS (2006) Phys Rev Lett 96:096104
28. Cho Y, Kim WY, Kim KS (2009) J Phys Chem A 113:4100
29. Ulman A (1996) Chem Rev 96:1533
30. Tulevski GS, Myers MB, Hybertsen MS, Steigerwald ML, Nuckolls C (2005) Science 309:591
31. Guo X, Small JP, Klare JE, Wang Y, Purewal MS, Tam IW, Hong BH, Caldwell R, Huang L, O'Brien S, Yan J, Breslow R, Wind SJ, Hone J, Kim P, Nuckolls C (2006) Science 311:356
32. Novoselov KS, Geim AK, Morozov SV, Jiang D, Dubonos SV, Grigorieva IV, Firsov AA (2004) Science 306:666
33. Kim KS, Zhao Y, Jang H, Lee SY, Kim JM, Kim KS, Ahn J-H, Kim P, Choi J-H, Hong BH (2009) Nature 457:706
34. Bae S, Kim H, Lee Y, Xu X, Park J-S, Zheng Y, Balakrishnan J, Lei T, Kim HR, Song YI, Kim Y-J, Kim KS, Ozyilmaz B, Ahn J-H, Hong BH, Iijima S (2010) Nat Nanotechnol 5:574
35. Kim N, Kim KS, Jung N, Brus L, Kim P (2011) Nano Lett 11:860
36. Yi JW, Park J, Singh NJ, Lee IJ, Kim KS, Kim BH (2011) Bioorg Med Chem Lett 21:704
37. Lee EC, Choi YC, Kim WY, Singh NJ, Lee S, Shim JH, Kim KS (2010) Chem Eur J 16:12141
38. Chandra V, Park J, Chun Y, Lee JW, Hwang I-C, Kim KS (2010) ACS Nano 4:3979
39. Chandra V, Kim KS (2011) Chem Commun 47:3942
40. Myung S, Park J, Lee H, Kim KS, Hong S (2010) Adv Mater 22:2045
41. Yu Y-J, Zhao Y, Ryu S, Brus LE, Kim KS, Kim P (2009) Nano Lett 9:3430

42. Cho Y, Choi YC, Kim KS (2011) J Phys Chem C 115:6019
43. Lee WH, Park J, Sim SH, Lim S, Kim KS, Hong BH, Cho K (2011) J Am Chem Soc 133:4447
44. Fisher DS, Lee PA (1981) Phys Rev B 23:6851
45. Chellappan K, Singh NJ, Hwang I-C, Lee JW, Kim KS (2005) Angew Chem Int Ed 44:2899
46. Kwon JY, Singh NJ, Kim NH, Kim SK, Kim KS, Yoon J (2004) J Am Chem Soc 126:8892
47. Xu Z, Singh NJ, Pan J, Kim H, Kim KS, Yoon J (2009) J Am Chem Soc 131:15528
48. Mardis ER (2011) Nature 470:198
49. Shendure J, Mitra RD, Varma C, Church GM (2004) Nat Rev Gen 5:335
50. Shendure J, Ji H (2008) Nat Biotechnol 26:1135
51. Kahvejian A, Quackenbush J, Thompson JF (2008) Nat Biotechnol 26:1125

Chapter 12
Computational Molecular Engineering for Nanodevices and Nanosystems

Norma L. Rangel, Paola A. Leon-Plata, and Jorge M. Seminario

Abstract Molecular electrostatic potentials (MEPs), electronics (moletronics), and vibrational electronics (vibronics) are novel scenarios to process information at the molecular level. These, along with the traditional current-voltage scenario can be used to design and develop molecular devices and systems for even more extended applications than traditional electronics. Successful control and communication features between scenarios would yield "smart" devices able to take decisions and act under difficult conditions. The design of molecular devices is a primordial step in the development of devices at the nanometer scale, enabling the next generation of sensors of chemical and biological agents molecularly sensitive, selective, and intelligent.

Keywords Moletronics • Vibronics • Molecular electrostatic potentials • Molecular electronics • Nanotechnology

12.1 Introduction

Molecular scenarios, along with the traditional current-voltage scenario, can be used to design and develop molecular devices for the next generation electronics. Control and communication features of these scenarios strongly help in the production of

N.L. Rangel • P.A. Leon-Plata • J.M. Seminario (✉)
Department of Chemical Engineering, Texas A&M University, College Station, TX, USA
e-mail: normalucre@gmail.com; paola.leon@tamu.com; seminario@tamu.edu

N.L. Rangel • J.M. Seminario
Materials Science and Engineering, Texas A&M University, College Station, TX, USA

J.M. Seminario
Department of Electrical and Computer Engineering, Texas A&M University, College Station, TX, USA

J. Leszczynski and M.K. Shukla (eds.), *Practical Aspects of Computational Chemistry I:* 347
An Overview of the Last Two Decades and Current Trends,
DOI 10.1007/978-94-007-0919-5_12, © Springer Science+Business Media B.V. 2012

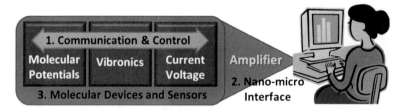

Fig. 12.1 Three major approaches to develop a new electronics

"smart" devices able to take decisions and act autonomously in aggressive environments. However, molecular potentials are still unreadable and non-approachable by any present technology. It is well-known that the proper assembly of molecules can result in a numerical processing system based on digital or even analogical computation. Thus, the outputs of molecular processing units need to be read and amplified in order to be interfaced to standard electronics; thus, the results of the most complex data processing can be successfully use in recent technologies.

Along with the extremely small size ($\sim$1 nm) of molecular devices, making devices of this size for other applications appears as a potential possibility to develop further areas of technology related with systems that were practically impossible to imagine just a few years ago, such as, for example, the detection and analysis of single molecules. In our particular case, the size of the device is of important consideration in sensor science; the ultimate detector of a molecule is another molecule that can respond quickly and selectively to several agents. This possibility directly implies a nano-micro interface to interconnect the output from a molecular device to standard microtechnologies.

In this review we show our advances in molecular engineering to develop sensors and nanosystems; their progress is characterized by three well-defined interconnected scenarios (Fig. 12.1), which requires, first, an understanding of how these scenarios for processing information can be implemented to operate, control and communicate in a device, in such a way that allows the development of smart devices. The second requirement is to design and develop a nano-micro interface that is able to read and amplify information and signals encoded and processed at the molecular level to current microtechnologies. Thus, this interface amplifies signals from nanoscale devices to levels compatible with standard microelectronics. And third, define a set of devices and sensors with high selectivity and sensitivity that will contribute to the scaling-down and performance in nanotechnology and to the improvement of conventional engineering applications.

12.1.1 Vibrational Electronics "Vibronics"

When a signal is injected into a molecule, the vibrational frequencies of the atoms around the injection point changes, affecting the atoms in the molecule through their bonds; frequency changes in vibrational frequencies of the molecule trigger displacements of atoms whose frequencies are transferred to their neighbor atoms by means of bond bending, bond stretching, van der Waals and Coulombic interactions between charges, etc. [1]. Thus, using vibronics, signals, information can be transferred through molecules [2–4]. When an atom in a molecule is displaced, a displacement signal is transmitted through the molecule by the vibrational modes. This is reminiscent of a mass-and-spring system, where a displacement from equilibrium experiences a proportional restoring force. This effect also occurs when an external stimulus is applied by another molecular vibration or potential in the neighborhood. Furthermore, an atom displacement induces a change in the field of the molecular electrostatic potential.

12.1.2 Molecular Electrostatic Potentials

Molecular potentials offer the opportunity to apply the art and science of chemistry to the implementation of molecules as electronic devices. As it stands, until a new technique to fabricate molecular circuits and transmit information between molecules is discovered, perhaps there is no other alternative than using the molecular potentials for the encoding and processing of information in of molecular electronic systems. Currently, we must rely on molecular potentials to develop systems that can continue the tremendous scaling down of photolithography in the massive fabrication and wiring of silicon based circuits.

Molecular potentials are calculated from the nuclei and electron density contributions. The electron density $\rho(r)$ is obtained from the wavefunction ψ by integrating on all electrons except one. Since the wavefunction is antisymmetric and the integral is quadratic on ψ, it does not matter what specific electron is left after the integrations.

$$\rho(\vec{r}_1) = N \int \psi^*(\vec{r}_1, \vec{r}_2, \cdots \vec{r}_N)\psi(\vec{r}_1, \vec{r}_2, \cdots \vec{r}_N)d\,\tau_2 \cdots d\,\tau_N \qquad (12.1)$$

where $\psi(\vec{r}_1, \vec{r}_2, \cdots \vec{r}_N)$ and $\psi^*(\vec{r}_1, \vec{r}_2, \cdots \vec{r}_N)$ are the wavefunction and its complex conjugate and N is the total number of electrons; substituting ψ by a single determinant wavefunction, the electron density reduces to

$$\rho(\vec{r}) = \sum_1^N |\phi_i(r)|^2 \qquad (12.2)$$

where ϕ_i are the molecular orbitals and the sum runs up to all N occupied ones, assuming one electron in each molecular orbital.

The MEP, $V(\vec{r})$, is calculated from the nuclei and electron density contributions [5],

$$V(\vec{r}) = \sum_i \frac{Z_i e}{|\vec{R}_i - \vec{r}|} - \int \frac{\rho(\vec{r}')}{|\vec{r} - \vec{r}'|} d\tau' \qquad (12.3)$$

where Z_i is the atomic number of atom i located at $\vec{R}_i$.

It has been demonstrated using reliable ab initio techniques that information can be encoded using molecular electrostatic potentials (MEP) [6–14]. This represents a new alternative to the standard charge-current approach used in standard electronics. Few molecules were found to function as molecular OR- [15, 16] and AND logical gates [3]. Present computers perform calculations reducing the information to binary notation whereby the only two binary digits (bits) are 1 and 0, which can be also assigned to pairs of logical variables or states such as TRUE or FALSE, ON or OFF, HIGH or LOW, POSITIVE or NEGATIVE, respectively. In binary notation the number nine is written as 1001, requiring at least four binary digits (bits). Operations between binary variables are performed using logical gates such as the NOT, AND, OR, etc. For instance the simplest one is the NOT which outputs a 0 if the input is 1 and yields 1 if the input is 0. The AND gate has two or more inputs and the output is 1 if all inputs are 1 and 0 if any input is 0. The OR gate yield a 0 when all inputs are zero and 1 when any input is 1. Interestingly, these gates can be combined to make adders, multipliers, integrators, etc. In principle we can numerically solve any operation that we can imagine. Recently, it has also been demonstrated that the output of two molecular OR-gates can be used as the input of a third molecular OR-gate, thus opening the possibility to implement complex molecular electrostatic potential circuits using simpler molecular logic gates [16]. These earlier calculations were performed with a molecule at vacuum and triggered by the electrostatic potential of small molecules as inputs to the gate. We showed how the substrate affects the molecular electrostatic potential device, by testing substrates on a molecule, which had been demonstrated to be an OR-gate in vacuum [3]. The effect of the substrate on the molecular devices is of paramount importance for the theoretical proof-of-concept of this new scenario for molecular computing.

12.1.3 Molecular Orbital Theory

The highest occupied molecular orbital (HOMO) and the lowest unoccupied molecular orbital (LUMO) are used to define the HOMO-LUMO gap (HLG), which is an indicator of the stability of wavefunctions. Short gaps (i.e., low hardness, high softness) are indicative of poor stability of the wavefunction, which worsens in stability when an external field is applied to the molecule. The localization of molecular orbitals (MOs) and the enlargement of HOMO-LUMO gaps suggest poor conduction in determined transport paths [17, 18].

12.1.4 Sensor Devices

A sensor device is a transducer of a physical or chemical quantity into, mainly, an electrical or optical signal. Sensors of physical parameters such as temperature, pressure, magnetic fields, forces, etc., can be encapsulated to avoid the effects of the environment; however, parameters of chemical sensors are concentrations of substances, of which there can be several in a single sample (e.g. air, blood). Thus the sensitivity of a sensor is a technological challenge along with its durability due to interfering substances and environmental effects of several factors such as light, corrosion and reactivity. Sensing science is a multidisciplinary field that pursues the development of new devices with high selectivity and sensitivity, enabling rapid analysis, durability and reduced costs.

There are several types of chemical sensors, including: conductive, electro-chemical and calorimetric. Conductive type sensors are based on measurements of material resistivity. They are usually employed to sense gases, though the reactivity of the materials with the environment leads to poor selectivity. Electrochemical sensors are used for pH measurements and gas detection, but there are limitations in fabrication, costs and usability time of the device due to the consumption of the electrodes; they can be fabricated in batches and allows the scaling down of sensor devices; however, for the conductivity sensors, electrochemical sensors are not suitable for aggressive environments. High sensitivity sensors have been developed using mass sensitive devices such as nano-electromechanical and resonator devices. For harsh environments absorption and emission of electromagnetic radiation have been used in optical sensors, among them, fiber optic, photo-acoustic sensors, luminescence, etc. Also, calorimetric sensors have been used to detect exothermic biochemical reactions based on the released thermal energy.

12.2 Molecular Engineering Theory

Integration of novel and traditional scenarios to develop nano-devices allows an improvement of conventional engineering applications with modern nanotechnologies such as molecular-electronics, which can be applied to develop devices for the post CMOS era, such as: solar cells, catalysis, sensing, chemical and biological defense and security, biotechnology, medicine and photo-electronics.

First, quantum mechanics methods and the formalism of the theoretical calculation of electrical properties for finite and infinite systems, level of theory implemented such as methods and basis sets, and single molecule conductance of molecular junctions are described.

Two types of molecular systems are used in the theoretical calculations: single molecules with a finite number of atoms and junctions combining finite systems (molecules) absorbed on metallic contact tips (bulk).

Electronic properties of finite systems are obtained using quantum chemistry by solving the Schrödinger equation, applying appropriate correlation and approximations factors as is done for example using density functional theory (DFT). Combined Green function theory (GF) and DFT are used to calculate the conductance of electrode-molecule-electrode junctions. All finite system calculations are performed with the program GAUSSIAN-09 [19] and earlier versions.

12.2.1 Ab Initio Molecular Orbital Theory

The energy and electrical properties of a molecule can be predicted by solving the nonrelativistic and time independent Schrodinger equation,

$$H \Psi = E \Psi, \tag{12.4}$$

which describes the wavefunction Ψ of the particles in the system, the energy of the molecule E, and the Hamiltonian operator H. The full Hamiltonian includes the kinetic and potential energy terms.

$$\hat{H} = \hat{T} + \hat{V} \tag{12.5}$$

The nuclear kinetic energy term can be neglected using the Born-Oppenheimer approximation which considers that the mass of the nuclei is greater than the electron mass and thus moves very slowly with respect to the electrons; therefore,

$$\hat{H} = \sum_i -\frac{\hbar^2}{2m_e}\nabla_i^2 - \sum_{i,A}\frac{Z_A e^2}{|\vec{r_i} - \vec{R_A}|} + \frac{1}{2}\sum_{i \neq j}\frac{e^2}{|\vec{r_i} - \vec{r_j}|} + \frac{1}{2}\sum_{A \neq B}\frac{Z_A Z_B e^2}{|\vec{R_A} - \vec{R_B}|} \tag{12.6}$$

Using proper boundary conditions and approximations, each possible solution of the time-independent Schrodinger equation corresponds to a stationary state of the system; the one with the lowest energy is considered the ground state. Also, the spatial wavefunction is independent of time and function of the positions of the electrons, and the nuclei should be normalized and anti-symmetrized. Thus,

$$E = \langle \Psi | H | \Psi \rangle \quad \langle \Psi | \Psi \rangle = 1 \tag{12.7}$$

To calculate the electronic properties of a system, the wavefunction is represented by a determinant of molecular orbitals (MOs) and spin functions "spin orbital". MOs can be expressed as linear combinations of the atomic orbitals χ_μ that resemble one-electron hydrogen-like wavefunctions known as the basis functions. Thus, each molecular orbital is expanded into a linear combination of atomic orbitals

(MO-LCAO), accepting that these atomic orbitals are actually pseudo atomic orbitals obtained by fitting one or more of the special functions to one hydrogen-like function.

$$\psi_i = \sum_\mu \mu C^i_\mu \chi_\mu \tag{12.8}$$

In turn, each pseudo atomic orbital is expanded into a sum of Gaussian functions gi, whose coefficients k_i and their corresponding exponents a are chosen to fit the best chemical properties, usually energies.

$$\chi_\mu = \sum_i k_i g_i \tag{12.9}$$

12.2.2 Basis Sets

The Gaussian functions introduced in a general manner in Eq. 12.9, also called primitives, are used to form a complete set of functions to describe the main features of the molecular orbitals. They are centered in each of the atoms of the molecular system and have the general Cartesian form,

$$g_i(\alpha_i, \vec{r}) = c_i x^{n_i} y^{m_i} z^{l_i} e^{-\alpha_i r^2} \tag{12.10}$$

where n_i, m_i and l_i are nonnegative integers, α_i is a positive orbital exponent, c_i is a normalization constant, r is the radial coordinate of the electron and x, y and z are its corresponding Cartesian coordinates, all with respect to the nucleus location of g_i. The Gaussian basis functions usually include additional primitives such as polarization and diffusion functions. Most of our work develops systematically increasing the quality of the basis set in trade-off with the cost of the calculation. Thus, we use basis set such as: 3-21G [20], 6-31G(d) [21, 22], 6-31G(d,p) [21, 22], and cc-pVTZ [23], which are used for first and second (and some third) row atoms; when heavier atoms are part of the calculation, Los Alamos National Laboratory (LANL2DZ) basis set and effective core potentials are used [24, 25].

12.2.3 Hartree-Fock Theory

Hartree Fock theory searches for a local minimum by finding the set of coefficients of the wave function that minimizes the total energy,

$$FC = SC\varepsilon_i \tag{12.11}$$

where ε_i is a diagonal matrix of one-electron energies; F is the Fock matrix, which includes the Hamiltonian for a single electron interacting with nuclei and a self-consistent field of other electrons; and S is an atomic-orbital overlap matrix.

The Fock matrix represents the average effects of the field of all the electrons and nuclei in each orbital; as the orbitals depend on the molecular orbital expansion coefficients, thus the self-consistent method (SCF) is used to solve

$$[F(C]_i)C_{i+1} = S_{i+1}C_{i+1}\varepsilon_{i+1}$$

until

$$C_i \cong C_{i+1}$$

At convergence, the energy is at a minimum and the electrons and nuclei yield a stationary field.

However, the Hartree-Fock method does not treat (by definition) what is called electron correlation, which physically represents instantaneous interactions between individual electrons rather than the average as done in HF. Electron correlation is a key factor to find important properties of chemical interest such as binding energies. One way to solve this problem is to include additional determinants to the wavefunction. Unfortunately, improving the single-determinant wavefunction, by adding additional determinants, yields methods that are extremely expensive in computational resources.

12.2.4 Density Functional Theory

The easier way to solve the problem of correlation is by using Density Functional Theory (DFT). DFT is less expensive than any of the correlated methods, and it is much more precise in many situations. It is our only possibility to include electron-correlation of large systems. DFT is based on the first Hohenberg-Kohn (1964) theorem, which establishes that properties in the ground state are functionals of the electron density (12.1). In 1965, Kohn-Sham demonstrated that the electron density of a molecular system of interacting electrons can be represented with the electron density of an equivalent system of non-interactive electrons subjected to an effective potential. Exact functionals for exchange and correlation are unknown and, thus, approximations found in the literature are needed to perform calculations using DFT.

Geometry optimizations are systematically carried out from the less expensive *ab initio* methods such as Hartree-Fock (HF), to density functional theory (DFT) [26] methods, which include electron correlation in a very efficient way. Second derivatives of the energy yielding the Hessian are also needed to obtain molecular vibrations.

12.2.5 Hybrid Functionals

The DFT B3PW91 hybrid functional includes a combination of the Perdew-Wang-91 [27, 28] and has shown good energetics predictions [29, 30], but the M05-2X meta functional has a larger range for electron correlation that improves the performance for nonbonded interactions and π-π stacking [31], showing good binding energies and geometry optimizations of molecules non-covalently bonded, as well as stacking order. Therefore, M05-2X is used for non-bonded interactions and geometries and B3PW91 for bonded systems and for the calculation of total energies.

12.2.6 Single Molecule Conductance

Extensive theoretical and experimental effort has been invested in the understanding of electron transfer mechanisms in single organic molecules due to their potential applications in nano-scale electronic systems [18]. It has been found that the current running through a molecule is the sum of the contributions from all molecular orbitals, each presenting a barrier to electron transport equal to their energy difference from the Fermi level of the contacts. Since a molecule is the fundamental component of a programmable molecular array, the electrical characteristics of the molecule, especially the current-voltage characteristic, are the major design parameters for the construction of a programmable molecular array [32].

The continuous electronic states of macroscopic contacts affect the discrete electronic states of isolated molecules calculated by solving the Schrödinger equation; thus, a combination of Density Functional Theory and Green function theory (DFT-GF) is used to obtain the electrical properties of the metal-molecule-metal junction.

After a geometry optimization of the extended molecule, using quantum chemistry techniques, is complete, a bias voltage is applied to determine the electronic properties of the extended molecule. During the field calculation, the geometry of the molecule may be kept the same or let it optimize.

The partial density of state (DOS) for the nanoelectrodes is calculated using Crystal 06 program, and the sets for each bias voltage of the Hamiltonian and overlap matrices, obtained using Gaussian 09, are entered in our in situ developed program, GENIP, which applies the combined DFT-GF formalism to obtain the current-voltage characteristics of the junction by considering the local nature of the molecule as well as the non-local features of the contacts.

12.3 Optimum Fit Material for a Nano-Micro Interface

A study of the best fit materials and molecules constituting a sensor system to achieve sensitivity, selectivity and efficiency of the active material is needed for the development of nanodevices.

High sensitivity and selectivity of the chosen material is needed, as well as appropriate lengths (nano-micro sizes), to achieve molecular sensitivity and facilitate the assembly on micro-sized chip gaps. Therefore, a screen of materials has been performed and summarized in this chapter. Materials considered include: large chain molecules such as DNA [33], DNA origami [34, 35], and alpha glycine [36]; carbon based materials such as carbon nanotubes [37] and graphene [38–40]; organic molecules such as the 4,4'-(diethynylphenyl)-2'-nitro-1-benzenethiolate [41], octyltrichlorosilane, and the 7-dehydrocholesterol (precursor of vitamin D_3) [42]; and thin films of metal nanoparticles [43]. Special emphasis is made on graphene due to its exceptional performance and capabilities to achieve the proposed goals, consequently, it is the focus and main material for the developed sensors in this chapter.

Despite the great findings, DNA based materials are not feasible for the development of sensors due to their environment reactivity, difficulties in the assembly, poor addressability and manipulation of single structures, and difficulties removing the structures from solutions. Thus, graphene sheets are considered as the main fabrication material.

Graphene has been demonstrated to act as a resonator, transducer, signal mixer, and high sensitivity and selectivity sensors, thanks to the exceptional intrinsic electrical, mechanical and thermal properties for applications in sensing science as an optimum material for each different type of sensor; the convenient fabrication of the devices, and atomic cross section allow graphene to be the perfect material for measurement of chemical and physical parameters.

12.4   Graphene Based Sensors

Graphene is a "novel" material recently proposed as one of the main alternatives to overcome the performance limitations of materials such as silicon and carbon nanotubes.

The number of layers, types of edges such as zigzag or arm-chair, and topological defects such as vacancies, impurities, ripples, non-hexagonal polygons, etc., affect the properties and behavior of the graphene crystals [44, 45]. We have calculated the electronic properties of graphene ribbons with non-hexagonal polygons defect types (Fig. 12.2) and graphane, a two-dimensional graphitic material based on sp^3 carbon atoms instead of sp^2.

The conductivity of graphene can significantly be affected by the presence of defects on the surface. The HLG of a pristine graphene ribbon with no defects is 2.28 eV and can be decreased down to 0.69 eV by introducing combinations of pentagons, heptagons and octagons on the surface (Fig. 12.2). More interestedly is the electronic properties of graphene; the HLG is more than three times larger than the graphene, with delocalized molecular orbitals but with very large barrier (7.64 eV) make graphene a graphitic insulator material.

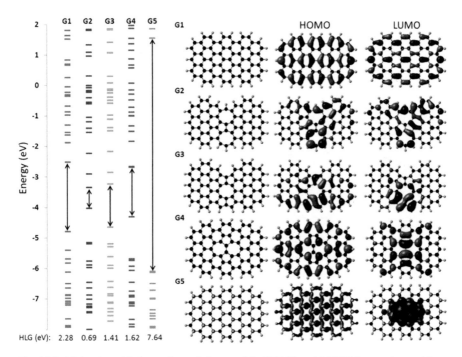

Fig. 12.2 Molecular orbital energies and shapes of the HOMO and LUMO for graphene without defects (*G1*), pentagon-octagon defects (*G2* and *G3*), and pentagon-heptagon defect (*G4*) and graphane (*G5*)

Using GENIP, we calculate the current-voltage characteristics of the graphene ribbons with defects. As expected from the HLG values and the shape of the molecular orbitals, when non-hexagon defects are introduced on graphene the conductivity increases (Fig. 12.3). It is also observed that the sheet of sp^3 carbon atoms is an insulator material.

Graphene defects can be used to modify the electronic properties of a ribbon, and the versatile characteristics of graphene allow the development of graphitic materials with different properties, suggesting that it is a promising successor material of the well-integrated silicon technologies.

Different stacking order, geometries, edge type, and passivation are considered in the study of the graphene sensors. The results show that graphene ribbons (actually molecules) resemble semiconductors but their HOMO-LUMO gaps are edge and passivation dependent.

Therefore, we add the hydrogen atoms to the ribbons obtained from Fig. 12.4 to warrant delocalized orbitals independent of the dominant edge. Despite of passivation effects, the HOMO-LUMO gaps of graphene ribbons affect the conductivity. Thus, we investigate various graphene molecules of different sizes starting from biggest one, the ribbon shown in Fig. 12.4; we calculate their HOMO-LUMO gaps as shown in Fig. 12.5.

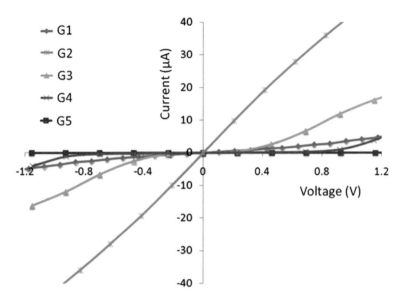

Fig. 12.3 Current-voltage characteristics of graphene, with and without defects, and graphane. The conductivity of graphene (*G1*) is improved by adding defects (*G2, G3* and *G4*) to the structure and significantly decreased in graphane (*G5*)

The dominant edge can influence the conductivity of the ribbons, as seen in Fig. 12.5. The shape of the molecular orbitals affects the transport properties but the calculations show that both, armchair and zigzag ribbons feature delocalized molecular orbitals when they are passivated; if the ribbon is not passivated, localized molecular orbitals are found especially along the zigzag edges.

Current-voltage characteristics of armchair (Fig. 12.6a) and zigzag (Fig. 12.6b) edged graphene ribbons, with and without hydrogen passivation are calculated to observe the effect of passivation on graphene. The results obtained with GENIP show that zigzag edged ribbons are more conductive than armchair edged ribbons, as is expected from the HLGs shown in Fig. 12.5. The effect on both zigzag and armchair ribbons is the same; when the ribbons are not passivated with hydrogen, the electronic density is localized at the edges, making the carbon at the edges more reactive and decreasing the conductivity of the ribbons (Fig. 12.6). When the ribbons are passivated, the electronic density is fully delocalized on the surface as plasmons and the conductivity is larger than the non-passivated.

Although for infinite and pristine graphene the theoretical band gap is zero, Fig. 12.4 suggests a finite HOMO-LUMO gap (HLG) for our small graphene molecules. It has been reported that armchair ribbons are metallic or semiconductors depending of their width and zigzag nanoribbons are mostly metallic [46]. Experiment has been reported that by decreasing the width of the ribbon the gaps increase [47], but there are not reported values for small graphene molecules with specific edge types.

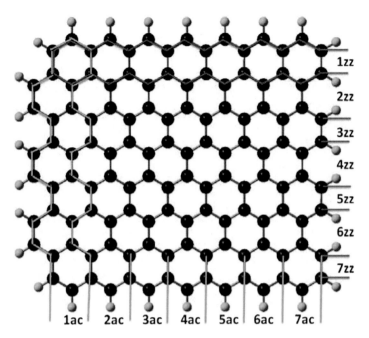

Fig. 12.4 Square graphene ribbon, the number of rings along each edge is used to label and characterize the size of the graphene ribbons used in this review. Shown with *red* is a unit with zigzag (zz) edge "1zz" and with green a unit of armchair (ac) edge "1 ac". The size of the ribbon shown is 7 ac × 7zz (Reprinted from Ref. [38] with kind permission of The American Institute of Physics)

Fig. 12.5 HOMO-LUMO gap (*HLG*) with respect to the number of edged size of the graphene ribbon shown in Fig. 12.4 (Reprinted from Ref. [38] with kind permission of The American Institute of Physics)

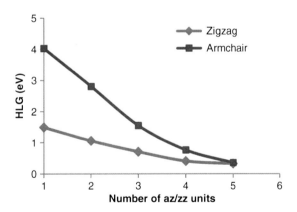

For small systems, the static conditions are calculated very precisely and can be taken to the highest level of precision simply by increasing the level of theory and increasing the size of the basis set until a satisfactory solution to the time-independent Schrödinger equation is found, or as in most cases, until the computational resources allow it.

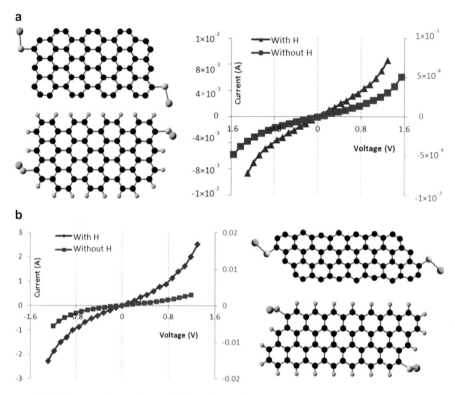

Fig. 12.6 Current-voltage characteristics of armchair (**a**) and zigzag (**b**) edged graphene ribbons and the effect of the hydrogen passivation on the conductivity

A high sensitivity and selectivity network can be build using graphene ribbons, which couples molecular vibrations and molecular electrostatic potentials, acting as amplifiers, able to even reach superconductivities, or be used as transducer converting signals into the usual current-voltage characteristics. Also, the mechanical and electrical properties of graphene lead to devices with exceptional performance, flexibility and scalability.

12.5 Molecular Interface to Read Molecular Electrostatic Potentials Based Electronics

Molecular potentials are unreadable and un-addressable by any present technology. It is known that the proper assembly of molecules can implement an entire numerical processing system based on digital or even analogical computation. In turn, the outputs of this molecular processing unit need to be amplified in order to be useful.

There are methods to strategically arrange molecules on a surface. One of the earliest developed method to manipulate molecules on a surface is by the use of a scanning tunneling microscope (STM) [48]. Other methods include several variants of controlled self-assembly monolayers [49] and the use of replicative molecules, such as DNA, to fabricate and locate molecules such as amino acids on a surface [50]. These techniques are still being developed and will hopefully provide new ways to utilize molecules as electronic devices.

However, no matter how well chemistry can assemble molecules into systems, the molecules need to be linked or interconnected to build electronic systems. The solution to this problem is one that may take an exponentially more complicated turn than the one created by the assembly of the molecules on a surface. It is our view that interconnection between neighboring molecules, as opposed to random molecular connection, would be the best resolution, with ideas similar to a cellular automata [51–53] or the like.

A technique that fulfills the requirements needed for the development of molecular circuits is the use of molecular potentials to encode and process information. The molecular potentials "outside of a molecule" vary between $+3$ and -3 V [14]. They were first used to determine the reactivity of molecules [9, 14, 54]. Later, they were used to create indicators or descriptors [55, 56] to determine several properties of the molecules. These indicators and descriptors have also been used in bulk materials with extrapolation to different phases [57]. As such, positive potentials outside a molecule or in the space where it interacts with others imply a shortage of electrons while negative potentials imply an excess. The practical importance of molecular potentials is the possibility to use them to act on and modify the potentials of neighbor molecules.

With this in mind, several molecular gates have been designed and studied. Consequently, these type works have made it possible to determine properties of materials such as energetic materials, biological and chemical agents that are very difficult to determine experimentally [8, 58–62]. Another important application of molecular potentials is their use to determine a set of point charges that better reproduce the potentials on a surface around a molecule. This provides much better consistency with concepts of charge and bonding used in chemistry [63, 64].

The use of molecular potentials also includes applications to hydrogen bonding interactions [65]. It has been demonstrated that the topography of the molecular electrostatic potential (MEP) provides a measure of the cavity dimensions and an understanding of the hydrogen-bonded interactions involving primary and secondary hydroxyl groups. The MEP topography qualitatively explains the binding patterns of the guest molecule with the host.

The strategic use of the MEP_s of several molecules have demonstrated the ability to perform logical operations. One of these included an OR operation [6, 7] to yield an output of 1 if at least one of its inputs is 1, otherwise it yields 0. It has also been demonstrated that other molecular single gates, such as organic molecules of less than 1 nm^2 in size that consist of two or more isolated functional groups such as ethynyl and butyl [12], are able to successfully interconnect to each other. This opens the door to extremely low-power consumption molecular devices [66]. The dihydroxybenzene molecule is an OR logic gate [15]. The 1,3,5-trifluorobenzene (TFB) molecule can function as a AND gate when input signals are applied to the 3- and 5-fluorine atoms and the output signal is taken from a buffer Be atom at 3.2 Angstroms from the 1-fluorine [3]

There are two general approaches for the use of a single molecule to perform computation [67] and other so called electronic functions [6, 68]. One is to continue the standard current-voltage approach: simply introduce molecules to continue the successful scaling-down of electronic devices based on silicon to sizes that still allows fabricating them and interconnecting them in a small integrated circuit [18, 67, 69–73]. The second approach involves using other alternatives to encode and transport information [66, 74]. One of the attractive approaches involves the concurrent use of molecular potentials and vibronics to perform logic, computation processing, and several other functions now performed by standard complementary metal-oxide semiconductor (CMOS) integrated circuits [75, 76].

However, no matter which approach is to be utilized in the future, there is one important issue that is common to both major scenarios for nano and molecular electronics: the nano-micro interface. As the nano and molecular world cannot be directly addressed presently, any data or signal extraction of the nano-micro interface has to be done by using the standard CMOS technology.

The introduction of data is not as difficult as the extraction, as the latter requires amplification of signals from individual molecules. A few years ago, the idea of a molecular amplifier was introduced [68, 77]. In this amplifier, the rotation of a ring in a dithiotolane molecule was used to control the current through the molecule. The rotation of one of the phenyl rings was able to increase the current through the molecule more than 60 times for the planar conformation over the perpendicular conformation and much more still for larger voltages. Although the calculations were performed using a simple molecule of the oligo (phenylene–ethynylene) (OPE) family, whereby the rotation of one ring was constrained, turning the control ring exclusively requires further analysis.

A modified OPE molecule that contains a local dipole moment in the central ring is implemented. Thus, two of the central carbon atoms are changed by adding polar substituents such as nitrogen, yielding the aza-OPE molecule (Fig. 12.7). The molecule can then be rotated by the effect of an external electric field. The effects of this rotation, and the changes in the molecular potential it produces, are tested. With the aza-OPE connected externally to a power source and to standard electronic amplifiers, it was discovered that the results can be used to design a specific amplifiers able to transform a small signal interacting with the dipole moment of the rings to a much larger signal through the aza-OPE, as shown in Fig. 12.7.

The aza-OPE molecule has the necessary properties to act as an amplifier such as large discrimination for different torsional angles. However, this is not only limited to get a current change due to the rotation of the angle but also due to just the value of the molecular potential, which strongly affects the current through the molecule even if the torsional angle has slightly or not changed. Our setting can read molecular potentials from molecular devices and convert (amplify in terms of power) their signal to be compatible with the standard microelectronics. Thus,

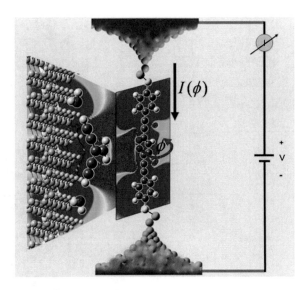

Fig. 12.7 Proposed nano-micro interface to read signals from molecular circuits operation using molecular potentials. Variations of the MEP are able to slightly rotate the central ring by an angle ϕ of the aza-OPE molecule, this in turn varies the current $I(\phi)$ through the external circuit (Reprinted from Ref. [41] with kind permission of The American Institute of Physics)

the molecular potential gates attached to the interface amplifier can also be used as sensors with built-in logic for molecular discrimination purposes. The output current through the aza-OPE molecule is the amplified signal from a chain of logical gates, representing the results of specific logical functions.

12.5.1 Graphene MEP Amplifier

The delocalized electronic density of graphene is highly sensitive to moieties adsorbed on the surface, allowing the development of sensors [38] and transducers [39] using the vibronics scenario. It has been demonstrated that information can be coded and processed using molecular electrostatic potentials, and we would like to amplify these signals to be readable in the conventional microelectronics.

We carried out accurate computational chemistry techniques, ab-initio calculations and used our GENIP program to calculate the conductivity along a graphene ribbon, when molecules with a strong dipole moment and either negative or positive molecular electrostatic potential are interacting in the two neighborhood sides of the graphene surface. Changes in the current-voltage characteristics calculated longitudinally on the graphene ribbon show the effect of the MEP on the electron transport; when a negative potential is applied using the molecules on the surface, the conductivity of the graphene ribbon increases. At the same time when a positive

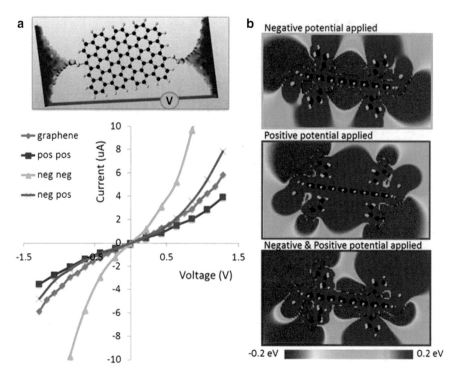

Fig. 12.8 Influence on graphene ribbons conductivity to the application of a MEP's on their surface (**a**) current response when highly polarized molecules are located at the ends and both sides of the graphene ribbon, and (**b**) the MEP for configurations of graphene ribbon and several potentials applied; negative potential (*top*) (*green*), positive potential (*middle*) (*red*), and, negative and positive potential applied to each side of graphene ribbon (*bottom*) (purple)

potential is applied, the conductivity of the ribbon decreases. When the ribbon is polarized by applying a negative potential at the left of the ribbon and a positive potential at the right side, a diode-like behavior is shown in the conductivity calculations. These results suggest a strong sensitivity of the graphene ribbons under the presence of a molecular electrostatic potential on the surface.

Graphene surface sensitivity to a negative or positive electrostatic potential on the near neighborhood has been shown using ab-initio calculations and Green function theory. Our calculations (Fig. 12.8) have shown that signals processed at the molecular level using MEPs can be amplified using graphene ribbons through changes in the conductivity of the graphene ribbons. Experimentally, we have shown that an electric field is digitally detected and amplified by graphene ribbons through changes in the conductivity.

12.6 Communication Between Molecular Scenarios: Single Molecule Detection Using Graphene Electrodes

Within the margins of what is allowed by quantum mechanical rules, the ability to read small perturbations of molecules, such as vibrations or molecular potentials, is a key point to implement the use of molecules as sophisticated molecular/electronic devices. Vibronics and molecular potentials are key scenarios for a new era of electronics [6, 78]. However, amplifiers and transducers of signals for these two scenarios are required to detect, transport, and encode information at the molecular level, as well as to facilitate the delivery of the new molecular technology. We are proposing graphene molecules, used as terahertz generators [23], as the base molecules to elaborate the reading/writing of information at the molecular level due to its atomic thickness. Graphene systems are considered the perfect materials to serve as interfaces between molecular/nano electronics and current silicon electronics.

We achieve communication between MEPs, vibronics and current-voltage characteristics by using the sensitive surface of graphene ribbons and a water molecule as a simple example. The molecular potential of a molecule is affected by the vibrational movement of atoms and the changes are then amplified through changes in the current-voltage characteristics.

It is shown, using density functional theory that the trapping of molecules between graphene electrode plates can be used to sense molecules through their vibrational fluctuations. This hypothesis is tested using water trapped on two graphene molecules connected to a potential difference. The electric current fluctuations generated through the junction correspond to the fluctuations of the vibrational modes. Since this system yields currents in a range workable by present electronic devices, the need of further "molecular amplification" is not required. Fluctuations of the three modes of water yield similar changes of potentials in the neighborhood accessible by other molecules; therefore, vibrations from a single water molecule, as an example, or vibrations from any other molecule can be transduced into electrical currents of magnitude compatible with present silicon-technology. In the particular case of the water molecule, a rectified potential signal is obtained from the fluctuations of the antisymmetric stretching mode, and a simple transduction is obtained from the symmetric stretching and bending modes. It is argued that the high sensitivity is due to the strong delocalization of the frontier molecular orbitals or molecular plasmons on graphene electrodes, which guarantees the detection based on molecular potentials or molecular vibrations; these plasmon-like molecules are of major importance for the development of molecular and nano electronics.

Molecular vibrations of the water molecule yield oscillations of the molecular potentials, which are calculated at 1.8 Å above the oxygen atom (Fig. 12.9a). This is a suitable distance to observe the MEPs because most of the important intermolecular interactions take place around that distance [11, 14]. Usually, vibrations of the agent (water in this case) are far beyond those of the detector and therefore, are not able to follow the vibrations of the agent; however, the concerted motion of delocalized electrons in the detector may allow us to follow those vibrations.

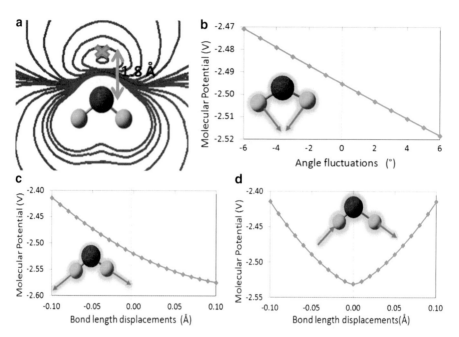

Fig. 12.9 (a) Molecular potential contours of the optimized water molecule in vacuum. Red lines are negative and blue lines are positive potential contours of values ranging from −2.7 to 2.7 V. All potential fluctuations of the water molecule are due to its vibrational modes and are calculated at the point × located at 1.8 Å from the oxygen atom. (b) Bending mode and (c) symmetric stretching: The MEP is linear for bending and symmetric stretching displacements; therefore, the potential (MEP) corresponds to the variations in geometry at the point x (transduction). This correspondence is highly linear for the angle fluctuations and quadratic for the bond lengths. (d) Antisymmetric stretching mode: positive or negative displacements in either of the O–H bond lengths with respect to the equilibrium geometry yield the same value of the molecular potential, resembling the behavior of a full rectifier (Reprinted from Ref. [39] with kind permission of The Institute of Physics)

Classically, a transducer is a device able to convert energy of one type into energy of other type; for instance, a microphone converts pressure vibrations in air into an electrical current. In this case, we extend the use of the term transducer to include, for instance, changes in molecular potentials due to the vibrational movement of the atoms. When comparing the molecular potentials versus the movement of atoms due to their vibrational modes, we find that a linear relation (transduction process) takes place due to the bending mode of the water molecule (Fig. 12.9b). Another transduction is observed, at least for small displacements, from the antisymmetric stretching mode (Fig. 12.9c), and a full rectification can be observed from the symmetric stretching mode (Fig. 12.9d).

Therefore, changes in molecular potentials due to molecular vibrations can be transduced and amplified into current-voltage characteristics on the delocalized electronic surface of graphene molecules [40]. A current response from each

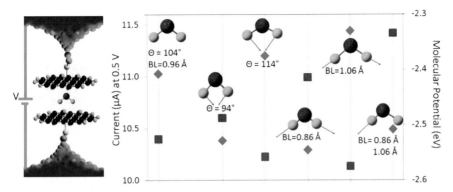

Fig. 12.10 Current response when a bias of 0.5 V is applied through two graphene ribbons acting as electrode plates and a water molecule (with several geometries associated to their vibrational modes) is placed in between the plates. From *left* to *right* the optimized geometry is shown (HOH angle 104° and OH bond lengths of 0.96 Å) followed by variations of this geometry due to the normal vibrational modes. For each geometry, the bond length and angle fluctuations affect the molecular potentials (*squares* and *right vertical axis*) and are detected through their effects on the current (*diamonds* and *left vertical axis*) across the junction (Reprinted from Ref. [39] with kind permission of The Institute of Physics)

vibrational mode is obtained when a constant voltage of 0.5 V is applied through a couple of graphene plates as shown in Fig. 12.10. These changes in the molecular geometry due to vibrational modes produce current fluctuations in the two-layer graphene junction.

The presence of a molecule between the graphene plates (sensor) produces changes in the current response due to perturbations of the fully delocalized electronic density of the graphene molecules (plasmons), which are sensitive to not only the trapped molecule (probe) but also to changes in the molecular electrostatic potentials due to the movement of the atoms. The contribution of the electrons can be either constructive or destructive to the conductivity, producing changes in the current response for each change in the molecular electrostatic potentials.

Plasmons on the graphene surface enhance transduction of molecular character-istics into signals readable by standard electronics. Thus, changes in both molecular electrostatic potentials and vibrations of an arbitrary molecule can be transduced and amplified into current-voltage characteristics.

12.7 Vibronics and Plasmonic Graphene Sensors

A high sensitivity and selectivity sensor is proposed using graphene ribbons that are able to read molecular vibrations and molecular electrostatic potentials, acting as an amplifier and as a transducer in order to convert molecular signals into current-voltage quantities of standard electronics. Two sensing mechanisms are used to demonstrate the concept using ab-initio density functional methods. The terahertz

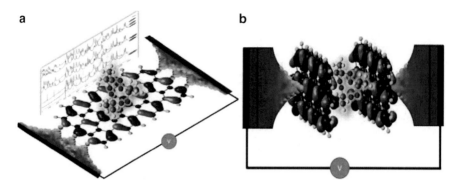

Fig. 12.11 Proposed scenarios using graphene ribbons as detectors. (**a**) Characteristic vibrations under the presence of an agent molecule on the surface of a graphene ribbon. Terahertz signals are transduced and amplified in terms of current-voltage characteristics. (**b**) Graphene plasmonic electrodes: the fully delocalized molecular orbitals of graphene ribbons act as plasmons on the surface, and are able to detect molecular vibrations by changes in the current flowing through the two ribbons acting as electrodes. Both *panels* show the fully delocalized highest occupied molecular orbital (HOMO) of graphene (Reprinted from Ref. [38] with kind permission of The American Institute of Physics)

region of the graphene spectrum can be used to characterize generated modes when single molecules are adsorbed on the ribbon surface. Characteristic modes can be obtained and used as fingerprints, which can be transduced into current by applying a voltage along the ribbons. On the other hand, the fully delocalized frontier molecular orbitals of graphene ribbons, commonly denominated plasmons in larger solid state structures, are extremely sensitive to any moiety approach; once plasmons are in contact with an "agent" (actually its molecular potential), the transport through the ribbons, that act as electrodes reading the signals (characteristics) from the agent, (Fig. 12.11), is strongly affected.

One possible application of graphene is as a molecular sensor using molecular vibrations (vibronics) [3]. Vibronics can be used to sense or transport signals, and theoretical simulations have shown the possible use for sensors to identify single molecules with modes in the terahertz (THz) region.

Raman intensities for low frequencies are shown in Fig. 12.12, for both passivated (blue lines) and non-passivated (pink lines) structures of the monolayer, bilayer and trilayer GNR. The presence of hydrogen passivation and extra layers of graphene increase the number of vibrational modes in the low frequency region.

12.8 Graphene Vibronics Sensor

The appearance of THz modes in the graphene spectrum occurs when molecules are adsorbed in the surface, even if the molecules do not have a signature in such region. These characteristic peaks can be used as THz-fingerprints of single molecules.

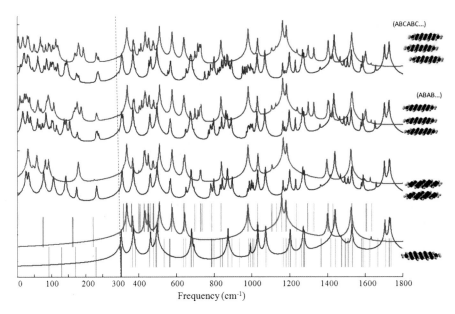

Fig. 12.12 Graphene nanoribbons as source of terahertz signals. Both hydrogen passivated and non-passivated structures show that the frequency spectrum for single graphene layer starts at frequencies greater than 300 cm^{-1}, while the second and third layered cluster structures show Raman intensities in the terahertz region (less than 100 cm^{-1}). *Vertical lines* in the two lower spectra show all, the non-active and active Raman vibrational modes (Reprinted from Ref. [40] with kind permission of The American Institute of Physics)

Even though the effect of temperature and the substrate on the graphene Raman spectroscopy is still unclear [79], its room temperature vibrational modes could be used as sensors of single molecules and as a generator of THz signals. Certainly, vibronics is only one of the several possible graphene applications; as ideas emerge, on a daily basis, our research develops and integrates new ideas into our current focus.

The optimized GNR-1, its bilayer and trilayer (Bernal and rhombohedral) clusters, are used as sensors of molecules. For the sake of proof-of-concept experiment, O_2 and N_2 are adsorbed in the surface, their small size facilitates the calculations, but this approach can be extended to any molecule (or explosive). Each cluster is optimized with the adsorbed molecule.

The surface sensitivity of graphene ribbons is studied using the molecules shown in Fig. 12.13 (the dimer Fig. 12.13a top) and trimer (Fig. 12.13b bottom) of acetone peroxide explosive molecules. They can be used as models to develop sensors for other explosives because they are very challenging to be detected by current sensors, due to their minimal absorbance, fluorescence, and molecular recognition patterns. However, the same concept of novel signature mechanisms using graphene ribbons, as mentioned earlier in this chapter can be used with other chemical or biological agents.

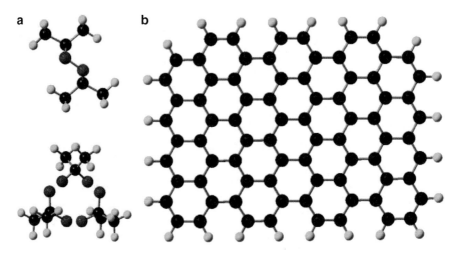

Fig. 12.13 (a) Molecules used to show the sensing application: the dimer (*top*) and trimer (*bottom*) of acetone peroxide. (b) Graphene ribbon used as sensor material of acetone peroxide based explosives (Reprinted from Ref. [38] with kind permission of The American Institute of Physics)

The characteristic modes that can be used as fingerprints of the triacetone triperoxide dimer for the detection using graphene ribbons are 0.7, 0.75 and 0.95 THz, and for the trimer: 0.3 and 0.35 THz (Fig. 12.14). These modes are attributed to the adsorption of the molecules on the graphene surface; however, it is difficult to distinguish differences between spectra taken with and without the agent because the observed modes of the agent can be mixed with the intrinsic modes of graphene. Therefore, spectroscopic detectors that could handle intensities in the THz region, different than those used in IR or Raman spectroscopies are needed. For example, inelastic electron tunneling spectroscopy [80] could be used for the for detection purposes.

12.9   Plasmonic Graphene Sensors

Graphene ribbons are used as sensors of dimer and trimer of acetone peroxide by calculating the current transversally and longitudinally. When the agent molecule is adsorbed on the surface of the graphene ribbon, new vibrational modes are generated in the terahertz region as previously shown in Fig. 12.14. The frequency calculations for multiple graphene layers show that vibrations between layers are in the same region as vibrations layer-molecule; therefore multiple layers of graphene are not convenient for the spectroscopic sensing in the terahertz region because the frequencies of modes can be misleading.

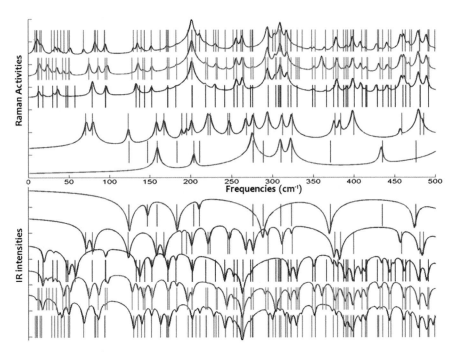

Fig. 12.14 Raman and infrared vibrational frequency spectrum of diacetone diperoxide (*pink line*), triacetone triperoxide (*blue line*), graphene ribbon (*red line*) diacetone diperoxide adsorbed in graphene (*green*) and triacetone triperoxide adsorbed on graphene (*gray line*). Vibrational modes generated in the low region of the spectrum (below 50 cm^{-1}) (Reprinted from Ref. [38] with kind permission of The American Institute of Physics)

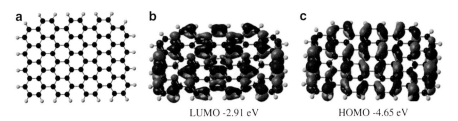

Fig. 12.15 Graphene ribbon (**a**) chosen for the sensing calculations; shape and energies of the molecular orbitals HOMO (**c**) and LUMO (**b**), following Fig. 12.4 the size is 7ac × 4zz (Reprinted from Ref. [38] with kind permission of The American Institute of Physics)

The electronic density on the graphene ribbons is completely delocalized on the surface, as shown with the molecular orbitals in Fig. 12.15.

In Fig. 12.16, we show an attempt to amplify the vibrations between the ribbon and the adsorbed agent by measuring the current longitudinally through the ribbon; however, the changes in conductivity are too small with and without the dimer and trimer molecules to be reliable and used as a sensing mechanism. The absorption

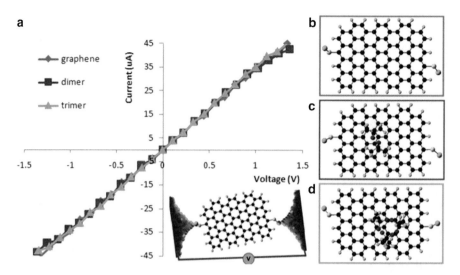

Fig. 12.16 (a) Current-voltage characteristics calculated longitudinally on the graphene ribbon isolated (**b**) (*blue*) when the dimer (**c**) (*red*) and trimer (**d**) (*green*) of acetone peroxide are adsorbed on the surface. A thio group is used in each case as an "alligator" to bond chemically the ribbons to the gold electrodes. The changes are not distinguishable for sensing or amplification (Reprinted from Ref. [38] with kind permission of The American Institute of Physics)

of the agent molecules has a weak effect on the electronic density of the graphene ribbons; the frontier molecular orbitals are always delocalized in the surface of the graphene molecules, and therefore the conductivity of any adsorbed agent does not affect the graphene conductivity.

Better results are obtained when the graphene ribbons are used as a sandwich-like arrangement, trapping the adsorbed molecule in-between. The optimized distance of the dimer sandwich is 9.15 Å and the trimer is 10.4 Å; for comparison purposes, we used 14.0 Å in both cases for the transport calculations. Therefore, there is no distance effect in the conductivity results.

When the current-voltage characteristics are calculated transversally along a couple of graphene ribbons, the changes in current are noticeable and can be used to amplify a signal that is indicative of the presence of agent molecules such as the dimer and trimer of acetone peroxide (Fig. 12.17).

The conductivity calculated transversally through the ribbons, shows changes in conductivity under the presence of the adsorbed molecules in between; even though the dimer molecule is smaller than the trimer, the conductivity is larger; which we attribute to the most outer atoms that in the dimer case are oxygen atoms, which are more electronically dense and contribute to the transport across ribbons, contrasting the trimer agent where the outer atoms are hydrogen yielding less channels for the transport of electrons and therefore smaller conductivity than the dimer.

The interaction between graphene layers yields vibrational modes in the terahertz region of the spectrum, independent of the type of edges, that are due the presence of

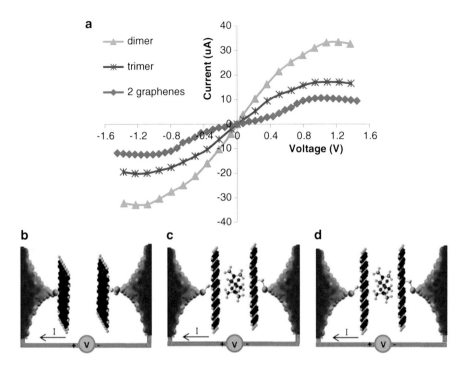

Fig. 12.17 (**a**) Current-voltage (I–V) characteristics calculated transversally between two graphene ribbons located at the same distance when (**b**) there is not molecule in between, (**c**) a acetone peroxide dimer is adsorbed in the middle of the ribbons and (**d**) when a acetone peroxide trimer is adsorbed (Reprinted from Ref. [38] with kind permission of The American Institute of Physics)

hydrogen pasivating the layers, or the number of layers. The presence of molecules on a graphene membrane can be detected from its vibrational modes in the terahertz spectrum. The interlayer distance of ribbons with zigzag edges is shorter. However, binding energies are slightly affected by the edge passivation with hydrogen; when there are not hydrogens, binding is 2.5 kcal/mol stronger. Then binding energy is also slightly affected stacking order; the Bernal stacking is 0.5 kcal/mol stronger than rhombohedral.

The electronic density of graphene molecules is delocalized on their external surfaces; they are plasmons, which can be used as highly selective and sensitive sensors as well as amplifiers of molecular potentials and vibrations. Two sensing mechanisms are used to demonstrate the concept in this review by *ab initio* density functional methods. Using a single graphene layer, a molecule adsorbed on the surface can be detected by spectroscopic methods as new vibrational modes are generated on the terahertz region. A sandwich-like sensor can be used to detect and amplify signals from molecules by measuring the current though the ribbons.

12.10   Graphene Mixer

Using a combination of density functional theory and Green's function theories, current-voltage fluctuations are found across three layers graphene clusters. These fluctuations come from the nuclei vibrations and displacement of layers; instantaneous potential in the neighborhood of graphene affects the intrinsic behavior of the electron density, corresponding to the HOMO and other energetically close molecular orbitals. Applications of nonlinear electronics include: RF signal mixing, detection of very weak forces and displacements [81, 82]. Carbon nanotubes have been used as electromechanical oscillators able to act as transducers of small forces [82] and as sensors of their own motion [83].

It has been proposed that nanotube-based transistors can operate at frequencies in the terahertz region as generators, frequency multipliers and detectors [84]; though carbon nanotube transistors are good candidates for RF and opto electronics [85], fabrication of nanotube arrays with controllable chirality and diameters is still a challenge for large scale fabrication of the devices.

The cross section of graphene, just one atom wide, and its electrical properties allow us to detect and amplify signals encoded at the molecular level as vibrations or potentials (vibronics or the molecular electrostatic potential scenarios [38]), as well as to amplify them into present electronic technologies. Graphene mechanical properties, such as stiffness, allow the implementation of electromechanical resonators that, when integrated with a graphene based transducer device [39] can be used as displacement sensors and detectors of very weak forces. Other examples of nonlinear devices are the graphene frequency multipliers [86, 87], whereby frequency doubling is achieved by biasing the gate of a single layer graphene transistor, and the ambipolar frequency mixer [88], which has been shown using a single graphene transistor acting as a RF mixer device.

We have previously proposed two sensing scenarios using graphene ribbons. One is based on the generation of characteristic terahertz signals from vibrations between an absorbed molecule on a single layer of graphene [38, 40]; the generated signals are characteristic for each molecule being detected. The second scenario is based on the electrical transport through graphene electrodes [39]; when the agent molecule is absorbed between graphene layers, different states of conductance are reached due to the interaction of the absorbed molecule with the delocalized electron density of graphene. The high sensitivity of graphene ribbons is due to their full delocalization of frontier molecular orbitals (plasmonic), which improve the detection of very small species from the effect on large substrates, rather than detecting the actual small molecules [38].

The restricted molecule (model under study itself), i.e. three graphene layers (Fig. 12.18) is fully optimized without any constraints; sulfur and an interfacial gold atoms are then added to each outer graphene layer, and single point calculations are performed. The optimized geometry of a C–S–Au group is used to chemically attach the graphene layers to the gold nano-electrodes (Fig. 12.18).

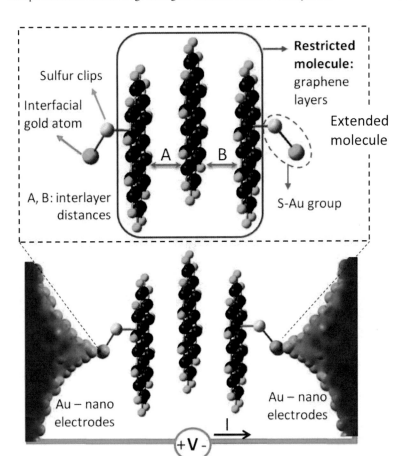

Fig. 12.18 Optimized three-graphene layers (*restricted molecule*) with sulfur clips and gold interfacial atoms to chemically bond the graphene ribbon to gold nanoelectrodes. The *extended molecule* is composed of the restricted molecule, sulfur clips and interfacial atoms. The three-graphene layers are separated by distances A and B (Reprinted from Ref. [89] with kind permission of The American Institute of Physics)

A geometry optimization of the three-layer graphene molecules (restricted molecule) yields an optimized average distance between layers of A = 3.42 Å and B = 3.44 Å (Fig. 12.18). To simulate the vibrational movement of graphene layers, subsequent optimizations are carried out while keeping the distance between layers (A and B) constant at 3.32, 2.42 and 3.52 Å. A resemblance to the vibrational "breathing" movement between layers is performed, setting "A" and "B" to show compressing (A = B = 3.32 Å), stretching (A = B = 3.52 Å) and anti-symmetric (A = 3.32 and B = 3.52 Å) vibrations between layers. The largest barrier to keep the layers at a constant distance is 0.52 kcal/mol (Table 12.1), corresponding to the relative energy of the layers separated by A = B = 3.52 Å with respect to the freely optimized system.

Table 12.1 Total and relative energies for the three-layer graphene molecules (restricted molecule Fig. 12.18) when freely optimized (opt) and when A and B are frozen (opt modR)

Calculation type	A (Å)	B (Å)	Total energy (Ha)	Relative energy (kcal/mol)
Opt	3.424	3.424	−2765.562715	0.00
Opt modR	3.42	3.42	−2765.562699	0.01
Opt modR	3.32	3.52	−2765.562231	0.30
Opt modR	3.32	3.32	−2765.561983	0.45
Opt modR	3.52	3.52	−2765.561875	0.52

Reprinted from Ref. [89] with kind permission of The American Institute of Physics

Energies from single-point calculations for the extended molecule (Fig. 12.18) are shown in Table 12.2. The largest barrier is for the graphene layers at 3.32 Å, and the shortest barrier corresponds to the stretched complex. The gold-gold distance varies from 12.82 to 13.05 Å, and the effect can be observed on the current-voltage curves, where the compressed case is more conductive than the stretched one.

Calculations of electron transport for the three-layer graphene extended molecule show oscillations (Fig. 12.19a) when the structure is freely optimized. In the nonlinear behavior, the response (current) is not proportional to its input (voltage). This nonlinearity is attributed to the fluctuations due to the vibrational modes between layers and is corroborated by the calculations where A is shorter than B (Fig. 12.19b). The two cases showing oscillations are embedded into the stretching (purple line) and compressing (green line) curves of Fig. 12.19; as expected when compression occurs, the Au–Au distance is shorter and thus larger conductivity is shown opposite to the stretching case where the conductivity slightly changes down. The current fluctuations diminish when the graphene layers are kept at the same distance.

This nonlinearity conductivity suggests multilayer graphene may act as signal mixer; therefore, we performed an experimental study to apply and recover the signals of a sample of few graphene layers.

Second and third order frequency signals are recovered in the output, and the results shown in this review are representative of all of our measured devices. When we apply a large-frequency signal of 2,500 kHz (F1) and a low-frequency signal of 300 kHz (F2) to the graphene sample, we recover, at the output, the two introduced inputs, as well as second-order modulated signals corresponding to the peaks at 2,200 kHz (|F2-F1|), 2,800 kHz (F1 + F2) and other harmonics (Fig. 12.20a). In addition, when two close frequency signals, F1 = 2,500 and F2 = 2,800 kHz, are introduced using the input electrodes, the graphene sample is able to demodulate the signal by subtracting the two inputs (|F2-F1|), showing a peak at 300 kHz as is shown in Fig. 12.20b.

The nonlinear behavior of the conductivity transversally calculated for few graphene layers comes from molecular movements, due to vibrations, which creates a nonlinear electrical component that allows the detection of signals with different

Table 12.2 Total, relative HOMO, LUMO and HLG energies for the single point calculations for three layers graphene molecules with sulfur clips and gold interfacial atoms (extended molecule)

A (Å)	B (Å)	Total energy (Ha)	Relative energy (kcal/mol)	HOMO (eV)	LUMO (eV)	HLG (eV)	Au–Au distance (Å)
–	–	–3832.190621	0.00	–3.65	–2.76	0.88	12.82
3.42	3.42	–3832.189789	0.52	–3.65	–2.76	0.89	12.76
3.32	3.52	–3832.188905	1.06	–3.64	–2.76	0.89	12.76
3.32	3.32	–3832.187500	1.93	–3.64	–2.72	0.92	12.51
3.52	3.52	–3832.189885	0.46	–3.65	–2.79	0.86	13.00

Au–Au distance is shown for each studied case
Reprinted from Ref. [89] with kind permission of The American Institute of Physics

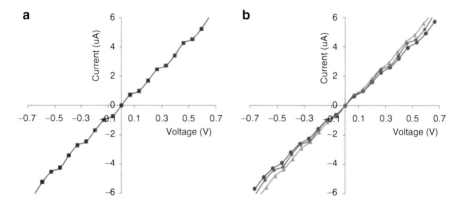

Fig. 12.19 Current-voltage characteristics using GENIP (**a**) Current voltage characteristics at the optimized distances A = 3.42 Å and B = 3.44 Å (**b**) current voltage characteristics when the distance between layers are kept constant at A = B = 3.32 Å (*green*), A = B = 3.52 Å (*purple*) and A = 3.32 Å and B = 3.52 (*blue*) (Reprinted from Ref. [89] with kind permission of The American Institute of Physics)

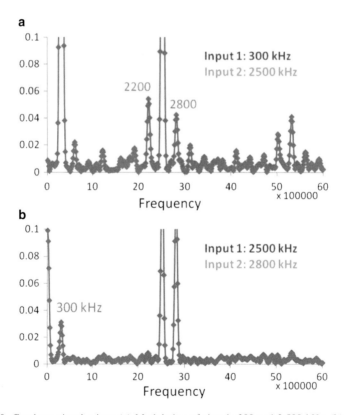

Fig. 12.20 Graphene signal mixer. (**a**) Modulation of signals 300 and 2,500 kHz. (**b**) Demodulation of 2,500 and 2,800 kHz (Reprinted from Ref. [89] with kind permission of The American Institute of Physics)

frequencies. The input signals from the agent are terahertz fingerprints generated by the adsorption of the agent molecule on the surface of graphene, which are mixed with the intrinsic vibrations of the detector (few layers graphene), modulated and demodulated by the graphene mixer, and amplified by changes in the electronic characteristics of the device.

12.11   Conclusions – Summary

We have designed, developed, and proposed nanodevices operating at the molecular level by a theoretical-experimental approach. Novel scenarios such as molecular electrostatic potentials and vibronics are complemented with the traditional current-voltage to perform operations, communicate, and process information at the molecular level.

A nano-micro interface to read information encoded in molecular level potentials and to amplify this signal to microelectronics levels has been shown. The amplification is performed by making the output molecular potential slightly twist the torsional angle between two rings of a pyridazine, 3,6-bis(phenylethynyl)(aza-OPE) molecule, requiring only fractions of kcal/mol energy. In addition, even if the signal from the molecular potentials is not enough to turn the ring, or even if the angles are the same for different combinations of outputs, still the current output yields results that resemble the device as a field effect transistor, providing the possibility to reduce channel lengths to the range of just 1 or 2 nm. The slight change in the torsional angle yields readable changes in the current through the aza-OPE biased by an external applied voltage.

The monoatomic thickness of the graphene and plasmonic surface material provides the best sensitivity available for a molecular sensor, and the absorption of single molecules can be detected by measuring changes in conductivity. Due to the fully delocalized frontier molecular orbitals of the graphene ribbons commonly denominated plasmons in larger solid state structures, the ribbons are extremely sensitive to any moiety approach. Plasmons on the graphene surface enhance transduction of molecular characteristics into signals readable by standard electronics. Therefore, the two dimensional structure of graphene has been suggested as an ideal material for sensing science and amplification of signals at the molecular level.

Changes in both MEPs and vibrations of an arbitrary molecule can be transduced and amplified into current–voltage characteristics using graphene. Under the presence of an agent molecule, the delocalized molecular orbitals of graphene are disturbed by molecular potentials, as well as vibrations affecting electron transport and vibrational spectrum of graphene.

Using molecular engineering, nanodevices with promising capabilities and applications in conventional engineering, nanotechnology and defense security as sensors of chemical and biological agents, among others, have been shown.

Some of the novel achievements of this research are graphene terahertz generators for molecular circuits and sensors, nano-micro interfaces to read molecular potentials into current-voltage based electronics using single layer graphene ribbons, vibronics and plasmonic graphene sensors, and single molecule detection using adsorption of molecules between graphene layers.

Acknowledgements We acknowledge financial support from the U. S. Defense Threat Reduction Agency DTRA through the U. S. Army Research Office, Project No. W91NF-06-1-0231; from the ARO/DURINT project # W91NF-07-1-0199, and the ARO/MURI project # W911NF-11-1-0024. We also thanks Jeremy Katusak for a thorough check of the final manuscript.

References

1. Allen MP, Tildesley DJ (1990) Computer simulation of liquids. Clarendon, Oxford
2. Seminario JM, Derosa PA, Bozard BH, Chagarlamudi K (2005) Vibrational study of a molecular device using molecular dynamics simulations. J Nanosci Nanotechnol 5:1–11
3. Seminario JM, Yan L, Ma Y (2005) Scenarios for molecular-level signal processing. Proc IEEE 93:1753–1764
4. Yan L, Ma Y, Seminario JM (2006) Terahertz signal transmission in molecular systems. Int J High Speed Electron Syst 16:669–675
5. Politzer P, Truhlar DG (1981) Chemical applications of atomic and molecular electrostatic potentials. Plenum Press, New York
6. Tour JM, Kosaki M, Seminario JM (1998) Molecular scale electronics: a synthetic/computational approach to digital computing. J Am Chem Soc 120:8486–8493
7. Tour JM, Kozaki M, Seminario JM (2001) Use of molecular electrostatic potential for molecular scale computation. U.S. Patent 6259277
8. Politzer P, Seminario JM (1989) Computational analysis of the structures, bond properties, and electrostatic potentials of some nitrotetrahedranes and nitroazatetrahedranes. J Phys Chem 93:4742–4745
9. Scrocco E, Tomasi J (1973) The electrostatic molecular potential as a tool for the interpretation of molecular properties. Top Curr Chem 42:95–170
10. Jeffrey GA (1991) The application of charge-density research to chemistry and drug design. Plenum Press, New York
11. Murray JS, Sen K (eds) (1996) Molecular electrostatic potentials. Concepts and applications. Theoretical and computational chemistry. Elsevier, Amsterdam, p 665
12. Yan L, Seminario JM (2006) Moletronics modeling towards molecular potentials. Int J Quantum Chem 106:1964–1969
13. Naray-Szabo G, Ferenczy GG (1995) Molecular electrostatics. Chem Rev 95:829–847
14. Politzer P, Murray J (1991) Molecular electrostatic potentials and chemical reactivity. In: Lipkowitz KB, Boyd DB (eds) Reviews in computational chemistry, vol 2. VCH Publishers, New York, pp 273–312
15. Seminario JM, Yan L, Ma Y (2005) Encoding and transport of information in molecular and biomolecular systems. Proc IEEE Nanotechnol Conf 5:65–68
16. Seminario JM, Yan L (2005) Molecular logical devices in cascade configuration with information encoded as electrostic potentials. J Am Chem Soc vol. Submitted
17. Hehre WJ, Radom L, Schleyer PvR, Pople JA (1986) Ab initio molecular orbital theory. Wiley, New York
18. Seminario JM, Yan L (2005) Ab initio analysis of electron currents in thioalkanes. Int J Quantum Chem 102:711–723
19. MJT Frisch GW, Schlegel HB, Scuseria GE, Robb MA, Cheeseman JR, Scalmani G, Barone V, Mennucci B, Petersson GA, Nakatsuji H, Caricato M, Li X, Hratchian HP, Izmaylov AF, Bloino J, Zheng G, Sonnenberg JL, Hada M, Ehara M, Toyota K, Fukuda R, Hasegawa J,

Ishida M, Nakajima T, Honda Y, Kitao O, Nakai H, Vreven T, Montgomery JA Jr, Peralta JE, Ogliaro F, Bearpark M, Heyd JJ, Brothers E, Kudin KN, Staroverov VN, Kobayashi R, Normand J, Raghavachari K, Rendell A, Burant JC, Iyengar SS, Tomasi J, Cossi M, Rega N, Millam NJ, Klene M, Knox JE, Cross JB, Bakken V, Adamo C, Jaramillo J, Gomperts R, Stratmann RE, Yazyev O, Austin AJ, Cammi R, Pomelli C, Ochterski JW, Martin RL, Morokuma K, Zakrzewski VG, Voth GA, Salvador P, Dannenberg JJ, Dapprich S, Daniels AD, Farkas Ö, Foresman JB, Ortiz JV, Cioslowski J, Fox DJ (2009) Gaussian-09, Revision A.01. Gaussian, Inc, Wallingford

20. McWeeny R, Diercksen G (1968) Self-consistent perturbation theory. II. Extension to open shells. J Chem Phys 49:4852–4856

21. Petersson GA, Al-Laham MA (1991) A complete basis set model chemistry. II. Open-shell systems and the total energies of the first-row atoms. J Chem Phys 94:6081–6090

22. Petersson GA, Bennett A, Tensfeldt TG, Al-Laham MA, Shirley WA, Mantzaris J (1988) A complete basis set model chemistry. I. The total energies of closed-shell atoms and hydrides of the first-row elements. J Chem Phys 89:2193–2218

23. Wang P, Moorefield CN, Lic S, Hwang S-H, Shreiner CD, Newkome GR (2006) TerpyridineCuII-mediated reversible nanocomposites of single-wall carbon nanotubes: towards metallo-nanoscale architectures. Chem Commun 10:1091–1093

24. Hay PJ, Wadt WR (1985) Ab initio effective core potentials for molecular calculations. Potentials for K to Au including the outermost core orbitals. J Chem Phys 82:299–310

25. Wadt WR, Hay PJ (1985) Ab initio effective core potentials for molecular calculations. Potentials for main group elements Na to Bi. J Chem Phys 82:284–298

26. Kohn W, Sham LJ (1965) Self-consistent equations including exchange and correlation effects. Phys Rev A 140:1133–1138

27. Perdew JP, Chevary JA, Vosko SH, Jackson KA, Pederson MR, Singh DJ, Fiolhais C (1992) Atoms, molecules, solids, and surfaces: applications of the generalized gradient approximation for exchange and correlation. Phys Rev B 46:6671–6687

28. Perdew JP, Wang Y (1992) Accurate and simple analytic representation of the electron-gas correlation energy. Phys Rev B 45:13244–13249

29. Seminario JM, Maffei MG, Agapito LA, Salazar PF (2006) Energy correctors for accurate prediction of molecular energies. J Phys Chem A 110:1060–1064

30. Seminario JM (1993) Energetics using DFT: comparisons to precise ab initio and experiment. Chem Phys Lett 206:547–554

31. Zhao Y, Schultz NE, Truhlar DG (2006) Design of density functionals by combining the method of constraint satisfaction with parametrization for thermochemistry, thermochemical kinetics, and noncovalent interactions. J Chem Theor Comput 2:364–382

32. Seminario JM, Ma Y, Tarigopula V (2006) The NanoCell: a chemically assembled molecular electronic circuit. IEEE Sensors 6:1614–1626

33. Hong S, Jauregui LA, Rangel NL, Cao H, Day BS, Norton ML, Sinitskii AS, Seminario JM (2008) Impedance measurements on a DNA junction. J Chem Phys 128:201103

34. Bellido EP, Bobadilla AD, Rangel NL, Zhong H, Norton ML, Sinitskii A, Seminario JM (2009) Current-voltage-temperature characteristics of DNA origami. Nanotechnology 20:175102

35. Bobadilla AD, Bellido EP, Rangel NL, Zhong H, Norton ML, Sinitskii A, Seminario JM (2009) DNA origami impedance measurement at room temperature. J Chem Phys 130:171101

36. Cristancho D, Seminario JM (2010) Polypeptides in alpha-helix conformation perform as diodes. J Chem Phys 132:065102

37. Rangel NL, Sotelo JC, Seminario JM (2009) Mechanism of carbon nanotubes unzipping into graphene ribbons. J Chem Phys 131:031105

38. Rangel NL, Seminario JM (2010) Vibronics and plasmonics based graphene sensors. J Chem Phys 132:125102

39. Rangel NL, Seminario JM (2010) Single molecule detection using graphene electrodes. J Physics B 43:155101

40. Rangel NL, Seminario JM (2008) Graphene terahertz generators for molecular circuits and sensors. J Phys Chem A 112:13699–13705

41. Rangel NL, Seminario JM (2008) Nanomicrointerface to read molecular potentials into current-voltage based electronics. J Chem Phys 128:114711
42. Rangel NL, Williams KS, Seminario JM (2009) Light-activated molecular conductivity in the photoreactions of vitamin D3. J Phys Chem A 113:6740–6744
43. Wang K, Rangel NL, Kundu S, Sotelo JC, Tovar RM, Seminario JM, Liang H (2009) Switchable molecular conductivity. J Am Chem Soc 131:10447–10451
44. Hashimoto A, Suenaga K, Gloter A, Urita K, Iijima S (2004) Direct evidence for atomic defects in graphene layers. Nature 430:870–873
45. Wakabayashi K, Fujita M, Ajiki H, Sigrist M (1999) Electronic and magnetic properties of nanographite ribbons. Phys Rev B 59:8271
46. Barone V, Hod O, Scuseria GE (2006) Electronic structure and stability of semiconducting graphene nanoribbon. Nano Lett 6:2748–2754
47. Han MY, Özyilmaz B, Zhang Y, Kim P (2007) Energy band-gap engineering of graphene nanoribbons. Phys Rev Lett 98:206805
48. Binning G, Rohrer H (1982) Scanning tunneling microscopy. Helv Phys Acta 55:726–735
49. Blum AS, Kushmerick JG, Long DP, Patterson CH, Yang JC, Henderson JC, Yao Y, Tour JM, Shashidhar R, Ratna BR (2005) Molecularly inherent voltage-controlled conductance switching. Nature 4:167–172
50. Ariga K, Hill JP, Endo H (2007) Developments in molecular recognition and sensing at interfaces. Int J Mol Sci 8:864–883
51. Von Neumann J (1966) Theory of self-reproduction automata. In: Burks A (ed) La vie artificielle. University of Illinois Press, Paris
52. Wolfram S (1984) Cellular automata as models of complexity. Nature 311:419–424
53. Wolfram S (1984) Universality and complexity in cellular automata. Phys D 10:1–35
54. Scrocco E, Tomasi J (1978) Electronic molecular structure, reactivity and intermolecular forces: a heuristic interpretation by means of electrostatic molecular potentials. Adv Quantum Chem 11:115
55. Valia Dimitrova SI, Galabov B (2002) Electrostatic potential at atomic sites as a reactivity descriptor for hydrogen bonding. Complexes of monosubstituted acetylenes and ammonia. J Phys Chem A 106:11801–11805
56. Geerlings PL, Langenaeker W, De Proft F, Baeten A (1996) Molecular electrostatic potentials vs. DFT [density-functional theory] descriptors of reactivity. General Review, pp 587–617
57. Robbins AM, Jin P, Brinck T, Murray JS, Politzer P (2006) Electrostatic potential as a measure of gas phase carbocation stability. Int J Quantum Chem 106:2904–2909
58. Dhumal NR, Patil UN, Gejji SP (2004) Molecular electrostatic potentials and electron densities in nitroazacubanes. J Chem Phys 120:749–755
59. Murray JS, Lane P, Politzer P (1998) Effects of strongly electron-attracting components on molecular surface electrostatic potentials; application to predicting impact sensitivities of energetic molecules. Mol Phys 93:187–194
60. Cheng X-l, Wang K-m, Zhang H, Yang X-d (2002) Relationships between impact sensitivities and the electrostatic potentials for five nitroaniline explosives. Inst Atomic Mol Phys 19:94–100
61. Politzer P, Murray JS (1996) Relationships between dissociation energies and electrostatic potentials of C–NO2 bonds: applications to impact sensitivities. J Mol Struct 376:419–424
62. Seminario JM, Yan L, Ma Y (2005) Nano-detectors using molecular circuits operating at THz frequencies. In: Jensen JO, Theriault JM (eds) Chemical and biological standoff detection III, vol 5995. SPIE, Bellingham, pp 230–244
63. Hao Hu ZL, Weitao Y (2007) Fitting molecular electrostatic potentials from quantum mechanical calculations. J Chem Theor Comput 3:1004–1013
64. Hall CMSGG (1984) Fitting electron densities of molecules. Int J Quantum Chem 25:881–890
65. Pinjari RV, Joshi KA, Gejji SP (2006) Molecular electrostatic potentials and hydrogen bonding in alpha-, beta-, and ç-cyclodextrins. J Phys Chem A 110:13073–13080
66. Seminario JM, Yan L (2007) Cascade configuration of logical gates processing information encoded in molecular potentials. Int J Quantum Chem 107:754–761

67. Seminario JM, Cordova LE, Derosa PA (2003) An Ab initio approach to the calculation of current-voltage characteristics of programmable molecular devices. Proc IEEE 91:1958–1975
68. Seminario JM, Zacarias AG, Tour JM (1998) Theoretical interpretation of conductivity measurements of thiotolane sandwich. A molecular scale electronic controller. Am Chem Soc 120:3970–3974
69. Seminario JM (2007) Quantum current-voltage relation for a single electron. J Phys B 40:F275–F276
70. Seminario JM, Zacarias AG, Tour JM (1999) Molecular current-voltage characteristics. J Phys Chem A 103:7883–7887
71. Xu G-b, Xu Q-x (2007) Development of advanced Hf based high-k gate dielectrics. Dianzi Qijian Inst Microelectron 30:1194–1199
72. Lee C, Choi DS, Park HR, Kim CS, Wang KL (2001) Side contact single electron devices for integrated circuit. J Korean Phys Soc 39:S442–S446
73. Wada Y (1997) Atom electronics: a proposal of atom/molecule switching devices. Surf Sci 30:265–278
74. Yang C, Zhong Z, Lieber CM (2005) Encoding electronic properties by synthesis of axial modulation-doped silicon nanowires. Science 310:1304–1307
75. Yuan Taur DAB, W Chen, Frank DJ, Ismail KE, Shih-Hsien Lo, Sai-Halasz GA, Viswanathan RG, Wann HC, Wind SJ, Wong H (1997) CMOS scaling into the nanometer regime. Proc IEEE 85:486–504
76. Moore GE (1965) Cramming more components onto integrated circuits. Electronics 38: 114–117
77. Seminario JM, Derosa PA (2001) Molecular gain in a thiotolane system. J Am Chem Soc 123:12418–12419
78. Seminario JM, Yan L, Ma Y (2006) Encoding and transport of information in molecular and biomolecular systems. Trans IEEE Nanotechnol 5:436–440
79. Calizo I, Balandin AA, Bao W, Miao F, Lau CN (2007) Temperature dependence of the Raman spectra of graphene and graphene multilayers. Nano Lett 7:2645–2649
80. Song H, Kim Y, Jang YH, Jeong H, Reed MA, Lee T (2009) Observation of molecular orbital gating. Nature 462:1039–1043
81. Knobel RG, Cleland AN (2003) Nanometre-scale displacement sensing using a single electron transistor. Nature 424:291–293
82. Sazonova V, Yaish Y, Ustunel H, Roundy D, Arias TA, McEuen PL (2004) A tunable carbon nanotube electromechanical oscillator. Nature 431:284–287
83. Witkamp B, Poot M, van der Zant HSJ (2006) Bending-mode vibration of a suspended nanotube resonator. Nano Lett 6:2904–2908
84. Portnoi ME, Kibis OV, Rosenau da Costa M (2007) Terahertz applications of carbon nanotubes. Superlattices Microstruct 43:399–407
85. Rutherglen C, Jain D, Burke P (2009) Nanotube electronics for radiofrequency applications. Nat Nanotechnol 4:811–819
86. Han W, Nezich D, Jing K, Palacios T (2009) Graphene frequency multipliers. Electron Device Lett IEEE 30:547–549
87. Wang Z, Zhang Z, Xu H, Ding L, Wang S, Peng L-M (2010) A high-performance top-gate graphene field-effect transistor based frequency doubler. Appl Phys Lett 96:173104–173104-3
88. Han W, Hsu A, Wu J, Jing K, Palacios T (2010) Graphene-based ambipolar RF mixers. Electron Device Lett IEEE 31:906–908
89. Rangel N, Gimenez A, Sinitskii A, Seminario JM (2011) Graphene signal mixer for sensing applications. J Phys Chem 115(24):12128–12134

Chapter 13
Theoretical Studies of Thymine–Thymine Photodimerization: Using Ground State Dynamics to Model Photoreaction

Martin McCullagh and George C. Schatz

Abstract Excited electronic states of DNA are extremely important in biology and yet most of these states have lifetimes of ~1 ps. This short lifetime allows for very little nuclear rearrangement during dissipation. In particular, photoinduced thymine-thymine (TT) dimer formation has been found to be a picosecond process. The most prevalent TT dimer is formed by a [2+2] addition of the C5–C6 double bonds of the dimerizing thymines. Given the topochemical rules known for photoinduced [2+2] addition of organic compounds in the solid state, a similar set of rules is presented for TT dimerization in solution phase DNA. It is found that a single ground state geometric parameter (the distance, d, between the C5–C6 double bonds) is sufficient as a constraint on when dimers can form such that accurate TT dimer quantum yields can be predicted. The electronic basis of such a model is examined along with calibration of the model for dT_{20} and $dA_{20}dT_{20}$. The application and validity of this model to a variety of double and single stranded DNA systems is then discussed.

Keywords Thymine-thymine dimmer • Photodimerization • Photoreaction • DNA damage • Mutation

13.1 Introduction

The ability to survive electronic excitation is a fundamental property of DNA [1]. While excited states of DNA nucleobases are important in mutation and repair processes, the average lifetime of these states is ~1 ps [2]. DNA excited states dissipate in a variety of ways with the dominant mechanism being non-radiative decay. Other decay pathways are important due to their health implications;

M. McCullagh • G.C. Schatz (✉)
Department of Chemistry, Northwestern University, Evanston, IL, 60208-3113, United States
e-mail: schatz@chem.northwestern.edu

J. Leszczynski and M.K. Shukla (eds.), *Practical Aspects of Computational Chemistry I: An Overview of the Last Two Decades and Current Trends*, DOI 10.1007/978-94-007-0919-5_13, © Springer Science+Business Media B.V. 2012

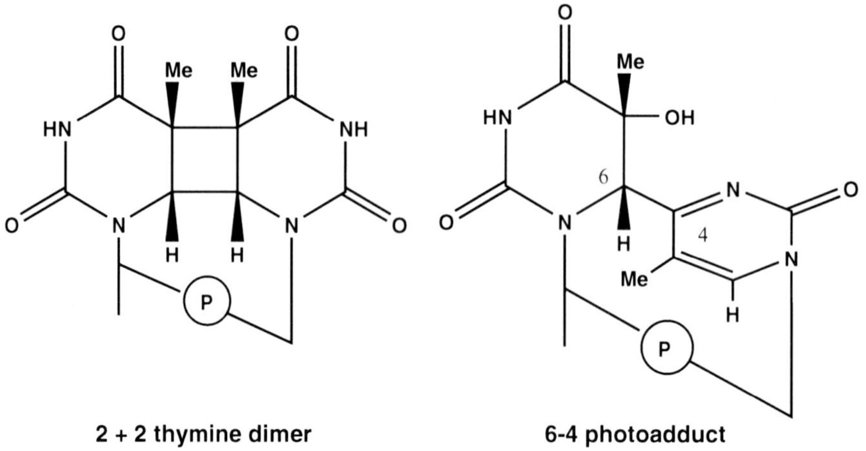

2 + 2 thymine dimer   6-4 photoadduct

Fig. 13.1 The two thymine–thymine dimers formed in DNA

these include pyrimidine dimer formation and strand cleavage, two of the most prevalent DNA mutations. Due to the chromophoric nature of the individual nucleobases and the highly π-stacked structure of the double helix, DNA excited states are inherently complicated and non-local. Theoretical investigation, therefore, into DNA mutation is exceedingly difficult. Important work on simplified DNA systems using complicated electronic structure methods have made some important breakthroughs (see for example reference [3]), some of which are addressed in other chapters of this book. In this chapter, we present a different approach. A simple ground state geometric model is suggested which can be coupled with molecular dynamics simulations to predict the amount of pyrimidine dimer formation in fairly large DNA strands.

One of the most prevalent examples of reaction involving DNA excited states is pyrimidine–pyrimidine dimer formation. Thymine and cytosine are the two pyrimidine bases present in DNA, and pyrimidine–pyrimidine dimers can form between any combination of these two bases. The most common of these is the thymine–thymine (TT) dimer [4–7]. Two types of TT dimers are known (shown in Fig. 13.1). The first, and sole focus of this chapter due to its prevalence, is called cyclobutane pyrimidine dimer (CPD) and is formed by the [2+2] addition of the C5–C6 double bonds. The second is called the 6-4 photoadduct and is formed by the addition of the C5–C6 double bond on one thymine to the C4–O4 double bond on the other. This leads to an oxetane intermediate that subsequently rearranges to form the 6-4 product. Both of these photoproducts are thought to form starting with initial excitation to a $^1(\pi\pi^*)$ state. There is some debate in the literature whether the subsequent excited state process goes through a singlet or triplet channel [8, 9]. Despite this controversy, it was recently determined, that TT dimers form within 1 ps giving little time for nuclear rearrangement [10]. Thus the ground state conformation prior to excitation greatly affects the ability of the thymines to dimerize.

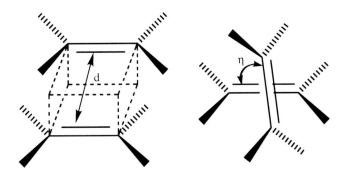

Fig. 13.2 The basic model for cyclobutane pyrimidine dimer formation is the addition of the fusing of the two C5=C6 double bonds. Here the distance, d, and dihedral angle, η, which define the relative C=C double bond geometry in depicted

In an attempt to build on the recent spectroscopic data of Schreier et al. [10], there have been a number of studies on the effect of ground state conformation on TT dimer formation [11–14]. The first of these studies looked at TT dimer formation in single stranded dT_{18} [11] and the second considered the dinucleotide dT_2 [12]. Two more recent studies attempted to develop a unified theory of both single stranded and double stranded systems [13, 14]. Double stranded systems are much more complicated because the added stability of the duplex creates better electronic coupling between bases. Additional electronic coupling between bases complicates the already complex excited state picture of the two thymines thus invalidating the basic model used in all of these studies.

Despite the complicated excited state nature of these reactions, ground state topochemical rules for the solid state photoinduced [2+2] addition of various organic compounds were developed over 50 years ago by Cohen et al. [15, 16]. The rules require that the distance, d, and the relative orientation, η, of the C=C double bonds be less than some cutoff (see Fig. 13.2). These values for different solid state structures are now relatively easy to determine using X-ray crystallography. Despite the general applicability of these topochemical rules, there are a few exceptions as noted by Ramamurthy et al. [16]. These include crystals in which [2+2] addition occurs with either the orientation or the distance between the C=C double bonds outside the accepted cutoffs.

Given this wealth of knowledge on solid state photoinduced [2+2] addition, the following question arises: are there a similar set of topochemical rules that apply to the [2+2] addition of two thymines in DNA? The answer to this is unclear for many reasons including that DNA is often in solution and the closely packed array of π-stacked chromophores make the excited states of duplex DNA much more complicated that the organic compounds that the model was originally designed for. In this chapter we address this question and present a simplified one parameter rule for [2+2] addition of thymines in both single and double stranded DNA. The model is presented in the Sect. 13.2 and calibrated in Sect. 13.3. Following calibration, three sets of applications are presented in order to determine whether ground state conformation is the dominant predictor of TT dimer formation in each set. The first of these, Sect. 13.4, involves probing the context dependence of TT dimer formation using three small DNA hairpins. The second application, Sect. 13.5, uses locked nucleic acids to both computationally and experimentally probe the role of ground state conformation. The final application, Sect. 13.6, probes the limits of the ground state conformation model by looking at the role of guanine and adenine on quenching TT dimer formation in small trinucleotides.

13.2 A Ground State Model for TT Dimerization in DNA

The method used to estimate dimerization quantum yields is related to methods used in previous studies by Law et al. [12] and Johnson et al. [11]. TT dimers are assumed to be formed by the cycloaddition between the C5 and C6 atoms on adjacent thymines while 6−4 photoproducts are formed by fusing O4–C5 bonds with C6–C4 bonds. In both cases it is assumed that the reactants need to be "close enough" to the conical intersection that connects the initially excited state and the ground state of the photoproduct for reaction to occur. Two geometrical parameters were proposed as being vital in the [2+2] TT formation. These parameters are the distance between the midpoints of the C5–C6 bonds, d, and the absolute value of the C5–C6–C6–C5 dihedral angle, η. Assuming that the thymines must be in a given range of d and η to dimerize and that in this range the dimerization yield is one, the overall quantum yield can be written as:

$$\phi = \int_{\eta_{min}}^{\eta_{max}} \int_{d_{min}}^{d_{max}} P(\eta, d) \, d\eta \, dd \tag{13.1}$$

where $P(\eta, d)$ is the normalized joint probability distribution of η and d. Law et al. suggest that $d_{min} = 0.0$ Å, $d_{max} = 3.7$ Å, $\eta_{min} = -48°$, and $\eta_{max} = 48°$ [12] are appropriate cutoffs for dimer formation, while Johnson et al. suggest that $d_{min} = 3.2$ Å, $d_{max} = 3.6$ Å, $\eta_{min} = 24°$, and $\eta_{max} = 30°$ are appropriate cutoffs for dimer formation [11]. For reference, the average d and η values obtained for dT$_{20}$ in our B-DNA starting structure (Fig. 13.7) are 4.44 Å and 47.6°, respectively and the values for canonical duplex DNA are 3.37 Å and 36.0°, respectively [17].

Both previously published models assume the need for two geometrical parameters to determine [2+2] photodimerization yields [11, 12]. To test this assumption, CASSCF calculations were carried out for ethylene+ethylene as a model [2+2] photodimerization system. Triplet mechanisms have been proposed for CPD formation [18, 19], however recent experiments have shown that the singlet pathway is dominant [9, 10]. This CASSCF study is therefore only done for singlet surfaces. Assuming initial excitation of the separated ethylenes to a singly excited π^* state, an eventual transition to a doubly excited π^* state will lead to the formation of cyclobutane. The topology of the $(\pi)^3(\pi^*)^1$ surface and the crossing between the $(\pi)^3(\pi^*)^1$ and $(\pi)^2(\pi^*)^2$ surfaces as a function of d and η will elucidate the [2+2]

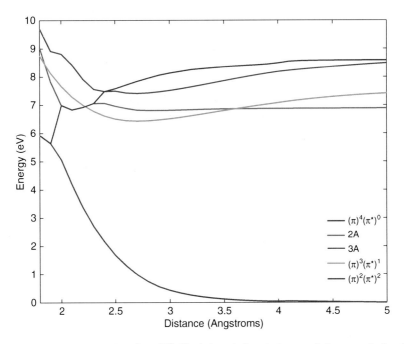

Fig. 13.3 Potential energy plots for a 25° dihedral angle for ethylene + ethylene as calculated by CASSCF(4,12)/aug-cc-pVDZ in D_2 symmetry. Surfaces 2A and 3A are all singly excited states. Surface $(\pi)^2(\pi^*)^2$ depicts where the CI vectors indicate significant population of the π^* orbital appropriate for cyclobutane formation

mechanisms dependency on d and η. Previous work by Bernardi et al. studied this reaction in detail but the surface dependence on η was never discussed [20, 21].

The dependence of the excited state surfaces on distance is investigated first. Some of the most important surfaces are plotted as a function of distance for a fixed dihedral angle of 25° in Fig. 13.3. At infinite separation, there are four degenerate $(\pi)^3(\pi^*)^1$ surfaces; for simplicity we only show one. The surfaces labeled 2A and 3A are ^{1}A states in D_2 symmetry while the surface labeled $(\pi)^3(\pi^*)^1$ is $^1B_2$ (and thus it could have been labeled $^1B_2$). There is a diabatic surface with significant $(\pi)^2(\pi^*)^2$ nature that we label in purple in Fig. 13.3. Unlike the other surfaces, which are adiabats, this surface is derived by examining the dominant configurations in the CASSCF calculations. At long distances this state is high in energy but rapidly decreases in energy as a function of d, eventually becoming the ground state at 2.0 Å. There is a crossing between the $(\pi)^2(\pi^*)^2$ state and the $(\pi)^3(\pi^*)^1$ surface at approximately 2.4 Å. This provides a mechanism whereby photoexcitation of the $(\pi)^3(\pi^*)^1$ state can internally convert to $(\pi)^2(\pi^*)^2$, leading to either to photodimerization or back to the reactants.

In Fig. 13.3, the ground state barrier between the separated ethylenes ($d>4$ Å) and cyclobutane ($d<2$ Å) is exaggerated due to the rigidity of the potential

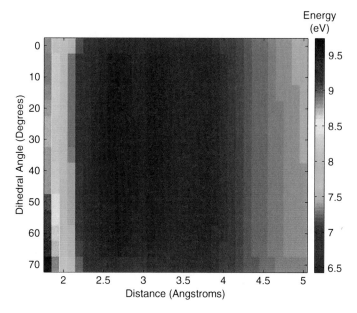

Fig. 13.4 Contour plot of the $(\pi)^3(\pi^*)^1$ surface of ethylene + ethylene as calculated by CASSCF(4,12)/aug-cc-pVDZ in D_2 symmetry

energy scan performed. Allowing for a change in the carbon–carbon and carbon–hydrogen bond lengths as well as a for out of plane bending on the hydrogen atoms would reduce the ground state barrier. A reduction in symmetry also plays a role in the excited state crossing as was determined by Dallos et al. [22]. Specifically, geometric relaxation is important to the energy of the crossing between the $(\pi)^2(\pi^*)^2$ and $(\pi)^3(\pi^*)^1$ states, which will move lower, making this intersection more favorable (i.e., downhill in energy starting from almost any geometry accessible from $(\pi)^3(\pi^*)^1$).

Note that the 2A state crosses $(\pi)^3(\pi^*)^1$ at $d < 3.6$ Å. This means that there is a conical intersection between these states at that point. This could provide a pathway for the initially excited $(\pi)^3(\pi^*)^1$ state to decay back to the reactants if formed at distances larger than this. However in reality photoexcitation produces both states with some probability, so the maximum distance that would allow access to the $(\pi)^2(\pi^*)^2$ state is difficult to estimate. In addition, in a solution reaction, the dependence of the dimerization probability on d is likely governed by additional issues, including solvent effects. In the analysis below we will vary d to optimize the agreement between theory and experiment.

The dependence of the $(\pi)^3(\pi^*)^1$ and $(\pi)^2(\pi^*)^2$ surfaces on distance and dihedral angle are displayed in the contour plots shown in Figs. 13.4 and 13.5 respectively. The contour plot in Fig. 13.4 shows a shallow well excimer on the $(\pi)^3(\pi^*)^1$ surface at 2.75 Å for a wide range of dihedral angles. Indeed this surface

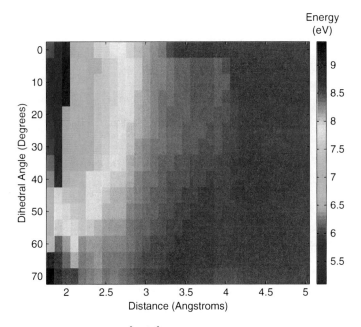

Fig. 13.5 Contour plot of the $(\pi)^2(\pi^*)^2$ surface of ethylene + ethylene as calculated in CASSCF(4,12)/aug-cc-pVDZ in D_2 symmetry. Energy is relative to the ground state at infinite separation

does not display significant dependence on dihedral angle. Figure 13.5 displays a gradient towards zero dihedral angle at short distance for the $(\pi)^2(\pi^*)^2$ state, however more important than the topology of the separate states is the nature of the crossing between the $(\pi)^3(\pi^*)^1$ and $(\pi)^2(\pi^*)^2$ states as it is this crossing that results in the formation of the ground state of cyclobutane. The difference between these two states is plotted in Fig. 13.6. As can be seen by the yellow contour (0.0 difference), the doubly excited $(\pi)^2(\pi^*)^2$ state crosses the singly excited $(\pi)^3(\pi^*)^1$ state for all dihedral values plotted. The crossing ranges from 2.2 Å at small dihedral angles to 2.0 Å at large dihedral angles.

Previous studies [11, 12] chose to limit the dihedral angle and the distance due to the necessary alignment of the carbon–carbon double bonds both in the products and conical intersection [23]. The excited state topology presented here, however, suggests that there are pathways for the $(\pi)^3(\pi^*)^1$ surface to cross the $(\pi)^2(\pi^*)^2$ surface over a large range of dihedral angles (Fig. 13.6). The topology of the $(\pi)^2(\pi^*)^2$ surface, presented in Fig. 13.5, displays a large gradient towards zero dihedral angle, however this is only accessed after the surface crossing, at a point where cyclobutane formation is guaranteed. It is thus possible for an ethylene + ethylene system at large dihedral angles to undergo excitation to the $(\pi)^3(\pi^*)^1$ surface and rapidly form cyclobutane.

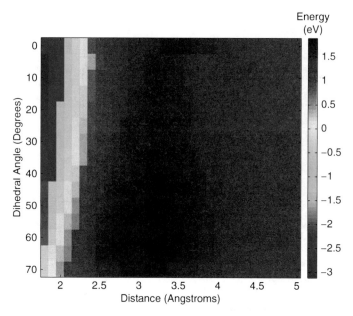

Fig. 13.6 Difference between $(\pi)^2(\pi^*)^2$ and $(\pi)^3(\pi^*)^1$ states of ethylene + ethylene as calculated by CASSCF(4,12)/aug-cc-pVDZ in D_2 symmetry

13.3 Model Calibration with dT_{20} and $dA_{20}dT_{20}$

13.3.1 Background

The dT_{20} and $dA_{20}dT_{20}$ systems are used as benchmarks for our studies due to their large size and high frequency of TT steps (see Fig. 13.7). Having both a single and double stranded system to calibrate our model is necessary due to the added complexity of a double stranded system and highly fluxional nature of the single stranded system. The difference in quantum yields between dT_{20} and $dA_{20}dT_{20}$ may reflect competitive absorption by dA and energy transfer between dA and dT as well as differences in the ground state conformation between the single strand and duplex. Thymine absorbs more strongly than adenine at the excitation wavelengths used in our quantum yield measurements (280–290 nm) [24]. Furthermore, fluorescence quantum yields are independent of excitation wavelength over the range 245–285 nm, suggesting that energy transfer between adenine and thymine takes place [25]. However, these factors appear to be insufficient to account for the lower quantum yields for $dA_{20}dT_{20}$ vs. dT_{20}.

Fig. 13.7 Starting structures
for (**a**) dT_{20} and (**b**) $dA_{20}dT_{20}$

13.3.2   Computational Details

In order to effectively sample the ground state conformations achieved by these systems molecular dynamics (MD) is employed. Starting structures for both systems were generated using the nucgen program in Amber 8 [26]. The systems were then neutralized with sodium ions and solvated with 8 Å buffers of TIP3 water. Short minimizations were carried out followed by NPT MD simulations. The simulations were all done using the NAMD [27] program with the CHARMM27 forcefield [28, 29]. A Langevin barostat was employed to keep the pressure at 1 bar with a piston period of 200 fs and a decay of 100 fs. The temperature was also kept constant with a Langevin thermostat which had a damping coefficient of $5.0 \, \text{ps}^{-1}$. Periodic boundary conditions were employed with a real space nonbonded cutoff of 12 Å and particle mesh Ewald summation for the long range electrostatics. An initial simulation of each system at 350 K was carried out in order to generate ten independent starting geometries for simulations at 300 K. Each of these ten runs consisted of a short minimization followed by 2 ns equilibration and 6 ns of production time. Geometries were sampled every 2 ps.

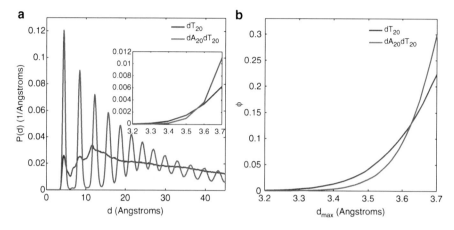

Fig. 13.8 (**a**) Normalized probability distributions of d for dT_{20} and $dA_{20}dT_{20}$ calculated from MD simulations. (**b**) Calculated quantum yield (ϕ) for TT dimer formation vs. d_{max} for dT_{20} and $dA_{20}dT_{20}$. The quantum yield is calculated using the normalized probability distribution function $P(d)$ from MD simulations and the formula $\phi = \int_0^{d_{max}} P(d)\,dd$

13.3.3 Results

Given the topologies of the excited states discussed in the previous section, limits on the C5–C6–C6–C5 dihedral angle are ignored and a distance cutoff giving agreement with the experimental dT_{20} result is sought. This simplifies Eq. 13.1 to:

$$\phi = \int_{d_{min}}^{d_{max}} P(d)\,dd \tag{13.2}$$

where $P(d)$ is the normalized probability distribution function of d. The probability distributions of d for all possible pairs of thymines in dT_{20} and $dA_{20}dT_{20}$ are plotted in Fig. 13.8a. A clear periodicity is seen for $dA_{20}dT_{20}$ due to the rigidity and symmetry of the double helix. This same periodicity is not as clear in $P(d)$ for dT_{20} due to the more random nature of the single stranded structure. There are, however, clear peaks at 4.2 Å, 8.2 Å, and 12.2 Å for both species. The inset in Fig. 13.8a shows a close-up of the range of d which we are most interested in.

The probability distribution functions of d in Fig. 13.8a show no population below $\sim$3.2 Å suggesting that a seeking a lower limit for d in Eq. 13.2 is unnecessary. This leaves our quantum yield as a simple function of d_{max} which is plotted in Fig. 13.8b. Values of d_{max} yielding experimentally relevant numbers are tabulated in Table 13.1 for both dT_{20} and $dA_{20}dT_{20}$. The MD cutoff yielding the smallest percent error from experiment for dT_{20} is $d < 3.52$ Å. This same cutoff value gives a quantum yield of 29×10^{-3} for $dA_{20}dT_{20}$ which is in good agreement with the experimental yield of 22×10^{-3}. Thus, geometrical arguments can explain the [2+2]

Table 13.1 [2+2] quantum yields for MD simulations of dT_{20} and $dA_{20}dT_{20}$ for four different cutoffs

	$10^3 \Phi_{TT}$	
	dT_{20}	$dA_{20}dT_{20}$
$d < 3.51$ Å	47	27
$d < 3.52$ Å	52	29
$d < 3.53$ Å	57	34
$d < 3.54$ Å	63	40
exp^a	50	22

[a]From reference [13]

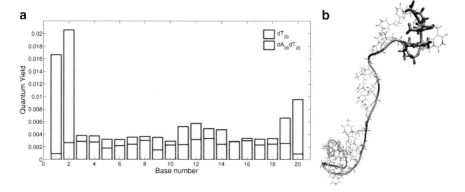

Fig. 13.9 (**a**) Quantum yields per base calculated from MD trajectories with the dimerizable cutoff of $d < 3.52$ Å. (**b**) Dimerizable snapshot of dT_{20}. Bases 2 and 4 are highlighted because they are the bases in the dimerizable geometry. The backbone is shown as a gold tube to show the hairpin region

quantum yield discrepancy between dT_{20} and $dA_{20}dT_{20}$. A separation of 3.52 Å for ethylene + ethylene on the $(\pi)^3(\pi^*)^1$ surface shows a negative gradient that would allow for excimer formation based on the CASSCF results (Fig. 13.4), and it is inside the conical intersection that would lead to nonreactive decay on the 2A surface (Fig. 13.3).

Johnson and Wiest concluded on the basis of their MD simulations of TT dimer formation in dT_{18} that dimer formation is more efficient in transiently-formed hairpin conformations having intra-strand T-T base pairing than in extended regions resembling the poly(dT) strand of a duplex (Fig. 13.7a) [11]. Analysis of our results for dT_{20} indicates that the terminal base pairs T1–T2 and, to a lesser extent, T19–T20 have the highest predicted dimerization quantum yields (Fig. 13.9a). The terminal bases can insert themselves between adjacent bases forming mini-hairpins (Fig. 13.9b) which allow for non-canonical stacked geometries to be sampled. In contrast to the results of our analysis of dT_{20}, we find that probability of dimer formation is similar for each of the thymines in $dA_{20}dT_{20}$ except for T1 and T20 which have only one neighboring T and thus a lower probability of dimer formation (Fig. 13.9a). Unfortunately there is no experimental data similar to Fig. 13.9a that we can use to text this prediction.

13.4 Sequence Dependence of TT Dimerization in DNA Hairpins

13.4.1 Background

Given the success of such a simple model at predicting the [2+2] TT dimer yields in double and single stranded DNA, a number of interesting questions can be addressed. One such question involves the sequence specificity of [2+2] TT dimer formation. Early studies on nucleosomal DNA formation found a 10.3 base pair periodicity in TT dimer hot spots [30–32]. This phenomenon is likely due to the nearly equivalent base pair periodicity found for TT steps in nucleosomal DNA (found in X-Ray results) which is thought to be due to a minimization of DNA bending free energy [33]. Other early studies on TT dimer formation have shown that individual neighboring bases can affect the quantum yield of TT dimer formation for a specific TT step [34,35]. In particular, it was found that neighboring purines reduced the quantum yield of dimerization for a TT step.

Recently, Hariharan et al. have synthesized and quantified the TT dimer yields of short DNA hairpins with different sequences [13,36]. These small systems are useful tools to probe the exact cause of sequence specificity of TT dimer yields. The three hairpins we will investigate are denoted St1, St3 and St5 all of which are depicted in Fig. 13.10. Each hairpin has six base pairs that are capped on one end by a twelve carbon C12 linker. The sequences are designed so that each hairpin only has one TT step. The TT step in St1 is on the 5' terminus, in St3 is in the middle of the sequence and in St5 is in the C12 linker adjacent site. Using our molecular dynamics model to compute the TT dimer yields of these sequences will allow us to determine the role of ground state geometry on sequence specificity in TT dimer yields.

13.4.2 Computational Details

Starting structures for the DNA bases were built using the nucgen program in Amber 8 [26]. The C12 linker (pictured at the top of Fig. 13.10) was built separately and manually positioned on the DNA bases and then minimized with restraints on the base pairs. Atom type parameters for the C12 linker were taken from standard CHARMM27 atom types [28, 29]. The partial charges were derived using HF/6-31G* calculations in conjunction with the resp module of Amber 8 [26]. The rest of the simulations were carried out as described above for the dT_{20} and $dA_{20}dT_{20}$ systems.

13.4.3 Results

The distance between the C5–C6 double bonds (d) of the thymines in all three of the hairpin structures was monitored. The probability distributions are plotted in

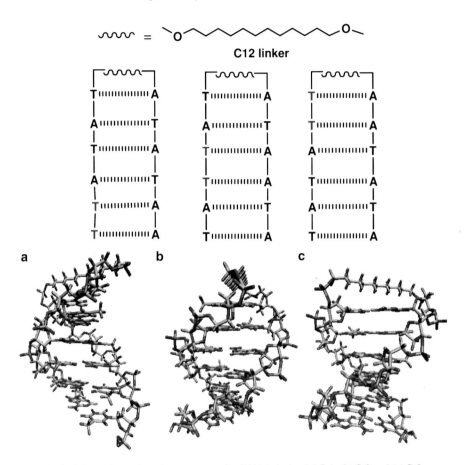

Fig. 13.10 Schematics and starting structures for DNA hairpins (**a**) St1, (**b**) St3 and (**c**) St5

Fig. 13.11a. The probability distributions are integrated as a function of d_{max} which is plotted in Fig. 13.11b. Using the distance cutoff for [2+2] dimerization found in section 13.3 ($d < 3.52\,\text{Å}$), the quantum yields for St1, St3 and St5 are shown in Table 13.2. Looking at the three curves in Fig. 13.11b, we can see that setting d_{max} to values in the range $d_{max} < 3.65\,\text{Å}$ would not alter the resulting trends in ϕ significantly. No clear trend in ϕ is seen until $d_{max} > 3.7\,\text{Å}$ at which point the experimental trend is upheld.

Upon comparison of the $d < 3.52\,\text{Å}$ results and the experimental results given in Table 13.2, two things are immediately apparent. First, the order of magnitude predicted by our one parameter model is correct. The percent error is only 14% for St5. The second major observation is that the experiment predicts a trend of increasing [2+2] quantum yield from St1 to St3 to St5, while the MD results predict the opposite trend. There are a couple of possible reasons for this, all stemming from the limitations of the model. First, the model depends on the accuracy of the force

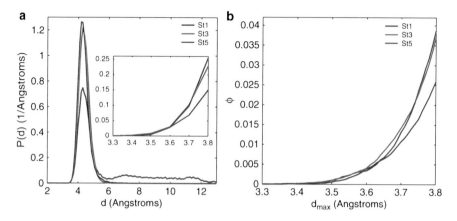

Fig. 13.11 (a) Normalized probability distributions of d for St1, St3, and St5 calculated from MD simulations. (b) Calculated quantum yield (ϕ) for TT dimer formation vs. d_{max} for St1, St3 and St5. The quantum yield is calculated using the normalized probability distribution function $P(d)$ from MD simulations and the formula $\phi = \int_0^{d_{max}} P(d)\,dd$

Table 13.2 [2+2] quantum yields for St1, St3, and St5 calculated from MD simulations using a distance cutoff of $d < 3.52$ Å

	$10^{-3}\Phi_{TT}$		
	St1	St3	St5
$d < 3.52$ Å	1.5	1.5	0.9
exp[a]	0.39	0.92	1.05

[a]From reference [13]

field employed. The geometries leading to dimer formation are rarely populated, a situation which can be especially difficult for a force field to sample accurately. Second, energy flow, excited state quenching via other pathways, and preferential excitation of certain bases are completely ignored. While these factors are not thought to be hugely important, hence the correctly predicted order of magnitude, they may be necessary to explain the subtle differences between St1, St3 and St5. Interestingly, it is suggested that end fraying of the DNA duplex is a possible cause for the lower experimental quantum yield of St1 as compared to St3 and St5 [13,36]. Our model explicitly incorporates end fraying with the use of molecular dynamics and yet is unable to accurately reproduce the experimental trend suggesting that end fraying does not play a significant role in differentiating the TT dimer quantum yields of St1, St3 and St5.

Comparison between the calculated results in Tables 13.1 ($d < 3.52$ Å) and 13.2 can be made by computing an average quantum yield per TT step. The values for dT_{20} and $dA_{20}dT_{20}$ are 2.7×10^{-3} and 1.5×10^{-3}, respectively. St1, St3 and St5 have only one TT step so the results in Table 13.2 do not require normalization. The calculated results suggest that the TT steps in St1 and St3 are as likely to be in dimerizable geometries as the average TT step in the longer $dA_{20}dT_{20}$ system.

However note that the calculated St5 TT step deviates from the average $dA_{20}dT_{20}$ result suggesting a special effect for constraints associated with the C12 linker. The dT_{20} system has a higher quantum yield per TT step than the four base paired systems demonstrating the importance of a more flexible structure in achieving $d < 3.52$ Å.

The ground state geometries of the TT steps in the three hairpins St1, St3 and St5 play important roles in the formation of the TT steps as evidenced by the order of magnitude agreement between predictions from our model and experimental values. The subtle differences between the TT dimer quantum yields of St1, St3 and St5 seen in experiment are not reproduced with our model. This implies our model is not capturing some of the subtle and yet important behavior of these systems. This could be due to anything from a lack of a refined ground state model to the need to incorporate an electronic effect.

13.5   Application to Locked Nucleic Acids

13.5.1   Background

Another way to probe the effect of ground state conformation on TT dimer formation is to synthetically change the conformational freedom of a TT step. This has been done successfully using locked nucleic acids (LNAs) [9, 14, 37]. Recently experimental evidence for ultrafast T–T dimerization based on femtosecond time-resolved IR spectroscopy has been presented by Gilch and co-workers for the oligonucleotide dT_{18} and the dinucleotides TT and T_LT_L, where T_L is the locked nucleic acid analog of dT in which the furanose ring is locked in the C3'-endo conformation (see bottom of Fig. 13.12) [9]. A recent report of higher dimerization efficiency and selectivity for formation of the major 2+2 dimer from T_LT_L vs. TT provides additional evidence for conformational control of T–T dimerization in dinucleotides [37]. This intriguing observation was extended to the more biologically relevant single strand and duplex systems by Hariharan et al. [14].

In this section we will discuss the results from our work in the paper by Hariharan et al. [14]. Five species were simulated and synthesized with TT steps having both a locked and non-locked sugar backbone. These species are depicted in Fig. 13.12. We will use the following nomenclature throughout the section, species **1** refers to the non-locked dinucleotide while species **1L** is the locked version. **1** and **1L** are single stranded dinucleotides, **2** and **2L** are single stranded six base species and **5–7** and **5L–7L** are double stranded DNA hairpins. Investigating the conformational freedom and TT dimer formation of all of these species allows us to probe the effect of locking the sugar on both single and double stranded systems.

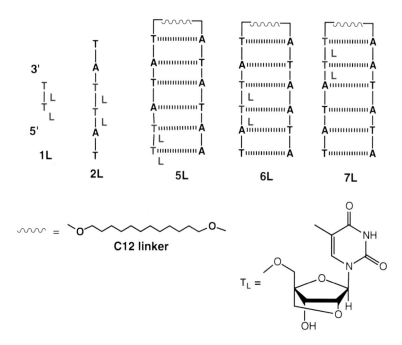

Fig. 13.12 Structures of the dinucleotide **1L**, single strand oligo **2L** and C12-linked hairpins **3L–7L** containing an LNA $T_L T_L$ step. Corresponding sequences with unmodified TT steps are designated **1–7**

13.5.2 *Computational Details*

Conformational modeling of **1L**, **2L** and **5L–7L** was carried out using the CHARMM27 force field [28, 29] with the addition of the locked nucleic acid parameters from Pande et al. [38]. Other than this modification, all MD simulations were carried out in the same manner as the initial dT_{20} and $dA_{20}dT_{20}$ systems above.

The resulting trajectories were all analyzed by monitoring the distance, d, between the C5–C6 double bonds and the C5–C6–C6'–C5' dihedral angle, η, of the TT step. The dimerization cutoff used was $d < 3.52\,\text{Å}$. All possible pairs of thymines were considered in determining the final quantum yields. Non-neighboring TT dimers were only found in structures **2** and **2L** and were minor in comparison to the neighboring TT dimers. The trajectories were also analyzed using 3DNA to determine their helical parameters [39]. For the inter base pair parameters (rise, shift, slide, roll, twist and tilt), only the strand containing the TT steps was analyzed as if it were single stranded DNA. This was done to remain consistent between single and double stranded systems.

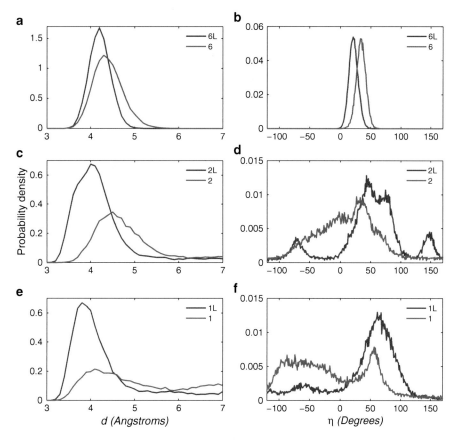

Fig. 13.13 Probability densities for the C5–C6 bond separation d (**a,c,e**) and the C5–C6–C6'–C5' dihedral angle η (**b,d,f**) for $T_L T_L$ (*blue lines*) vs TT steps (*green lines*). (**a**) and (**b**) are for **6L** vs. **6**, (**c**) and (**d**) are for **2L** vs. **2** and (**e**) and (**f**) are for **1L** vs. **1**

13.5.3 Results

Probability densities for the distance d separating the mid-points of the $T_L T_L$ C5–C6 double bonds in **1L**, **2L**, and **6L** are shown on the left side of Fig. 13.13 along with results for the unmodified sequences. Significantly narrower distributions of d were seen for all of the locked systems in comparison to the non-locked systems, the difference being more pronounced for **1L** and **2L** than for **6L**. The peak in the probability density for the sequences **1L** and **2L** lies at shorter distances (ca. 3.8 and 4.0 Å) as compared to the 4.2 Å separation in hairpin **6L**. Narrower distributions of the C5–C6–C6'–C5' dihedral angle, η, were also observed for the locked vs. unmodified TT steps (right side of Fig. 13.13).

Quantum yields were calculated using Eq. 13.2 with $d_{max} = 3.52$ Å as determined in Sect. 13.3. These values are given in Table 13.3. They are not corrected

Table 13.3 Calculated CPD quantum yields of single strand (**1–2**) and hairpin sequences (**5–7**) using a single parameter cutoff of $d < 3.52$ Å. Quantum yields for all five systems were calculated for both normal TT and locked $T_L T_L$ steps

	1	2	5	6	7
$10^3 \Phi_{TT}$	9.0	2.3	1.5	1.5	0.9
$10^3 \Phi_{T_L T_L}$	62	77	32	1.8	2.5
$\Phi_{T_L T_L} / \Phi_{TT}$	6.9	33	21	1.2	2.8

Table 13.4 Experimental CPD quantum yields of single strand (**1–2**) and hairpin sequences (**5–7**) from reference [14]. Quantum yields for all five systems were measured with a normal TT step and replacing that with the locked $T_L T_L$ analog

	1	2	5	6	7
$10^3 \Phi_{TT}$	0.95	0.40	0.39	0.92	1.1
$10^3 \Phi_{T_L T_L}$	8.3	6.1	4.1	3.4	4.4
$\Phi_{T_L T_L} / \Phi_{TT}$	9	15	11	4	4

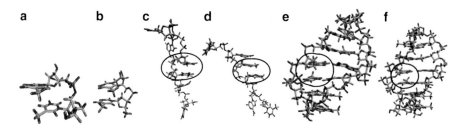

Fig. 13.14 Snapshots of TT stacked conformations from molecular dynamics simulations of (**a**) **1L**; (**b**) **1**; (**c**) **2L**; (**d**) **2**; (**e**) **6L**; and (**f**) **6**. Ovals indicate the location of the TT step in structures (**c–f**)

for competitive absorption of light by non-reactive bases or energy transfer between reactive and non-reactive bases and thus are not expected to reproduce the experimental quantum yields (Table 13.4). However, the calculated values for the ratios of $T_L T_L$ vs. T–T dimerization replicate the important trends in the experimental data; namely the large ratio for **1**, **2**, and **5** and the much smaller ratios for **6** and **7**.

Randomly selected snapshots of reactive conformations of **1L**, **2L**, and **6L** and their unmodified analogs are shown in Fig. 13.14. Other than the obvious difference in sugar conformation, there are no noticeable differences between the structures of **1L** vs. **1** or **2L** vs. **2** (Fig. 13.14a–d). Differences between these structures arise upon analysis of probability distributions of helical parameters (rise, shift, slide, tilt, roll and twist) between adjacent thymines for **1L** and **1** (Fig. 13.15) and for **2L** vs. **2** (Fig. 13.16). The locked TT pairs sample a much narrow range of values than

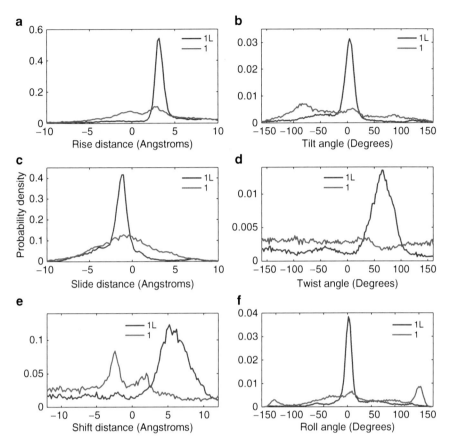

Fig. 13.15 Probability density distributions of inter base pair helical parameters of **1L** and **1** from molecular dynamics simulations. (**a**) rise distance, (**b**) tilt angle, (**c**) slide distance, (**d**) twist angle, (**e**) shift distance and (**f**) roll angle in **1L** (*blue line*) and **1** (*green line*)

the non-locked. Differences in the structures of **6L** and **6** (Fig. 13.14e, f) are more obvious and include larger buckle angles for T_L3-A and T_L4-A base pairs ($-8°$ and $-16°$) than for the corresponding T–A base pairs ($0°$ and $-4°$) (Fig. 13.17). Similar large buckle angles are observed in the NMR structure of a 10-base pair duplex having a T_LT_L step at mid-strand [40]. Smaller differences are calculated for the propeller angle and opening angle (Fig. 13.17) and the rise, roll, twist, and tilt distributions (Fig. 13.16) for the T_LT_L vs. TT steps of **6L** vs. **6**. The A–T base pairs adjacent to the T_LT_L step in **6L** appear to have normal B-DNA geometries, in accord with previous NMR, [40] X-ray crystal structure [41] and molecular dynamics studies [28, 38] which show that the structural perturbation introduced by a single locked nucleotide is localized [42].

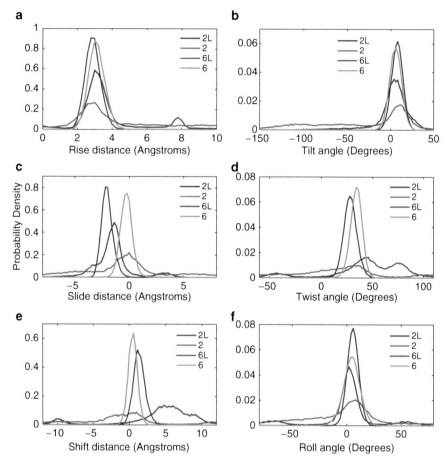

Fig. 13.16 Probability density distributions of inter base pair helical parameters of **6L** and **6** from molecular dynamics simulations. (**a**) rise distance, (**b**) tilt angle, (**c**) slide distance, (**d**) twist angle, (**e**) shift distance and (**f**) roll angle for TT steps in single stranded systems **2L** (*blue line*) and **2** (*green line*) and double stranded systems **6L** (*red line*) and **6** (*teal line*)

In summary, we observe increases in the quantum yields for dimerization at $T_L T_L$ vs. TT steps in a single strand (Table 13.3) which are even larger than that previously reported by Desnous et al. [37] for the dinucleotides. Molecular dynamics simulations indicate that the increased quantum yield for the single strand sequence **2L** is a consequence of a marked increase in the population of ground state conformations having geometries appropriate for dimerization of $T_L T_L$ vs. TT steps. More modest increases in quantum yields are observed for the $T_L T_L$ vs. TT steps in hairpins **6L** and **7L**. Molecular dynamics simulations indicate that the more rigid duplex structure of hairpin **6L** prevents the close approach of the reactive double bonds observed in **1L** and **2L** and thus allows only a minor increase in the population of reactive conformations for the $T_L T_L$ vs. TT step (Fig. 13.14e, f). The increase in

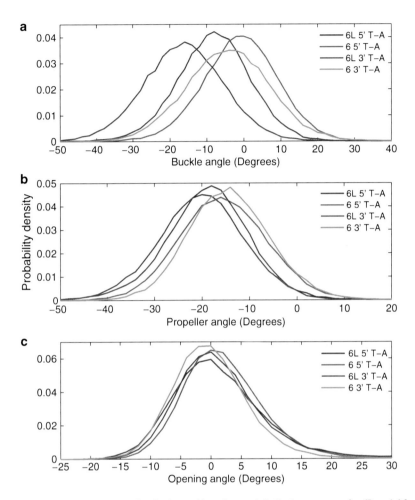

Fig. 13.17 Probability density distributions of intra base pair helical parameters for **6L** and **6** from molecular dynamics simulations (**a**) buckle , (**b**) propeller, and (**c**) opening angles for the 5' and 3' A–T base pair of the TT step in **6** (*blue line* and *red line*) and the 5' and 3' A–T base pair of the TT step in **6L** (*green line* and *teal line*)

quantum yield for **5L** vs. **5** is intermediate between those for the duplex interior (**6L** vs. **6**) and for the single and double overhangs, suggestive of intermediate conformational populations for a terminal $T_L T_L$ step. The more rigid structures of the $T_L T_L$ vs. TT steps are inappropriate for the formation of 6–4 adducts or other minor T–T photoadducts, resulting in highly selective formation of the syn 2+2 photoadducts. These results serve to elucidate the photochemical behavior of $T_L T_L$ steps in single strand and duplex DNA and to further establish the importance of ground state conformation in determining the efficiency and selectivity of T–T dimerization, particularly in single strand sequences.

13.6 Quenching of TT Dimer Formation in Trinucleotides by Purines

13.6.1 Background

So far we have seen two sets of systems in which our ground state model for predicting photoinduced TT dimer yields works well (Sects. 13.3 and 13.5). The model failed to predict the subtle trend in TT dimer yields of hairpins St1, St3 and St5 in Sect. 13.4 and yet was able to predict quantum yields for each species that were within an order of magnitude of experimental results. In this section we further test the limits of our ground state model by investigating sequence specificity in TT dimer yields more closely.

Specifically, we investigate the known TT dimer formation quenching ability of neighboring purine bases [43–46]. The ability of a neighboring purine base to quench the formation of a TT dimer depends on two variables. The first is whether the purine is 3' or 5' to the TT step. It has been shown that a purine only on the 5' side of a TT step quenches TT dimer formation more effectively than does the equivalent purine only on the 3' side of the TT step [44, 46]. The second variable is the type of purine base. Guanine and other purines with lower ionization potential than adenine have been shown to be more effective at quenching TT dimer formation than adenine [45, 46].

The smallest systems that can be used to test both of the TT dimer quenching variables are trinucleotides. Four of these have been synthesized and modeled in order to test their TT dimer formation ability [46]. These sequences are depicted in Fig. 13.18. The sequences consist of TT steps with either a guanine or adenine 3' or 5' and are labeled A–T–T, T–T–A, G–T–T and T–T–G. Note that the nomenclature here starts with the 5' base. The difference in ground state geometry of these four sequences is investigated in the following sections.

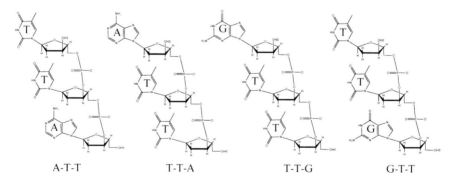

A-T-T   T-T-A   T-T-G   G-T-T

Fig. 13.18 The four trinucleotide systems used to investigate the TT dimer quenching ability of neighboring purine bases. A–T–T denotes a 5'adenine terminus and a 3' thymine terminus

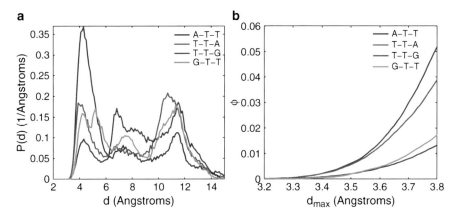

Fig. 13.19 (**a**) Normalized probability distribution of d for A–T–T, T–T–A, T–T–G, and G–T–T calculated from MD trajectories (**b**) Quantum yield vs. d_{max} for A–T–T, T–T–A, T–T–G, and G–T–T

13.6.2 Computational Details

The trinucleotide systems were all simulated in a similar fashion to the dT_{20} and $dA_{20}dT_{20}$ with the exception of having 12 ns production runs as opposed to 6 ns. This yielded a total of 120 ns of analyzed simulation time for each system.

13.6.3 Results

The distance between the midpoints of the C5–C6 double bonds in the two thymines in each of the four trinucleotides was monitored. The normalized probability distributions of this value are plotted in Fig. 13.19a. Unlike the double stranded systems reported in previous sections, the trinucleotides all have broad distributions of d values. In comparison to the standard dinucleotide, **1**, reported in the previous section, most of the trinucleotides show a stronger preference for short d values. All four trinucleotide species also show significant population for $d \sim 11$ Å. The origin of these peaks will be discussed in more detail.

Integrating the probability distributions of d in accord with Eq. 13.2 allows us to calculate the quantum yield of [2+2] TT dimer formation as a function of d_{max}. This is plotted in Fig. 13.19b for all four trinucleotide systems. Using the previously determined value, $d_{max} = 3.52$ Å, the calculated quantum yields are tabulated along with experimental values in Table 13.5. Calculated values agree with experiment not only in absolute values but also in trends. Both adenine containing species are found to have higher quantum yields than the corresponding guanine containing species. Also, both 5' purine species are found to have higher quantum yields than their 3'

Table 13.5 Quantum yields
for four trinucleotide systems.
Calculated results are from
MD simulation with
$d < 3.52\,\text{Å}$

Species	$\phi_{calc} \times 10^3$	$\phi_{exp}^a \times 10^3$
A–T–T	7.2	6.1
T–T–A	6.7	3.5
T–T–G	2.4	2.7
G–T–T	2.5	3.3

[a]From reference [46]

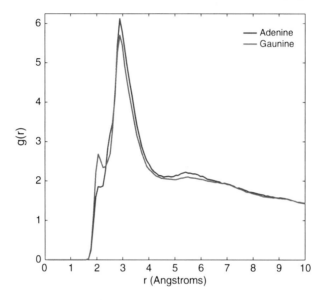

Fig. 13.20 The pair distribution function between either adenine or guanine and water from
A–T–T or G–T–T respectively

counterparts. We can see from Fig. 13.19b that these trends are robust with respect
to the specific choice of d_{max} value.

Both species containing guanine have significantly lower quantum yields than
the species containing adenine. This is due, in part, to guanine having a lower free
energy of solvation [47]. Because guanine is more readily solvated, the stacking
of the three bases is more easily perturbed in the guanine trinucleotides. The
preferential guanine solvation can be seen in our trajectories by monitoring the
pair distribution function between waters and either adenine or guanine. This is
plotted for A–T–T and G–T–T in Figure 13.20. Guanine has a much larger peak at
$r \sim 2\,\text{Å}$ representing its additional hydrogen bond available to water.

Guanine's higher affinity for water can also be seen in the probability distribution
of the base–base stacking energies. In order to determine if the purine–thymine
stacking energy correlates to the thymine–thymine stacking energy we monitor
the joint probability of these two variables. These are plotted in Fig. 13.21 for all
four trinucleotide systems. Looking at the distributions for A–T–T and G–T–T (top
row of Fig. 13.21) we can see that both species spend significant time with no

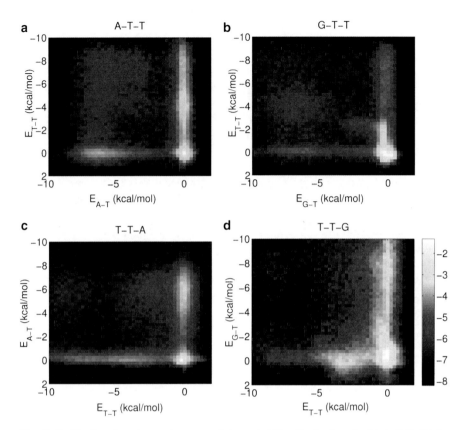

Fig. 13.21 The base-base interaction energy joint probability distributions for (**a**) A–T–T, (**b**) G–T–T, (**c**) T–T–A, and (**d**) T–T–G. The population is given in the log scale

stacked bases. Comparing these two plots, we see that the A–T–T system spends significantly more time than the G–T–T system with two of the bases stacked together. In particular, there is significant population with $E_{A-T} = 0$ kcal/mol and $E_{T-T} < -2$ kcal/mol as well as $E_{T-T} = 0$ kcal/mol and $E_{A-T} < -2$ kcal/mol for the A–T–T system. The G–T–T system shows little stacking of either G–T or T–T with an overwhelming population at $E_{G-T} = E_{T-T} = 0$ kcal/mol. While neither species shows major population with all three bases stacked, A–T–T has more than G–T–T. The additional stacking of adenine with the neighboring thymine does seem to lead to additional T–T stacking, increasing the potential for dimer formation.

Also of interest is the difference seen in experimental quantum yields for 5' vs. 3' purines. A decrease by a factor of approximately two is seen in the experimental quantum yields when moving adenine from the 5' side of the TT step to the 3' side of the TT step in these trinucleotides (see Table 13.5). The calculated decrease is not as dramatic being only a factor of ∼1.1. By comparing the base–base interaction energy joint probability distributions of Fig. 13.21a and c we can see the effect of having the 5' vs 3' adenine. Both A–T–T and T–T–A have significant population with $E_{T-T} < -2$ kcal/mol and $E_{A-T} = 0$ kcal/mol. A–T–T, however, has significantly more population in which both bases are stacked ($E_{T-T} < -2$ kcal/mol and $E_{A-T} < -2$ kcal/mol). It is clear that the added stacking ability of adenine as compared to guanine provides stability of T–T stacking and thus increases the TT dimer yields. Stacking of the 5' adenine with the central thymine seems to induce more population in which all three bases are stacked than stacking of the 3' adenine thus slightly increasing the TT dimer quantum yield for A–T–T as compared to T–T–A.

The discrepancy between simulation and experiment in the [2+2] TT dimer quantum yield for T–T–A is likely due to small inaccuracies in the force field as well as insufficiencies in our dimerization model. The overall agreement between theory and experiment for the trinucleotide systems, however, is remarkably good. Given that the TT dimer quenching ability of neighboring purines likely has a significant electronic mechanism, a ground state model for dimerization should be unable to grasp the experimental trends. The agreement between experiment and simulation is thus suggesting that ground state conformation is the overwhelming factor determining TT dimer yields in trinucleotides. The highly random nature of these systems limits the electronic effect of neighboring purines.

13.7 Concluding Remarks

A single parameter ground state model to determine [2+2] thymine–thymine dimer quantum yields in single and double stranded DNA was presented. A similar two parameter model was presented over 50 years ago to predict [2+2] addition yields in solid state organic molecules [15]. The reduction to one parameter was validated by CASSCF results for the ethylene+ethylene system and the distance parameter in the model was calibrated using data which compared results for dT_{20} and $dA_{20}dT_{20}$. By comparing to experimental quantum yields, it was determined that two thymines which have their C5–C6 double bonds within 3.52 Å of each other will form a dimer.

Following calibration, the model was applied to determine [2+2] TT dimer yields in a variety of systems. The first set of systems were three six base pair DNA hairpins capped by a C12 linker. These hairpins were designed to test the sequence specificity of TT dimer formation. Each hairpin contains one TT step, the location of which is varied between the hairpins. The ground state geometry of the TT steps was found to be the major factor in determining TT dimer yields given that quantum yield predictions from our model agree with the order of magnitude found in experiment. The experimental quantum yield trend within these three hairpins was not, however, reproduced by our model. This is likely due to electronic coupling between the TT steps and neighboring adenines which is in no way incorporated into our model.

The next set of systems investigated were DNA sequences containing locked nucleic acids (LNAs) at the TT steps. Comparing these sequences to the unlocked sequences allowed both simulation and experiment to probe the effect of limiting ground state conformation of the TT step. A large increase in TT dimer quantum yield was found for all locked TT steps in both theory and experiment. This effect was amplified in single stranded DNA (ssDNA) as compared to double stranded DNA (dsDNA) because rigidity is already present in the non-locked TT steps of dsDNA.

The final set of systems investigated were four trinucleotides. Each trinucleotide contained a TT step with a purine base on either the 3' or 5' side. Adenine and guanine were the two purine bases investigated. These systems were of interest due to the TT dimer quenching ability of neighboring purine bases. This effect is thought to be electronic in nature thus suggesting that our model would be insufficient to reproduce experimental TT dimer quantum yields. In fact, our model accurately predicted not only the absolute value of the quantum yields but also the relative trend in quantum yields among the four trinucleotides. Ground state conformation is likely the most important aspect in these systems due to the large fluctuations seen in each trinucleotide structure preventing significant electronic coupling between the TT step and the neighboring purine.

Ground state conformation plays a huge role in determining the amount of TT dimer formed at a particular TT step. This model is obviously simplistic and has a variety of failings. One of these is that given a geometric cutoff (or set of cutoffs) we assume that the dimer forms with a quantum yield of one. Realistically, the T+T excited state must go through a conical intersection to form the TT dimer. As the excited T+T passes through the conical intersection it has some probability of forming the TT dimer and some probability of going back to T+T. This is a fairly minor issue because recalibration will only change the cutoff value(s) slightly. Another important failing of this model is suggested by the sequence specificity section. When there is significant electronic coupling between the TT step and the neighboring base a variety of excited state processes can occur which are not taken into account in the ground geometry model. Despite these failings, our model has been shown to accurately predict TT dimer yields in a variety of systems, thus proving that the ground state geometry of the TT step prior to excitation is of pivotal importance to the quantum yield of dimerization.

Acknowledgements The authors would like to thank Fred Lewis, Mahesh Hariharan and Zhengzheng Pan for performing almost all of the experiments referred to in this chapter and numerous useful discussions. Funding for this research was provided by the National Science Foundation (NSF-CRC Grant CHE-0628130).

References

1. Shukla MK, Leszczynski J (2008) Radiation induced molecular phenomena in nucleic acids: a brief introduction. In: Shukla MK, Leszczynski J (eds) Radiation induced molecular phenomena in nucleic acid. Springer, Dordrecht/London, pp 1–14

2. Crespo-Hernandez CE, Cohen B, Hare PM, Kohler B (2004) Ultrafast excited-state dynamics in nucleic acids. Chem Rev 104:1977
3. Shukla MK, Leszczynski J (2008) Radiation induced molecular phenomena in nucleic acids. Springer, Dordrecht/London
4. Beukers R, Berends W (1960) Isolation and identification of the irradiation product of thymine. Biochim Biophys Acta 41:550
5. Beukers R, Eker APM, Lohman PHM (2008) 50 years thymine dimer. DNA Repair 7:530
6. Setlow RB (1966) Cyclobutane-type pyrimidine dimers in polynucleotides. Science 153:379
7. Taylor JS, Brockie IR, O'Day CL (1987) A building block for the sequence-specific introduction of cis-syn thymine dimers into oligonucleotides. solid-phase synthesis of TpT[c,s]pTpT. J Am Chem Soc 109:6735
8. Marguet S, Markovitsi D (2005) Time-resolved study of thymine dimer formation. J Am Chem Soc 127:5780–5781
9. Schreier WJ, Kubon J, Regner N, Haiser K, Schrader TE, Zinth W Clivio P, Gilch P (2009) Thymine dimerization in DNA model systems: cyclobutane photolesion is predominantly formed via the singlet channel. J Am Chem Soc 131:5038
10. Schreier WJ, Schrader TE, Koller FO, Gilch P, Crespo-Hernandez CE, Swaminathan VN, Carell T, Zinth W, Kohler B (2007) Thymine dimerization in DNA is an ultrafast photoreaction. Science 315:625–629
11. Johnson AT, Wiest O (2007) Structure and dynamics of poly(T) single-strand DNA: implications toward CPD formation. J Phys Chem B 111:14398–14404
12. Law YK, Azadi J, Crespo-Hernandez CE, Olmon E, Kohler B (2008) Predicting thymine dimerization yields from molecular dynamics simulations. Biophys J 94:3590–3600
13. McCullagh M, Hariharan M, Lewis FD, Markovitsi D, Douki T, Schatz GC (2010) Conformational control of TT dimerization in DNA conjugates. A molecular dynamics study. J Phys Chem B 114:5215–5221
14. Hariharan M, McCullagh M, Schatz GC, Lewis FD (2010) Conformational control of thymine photodimerization in single-strand and duplex DNA containing locked nucleic acid TT steps. J Am Chem Soc 132:12856–12858
15. Cohen MD, Schmidt GMJ (1964) Topochemistry. Part I. A Survey. 1996–2000
16. Ramamurthy V, Venkatesan K (1987) Photochemical-reactions of organic-crystals. Chem Rev 87:433
17. Neidle S (1999) Oxford handbook of nucleic acid structure. Oxford University Press, Oxford/New York
18. Zhang RB, Eriksson LA (2006) A triplet mechanism for the formation of cyclobutane pyrimidine dimers in UV-irradiated DNA. J Phys Chem B 110:7556–7562
19. Kwok W-M, Ma C, Phillips DL (2008) A doorway state leads to photostability or triplet photodamage in thymine DNA. J Am Chem Soc 130:5131
20. Bernardi F, Olivucci M, Robb MA (1990) Predicting forbidden and allowed cycloaddition reactions – potential surface-topology and its rationalization. Acc Chem Res 23:405
21. Bernardi F, De S, Olivucci M, Robb MA (1990) Mechanism of ground-state-forbidden photochemical pericyclic-reactions – evidence for real conical intersections. J Am Chem Soc 112:1737
22. Dallos M, Lischka H, Shepard R, Yarkony, DR, Szalay PG (2004) Analytic evaluation of nonadiabatic coupling terms at the mr-ci level. ii. minima on the crossing seam: formaldehyde and the photodimerization of ethylene. J Chem Phys 120:7330
23. Boggio-Pasqua M, Groenhof G, Schaefer LV, Grubmuller H, Robb MA (2007) Ultrafast deactivation channel for thymine dimerization. J Am Chem Soc 129:10996
24. Du H, Fuh RA, Li AC, Lindsey JS (1998) Photochemcad: a computer-aided design and research tool in photochemistry. Photochem Photbiol 68:141
25. Markovitsi D, Gustavsson T, Talbot F (2007) Excited states and energy transfer among DNA bases in double helices. Photochem Photobiol Sci 6:717
26. Case DA et al (2004) Amber 8. University of California, San Francisco

27. Kale L, Skeel R, Bhandarkar M, Brunner R, Gursoy A, Krawetz N, Phillips J, Shinozaki A, Varadarajan K, Schulten K (1999) NAMD2: greater scalability for parallel molecular dynamics. J Comput Phys 151:283–312
28. Foloppe N, MacKerell AD (2000) All-atom empirical force field for nucleic acids: I. Parameter optimization based on small molecule and condensed phase macromolecular target data. J Comput Chem 21:86–104
29. MacKerell AD, Banavali NK (2000) All-atom empirical force field for nucleic acids: II. Application to molecular dynamics simulations of DNA and RNA in solution. J Comput Chem 21:105–120
30. Gale JM, Nissen KA, Smerdon MJ (1987) UV-induced formation of pyrimidine dimers in nucleosome core DNA is strongly modulated with a period of 10.3 bases. Proc Natl Acad Sci U S A 84:6644–6648
31. Gale JM, Smerdon MJ (1990) UV induced (6-4) photoproducts are distributed differently than cyclobutane dimers in nucleosomes. Photochem Photobiol 51:411–417
32. Mitchell DL, Nguyen TD, Cleave JE (1990) Nonrandom induction of pyrimidine-pyrimidone (6–4) photoproducts in ultraviolet-irradiated human chromatin. J Biol Chem 265:5353–5356
33. Segal E, Fondufe-Mittendorf Y, Chen L, Thastrom A, Field Y, Moore IK, Wang J-PZ, Widom J (2006) A genomic code for nucleosome positioning. Science 442:772
34. Bourre F, Renault G, Seawell PC, Sarasin A (1985) Distribution of ultraviolet-induced lesions in simian virus 40 DNA. Biochimie 67:293–299
35. Becker MM, Wang Z (1989) Origin of ultraviolet damage in DNA. J Mol Biol 210:429–438
36. Hariharan M, Lewis FD (2008) Context-dependent photodimerization in isolated thymine–thymine steps in DNA. J Am Chem Soc 130:11870–11871
37. Desnous C, Babu BR, Moriou C, Mayo JUO, Favre A, Wengel J, Clivio P (2008) The sugar conformation governs (6–4) photoproduct formation at the dinucleotide level. J Am Chem Soc 130:30–31
38. Pande V, Nilsson L (2008) Insights into structure, dynamics and hydration of locked nucleic acid (LNA) strand-based duplexes from molecular dynamics simulations. Nucleic Acids Res 36:1508–1516
39. Lu X-J, Olson WK (2003) 3DNA: a software package for the analysis, rebuilding and visualization of three-dimensional nucleic acid structures. Nucleic Acids Res 31:5108–5121
40. Nielsen KE, Singh SK, Wengel J, Jacobsen JP (2000) Solution structure of an LNA hybridized to DNA: NMR study of the d(CT^{L}GCT^{L}T^{L}CT^{L}GC):d(GCAGAAGCAG) duplex containing four locked nucleotides. Bioconjug Chem 11:228–238
41. Egli M, Minasov G, Teplova M, Kumar R, Wengel J (2001) X-ray crystal structure of a locked nucleic acid (LNA) duplex composed of a palindromic 10-mer DNA strand containing one LNA thymine monomer. J Chem Commun 7:651–652
42. Ivanova A, Rosch N (2007) The structure of LNA:DNA hybrids from molecular dynamics simulations:the effect of locked nucleotides. J Phys Chem A 111:9307–9319
43. Crespo-Hernandez CE, Cohen B, Kohler B (2005) Base stacking controls excited-state dynamics in A-T containing DNA. Nature 436:1141–1144
44. Chinnapen DJF, Sen D (2004) A deoxyribozyme that harnesses light to repair thymine dimers in DNA. Proc Natl Acad Sci U S A 101:65–69
45. Cannistraro VJ, Taylor J-S (2009) Acceleration of the 5-methylcytosine deamination in cyclobutane dimers by G and its implications for UV-induced C-to-T muataion hotspots. J Mol Biol 392:1145–1157
46. Pan Z, McCullagh M, Schatz GC, Lewis FD (2011) Conformational control of thymine photodimerization in purine-containing trinucleotides. J Phys Chem Lett 2:1432–1438
47. Monajjemi M, Ketabi S, Zadeh MH, Amiri A (2006) Simulation of DNA bases in water: comparison of the monte carlo algorithm with molecular mechanics force fields. Biochemistry (Moscow) 71:S1–S8

Chapter 14
Excited State Structural Analysis: TDDFT and Related Models

A.V. Luzanov and O.A. Zhikol

Abstract We review and further develop the excited state structural analysis (ESSA) which was proposed many years ago [Luzanov AV (1980) Russ Chem Rev 49: 1033] for semiempirical models of $\pi\pi^*$-transitions and which was extended quite recently to the time-dependent density functional theory. Herein we discuss ESSA with some new features (generalized bond orders, similarity measures etc.) and provide additional applications of the ESSA to various topics of spectrochemistry and photochemistry. The illustrations focus primarily on the visualization of electronic transitions by portraying the excitation localization on atoms and molecular fragments and by detailing excited state structure using specialized charge transfer numbers. An extension of ESSA to general-type wave functions is briefly considered.

Keywords Charge transfer • Conjugated systems • Density functional theory • Localization • Photochemistry

14.1 Introduction

Electronic excited states of molecules are fundamental objects of the current photophysical and photochemical techniques which make it possible to study them in many fine details [1–4]. Regarding the possibilities of applications of computational chemistry in this field, we refer the reader to reviews [5–10]. From them one can infer that the modern quantum chemistry is capable of describing low-lying electronic excitations with a satisfactory accuracy, provided suitable

A.V. Luzanov (✉) • O.A. Zhikol
STC "Institute for Single Crystals" of National Academy of Sciences of Ukraine,
60 Lenin ave, Kharkiv 61001, Ukraine
e-mail: luzanov@xray.isc.kharkov.com; zhikol@xray.isc.kharkov.com

J. Leszczynski and M.K. Shukla (eds.), *Practical Aspects of Computational Chemistry I: An Overview of the Last Two Decades and Current Trends,*
DOI 10.1007/978-94-007-0919-5_14, © Springer Science+Business Media B.V. 2012

configuration interaction (CI) schemes are used. Many efforts were undertaken to improve reliability and applicability of low-level and moderate-level methods based on the truncated CI schemes such as singles CI (CIS) [11], singles and doubles CI (CISD) methods etc. Within such methodology, CIS(D), that is, CIS with perturbative doubles correction [12], is especially tractable for large-scale systems. More efficient, but computationally more demanding, is the so-called SAC-CI (symmetry-adapted cluster CI method) scheme [5]. SAC-CI can be viewed as an alternative way in excited state theory within the coupled-cluster singles and doubles (CCSD) approximation based on the equation-of-motion (EOM) approach (EOM-CCSD and related techniques [9]).

In the last two decades the density functional theory (DFT) techniques became an effective and seemingly simple philosophy for an approximate treatment of a very wide range of theoretical chemistry and physical chemistry problems [13–15]. Now a 100 or more papers appear annually on the DFT theory of excited states. Most of them are made within the conventional time-dependent DFT (TDDFT) [8]. Notice that the formal (algebraic) grounds of TDDFT are very similar to those of RPA (random phase approximation) [16], the latter being based on the Hartree-Fock linear response theory. At the same time, RPA equations can be somehow connected to the CIS theory [17]. Roughly speaking, RPA is a certain transformation of CIS in a hyperbolic metric. More specifically, RPA and TDDFT solutions can be presented in the form of a doubled set of CIS-like solutions in the Euclidian metric (see the next section). We will see that for many practical approaches (CIS, RPA, CIS(D), and TDDFT) their algebraic structure holds similarity in many respects. Therefore, the widely used TDDFT approach may be, in a sense, transformed into the general CIS theory and its application-oriented techniques.

Now we turn to another side of computational studies. Let excitation energies and spectral line intensities be computed more or less satisfactorily. Usually, however this is not sufficient because it seldom gives the solution which is clearly interpreted in chemical or chemistry-related terms. Really, it is not easy to grasp the corresponding wave functions by inspecting very huge data sets (in particular, expansion coefficients of different kinds). In other words, "to calculate is not in itself to analyze" [18]. For this reason, in quantum chemistry, various interpretive and visualization tools were designed to provide efficient ways to understand the computed electronic states (see [19–21] and references therein). As to the CIS model proper, the first internally consistent interpretive tool was apparently proposed in [22] and then extended in [23]. This tool, named also the excited state structural analysis (ESSA) [17] was frequently applied in semiempirical theories of optical and photochemical properties [24–35]. Recently it was extended to the ab initio CIS [36] and to RPA and TDDFT descriptions [17]. The aim of this chapter is to present typical formal results and practical applications of ESSA and the related schemes. In this exposition we will try to suppress much of matrix machinery, yet retaining and generalizing the most important constructs and adding some new features to the ESSA.

The plan of the chapter is as follows. The next section provides a very brief overview of the CIS and RPA theories in the matrix-covariant representation ensuring basis-independent formulation (for the full AO or MO sets). Section 14.3

introduces the principal machinery of our ESSA analysis. In Sect. 14.4 we consider additional physical aspects of the introduced indices and also supplement the ESSA by similarity measures (statistical overlaps) and generalized Wiberg bond indices for excites states. Sections 14.5 and 14.6 are more oriented to practitioners in the area of real computational chemistry problems, and they can be read almost independently of the preceding ones. In Sect. 14.5 various molecular systems are examined, and in Sect. 14.6 ESSA in the close context of typical photochemical issues is given. Section 14.7 presents an extension of ESSA beyond the CIS-like models. Finally, in the last section, we conclude the chapter with some remarks concerning a further progress in the TDDFT technique and its interpretation. In Appendix we give details on the tight-binding model of interacting chromophores in a trimer.

14.2   CIS and Related RPA and TDDFT Methods

In this work CIS serves mainly as a basic pattern for deriving electronic properties and structural indices of the RPA and TDDFT excited states. In what follows we will sketch the results of the cited paper [17] which itself is based on previous works [22, 23]. The matter below is also necessary to understand difficulties in interpreting TDDFT solutions and possible ways for overcoming them.

In a sense, CIS is a certain counterpart of the (1-electron) Hartree-Fock method for excited state. Indeed, the Hartree-Fock (HF) variational parameters can be packed into an 1-electron matrix, C, of the conventional occupied MO expansion coefficients. In turn, the CIS configurational coefficients comprise the matrix which is just the first variational derivative of the C-matrix. More exactly, given the ground state N-electron Slater determinant $|\Phi\rangle \equiv |\Phi(1\ldots N)\rangle$. This $|\Phi\rangle$ is the antisymmetrized product of the spin-orbitals, the latter being the standard spinless spatial MOs, $|\phi_i^\circ\rangle$, equipped with spin variables. As usually, MOs $|\phi_i^\circ\rangle$ are constructed as LCAO (linear combinations of atomic orbitals (AOs) $|\chi_\mu^\circ\rangle$):

$$|\phi_i^\circ\rangle = \sum_{\mu=1}^{\dim} |\chi_\mu^\circ\rangle c_{\mu i}^\circ \tag{14.1}$$

where dim is the size of an AO basis set. For simplicity, these AOs are assumed to be orthonormal. Furthermore, we will use similar notation for spin-orbitals, so that $|\phi_i\rangle$ is the ith occupied spin-orbital. Accordingly, $|\phi_a\rangle$ and $|\phi_a^\circ\rangle$ will denote the ath vacant spin-orbital and ath vacant spinless orbital, respectively. For the general spin-orbitals $|\phi_i\rangle$ we imply the validity of LCAO representation as in Eq. 14.1, so the above matrix C is in fact

$$C = \| C_{\mu i} \| \tag{14.2}$$

where index μ in $C_{\mu i}$ is related to the atomic spin orbital $|\chi_\mu\rangle$.

Now we accept this $|\Phi\rangle$ as an admissible model of the ground state. Then we can easily construct a model of low-lying excited states. Let the molecule under consideration be perturbed by arbitrary 1-electron perturbation. Then $|\Phi\rangle$ is varied, and in the first order in the perturbation the correction term takes the form of a superposition of singly excited configurations $|\Phi_{i \to a}\rangle$. This is a rather general pattern for the excited state wave functions deviating least from the ground state determinant $|\Phi\rangle$. We conventionally define the corresponding excited state vector $|\Phi^*\rangle$, that is the CIS state vector, as follows:

$$|\Phi^*\rangle = \sum_{i=1}^{N_{occ}} \sum_{a=1}^{N_{vac}} |\Phi_{i \to a}\rangle \, \tau_{ai}, \tag{14.3}$$

where N_{occ} and N_{vac} are numbers of active occupied and vacant spin-orbitals. The configurational coefficients τ_{ai} constitute the key 1-electron matrix of the CIS theory, namely

$$\tau = \| \tau_{ai} \|. \tag{14.4}$$

With this matrix, the 1-electron operator can be commonly associated by the rule

$$\tau = \sum_{i=1}^{N_{occ}} \sum_{a=1}^{N_{vac}} \tau_{ai} \, |\phi_a\rangle \langle \phi_i|. \tag{14.5}$$

As usual, we will use the same symbols for matrices and relevant operators.

The condition for $|\Phi^*\rangle$ to be normalized is the corresponding normalization condition

$$\| \tau \|^2 = 1, \tag{14.6}$$

where for a given operator Z the squared norm is generally defined as follows:

$$\| Z \|^2 = \mathrm{Tr} \, Z^+ Z = \sum_{k,l} |Z_{kl}|^2. \tag{14.7}$$

On the other side, $\{\tau_{ai}\}$ is a set of linear variational parameters, and the standard variational procedure leads to the special eigenvalue problem that determines excitation energies and associated τ−matrices. Owing to (14.6), the individual contributions $|\tau_{ai}|^2$ can be treated as probabilities that is important for our interpretive models.

For completeness, we introduce the main operator ρ needed in the HF theory. It is produced, from N_{occ} given occupied spin-orbitals $|\phi_i\rangle$, as the projector operator of the form

$$\rho = \rho^2 = \sum_{i=1}^{N_{occ}} |\phi_i\rangle \langle\phi_i|. \tag{14.8}$$

This ρ is exactly the 1-electron density matrix of determinant $|\Phi\rangle$ (it is usually named the Dirac or Dirac-Fock matrix). Following Löwdin [37], we can term it the fundamental idempotent. The τ−matrices basically differ from the Dirac-Fock matrix in its algebraic nature since

$$\tau^2 = 0 \tag{14.9}$$

in virtue of orthogonality $\langle\varphi_i|\varphi_a\rangle = 0$. For this reason, τ will be termed the (fundamental) nilpotent.

What is important here is that similar objects can be constructed for RPA and TDDFT. At this stage we need only one basic result of the reduced density matrix (RDM) theory for determinant states. Let ρ and $\tilde{\rho}$ be two possible Dirac-Fock matrices of the respective nonorthogonal determinants, $|\Phi\rangle$ and $|\tilde{\Phi}\rangle$. Then it is possible to present

$$\tilde{\rho} = (\rho + \tilde{\tau})\frac{I}{I + \tilde{\tau}^+\tilde{\tau}}(\rho + \tilde{\tau}^+) \tag{14.10}$$

where $\tilde{\tau}$ is an admissible nilpotent of type (14.5). This is the remarkable McWeeny formula [38] which is sometimes rediscovered [16]. If τ is small then

$$\tilde{\rho} \approx \rho + \tilde{\tau} + \tilde{\tau}^+, \tag{14.11}$$

and $|\tilde{\Phi}\rangle$ is a small variation of $|\Phi\rangle$. More exactly, τ can be viewed as a generator of transformation $|\Phi\rangle \rightarrow |\tilde{\Phi}\rangle$, producing the first variation $|\delta\,\Phi\rangle = |\tilde{\Phi}\rangle - |\Phi\rangle$, namely

$$|\delta\,\Phi\rangle = \sum_{1\leqslant k\leqslant N} \tilde{\tau}(k)\,|\Phi\rangle, \tag{14.12}$$

where $\tilde{\tau}(k)$ is the operator (14.5) acting on the kth electron. Likewise, CIS state vector (14.3) is

$$|\Phi^*\rangle = \sum_{1\leqslant k\leqslant N} \tau(k)\,|\Phi\rangle. \tag{14.13}$$

Hence, these ρ and τ can be regarded as the principal algebraic quantities containing all needed information about $|\Phi\rangle$ and $|\Phi^*\rangle$.

The above situation is aggravated when one goes to RPA. Indeed, RPA equations can be deduced within the linear response HF theory for the ground state, and the final result is equivalent to that involving two coupled correction wave functions of the type (14.13). Accordingly, a RPA excited state is described by two special nilpotents. Let these matrices be denoted by X and Y, viz.

$$X = \| X_{ai} \|, \ Y = \| Y_{ai} \|. \tag{14.14}$$

They can be determined using the special variational principle due to Thouless [39]. The principal distinction between RPA and CIS solutions is that the RPA fundamental nilpotents (14.14) are not usually normalized. Instead of Eq. 14.6 one must impose the indefinite metric condition

$$\| X \|^2 - \| Y \|^2 = 1. \tag{14.15}$$

This unfavorable condition precludes us from defining correctly the probabilities analogous to $|\tau_{ai}|^2$. More than that, for RPA solutions we are not able, in principle, to construct the well defined RDMs, and the same is true for TDDFT. These difficulties can be formally overcome by mimicking the CIS method [17].

The trick we have used rests on the following observation. The RPA eigenvalue problem is non-Hermitian, of the form

$$\left[\begin{pmatrix} \hat{A} & \hat{B} \\ \hat{B} & \hat{A} \end{pmatrix} - \omega \begin{pmatrix} I & 0 \\ 0 & -I \end{pmatrix} \right] \begin{pmatrix} X \\ Y \end{pmatrix} = 0, \tag{14.16}$$

where ω is an excitation energy, and $\hat{A}$ and $\hat{B}$ are superoperators acting in a linear space of nilpotents of type (14.5) (see [17] for more detail). The non-Hermitian problem (14.16) can be replaced by the equivalent Hermitian one:

$$\begin{pmatrix} \hat{A} & \hat{B} \\ \hat{B} & \hat{A} \end{pmatrix}^{1/2} \begin{pmatrix} I & 0 \\ 0 & -I \end{pmatrix} \begin{pmatrix} \hat{A} & \hat{B} \\ \hat{B} & \hat{A} \end{pmatrix}^{1/2} \begin{pmatrix} X_0 \\ Y_0 \end{pmatrix} = \omega \begin{pmatrix} X_0 \\ Y_0 \end{pmatrix}, \tag{14.17}$$

where the new coupled nilpotents X_0, Y_0 are now subjected to the Euclidean normalization condition

$$\| X_0 \|^2 + \| Y_0 \|^2 = 1. \tag{14.18}$$

The such formulation is quite suitable for reinterpreting RPA and TDDFT in CIS-like terms. The above nilpotents X_0 and Y_0 are explicitly expressed via nilpotents X and Y by Eqs. 4.9, 4.12 and 4.18 in [17].

14.3 Main Structural Indices

At this point, we elucidate what one can gain from the RDM study of CIS and related approaches. A rather exhaustive formal examination of the CIS density matrices was given long ago in [40]. One can also find some related results in [11]. On the interpretive side of the CIS and related models, see reviews [6, 23, 41]. In this section we outline in more detail the notions and methods that have proved most useful for ESSA.

We start with definitions and focus on the most important RDMs, that is the first order RDM, D_1, and the second order RDM, D_2. These RDMs are now the standard entities widely used in modern quantum chemistry [42, 43]. The main relations are 1-electron averages

$$\left\langle \sum_{1 \leqslant k \leqslant N} h(k) \right\rangle = \text{Tr} \, h D_1, \tag{14.19}$$

and 2-electron averages

$$\left\langle \sum_{1 \leqslant k < l \leqslant N} g(kl) \right\rangle = \text{Tr} \, g D_2 \tag{14.20}$$

In above, $\langle \rangle$ stands for averaging over the electronic state under study, $h = h(1)$ is arbitrary 1-electron operator, and $g = g(12)$ is a 2-electron operator.

Having at disposal D_1 and D_2 of the given state vector, one straightforwardly yields the corresponding energy functional and physical observables. For instance, in case of the HF model we have the well known representations $D_1^{HF} = \rho$ and $D_2^{HF} = A_2\rho(1)\rho(2)$ where 2-electron antisymmetrizer $A_2 = (I - P_{12})/2$ with P_{12} being the usual transposition operator. The results for CIS can be obtained fairly readily. In terms of the fundamental nilpotent (14.5) we have

$$D_1^{CIS} = \rho + \tau\tau^+ - \tau^+\tau, \tag{14.21}$$

Somewhat more complicated expression is derived for D_2^{CIS} (e.g., see Eqs. 24 in [40] or 3.14 in [17]).

It is well known that the interpretation in terms of natural spin orbitals (or simply, natural orbitals) is very efficient [37, 42, 43]. We observe that eigenvectors of D_1^{CIS} in (14.21), that is the CIS natural orbitals, arise from two independent contributions, $\tau\tau^+$ and $\tau^+\tau$. Let us designate these natural orbitals by $|\alpha^*\rangle$ and $|\alpha\rangle$, so $|\alpha^*\rangle$ and $|\alpha\rangle$ belong to the same eigenvalue μ_α. It means that D_1^{CIS} allows the spectral resolution

$$D_1^{CIS} = \sum_{\alpha=1}^{N} \{(1 - \mu_\alpha) \, |\alpha\rangle \, \langle\alpha| + \mu_\alpha \, |\alpha^*\rangle \, \langle\alpha^*|\}, \tag{14.22}$$

that is long ago established in more general form [44, 45]. Furthermore, in [22, 45] the so-called singular value decomposition for τ was in fact given (it was termed in [45] the representation of singular numbers). It is tantamount, in the above notation, to the following expression:

$$\tau = \sum_{\alpha=1}^{N} \sqrt{\mu_\alpha} \, |\alpha^*\rangle \, \langle\alpha| \,. \tag{14.23}$$

It is interesting that these results were recently rediscovered [46, 47]. At present, such CIS or CIS-like natural orbitals are frequently termed "transition orbitals" (see [46, 48] and similar works on TDDFT calculations).

We consider also a consequence of the above decomposition for the CIS state vector. Substituting (14.23) into (14.13) comes to the compact representation

$$|\Phi^*\rangle = \sum_{\alpha=1}^{N} \sqrt{\mu_\alpha} \, |\Phi_{\alpha\to\alpha^*}\rangle, \qquad (14.24)$$

where $|\Phi_{\alpha\to\alpha^*}\rangle$ is the 'natural' singly excited configuration. It is constructed by the substitution $|\alpha\rangle \to |\alpha^*\rangle$ in the reference determinant $|\Phi\rangle$ which can be re-expressed in terms of natural occupied states $\{|\alpha\rangle\}$ (see also a direct derivation in [22]).

This topic is related to the problem how many significant terms $|\Phi_{\alpha\to\alpha^*}\rangle$ in (14.24) can effectively characterize $|\Phi^*\rangle$. The corresponding effective number of CIS configurations is closely connected with the notion of average rank of matrices. The first solution of the problem was given in [49] where the so-called collectivity number at CIS level, κ, was defined. Then the idea of the average rank was extended to the general CI-type wave functions [50, 51]. The CIS collectivity number is readily computed as follows

$$\kappa = 1/\mathrm{Tr}(\tau^+\tau)^2. \qquad (14.25)$$

This matrix invariant suitably presents a many-configurational nature of the CIS state. In particular, it permits us to discern a false collectivity which might be suggested by a superficial inspection of configurational coefficients. The example of the state with all identical amplitudes ($\tau_{ai} \equiv const$) [22] clearly demonstrates the point. In this case, $\kappa = 1$ whereas the corresponding set $\{\tau_{ai}\}$ tells us about seemingly the most involved CI state. This shows that a straightforward use of some non-invariant entities, such as MOs or configuration coefficients, should be avoided, and the ESSA is just a route, at least at formal level, to make the analysis in invariant terms which are produced from the correctly defined matrix entities. Of course, in practice, the old 'basis-dependence curse' remains unbroken with most available interpretive schemes, and here we cannot discuss this morbid issue in detail.

The excitation localization indices are the main quantities in ESSA, and we describe them more completely. Before giving some specific relations, we briefly notice that the structural-chemistry interpretation of excited states is in conformity with the rich chemical and spectrochemical experience. Really, the latter conclusively shows that molecular systems can possess separated fragments (subunits) even in excited states (see e.g. [52–54]). Therefore, it was practically important to estimate a measure of excitation localization in one or another way. The technique of excitation localization indices [22] and charge transfer numbers [23, 55] opened a possibility for an internally consistent quantum description of localization phenomena in spectrochemistry. Initially this was applied to the CIS π-electron model. Notice that more elementary, but not invariant, scheme was earlier proposed in [56].

Within CIS the localization problem is directly solved by using the above RDM results. We see from inspecting (14.21) that

$$\Delta D \equiv \tau\tau^+ - \tau^+\tau \qquad (14.26)$$

is a deviation of 1-electron density matrix from that of the ground state. Thus, to extract the desired information from this ΔD matrix, all one needs is to form the positive definite matrix $|\Delta D|$ instead of ΔD. Then we arrive at the simple result

$$|\Delta D| = \tau\tau^+ + \tau^+\tau, \qquad (14.27)$$

that is correct in the usual matrix algebraic sense. It is important that this $|\Delta D|$ automatically possesses the size-consistent normalization to two. Therefore the permissible normalized localization operator is

$$\hat{L}^* = (\tau\tau^+ + \tau^+\tau)/2. \qquad (14.28)$$

Its diagonal matrix element, in the AO basis set,

$$L^*_{\mu\mu} = \langle\chi_\mu| \tau\tau^+ + \tau^+\tau |\chi_\mu\rangle/2, \qquad (14.29)$$

provides a suitable measure of localizing excitation on the given spin-AO $|\chi_\mu\rangle$. Then the index

$$L^*_A = \sum_{\mu\in A} L^*_{\mu\mu} \qquad (14.30)$$

is the gross measure of the excitation localization on the molecular fragment A. These excitation indices are correctly normalized, viz.

$$\sum_A L^*_A = 1, \qquad (14.31)$$

that permits interpreting L^*_A in probability terms.

At this point, it is pertinent to compare the above excitation index method with the approaches based on visualizing the transition orbitals [46]. When using L^*_A-indices for the given excitation one need displaying only the full $\{L^*_A\}$ excitation distribution. In the transition orbitals method, to be consistent, one must display all the significant orbitals, which makes it difficult to get the united picture. In fact, such situations arise whenever collectivity number (14.25) is markedly greater than one. From the results of Sect. 14.5 we can conclude that these situations occur rather frequently (see in particular the instructive analysis of triplet states of terphenyl in Sect. 14.5.4).

The next useful indices are charge transfer (CT) indices. The idea here is to think of the AO matrix element,

$$\tau_{\mu\nu} = \langle \chi_\mu | \tau | \chi_\mu \rangle, \tag{14.32}$$

as being a probability amplitude for the electron transfer from $|\chi_\nu\rangle$ to $|\chi_\mu\rangle$. Explicitly

$$\tau_{\mu\nu} = \sum_{i=1}^{N_{occ}} \sum_{a=1}^{N_{vac}} \tau_{ai} C_{\mu a} C_{\nu i}^*, \tag{14.33}$$

where $C_{\mu a}$ and $C_{\mu i}$ are the LCAO coefficients as in (14.2). Then, the number

$$l_{A\rightarrow B} = \sum_{\mu\in A} \sum_{\nu\in B} |\langle \chi_\nu | \tau | \chi_\mu \rangle|^2. \tag{14.34}$$

is identified as a desired CT number giving a probability of electron transfer from A to B. The interrelation between the full localization indices L_A^* and the CT numbers $l_{A\rightarrow B}$ are easily elucidated [23, 51]:

$$L_A^* = l_A + \frac{1}{2} \sum_{B\neq A} l_{A\rightarrow B}, \tag{14.35}$$

where $l_A = l_{A\rightarrow A}$.

Indices l_A and $l_{A\rightarrow B}$ give also a simple way to estimate the relative weights of locally excited and CT configurations. The total weight of local excitations, symbolized by $|\Psi_A^* \Psi_B \ldots\rangle$, is identified with l_A, and the weight of CT excitations, that is $|\Psi_A^+ \Psi_B^- \ldots\rangle$, with $l_{A\rightarrow B}$. This interpretation is in concordance with the interrelation, proven rigorously [55], between $l_{A\rightarrow B}$ and the change, ΔD_A, of the electronic population on the fragment A:

$$\Delta D_A = \sum_{B\neq A} (l_{B\rightarrow A} - l_{A\rightarrow B}). \tag{14.36}$$

All these indices can be readily computed within RPA and TDDFT by making a use of the coupled nilpotents X_0, Y_0 described above. For instance,

$$l_{A\rightarrow B} = \sum_{\mu\in A} \sum_{\nu\in B} (|\langle \chi_\nu | X_0 | \chi_\mu \rangle|^2 + |\langle \chi_\nu | Y_0 | \chi_\mu \rangle|^2), \tag{14.37}$$

$$\sum_{A,B} l_{A\rightarrow B} = 1, \tag{14.38}$$

are the TDDFT counterparts of CT numbers (14.34). Similarly, the other CIS quantities can be consistently reformulated for TDDFT [17]. It is worth mentioning that almost at the same time [15] the related problem of reconstructing (or rather mimicking) RDMs for RPA and TDDFT schemes was discussed as the "assignment problem". In general, the expression for ΔD, proposed in [15] (equations (II.83) and (II.84) in *loc cit*), cannot provide the desirable localization operators which would be naturally endowed with the unity norm, like $\hat{L}^*$ in (14.28). At the same time, the Y norm is typically small in practical TDDFT calculations ($\|Y\| = 10^{-2} \div 10^{-4}$ [17]), and the inaccuracy from using more simple formulations is not serious. However, in order to prevent one from possible difficulties with $\|Y\| \approx \|X\|$, the technique [17] is recommended. In passing, notice that CIS cannot treat the double CT states such as $\left|\Psi_A^{**} \Psi_B \ldots\right\rangle$ or $\left|\Psi_A^{++} \Psi_B^{--} \ldots\right\rangle$, but RPA and TDDFT implicitly involve them to some extent [6], and so do the corresponding indices L_A^* and $l_{A \to B}$.

14.4 CT and Hole-Particle Interpretation; Other Structural Indices

The nature of the above described localization indices may be additionally discussed in a more wide context which treats $|\Phi^*\rangle$ and corresponding RDMs in hole-particle terms. For arbitrary CI wave functions, a general hole-particle analysis involving high-order RDMs was given in [57]. Following this analysis we reveal the connection of $\hat{L}^*$ with hole and particle density matrices. More specifically, the hole density matrix turns out to be $D_1^h = \tau^+ \tau$, and, respectively, the particle density matrix $D_1^p = \tau \tau^+$ (see Eq. 7.1 and 7.2 in [57]), that is just two terms which constitute excitation operator $\hat{L}^*$ (14.28). In so doing we can also employ a mixed hole-particle density which proves to be $D_2^{hp} = P_{12} \tau(1) \tau(2)$ (as in Eq. 7.4 in [57]). It leads to identifying the hole-particle densities with the CT indices, that is

$$(D_2^{hp})_{\mu\nu,\mu\nu} = \left|\langle \chi_\nu | \tau | \chi_\mu \rangle\right|^2. \tag{14.39}$$

Furthermore, the problem of comparing excited states sometimes arises, and quantities familiar from probability theory can be exploited for this purpose. When needed, we apply the so-called classical fidelity measure, or statistical overlap, computed in our case from the CT probability distribution (14.37) as follows:

$$\wp_{CT}[I, J] = \sum_{A,B} \sqrt{l_{A \to B}[I] \, l_{A \to B}[J]}, \tag{14.40}$$

where $l_{A \to B}[I]$ are CT numbers (14.34) for the Ith CIS state $|\Phi_I^*\rangle$. This quantity quantitatively describes a similarity between the two excited states with wave functions $|\Phi_I^*\rangle$ and $|\Phi_J^*\rangle$. Evidently, all $\wp[I, I] \equiv 1$. Other distributions, for instance (14.39), can be used for computing statistical overlaps, but this topic is beyond the main scope of the chapter.

Now, we look again at the above hole-particle analysis by invoking another currently used approach. It is based on the quantum chemical notion termed frequently 'effectively unpaired electrons' [58–60]. The very idea of such "odd" electrons was pioneered by work [61]. In the present chapter we cannot go into much detail. The problem is how to describe correlated electrons in terms of properly defined distributions. From the viewpoint of the hole-particle analysis, the Head-Gordon approach [60] seem to be more appropriate and simple solution (see [57], Section 7a). Therefore, here we apply this approach to CIS-like models. Notice that previously this problem was not formulated for CIS explicitly, including our own studies. But in the context of the TDDFT and above given CIS-like reformulation, the inference from the such analysis becomes fairly valuable. It gives a simple conclusion that the orbital and atomic excitation indices (14.29) and (14.30) correspond precisely to the distributions of effectively unpaired electron in the given excited state.

To be more specific, we restate the approach [60] for the full (spin orbital) density matrix D_1 what is not principal for our purposes. Then the approach is in fact assumed that the special density matrix, D_1^{unpair}, associated with effectively unpaired electrons, has common natural orbitals $\{|\varphi_k\rangle\}$ with D_1. In doing so, the associated occupation numbers, $\lambda_k^{\text{unpair}}$, are to be recalculated from the D_1 natural occupation numbers, λ_k, by the rule

$$\lambda_k^{\text{unpair}} = \min[\lambda_k, \, 1 - \lambda_k]. \tag{14.41}$$

Accordingly, the spectral resolution

$$D_1^{\text{unpair}} = \sum_{k=1}^{N} \lambda_k^{\text{unpair}} |\varphi_k\rangle \langle\varphi_k|, \tag{14.42}$$

can be taken as a possible definition of D_1^{unpair} in the Head-Gordon scheme. Passing to the CIS density matrix (14.22) we observe the identity $D_1^{\text{unpair}} = |\Delta D| = 2 \hat{L}^*$ due to the specific structure of the D_1^{CIS} spectrum. It means that the excitation localization indices (14.29) and (14.30) can be indeed regarded as localization indices of the unpaired electrons. In case of the conventional CIS triplet state, this connection conforms with the fact that $\hat{L}^*$ coincides strictly with the corresponding spin-density matrix. These reasonings justify using the indices not only for the Franck-Condon transitions. Merely, the interpretation of the indices for arbitrary points on the excited-state potential energy surface is somewhat different. Namely, by these indices we detect photo-activated valence centers for any reasonable configuration of the excited molecule.

Another useful notion is the so-called generalized bond index K_{AB} referred to the given atoms A and B. According to [36], in case of CIS excited states, K_{AB}, more exactly, $K_{AB}[CIS]$, turns out to be connected with CT numbers $l_{A \to B}$ and some additional quantities. Before giving expressions applicable also to RPA and related models, we briefly clarify the meaning of the generalized bond index. Even within the elementary MO theory, such as the π-electron model, the corresponding quantity, that is $K_{\mu\nu}[MO]$, is not the same as the usual bond order $P_{\mu\nu}$. Instead, the squared bond order makes its appearance, namely, by definition the orbital index

$K_{\mu v}[\text{MO}] = (P_{\mu v})^2$. From these $K_{\mu v}[\text{MO}]$ the conventional bond index $K_{AB}[\text{MO}]$, that is the Wiberg index [62], is computed. With minor complications, similar MO quantities can be defined within ab initio MO schemes [63].

To define K_{AB} and more general multicenter indices at a post Hartree-Fock level, one must invoke more sophisticated techniques [64–67]. We follow our work [36] in which the full CI and some approximated models were considered in detail. The conventional definition of the generalized bond index is based on identification of K_{AB} with a charge density fluctuation measured via the second-order joint statistical moment [64] (generally, the joint cumulant):

$$K_{AB} = 2 \left(\langle U_A U_B \rangle - \langle U_A \rangle \langle U_A \rangle \right). \tag{14.43}$$

Here the operator $U_A = \sum_{1 \leq i \leq N} u_A(i)$ with u_A being the AO projector associated with the given atom A. In particular, $\langle U_A \rangle = (D_1)_{AA} \equiv D_A$, that is the gross electronic population on A. Manipulating (14.43) over CIS wave functions gives the compact result which follows from Eqs. 84 and 85 in [36]:

$$K_{AB}[CIS] = K_{AB}[\text{MO}] + \Delta K_{AB}[\tau], \tag{14.44}$$

$$\Delta K_{AB}[\tau] = 2 \{ l_{A \to B} + l_{B \to A} + \Delta_A \Delta_B + \partial_{AB} - (1 - \sigma) T_A T_B \}, \tag{14.45}$$

Here $\sigma = 0$ for singlet states and $\sigma = 1$ for triplet states, and furthermore

$$\Delta_A = \sum_{\mu \in A} \Delta D_{\mu\mu}, \quad \partial_{AB} = \sum_{\mu \in A} \sum_{v \in B} \Delta D_{\mu v} P_{\mu v}, \quad T_A = 4 \sum_{\mu \in A} \tau_{\mu\mu}. \tag{14.46}$$

In these relations we imply that a spin free component of τ is used (for more detail see [17]).

The extension of the above expressions to RPA and TDDFT is obvious:

$$K_{AB}[\text{RPA}] = K_{AB}[\text{TDDFT}] = K_{AB}[\text{MO}] + \Delta K_{AB}[X_0] + \Delta K_{AB}[Y_0], \tag{14.47}$$

and nilpotents X_0, Y_0 from Eq. 14.17 are now used as arguments in (14.44)–(14.46). More general open-shell electronic states with non-zero spin values are not considered in this work, but no principal difficulties are presented in such cases (if ignoring spin purity problems of the appropriate UHF extension of the theory).

14.5 ESSA of Some Generic Systems

Now we examine the practical applications of ESSA to various molecular problems. All of the TDDFT illustrations are obtained by employing the special program code [17]. In the cited work some specific details are given, and here we also exploit the same B3LYP scheme and the AO basis set of aug-cc-pvdz quality along with a slight simplification of this aug-cc-pvdz for ESSA. We can refer to this scheme as ESSA/TDDFT/B3LYP/aug-cc-pvdz. In what follows, we discuss some typical results obtained in [17] and in spectrochemical applications [68–70]. Besides, many extra molecular systems are specially examined to be added to this section. Previously the usefulness of the π-electron ESSA for solving practical spectroscopy problems was repeatedly demonstrated (see references [23–35] which are only a small part of the papers on this subject). Now, we present a rather wide range of chemically interesting examples of ESSA, adhering, at this stage, to the above defined ESSA/TDDFT/B3LYP/aug-cc-pvdz technique. It is this scheme we will imply when considering ESSA/TDDFT results below.

14.5.1 $\pi\pi^*$-Transitions

As a typical example let us consider singlet and triplet excited states in anthracene, tetracene, pyrene, and perylene, all being generic photophysical systems within a class of polycyclic aromatic hydrocarbons (PAHs). In these systems, the lowest singlet and triplet excited states have, as a rule, the same spatial symmetry B_{2u} (polarization vector along the short molecular axis). The spectral data (experiment from [71–75]) and excitation distribution $\{L_A^*\}$ diagrams are presented in Table 14.1. We see that in anthracene the two distributions are markedly similar, and the main difference is in L_A^* values on the positions 9 and 10 in the central ring: the triplet state is somewhat more localized on these sites. As a rule, the lowest excitations of linearly fused PAHs are more localized on the central aromatic rings (e.g., see results for tetracene in Table 14.1). The indicated features are in agreement with the enhanced reactivity of the central regions in the linear acene molecules when

Table 14.1 Excitation energy λ (in eV; experimental values after slash), collectivity number κ and excitation localization distribution $\{L_A^*\}$ for the lowest singlet and triplet transitions in some typical PAHs

Molecule	$^1A_{1g} \rightarrow {}^1B_{2u}$			$^1A_{1g} \rightarrow {}^3B_{2u}$		
	λ	κ	$\{L_A^*\}$	λ	κ	$\{L_A^*\}$
Anthracene	3.16 /3.27	1.03		1.80 /1.82	1.04	
Tetracene	2.39 /2.71	1.06		1.13 /1.27	1.13	
Pyrene	3.62 /3.71	1.27		2.12 /2.08	1.22	
Perylene	2.74 /2.85	1.07		1.50 /1.56	1.15	

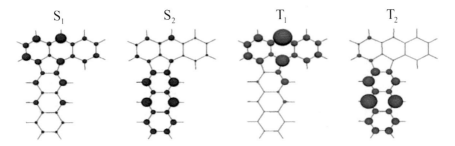

Fig. 14.1 Excitation localization distribution $\{L_A^*\}$ for low excitations in 2, 3-benznaphthfluorantene

photoaddition and photooxidation reactions take places [76]. It seems likely that for such reactions excitation localization indices L_A^* can serve as good photoreactivity indices (see also Sect. 14.6).

We briefly touch on another side of the same topic. What are the main factors which are favorable in forming a local structure of low-lying electronic excitations? A simple analysis along with π-electron calculations [23] and spectrochemical experience tell us that molecular asymmetry is a natural factor giving rise, as a rule, to a preferable localizability (even in conjugated systems). With this, some highly excited states can be notably delocalized. The ESSA within TDDFT confirms such observations. In Fig. 14.1 we present localization diagrams for two lowest transitions in 2,3-benznaphthfluorantene computed within TDDFT/B3LYP. These diagrams are rather similar to ones obtained in [17] for the same system within a more rough semiempirical π-electron approximation.

Similar excitation distributions and CT diagrams are obtained for $\pi\pi^*$ excitations in heteroaromatic systems. Some interesting systems are treated in [70] by using the ESSA/TDDFT analysis for proton phototransfer processes. In particular, for the first excited $\pi\pi^*$ state in the quinoline-substituted 3-hydroxychromone one observes the pronounced localization picture which can be revealed in the corresponding CT diagrams (see Table 2 in [69]).

14.5.2 $n\pi^*$- and $\sigma\sigma^*$-Transitions

Unlike the conjugated hydrocarbons properly, heteroatom conjugated systems allow one to observe, in addition, low-lying excited $\sigma\pi^*$ states usually referred to $n\pi^*$ states (σ−HOMO is well approximated by a lone-pair AO, that is n−AO of the corresponding heteroatom). We discuss them in more detail by considering the benzaldehyde molecule. For this molecule the lowest-lying excited state is of $n\pi^*$ type and the next one is of $\pi\pi^*$ type, and it is interesting to show the difference between them (see Fig. 14.2).

Table 14.2 Excitation localization distribution $\{L_A^*\}$ for low-lying triplet states in **I–III**

Molecule	$S_0 \rightarrow T_1$	$S_0 \rightarrow T_2$	$S_0 \rightarrow T_3$
I			
	B_1, 3.613, 2.51	B_1, 3.703 , 4.97	B_2, 4.185, 1.10
II			
	B_2, 3.279, 1.59	A_2, 3.581, 3.38	B_2, 3.657, 4.92
III			
	B_1, 2.526, 1.43	B_3, 3.123, 3.11	B_1, 3.179, 3.79

The electronic state symmetry, transition energy (in eV), and collectivity number are given at the bottom of each image

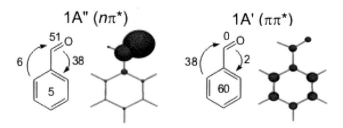

Fig. 14.2 Localization and CT numbers for the lowest $n\pi^*$ and $\pi\pi^*$ singlet excitations in formaldehyde. CT numbers are placed at arrows, and partial atomic localizations at or in the selected fragments. All values are in %

In this rather typical case the clear distinction between $n\pi^*-$ and $\pi\pi^*-$ excitations is observed. These peculiarities are easily understood in term of natural orbitals within the $\pi-$ electron CIS model [23], and it is instructive to discuss this example in more detail.

For $n\pi^*-$ transitions in carbonyl-containing molecules, $n-$ orbital is in fact the only active natural "occupied" orbital of the excited state. Among vacant orbitals a certain set of the low-lying delocalized π_a^*- orbitals are most important. When passing to the natural vacant orbitals we observe that this set of π_a^*- orbitals is just replaced by the single 'excited' natural orbital which we signify as π_{no}^*. It is an obvious result implying from the rank-1 structure of the corresponding nilpotent (14.5) (exactly its spinless counterpart):

Fig. 14.3 Transformation of
the lowest $n\pi^*$ transition to
the single configuration in the
natural orbital representation

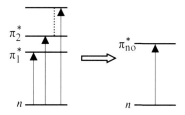

Table 14.3 Excitation distribution for low-lying singlet states in **I–III**

Molecule	$S_0 \to S_1$	$S_0 \to S_2$	$S_0 \to S_3$
I	B_2, 4.827 (0.062), 1.39	B_2, 5.081 (0.039), 2.49	A_1, 5.091 (0.010), 2.84
II	B_1, 4.425 (0.151), 1.25	B_2, 4.448 (0.000), 1.06	A_2, 4.873 (0.000), 1.03
III	B_3, 3.712 (0.002), 1.80	B_1, 3.905 (0.009), 1.10	B_3, 4.133 (0.008), 1.80

The electronic state symmetry, transition energy (in eV), oscillator strength (in parentheses), and collectivity number are given at the bottom of each image

$$\tau = \sum_{a=1}^{N_{vac}} \tau_{ai} |\pi_a^*\rangle \langle n| = |\pi_{no}^*\rangle \langle n|, \tag{14.48}$$

where

$$|\pi_{no}^*\rangle = \sum_{a=1}^{N_{vac}} \tau_{ai} |\pi_a^*\rangle. \tag{14.49}$$

Thus, instead of the above CI scheme involving N_{vac} configurations $i \to a$, we have the sole natural configuration as in Fig. 14.3.

In addition, the excitation energy consideration suggests that coulomb integral, $J(n, \pi_{no}^*)$, between this π_{no}^* and the fixed $n-$ orbital should be maximal, and this causes the π_{no}^* to be localized as possible on the same oxygen atom (we recall that for one-configuration transition $i \to a$ the excitation energy equals to $\varepsilon_a - \varepsilon_i - J(i,a)$ with $\varepsilon_i, \varepsilon_a$ being orbital energies). If we turn to $\pi\pi^*-$ transition in Fig. 14.2, we can see that this transition is fairly delocalized, and at the same time it involves significant CT processes. The fact that the lowest $n\pi^*-$ transition is strongly localized in the region nearest to the heteroatom is in agreement with

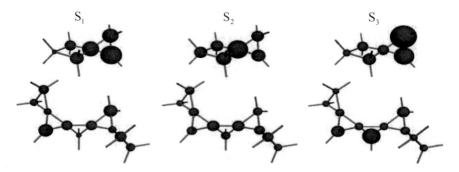

Fig. 14.4 Excitation localization distribution $\{L_A^*\}$ in spiro[2.3]hexane and [5]triangulane

molecular spectroscopy data [52]. ESSA/TDDFT for other carbonyl-containing systems confirms that the electronic transitions in such compounds share similar distinctive features. This fact is well demonstrated by the recent computations on $n\pi^*-$ states of hydroxychromones (see Table 3 in [69]).

It is also worth mentioning an alternative approach [77] to describe the localized character of $n\pi^*-$ excitations. It based on the PCILO (Perturbative Configuration Interaction using Localized Orbitals) method where CI calculations are carried out in a basis of locally excited determinants. Accordingly, the local structure of $n\pi^*-$ excitations is detected by the weights of the appropriate determinant. At the same time, our approach includes no localization procedure for MOs, and excited state local properties arise directly from the main indices (14.35), (14.37), and (14.38). Moreover, the PCILO method is somewhat outdated; at least its extension to the DFT methodology seems not reasonable. We cite work [77] as the first paper where the localization phenomena of $n\pi^*-$ transitions were discussed from a quantum-chemical viewpoint.

Rather interesting are excitation distributions in spatial (non-planar) molecular structures. Among them, spiranes present a peculiar class of hydrocarbons, and as examples we take spiro[2.3]hexane and the special spiro-linked cyclopropane aggregate (Fig. 14.4) named [5]triangulane [78].

We see that carbon spiroatoms are markedly involved in forming excitations, especially for the second transitions. In [5]triangulane the low-lying electronic transitions have a very small intensity, and in fact they are quasi-degenerated, with transition energies (in eV) calculated to be 6.35, 6.38, and 6.44, respectively. At the same time, these transitions more or less uniformly spread over cyclo-propyl moieties due to a strong topological connectivity between the localized $\sigma-$orbitals of neighbouring subunits (in organic chemistry terms one would say of σ-conjugation effects). But there are many other systems where inter-fragment interactions are weak thereby leading to specific effects considered below.

Fig. 14.5 Splitting excited states in the linear trimer

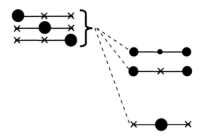

Table 14.4 The excitation distribution for low-lying singlet states in squalene

j	$S_0 \rightarrow S_j$	$S_0 \rightarrow T_j$
1	5.265 (0.001), 1.33	3.905, 2.14
2	5.325 (0.001), 1.74	3.929, 2.21
3	5.381 (0.000), 3.05	3.941, 4.21

Transition energy (in eV), oscillator strength (in parentheses), and collectivity number are given at the bottom of each image

14.5.3   Weakly Coupled Subsystems

In many realistic cases, we can regard the given system as being composed of several weakly coupled subsystems. Below we present ESSA results for three systems of this kind, namely, for 5,7,12,14-tetrahydropentacene (**I**), 5,7,12,14-tetrahydro-5,14:7,12-dimethanopentacene (**II**), and three-layered [2.2]paracyclophane (**III**), all displayed in Table 14.2. But before that, a simple reasoning can be produced when the subsystems, that is fragments or monomers (which frequently are actually chromophores capable of selective light absorption), are the same, and a counterpart of tight-binding approximation is applicable [79, 80]. For instance, consider low-lying triplet states of a linear trimer $A - A - A$. The main phenomenological parameters are the effective local shift parameter, V_*, and the nearest-neighbour exchange integral K_* (see the Appendix). From expressions given there we see that for a small K_* the lowest excitation is preferably localized on the central fragment, that is opposite to the localization of the rest excitations (Fig. 14.5).

The requirement of K_* to be small looks quite realistic for the exchange integrals with their short-distance vanishing. In fact, this model with the parameters (all in eV) $\lambda_0=3.8$, $V_*=0.25$, and $K_* =0.0675$ is in a semiquantitative agreement with TDDFT results for the lowest triplet states in **II**: compare the model energies 3.27;

3.55; 3.58 (the λ_i from the Appendix) with the relevant TDDFT data of Table 14.2. Clearly, intermediate cases exhibit when $K_*/V_* \approx 1$, but they share many of the above features. Nevertheless, the tight-binding model cannot reproduce all peculiarities of the systems in question, and in practice TDDFT calculations demonstrate more variable behaviour. It is confirmed by other results from Tables 14.2 and 14.3, along with Table 14.4 for the squalene known as an important physiological substance. Additionally, usual differences between singlet and triplet states should be also noticed. Even in systems with weakly coupled subsystems, $S_0 \rightarrow S_j$ transitions are typically more delocalized and have more significant CT components than $S_0 \rightarrow T_j$ transitions. We illustrate this fact for $S_0 \rightarrow S_1$, $S_0 \rightarrow S_2$, and $S_0 \rightarrow S_3$ in **I** by the summarized CT index values for nearest benzenoid moieties, namely by the quantity $l_{1\rightarrow 2} + l_{2\rightarrow 1}$ which equals (in %) 11, 9, and 8, respectively, that can be compared with the values 2, 1, and 4 for $S_0 \rightarrow T_1$, $S_0 \rightarrow T_2$, and $S_0 \rightarrow T_3$, respectively.

14.5.4   Intramolecular Mixing of Local and CT Excitations

Mixing local and CT excitations is a well-known effect [81] arising when inter-fragment interactions become moderate or rather strong, and the simple model consideration given in the previous subsection is no longer valid. To understand excited states more clearly, one can explicitly invoke CT numbers (14.37). Here we study how TDDFT describes the systems with the intermediate bonding of monomeric subunits. For illustrations we take the molecules composed entirely of phenyl or phenylene groups. The *para*-polyphenyls are particularly appropriate for these purposes since they are linearly arranged. Each of them has a common conjugated π-system (without significant irregularities in the $\pi - \pi$-conjugation), even though π-conjugation in polyphenyls is weakened by significant relative twisting of phenyl rings: torsion angles are about 40° in average. And yet, the inter-ring interactions suffice to cause a pronounced delocalization which manifests as well through CT components of excited states. For composite molecules such as the polyphenyls, the chemical fragments (benzenoid rings) are almost the same, and a CT character of the given excited state is conveniently captured by the total CT index

$$\mathrm{CT}_{\mathrm{tot}} = 1 - \sum_A l_A, \qquad (14.50)$$

(recall identities (14.35) and (14.38)), where summation is over all fragments (benzenoid rings in the present case). From Tables 14.5–14.7 we see that CT states contribute significantly to the singlet states, especially for long polyphenyls, and similar effects take place for the triplet states.

We also observe that in many cases of oligomeric structures, collectivity numbers (14.25) can amount to large values. Therefore, some additional words are called for

Table 14.5 Transition energy λ (in eV), oscillator strength (in parentheses), collectivity number κ, excitation distribution $\{L_A{}^*\}$, CT numbers (14.37) for intra-fragment ($l_I \equiv l_{I \to I}$) and inter-fragment ($l_{I \to II}$) interactions, and CT_{tot} (in %) in the lowest p-transitions of biphenyl

State	λ	κ	$\{L_A^*\}$	l_I	$l_{I \to II}$	CT_{tot}
$^1B_2$	4.834 (0.437)	1.10		0.24	0.26	52
$^3B_2$	3.278	1.71		0.32	0.18	36

Table 14.6 Transition energy λ (in eV), oscillator strength (in parentheses), collectivity number κ, excitation distribution $\{L_A{}^*\}$, CT numbers for intra-fragment ($l_A \equiv l_{A \to A}$) and inter-fragment ($l_{A \to B}$) interactions, and CT_{tot} (in %) in the lowest transitions of p-terphenyl

State	λ	κ	$\{L_A^*\}$	l_I	l_{II}	$l_{I \to II}$	$l_{II \to I}$	$l_{I \to III}$	CT_{tot}
$^1B_2$	4. 247 (0.900)	1.05		0.09	0.18	0.13	0.13	0.07	64
$^1B_3$	4.591 (0.000)	1.82		0.04	0.44	0.14	0.09	0.01	48
$^1B_1$	4.763 (0.000)	2.25		0.25	0.01	0.07	0.12	0.06	49
$^3B_2$	2.990	1.51		0.14	0.33	0.08	0.08	0.02	39
$^3A_1$	3. 589	3.53		0.40	0.02	0.04	0.04	0.01	18
$^3B_2$	3. 936	5.88		0.29	0.34	0.02	0.02	0.00	8

here to elucidate how collectivity numbers reflect a many-configurational nature of the excited states. As we noted in Sect. 14.3, κ can be identified with the average rank of the fundamental nilpotent, that is the average number of most important natural configurations in (14.24). The weight with which the given vector $|\Phi_{\alpha \to \alpha^*}\rangle$ contributes to the total transition, is, evidently, equal to the squared singular value, μ_α. Therefore, by presenting a set $\{\mu_\alpha\}$ of the most important μ_α, we can detail, if needed, the relative contributions of the chief natural configurations. The example of terphenyl triplet states suitably demonstrates this point. For the three triplet states in Table 14.6, the significant $\{\mu_\alpha\}$ are, respectively $\{0.809, 0.055, 0.054\}$, $\{0.357, 0.342, 0.140, 0.139\}$, and $\{0.225, 0.188, 0.185, 0.162, 0.113, 0.103\}$.

Table 14.7 Transition energy λ (in eV), oscillator strength (in parentheses), collectivity number κ, excitation distribution $\{L_A{}^*\}$, CT numbers for the intra-fragment (l_A) interactions and CT$_{tot}$ (in %) in the lowest transitions of p-quaterphenyl

State	λ	κ	$\{L_A^*\}$	l_I	l_{II}	CT$_{tot}$
$^1B_2$	3.927 (1.327)	1.04		0.04	0.11	70
$^1B_1$	4.471 (0.000)	1.73		0.01	0.21	56
$^1B_3$	4.585 (0.000)	2.13		0.02	0.22	52
$^1A_1$	4.586 (0.000)	1.98		0.01	0.03	92
$^3B_2$	2.842	1.53		0.06	0.21	46
$^3A_1$	3.310	3.20		0.12	0.22	32
$^3B_2$	3.740	5.68		0.35	0.07	16
$^3A_1$	3.959	7.53		0.19	0.26	10

These values are in conformity with the κ-values in the table. For instance, the last set tells us that the third excitation (the second $^1A_1 \rightarrow {}^3B_2$ transition in the table) requires at least six configuration, and it agrees with the value $\kappa = 5.88$. However, unlike the transition orbital method, we do not need to visualize the corresponding 12 natural orbitals (six $\{\alpha\}$ and six $\{\alpha^*\}$ orbitals). The distribution diagrams in Table 14.6 take properly into account the whole set of natural contributions. Notice that ESSA for one-dimensional polimer systems, treated at a semiempirical level, was given in [33].

Consider now an interesting class of macrocyclic polyphenylenes which were synthesized [82]. For the simplest ones, some results of ESSA/TDDFT are displayed in Table 14.8. They demonstrate various patterns of spreading excitation over molecules with a peculiar benzenoid structure.

14.6   Some Photochemical Applications

We previously mentioned that excitation localization indices L_A^* point to photoactive sites in excited molecule, and as such they can serve as suitable photochemical indices. For instance, anthracene and tetracene undergo photoaddition and photooxidation primarily at the peripheral atoms of central rings, and the diagrams in Table 14.1 agree with this fact. In peri-condensed PAHs, such as pyrene and

Table 14.8 Excitation distributions { L_A^* } and CT_{tot} (in %, at the bottom of each image) for the three lowest $S_0 \rightarrow S_j$ and $S_0 \rightarrow T_j$ transitions in m,m,o,p,o-pentaphenylene and m,m,m,o,m,o-hexaphenylene

	m,m,o,p,o-pentaphenylene		m,m,m,o,m,o-hexaphenylene	
j	$S_0 \rightarrow S_j$	$S_0 \rightarrow T_j$	$S_0 \rightarrow S_j$	$S_0 \rightarrow T_j$
1	79	32	74	34
2	72	32	80	34
3	67	24	78	34

perylene, the photoactive sites are positions 3 and 1, respectively, and these experimental facts are also in agreement with the $\{L_A^*\}$ diagrams displayed in the same table.

Another interesting example of this kind is the photodimerization of acenaphthylene previously discussed in many papers (see [83] and references therein). It is not so surprising that [2 + 2] photodimerization takes place at the ethylenic moiety, but free valence indices, unlike the frontier orbital analysis [83], favor the opposite positions in the naphthalene moiety. In correspondence with the experiment the excitation distributions in Fig. 14.6 unambiguously indicate the double bond sites as most photoactive. Moreover, one can expect more efficient dimerization from the triplet state rather than from the singlet state, and this agrees with [84]. Such examples can be multiplied easily.

Additional complications may arise in some cases when $n\pi^*$ excitations are involved in photoreactions, and the main mechanism is not so clear. For definiteness, take two rather popular molecular systems, thymine and coumarin. In the former, $S_0 \rightarrow S_1$ transition is of $n\pi^*$ nature, whereas in the latter is of $\pi\pi^*$ nature. With this, $S_0 \rightarrow T_1$ is the $\pi\pi^*$ transition for both molecules. The photodimerization is possible for both the systems [54, 85]. There are various viewpoints on the mechanism of the photoreactions (compare [85] and [86]), but, in our opinion,

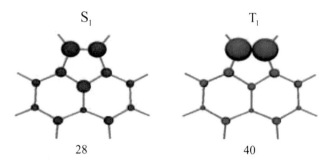

Fig. 14.6 $S_0 \rightarrow S_1$ and $S_0 \rightarrow T_1$ excitation distributions in acenapthylene. The values of L^*_{CH-CH} (in %) are given at the bottom of each image

the heavy atom effect and other details support the triplet mechanism as being the more likely to represent the situation. Looking at pictures in Table 14.9 and the values of the excitation localization index for the relevant double bond, one can observe that the TDDFT lowest transition $S_0 \rightarrow T_1$ notably concentrates just on the double bond. Strictly speaking, we cannot ignore the singlet mechanism of the photodimerization in thymine and coumarin, implying yet the related probabilities to be minor.

Much more difficult are the ESSA applications to valence photoisomerizations. Below we will try to understand two interesting type of photoreactions, namely photocyclizations in short polyenes and di-π-methane rearrangements. To make a picture more clear the generalized bond orders (14.47), or simply, bond indices, will be added to the usual ESSA analysis. We are quite aware that the very idea to understand possible ways of molecular photo processes via Franck-Condon transitions is under question. Indeed, well elaborated nonadiabatic theories and corresponding techniques (in particular, conical section methods) [7, 87] are to be much more appropriate tools. However, these techniques are very computationally demanding for complex systems (the need for the extended CI approaches etc.). Furthermore, before studying elementary photochemical acts one must explore the most important regions on the potential energy surface (PES) and make additional assumptions on account of a very complicated structure of the PESs. For making presumptions, chemical intuition and properties of Franck-Condon transitions, as well as corresponding structural-chemistry interpretation, can be helpful too. The semiquantitative schemes of this kind are well known in photochemistry starting from Zimmerman's papers [88, 89] where the difference charge density matrix, analogous to (14.26), and related bond orders are used (see also [90–92]). Recently the work in this direction was revived [93, 94], and in the present chapter we modify the Zimmerman approach in our own manner by invoking excitation indices (14.30) and bond indices (14.47).

Table 14.9 Excitation distributions $\{L_A{}^*\}$ and $L^*_{C=CH}$ (or $L^*_{CH=CH}$) (all in %) for low-lying singlet and triplet states in thymine and coumarin

j	$S_0 \rightarrow S_j$	$L^*_{C=CH}$	$S_0 \rightarrow T_j$	$L^*_{C=CH}$
1	4.680 (0.000), 1.01	23	3.247, 1.02	53
2	4.903 (0.128), 1.12	39	4.295, 1.01	22
1	4.083 (0.116), 1.12	23	2.753, 1.12	38
2	4.387 (0.000), 1.00	24	3.510, 1.10	14

Transition energy (in eV), oscillator strength (in parentheses) and collectivity number are given at the bottom of each image

We start with hexatriene and cycloheptatriene molecules for which the bond indices related to photocyclizations are displayed in Table 14.10. For these molecules 1,6-interaction is well known as the most important one. Some other nonbonded interactions are also interesting *per se*, while they are not manifested explicitly in photochemical experiments. In Table 14.10 we observe that in the case of cycloheptatriene, 1,6-interaction is well developed in all the studied states whereas for hexatriene the cyclization, leading to cyclohexadiene, is most probable in S_1 only. The photochemical data [95] more or less support these expectations. At least, cycloheptatriene undergoes 1,6-cyclization both thermally and (in gas phase) photochemically, leading to norcaradiene [96]. The products resulting from 1,4- or 3,6-photocyclizations are also observed [97, 98]. However, no clear evidences of the triplet mechanism have been described for these rearrangements. In case of cycloheptatriene we additionally computed the bond indices for S_1 in its optimized geometry, and the results for this relaxed state are mainly the same as for the Franck-Condon state in Table 14.10. While these results seem to be rather plausible, we can hardly believe that this situation is typical because most current DFT and TDDFT schemes are not able to correctly predict bond cleavage and biradical processes. At the same time, the Franck-Condon region can be described by TDDFT more or less satisfactorily.

Table 14.10 Excitation distributions and bond indices (14.47) for hexatriene (the first part of the table) and for cycloheptatriene (the second part of the table) in the lowest electronic states

Bond	S_0	S_1	T_1
1-6	0.048	0.235	0.070
1-5	0.046	0.137	0.065
1-4	0.101	0.164	0.100
1-3	0.444	0.449	0.461
2-5	0.068	0.119	0.075
1-6	0.319	0.432	0.324
1-5	0.095	0.129	0.085
1-4	0.116	0.144	0.096
1-3	0.413	0.417	0.437
2-5	0.091	0.119	0.090

Considering the di-π-methane rearrangement, we take three generic examples (see Table 14.11). In particular, for 3,3-dimethyl-penta-1,4-diene from the table one deduces that in the lowest singlet excited state 2,4-interaction becomes the most important (among nonbonded interactions), and this is in accordance with the first stage in Zimmerman's mechanism of the di-π-methane rearrangement [54, 99]. We also see that in the first triplet state the 2,4-bonding seems not to be so important, although the triplet channel of the di-π-methane rearrangement is usually accepted as essential too [100]. This distinction of the S_1 and T_1 behavior (in the Franck-Condon region) is rather typical for TDDFT models we use here (see other systems in Table 14.11). For the present we cannot take this result as concluding because generalized bond indices (14.47) seem not good quantities to be sufficiently discriminative for predicting photoreaction paths (which are very complex and convoluted as usually). From our experience (unlike other viewpoints in the literature), the current TDDFT approach is not a sufficiently reliable basis for solving nontrivial photochemical problems. Moreover, in many cases one cannot be quite sure that photoreactions go through the lowest singlet or triplet excited states. Nevertheless, the above nonbonded interaction analysis may be in use when the system under study is too complex, and a preliminary picture of possible nonbonded interactions is desirable. We should also keep in mind that excitation localization indices L_A^* serve as important independent indicators pointing to the

Table 14.11 Excitation distributions and bond indices (14.47) in the lowest states of 3,3-dimethylpenta-1,4-diene, norbornadiene and benzobarrelene

Bond	S_0	S_1	T_1
1-3	0.358	0.345	0.329
1-5	0.075	0.409	0.088
2-4	0.223	0.413	0.229
2-5	0.078	0.289	0.085
2-6	0.245	0.424	0.278
2-5	0.135	0.245	0.175
1-4	0.351	0.351	0.347
2-7	0.317	0.318	0.314
3-11	0.247	0.352	0.273
3-5	0.241	0.263	0.250

most significant photoreactivity centers. In conjunction with bond indices (14.47), these L_A^*, being densities of effectively unpaired electrons (see Sect. 14.4), can provide more informative results.

14.7 Extension of ESSA to the General CI Case

We address here the important issue how to interpret transitions between multiconfigurational state vectors which are distinct from the Hartree-Fock and associated CIS states. In the preceding sections we demonstrated ESSA for the special CIS-like wave functions produced from RPA or TDDFT. We recall that the core idea of the method is constituted by the possibility of interpreting the basic 1-electron nilpotent τ (14.5) as a matrix of transition probability amplitudes. The τ is naturally normalized owing to the specific structure of the CIS wave-function (14.13) (no high-order excitation operators are involved), and, along with simplicity, this normalization is an advanced property for making the correctly defined interatomic CT probabilities. At this stage, our previous ESSA schemes are to be referred as ESSA/CIS, ESSA/RPA or ESSA/TDDFT, depending on the specific scheme used for computing excited states.

Now we realize that for extension of ESSA to the general CI (GCI) wave functions, only a small generalization is required, thus leading us to the corresponding ESSA/GCI approach. Really, instead of τ we must employ the corresponding 1-electron transition RDM (in the Löwdin sense) which we signify here as τ_{GCI}. Clearly, we cannot generally expect from τ_{GCI} neither the natural normalization (14.6) nor nilpotency (14.9). Nevertheless, the τ_{GCI} matrix elements can be envisaged as related to transition probabilities. This is rather evident from the standard expression for the transition momentum formula $\vec{Q} = \text{Tr}\,\vec{r}\,\tau_{GCI}$, with $\vec{r}$ being the position operator.

Because of an additive separability of τ_{GCI} (this follows from McWeeny's theory of generalized group functions [42]), renormalizing this matrix is a permissible operation. Thus, after renormalizing to unity, the τ_{GCI} becomes the main object to be analyzed within the ESSA/GCI scheme we propose here. Hence, in most relations of Sect. 14.3 we must make the replacement

$$\tau \;\rightarrow\; \tau_{GCI}/\;\|\tau_{GCI}\|, \qquad\qquad (14.51)$$

in order to compute, at the ESSA/GCI level, excitation localization indices (14.30), CT numbers (14.34), and the electronic transition collectivity number (14.25). Such a simple way leads to quite reasonable results.

However, one unsolved issue arises with the generalized ESSA of this kind. In the original ESSA [23] we ensured the intrinsic consistency of CT numbers what is expressed in identity (14.38). For CT numbers in ESSA/GCI, such charge density relations can be only approximately valid (if at all). This is a consequence of the fact that (14.36) is no longer true. Thus, passing to more general models we inevitably lose some attractive features of the original CT numbers (14.34) while acquiring a more correct method in whole.

This chapter affords only a preliminary view on the subject matter. In the case study below, we simplify computations by employing the conventional π-electron approximation for the full CI description of π-electron subshells of conjugated hydrocarbons. More specifically, we make using the π- parameterization scheme [101] for the so-called Pariser-Parr-Pople (PPP) Hamiltonian, and consider the phenyl-pentafulvene molecule known in aromatic compound chemistry [102]. In Table 14.12 we compare the ESSA results, thus obtained from CIS/PPP and FCI/PPP (and TDDFT/B3LYP, for completeness). The FCI solutions and corresponding RDMs were computed following the wave-function operator technique [36, 103]. We see that both approaches furnish the similar description, in particular the same assignment of $S_0 \rightarrow S_1$ to the transition basically localized on the fulvene moiety, and besides, small CT effects are observed. The same picture is given by the conventional TDDFT/B3LYP method.

At the same time, an apparent similarity of the above CIS and FCI solutions is illusive. We can estimate quantitatively their similarity by the fidelity measures (14.40). The corresponding index

Table 14.12 Excitation distribution and CT numbers (in %) for the lowest singlet $\pi\pi^*$ transition in phenyl-pentafulvene at CIS/PPP, FCI/PPP and TDDFT/B3LYP levels

CIS/PPP	FCI/PPP	TDDFT/B3LYP
3.263 (0.042), 1.02	2.799 (0.004), 1.03	2.803 (0.002), 1.01

Excitation energies (in eV), oscillator strengths (in parentheses), and collectivity numbers are given at the bottom of each image

$$\wp_{CT}[\mathrm{CIS, FCI}] = 0.56,$$

and it seems that internal structure of excited states is more varied and more complex than that obtained merely by inspecting localization diagrams. Notice that more consistent approaches for measuring similarity between two electronic states are much involved, making use of the RDMs of appropriate order [20, 104, 105]. However, in practice the simplified (based on (14.40)) technique often appears quite reasonable.

In the whole, ESSA/FCI or more practical ESSA/GCI schemes, as going beyond the simplified DFT models, provide us with the suitable interpretive tool which is, in many general respects, similar to the ESSA/CIS and ESSA/TDDFT schemes we examined in the chapter.

14.8 Concluding Remarks

An excited state structural analysis [23] was conceived within π-electron model as a response to the experimental evidence of localization phenomena in excited molecules. Thirty years after it was revived in a new (TDDFT) clothing [17] that allows one to extend the approach to a more realistic quantum-chemical description. Now equipped with this revitalized technique, we tried here to elucidate how ESSA, more exactly ESSA/TDDFT, works on complex molecular systems. Before summing up, several critical remarks need to be made. They are related to the current status of DFT and TDDFT in particular, and encompass issues too difficult to permit anything but a preliminary and inevitably very sketchy consideration.

We begin with the most important issue. It is assumed that in principle it is possible to obtain a full description of many-electron systems by DFT if the exact (yet unknown) density-dependent exchange-correlation functional is employed. This seemingly undisputed tenet, based on the Hohenberg-Kohn theorem, was recently called into question [106, 107]. In his works Kaplan shows that the conventional Kohn-Sham equations are invariant with the respect to the total spin, *s,* due to the invariance of electron density with respect to *s* (the Kaplan theorem). The analysis [107] of the existing DFT procedures for spin multiplets has shown that these modify only the expressions for the exchange energy, and in all such procedures the correlation functionals do not correspond to a definite total spin of the state. As a consequence, no explicit dependence on the total spin is specified, and it is a fundamental drawback of the whole DFT methodology. To our knowledge, this viewpoint is not challenged, and up to now the crucial issue about spin within DFT has been open.

Another issue concerns the difficulties of the foundations of TDDFT *per se* [108–110]. From these works we infer that TDDFT schemes of practical importance are rather semi-ab initio (in terms of Dewar [111]) than ab initio approaches, and therefore they can suffer from unpredictable inaccuracies. An extensive comparison of TDDFT with other schemes was given in [112]. The work confirms that TDDFT provides rather modest possibilities for treating dyes that have chromophores with low transition energies.

Many efforts were made to improve the TDDFT description of the excited states with significant CT components (see the recent papers [113–116] and references therein). Unfortunately, some expectations on the range-separated exchange-correlation functionals did not fully come true, as it is shown in [117] where the singlet-triplet splitting is examined. That work and papers [118–120] tell us that the M06 family of Truhlar's functionals seems to be more attractive and promising for solving molecular spectroscopy problems.

Now, in the light of the above discussion, we can estimate the matter-of-fact applicability of the proposed ESSA in an attainable perspective. We must acknowledge that TDDFT/B3LYP, and hence ESSA/TDDFT/B3LYP, offers no more than approximate characterization of the electronic excited states. More reliable would be the similar treatment within ESSA/TDDFT/M06-2X or related schemes. Moreover, the usage of the well-defined natural AOs technique [21] might alleviate the basis-dependence problem. At the same time, the key principles of the ESSA/CIS and ESSA/TDDFT approaches can be extended to bona-fide many-electron wave functions at a proper ab initio level. Therefore, ESSA at any level of the excited state theory enables us to reveal, in pictorial form, an internal structure of the excited states under study. The proposed analysis includes portraying distributions which are produced by excitation localization indices (14.30), and, additionally, the analysis provides more detailed characterization via charge transfer numbers (14.37) and generalized bond indices (14.47). The usefulness of this interpretive tool has been repeatedly demonstrated in this chapter, and we believe that more reliable DFT schemes or genuine good-quality many-electron models, in conjunction with the present tool, will produce qualitatively comparable results.

Acknowledgments Authors are very grateful to A. O. Doroshenko, I. G. Kaplan, and O. V. Shishkin for valuable discussions.

Appendix

We consider the conventional exciton-like model [79, 80], slightly modified here for triplet states of a finite-size system (assembly) of m weakly interacting identical subunits (fragments) $\{A_k\}_{1 \leq k \leq m}$. Let Ξ_k be the the kth basis vector which corresponds to the local excitation of the kth site of the assembly, viz. $\Xi_k = |A_1^0 \ldots A_{k-1}^0 A_k^* A_{k-1}^0 \ldots A_m^0\rangle$. Then in a tight-binding approximation we can approximate the exciton-like Hamiltonian matrix $\Lambda = \|\Lambda_{kl}\|$, as follows

$$\Lambda_{kl} = (\lambda_0 - \upsilon_k V_*)\delta_{kl} + \{K_* \text{ if } A_k \text{ and } A_l \text{ are neighbors, } 0 \text{ otherwise}\}$$

where λ_0 is an excitation energy of the isolated fragment; the positive V_* and K_* are effective local level shift and exchange parameters, respectively. The factor υ_k in term $\upsilon_k V_*$ is equal to a number of neighbours of site A_k, which takes into account the local environment effects.

Eigenvalues, λ_k, of Λ are excitation energies of the whole system. For instance, in a linear trimer we have

$$\Lambda = \begin{Vmatrix} \lambda_0 - V_* & K_* & 0 \\ K_* & \lambda_0 - 2V_* & K_* \\ 0 & K_* & \lambda_0 - V_* \end{Vmatrix}$$

with simply computed excitation energies. They take the form

$$\lambda_1 = \lambda_0 - V_*\left(\sqrt{1 + 2q_*^2} + 3\right)/2, \ \lambda_2 = \lambda_0 - J_*,$$

$$\lambda_3 = \lambda_0 + V_*\left(\sqrt{1 + 2q_*^2} - 3\right)/2$$

where the dimensionless parameter $q_* = 2K_*/V_*$ is used. The corresponding (non-normalized) eigenvectors are of the form

$$\begin{Vmatrix} -q_* \\ \sqrt{1 + 2q_*^2} + 1 \\ -q_* \end{Vmatrix} \begin{Vmatrix} 1 \\ 0 \\ -1 \end{Vmatrix}, \begin{Vmatrix} q_* \\ \sqrt{1 + 2q_*^2} - 1 \\ q_* \end{Vmatrix}$$

From them we obtain the distribution localization indices L_A^* by merely squaring the corresponding "coordinates" in the above eigenvectors. The results become more clear for very weakly coupled fragments (chromophores) when $q_* \ll 1$, and we have

$$\lambda_1 = \lambda_0 - J_* - 2K_*^2/V_*, \ \{L_A^*\} = \{1/2 - (K_*/V_*)^2, \ 2(K_*/V_*)^2,$$

$$1/2 - (K_*/V_*)^2\};$$

$$\lambda_2 = \lambda_0 - V_*, \ \{L_A^*\} = \{1/2, \ 0, \ 1/2\};$$

$$\lambda_3 = \lambda_0 - 2J_* + 2K_*{}^2/V_*, \ \{L_A^*\} = \{(K_*/V_*)^2, \ 1 - 2\,(K_*/V_*)^2, \ (K_*/V_*)^2\}.$$

We see that the lowest excitation is predominantly localized on the internal fragment A whereas the rest are shared between the two terminal fragments of the trimer. The qualitatively similar picture is appropriate in the case of moderate values of q_* as well.

References

1. Ramamurthy V, Schanze KS (eds) (2001) Understanding and manipulating excited-state processes. Molecular and supramolecular photochemistry, vol 8. Marcel Dekker, New York
2. Lanzani G (ed) (2006) Photophysics of molecular materials. From single molecules to single crystals. Wiley, Berlin
3. Gell C, Brockwell D, Smith A (2006) Handbook of single molecule fluorescence spectroscopy. Oxford University Press, USA
4. Gräslund A, Rigler R, Widengren J (eds) (2010) Single molecule spectroscopy in chemistry, physics and biology. Springer, New York
5. Ehara M, Nakatsuji H (2010) In: Cársky P, Paldus J, Pittner J (eds) Recent progress in coupled cluster methods, vol 11, Challenges and advances in computational chemistry and physics. Springer, Dordrecht/Heidelberg/London/New York, p 79
6. Dreuw A, Head-Gordon M (2005) Chem Rev 105:4009
7. Olivucci M (ed) (2005) Computational photochemistry. Elsevier, Amsterdam
8. Marques MAL, Ullrich C, Nogueira F, Rubio A, Gross EKU (eds) (2006) Time-dependent density-functional theory, vol 706, Lecture notes in physics. Springer, Berlin
9. Shavitt I, Bartlett RJ (2009) Many-body methods in chemistry and physics. Cambridge University Press, Cambridge
10. Fedorov DG, Kitaura K (eds) (2009) The fragment molecular orbital method: practical applications to large molecular systems. CRC, Boca Raton
11. Foresman JB, Head-Gordon M, Pople JA, Frisch MJ (1992) J Phys Chem 96:135
12. Head-Gordon M, Rico RJ, Oumi M, Lee TJ (1994) Chem Phys Lett 219:21
13. Koch W, Holthausen MC (2002) A chemist's guide to density functional theory, 2nd edn. Wiley, Weinheim
14. Sholl D, Steckel JA (2009) Density functional theory: a practical introduction. Wiley, Hoboken
15. Casida ME (2009) J Mol Struct Theochem 914:3
16. Blaisot JP, Ripka G (1986) Quantum theory of finite systems. Massachusetts Institute of Technology, Massachusetts
17. Luzanov AV, Zhikol OA (2010) Int J Quant Chem 110:902
18. Edgar AP (1902) Tales, vol 4, The complete works of Edgar Allan Poe. Knickerbocker, New York, p 175
19. Bader RFW (1990) Atoms in molecules – a quantum theory. Oxford University Press, Oxford
20. Carbó-Dorca R, Girones X, Mezey PG (eds) (2001) Fundamentals of molecular similarity, Mathematical and computational chemistry. Kluwer, New York
21. Weinhold F, Landis CR (2005) Valency and bonding: a natural bond orbital donor-acceptor perspective. Cambridge University Press, Cambridge

22. Luzanov AV, Sukhorukov AA, Umanski VE (1974) Theor Exp Chem 10:354
23. Luzanov AV (1980) Russ Chem Rev 49:1033
24. Blagoy Yu P, Sheina GG, Luzanov AV, Silina LK, Pedash VF, Rubin Yu V, Leibina EA (1980) Int J Quant Chem 17:913
25. Fabian J, Mehlhorn A, Fratev F (1980) Int J Quant Chem 17:235
26. Czuchajowski L, Wisor AK (1986) J Phys Chem 90:1541
27. Kachkovski AD (1994) Dyes Pigments 24:171
28. Reinhardt M, Kirschke K, Baumann H (1995) J Mol Struct 348:417
29. Mitina VG, Ivanov VV, Ponomarev OA, Sleta LA, Shershukov VM (1996) Mol Engin 6:249
30. Doroshenko AO, Kirichenko AV, Mitina VG, Ponomarev OA (1996) J Photochem Photobiol A Chem 94:15
31. Doroshenko AO, Grigorovich AV, Posokhov EA, Pivovarenko VG, Demchenko AP (1999) Mol Engin 8:199
32. Doroshenko AO, Posokhov EA, Verezubova AA, Ptyagina LM (2000) J Phys Org Chem 13:253
33. Luzanov AV (2002) J Struct Chem 43:711
34. Doroshenko AO, Posokhov EA, Verezubova AA, Ptyagina LM, Skripkin VT, Shershukov VM (2002) Photochem Photobiol Sci 1:92
35. Nijegorodov N, Zvolinski V, Luhanga PVC (2008) J Photochem Photobiol A Chem 196:219
36. Luzanov AV, Prezhdo OV (2005) Int J Quant Chem 102:582
37. Löwdin P-O (1955) Phys Rev 97:1490
38. McWeeny R (1956) Proc Roy Soc Ser A 237:355
39. Thouless DJ (1961) Nucl Phys 22:78
40. Luzanov AV (1975) Theor Exp Chem 9:567
41. Mukamel S, Tretiak S (2002) Chem Rev 102:3171
42. McWeeny R (1992) Methods of molecular quantum mechanics, 2nd edn. Academic, London
43. Davidson ER (1976) Reduced density matrices in quantum chemistry. Academic, New York
44. Luzanov AV (1978) Theor Exp Chem 13:433
45. Luzanov AV, Mestechkin MM (1981) Theor Math Phys 48:740
46. Martin RL (2003) J Chem Phys 118:4775
47. Mayer I (2007) Chem Phys Lett 437:284
48. Batista ER, Martin RL (2005) J Phys Chem A 109:3128
49. Luzanov AV, Umanski VE (1977) Theor Exp Chem 13:162
50. Luzanov AV, Pedash Yu F, Mohamad S (1990) Theor Exp Chem 26:485
51. Luzanov AV, Zhikol OA (2005) Int J Quant Chem 104:167
52. Ermolaev VL, Terenin AN (1960) Sov Phys Usp 3:423
53. Shigorin DN (1977) Zh Fiz Khim 51:1894
54. Turro NJ (1991) Modern molecular photochemistry. University Science, Mill Valley
55. Luzanov AV, Pedash VF (1979) Theor Exp Chem 15:338
56. Ohta T, Kuroda H, Kunii TL (1970) Theor Chim Acta 19:167
57. Luzanov AV, Prezhdo OV (2006) J Chem Phys 124:224109
58. Staroverov VN, Davidson ER (2000) Chem Phys Lett 330:161
59. Alcoba DR, Bochicchio RC, Lain L, Torre A (2006) Chem Phys Lett 429:286
60. Head-Gordon M (2003) Chem Phys Lett 380:488
61. Takatsuka K, Fueno T, Yamaguchi K (1978) Theor Chim Acta 48:175
62. Wiberg K (1968) Tetrahedron 24:1083
63. Mayer I (1983) Chem Phys Lett 97:270
64. de Giambiagi MS, Giambiagi M, Jorge FE (1985) Theor Chim Acta 68:337
65. Ponec R, Uhlík F, Cooper DL, Jug K (1996) Croat Chem Acta 69:933
66. Yamasaki T, Goddard WA III (1998) J Phys Chem A 102:2919
67. Alcoba DR, Bochicchio RC, Lain L, Torre A (2008) Phys Chem Chem Phys 10:5144
68. Doroshenko AO (2009) J Mol Struct 933:169
69. Chepeleva LV, Matsakov AY, Kondratyuk ZA, Yaremenko FG, Doroshenko AO (2010) J Photochem Photobiol A Chem 209:163

70. Svechkarev D, Doroshenko A, Baumer V, Dereka B (2011) J Luminescence 131:253
71. Kawashima Y, Hashimoto T, Nakano H, Hirao K (1999) Theor Chem Acc 102:49
72. Parac M, Grimme S (2003) Chem Phys 292:11
73. Halasinski TM, Weisman JL, Ruiterkamp R, Lee TJ, Salama F, Head-Gordon M (2003) J Phys Chem A 107:3660
74. Malloci G, Mulas G, Cappellini G, Joblin C (2007) Chem Phys 340:43
75. Kadantsev ES, Stott MJ, Rubio A (2006) J Chem Phys 124:134901
76. Clar E (1964) Polycyclic hydrocarbons, vol 1. Academic, London
77. Langlet J (1975) Theor Chim Acta 38:199
78. de Meijere A, Kozhushkov SI, Fokin AA, Emme I, Redlich S, Schreiner PR (2003) Pure Appl Chem 75:549
79. Kasha M, Rawls HR, El-Bayoumi MA (1965) Pure Appl Chem 11:371
80. Hochstrasser RM, Whiteman JD (1972) J Chem Phys 56:5945
81. Parson WW (2007) Modern optical spectroscopy. Springer, New York
82. Fujioka Y (1984) Bull Chem Soc Jap 57:3494
83. Fukui K, Morokuma K, Yonezawa T (1961) Bull Chem Soc Jap 34:1178
84. Cowan DO, Koziar JC (1975) J Am Chem Soc 97:249
85. Lamola AA (1970) Pure Appl Chem 24:599
86. Boggio-Pasqua M, Groenhof G, Schafer LV, Grubmuller H, Robb MA (2007) J Am Chem Soc 129:10996
87. Domcke W, Yarkony DR, Köppel H (eds) (2004) Conical intersections: electronic structure, dynamics and spectroscopy. World Scientific, Singapore
88. Zimmerman HE, Swenton JS (1967) J Am Chem Soc 89:906
89. Zimmerman HE, Binkley RW, McCullough JJ, Zimmerman GA (1967) J Am Chem Soc 89:6589
90. Weltin EE (1973) J Am Chem Soc 95:7650
91. Vysotskii YB, Sivyakova LN (1986) Khim Geterotsikl Soedin (2): 173
92. Lee GH, Park Y-T (1994) Bull Korean Chem Soc 15:857
93. Zimmerman HE, Alabugin IV (2000) J Am Chem Soc 122:952
94. Zimmerman HE (2005) In: Olivucci M (ed) Computational photochemistry. Elsevier, London, p 255
95. Srinivasan R (1961) J Am Chem Soc 83:2806
96. Samuni U, Kahana S, Haas Y (1998) J Phys Chem 102:4758
97. Dauben WG, Cargill RL (1961) Tetrahedron 12:186
98. Jones LB, Jones VK (1968) J Am Chem Soc 90:1540
99. Zimmerman HE, Armesto D (1996) Chem Rev 96:3065
100. Frutos LM, Sancho U, Castano O (2004) Org Lett 6:1229
101. Schulten K, Ohmine I, Karplus M (1976) J Chem Phys 64:4422
102. Neuenschwander M (1986) Pure Appl Chem 58:55
103. Luzanov AV, Wulfov AL, Krouglov VO (1992) Chem Phys Lett 197:614
104. Nielsen MA, Chuang IL (2000) Quantum computation and quantum information. Cambridge University Press, Cambridge
105. Luzanov AV (2011) Int J Quant Chem 111:2196
106. Kaplan IG (2007) J Mol Struct 838:39
107. Kaplan IG (2007) Int J Quant Chem 107:2595
108. Schirmer J, Dreuw A (2007) Phys Rev A 75:022513
109. Schirmer J, Dreuw A (2008) Phys Rev A 78:056502
110. Niehaus TA, March NH (2009) Theor Chem Acc 125:427
111. Dewar MJS (1993) Org Mass Spectrom 28:305
112. Fabian J (2010) Dyes Pigments 84:36
113. Rohrdanz MA, Martins KM, Herbert JM (2009) J Chem Phys 130:054112
114. Stein T, Kronik L, Baer R (2009) J Am Chem Soc 131:2818
115. Ipatov A, Hesselmann A, Gorling A (2010) Int J Quant Chem 110:2202

116. Jacquemin D, Perpète EA, Ciofini I, Adamo C, Valero R, Zhao Y, Truhlar DG (2010) J Chem Theor Comput 6:2071
117. Cui G, Yang W (2010) Mol Phys 108:2745
118. Jacquemin D, Perpète EA, Ciofini I, Adamo C (2010) J Chem Theor Comput 6:1532
119. Liu X, Yang D, Ju H, Teng F, Hou Y, Lou Z (2011) Chem Phys Lett 503:75
120. Liu X, Zhang X, Hou Y, Teng F, Lou Z (2011) Chem Phys 381:100

Chapter 15
VCD Chirality Transfer: A New Insight into the Intermolecular Interactions

Jan Cz. Dobrowolski, Joanna E. Rode, and Joanna Sadlej

Abstract The Vibrational Circular Dichroism (VCD) spectroscopy has been developing rapidly in both experimental and theoretical aspects. Currently, the VCD has become one of the most effective and reliable spectroscopic technique to determine the absolute configuration of chiral molecules. Its success is related to the availability of instrumentation and software for quantum-chemical calculation of the spectra. Nowadays, large parts of the VCD spectra can be trustfully predicted by theory and critically verified by confiding experiment, and vice versa. In the last decade, several theoretical and experimental VCD studies reported on VCD chirality transfer phenomenon occurring when an achiral molecule becomes VCD active as a result of intermolecular interactions with a chiral one. There are still some theoretical and experimental uncertainties about the VCD chirality transfer, however, benefits from an comprehensive use of the phenomenon can push our ability to diversify the intermolecular complexes and deepen our understanding of intermolecular interactions. This chapter is a review of the computational studies on VCD chirality transfer phenomenon supported by the experimental references, and ended by perspectives.

Keywords Chirality transfer • Chiroptical methods • DFT • Intermolecular interactions • VCD spectra

J.C. Dobrowolski • J. Sadlej
National Medicines Institute, 30/34 Chełmska Street, 00-725 Warsaw, Poland

J.C. Dobrowolski • J.E. Rode
Industrial Chemistry Research Institute, 8 Rydygiera Street, 01-793 Warsaw, Poland

J. Sadlej (✉)
Faculty of Chemistry, University of Warsaw, 1 Pasteura Street, 02-093 Warsaw, Poland

J. Leszczynski and M.K. Shukla (eds.), *Practical Aspects of Computational Chemistry I:* 451
An Overview of the Last Two Decades and Current Trends,
DOI 10.1007/978-94-007-0919-5_15, © Springer Science+Business Media B.V. 2012

15.1 Introduction

Chirality was discovered by Pasteur as *"l'hémiédrie"* of crystals mechanically separated into two enantiomeric forms in 1848 [1].[1] Moreover, Pasteur did correctly interpreted *"l'hémiédrie"* as a manifestation of *"la dissymétrie"* of the smallest elements forming the crystal [2]. The term *chirality* was introduced in a footnote in the *Second Robert Boyle Lecture*, *"The Molecular Tactics of a Crystal"* delivered by Kelvin to the Oxford University Junior Scientific in 1893, and published a year later [3, 4]. Yet, consciousness of incongruency between the left and right handedness existed much before. In 1783 in *Prolegomena*, Kant wrote *"what can be more similar and in every part more alike to my hand or to my ear, than their images in the mirror? And yet I cannot put such a hand as is seen in the mirror in the place of its original; for if this is a right hand, that in the mirror is a left one … (they cannot be made congruent); the glove of one hand cannot be employed for the other"* [5a].[2] Already in ancient Greece, in ca. 350 B.C.E Aristotle mentioned that for Pythagoreans *right and left* represented one of the principle contrarieties [5b].[3]

Nowadays, the ruling IUPAC definition of chirality is the following: *"The geometric property of a rigid object (or spatial arrangement of points or atoms) of being non-superposable on its mirror image; such an object has no symmetry elements of the second kind (a mirror plane, $\sigma = S_1$, a centre of inversion, $i = S_2$, a rotation-reflection axis, S_{2n}). If the object is superposable on its mirror image the object is described as being achiral."* [6]. Thus, a molecule is chiral if and

---

[1]*"Je montre, en effet, que l'hémiédrie est liée avec le sens de la polarisation rotatoire. Or, ce dernier phénomène étant moléculaire et accusant une dissymétrie dans les molécules, l'hémiédrie, à son tour, se trouve donc en étroite connexion avec la dissymétrie des derniers éléments qui composent le cristal."*

[2]"Was kann wohl meiner Hand oder meinem Ohr ähnlicher, und in allen Stücken gleicher sein, als ihr Bild im Spiegel? Und dennoch kann ich eine solche Hand, als im Spiegel gesehen wird, nicht an die Stelle ihres Urbildes setzen; denn wenn dieses eine rechte Hand war, so ist jene im Spiegel eine linke, und das Bild des rechten Ohres ist ein linkes, das nimmermehr die Stelle des ersteren vertreten kann. Nun sind hier keine innre Unterschiede, die irgendein Verstand nur denken könnte; und dennoch sind die Unterschiede innerlich, soweit die Sinne lehren, denn die linke Hand kann mit der rechten, ohnerachtet aller beiderseitigen Gleichheit und Ähnlichkeit, doch nicht zwischen denselben Grenzen eingeschlossen sein, (sie können nicht kongruieren) der Handschuh der einen Hand kann nicht auf der andern gebraucht werden. Was ist nun die Auflösung? Diese Gegenstände sind nicht etwa Vorstellungen der Dinge, wie sie an sich selbst sind, und wie sie der pure Verstand erkennen würde, sondern es sind sinnliche Anschauungen, d. i. Erscheinungen, deren Möglichkeit auf dem Verhältnisse gewisser an sich unbekannten Dinge zu etwas anderem, nämlich unserer Sinnlichkeit beruht. Von dieser ist nun der Raum die Form der äußern Anschauung, und die innere Bestimmung eines jeden Raumes ist nur durch die Bestimmung des äußeren Verhältnisses zu dem ganzen Raume, davon jener ein Teil ist, (dem Verhältnisse zum äußeren Sinne) d. i. der Teil ist nur durchs Ganze möglich, welches bei Dingen an sich selbst, als Gegenständen des bloßen Verstandes niemals, wohl aber bei bloßen Erscheinungen stattfindet. Wir können daher auch den Unterschied ähnlicher und gleicher, aber doch inkongruenter Dinge (z. B. widersinnig gewundener Schnecken) durch keinen einzigen Begriff verständlich machen, sondern nur durch das Verhältnis zur rechten und linken Hand, welches unmittelbar auf Anschauung geht."

only if it belongs to the point symmetry group missing out all orientation-reversing isometries, namely C_1, C_n, D_n, and, very rare in molecules, tetrahedral, octahedral, and icosahedral point groups of symmetry, T, O, and I, respectively [7]. The rigidness required by the IUPAC definition is not satisfied for majority of molecules. However, IUPAC defines also *chirality centre, chirality axis, and chirality plane* [8], which, if conserved over conformational changes, guarantees the molecular chirality to be preserved. So far, IUPAC has not included chirality originating from topological entangling of the molecules and biomolecules [9, 10], knotted [11], linked in catenanes and rotaxanes [12], turned in moebius strips [13] *etc.* which in most cases generates molecular chirality. Yet, topological chemistry which is relatively new field, needs more work on classification and naming topological structures which probably cannot be done without help of mathematicians and physicists [14–16].

Not only rigidness can be important for chirality. Indeed, the definition pass over observation time scale, physical conditions (temperature, pressure), state of aggregation, solvation, isotopic composition *etc.* which are, for obvious reasons, important for chirality of molecules. The time scale is especially crucial for spectroscopic observations of non-rigid chiral molecules, because, the effects observed,

---

"What can be more similar in every respect and in every part more alike to my hand and to my ear, than their images in a mirror? And yet I cannot put such a hand as is seen in the glass in the place of its archetype; for if this is a right hand, that in the glass is a left one, and the image or reflection of the right ear is a left one which never can serve as a substitute for the other. There are in this case no internal differences which our understanding could determine by thinking alone. Yet the differences are internal as the senses teach, for, notwithstanding their complete equality and similarity, the left hand cannot be enclosed in the same bounds as the right one (they are not congruent); the glove of one hand cannot be used for the other. What is the solution? These objects are not representations of things as they are in themselves, and as the pure understanding would know them, but sensuous intuitions, that is, appearances, the possibility of which rests upon the relation of certain things unknown in themselves to something else, viz., to our sensibility. Space is the form of the external intuition of this sensibility, and the internal determination of every space is only possible by the determination of its external relation to the whole space, of which it is a part (in other words, by its relation to the external sense). That is to say, the part is only possible through the whole, which is never the case with things in themselves, as objects of the mere understanding, but with appearances only. Hence the difference between similar and equal things, which are yet not congruent(for instance, two symmetric helices), cannot be made intelligible by any concept, but only by the relation to the right and the left hands which immediately refers to intuition."

[3]"Other members of this same school say there are ten principles, which they arrange in two columns of cognates-limit and unlimited, odd and even, one and plurality, right and left, male and female, resting and moving, straight and curved, light and darkness, good and bad, square and oblong. In this way Alcmaeon of Croton seems also to have conceived the matter, and either he got this view from them or they got it from him; for he expressed himself similarly to them. For he says most human affairs go in pairs, meaning not definite contrarieties such as the Pythagoreans speak of, but any chance contrarieties, e.g. white and black, sweet and bitter, good and bad, great and small. He threw out indefinite suggestions about the other contrarieties, but the Pythagoreans declared both how many and which their contrarieties are."

for instance, by electronic, vibrational and NMR spectroscopies can be detectable by one of the method whereas undetectable by the other. The same holds true for the other parameters precising state of molecules.

Chiral molecules play a fundamental role in many aspects of live science, chemistry and physics, whereas chirality is responsible for structural and functional diversity of biological macromolecules. Homochirality in the monomeric left-handed amino acid of proteins leads to homochirality in higher order structures such as the right-handed α-helix (secondary structure), and the fold (tertiary structure), unique to different protein. Nuclear acids consist of chains of DNA or RNA and are connected with monomeric D-deoxyribose or D-ribose sugar rings [17]. In enzyme catalysis the Fischer's "lock and key" principle governs the stereochemical selection. The absolute configuration of chiral system is very important in understanding the activity-structure relations in chemistry and pharmacology, at the supramolecular level in coordination complexes and material science [18–23]. There is no doubt, that Vibrational Circular Dichroism spectroscopy is a powerful method to study a variety of fundamental actions of chiral compounds [24]. We are sure, that use of the VCD chirality transfer effects can help in better discriminating among different important physico-chemical phenomena essential for life sciences, chemistry, and materials science.

One of the focuses of this review is to show the potential in using VCD technique in studies of the molecular interactions. The characterization of inter-molecular interactions is one of the basic goals of experimental and theoretical physical chemistry. The hydrogen bond, which is often a dominant intermolecular interaction, is particularly interesting and intensively investigated on account of its importance in biochemistry. IR and NMR spectroscopies have been extensively used for explaining and predicting spectroscopic features of hydrogen bonds such as vibrational frequencies and intensities, NMR chemical shifts, and also spin-spin coupling constants [25]. A thorough understanding of the experimental spectra is impossible without the reference to quantum mechanics models. Theoretical models allow us to better understand the relationship between the spectroscopic parameters and the structure of the samples and they are indispensable for a realistic modeling of the parameters as a function of structure.

In recent years few groups started to use VCD spectroscopy as a new method providing an evidence for hydrogen bond formation and its conformational characteristic [26–29]. Such an approach is complementary to the one obtainable from vibrational absorption (VA) spectra. Therefore, we refer to our experience in computational spectroscopy and we will pay special attention to the VCD spectra as a tool for predictions of the conformational characteristic of the interacting molecules.

The main purpose of this review is to explain how the VCD spectra can potentially be used to better characterize the intermolecular interactions and to document the advance of this aspect of VCD spectroscopy to the present data. We overview recent theoretical predictions and the innovative VCD observations of chirality transfer (Sect. 15.2.2) from a chiral molecule to an achiral one as a result of hydrogen bond interactions between them. Of particular interest is the hydrogen bonding interaction between chiral molecules with water. Throughout this review, illustrative examples highlight the main aspects of chiral transfer in molecular complexes. In our opinion, the VCD technique has opened new horizons for both understanding and monitoring the intermolecular actions and applicability of relatively new and powerful physicochemical method.

The structure of this review is as follows: first we give a short presentation of the theory of VCD and of contemporary *ab initio* methods used for the VCD spectra calculations. We center on theoretical investigations of the problems that affect the calculated results, such as basis set, the effects of molecular vibrations and the solvent effects. Next, we focus on the calculations of the VCD spectra of hydrogen-bonded systems and the possibility to use this method for the investigation of the interacting systems and the spectral fingerprint of chirality transfer upon complexation. The last part of this review presents a survey on the perspective of this field.

15.2   The Physical Manifestation of Optical Activity in Chiroptical Spectroscopic Methods

15.2.1   Chiroptical Methods

The interaction of chiral molecules with electromagnetic fields has attracted interest for decades. There are two classical methods to measure optical activity of chiral molecules. The simplest and the oldest one is optical rotation of linearly polarized light (OR) in the transparent region (measured at a single frequency, usually using the sodium D-line). The other one is circular dichroism (CD) of circularly polarized light in absorbing region of UV–vis region [30, 31]. The two methods have the same origin because the enantiomers have different indices of refraction in their real and imaginary parts. The real part of difference between the refraction indices measures the rotation of the polarization vector of an incident linearly polarized field (ORD), whereas the imaginary part accounts for the difference in the absorption (CD).

More precisely, if a beam of linearly polarized light traverses a sample of chiral molecules, the plane of polarization is rotated by a certain angle κ. Such light can be decomposed into left and right circular polarizations (LCD and RCD, respectively). A net rotation of the plane of polarization appears, because, for circular waves of opposite rotations, the real parts of the index of refraction are different ($n_L \neq n_R$). The spectrum of optical rotation (OR) as a function of wavelength of the incident light is called the optical rotatory dispersion (ORD). Measurement of the optical rotation, carried out for a given wavelength, is still used as a simple method for a determination of the optical purity of a chiral sample [30].

The second phenomenon, circular dichroism, can be rationalized by the fact that the rate of decay of the imaginary component of the refractive index is different for the left- and right circular polarized light. Circular dichroism is defined as the difference in the absorbance $A = \log_{10}(I/I_0)$ of a molecule for left versus right

circular polarized radiation in the infrared region of the electromagnetic spectrum
$\Delta A = A_L - A_R$. Under the assumption that the Beer-Lambert's law is satisfied,
$A = \varepsilon(\lambda)cl$ (where $\varepsilon(\lambda)$ is the molar extinction coefficient, c is the solute molarity,
l is the sample path-length) the difference $\Delta\varepsilon(\lambda) = \varepsilon_L(\lambda) - \varepsilon_R(\lambda)$ determines the
spectrum. For two enantiomers L and R, the intensities of the spectra are of equal
magnitudes but of opposite signs, $\Delta\varepsilon_L(\lambda) = -\Delta\varepsilon_R(\lambda)$. This is why the spectra
contain information about the absolute configuration of a chiral molecule. Electronic
CD (ECD) is the original form of CD, which probes electronic transitions using UV–
vis light [30–32]. These measurements are still extensively used for determination
of the absolute configuration (AC) of chiral compounds.

Yet, two other, very promising, techniques belong to the group of chiroptical
methods. They are known under the general name of vibrational optical activity
(VOA) [30]. The first is vibrational circular dichroism (VCD) [33, 34] and the
second is vibrational Raman optical dichroism (VROA) [34]. As ECD probes the
effect of chirality on electronic transitions, VCD probes the chirality of vibrational
transitions. One advantage of VCD over ECD is that high quality vibrational spectra
can be obtained without treating excited electronic states. Unfortunately, VCD
has a disadvantage compared to ECD: VCD is about four orders of magnitude
weaker than the parent IR vibrational absorption (VA). This explains the difficulties
with observation of the VCD spectra before the era of the Fourier-transform IR
spectrometers. The rapid advancements in laser technology also have made possible
measurements of the vibrational Raman optical activity (VROA) spectra. Like the
other chiroptic methods, VROA measures scattering cross-section of the chiral
sample for left- and right-hand circular polarized light. In spite of the very low
intrinsic intensities, the VCD and VROA spectra have been proved to be good
sources of information on molecular conformation, which is hardly accessible
by other spectroscopic techniques like NMR or X-ray [34–37]. CD and ROA
are capable of distinguishing left and right enantiomers, but are also sensitive
to molecular stereochemistry. Usually only fundamental vibrational transitions
are considered in the VCD spectra. Similarly to the VA spectra, the VCD band
frequency yields information on energy of particular vibrational mode, but unlike
the VA method, the VCD band intensity exhibits either positive or negative sign.

First chiroptical observations of molecular vibrations were done in early 1960s
for vibrational fine structure of the ECD and ORD spectra observed in the UV–vis
region [38–41]. In late 1960s, Deutsche and Moscowitz published two theoretical
and computational papers arguing for need and usefulness of studies on optical
activity of vibrational origin in the field of polymer science [42, 43]. Probably
first VCD measurements were published in 1972 by Dudleys, Mason, and Peacock
for cholesteric mesophase [44] and Chabay, Hsu, and Holzwarth for tartarate salt
and inorganic salts [45–47]. Soon, Schellman published formulas for calculation
of the VCD rotatory strengths [48], and in 1975 the theoretical fundament for
VCD theory was established in Barron, Buckingham, and Raab papers [49, 50].
The same year, Nafie, Cheng, and Stephen presented VCD spectra of improved
quality [51] than those measured earlier [52], and one year later they presented
good quality vibrational circular dichroism spectra of a number of chiral molecules

in the liquid phase [53]. In 1979, Nafie and Diem designed and built the first FTIR-VCD spectrometer at Syracuse University [54]. Commercial attempts to sell the VCD instruments yet failed in the 1980s [55], and in mid 1990s first commercially available VCD Chiral*IR* spectrometer, from Bomem/BioTools, appeared at the market [56].

There are three principal areas of application of the VCD technique: (1) determination of the absolute configuration, (2) determination of the optical purity, and (3) determination of the conformational equilibria of chiral molecules in solution. Therefore now, this method is recognized as an excellent tool for structural studies on chiral molecules, which is especially relevant for studies of biomolecules in their native aqueous environment [33, 37].

15.2.2   Chirality Transfer

We say that the *chirality transfer* occurs from a chiral to an achiral molecule if, in presence of the chiral molecule, the achiral one gains a property observable by a *chirality-sensitive-method* (CSM). The origin of this phenomenon is a symmetry breaking incident induced by intermoleculecular interactions between the two individua which shifts the achiral molecule from the achiral point symmetry group to the chiral one. This means that achiral molecule possessing either a symmetry plane, or symmetry center, or *2k*-fold inversion axis looses one or more such symmetry elements as a result of intermolecular interactions, and thus it falls into a chiral point symmetry group.

We call the chirality-sensitive-method any physico-chemical method yielding a response selective to the chiral systems. Obviously, the magnitude of the response of the method used for observation of chirality transfer phenomena does matter. In this paper we focus on chiroptical methods with special attention to the VCD technique. Clearly, chirality transfer observed by the VCD method may be hardly observable by the ECD method, as for example, the non-aromatic amino acids, which mainly have only week and unspecific spectra in the UV–vis range, exhibit quite specific ones in the IR region. Also, some molecules can be observed easier by ROA than VCD measurements, and vice versa.

As far as we know, the first VCD study on chirality transfer phenomenon was done by Bürgi, Vargas, and Backer in 2002 [57]. They measured and calculated the VCD spectra of H-bond complexes of a cinchonidine derivative and its modified enantiomer with trifluoroacetic acid. As a result, they demonstrated that the chiral cinchonidine induced, through the H-bonding with the achiral CF_3COOH, the VCD bands of the achiral acid. Also, the sign of the $\nu(C = O)$ band was sensitive to the absolute configuration of the complex partner. The authors suggested the "induced VCD" effect to allow probing of the chiral binding site. One year later, Ellzy *et al.* observed and calculated the VCD spectra of L-alanine dissolved in water [58]. The calculated spectra clearly demonstrated the chirality transfer phenomenon; yet, the authors focused their attention on alanine bands. Also in 2003, we published a study on VCD spectra of intermolecular H-bonding complexes of model chiral β-lactams with the HX molecules (X = F, Cl, Br) [59]. Two stable types of the complex were shown to be competitive: a cyclic one, in which the HX molecule interacted with both C=O and NH moieties; and a bent one, where the HX was hydrogen bonded solely to the C=O group. We stressed the achiral HX molecules to gain significant VCD rotational strength and which sign revealed the geometry of the hydrogen bond. In our further studies, the geometrical arrangement of the H-bond partners was argued to be determinable based on the VCD spectra. However, a difficulty arose for the D-lactic–water [60] and cysteine-water [61] complexes which have exhibited a rich conformational changeability and their VCD spectra were found to be very sensitive to the conformational alterations. Again, the vibrational modes of the achiral water molecule become VCD active after the complex formation, and the sign of some of them followed the geometry changes [60, 61]. In the same time experimental and theoretical studies demonstrated the VCD chirality transfer phenomenon to be promising tool in investigations of intermolecular interactions [29, 62–77].

15.2.3   Mode Robustness

To evaluate whether the calculated vibrational mode can be assumed as a reliable VCD manifestation of molecular arrangement, which can be applied to establish an absolute configuration of a molecule, or not, Baerends group introduced the concept of the *VCD mode robustness* [28, 78–80]. Indeed, using an inappropriate computational method, accidental basis set, choosing an arbitrary mode, or unfortunate solvent one can obtain false conclusions from the comparison between calculated and experimental VCD spectra. Nicu and Baerends [28, 79], Nicu, Neugebauer, and Baerends [78] and Nicu *et al.* [80] performed detailed analyses of three components on which the VCD intensity is dependent: the nuclear displacement vectors of the normal modes and atomic axial and atomic polar tensors: AATs, APTs. The VCD intensity of the i-th vibrational mode is determined by the rotational strength R(i). The rotational strength is the scalar product of the Electric and Magnetic Dipole Transition Moments (EDTM and MDTM, respectively). The sign of the scalar product is determined by the angle ξ between the vectors. Thus when $\xi < 90°$ the sign of the VCD intensity is positive (R(i) > 0), whereas when $\xi > 90°$ it is negative (R(i) < 0). In achiral molecules, the two vectors (generally different from zero) are perpendicular and yield the scalar product equal to zero (R(i) = 0). For ξ close to 90°, even small perturbation produced e.g. by solvent, interactions, molecular conformation, or, in computations, by an inadequate theory level or basis set, may change ξ across the 90° and thus may induce an erratic change of the sign.

The congruity between experimental and theoretical VCD spectra was studied for numerous molecules by Nicu and Baerends [28]; Nicu, Neugebauer, and Baerends [78]; Nicu [79], and Nicu *et al.* [80]. They varied DFT functionals, basis sets, and solvents, and claim to the distinction of six categories of VCD bands behavior. The modes belong to the class A if they are insensitive to any perturbations. If the sign of the rotational strength of the mode varies upon a change of the mode, it is classified to the group B, whereas if the rotational strength just changes its magnitude upon a stress it falls into the category C. The modes newly appearing as a consequence of "chirality transfer" are grouped into class D, whereas the modes which undergo large changes in both: position and intensity (IR as well as VCD), are grouped into class E. The last group contains, in vast majority, the modes engaged into a strong hydrogen bonding. New bands, resulting from interaction of a chiral solute with solvent, unpredictable by direct calculation of solute molecule, form the class F.

At the end of such a categorization and analysis of factors determining the VCD intensity they defined the *VCD mode robustness*. The VCD signals can be classified as *robust* if and only if they are intense and whose sign is insensitive to a (slight) perturbation. Oppositely, the VCD signals are *non-robust* if and only if they are either weak and or whose sign is sensitive to a (slight) perturbation. Thus, the mode is robust if the two transition moments, electric dipole and magnetic dipole, are large and the angle between them is far from the right angle (not less than 30°). Additionally, the robust modes are required to be "unmixed" (in terms of local modes). The *non-robust* modes are those whose either ξ is close to 90° or at least one of the two transition moments is small. As a conclusion, Nicu and Baerends have recommended indicating the mode robustness when calculating the VCD spectra. This is especially important for determination of the absolute configuration [28, 78–80]. Thus, the calculated data should be accompanied by the ξ angles and values of the electric and magnetic dipole transition moments.

Quite recently Nicu, Neugebauer, and Baerends [78]; Nicu [79]; Debie *et al.* [80] have been studied the chirality transfer for the pulegone molecule in $CDCl_3$ solvent [80]. They studied also the 1:1 pulegone-$CDCl_3$ complex computationally (in vacuum and in solvent simulated by the COSMO model) by using the OLYP and BP86 DFT functionals combined with TZP basis sets and using different geometry convergence criteria. Additionally, different stable arrangements of the complex partners were considered. They also took into consideration change of the normal mode with the change of the computational method by introducing a parameter revealing similarity between two movements called the *overlap between two modes*. The parameter is ranging from 0 to 1 as the similarity ranges from none to complete. For the focus of this chapter, the most important conclusion of the Nicu *et al.* article is that the transfer of chirality calculated for the $\nu(C–D)$ stretching vibration VCD band of $CDCl_3$ is non-robust and that the computational methods used cannot reliable predict their experimental chirality transfer findings. Indeed, change of each of the above mentioned parameters can change sign of the $\nu(C–D)$ VCD intensity. Figure 15.1a, b shows variation of the ξ angle (determining the sign) with the change of 16 diverse computational criteria and nine diverse arrangements of the 1:1 pulegone-$CDCl_3$ complex, respectively.

Nicu *et al.* paper shows that predictability of the chirality transfer phenomenon by standard DFT methods approached serious limitations [80], however, it leaves reader with several questions that have to be answered in future studies. The most

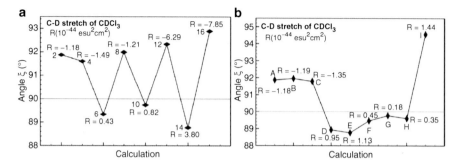

Fig. 15.1 Comparison of the ξ angle between electric and magnetic dipole transition moments of the chirality transfer ν(C–D) mode in the 1:1 pulegone-CDCl₃ complexes calculated by using diverse computational criteria (**a**) and 9 diverse complex arrangements (**b**) (Reproduced from Ref. [80] with kind permission of Wiley-Liss, Inc)

important are about choice of the best: (i) functional (*ab initio* method), (ii) basis set, (iii) solvent model, (iv) geometry convergence criteria, (v) PES shape, to best reproduce the VCD chirality transfer phenomena. We agree with one of the reviewers of this chapter that the VCD sign of a "truly non-robust" mode cannot be predicted safely by any computations, since even a very small perturbation during the experiment can result in a sign change. This does not mean that "truly non-robust" mode cannot be observed experimentally. Also, that an important question is whether there are cases when the chirality transfer effect is large enough to yield a robust mode of the complexed achiral molecule. Thus still, it seems that more discussion of connections between perturbation slightness, accuracy of the computational methods and robustness of the VCD intensities is needed to better understand various aspects of the VCD chirality transfer effect.

15.3   Methods of Calculations of the VCD Spectra

Theoretical models allow us to better understand the relationship between the spectroscopic parameters and the structure of the samples and they are indispensable for a realistic modeling of the parameters as a function of structure. Therefore, we present here the essential features of the theory underlying the calculations of VCD spectra. The first principle calculations of the chiroptic spectroscopic parameters have appeared over the years. Theoretical foundations of the models including their mathematical formulation and numerical implementation have been discussed in the articles [30–32], while the *ab initio* most successful – in terms of its applications-the perturbative approaches by Stephens calculations are presented in Refs. 81–87. We shall refer to some of them in what follows. However, we directing the reader to the review articles for a more detailed description [81–83].

15.3.1 The Molecular Origin of Vibrational Circular Dichroism

Here an outline of the derivation of the quantum mechanical expression for the vibrational circular dichroism will be presented. The chiroptical analog of the vibrational excitation spectrum (VA) is the VCD spectrum. The chiroptical analog of the band strength (band intensity) of VA is the transition moment called the rotatory strength $R_{g1,g0}(i)$. Let us remind: the intensity of the VA band for each mode is proportional to the absolute square of the electric-dipole transition moment (EDTM) of the molecule, called dipole strength at the molecular level. It is denoted $D_{g1,g0}(i)$ for the i-th mode and it is connected with the transition from the ground electronic (g) – ground vibrational state (0) described by the Ψ_{g0} wave function to the ground electronic (g) – excited vibrational state (1) denoted as Ψ_{g1}, when Born-Oppenheimer separation of electron and nuclear motion is accepted,

$$\Psi_{g0} = \psi_g\,(\mathbf{r}; \mathbf{R})\,\chi_0\,(\mathbf{R})$$

$$\Psi_{g1} = \psi_g\,(\mathbf{r}; \mathbf{R})\,\chi_1\,(\mathbf{R}) \tag{15.1}$$

where $\mathbf{r}$ and $\mathbf{R}$ denotes the set of electronic and nuclear coordinates, respectively.

The dipole strength expression contains also the absorption operator proportional to the dipole moment change in the i-th mode multiplied by the geometrical change of molecule in the mode i.e., the normal mode coordinate. The dipole strength is always positive as an absolute square. The corresponding intensity of the VCD band of the i-th mode is connected to the rotatory strength $R_{g1,g0}(i)$, which is the imaginary part of the scalar product of the electric-dipole transition moment (EDTM) and the magnetic-dipole transition moment (MDTM).

The equations below define these two quantities at the molecular level for the i-th vibrational mode:

$$D_{g1,g0}(i) = \left|\langle \Psi_{g0}\,|\mu_{el}\,|\Psi_{g1}\,\rangle\right|^2 = E_{g1,g0}^2(i) \tag{15.2}$$

$$R_{g1,g0}(i) = \mathrm{Im}\left[\langle \Psi_{g0}\,|\mu_{el}\,|\Psi_{g1}\rangle\langle \Psi_{g1}\,|\mu_{magn}\,|\Psi_{g0}\rangle\right] = E_{g1,g0}(i)\mathrm{Im}\left[M_{g1,g0}(i)\right] \tag{15.3}$$

The sign of the rotatory strength for the i-th mode is thus determined by the angle between the vector $E_{g1,g0}(i)$ of the EDTM and the vector $M_{g1,g0}(i)$ of the MDTM, which can be seen from Eq. 15.3. If two vectors are orthogonal with respect to each other, as in achiral molecules, the rotatory strength is zero.

Molecular theory of the VCD by Stephens involves the determination of wave-function parameters perturbed with respect to the nuclear displacements and the magnetic perturbation. It means, it requires solution of $3N+3$ response equations (N is the number of atoms in the system). Theory of the VCD spectra is constructed within the following assumptions of: (a) Born-Oppenheimer approximation (BO) (15.1), in which total wavefunction is a product of electronic and nuclear wave functions for an electronic reference state "g0" and vibrational state "g1";

(b) exclusive population of the lowest vibrational level; (c) double harmonic (electrically and mechanically harmonic) approximation and use of harmonic force field (HFF); (d) since we are concerned with nuclear motion, including vibronic coupling is essential for vibrational CD to obtain the electronic contributions to the rotatory strength (MDTM part), because within BO approximation $\langle \psi_{g0} | \mu_{magn}^n | \psi_{g0} \rangle = 0$ at all molecular geometries and only nuclear contribution survives [81, 82] (for closed shell molecules the contribution from the electronic part of the magnetic moment vanishes, irrespective of geometry), (e) separability of the electric and magnetic dipole transition moment operators into electronic and nuclear parts, Eq. 15.4.

$$\mu_{el} = \mu_{el}^e + \mu_{el}^n \text{ and } \mu_{magn} = \mu_{magn}^e + \mu_{magn}^n \tag{15.4}$$

Under these assumptions the atomic contributions of the EDTM and MDTM are given by two important tensors defined per atom γ: $P_{\alpha\beta}^\gamma$ are the derivatives of molecular electric-dipole moments with respect to Cartesian displacements of nuclei from equilibrium and is known as an atomic polar tensor (APT). The APT tensors represent the derivative of molecular EDTM of the entire molecule with respect to the Cartesian displacement of the γ–th atom from equilibrium. Hence it quantifies the change in molecular charge distribution during nuclear motion. This tensor is known in the description of the VA intensity. Computationally, the APT are obtained as the geometric gradient of the electric dipole moment for the electronic ground state. The second tensor – $M_{\alpha\beta}^\gamma$ is called atomic axial tensor (AAT). It is the derivative of the ground state magnetic moment with respect to the velocity of the nuclei. Usually, in practice, the derivatives of the EDTM and MDTM are obtained in terms of the Cartesian coordinate of γ atom, X_α^γ, $\alpha = (x,y,z)$ and then are converted to normal coordinates $Q(i)$, where $S_\alpha^\gamma(i)$ is the transformation matrix from the Cartesian X to the normal coordinates $Q(i)$. The so-called nuclear displacement S-vector components are obtained in a vibrational frequency calculation.

$$X_\alpha^\gamma = \sum_i S_\alpha^\gamma(i) Q(i) \tag{15.5}$$

Within these three tensors final Stephens' and Devlins' equations for the rotatory strength $R_{g1,g0}(i)$ of the fundamental excitation of i-th mode is the following [82]:

$$R_{g1,g0}(i) = \hbar^2 \sum_\beta \sum_{\alpha\alpha'\gamma\gamma'} \left\{ S_\alpha^\gamma(i) P_{\alpha\beta}^\gamma \right\} \left\{ S_{\alpha'}^{\gamma'}(i) M_{\alpha'\beta}^{\gamma'} \right\} \tag{15.6}$$

Both tensors APT ($P_{\alpha\beta}^\gamma$) and AAT ($M_{\alpha\beta}^\gamma$) are composed of the electronic $E_{\alpha\beta}^\gamma$, $N_{\alpha\beta}^\gamma$ and nuclear parts $I_{\alpha\beta}^\gamma$, $J_{\alpha\beta}^\gamma$ respectively [82–84].

$$P_{\alpha\beta}^\gamma = \left(\frac{\partial \mu_\beta}{\partial X_\alpha^\gamma} \right)_0 = E_{\alpha\beta}^\gamma + I_{\alpha\beta}^\gamma \tag{15.7}$$

$$M_{\alpha\beta}^{\gamma} = \left(\frac{\partial \mu_{magn}}{\partial \overset{\bullet}{X}_{\alpha}^{\gamma}} \right)_0 = N_{\alpha\beta}^{\gamma} + J_{\alpha\beta}^{\gamma} \qquad (15.8)$$

The pairs: electronic and nuclear contributions are, as below:

$$E_{\alpha\beta}^{\gamma} = 2 \left\langle \left(\frac{\partial \psi_g(X)}{\partial X_{\alpha}^{\gamma}} \right)_{X_0} \left| (\mu_{el}^e)_\beta \right| \psi_g(X_0) \right\rangle \qquad (15.9)$$

$$I_{\alpha\beta}^{\gamma} = (Z^{\gamma} e) \delta_{\alpha\beta} \qquad (15.10)$$

$$N_{\alpha\beta}^{\gamma} = \left\langle \left(\frac{\partial \psi_g(X)}{\partial X_{\alpha}^{\gamma}} \right)_{X_0} \left| \left(\frac{\partial \psi_g(X_0, H)}{\partial H_\beta} \right)_{H=0} \right\rangle \qquad (15.11)$$

$$J_{\alpha\beta}^{\gamma} = \frac{ieZ^{\gamma}}{4\hbar c} \sum_{\delta} X_0^{\gamma} \varepsilon_{\alpha\beta\delta} \qquad (15.12)$$

The meaning of the symbols are the following: $\hbar$ is the reduced Planck constant, ν_i is the frequency of the i-th vibration, H is the static magnetic field, e and c are the charge and speed of light in vacuum, μ_{el}^e is the electronic contribution to μ_{el}, eZ^{γ} is the charge of nucleus γ, respectively, $\delta_{\alpha\beta}$ is the Kronecker delta, and $\varepsilon_{\alpha\beta\delta}$ is the Levi-Civita tensor. The wavefunction $\psi_g(X)$ depends on the not-fixed coordinates of the nuclei, while $\psi_g(X_0, H)$ is the wavefunction for the equilibrium positions of the nuclei in the presence of a magnetic field. The electronic part of AAT tensor $N_{\alpha\beta}^{\gamma}$ is the overlap integral containing the derivatives of the electronic wavefunction of the ground state g with respect to the external magnetic field.

The electric- and magnetic-dipole moment operators entering the rotatory strength tensor are dependent on the origin. The EDTM causes no problems, but in computations with a finite atomic orbital basis set the response properties involving MDTM are origin dependent, i.e. they may change if an origin shift is applied to the coordinates of the molecule. The Gauge Invariant/Including Atomic Orbital (GIAO) basis, where each individual basis function is made to depend explicitly on the external magnetic field induction, eliminates this origin dependence even for finite basis set [85, 86] (see *The basis set requirements*).

To summarize this part, the prediction of VA and VCD spectra within the harmonic approximation requires:

(a) the normal modes and the frequency calculations at the harmonic force field
(HFF), i.e., the Hessian matrix with the elements: $\left(\frac{\partial^2 E}{\partial X_{\alpha}^{\gamma} \partial X_{\alpha'}^{\gamma'}} \right)$

(b) the intensity of VA spectra, i.e. the APT tensor, with the elements $\left(\frac{\partial \psi_g}{\partial X_{\alpha}^{\gamma}} \right)_0$

(c) the intensity of VCD spectra, i.e. in addition to APT also the AAP tensor, with
the elements $\left(\frac{\partial \psi_g}{\partial H_\beta} \right)_0$.

15.3.2   Practical Aspects of the Calculations: Methods, Basis Set, Software

The previous section has outlined the general computational schemes for the calculations of VCD spectra. In the following, we will shortly discuss the computational methods used for calculation of the spectra.

15.3.2.1   Electronic Structure Methods

Spectroscopic properties can be calculated from an approximate electronic wave function or from approximate electronic density (density functional theory, DFT). VCD requires the knowledge of geometrically and magnetically perturbed wave functions, but VCD does not require the calculation of frequency dependent response. The Hartee-Fock approximation (HF) lacks of treatment of the electron correlation, yet, it yields some properties at acceptable accuracy level. However, the electron correlation energy is necessary for proper description of the AAT tensor [87]. Therefore, the HF method is generally not recommended for computing chiroptical properties. The static correlation is well described by the multiconfigurational self-consistent field (MC SCF) model and this method was successfully implemented for the calculations of $R_{g1,g0}$ [88, 89]. Nevertheless, it becomes difficult to perform calculations with explicit inclusion of the electron correlation effects for larger chemical systems.

The most widespread method, which can be applied to large systems, where the use of MC SCF or Coupled Clusters (an approximate singles-doubles CC2) methods is not possible, is density functional method (DFT). DFT is now an acronym for a variety of different approximations of the functional [90]. It has a comparatively low scaling with the system size and typically also has lower demands on the basis set. Its cost is similar to that of RHF model and nowadays it plays a dominant role in the theoretical investigations of the vibrational rotatory strength $R_{g1,g0}$. The introduction of the DFT methods caused that the evaluation of rotatory strength became a routine procedure, now used to support experimental works.

The exact form of the universal exchange-correlation functional in DFT method remains unknown. Therefore, over the years, several approximate exchange-correlation functionals have been developed and tested. Functionals used for calculations of different molecular properties can be grouped into three classes [90]: (a) local-density approximation (LDA) successfully used in solid-state physics [91], (b) non-local, gradient approximation (GGA) [94, 95] or Keal-Tozer (KT1) [96], and (c) hybrid functional Becke-3-Lee-Yang-Parr (B3LYP, [92, 93]) (based on the BLYP functional with 20% exact exchange and fitted to thermochemical data) [97] or Becke 97 (B97) or Coulomb-attenuated (CAM-B3LYP) functional, where exchange included in the functional depends on the inter-electronic distance [98, 99]. Some of these functionals are better suited than others to the calculation of spectroscopic properties. The most popular for the rotatory strength calculations is now the Becke-3-Lee-Yang-Parr (B3LYP) hybrid functional [97]. For some molecules containing peptide groups, the BPW91 functional has shown better agreement with experiment [100].

15.3.2.2 The Basis Set Requirements

In the *ab initio* calculations, except the choice of the computational method, the choice of the proper basis set for determination of any molecular property (VCD intensity included) is an additional problem. A general rule for basis set choice is hardly to be given, yet, some suggestions can be provided. As for the energy, larger basis set at correlated level are required than in SCF for the energy calculations, aiming at a representation of the Coulomb hole. Calculations of the magnetic properties make the calculations more difficult than those of the energy alone. The standard basis sets are not adequate to predict them correctly. Basis set with flexibility, such as with correlation-consistent basis set aug-cc-pVXZ [101] appear to offer a good compromise between accuracy and computational cost for medium size molecules. However, it should be kept in mind, that the description of the interacting systems (by H-bonds or van der Waals interactions), like those discussed in this review, requires economic polarized basis with an addition of diffuse functions.

The problem of lack of origin independence in *ab initio* calculations of circular dichroism in finite basis sets has been dealt through the introduction of London atomic orbitals (LAO) or the gauge invariant (independent) atomic orbitals (GIAO) in calculation of AATs. Accordingly, AATs are gauge-independent, and yield origin-independent rotational strengths. Each individual basis function is made to depend explicitly on **B**-the external magnetic field.

$$\Phi_\mu \left(\mathbf{B}, \mathbf{P}_M \right) = \exp \left(-i\, \mathbf{A}_{OM}\mathbf{r} \right) \chi_\mu \left(\mathbf{P}_M \right) \tag{15.13}$$

where $\chi_\mu \left(\mathbf{P}_M \right)$ is a Gaussian basis function, **A** is the magnetic vector potential. The effect of this phase factor is to move the global gauge origin **O** to the best local gauge origin for each basis function, **M** – the atomic center to which the basis function is attached. GIAO improve the basis set convergence in addition to providing gauge origin independent results for nuclear magnetic shielding, magnetizabilities and VCD properties.

15.3.2.3 Implementation Procedure and Program Packages

A first step, before to compute rotatory strength, is to determine the molecular structure. This point involves geometry optimization by proper method with a proper quality basis set (with polarization functions), followed by a confirmation of the stationary point characteristics on the potential energy surface (PES), i.e.

calculations of the harmonic frequencies. The minimum nature of conformers is to be confirmed by the harmonic frequency calculations, which must show no imaginary frequencies. Next step is to compute the chiroptical response property [102]. Even when a molecule of interest is rigid in the gas phase, averaging over the nuclear vibrational motions (zero point) is desirable, as in calculations of NMR parameters and some other optical ones too (which at present are not performed routinely). When a molecule is flexible and it have a few different conformers, they could have different chiroptical responses (and even the sign). If it is possible, the Boltzmann-population-weighted spectra could be generated and then compared with the experimental one.

The calculation of the $R_{g1,g0}$ has been implemented at different levels of theory in several computer codes of widespread use. The HFF, APT and AAP tensors, as derivatives of energies and wave functions with respect to the proper perturbations, can be evaluated using either numerically (finite differences of gradients) or directly, analytically. Software packages that are capable of VCD spectra calculations are available commercially. Here, we present, in alphabetical order, the most popular software packages implementing analytical derivatives in the calculations of the HHF, APT and AAP tensors: (a) Amsterdam Density Functional, ADF [103]; (b) CADPAC [104]; (c) DALTON [105]; and (d) GAUSSIAN, G03, G09 release [106, 107].

Recently, due to both availability of the Fourier transform VCD spectrometers (FT-VCD) and DFT software for predicting the VCD spectra, the VCD technique has become widely recognized and used [108–110]. DFT has been accepted by the *ab initio* quantum chemistry community as a cost-effective approach to computations of molecular structures and spectra (vibrational and NMR) of molecules of chemical interest. Many studies have shown that vibrational frequencies and VCD intensities calculated by means of DFT methods are more reliable than those obtained at the MP2 level [87].

15.4 Applications of VCD to Study Chirality Transfer

Obtaining comparable conditions for a system studied theoretically and experimentally has been always fundamental for spectroscopy. The more system is simplified the easier and exacter it is modeled theoretically. However, a simplification of the system is usually difficult to satisfy experimentally. In case of the vibrational spectroscopies intermolecular interactions, especially hydrogen bondings, play a primary role in changing the spectral patterns. Therefore, simulating the surrounding of the molecule, the most exactly reproducing the experimental conditions of the studied molecule, is the theoretical challenge, whereas efforts to limit the undesired intermolecular interactions between surrounding (solvent) and the molecule simultaneously preserving a detectable concentration of a solute molecule, is a challenge for experiments. This is also the truth for the VCD spectroscopy.

To solve the dilemma, there are two common ways to reduce the uncertainties generated by the experiment: (1) *use an inert solvent*; (2) *apply deposition in a low-temperature noble gas matrix*. However, often, the more inert is the solvent the lower is concentration of the saturated solutions. Also, a lot of substances decompose when being heated to depose them in the matrix.

On the other hand, theory offers three main approaches: (1) *the supermolecular calculations*, where some neighbouring solvent molecules are explicitly included in the quantum mechanics calculations; (2) *continuum solvent models* in which solvent is modelled as a macroscopic continuum dielectric medium and solute is described quantum-mechanically; (3) *the combined model*, where cluster of molecules of interest is surrounded by solvent molecules, which are generated by Monte Carlo (MC) or by a classical simulation trajectory using molecular dynamics methods. The supermolecular strategy allows reproducing relatively correctly intermolecular interactions within the closest neighbours of the studied molecule; still, the costs of computations increases rapidly as the number of solvent molecules is increased. Moreover, the number of complex arrangements increases rapidly as well. On the other hand, the two most widely used continuum solvent models, the polarizable continuum models (PCMs) and the conductor-like screening model (COSMO), help in imitating the non-directional forces and interactions without a rapid increase of computational time; yet, important parts of vibrational spectra are not predicted accurately [111]. Finally, the combined model, reduces (to some respect) some defects of the former two ones, introducing another inaccuracies being a consequence of the methods used for MC or dynamic calculations.

The VCD spectra are the most often recorded for molecules dissolved in variety of solvents, therefore the solvent influence on the spectra was very seriously considered by the VCD spectroscopists. However, the principal issue was to understand the changes induced by a solvent in the VCD bands of solute molecules. Therefore, the principal was to deduce both the conformational changes of the solute, and to best reproduce its spectrum by different theoretical approaches [112–128]. For amino acids, peptides and proteins in aqueous solutions it was advocated to use both the explicit water molecules and continuum solvent treatment, as even the proper conformation often could not be found without such a methodology [123].

The number of studies considering activation of the vibrational modes of an achiral solvent by a chiral solute increases. Thanks to quantum chemical calculations of the VCD spectra of the interacting systems it is possible to associate some of the VCD signals to specific solute-solvent interactions. This was the case for chinchonidine interacting with trifluoroacetic acid [57], L-alanine in aqueous solution [58, 112], tetrahydrofuran-2-carboxylic acid in $CDCl_3$ [62], camphor in $CDCl_3$ and CD_2Cl_2 [63], and pulegone in $CDCl_3$ [64]. Also the interactions of chiral methyl lactate [65], (S)-glicidol [66], lactic acid [67], propylene oxide [68], and glicidol [69] with water solvent as well as methyl lactate in methanol [29] revealed unequivocally the VCD chirality transfer from solute to achiral solvent. For example, the $\beta(HOH)$ bending vibrations band located at $1,640\,cm^{-1}$, present in VCD spectra of the lactic acid in water and absent in methanol, cannot be assigned to anything else than water H-bonded to lactic acid [67].

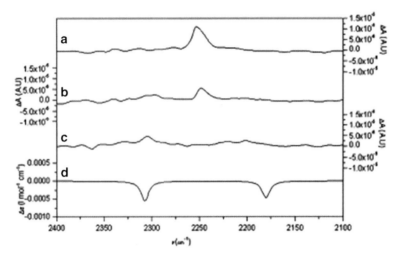

Fig. 15.2 The experimental VCD spectra of (+)-pulegone in binary mixtures of CDCl$_3$ and CD$_2$Cl$_2$: (**a**) 100:0, (**b**) 50:50, (**c**) 0:100, and (**d**) the calculated spectrum of the R-pulegone-CD$_2$Cl$_2$ association (Reproduced from Ref. [64] with kind permission of PCCP Owner Societies)

The low-temperature matrix-isolation conditions are the excellent reference systems for calculations of the intermolecular interacting molecules. In the 1980s the pioneering work by Schlosser *et al.* [70] and Henderson and Polavarapu [71] opened new possibilities for experimental investigations using matrix isolation vibrational circular dichroism spectra (MI–VCD). Recently, Tarczay, Magyarfalvi, and Vass performed modern MI–VCD experiments for (R)-2-amino-1-propanol [72] and shown the MI–VCD to be useful technique for study conformation of floppy molecules interacting by H-bonding. The VCD spectra of Ac-Gly–NHMe and Ac-L-Ala–NHMe molecules were registered in matrices and the VCD signals of water molecules interacting with peptides were detected [73, 74]. Moreover, it was also shown that the signs of these transitions were sensitive to the alignment of the complexed water and it could be used for structural identification of the complex geometry [74]. However very recently, Góbi *et al.* studied conformational landscape of β-homo-proline derivative in both cryogenic matrices and dissolved state in several solutions by using IR, VCD, and computational methods, they did not reported on chirality transfer phenomena, which could be covered by very complex conformational equilibria [75].

In the immersing article, Debie *et al.* showed that in the VCD spectra of chiral pulegone dissolved in CDCl$_3$ the solvent C–D stretching vibration band has been observed (Fig. 15.2). Moreover, the spectrum was well reproduced by quantum chemical calculations assuming 1:1 H-bond complex between pulegone and the solvent [64]. On the other hand, despite the B3LYP/6 − 311 + G** calculations predicted two CD$_2$ stretching vibration bands (regardless the sign) to be manifested in the VCD spectra of pulegone dissolved in CD$_2$Cl$_2$, they were not observed [64]. This was interpreted as either experimental evidence of weak solute-solvent

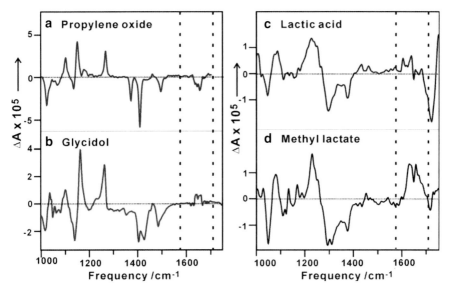

Fig. 15.3 Comparison of the induced solvent chirality features in the water bending vibrational region for the S form of (**a**) propylene oxide, (**b**) glycidol, (**c**) lactic acid, and (**d**) methyl lactate in aqueous solutions (Reproduced from Ref. [69] with kind permission of The American Institute of Physics)

interactions or result of more complicated solvation phenomena present in the system, where one pulegone molecule is surrounded by nine methylene chloride molecules.

Quite recently, Merten, Amkreutz, and Hartwig have studied the VCD spectra of α-phenylethyl isocyanide (PENC) in CDCl₃ by DFT method [76]. They found the calculated VCD spectra of PENC and its dimer to be almost unaffected by the solvation, except of the ν(C–D) stretching vibration band of chloroform that was activated at ca 2,220 cm^{-1} [76]. On the other hand, Aviles-Moreno *et al.* have reported interesting behavior of the C=O and C=C stretching vibration VCD bands of pulegone, which changed their intensities and signs from solvent to solvent (CCl₄, CH₂Cl₂, CHCl₃, DMSO) [77]. Thus, the bands are evidently non-robust [28, 78, 79], and moreover, the B3LYP/IEF–PCM predictions of the VCD intensities varies with the change of the basis sets used, even if a correct vibrational assignment can be achieved for the IR and Raman spectra [77].

The just studied VCD spectra of (S)-(−)-glycidol in water by Yang and Xu exhibited prominent VCD chirality transfer at the bending bands of water that become optically active due to hydrogen bonding to glycidol molecules [69]. Over 30 small glycidol-(water)$_N$ clusters (N = 1, 2, 3, 4) studied at the B3LYP/6 − 311++G** level revealed strong cooperative hydrogen bonding effects and enabled for satisfactory reproduction of the observed VCD chirality transfer (Fig. 15.3).

Finally, the additional problems arise when a solute molecule is flexible. The VCD chirality transfer phenomenon is very sensitive to conformational behavior of the chiral molecule and the location of the interacting achiral partner. In our earliest computational study we have chosen a model rigid chiral β-lactam and demonstrated that the VCD band of the H-X stretching vibrations of the hydrogen halide interacting with the lactam, changes its sign with the change of the complex geometry [59]. Next, we have studied non-rigid small biomolecules such as lactic acid [60], cysteine [61], and quinine [129] with increasing complexity of the conformational landscape. The knottiness of the spectra increases very quickly with the number of conformers necessary to be considered and number of interacting sites between the complex partners. Nevertheless, for the systems with a few dominating complexes it is possible to indicate either the complex geometry or the complexation site. Therefore, the VCD chirality transfer phenomenon seems to conserve the predictive power if and only if the number of species to be distinguished is relatively small. However, more study is needed to understand scopes and limitations of the VCD chirality transfer bound by the mode robustness conditions.

15.5 Perspectives

In this review, we have briefly presented the basic concepts associated with the theoretical interpretation of the VCD intensities, the computational requirements needed for their accurate theoretical studies, and an overview of recent predictions of VCD observations of chirality transfer from a chiral molecule to an achiral one as a result of hydrogen bonding interactions between them. With the recent developments of experimental and theoretical methods and computer software, the calculations of VCD intensities have become a routine task, permitting a better interpretation of large parts of the spectra, yet; important features, such as chirality transfer bands, are still unreliably reproduced. Therefore, measuring and interpreting VCD phenomenon is still a challenging task for both experimental and theoretical investigations.

Most calculations of VCD spectra are nowadays carried out with DFT methods. In these calculations the most important is choice of a proper functional. Unfortunately, no single functional performs the best for different classes of compounds and different classes of problems. Construction of a functional optimal for calculations of VCD spectra is needed. Second, an alternative approach to speed up the calculations of the rotatory strength $R_{g1,g0}(i)$ would be desirable, as has very recently been shown in Ref. [130]. Third, the VCD chirality transfer effect is well confirmed to occur in achiral solvents and in low-temperature matrices; yet, its predictability by the quantum chemical methods is still being questioned because most of the chirality transfer bands seem to be non-robust. Therefore, we see an urgent need for development of both: experimental findings of the VCD chirality transfer effects and deepening of the theoretical understanding of the mode robustness criteria [28, 78–80]. To trust the calculated VCD intensity, the strict criteria should define, *inter alia*, the role of the following factors on the correctness of prediction of the VCD intensities: (i) electron correlation contribution; (ii) mechanical, electrical, and magnetical anharmonicities; (iii) basis set; (iv) solvent model; (v) geometry convergence criteria; (vi) potential energy surface shape, (vii) stabilization energy of the complex. Thus still, it seems that more discussion of connections between perturbation slightness, accuracy of the computational methods and robustness of the VCD intensities is needed to better comprehend various aspects of the VCD chirality transfer effect.

As for experimental studies, the symmetry nature of the chirality phenomenon indicates that the effect that could be called the (VCD) *chirality quenching* can probably be measured soon. It should take place when the spectrum disappears as an effect of intermolecular interactions shifting system from a chiral group to an achiral one. The simplest case would be the chiral acid dimer with a symmetry center. In fact, we expected to see the chirality quenching when reading Losada, Tran, and Xu article [67] on the VCD spectra of lactic acid self-aggregation in $CDCl_3$ solution. The L-lactic acid molecule belongs to the C_1 point symmetry group whereas some of its dimers, for instance two L-lactic acid molecules forming the AA dimer in Ref. 67, fall into the achiral C_i group and therefore are expected to be not VCD active [26]. In such a case, detectability of the asymmetric dimers should be possible in presence of the more stable, but symmetric, complex. Probably, in the case studied in Ref. 67, either chloroform or more complicated complexation equilibria enforced asymmetrization of the C_i dimers, and caused them to be active in the VCD spectra.

The other interesting chiral systems have been reported recently by Bouchet *et al.* [131, 132]. For the water-soluble chiral cryptophanol they have showed that for pH > 12 the VCD (and ECD) spectra strongly depended on the nature of the alkali metal ions (Li^+, Na^+, K^+, Cs^+) surrounding the cryptophane. Moreover, presence of the guest molecule inside the cavity of the cryptophanol modified the spectra. Although, the VCD bands of the guest molecules have not been analyzed, they might reveal additional features of the host-guest-counterions interactions. Thus, there is possibility to observe the chiroptical properties transferred to the achiral guest molecule from a host and modified by counterions. Also, one can easily imagine that the opposite VCD chirality transfer, from a chiral guest to an achiral host, modulated by surrounding counterions, can be observed as well.

In summary, although the theoretical calculations of the VCD spectra are technically less and less difficult, there are still challenging theoretical problems to be solved. The reliable prediction of the VCD chirality transfer phenomenon is one of such problems. Once this problem is overcome, the VCD technique supported by theoretical calculations will accelerate the current understanding of the role o chiral substances and their intermolecular interactions.

Note added in proof After submission of this Chapter, Góbi and Magyarfalvi [133] and Góbi *et al.* [134] have published a mode robustness measure alternative to the Nicu and Baerends one. It is defined as the ratio $\zeta = R/D$ between the rotational and dipole strengths of the mode, which is dimensionless and gauge independent. The authors defined the mode to be the ζ-robust when ζ exceeds 10 ppm.

Acknowledgements This work was supported by Ministry of Science and Higher Education in Poland Grants No. NN204 242034 and N N204 443140. The computational Grants G19-4 and G31-13 from the Interdisciplinary Center of Mathematical and Computer Modeling (ICM) at the University of Warsaw are gratefully acknowledged.

References

1. Pasteur L (1848) Recherches sur les relations qui peuvent exister entre la forme cristalline, la composition chimique et le sens de la polarisation rotatoire. Annales de chimie et de physique 24:442–459, 3e série
2. Flack HD (2009) Louis Pasteur's discovery of molecular chirality and spontaneous resolution in 1848, together with a complete review of his crystallographic and chemical work. Acta Crystallogr A 65:371–389
3. Kelvin L (1894) The molecular tactics of a crystal. J Oxf Univ Junior Sci Club 18:3–57
4. Bentley R (2010) Chiral: a confusing etymology. Chirality 22:1–2
5. (a) Kant I (1783) Prolegomena zu einer jeden künftigen Metaphysik die als Wissenschaft wird auftreten können. Hamburg (Meiner) 2001 http://www.uni-potsdam.de/u/philosophie/texte/prolegom/!start.htm (§13) (English translation: http://philosophy.eserver.org/kant-prolegomena.txt (cf. first part, section 13)); (b) Aristotle (350 B.C.E) Metaphysics (1. 5) (τἀμετἀτἀφυσικἀ) trans: Ross WD http://classics.mit.edu/Aristotle/metaphysics.1.i.html
6. Moss GP (1996) Basic terminology of stereochemistry (IUPAC recommendations 1996). Pure Appl Chem 68:2193–2222, p. 2203, Blue Book, p. 479
7. Mizerski W (2008) Chemical tables, (in polish). Adamantan Publishing House, Warsaw
8. Moss GP (1996) Basic terminology of stereochemistry (IUPAC recommendations 1996). Pure Appl Chem 68:2203
9. Sauvage J-P (1993) Topology in molecular chemistry. New J Chem 17:618–763 (whole issue)
10. Dobrowolski JCz (2003) DNA knots and links (węzły i sploty DNA). Polimery 48:3–15
11. Dietrich-Buchecker CO, Rapenne G, Sauvage J-P, De Cian A, Fisher J (1999) A dicopper(I) trefoil knot with *m*-phenylene bridges between the ligand subunits: synthesis, resolution, and absolute configuration. Chem Eur J 5:1432–1439
12. Amabilino DB, Stoddart JF (1995) Interlocked and interwined structures and superstructures. Chem Rev 95:2725–2828
13. Herges R (2006) Topology in chemistry: designing möbius molecules. Chem Rev 106:4820–4842
14. Murasugi K (1996) Knot theory and its applications. Birkhäuser, Boston-Bazylea-Berlin
15. Pieranski P (1998) In search of ideal knots. In: Stasiak A, Katritch V, Kauffman LH (eds) Ideal knots, vol 19, Series on knots and everything. World Scientific, Singapore/New Jersey/London/Hong Kong
16. Dobrowolski JCz (2003) On the classification of topological isomers: Knots, Links, Rotaxanes, etc. Croat Chem Acta 76:145–152.
17. Barron LD (2009) An introduction to chirality at the nanoscale. In: Amabilino DB (ed) Chirailty at the nanoscale: nanoparticles, surface, materials and more. Wiley, Weinheim
18. Zehnacker A, Suhm MA (2008) Chirality recognition between neutral molecules in the gas phase. Angew Chem Int Ed 47:6970–6992
19. Crassous J (2009) Chiral transfer in coordination complexes: towards molecular materials. Chem Soc Rev 38:830–845
20. Berova N, Di Bari L, Pescitelli G (2007) Application of electronic circular dichroism in configurational and conformational analysis of organic compounds. Chem Soc Rev 36:914–931

21. Feringa BL, van Delden RA (1999) Absolute asymmetric synthesis: the origin, control and amplification of chirality. Angew Chem Int Ed 38:3418–3438
22. Hembury GA, Borovkov VV, Inoue Y (2008) Chirality-sensing supermolecular systems. Chem Rev 108:1–70
23. Alkorta I, Elguero J (2009) Chirality and chiral recognition. In: Leszczynski J, Shukla MK (eds) Practical aspects of computational chemistry methods, concepts and applications. Springer, Heidelberg/Dordrecht/London/New York
24. Keiderling TA (1996) In: Fasman GD (ed) Circular dichroism and the conformational analysis of biomolecules. Plenum Press, New York
25. Pecul M, Sadlej J (2003) Ab initio calculations of the intermolecular nuclear spin-spin coupling constants. In: Leszczynski J (ed) Computational chemistry, vol 8, Review of current trends. World Scientific, New York, pp 131–160
26. Sadlej J, Dobrowolski JCz, Rode JE (2010) VCD spectroscopy as a novel probe for chirality transfer in molecular interactions. Chem Soc Rev 39:1478–1488
27. Polavarapu PL (2007) Renaissance in chiroptical spectroscopic methods for molecular structure determination. Chem Rec 7:125–132
28. Nicu VP, Baerends EJ (2009) Robust normal modes in vibrational circular dichroism spectra. Phys Chem Chem Phys 11:6107–6118
29. Liu Y, Yang G, Losada M, Xu Y (2010) Vibrational absorption, vibrational circular dichroism, and theoretical studies of methyl lactate self-aggregation and methyl lactate-methanol intermolecular interactions. J Chem Phys 132:234513
30. Barron LD (2004) Molecular light scattering and optical activity, 2nd edn. Cambridge University Press, Cambridge
31. Buckingham AD (1994) Faraday Discuss 99:1, whole issue: Vibrational Optical Activity: from fundamentals to biological applications
32. Autschbach J (2009) Computing chiroptical properties with first-principles theoretical methods: background and illustrative examples. Chirality 21:E116–E152
33. Nafie LA, Dukor RK, Freedman TB (2002) Vibrational circular dichroism. In: Chalmers JM, Griffiths PR (eds) Handbook of vibrational spectroscopy. Wiley, Chichester, pp 731–744
34. Polavarapu PL (1998) Vibrational spectra: principles and applications with emphasis on optical activity. Elsevier, New York, pp 143–182
35. Polavarapu PL (1990) Ab initio vibrational raman and raman optical activity spectra. J Phys Chem 94:8106–8112
36. Polavarapu PL (2006) Quantum mechanical predictions of chirooptical vibrational properties. Int J Quantum Chem 106:1809–1814
37. Polavarapu PL (2008) Why is it important to simultaneously use more than one chiroptical spectroscopic method for determining the structures of chiral molecules. Chirality 20:664–672
38. Ballard RF, Mason SF, Vane GW (1963) Circular dichroism of dissymmetric α, β-unsaturated ketones. Discuss Faraday Soc 35:43–47
39. Moscowitz A, Wellman KM, Djerassi C (1963) Optical rotatory dispersion studies. XCIV. Some effects of solvation upon optically active molecules. Proc Natl Acad Sci USA 50:799–804
40. Wellman KM, Records R, Bunsenberg E, Djerassi C (1964) Optical rotatory dispersion studies. XCI. The Use of Low-temperature circular dichroism measurements for "fingerprinting" of steroidal ketones. J Am Chem Soc 86:492–498
41. Weigang OE Jr (1965) Vibrational structuring in circular dichroism. J Chem Phys 42:2244–2245
42. Deutsche CW, Moscowitz A (1968) Optical activity of vibrational origin. I. A model helical polymer. J Chem Phys 49:3257–3272
43. Deutsche CW, Moscowitz A (1970) Optical activity of vibrational origin. II. Consequences of polymer conformation. J Chem Phys 53:2630–2644
44. Dudleys RJ, Mason F, Peacock RD (1972) Infrared vibrational circular dichroism. J Chem Soc Chem Commun 1084–1085

45. Chabay I, Hsu EC, Holzwarth G (1972) Infrared circular dichroism measurement between 2000 and 5000 cm^{-1}: Pr^{+3} tartrate complexes. Chem Phys Lett 15:211–214

46. Holzwarth G, Chabay I (1972) Optical activity in molecular vibrations: a coupled oscillator model. J Chem Phys 57:1632–1635

47. Hsu EC, Holzwarth G (1973) Vibrational circular dichroism in crystalline alpha-NiSO$_4$.6H$_2$O and alpha-ZnSeO$_4$.6H$_2$O Between 1900 and 5000 cm^{-1}. J Chem Phys 59:4678–4685

48. Schellman JA (1973) Vibrational optical activity. J Chem Phys 58:2882–2886

49. Barron LD, Buckingham AD (1975) Rayleigh and raman optical activity. Annu Rev Phys Chem 26:381

50. Buckingham AD, Raab RE (1975) Electric-field-induced differential scattering of right and left circularly polarized light. Proc R Soc Lond Ser A 345:365–377

51. Nafie LA, Cheng JC, Stephens PJ (1975) Vibrational circular dichroism of 2, 2, 2-trifluoro-1-phenylethanol. J Am Chem Soc 97:3842–3843

52. Holzwarth G, Hsu EC et al (1974) Infrared circular dichroism of carbon-hydrogen and carbon-deuterium stretching modes. Observations. J Am Chem Soc 96:251–252

53. Nafie LA, Keiderling TA, Stephens PJ (1976) Vibrational circular dichroism. J Am Chem Soc 98:2715–2723

54. Nafie LA, Diem M (1979) Theory of high frequency differential interferometry: application to the measurement of infrared circular and linear dichroism via fourier transform spectroscopy. Appl Spectrosc 33:130–135

55. Bouř P (2009) Cross-polarization detection enables fast measurement of vibrational circular dichroism. Chem Phys Chem 10:1983–1985

56. Nafie LA (2008) Vibrational circular dichroism: a New tool for the solution-state determination of the structure and absolute configuration of chiral natural product molecules. Nat Prod Commun 3:451–466

57. Bürgi T, Vargas A, Baiker A (2002) VCD spectroscopy of chiral cinchona modifiers used in heterogeneous enantioselective hydrogenation: conformation and binding of non-chiral acids. J Chem Soc Perkin Trans 2:1596–1601

58. Ellzy MW, Jensen JO, Hameka HF, Kay JG (2003) Correlation of structure and vibrational spectra of the zwitterion L-alanine in the presence of water: an experimental and density functional analysis. Spectrochim Acta A Mol Biomol Spectrosc 59:2619–2633

59. Rode JE, Dobrowolski JCz (2003) VCD technique in determining intermolecular H-bond geometry: a DFT study. J Mol Struct Theochem 637:81–89

60. Sadlej J, Dobrowolski JCz, Rode JE, Jamróz MH (2006) DFT study of vibrational circular dichroism spectra of D-lactic acid–water complexes. Phys Chem Chem Phys 8:101–113

61. Sadlej J, Dobrowolski JCz, Rode JE, Jamróz MH (2007) Density functional theory study on vibrational circular dichroism as a tool for analysis of intermolecular systems: (1:1) cysteine-water complex conformations. J Phys Chem A 111:10703–10711

62. Kuppens T, Herrebout H, van der Veken B, Bultinck P (2006) Intermolecular association of tetrahydrofuran-2-carboxylic acid in solution: a vibrational circular dichroism study. J Phys Chem A 110:10191–10200

63. Debie E, Jaspers L, Bultinck P, Herrebout W, van der Veken B (2008) Induced solvent chirality: a VCD study of camphor in CDCl$_3$. Chem Phys Lett 450:426–430

64. Debie E, Bultinck P, Herrebout W, van der Veken B (2008) Solvent effects on IR and VCD spectra of natural products: an experimental and theoretical VCD study of pulegone. Phys Chem Chem Phys 10:3498–3508

65. Losada M, Xu Y (2007) Chirality transfer through hydrogen-bonding: experimental and ab initio analyses of vibrational circular dichroism spectra of methyl lactate in water. Phys Chem Chem Phys 9:3127–3135

66. Sun W, Wu J, Zheng B, Zhu Y, Liu C (2007) DFT study of vibrational circular dichroism spectra of (S)-glycidol–water complexes. J Mol Struct Theochem 809:161–169

67. Losada M, Tran H, Xu Y (2008) Lactic acid in solution: investigations of lactic acid self-aggregation and hydrogen bonding interactions with water and methanol using vibrational absorption and vibrational circular dichroism spectroscopies. J Chem Phys 128:014508-1–014508-11

68. Losada M, Nguyen P, Xu Y (2008) Solvation of propylene oxide in water: vibrational circular dichroism, optical rotation, and computer simulation studies. J Phys Chem A 112:5621–5627

69. Yang G, Xu Y (2009) Probing chiral solute-water hydrogen bonding networks by chirality transfer effects: a vibrational circular dichroism study of glycidol in water. J Chem Phys 130:164506-1–164506-9

70. Schlosser DW, Devlin F, Jalkanen K, Stephens PJ (1982) Vibrational circular dichroism of matrix-isolated molecules. Chem Phys Lett 88:286–291

71. Henderson DO, Polavarapu P (1986) Fourier transform infrared vibrational circular dichroism of matrix-isolated molecules. J Am Chem Soc 108:7110–7111

72. Tarczay G, Magyarfalvi G, Vass E (2006) Towards the determination of the absolute configuration of complex molecular systems: matrix isolation vibrational circular dichroism study of (R)-2-amino-1-propanol. Angew Chem Int Ed 45:1775–1777

73. Pohl G, Perczel A, Vass E, Magyarfalvi G, Tarczay G (2007) A matrix isolation study on Ac-Gly-NHMe and Ac-L-Ala-NHMe, the simplest chiral and achiral building blocks of peptides and proteins. Phys Chem Chem Phys 9:4698–4708

74. Tarczay G, Góbia S, Vass E, Magyarfalvi G (2009) Model peptide–water complexes in Ar matrix: complexation induced conformation change and chirality transfer. Vib Spectrosc 50:21–28

75. Góbi S, Knapp K, Vass E, Majer Z, Magyarfalvi G, Hollósi M, Tarczay G (2010) Is β-homo-proline a pseudo-γ-turn forming element of β-peptides? an IR and VCD spectroscopic study on Ac-β-HPro-NHMe in cryogenic matrices and solutions. Phys Chem Chem Phys 12:13603–13615

76. Merten C, Amkreutz M, Hartwig A (2010) Determining the structure of α-phenylethyl isocyanide in chloroform by VCD spectroscopy and DFT calculations-simple case or challenge? Phys Chem Chem Phys 12:11635–11641

77. Aviles-Moreno JR, Urena Horno E, Partal Urena F, López González JJ (2010) IR–raman–VCD study of R-(+)-pulegone: influence of the solvent. Spectrochim Acta A Mol Biomol Spectrosc. doi:10.1016/j.saa.2010.08.051

78. Nicu VP, Neugebauer J, Baerends EJ (2008) Effects of complex formation on vibrational circular dichroism spectra. J Phys Chem 111:6978–6991

79. Nicu VP (2009) Implementation, calculation and interpretation of vibrational circular dichroism spectra. PhD Dissertation, Vrije Universiteit

80. Nicu VP, Debie E, Herrebout W, van der Veken B, Bultinck P, Baerends EJ (2010) A VCD robust mode analysis of induced chirality: the case of pulegone in chloroform. Chirality 21:E287–E297

81. Stephens PJ, Lowe MA (1985) Vibrational circular dichroism. Annu Rev Phys Chem 36:213–241

82. Stephens PJ, Devlin FJ (2007) Vibrational circular dichroism. In: Mennucci B, Cammi R (eds) Continuum solvation models in chemical physics. From theory to application. Wiley, New York, p 180

83. Hug W (2007) Raman opital activity. In: Mennucci B, Cammi R (eds) Continuum solvation models in chemical physics. From theory to application. Wiley, New York, p 220

84. Stephens PJ (1985) Theory of vibrational circular dichroism. J Phys Chem 89:748–752

85. Stephens PJ (1987) Gauge dependence of vibrational magnetic dipole transition moments and rotational strengths. J Phys Chem 91:1712–1715

86. Ditchfield R (1974) Self-consistent perturbation theory of diamagnetism. Mol Phys 27:789–807

87. Devlin FJ, Stephens PJ (1994) Ab initio calculation of vibrational circular dichroism spectra of chiral natural products using MP2 force fields: camphor. J Am Chem Soc 116:5003–5004

88. Bak KL, Jørgensen P, Helgaker T, Ruud K, Jensen HJAa (1993) Gauge-origin independent multiconfigurational self-consistent-field theory for vibrational circular dichroism. J Chem Phys 98:8873–8887

89. Bak KL, Jürgensen P, Helgaker T, Ruud K (1994) Basis set convergence and correlation effects in vibrational circular dichroism calculations using London atomic orbitals. Faraday Discuss 99:121–129

90. Malkin VG, Malkina OL, Eriksson LA, Salahub DR (1995) In: Seminario JM, Politzer P (eds) Modern density functional theory: a tool for chemistry, vol 2. Elsevier, Amsterdam

91. Anisimov VI, Aryasetiawan F, Lichtenstein AI (1997) First-principles calculations of the electronic structure and spectra of strongly correlated systems: the LDA + U method. J Phys Condens Matter 9:767–808

92. Becke AD (1988) A multicenter numerical integration scheme for polyatomic molecules. J Chem Phys 88:2547–2553

93. Lee C, Yang W, Parr RG (1988) Development of the colle-salvetti correlation-energy formula into a functional of electron density. Phys Rev B 37:785–789

94. Perdew JP, Burke K, Wang Y (1996) Generalized gradient approximation for the exchange-correlation hole of a many-electron system. Phys Rev B 54:16533–16539

95. Burke K, Perdew JP, Wang Y (1998) In: Dobson JF, Vignale G, Das MP (eds) Electronic density functional theory: recent progress and new directions. Plenum, New York

96. Keal TW, Tozer DJ (2003) The exchange-correlation potential in Kohn–Sham nuclear magnetic resonance shielding calculations. J Chem Phys 119:3015–3024

97. Becke AD (1993) Density-functional thermochemistry. III. The role of exact exchange. J Chem Phys 98:5648–5652

98. Yanai T, Tew DP, Handy NC (2004) A new hybrid exchange-correlation functional using the coulomb-attenuating method (CAM-B3LYP). Chem Phys Lett 393:51–57

99. Peach MJG, Helgaker T, Salek P, Keal TW, Lutnaes OP, Tozer DJ, Handy NC (2006) Assessment of a coulomb-attenuated exchange-corelation energy functional. Phys Chem Chem Phys 5:558–562

100. Bouř P, Kubelka J, Keiderling TA (2002) Ab initio quantum mechanical model of peptide helices and their vibrational spectra. Biopolymers 65:45–59

101. Dunning TH Jr (1989) Gaussian basis sets for use in correlated molecular calculations. I. The atoms boron through neon and hydrogen. J Chem Phys 90:1007–1023

102. Jalkanen KJ, Jürgensen VW, Degtyarenko IM (2005) Linear response properties required to simulate vibrational spectra of biomolecules in various media: (R)-phenyloxirane (a comparative theoretical and spectroscopic vibrational study). Adv Quantum Chem 50:91–124

103. ADF Program System Release (2010) Scientific computing and modelling NV. Vrije Universiteit, Theoretical chemistry De Boelelaan 1083; 1081 HV Amsterdam; The Netherlands, Copyright © 1993–2010: Scientific Computing & Modelling NV Vrije Universiteit, Theoretical chemistry, Amsterdam, The Netherlands

104. Amos RD, Alberts IL, Andrews JS, Colwell SM, Handy NC, Jayatilaka D, Knowles PJ, Kobayashi R, Laidig KE, Laming G, Lee AM, Maslen PE, Murray CW, Rice JE (1995) In: Simandiras ED, Stone AJ, Su M-D, Tozer DJ (ed) CADPAC: the Cambridge analytic derivatives, package issue 6. Cambridge

105. Dalton J (2005) An ab initio electronic structure program. Release 2.0. Available at http://www.kjemi.uio.no/software/dalton/dalton.html. Accessed 1 Mar 2005

106. Gaussian 03, Revision E.01, Frisch MJ, Trucks GW, Schlegel HB, Scuseria GE, Robb MA, Cheeseman, JR, Montgomery JA Jr, Vreven T, Kudin KN, Burant JC, Millam JM, Iyengar SS, Tomasi J, Barone V, Mennucci B, Cossi M, Scalmani G, Rega N, Petersson GA, Nakatsuji H, Hada M, Ehara M, Toyota K, Fukuda R, Hasegawa J, Ishida M, Nakajima T, Honda Y, Kitao O, Nakai H, Klene M, Li X, Knox JE, Hratchian HP, Cross JB, Bakken V, Adamo C, Jaramillo J, Gomperts R, Stratmann RE, Yazyev O, Austin AJ, Cammi R, Pomelli C, Ochterski JW, Ayala PY, Morokuma K, Voth GA, Salvador P, Dannenberg JJ, Zakrzewski VG, Dapprich S, Daniels AD, Strain MC, Farkas O, Malick DK, Rabuck AD, Raghavachari K, Foresman JB, Ortiz JV, Cui Q, Baboul AG, Clifford S, Cioslowski J, Stefanov BB, Liu G,

Liashenko A, Piskorz P, Komaromi I, Martin RL, Fox DJ, Keith T, Al-Laham MA, Peng CY, Nanayakkara A, Challacombe M, Gill PMW, Johnson B, Chen W, Wong MW, Gonzalez C, Pople JA (2004). Gaussian Inc, Wallingford

107. Gaussian 09, Rev. A.1, Frisch MJ, Trucks GW, Schlegel HB, Scuseria GE, Robb MA, Cheeseman JR, Scalmani G, Barone V, Mennucci B, Petersson GA, Nakatsuji H, Caricato M, Li X, Hratchian HP, Izmaylov AF, Bloino J, Zheng G, Sonnenberg JL, Hada M, Ehara M, Toyota K, Fukuda R, Hasegawa J, Ishida M, Nakajima T, Honda Y, Kitao O, Nakai H, Vreven T, Montgomery JA Jr, Peralta JE, Ogliaro F, Bearpark M, Heyd JJ, Brothers E, Kudin KN, Staroverov VN, Kobayashi R, Normand J, Raghavachari K, Rendell A, Burant JC, Iyengar SS, Tomasi J, Cossi M, Rega N, Millam NJ, Klene M, Knox JE, Cross JB, Bakken V, Adamo C, Jaramillo J, Gomperts R, Stratmann RE, Yazyev O, Austin AJ, Cammi R, Pomelli C, Ochterski JW, Martin RL, Morokuma K, Zakrzewski VG, Voth GA, Salvador P, Dannenberg JJ, Dapprich S, Daniels AD, Farkas Ö, Foresman JB, Ortiz JV, Cioslowski J, Fox DJ (2009). Gaussian Inc, Wallingford

108. Cheeseman JR, Frisch MJ, Devlin FJ, Stephens PJ (1996) Ab initio calculation of atomic axial tensors and vibrational rotational strengths using density functional theory. Chem Phys Lett 252:211–220

109. Devlin FJ, Stephens PJ, Cheeseman JR, Frisch MJ (1997) *Ab initio* prediction of vibrational absorption and circular dichroism spectra of chiral natural products using density functional theory: camphor and fenchone. J Phys Chem A 101:6322–6333

110. Devlin FJ, Stephens PJ, Cheeseman JR, Frisch MJ (1996) Prediction of vibrational circular dichroism spectra using density functional theory: camphor and fenchone. J Am Chem Soc 118:6327–6328

111. Mennucci B, Cammi R (eds) (2007) Properties and spectroscopies – Continuum solvation models in chemical physics: from theory to applications. Wiley, Chichester/ Hoboken

112. Frimand K, Bohr H, Jalkanen KJ, Suhai S (2000) Structures, vibrational absorption and vibrational circular dichroism spectra of L-alanine in aqueous solution: a density functional theory and RHF study. Chem Phys 255:165–194

113. Jalkanen KJ, Suhai S (1996) N-acetyl-N-alanine N'-methylamide: a density functional analysis of the vibrational absorption and vibrational circular dichroism spectra. Chem Phys 208:81–116

114. Kim J, Kapitán J, Lakhani A, Bouř P, Keiderling TA (2008) Tight β-turns in peptides. DFT-based study of infrared absorption and vibrational circular dichroism for various conformers including solvent effects. Theor Chem Acc 119:81–97

115. Han WG, Jalkanen KJ, Elstner M, Suhai S (1998) Theoretical study of aqueous N-acetyl-L-alanine N-methylamide: structures and Raman, VCD, and ROA spectra. J Phys Chem B 102:2587–2602

116. Capelli C, Monti S, Rizzo A (2005) Effect of the environment on vibrational infrared and circular dichroism spectra of (S)-proline. Int J Quantum Chem 104:744–757

117. Tajkhorshid E, Jalkanen KJ, Suhai S (1998) Structure and vibrational spectra of the zwitterion L-alanine in the presence of explicit water molecules: a density functional analysis. J Phys Chem B 102:5899–5913

118. Kubelka J, Huang R, Keiderling TA (2005) Solvent effects on IR and VCD spectra of helical peptides: DFT-based static spectral simulations with explicit water. J Phys Chem B 109:8231–8243

119. Oh K-I, Han J, Lee K-K, Hahn S, Han H, Cho M (2006) Site-specific hydrogen-bonding interaction between N-acetylproline amide and protic solvent molecules: comparisons of IR and VCD measurements with MD simulations. J Phys Chem A 110:13355–13365

120. Vargas A, Bonalumi N, Ferri D, Baiker A (2006) Solvent-induced conformational changes of O-phenyl-cinchonidine: a theoretical and VCD spectroscopy study. J Phys Chem A 110:1118–1127

121. Jürgensen VW, Jalkanen KJ (2006) The VA, VCD, Raman and ROA spectra of tri-L-serine in aqueous solution. Phys Biol 3:S63–S79

122. Schweitzer-Stenner R, Measey T, Kakalis L, Jordan F, Pizzanelli S, Forte C, Griebenow K (2007) Conformations of alanine-based peptides in water probed by FTIR, Raman, vibrational circular dichroism, electronic circular dichroism, and NMR spectroscopy. Biochemistry 46:1587–1596

123. Jalkanen KJ, Degtyarenko IM, Nieminen RM, Cao X, Nafie LA, Zhu F, Barron LD (2008) Role of hydration in determining the structure and vibrational spectra of L-alanine and N-acetyl L-alanine N-methylamide in aqueous solution: a combined theoretical and experimental approach. Theor Chem Acc 119:191–210

124. Yang S, Cho M (2009) Direct calculations of vibrational absorption and circular dichroism spectra of alanine dipeptide analog in water: quantum mechanical/molecular mechanical molecular dynamics simulations. J Chem Phys 131:135102-1–135102-8

125. Hatfield MPD, Murphy RF, Lovas S (2010) VCD spectroscopic properties of the β-hairpin forming miniprotein CLN025 in various solvents. Biopolymers 93:442–450

126. Yang G, Xu Y, Hou J, Zhang H, Zhao Y (2010) Determination of the absolute configurations of the pentacoordinate chiral phosphorus compounds in solution using vibrational circular dichroism spectroscopy and density functional theory. Chem Eur J 16:2518–2527

127. Yang G, Xu Y, Hou J, Zhang H, Zhao Y (2010) Diastereomers of the pentacoordinate chiral phosphorus compounds in solution: absolute configurations and predominant conformations. Dalton Trans 39:6953–6959

128. Brizard A, Berthier D, Aimé C, Buffeteau T, Cavagnat D, Ducasse L, Huc I, Oda R (2009) Molecular and supramolecular chirality in gemini-tartrate amphiphiles studied by electronic and vibrational circular dichroisms. Chirality 21:S153–S162

129. Rode JE, Dobrowolski JCz On theoretical VCD spectra of quinine-BF$_3$ complexes (submitted)

130. Coriani S, Thorvaldsen AJ, Kristensen K, Joergensen P (2011) Variational response- function formulation of vibrational circular dichroism. Phys Chem Chem Phys. 13:4224–4229

131. Bouchet A, Brotin T, Cavagnat D, Buffeteau T (2010) Induced chiroptical changes of a water-soluble cryptophane by encapsulation of guest molecules and counterion effects. Chem Eur J 16:4507–4518

132. Bouchet A, Brotin T, Linares M, Ågren H, Cavagnat D, Buffeteau T (2011) Conformational effects induced by guest encapsulation in an enantiopure water-soluble Cryptophane. J Org Chem 76:1372–1383

133. Góbi S, Magyarfalvi G (2011) Reliability of computed signs and intensities for vibrational circular dichroism spectra. Phys Chem Chem Phys 13:16126–16129.

134. Góbi S, Vass E, Magyarfalvi G, Tarczay G (2011) Effects of strong and weak hydrogen bond formation on VCD spectra: a case study of 2-chloropropionic acid, Phys Chem Chem Phys 13:13972–13984

Chapter 16
Non-hydrogen-Bonding Intramolecular Interactions: Important but Often Overlooked

Peter Politzer and Jane S. Murray

Abstract Our focus in this chapter is upon intramolecular noncovalent interactions, electrostatically driven, but not including hydrogen bonding. They often involve a positive σ-hole or π-hole on a covalently-bonded Group IV – VII atom, in conjunction with a negative site in the molecule. Examples are given involving NO_2 groups, Si–O–N bond angles, and specific 1,3-Si—O, 1,4-S—O, 1,4-Se—O, 1,3-P—Cl and 1,4-C—O interactions. These examples demonstrate that intramolecular interactions can play significant roles in determining the structure of a molecule and also its reactive properties. This often involves stabilizing a particular conformation, but can also include markedly affecting bond lengths and/or angles. It is essential to take intramolecular interactions into account in trying to understand and predict molecular behavior, and furthermore to exploit them in designing new materials, in pharmacology, crystal engineering, etc.

Keywords Intramolecular interactions • Electrostatic potentials • σ-hole interactions • π-hole interactions

16.1 Noncovalent Interactions

The last decade has seen a surge of interest and activity in noncovalent interactions. Even in venerable hydrogen bonding, which has so long been studied and analyzed [1–4], there have been new developments [5–10]. The old concept of a hydrogen bond A-H—B involving an electronegative atom A and a basic site B has been expanded to include C–H—B systems and A–H—H-M "dihydrogen bonds;" A–H "blue shifting" is recognized as a possibility. Hydrogen bonding has even been described as a type of σ-hole interaction [11, 12].

P. Politzer (✉) • J.S. Murray
CleveTheoComp, 1951 W. 26th Street, Cleveland, OH 44113, USA
e-mail: ppolitze@uno.edu

J. Leszczynski and M.K. Shukla (eds.), *Practical Aspects of Computational Chemistry I:*
An Overview of the Last Two Decades and Current Trends,
DOI 10.1007/978-94-007-0919-5_16, © Springer Science+Business Media B.V. 2012

Dihydrogen bonding may sound like an anomaly until it is pointed out that M is typically an electron donor (e.g. Li or Na) so that the hydrogen in H–M has hydride character. Other types of noncovalent interactions may have also initially seemed puzzling. For instance, Guru Row et al. found short contacts between the sulfurs of identical R_1R_2S molecules in crystal lattices [13]. Equally enigmatic was "halogen bonding," R–X—B, in which X is a halogen and B is a basic site. Such interactions have long been known (see overview by Politzer et al. [14]), but the obvious question was: Why is an electronegative halogen attracted to a negative site? Furthermore, the aforementioned sulfurs and the halogens are often observed to also interact favorably with an electrophilic moiety [13, 15].

An explanation came out of the work of Brinck et al. in 1992 [16]. (See also Auffinger et al. [17].) They showed that a covalently-bonded halogen can have a region of positive electrostatic potential on its outer side, on the extension of the covalent bond to it. Through this positive potential the halogen can interact electrostatically with a negative site, such as the lone pair of a Lewis base; for example, H_3C–Br—NH_3. It has subsequently been demonstrated that there are frequently analogous positive regions on covalently-bonded atoms of Groups IV [18], V [19] and VI [20], again on the extensions of the bonds to these atoms. These positive regions, which can be seen for the molecule SeBrCl in Fig. 16.1, can also interact favorably with Lewis bases and other nucleophiles. The positive outer portion of the atom is usually (but not always) surrounded by a negative region, to which electrophiles can be attracted (Fig. 16.1). Extensive crystallographic surveys fully support this picture [13, 21–23]. These findings demonstrate the fallacy of assigning a single charge to an atom in a molecule, by whatever method [15, 17]. A single charge cannot account for the atom being able to interact electrostatically with both electrophiles and nucleophiles.

The origin of the positive potentials is the anisotropic charge distribution of a covalently-bonded atom [24–26]. This involves a diminished electron density, a "σ-hole" [27], on the side of the atom opposite to each bond. Within each Group of the periodic table, the σ-holes become more positive in going from the lighter to the heavier (more polarizable and less electronegative) atoms, and as the remainder of the molecule is more electron-withdrawing [11, 14, 18–20, 27]. Thus the three σ-holes on the phosphorus in PF_3 are less positive than those on the arsenic in AsF_3, but more positive than those in $As(CH_3)_3$ [19]. However $PF(CH_3)_2$ has a positive σ-hole only on the extension of the F–P bond; the other two σ-holes are negative (although less negative than their surroundings). Further examples of the trends in the relative strengths of σ-hole potentials are in Fig. 16.1. Negative σ-holes are often found on first-row atoms, i.e. C, N, O and F (and sometimes second-row), since these are less polarizable and more electronegative than the heavier members of their Groups; however when first-row atoms are bonded to strongly-electron-withdrawing substituents, positive σ-holes are observed (e.g. on the fluorines in F_2 and F–CN) [11, 15, 18–20].

The interactions of positive σ-holes with negative sites is called σ-hole bonding; halogen bonding is a subset of this. Since σ-holes are on the extensions of the

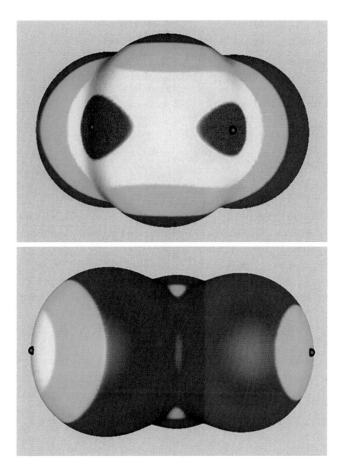

Fig. 16.1 Computed electrostatic potential on the 0.001 au molecular surface of SeBrCl. *Top view*: selenium is in the foreground, with bromine at the *left* rear and chlorine at the *right* rear. *Bottom view*: bromine is at the *left* front, chlorine at the *right* front; selenium is in the rear. Color ranges, in kcal/mol: *red*, greater than 24; *yellow*, between 24 and 12; *green*, between 12 and 0; *blue*, less than 0 (negative). The black hemispheres indicate the locations of the σ-hole $V_{S,max}$. The two on selenium (*top*) have magnitudes of 30.9 kcal/mol (*left*) and 28.2 kcal/mol (*right*). The more positive is on the extension of the Cl–Se bond. The bromine and chlorine $V_{S,max}$ are 19.7 and 8.5 kcal/mol, respectively

covalent bonds to the atoms, their interactions are highly directional, essentially colinear with these bonds. For example, in the complex **1** formed between $PF(CH_3)_2$ and NH_3,

$$\mathbf{1} \qquad F\!-\!\underset{\underset{CH_3}{|}}{\overset{\overset{CH_3}{|}}{P}}\!\cdots\!NH_3$$

the angle F–P–N will be very close to 180° (unless affected by secondary interactions). The strengths of σ-hole bonds have been shown to correlate with the magnitudes of the positive σ-hole potentials [12, 28, 29], which reinforces the interpretation of these interactions as electrostatically-driven. More detailed discussions of halogen and σ-hole bonding can be found elsewhere [11, 30, 31].

The significance of σ-hole interactions in molecular biology, pharmacology and the design of new materials is increasingly recognized [17, 30–33]. This is particularly true for Group VII, for which they are best known and most studied. A very recent development has been the expansion of the σ-hole concept to include π-holes [34]. These are regions of positive electrostatic potential that are perpendicular to a portion of a molecule. For example, SO_2 has π-holes above and below the molecular plane, over the sulfur atom. The characteristics of π-holes and their interactions with negative sites are analogous to those of σ-holes. (Both σ- and π-hole interactions do sometimes evolve into a stronger bonding that appears to have some degree of coordinate covalency [34, 35].)

It seems fair to say that, in general, intermolecular noncovalent interactions, of whatever type, receive more attention than do intramolecular ones. However the latter can also be very important. For instance, intramolecular attractions are responsible for chlorotrinitromethane, $Cl–C(NO_2)_3$, having the shortest $C(sp^3)$–Cl bond ever observed crystallographically [36]. Another intramolecular interaction is a key factor in the remarkable increase in detonation sensitivity that occurs when the central carbon in the explosive PETN (2) is replaced by a silicon (3) [37].

$$O_2NOH_2C-\underset{\underset{CH_2ONO_2}{|}}{\overset{\overset{CH_2ONO_2}{|}}{C}}-CH_2ONO_2 \qquad\qquad O_2NOH_2C-\underset{\underset{CH_2ONO_2}{|}}{\overset{\overset{CH_2ONO_2}{|}}{Si}}-CH_2ONO_2$$

$$\textbf{2} \qquad\qquad\qquad\qquad\qquad\qquad \textbf{3}$$

Our objective in this chapter is to draw attention to these and several other interesting examples of intramolecular noncovalent interactions. Our analyses shall be in terms of physical observables – structures and electrostatic potentials. We view these interactions as primarily electrostatic in nature, which of course includes the polarizing effects of the participants upon each other.

16.2   The Electrostatic Potential

The electrons and nuclei of any molecule create an electrostatic (Coulomb) potential in the surrounding space. Its sign and magnitude at any point $\mathbf{r}$, designated $V(\mathbf{r})$, are given by Eq. 16.1:

$$V(\mathbf{r}) = \sum_{A} \frac{Z_A}{|\mathbf{R}_A - \mathbf{r}|} - \int \frac{\rho(\mathbf{r}\prime)d\mathbf{r}\prime}{|\mathbf{r}\prime - \mathbf{r}|} \qquad (16.1)$$

Z_A is the charge on nucleus A, located at $\mathbf{R}_A$, and $\rho(\mathbf{r})$ is the molecule's electronic density.

$V(\mathbf{r})$ is a physical observable, which can be determined experimentally by diffraction methods [38, 39] as well as computationally. Its sign in any region depends upon whether the positive contribution of the nuclei or the negative one of the electrons is dominant there.

The electrostatic potential is related rigorously to the electronic density by Poisson's equation and is therefore, through the Hohenberg-Kohn theorem [40], a property of fundamental significance [41, 42]. For example, exact atomic and molecular energies can be formulated as functions of $V(\mathbf{r})$ [43].

As is evident already in Sect. 16.1, $V(\mathbf{r})$ provides an effective basis for interpreting and predicting noncovalent interactions [41, 42]. It has in fact been shown that condensed phase physical properties that depend upon noncovalent interactions can be expressed analytically in terms of the features of $V(\mathbf{r})$. For these purposes, $V(\mathbf{r})$ is commonly computed on the surface of the molecule, taking this to be the 0.001 au (electrons/bohr^3) contour of its electronic density, as proposed by Bader et al. [44]. Representing the surface as a contour of $\rho(\mathbf{r})$ has the advantage that it is specific to that particular molecule, and reflects its lone pairs, π electrons, etc. $V(\mathbf{r})$ computed on a molecular surface is labeled $V_S(\mathbf{r})$; its local most positive and most negative values, of which there may be several, are designated $V_{S,max}$ and $V_{S,min}$, respectively.

Figure 16.1 shows $V_S(\mathbf{r})$ and the sites of the σ-hole $V_{S,max}$ for SeBrCl, obtained with the Wave Function Analysis-Surface Analysis Suite [45]. The selenium has two σ-hole $V_{S,max}$; the one on the extension of the Cl–Se bond is stronger, due to the greater electronegativity of chlorine compared to bromine. The bromine $V_{S,max}$ is more positive than that of the chlorine, which is also in keeping with the trends summarized in Sect. 16.1.

16.3 Some Noncovalent Intramolecular Interactions

16.3.1 The Nitro Group

The nitro group, NO_2, offers an array of possibilities for intramolecular interactions [46], especially when there are two or more on the same carbon. The trinitromethyl group, $C(NO_2)_3$, is of particular interest as a component of energetic materials, e.g. explosives. Molecules of the type $Z–C(NO_2)_3$ have been prepared and studied extensively [47] because of their potential energetic performance. A noteworthy example is $[(O_2N)_3C–CH_2–O]_2C\!\!=\!\!O$, which has a desirably high density of 1.975 g/cm^3 [48].

Table 16.1 Some structural data[a] (Distances are in Ångstroms, angles in degrees. For nonbonded separations, sums of van der Waals radii are in parentheses[b])

Molecule	Property	Computed value[c]	Experimental
Cl–CH$_3$	C–Cl bond length	1.782	1.785[d]
Cl–CH$_2$NO$_2$	C–Cl bond length	1.744	
	Cl–C–N–O dihedral angles	−0.2, 179.8	
	O$_{upper}$—Cl separation	2.88 (3.27)	
	O$_{lower}$—H separations	2.52, 2.51 (2.72)	
	C–Cl bond length	1.722	
	Cl–C–N$_a$–O dihedral angles	3.2, −178.2	
	Cl–C–N$_b$–O dihedral angles	69.9, −108.9	
	O$_a$—Cl separation	2.86 (3.27)	
	O$_c$—N$_a$ separation	2.77 (3.07)	
	O$_b$—N$_b$ separation	2.74 (3.07)	
	O$_b$—H separation	2.54 (2.72)	
	O$_d$—H separation	2.38 (2.72)	
Cl–C(NO$_2$)$_3$	C–Cl bond length	1.702	1.694[e]
	Cl–C–N–O dihedral angle	41.9	37.4–45.9[e]
	O$_{upper}$—Cl separations	2.91, 2.91, 2.90 (3.27)	
	O$_{lower}$—N separations	2.58, 2.58, 2.58 (3.07)	
Cl–CH$_2$CN	C–Cl bond length	1.786	
Cl–C(CN)$_3$	C–Cl bond length	1.807	1.78[f]

[a]All data are from reference [46]
[b]Reference [49]
[c]Computational level: B3PW91/6-311 G(3d,2p)
[d]Reference [50]
[e]Reference [36]
[f]Reference [51]

We shall begin by considering methyl chloride, H$_3$C–Cl, and progressively introducing nitro groups. The relevant structural data are in Table 16.1 and electrostatic potentials are in Table 16.2. In H$_3$C–Cl, the chlorine is completely negative, including its σ-hole. The computed C–Cl bond length is 1.782 Å, in good agreement with the experimental 1.785 Å.

The introduction of just one NO$_2$ is enough to change both of these properties considerably. The strongly electron-withdrawing NO$_2$ makes the chlorine surface nearly completely positive, with a maximum ($V_{S,max}$) of 13.3 kcal/mol on the extension of the C–Cl bond (the chlorine σ-hole) (Table 16.2). The C–Cl distance decreases significantly from its value in Cl–CH$_3$, to 1.744 Å (Table 16.1).

What is the reason for this marked C–Cl shortening, by about 0.04 Å? It can be understood by looking at the orientation of the NO$_2$. It is coplanar with the C–Cl bond (Fig. 16.2); the Cl–C–N–O dihedral angles are 0° and 180°. This orientation is stabilized by three 1,4 intramolecular close contacts, all involving the two oxygens, which are entirely negative. The upper oxygen interacts with the positive chlorine, the lower one with the two hydrogens. The O—Cl and O—H separations are all

Table 16.2 Most positive ($V_{S,max}$) and most negative ($V_{S,min}$) electrostatic potentials on 0.001 au molecular surfaces, in kcal/mol[a,b]

Molecule	$V_{S,max}$	$V_{S,min}$
Cl–CH$_3$	H: 20.6	Cl: -16.6
Cl–CH$_2$NO$_2$	H: 35.7	O: -24.4 to-22.5
	Cl: 13.3[c]	
Cl–CH(NO$_2$)$_2$	H: 46.1	O: -18.4 to-11.9
	C: 45.9	
	Cl: 26.5[c]	
	N: 12.0, 24.1	
Cl–C(NO$_2$)$_3$	Cl: 35.5[c]	O: -12.4 to-11.6
	N: 11.9, 36.0	
	C: 11.3[d]	
Cl–CH$_2$CN	H: 35.7	N: -32.9
	Cl: 12.0[c]	Cl: -9.4
Cl–C(CN)$_3$	C: 42.4[d], 36.4[e]	N: -20.0
	Cl: 31.5[c]	

[a] All data are from reference [46]
[b] Computational level: B3PW91/6-31 G(d,p)
[c] Located on the surface of the chlorine, along the extension of the C–Cl bond
[d] Located on the surface of the carbon, along the extension of the Cl–C bond
[e] There are three of these $V_{S,max}$, located on the central carbon on the extension of each NC–C bond

Fig. 16.2 Ball-and-stick structure of Cl–CH$_2$NO$_2$. Chlorine is *green*, carbon is *gray*, nitrogen is *blue*, oxygen is *red* and hydrogen is *white*. The NO$_2$ group is coplanar with the C–Cl bond

substantially less than the sums of the respective van der Waals radii (Table 16.1). Note that the O—Cl interaction is not with the chlorine σ-hole, which is on the extension of the C–Cl bond, but rather with one of the lateral sides of the chlorine; these are also positive in this molecule.

We believe that the O—Cl attractive interaction is the cause of the decrease in the C–Cl bond length [46]. One piece of supporting evidence is that when the NO_2 is rotated so that the O—Cl distance becomes larger, the C–Cl bond lengthens. For example, a rotation of 60° increases the O—Cl separation from 2.88 to 3.05 Å, and the C–Cl bond goes from 1.744 to 1.754 Å. It might be argued that the shortening of this bond is due to hyperconjugation promoted by the electron-attracting NO_2:

$$ \mathbf{4} \qquad\qquad \mathbf{5} \qquad\qquad (16.2) $$

However a similar effect should then be observed in $Cl–CH_2–CN$, since the CN group is also strongly electron-attracting. Instead, the C–Cl bonds in $Cl–CH_3$ and $Cl–CH_2–CN$ are of essentially the same length [46]. Furthermore, the C–Cl bond was found to be weaker in $Cl–CH_2–NO_2$ than in $Cl–CH_3$, by about 7 kcal/mol, which is not consistent with its having acquired some double bond character (structure **5**).

Proceeding to $Cl–CH(NO_2)_2$, Table 16.2 shows that the chlorine has become more positive, and each nitrogen now has two local maxima ($V_{S,max}$). These positive regions are π-holes, perpendicular to the NO_2 plane. The fact that one NO_2 is rotated, and is not coplanar with the C–Cl bond (Fig. 16.3 and Table 16.1) allows a short contact between each nitrogen and an oxygen of the other NO_2, through a π-hole of the nitrogen. (This is why one of each nitrogen's $V_{S,max}$ and one of each oxygen's $V_{S,min}$ are smaller in magnitude than the other $V_{S,max}$ and $V_{S,min}$; they have been partially neutralized by this interaction.) There is again a short O—Cl contact, and two O—H. Thus $Cl–CH(NO_2)_2$ has a total of five intramolecular interactions: two O—N, two O—H and one O—Cl. Because the chlorine has become considerably more positive, the O—Cl attraction further reduces the C–Cl bond length, by 0.02 Å.

Finally we come to $Cl–C(NO_2)_3$. This has the distinction of possessing the shortest $C(sp^3)$–Cl bond found in the Cambridge Structural Database [36]. Its crystallographically-determined length is 1.694 Å; our computed value is 1.702 Å (Table 16.1). In $Cl–C(NO_2)_3$, as in other molecules of the type $Z–C(NO_2)_3$ [36, 46, 47], all three NO_2 have the same orientation relative to the C–Cl bond, the propeller-like arrangement shown in Fig. 16.4. Three of the oxygens, one from each NO_2, are in an upper plane near the chlorine while the other three are in a lower plane. The former are in close contact with the chlorine, the latter with the nitrogens of

Fig. 16.3 Ball-and-stick structure of Cl–CH(NO$_2$)$_2$. Chlorine is *green*, carbon is *gray*, nitrogen is *blue*, oxygen is *red* and hydrogen is *white*. The NO$_2$ group at the right is essentially coplanar with the C–Cl bond

Fig. 16.4 Ball-and-stick structure of Cl–C(NO$_2$)$_3$. Chlorine is *green*, carbon is *gray*, nitrogen is *blue* and oxygen is *red*. This shows the propeller-like arrangement of the NO$_2$ groups. Three of the oxygens are in an upper plane and three are in an essentially parallel lower one

neighboring NO$_2$ groups, through the nitrogen π-holes. Two noninteracting nitrogen π-holes as well as the positive chlorine and its σ-hole can be seen in Fig. 16.5. There are accordingly six intramolecular interactions, three O—Cl and three O—N. The three O—Cl attractions are responsible for the greatly diminished C–Cl distance. In contrast, Cl–C(CN)$_3$ also has a very positive chlorine (Table 16.2) but no possibilities for intramolecular interactions with it, and so the C–Cl bond has a normal length (Table 16.1).

Fig. 16.5 Computed electrostatic potential on the 0.001 au molecular surface of chlorotrini-tromethane, $Cl–C(NO_2)_3$. The chlorine is pointing to the *right*. Color ranges, in kcal/mol: *red*, greater than 24; *yellow*, between 24 and 12; *green*, between 12 and 0; *blue*, less than 0 (negative). The black hemispheres indicate the locations of the $V_{S,max}$ of the σ-hole on chlorine and a π-hole on each of two nitrogens. The magnitudes are 35.5 kcal/mol (chlorine) and 36.0 kcal/mol (nitrogens)

A point that might be raised is that if the three NO_2 groups in $Cl–C(NO_2)_3$ were all coplanar with the C–Cl bond, the O_{upper}—Cl separations would be less and the interactions stronger. This is true, as can be seen by comparing the O—Cl distances in $Cl–C(NO_2)_3$ with those in $Cl–CH_2NO_2$ and $Cl–CH(NO_2)_2$, each of which has an NO_2 coplanar with C–Cl. However Göbel et al. [36] showed that such a structure would increase the total energy of $Cl–C(NO_2)_3$, because it interferes with the favorable O—N close contacts and increases O_{lower}—O_{lower} and N—N repulsions. The beauty of the propeller-like structure is that it finds the optimum balance between all of these factors.

Another interesting $Z–C(NO_2)_3$ molecule is trinitromethane, $H–C(NO_2)_3$ (ni-troform), which can be used to prepare various energetic compounds [52]. The hydrogen in $H–C(NO_2)_3$ is very acidic due to the three electron-withdrawing NO_2 groups, although not as much so as might be anticipated. Furthermore, whereas the transfer of this hydrogen to an NO_2 oxygen to give the *aci* tautomer,

$$H-C(NO_2)_3 \quad \rightleftharpoons \quad \underset{O_2N}{\overset{O_2N}{>}}C=N\underset{O^-}{\overset{+OH}{<}}$$

6   **7**

(16.3)

could be expected to be quite facile, the tautomer **7** has, to our knowledge, not been isolated. These seemingly anomalous features of $H-C(NO_2)_3$ are believed to reflect the O—H interactions between the hydrogen and the three nearby upper-plane oxygens in the propeller-like arrangement of the NO_2 groups [53, 54].

The examples that have been discussed illustrate the remarkable capacity of the NO_2 group for intramolecular interactions, and the effects that these can have upon a molecule's structure and properties. As pointed out earlier [46], several features of NO_2 give it this exceptional propensity for attractive close contacts: (a) It has two centers of negative charge, the oxygens, plus sometimes positive π-holes associated with the nitrogen. (b) Its electron-attracting power can promote the formation of neighboring positive centers. (c) It can rotate to maximize favorable interactions and minimize unfavorable ones.

16.3.2   The Si–O–N Linkage

In the preceding section, we looked at 1,4 intramolecular interactions involving the NO_2 group. Our focus now shall be Si—N (i.e. 1,3) interactions within the Si–O–N linkage. Specifically, we will consider molecules of the type,

8
$$\underset{Y}{\overset{X}{>}}Si\underset{}{\overset{O}{\diagdown}}N(CH_3)_2$$
$$Z$$

where X, Y and Z represent any atoms or groups bonded to the silicon. X is essentially coplanar with Si, O and N, and is said to be *anti* to the nitrogen, while Y and Z are *gauche* to it.

Molecules in this category have been prepared and extensively characterized by Mitzel et al. [55–59]. They possess the interesting feature that the Si–O–N angle is relatively small, in the range from 75° to 110°, whereas it is approximately 125° in the isoelectronic $XYZSi-O-CH(CH_3)_2$ [55, 60]. Furthermore, the magnitude of the Si–O–N angle is very dependent upon the natures of the substituents X, Y and Z, and whether they are *anti* or *gauche* to the nitrogen.

All of this can be explained in terms of the positive σ-holes on the silicon, on the extensions of the X–Si, Y–Si and Z–Si bonds [61]. Particularly important is the σ-hole on the extension of the bond that is *anti* to the nitrogen (the X–Si bond in structure **8**), since this σ-hole is best positioned to interact effectively with the lone pair of the N(CH$_3$)$_2$ nitrogen. It is primarily this interaction that causes the contraction of the Si–O–N angle. In XYZSi–O–CH(CH$_3$)$_2$, on the other hand, there is no significant negative site on the CH(CH$_3$)$_2$ group.

Consider, for instance, the two conformers of ClH$_2$Si–O–N(CH$_3$)$_2$, **9** and **10**. In **9**, the chlorine is *anti* to the nitrogen; in **10** it is *gauche*. In order to understand the intramolecular interactions, it is necessary to look at the σ-holes on the silicon before these take place (since they have a neutralizing effect). Accordingly we evaluate the σ-holes on the silicon in ClH$_2$Si–H, using a hydrogen to fill the fourth position. The σ-hole V$_{S,max}$ are 31.0 kcal/mol on the extension of the Cl–Si bond and 17.7 kcal/mol on the extensions of the H–Si bonds [61]. The Si–O–N angle should therefore be contracted more in **9** than in **10**, because the nitrogen lone pair in **9** is facing the stronger positive region, as indicated by its V$_{S,max}$. This is indeed the case; the gas phase Si–O–N angle is 87.1° in **9** and 104.7° in **10** [56].

The role of secondary interactions must always be kept in mind. In **10**, the Si–O–N angle would be smaller if not for the repulsion between the nitrogen lone pair and the lateral side of the chlorine, which is negative (see Fig. 16.2 in Murray et al. [61]). The effect of this is evident in the dichloro analogue, **11**. The primary interaction is with the σ-hole on the extension of the *anti* H–Si bond, just as in **10**. However the Si–O–N angle in **11** is 111.1° [58], much greater than the 104.7° in **10**. This reflects the additional repulsion between the nitrogen lone pair and the lateral side of the *second* chlorine.

16.3.3   Some 1,3 Si—O Interactions

The molecules pentaerythritol tetranitrate (PETN, **2**) and its silicon analogue (Si-PETN, **3**) provide a striking example of the consequences that can follow from a 1,3 intramolecular interaction [37]. PETN is a well known explosive, quite sensitive to impact; dropping a 2.5 kg mass upon it from a height of just 14 cm has a 50% probability of producing detonation [62]. However Si-PETN, prepared only recently [63], is reported to be so remarkably more sensitive that impact is not even needed; touching with a spatula is sufficient to cause explosion!

In an effort to understand this difference in sensitivity between PETN and Si-PETN, Liu et al. analyzed computationally several possible initial steps in the explosive decompositions of these molecules [64]. For PETN, the energetically-preferred possibilities were O–NO$_2$ rupture and HONO elimination; these were both found to have energy barriers of about 39 kcal/mol. For Si-PETN, the analogous processes had barriers of 36 and 39 kcal/mol; these do not explain the enormous difference in sensitivities. However Liu et al. found that Si-PETN can undergo an internal rearrangement in which a linking oxygen interacts with the silicon:

$$\textbf{12} \qquad \text{transition state} \qquad \textbf{13} \qquad (16.4)$$

The computed activation energy for this was 32 kcal/mol [64], which is less than for any of the other processes investigated for either PETN or Si-PETN. In addition, this rearrangement is overall quite exothermic, with a predicted ΔE of about -45 kcal/mol, and therefore can provide energy for further decomposition. In the case of PETN, on the other hand, the corresponding process has a much higher activation barrier of about 80 kcal/mol.

The fact that the rearrangement in Eq. 16.4 is available for initiating the decomposition of Si-PETN but not PETN was cited by Liu et al. as the reason for the vastly greater sensitivity of the former [64]. But why does Si-PETN rearrange in this manner so much more readily than PETN? One factor, as Liu et al. pointed out, is likely to be the greater size of the silicon atom compared to carbon, which facilitates the expansion of its coordination sphere from four to five during the transition from **12** to **13**.

A second factor was brought out by a detailed analysis of the analogous rearrangements of the molecules **14** and **15**, used as models for PETN (**2**) and Si-PETN (**3**) to shorten the

$$\textbf{14} \qquad\qquad\qquad \textbf{15}$$

calculations [37]. Both **14** and **15** have been synthesized and characterized [65], and **15** is again by far the more sensitive. As can be seen in Eq. 16.4, the rearrangements involve the formation of a new bond between the linking oxygen and the silicon (or central carbon). This is significantly promoted in the case of **15** by an attractive interaction of a silicon σ-hole with the oxygen [37]. The σ-hole is on the extension of one of the H_3C–Si bonds. The central carbon in **14** does not have a positive σ-hole, so bonding to the linking oxygen does not receive this assistance. This beneficial 1,3 Si—O interaction should occur as well in Si-PETN, and – together with the larger size of the silicon atom – accounts for its lower activation energy for the rearrangement in Eq. 16.4.

16.3.4   Conformation Stabilization

The crystalline forms of the antitumor agents tiazofurin (**16**) and its selenium analogue selenazofurin (**17**) have 1,4 close contacts between the sulfur or selenium and the furanose oxygen [66, 67]; the S—O and Se—O separations are both less than the sums of the respective van der Waals radii. As pointed out by Burling and Goldstein [68], these interactions are biologically significant because they constrain rotation around the C–C bonds that link the two rings in **16** and **17**, and thereby influence the specificity of the binding of anabolites of **16** and **17** to cellular targets. Burling and Goldstein concluded that the S—O and Se—O interactions are electrostatic. Our subsequent analysis fully supports this interpretation, showing that the negative oxygens are attracted to positive σ-holes on the sulfur and selenium [68], which are on the extensions of C–S and C–Se bonds.

16   17

1,3-thiazole (**18**), 1,3-selenazole (**19**) and their derivatives, such as those that are components of **16** and **17**, provide good examples of the "tunability" of σ-holes [68]. By this we mean that both the strengths and the directions of σ-hole interactions involving the sulfur and selenium can be controlled by appropriate substitution on the rings. This is shown for 1,3-thiazole in Table 16.3.

18   19

The unsubstituted molecule has two positive sulfur σ-holes, on the extensions of the C_2–S and C_5–S bonds, with $V_{S,max}$ of 19.5 and 22.9 kcal/mol, respectively. Without the aza nitrogen in the ring, both $V_{S,max}$ would be 14.7 kcal/mol. Table 16.3 shows that by substituting an electron-withdrawing chlorine or CN at positions 2, 4 or 5 of **18**, the relative strengths of the two σ-holes can be varied and even reversed. On

Table 16.3 Computed σ-hole $V_{S,max}$ on the sulfur 0.001 au surfaces in some substituted 1,3-thiazoles[a,b]

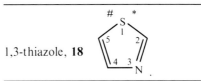

1,3-thiazole, **18**

Molecule	Sulfur $V_{S,max}$
1,3-thiazole, **18**	(#) 19.5, (*) 22.9
2-chloro-1,3-thiazole	(#) 25.8, (*) 12.2
4-chloro-1,3-thiazole	(#) 25.4, (*) 30.2
5-chloro-1,3-thiazole	(#) 11.7, (*) 29.4
2-cyano-1,3-thiazole	(#) 32.8, (*) 22.5
4-cyano-1,3-thiazole	(#) 32.3, (*) 36.9
5-cyano-1,3-thiazole	(#) 21.1, (*) 37.4
2-amino-1,3-thiazole	(#) 14.5
4-amino-1,3-thiazole	None
5-amino-1,3-thiazole	None
4-amido-1,3-thiazole	(#) 21.4, (*) 29.1

The $V_{S,max}$ are on the extensions of the C_2–S and C_5–S bonds; their approximate positions are indicated by the symbols # and *. The values are in kcal/mol
[a] All data are from reference [68]
[b] Computational level: B3PW91/6-31 G(d,p)

the other hand, the introduction of the electron-donating NH_2 weakens or eliminates the positive σ-holes. The amido group, $C(=O)NH_2$, is what is present in tiazofurin (**16**); Table 16.3 confirms that it strengthens particularly the σ-hole on the extension of the C_5–S bond, in agreement with the close contact found in tiazofurin (**16**).

The effects of Cl, CN, NH_2 and $C(=O)NH_2$ substitution in selenazole (**19**), with respect to the selenium σ-holes, are analogous to those obtained for thiazole (Table 16.3) [68]. The primary difference is that the 4-NH_2 derivative retains the two selenium positive σ-holes.

It should be noted that the substituted 1,3-thiazoles and 1,3-selenazoles typically contain $V_{S,max}$ other than the sulfur and selenium σ-holes (e.g. on the chlorines and hydrogens) as well as a number of $V_{S,min}$ [68]. Thus there can be opportunities for multiple interactions. It must be kept in mind, however, that when there are several sources of positive and negative molecular surface electrostatic potentials, these regions are likely to overlap – which may strengthen, weaken or eliminate some $V_{S,max}$ and/or $V_{S,min}$. This is further discussed by Murray et al. [68].

The examples that have been given here and elsewhere [28, 29, 68] illustrate how the interactive capacities of σ-holes can be enhanced or inhibited by modifying the molecular framework and/or appropriately choosing and positioning the substituents. This gives rise to the possibility of selectively strengthening certain σ-holes relative to others in order to achieve the directionalities and orientations of interactions that are required for specific purposes, such as drug-receptor binding or crystal formation.

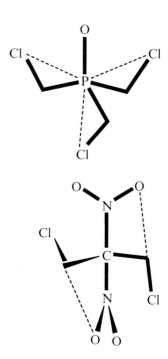

Fig. 16.6 Diagram showing the directionalities of the intramolecular interactions in O=P(CH$_2$Cl)$_3$. They involve positive σ-holes on the phosphorus, along the extensions of the O–P and C–P bonds, and the negative sides of the chlorines

Fig. 16.7 Diagram showing the directionalities of the intramolecular interactions in ClH$_2$C–C(NO$_2$)$_2$–CH$_2$Cl. They involve positive σ-holes on the terminal carbons, along the extensions of the Cl–C bonds, and a negative oxygen of each nitro group

Tiazofurin (**16**) and selenazofurin (**17**) are molecules that are conformationally stabilized through intramolecular interactions, with important biological conse-quences as mentioned above. We will just briefly touch upon two recently synthe-sized compounds that also display conformational stabilization, **20** and **21** [69]:

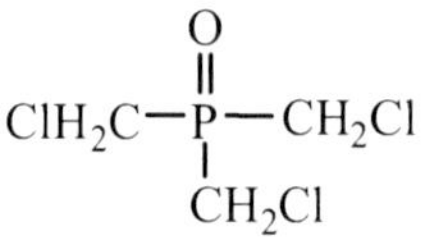

$$
\text{ClH}_2\text{C}-\overset{\overset{\textstyle O}{\|}}{\underset{\underset{\textstyle CH_2Cl}{|}}{P}}-\text{CH}_2\text{Cl}
$$

$$
\text{ClH}_2\text{C}-\overset{\overset{\textstyle NO_2}{|}}{\underset{\underset{\textstyle NO_2}{|}}{C}}-\text{CH}_2\text{Cl}
$$

20 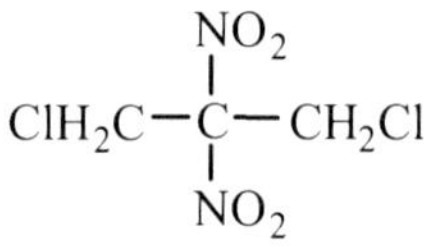   **21**

It was found that neither **20** nor **21** shows the expected reactivity toward nucleophiles, e.g. replacement of Cl$^-$ by a negative ion such as N$_3$$^-$ or C(NO$_2$)$_3$$^-$. In both instances, this can be attributed to intramolecular σ-hole interactions [69]. In **20**, these involve positive σ-holes on the phosphorus, on the extensions of the O–P and C–P bonds, which are attracted to the negative sides of the chlorines (Fig. 16.6). In **21**, σ-holes on the terminal carbons, on the extensions of the Cl–C bonds, are interacting with nitro oxygens (Fig. 16.7). The consequences are the freezing of the conformations of **20** and **21**, thereby interfering with the access of an approaching negative ion and preventing the anticipated substitution.

16.4 Summary

We have sought, through a series of examples, to demonstrate the important role that can be played by non-hydrogen-bonding intramolecular interactions. These often involve one or more positive σ-holes or π-holes in conjunction with some negative site(s) on the molecule. These interactions, which are electrostatically-driven, can significantly influence not only the structures of molecules, but also their reactive behavior, both noncovalent and covalent. They should accordingly be viewed as an essential factor to take into account and indeed to exploit in designing molecules for particular functions, e.g. drugs.

References

1. Joesten MD, Schaad LJ (1974) Hydrogen bonding. Marcel Dekker, New York
2. Schuster P, Zundel G, Sandorfy C (eds) (1976) The hydrogen bond. Recent developments in theory and experiments. North-Holland Publishing Co, Amsterdam
3. Jeffrey GA, Saenger W (1991) Hydrogen bonding in biological structures. Springer, Berlin
4. Scheiner S (1997) Hydrogen bonding. A theoretical perspective. Oxford University Press, New York
5. Hobza P, Havlas Z (2000) Chem Rev 100:4253
6. Belkova NV, Shubina ES, Epstein LM (2005) Acc Chem Res 38:624
7. Grabowski SJ, Sokalski WA, Leszczynski J (2006) Chem Phys Lett 432:33
8. Grabowski SJ (ed) (2006) Hydrogen bonding – new insights. Springer, Dordrecht
9. Gilli G, Gilli P (2009) The nature of the hydrogen bond. Oxford University Press, Oxford
10. de Oliveira BG, Ramos MN (2009) Int J Quantum Chem 110:307
11. Politzer P, Murray JS, Clark T (2010) Phys Chem Chem Phys 12:7748
12. Shields ZP, Murray JS, Politzer P (2010) Int J Quantum Chem 110:2823
13. Guru Row TN, Parthasarathy R, Dunitz JD (1977) J Am Chem Soc 99:4860
14. Politzer P, Lane P, Concha MC, Ma Y, Murray JS (2007) J Mol Model 13:305
15. Politzer P, Murray JS, Concha MC (2008) J Mol Model 14:659
16. Brinck T, Murray JS, Politzer P (1992) Int J Quantum Chem 44:57
17. Auffinger P, Hays FA, Westhof E, Shing Ho P (2004) Proc Natl Acad Sci USA 101:16789
18. Murray JS, Lane P, Politzer P (2009) J Mol Model 15:723
19. Murray JS, Lane P, Politzer P (2007) Int J Quantum Chem 107:2286
20. Murray JS, Lane P, Clark T, Politzer P (2007) J Mol Model 13:1033
21. Rosenfield RE Jr, Parthasarathy R, Dunitz JD (1977) J Am Chem Soc 99:4860
22. Murray-Rust P, Motherwell WDS (1979) J Am Chem Soc 101:4374
23. Murray-Rust P, Stallings WC, Monti CT, Preston RK, Glusker JP (1983) J Am Chem Soc 105:3206
24. Stevens ED (1979) Mol Phys 37:27
25. Lommerse JPM, Stone AJ, Taylor R, Allen FH (1996) J Am Chem Soc 118:3108
26. Ikuta S (1990) J Mol Struct (Theochem) 205:191
27. Clark T, Hennemann M, Murray JS, Politzer P (2007) J Mol Model 13:291
28. Riley KE, Murray JS, Politzer P, Concha MC, Hobza P (2009) J Chem Theor Comput 5:155
29. Riley KE, Murray JS, Fanfrlík J, Řezáč J, Solá RJ, Concha MC, Ramos FM, Politzer P (2011) J Mol Model. doi:10.1007/s00894-011-1015-6
30. Politzer P, Murray JS (2009) In: Leszczynski J, Shukla M (eds) Practical aspects of computational chemistry. Springer, Heidelberg, pp 149–163

31. Murray JS, Riley KE, Clark T, Politzer P (2010) Aust J Chem 63:1598
32. Metrangolo P, Neukirch H, Pilati T, Resnati G (2005) Acc Chem Res 38:386
33. Metrangolo P, Resnati G (eds) (2008) Halogen bonding: fundamentals and applications. Springer, Berlin
34. Murray JS, Lane P, Clark T, Riley KE, Politzer P (2011) J Mol Model doi:10.1007/s00894-011-1089-1
35. Del Bene JE, Alkorta I, Elguero J (2010) J Phys Chem A 114:12958
36. Göbel M, Tchitchanov BH, Murray JS, Politzer P, Klapötke TM (2009) Nat Chem 1:229
37. Murray JS, Lane P, Nieder A, Klapötke TM, Politzer P (2010) Theor Chem Acc 127:345
38. Stewart RF (1979) Chem Phys Lett 65:335
39. Politzer P, Truhlar DG (eds) (1981) Chemical applications of atomic and molecular electrostatic potentials. Plenum Press, New York
40. Hohenberg P, Kohn W (1964) Phys Rev 136:B864
41. Politzer P, Murray JS (2002) Theor Chem Acc 108:134
42. Murray JS, Politzer P (2011) Comput Mol Sci 1: 153–163. Wiley interdisciplinary reviews. doi:10.1002/wcms.19
43. Politzer P (2004) Theor Chem Acc 111:395
44. Bader RWF, Carroll MT, Cheeseman JR, Chang C (1987) J Am Chem Soc 109:7968
45. Bulat FA, Toro-Labbé A, Brinck T, Murray JS, Politzer P (2010) J Mol Model 16:1679
46. Macaveiu L, Göbel M, Klapötke TM, Murray JS, Politzer P (2010) Struct Chem 21:139
47. Göbel M, Klapötke TM (2009) Adv Funct Mater 19:347
48. Göbel M, Klapötke TM (2008) Acta Cryst C64:o58
49. Bondi A (1964) J Phys Chem 68:441
50. Lide DR (ed) (2006) Handbook of chemistry and physics, 87th edn. CRC Press/Taylor and Francis, Boca Raton
51. Witt JR, Britton D, Mahon C (1972) Acta Cryst B28:950
52. Urbański T (1984) Chemistry and technology of explosives, vol 5. Pergamon Press, New York, p 245
53. Murray JS, Lane P, Göbel M, Klapötke TM, Politzer P (2009) J Chem Phys 130:104304
54. Murray JS, Lane P, Göbel M, Klapötke TM, Politzer P (2009) Theor Chem Acc 124:355
55. Mitzel NW, Blake AJ, Rankin DWH (1997) J Am Chem Soc 119:4143
56. Mitzel NW, Losehand U (1998) J Am Chem Soc 120:7320
57. Mitzel NW, Losehand U, Wu A, Cremer D, Rankin DWH (2000) J Am Chem Soc 122:4471
58. Vojinovic K, Mitzel NW, Foerster T, Rankin DWH (2004) Z Naturforsch 59b:1505
59. Mitzel NW, Vojinovic K, Froehlich R, Foerster T, Robertson HE, Borisenko KB, Rankin DWH (2005) J Am Chem Soc 127:13705
60. Blake AJ, Dyrbush M, Ebsworth EAV, Henderson SGD (1988) Acta Cryst C44:1
61. Murray JS, Concha MC, Politzer P (2011) J Mol Model 17:2151 doi:10.007/s00894-010- 0836-z
62. Gibbs TR, Popolato A (eds) (1980) LASL explosive property data. University of California Press, Berkeley
63. Klapötke TM, Krumm B, Ilg R, Troegel D, Tacke R (2007) J Am Chem Soc 129:6908
64. Liu W-G, Zybin SV, Dasgupta S, Klapötke TM, Goddard WA III (2009) J Am Chem Soc 131:7490
65. Klapötke TM, Krumm B, Nieder A, Tacke R, Troegel D (2009) New trends in research of energetic materials, part 1. Institute of Energetic Materials, University of Pardubice, Czech Republic, Pardubice, pp 262–276
66. Goldstein BM, Takusagawa F, Berman HM, Scrivastava PC, Robins RK (1983) J Am Chem Soc 105:7416
67. Goldstein BM, Takusagawa F, Berman HM, Scrivastava PC, Robins RK (1985) J Am Chem Soc 107:1394
68. Murray JS, Lane P, Politzer P (2008) Int J Quantum Chem 108:2770
69. Evangelisti C, Kettner M, Penger A, Del Costello D, Klapötke TM, Murray JS, Politzer P, submitted

Chapter 17
X–H···π and X–H···σ Interactions – Hydrogen Bonds with Multicenter Proton Acceptors

Sławomir J. Grabowski

Abstract X–H···π interactions which are often classified as hydrogen bonds, are the subject of numerous studies since they are important in chemical and biological processes. For typical hydrogen bonds there is the one-center and localized proton acceptor while for X–H···π interactions π-electron systems act as the Lewis base. Hence the latter hydrogen bonds should be classified as those which possess multicenter proton acceptors. However X–H···σ interactions may be also classified as hydrogen bonds possessing multicenter proton acceptors. Ab initio calculations as well as Quantum Theory of 'Atoms in Molecules' (QTAIM) show unique properties of X–H···π and X-H···σ interactions.

Keywords Hydrogen bond • X–H···π interaction • X–H···σ interaction • Crystal structure • Ab initio calculations • Quantum theory of 'Atoms in Molecules' • Decomposition of interaction energy • Dispersive interactions • π-electrons • Multicenter proton acceptors

17.1 Classification of Hydrogen Bonds According to Properties of Proton Donor and Proton Acceptor

X–H···π interactions possess numerous characteristics usually attributed to hydrogen bonds but it is also possible to indicate features typical only for them [1]. These interactions are the subject of extensive studies since they play the key role in various chemical, physical and bio-chemical processes, similarly as the

S.J. Grabowski (✉)
Kimika Fakultatea, Euskal Herriko Unibertsitatea and Donostia International Physics Center (DIPC), P.K. 1072, 20080 Donostia, Euskadi, Spain

Ikerbasque, Basque Foundation for Science, 48011 Bilbao, Spain
e-mail: s.grabowski@ikerbasque.org

J. Leszczynski and M.K. Shukla (eds.), *Practical Aspects of Computational Chemistry I:*   497
An Overview of the Last Two Decades and Current Trends,
DOI 10.1007/978-94-007-0919-5_17, © Springer Science+Business Media B.V. 2012

Table 17.1 The classification of hydrogen bonds [6]

X–H···Y H-bond	More detailed characterization	Examples
One center X and one center Y	Pauling type H-bond	O–H···O, N–H···O, N–H···N
	Nonelectronegative X	C–H···O, C–H···N, C–H···S
	Nonelectronegative Y	O–H···C, N–H···C
	Nonelectronegative X and Y	C–H···C
	X–H···H–Y dihydrogen bond	N–H···H–Re, C–H···H–C, O–H···H–Be
Multicenter Y or/and X	Multicenter Y	X–H···π, X–H···σ
	Multicenter X and multicenter Y	π–H···π, π–H···σ, σ–H···σ
	Multicenter X	π–H···O

Reprinted from Ref. [6] with kind permission of The American Chemical Society

other hydrogen bonds possessing one-center electronegative proton donor and the same type of proton acceptor. The latter hydrogen bonds are often designated as X–H···Y. These are, for example, O–H···O, N–H···O or N–H···N [2, 3]. It is worth mentioning that C–H···π interactions, the sub-class of X–H···π hydrogen bonds, is well represented in crystal structures of organic, and organometallic compounds [4] since C–H bonds as well as π-electron systems are common for organic compounds. The existence of the multicenter proton acceptor is the unique characteristic of X–H···π hydrogen bonds. This is probably one of reasons why such interactions possess slightly different features than the other hydrogen bonds. The similar situation occurs if one considers X–H···σ interactions where σ-electrons act as the proton acceptor [5]. However it seems that in the latter case only H_2 molecule may act as the Lewis base (proton acceptor) for H-bond interaction. This is the different situation than for X–H···π hydrogen bonds where there is the wide spectrum of π-electron systems acting as the Lewis base; acetylene and its derivatives, ethylene and derivatives, benzene and the other aromatic systems, there are many systems containing π-electrons.

The classification of hydrogen bonds according to the kind of proton donor and proton acceptor was proposed [5, 6]. The following types of X–H···Y hydrogen bonds may be mentioned (Table 17.1); hydrogen bonds possessing the one center proton donor and the one center proton acceptor. If both centers are electronegative atoms thus the characteristics of interaction (O–H···O, N–H···O, O–H···Cl, etc.) are in accordance with the early Pauling definition of hydrogen bonding [7]. There are the other types of hydrogen bonding, for example those, where X and/or Y are not electronegative centers, such interactions as C–H···O, O–H···C and C–H···C. The latter ones are not in line with the Pauling definition but still they possess the one center X and Y, similarly as dihydrogen bond where there is the negatively charged hydrogen atom acting as the one center proton acceptor. Table 17.1 also contains examples of interactions where there are multicenter X and/or Y. Among them there are the mentioned above X–H···π and X–H···σ hydrogen bonds.

Fig. 17.1 Structure of
$[H_3O.3C_6H_6]^+[CHB_{11}Cl_{11}]^-$
C_6H_6 (Reprinted from Ref.
[8] with kind permission of
The American Chemical
Society)

There are numerous review articles and monographs on X–H···π hydrogen bonds, especially since such interactions occur commonly in crystal structures. Figure 17.1 presents an interesting case of such interactions existing in the crystal structure of $[H_3O.3C_6H_6]^+$ $[CHB_{11}Cl_{11}]^-C_6H_6$ [8] where H_3O^+ ion is surrounded by three benzene molecules. In such a way hydronium ion acts as the proton donor for O–H···π hydrogen bonds since π-electrons of benzene molecules are the proton acceptors. In the other words, all O–H bonds of H_3O^+ are saturated by H-bond interactions. Figure 17.1 also presents $[CHB_{11}Cl_{11}]^-$ion which directly interacts with the other benzene molecule through C–H···π hydrogen bonding. One can see that for this crystal structure there is the dominant influence X–H···π hydrogen bonds on the arrangement of molecules and ions.

The results on crystal structures are not the only experimental evidences of the existence of X–H···π interactions. There are numerous gas-phase experimental studies. For example, the high-resolution optical and microwave spectra on the benzene-ammonia complex were presented [9]. It was found that in the vibrationally averaged structure, the C_3 symmetry axis of ammonia is tilted by approximately 58° relative to the C_6 benzene axis. In such a way the N–H bonds interact with π-electrons of benzene through N–H···π hydrogen bonds. The resonance-enhanced two-photon ionization, the microwave spectroscopy and the other spectroscopic techniques were used to analyze such complexes as C_6H_6–H_2O [10], C_6H_6–HF [11], C_6H_6–HCl [12] and also T-shaped complexes where π-electrons of acetylene act as the proton acceptor while such moieties as HF, HCl or HCN are the proton donors [13–15].

The other situation is observed for X-H···σ interactions since σ-electrons are much weaker Lewis base than π-electrons [5]. The comparison of X–H···π and X–H···σ hydrogen bonds was performed showing similar dependencies, relationships and tendencies for both kinds of interactions. However, as one could expect, X–H···σ are much weaker hydrogen bonds than their X–H···π counterparts. For both kinds of interactions the same set of the proton donating moieties was chosen and two simple species, acetylene (for X–H···π) and molecular hydrogen (for X–H···σ) were taken into account as the Lewis bases [5]. There are numerous more or less simple π-electron systems which may play the role of proton acceptors in H-bond interactions. However in the case of σ-electrons practically only H_2 molecule may act as the proton acceptor. From time to time the other σ-bonds are considered as the proton acceptors [16] but this requires additional studies. For example, this is commonly accepted that for $X–H^{+\delta}···^{-\delta}H–Y$ dihydrogen bond there are two H-atoms being in contact where one of them is positively charged as for hydrogen bond interaction while the second one acting as the proton acceptor is negatively charged [17]. Hence there is the $H^{+\delta}···^{-\delta}H$ contact and the interaction of H-atoms of opposite charges is the main factor responsible for the stability of dihydrogen bonded complexes. However it was also claimed that for dihydrogen bonds the $^{-\delta}H–Y$ σ-bond may be considered as the proton acceptor and not the negatively charged H-atom center [16]. The latter idea may be supported by the fact that for dihydrogen bonded systems the main interaction energy part connected with the electron charge transfer from the proton acceptor to the proton donor may be classified as $σ → σ^*$ interaction [18], the same type of interaction exists for T-shaped complexes with H_2 molecule, $FH···H_2$, $ClH···H_2$, $NH_4^+···H_2$ etc. [5].

It is worth mentioning that H_2 molecule may act as Lewis acid and also as Lewis base. For example high level ab initio calculations have been performed recently on complexes where dihydrogen bonds and metal-sigma interactions exist between H_2 molecule and metal hydrides [19]. In the case of dihydrogen bonds molecular hydrogen acts as Lewis acid while for the latter interactions it plays the role of Lewis base. Besides the metal-sigma interactions are much stronger than dihydrogen bonds. The similar situation may be found for X–H···σ hydrogen bonds where H_2 is the Lewis base (proton acceptor). What is an experimental evidence of interactions with molecular hydrogen? The search through Cambridge Structural Database (CSD) [20] shows 37 crystal structures containing molecular hydrogen. Only in one case molecular hydrogen is encapsulated in fullerene [21]. For the remaining 36 structures there are metal-sigma interactions where H_2 molecule interacts as a ligand with transition metal elements (or ions), mainly with Ru, but also with Os, Fe, Mo and Cr.[1] This is not surprising since such interactions of H_2 with metals are much stronger than the other interactions of molecular hydrogen [19].

However the gas phase experiments were performed on X–H···σ hydrogen bonds, for example T-shaped structure for $FH···H_2$ complex was found [22]. The ab initio calculations on that complex were performed up to the MP2/aug-cc-pV5Z

---

[1]For the purpose of this chapter CSD was searched for molecular hydrogen, the following restrictions concerning the quality of the diffraction measurements and results were taken into account: only non-disordered structures, non-powder structures, non-polymeric structures, structures without errors.

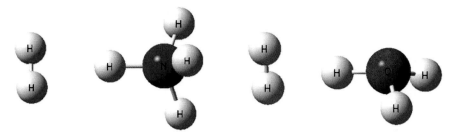

Fig. 17.2 The hydrogen – ammonia and hydrogen – hydronium complexes connected through N–H$\cdots\sigma$ and O–H$\cdots\sigma$ hydrogen bonds

level and the binding energy equal to 1 kcal/mol was found (corrected for BSSE) [23].[2] The MP2/6-311++G(3df,3pd) calculations performed on $NH_4^+\cdots H_2$ and $H_3O^+\cdots H_2$ complexes (see Fig. 17.2) show such energies equal to 2.3 and 5.2 kcal/mol, respectively [5]. The corresponding energies for complexes with acetylene, $NH_4^+\cdots C_2H_2$ and $H_3O^+\cdots C_2H_2$ amount to 10.6 and 19.2 kcal/mol respectively [5]. The mentioned above X–H$\cdots\sigma$ interactions may be important to resolve the problem of hydrogen storage [24]. The clusters of ammonia containing molecular hydrogen were analyzed, $(NH_4^+)\cdots(H_2)_n$ [25] as well as the other $(XH_4^+)\cdots(H_2)_n$ related systems (where X = N, P, As, Sb and Bi) [26]. It was found that there is the stable shell of four H_2 molecules connected with XH_4^+ ion (through four X–H$\cdots\sigma$ interactions). However there are also the other face-centered (X$\cdots\sigma$) interactions [25, 26]; Fig. 17.3 shows $(SbH_4^+)\cdots(H_2)_5$ cluster where one Sb–H$\cdots\sigma$ hydrogen bond exists and four face-centered interactions.

17.2   Energies of X–H$\cdots\pi$ and X–H$\cdots\sigma$ Interactions

It was stated early that π-electrons, especially aromatic systems, may act as proton acceptors for hydrogen bond interactions. Pimentel and McClellan have concluded [27] that "Since there is ample evidence that the proton of the acid is specifically involved in the interaction, it can properly be called a H bond. An important implication of the conclusion that the base B can be aromatic is that the base must be a good electron donor but not necessarily a highly electronegative group." The possibility of π-electron systems to act as proton acceptors in crystals was claimed by Leiserowitz for the crystal structure of propiolic acid [28]. For this structure centrosymmetric dimers exist since monomers are linked doubly through

---

[2]The hydrogen bond energy possesses a negative value and more negative value indicates stronger interaction. However, very often for the convenience of explanations the absolute values are given in texts. The same concerns energies given here, the positive values will be given hereafter through the whole text. The true negative values are included in Tables and Figures.

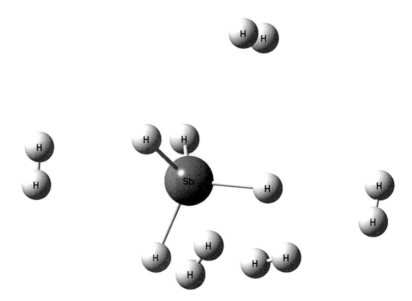

Fig. 17.3 The $(SbH_4{}^+)\cdots(H_2)_5$ complex, one Sb–H$\cdots\sigma$ hydrogen bond and four face-centered Sb$\cdots\sigma$ interactions

equivalent O–H$\cdots$O hydrogen bonds. Besides, C–H$\cdots$C$\equiv$C (midpoint) contacts of 3.8 Å (H$\cdots$C$\equiv$C of 2.8 Å) link the dimer units. The latter interactions may be classified as C–H$\cdots\pi$ hydrogen bonds. The other gas-phase and solid-phase examples of such interactions were presented in the previous section. It is usually pointed out that C–H$\cdots\pi$ interactions are rather weak if compared with the other, especially Pauling-type (see Table 17.1) hydrogen bonds.

In general, for X–H$\cdots\pi$ interactions the X–H bond is directed to the π-electron system [4]. However it is worth mentioning that such interactions are sometimes comparable in the strength with the Pauling-type hydrogen bonds, especially if X is not the carbon atom center. For example, complexes of benzene with different proton donating sub-units were analyzed and the X–H$\cdots\pi$ interaction energies were calculated at CCSD(T) level at the basis set limit [29, 30]. It was found that the interaction energy for benzene-methane complex amounts to 1.5 kcal/mol while such an energy for benzene-NH$_4{}^+$ complex is equal to 19.3 kcal/mol, like for the strong hydrogen bonds [31]. However for the latter complex there is the N–H$\cdots\pi$ positively charge assisted, CAHB(+), hydrogen bonding.

Table 17.2 presents binding energies, ΔE's, for a series of complexes of π-electron systems with molecules acting as proton donors, hydrogen fluoride, HF, and acetylene, [32]. Certainly for complexes with HF the interactions are stronger than the corresponding interactions for complexes with C_2H_2 since hydrogen fluoride is the stronger Lewis acid than acetylene. Except of two triplet cases the singlet systems are presented in Table 17.2. In two cases there are the meaningful

Table 17.2 Binding energies (in kcal/mol), calculated at MP2/6-311++G(d,p) level of approximation (corrected for BSSE, the deformation energy being the result of the complexation is taken into account) [32]

Proton acceptor	ΔE	
	C_2H_2 proton donor	HF proton donor
C_6H_6	−2.14	−3.25
$C_5H_5^-$	−9.84	−16.31
C_4H_4	−1.63	−3.72
C_4H_4[a]	−1.66	−2.82

Reprinted from Ref. [32] with kind permission of The American Chemical Society

[a] Triplet

Table 17.3 The interaction energy terms (in kcal/mol), calculated at MP2/6-311++G(d,p) level of approximation, for the convention of signs see footnote 2 [36]

Energy term	$C_2H_2\cdots H_3O^+$	$C_2H_2\cdots HF$	$C_2H_2\cdots HCl$	$C_2H_2\cdots CHF_3$	$C_2H_2\cdots C_2H_2$
EL	−19.12	−6.42	−4.20	−2.75	−2.15
EX	25.37	6.28	4.65	2.43	2.13
PL	−15.34	−1.51	−0.82	−0.46	−0.26
CT	−16.32	−2.16	−1.51	−0.53	−0.46
MIX	6.89	0.80	0.64	0.28	0.16
E_{SCF}	−18.52	−3.01	−1.23	−1.02	−0.58
CORR	−3.48	−1.43	−1.94	−1.27	−1.42
E_{MP2}	−22.00	−4.44	−3.17	−2.29	−2.00

Reprinted from Ref. [36] with kind permission of Elsevier

interactions energies. For $C_5H_5^-\cdots HF$ complex it amounts to 16.3 kcal/mol, even for $C_5H_5^-\cdots C_2H_2$ complex, where the C–H···π hydrogen bonding exists the interaction energy is close to 10 kcal/mol.

It was stated that for strong and very strong hydrogen bonds the delocalization (DEL) is the most important attractive interaction energy term [33], for medium in strength hydrogen bonds the electrostatic energy (EL) is the most important [33] while for weak such interactions the dispersive energy (DISP) dominates [34]. It is worth mentioning that the meaning of these interaction energy terms is different for different decomposition schemes. Briefly speaking, the delocalization term roughly corresponds to the sum of polarization (PL) and charge transfer (CT) attractive terms which appear in the Kitaura-Morokuma partitioning scheme [35].

The Kitaura-Morokuma decomposition scheme was applied recently to analyze the nature of intermolecular interactions for complexes of acetylene with different proton donors [36]. Table 17.3 shows the interaction energy terms for few of the complexes analyzed. These are so-called T-shaped systems since the proton donating bond for them is perpendicular, or nearly so, to the C≡C bond of acetylene. The interaction energy terms presented in the table may be briefly described in the following way. EL is the electrostatic interaction energy between the undistorted charge distributions of two interacting species (acetylene and the proton donating moiety for the sample presented here), EX is the exchange energy resulting from the electron clouds repulsion, PL is the polarization interaction energy defined as the energy of the distortion of charge distribution of monomers as an effect of complexation, CT is the energy of charge transfer from one monomer to the other one as an effect of the formation of the complex and MIX term is the energy difference between SCF interaction energy (E_{SCF}) and the sum of the previously mentioned terms. Table 17.3 contains also E_{MP2} energies since the calculations were performed at MP2/6-311++G(d,p) level. It is worth mentioning that the interaction energies presented in Table 17.3 (E_{MP2}'s) do not include the deformation energies being the result of the complexation. They were evaluated as the difference between the energy of complex and the sum of the corresponding monomers' energies. The energies of these monomers correspond to geometries taken from the complex.

The correlation energy, CORR, contains DISP – the dispersion interaction energy term. DISP is one of the most important attractive terms resulting from the decomposition of CORR. The results collected in Table 17.3 show that for all cases of complexes the exchange energy is comparable with the modulus of electrostatic energy, in the other words for these systems the other attractive interactions are very important. Even for $C_2H_2\cdots H_3O^+$ complex EX outweighs significantly EL term. Additionally, for the later complex, where O–H$\cdots\pi$ CAHB(+) exists, the modulus of sum of polarization and charge transfer terms is greater than the modulus of electrostatic energy term. This is well known phenomenon for strong hydrogen bonds, especially those interactions which possess the covalent character [31]. For all remaining complexes of Table 17.3 the electrostatic and correlation energy terms are the most important attractive contributions. Thus it is partly confirmed, that for weak X–H$\cdots\pi$, and especially C–H$\cdots\pi$ interactions, the correlation and consequently the dispersion energies are very important for the stabilization of complexes.

The other decomposition scheme [37] was applied to analyze $NH_4^+(H_2)_n$ clusters [25]. It was found that for the T-shaped conformation of $NH_4^+(H_2)_1$ complex there are the following interaction energy terms: EL(-1.6 kcal/mol), EX($+2.75$ kcal/mol), DEL(-2.62 kcal/mol) and DISP(-1.0 kcal/mol); calculations were performed at MP2/aug-cc-pVTZ level. It is difficult to point out the general conclusions on the nature of X–H$\cdots\sigma$ interactions taking into account only one example. However one can see that for the latter complex the exchange energy, EX, outweighs the electrostatic energy, ES, and the complex is stable due to the meaningful DEL and CORR contributions. Hence the distribution of the energy terms is similar here to those observed for the complexes linked through X–H$\cdots\pi$ interactions. The comparison of X–H$\cdots\pi$ and X–H$\cdots\sigma$ interactions was performed [5] for the complexes with C_2H_2 and H_2 proton acceptors, respectively. The following molecules were chosen as the proton donors: HF, C_2H_2, NH_4^+, H_3O^+ and $C_2H_3^+$. Figure 17.4 shows two complexes where $C_2H_3^+$ is the proton donating moiety and C_2H_2 and H_2 are the proton acceptors. Thus two types of hydrogen bonds are observed here, π-H$\cdots\pi$ and π-H$\cdots\sigma$ since there are multicenter proton donors and multicenter proton acceptors for these systems (see Table 17.1).

It is worth mentioning that π-H$\cdots\pi$ interactions, especially for $C_2H_3^+\cdots C_2H_2$ complex, were analyzed both theoretically [38] as well as experimentally [39].

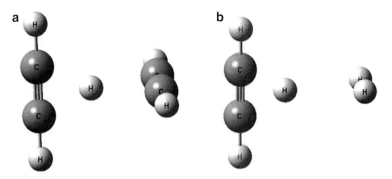

Fig. 17.4 $C_2H_3^+ \cdots C_2H_2$ (a) and $C_2H_3^+ \cdots H_2$ (b) complexes, where $C_2H_3^+$ is the proton donor and acetylene and molecular hydrogen are proton acceptors, respectively

Table 17.4 shows the characteristics of the mentioned above complexes linked through X–H···π and X–H···σ interactions [5]. One can see that σ-electrons are much weaker Lewis base than π-electrons since for the complexes with molecular hydrogen the binding energies are smaller than the corresponding values for complexes with acetylene. The other results confirm this observation since the electron transfer from the proton acceptor to the proton donor is greater for complexes with acetylene. There are the expected changes within sub-samples presented in Table 17.4. For the stronger interactions there are shorter proton···acceptor distances and the greater elongation of C≡C or H–H bond. There is the greatest elongation of the latter bonds for complexes with H^+, i.e. for $C_2H_2 \cdots H^+$ and $H_2 \cdots H^+$. However for those moieties proton is practically covalently bonded to acetylene and molecular hydrogen, respectively. The latter interactions may be classified as multicenter covalent bonds [40].

Generally, within the sub-sample characterized by the same proton acceptor, the stronger interaction is connected with the greater electron charge transfer, greater elongation of the acceptor's bond and shorter H···π(σ) contact. Hence one can see that both X–H···π and X–H···σ interactions posses the characteristics of hydrogen bonding since the mentioned changes are characteristic for the hydrogen bond formation.

17.3   The Use of Quantum Theory of 'Atoms in Molecules' to Characterize X–H···π and X–H···σ Interactions

The Quantum Theory of 'Atoms in Molecules' (QTAIM) [41, 42] is useful to characterize and classify inter- and intramolecular interactions [31] since it is possible to analyze the electron density of any system considered; molecule, ion, complex, more or less complicated cluster, even the analysis of the electron density of crystals is possible [42, 43]. Figure 17.5 presents the molecular graph of

Table 17.4 Characteristics of interactions for complexes linked through X–H$\cdots\pi$ and X–H$\cdots\sigma$ hydrogen bonds; MP2/6-311++G(3df,3pd) results [5]

Complex	C≡C (H–H) bond length	H$\cdots\pi(\sigma)$ distance	Electron transfer	Binding energy
C$_2$H$_2$ Lewis base				
H$^+\cdots$ C$_2$H$_2$	1.228	1.117	744	−154.5
FH$\cdots$ C$_2$H$_2$[a]	1.212	2.117	87	−3.9
C$_2$H$_2\cdots$ C$_2$H$_2$[a]	1.211	2.636	15	−1.4
NH$_4^+\cdots$ C$_2$H$_2$	1.214	2.074	182	−10.7
H$_3$O$^+\cdots$ C$_2$H$_2$	1.216	1.729	308	−19.4
C$_2$H$_3^+\cdots$ C$_2$H$_2$	1.218	1.665	368	−15.5
H$_2$ Lewis base				
H$^+\cdots$ H$_2$	0.871	0.754	667	−105.7
FH$\cdots$ H$_2$[a]	0.739	1.983	46	−0.8
C$_2$H$_2\cdots$ H$_2$[a]	0.737	2.573	1	−0.3
NH$_4^+\cdots$ H$_2$	0.741	1.971	62	−2.3
H$_3$O$^+\cdots$ H$_2$	0.747	1.590	139	−5.2
C$_2$H$_3^+\cdots$ H$_2$	0.743	1.899	99	−2.8

The following features are collected; C≡C and H–H bond lengths (in Å), H$\cdots\pi$ and H$\cdots\sigma$ distances (in Å, from H-atom to the mid-point of C≡C or H–H bond), electron transfer (in milielectrons) from the proton acceptor to the proton donating moiety, the binding energy (BSSE and the deformation energy as a result of complexation are taken into account) (Reprinted from Ref. [5] with kind permission of The American Chemical Society)
[a] T-shaped complex

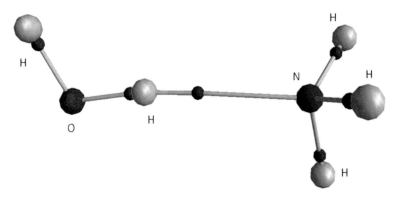

Fig. 17.5 The molecular graph of ammonia-water complex

ammonia-water complex; big circles correspond to local maxima of the electron density (attractors), there are also bond paths (BPs) – the lines of the minimal electron density which connect attractors. The bond critical point (BCP, designated in figure by red small circle) is the point of the minimal electron density on the bond path. There are the other critical points which do not occur for ammonia-water complex; ring critical points (RCPs) and cage critical points (CCPs).

The characteristics of BCP inform on the nature of atom-atom interaction since attractors are attributed to the positions of atoms; these are: electron density at BCP (ρ_{BCP}), laplacian of this electron density ($\nabla^2\rho_{BCP}$), the total electron energy density at BCP (H_{BCP}) and the components of the latter value: the kinetic electron energy density (G_{BCP} – always positive value) and the potential electron energy density (V_{BCP} – always negative value). There are the following relations between the mentioned above values; Eqs. 17.1 and 17.2.

$$(1/4)\nabla^2\rho_{BCP} = 2G_{BCP} + V_{BCP} \qquad (17.1)$$

$$H_{BCP} = G_{BCP} + V_{BCP} \qquad (17.2)$$

It is worth mentioning that the negative value of laplacian is attributed to the covalent interactions while its positive value is characteristic for van der Waals interactions, ionic interactions and hydrogen bonds. However for very strong hydrogen bonds H_{BCP} is negative and even in extreme cases laplacian is negative like in the case of covalent bonds [31, 44].

It was stated that; "The presence of the bond path linking a pair of atoms fulfils the sufficient and necessary conditions that the atoms are bonded to one another [45, 46]." It does not mean that there is the chemical bond between two atoms (exactly attractors) linked by the bond path. This means that there is the interaction which stabilizes the system considered, and it may be not only the covalent bond. It was found recently for selected complexes linked through X–H··· π hydrogen bonds that the bond paths exist for the preferable interactions [32]. Figure 17.5 shows bond paths between pairs of atoms linked through covalent bonds, however there is also the bond path linking H and N-atoms, i.e. the proton with the proton acceptor. The latter connection corresponds to hydrogen bonding.

Table 17.5 presents characteristics of the proton – proton acceptor BCPs for complexes analyzed in the previous section (linked through X–H··· π and X–H··· σ interactions). It is worth mentioning that in two cases $H^+\cdots C_2H_2$ and $H^+\cdots H_2$ complexes one can observe the negative value of the laplacian of electron density at BCP. These interactions may be classified as the multicenter covalent bonds. Additionally the electron density at BCP for $H^+\cdots C_2H_2$ and $H^+\cdots H_2$ complexes amounts to 0.21 and 0.14 au, respectively. These values are typical for covalent bonds, one can observe that ρ_{BCP}'s for the remaining systems are ten times smaller or even more, like it is usually observed for the closed-shell interactions. For few complexes presented in Table 17.5 $\nabla^2\rho_{BCP}$ is positive but H_{BCP} is negative. This means that hydrogen bonding for them is partially covalent in nature [31, 47]. Such a situation is observed for $NH_4^+\cdots C_2H_2$, $H_3O^+\cdots C_2H_2$ and $C_2H_3^+\cdots C_2H_2$ complexes where X–H··· π hydrogen bonds exist, but also for $H_3O^+\cdots H_2$ with X–H··· σ interaction. Thus one can see that even X–H··· σ interactions with weak Lewis base may posses covalent character.

Figure 17.6 presents two examples of molecular graphs of the complexes analyzed, $NH_4^+\cdots C_2H_2$ and $NH_4^+\cdots H_2$. The complexes are linked through

Table 17.5 Characteristics (in au) of BCPs corresponding to H···π and H···σ interactions for complexes linked through X–H···π and X–H···σ hydrogen bonds, respectively [32]

Complex	ρ_{BCP}	$\nabla^2\rho_{BCP}$	G_{BCP}	V_{BCP}	H_{BCP}
C_2H_2 Lewis base					
$H^+\cdots C_2H_2$	0.2059	−0.3220	0.0776	−0.2357	−0.1581
$FH\cdots C_2H_2{}^a$	0.0199	0.0532	0.0127	−0.0122	0.0005
$C_2H_2\cdots C_2H_2{}^a$	0.0077	0.0235	0.0047	−0.0036	0.0011
$NH_4^+\cdots C_2H_2$	0.0251	0.0554	0.0147	−0.0156	−0.0009
$H_3O^+\cdots C_2H_2$	0.0534	0.0278	0.0265	−0.0461	−0.0196
$C_2H_3^+\cdots C_2H_2$	0.0651	0.0126	0.0289	−0.0546	−0.0257
H_2 Lewis base					
$H^+\cdots H_2$	0.1380	−0.2458	0.0021	−0.0656	−0.0635
$FH\cdots H_2{}^a$	0.0109	0.0079	0.0381	−0.0062	0.0319
$C_2H_2\cdots H_2{}^a$	0.0033	0.0124	0.0024	−0.0016	0.0008
$NH_4^+\cdots H_2$	0.0131	0.0406	0.0088	−0.0074	0.0014
$H_3O^+\cdots H_2$	0.0307	0.0536	0.0185	−0.0235	−0.0050
$C_2H_3^+\cdots H_2$	0.0167	0.0428	0.0100	−0.0094	0.0006

Reprinted from Ref. [32] with kind permission of The American Chemical Society
^a T-shaped complex

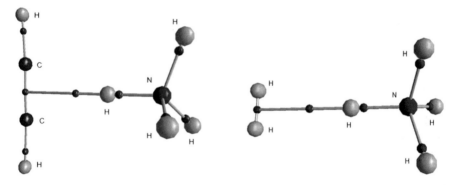

Fig. 17.6 The molecular graphs of $NH_4^+\cdots C_2H_2$ and $NH_4^+\cdots H_2$ complexes linked through X–H···π and X–H···σ interactions, respectively

X–H···π and X–H···σ interactions thus they may be classified as hydrogen bonds with multicenter proton acceptors (Table 17.1). H-attractor of the proton donating N–H bond is linked through the bond path with the BCP of C≡C or H–H bond. For the Pauling-type hydrogen bonds (see Fig. 17.5) the bond path links two attractors which correspond to H-atom (of the proton donating X–H bond) and one-centered Y acceptor. For $NH_4^+\cdots C_2H_2$ and $NH_4^+\cdots H_2$ complexes BCPs of C≡C and H–H bonds mimic the acceptor centers of π and σ-electrons. The similar molecular graphs are observed for all complexes collected in Table 17.5 where H – BCP bond paths exist which connect Lewis acid – Lewis base sub-units. In general,

Fig. 17.7 The molecular graph of the cluster containing four acetylene molecules; *big black* and *grey circles* correspond to C and H-attractors, respectively, *small red, yellow* and *green circles* designate, bond, ring and cage critical points, respectively

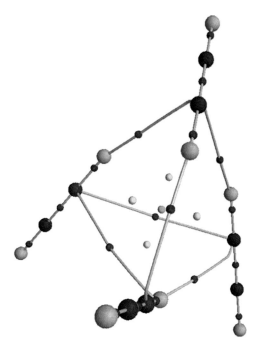

these are highly symmetrical species since almost for all cases the proton donating bond is perpendicular (or nearly so) to the C≡C or H–H bond. Any distortion from the symmetry may cause the non-symmetrical electron charge distribution and in consequence the preference of the single atomic center to act as an acceptor. For example, one of carbon atoms of acetylene may act as the Lewis base. Figure 17.7 presents such a situation for the cluster containing four acetylene molecules. The bond paths link here H and C-attractors what suggests the existence of C–H⋯C hydrogen bonds instead of C–H⋯π ones. There is additional C⋯C bond path (Fig. 17.7). what suggests the existence of the stabilizing interaction between the corresponding carbon atoms.

Table 17.1 contains also the case of hydrogen bonding with multicenter proton donor and multicenter proton acceptor. The complexes possessing such type of hydrogen bonds were analyzed previously [5, 6, 23, 31, 38, 39]. Figure 17.8 presents the molecular graph of $C_2H_3^+\cdots C_2H_2$ complex where there are two-center donor and acceptor since acetylene molecule donates proton and the other acetylene acts as a Lewis base. The proton is closer to one of acetylene molecules (left in Fig. 17.8) forming with it multicenter covalent bond since the laplacian of the electron density at BCP, and consequently H_{BCP}, are negative. The proton is slightly more far from the second acetylene molecule (right side of Fig. 17.8). However for the latter interaction there is also strong and partly covalent in nature interaction since H_{BCP}-value for the corresponding BCP is negative. The binding energy for this complex calculated at MP2/6-311++G(d,p) level amounts to 15.5 kcal/mol (Table 17.4). It

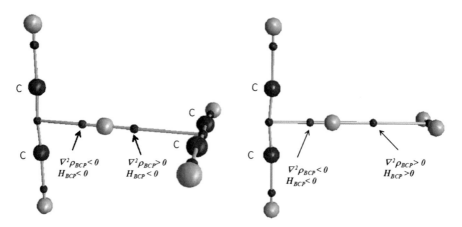

Fig. 17.8 The molecular graphs of the $C_2H_3^+\cdots C_2H_2$ and $C_2H_3^+\cdots H_2$ complexes. *Big circles* correspond to attractors (*black* – carbon, *grey* – hydrogen) (Reprinted from Ref. [5] with kind permission of The American Chemical Society)

means that the latter interaction belongs to the strong ones, this is supported by the QTAIM characteristics of critical point. The similar situation is observed for the $C_2H_3^+\cdots H_2$ (Fig. 17.8). However in such a case there is multicenter covalent bond within $C_2H_3^+$ sub-unit and the other multicenter interaction, $H^+\cdots\sigma$ (proton – molecular hydrogen) is much weaker since both, $\nabla^2\rho_{BCP}$ and H_{BCP} values at the corresponding BCP are positive. Table 17.4 shows binding energy for the latter complex amounting to 2.8 kcal/mol, much less if compared with $C_2H_3^+\cdots C_2H_2$.

17.4 The Case of Multicenter Proton Acceptors

The examples of X–H$\cdots\pi$ and X–H$\cdots\sigma$ hydrogen bonds were presented in the previous section. However for both types of interactions there were two-center proton acceptors, acetylene or molecular hydrogen. For the other complexes there is controversy if π and σ-electrons act as the proton acceptor or only one of atoms of the Lewis base sub-unit possessing the excess of negative charge. In the latter case it would be the one-center proton acceptor. Especially, the situation is complicated if such systems as benzene or more general the multi-atom π-electron systems act as the Lewis base in the intermolecular interactions. Different complexes possessing multicenter proton acceptors were studies recently [32]. Figure 17.9 presents molecular graphs of few examples of such complexes. Hydrogen fluoride (Fig. 17.9) or acetylene (Fig. 17.9) act as the proton donating species. There are two kinds of interactions for complexes with HF molecule. For $C_6H_6\cdots$HF and $C_4H_4\cdots$HF complexes the bond path links H-attractor of HF with the BCP corresponding to one of CC bonds of C_6H_6 or C_4H_4 molecule. This may suggest that in such a case there

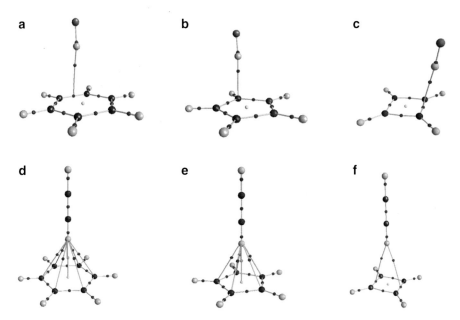

Fig. 17.9 The molecular graphs of the following complexes: (**a**) $C_6H_6\cdots HF$, (**b**) $C_5H_5^-\cdots HF$, (**c**) $C_4H_4\cdots HF$, (**d**) $C_6H_6\cdots HCCH$, (**e**) $C_5H_5^-\cdots HCCH$ and (**f**) $C_4H_4\cdots HCCH$. *Big circles* correspond to attractors (*black* – carbon, *grey* – hydrogen, *green* – fluorine) (Reprinted from Ref. [32] with kind permission of The American Chemical Society)

is the two-center proton acceptor and not six or four-center as one could expect. For $C_5H_5^-\cdots HF$ complex the bond path links H-attractor of HF with the single C-attractor of the ring ($C_5H_5^-$). This shows that probably F–H···C hydrogen bond exists here and it may be classified as the interaction possessing one center X and one center Y. For such an interaction Y is non-electronegative carbon atom (see Table 17.1).

There is the different situation for complexes with acetylene. For $C_6H_6\cdots HCCH$ and $C_5H_5^-\cdots HCCH$ complexes the systems are highly symmetrical since they possess C_{6v} and C_{5v} symmetries, respectively. For these complexes there are bond paths linking H-attractor of HCCH proton donating molecule with all C-attractors of the rings (C_6H_6 and $C_5H_5^-$). One may say that there are six-center and five-center proton acceptors here since it was stated that the bond paths show preferable interactions [45, 46] and for these two complexes the carbon atoms of the rings are equivalent. For these complexes there are also bond paths linking H-attractors with RCPs (the ring critical points) of C_6H_6 and $C_5H_5^-$, such pats (one for each of complexes) are perpendicular to the proton accepting ring and contain the cage critical point (CCP). For $C_4H_4\cdots HCCH$ of C_{2v} symmetry two bond paths link H-attractor of acetylene with two-mid points of CC bonds, one may say that there is four-center proton acceptor here.

Table 17.6 Characteristics of HF and C_2H_2 complexes, the type of bond path is given, the binding energy (kcal/mol) where the deformation as a result of complexation is included and it is corrected for BSSE, NBO energy (in kcal/mol) [32]

Complex	Bond path	Binding energy	NBO energy
$C_6H_6\cdots HF$	$H\cdots CC$ (BCP)	-3.25	2.6
$C_5H_5^-\cdots HF$	$H\cdots C$	-16.31	72.9
$C_4H_4\cdots HF$	$H\cdots CC$ (BCP)	-3.72	23.2
$C_6H_6\cdots C_2H_2$	$H\cdots\ C_6H_6$ (RCP)[a]	-2.14	8.7
$C_5H_5^-\cdots C_2H_2$	$H\cdots\ C_5H_5^-$ (RCP)[a]	-9.84	46.8
$C_4H_4\cdots C_2H_2$	$H\cdots CC$ (BCP)	-1.63	8.1

Reprinted from Ref. [32] with kind permission of The American Chemical Society
[a] There are also six (C_6H_6) and five ($C_5H_5^-$) $H\cdots C$ bond paths for these complexes

Why complexes with acetylene are characterized by the high symmetry while complexes with hydrogen fluoride do not possess high symmetry. In the latter case HF is not perpendicular to the plane of the Lewis base ring. It seems that in a case of HF complexes there is the electrostatic attraction between fluorine negatively charged center and the positively charged H-atoms of the Lewis base. This may be the reason of the distortion of the system from the high C_{nv} symmetry. In a case of complexes with acetylene there is the negligible repulsion between positively charged atoms of terminal H-atom of acetylene and H-atoms of the rings; this does not change the high symmetry of the systems.

Table 17.6 presents some additional characteristics of the complexes analyzed here. The types of bond paths are presented, the binding energy and the NBO energy [48] connected with the electron charge transfer from the Lewis base to the Lewis acid. One can see that in few cases, especially for the complexes with $C_5H_5^-$ there is the meaningful Lewis acid – Lewis base interaction. The binding energy for $C_5H_5^-\cdots HF$ complex amounts 16.3 kcal/mol and the NBO energy exceeds 70 kcal/mol. Also for the $C_5H_5^-\cdots C_2H_2$ there is the meaningful binding energy amounting 9.8 kcal/mol and the NBO energy is equal to 46.8 kcal/mol. However for latter complexes hydrogen bonding is enhanced by negative charge, they may be classified as CAHB($-$), for the complex with HF this is F–H$\cdots$C interaction while for the complex with acetylene this is C–H$\cdots\pi$. For $C_5H_5^-\cdots HF$ system, for BCP lying on the H$\cdots$C bond path the total electron energy density at BCP (H_{BCP}) is negative what means that such an interaction may be treated as partially covalent in nature [31, 47].

These results on complexes with multicenter proton acceptors show that the meaning of X-H$\cdots\pi$ should be verified. This is common that in the case of benzene and the other aromatic systems acting as the Lewis bases the whole rings are treated as proton acceptors since π-electrons are delocalized within them. However the QTAIM and NBO results show that in some of cases there is the accumulation of the electron charge, on single atomic center like for the $C_5H_5^-\cdots HF$ complex or on the single bond, like for the $C_6H_6\cdots HF$ complex (see Fig. 17.9).

Table 17.7 Characteristics of HF complexes, the H···C distance is presented between H-atom of HF and C-atom of the proton acceptor (in Å), the charge of the carbon atom corresponding to the H···C distance (in au) [32]

Carbon atom	H···C	C-atom charge
C_6H_6···HF		
1	2.929	−0.2038
2	2.927	−0.2039
3	2.726	−0.2097
4	2.505	−0.2246
5	2.503	−0.2249
6	2.723	−0.2099
$C_5H_5^-$···HF		
1	2.007	−0.4330
2	2.395	−0.3532
3	2.900	−0.3497
4	2.902	−0.3497
5	2.399	−0.3530
C_4H_4···HF		
1	2.245	−0.2172
2	2.248	−0.2178
3	3.063	−0.1927
4	3.061	−0.1927

Reprinted from Ref. [32] with kind permission of The American Chemical Society

The results for complexes with HF presented in Table 17.7 support the latter statements. The H···C distances between H-atom of HF molecule and C-atoms of Lewis base are presented. For the C_6H_6···HF complex there are two distinctly shortest H···C distances of 2.505 and 2.503 Å for C-atoms numbered as 4 and 5. The most negative charges are observed for these carbon atoms. In the other words the most important attractive electrostatic interactions exist for the shortest H···C contacts. This is very important that the bond path connecting two sub-units, HF and C_6H_6, links H-attractor with BCP lying on the bond path connecting 4 and 5 carbon attractors! One can see that the bond path shows the preferable interaction [32].

The similar situation is observed for the other complexes, for $C_5H_5^-$···HF the shortest H···C distance amounts to 2.007 Å and the most negative charge exists for the corresponding carbon atom. The preferable attractive electrostatic H···C interaction is reflected by the existence of the bond path (Fig. 17.9) connecting the corresponding attractors. For C_4H_4···HF complex two H···C distances are much shorter than the remaining ones, the corresponding C-atoms possess more negative charges than the other C-atoms. In this case the bond path connects H-attractor of Lewis acid with BCP of CC bond, where the latter C-atoms correspond to numbers 1 and 2 in Table 17.7.

The results on complexes with acetylene are not presented in Table 17.7. This is because all H···C distances within the same complex are equal to each other, also charges of C-atoms in the ring for the same complex are equivalent. This leads to the symmetrical systems, for complexes with benzene and cyclopentadienyl there are 6 and 5 $H \cdots C$ bond paths respectively. For $C_4H_4 \cdots HCCH$ complex two bond paths connect H-attractor of acetylene with two BCPs of CC bonds. It is worth mentioning that the latter CC bonds are formally double ones (1.350 Å) since two other are much longer (1.573 Å). Thus there are interactions with bonds where the accumulation of π-electron charge is observed.

It should be noted here that $X-H \cdots \sigma$ interactions with more than two-center acceptors formally should not exist since practically the delocalization of sigma electrons within any ring system is not so important to influence the creation of multicenter Lewis bases.

17.5 Summary

$X-H \cdots \pi$ and $X-H \cdots \sigma$ interactions possess the characteristics of hydrogen bonds. For the latter type of interaction practically only molecular hydrogen may play the role of the proton acceptor. For example, σ-electrons act as the Lewis base for T-shaped $FH \cdots H_2$ complex; it is observed experimentally as well as theoretical calculations on it were performed. Similarly the other T-shaped complexes with hydrogen molecule were analyzed and classified as $X-H \cdots \sigma$ hydrogen bonds.

The situation is more complicated for $X-H \cdots \pi$ hydrogen bonds where two-centered proton acceptors are possible as for example acetylene or ethylene molecules. However also more than two-center Lewis bases play the role of proton acceptors; benzene and the other aromatic, or more general, π-electron systems. The deeper analysis of the latter complexes, like for example, $C_6H_6 \cdots HF$, $C_5H_5^- \cdots HF$ and $C_6H_6 \cdots HCCH$ shows that $F-H \cdots \pi$, $F-H \cdots C$ and $C-H \cdots \pi$ hydrogen bonds are observed for them, respectively. It seems that for $C_6H_6 \cdots HF$ complex there is two-center proton acceptor while for $C_6H_6 \cdots HCCH$ this is six-center acceptor.

Acknowledgments Technical and human support provided by *IZO-SGI SGIker* (*UPV/EHU, MICINN, GV/EJ, ESF*) is gratefully acknowledged.

References

1. Desiraju GR, Steiner T (1999) The weak hydrogen bond in structural chemistry and biology. Oxford University Press Inc., New York
2. Jeffrey GA, Saenger W (1991) Hydrogen bonding in biological structures. Springer, Berlin
3. Grabowski SJ (ed) (2006) Hydrogen bonding – new insights. Springer, Dordrecht. Third part of the series: Leszczynski J (ed) Challenges and advances in computational chemistry and physics
4. Nishio M, Hirota M, Umezawa Y (1998) The CH/π interaction, evidence, nature, and consequences. Wiley-VCH, New York
5. Grabowski SJ (2007) Hydrogen bonds with π and σ electrons as the multicenter proton acceptors: high level ab initio calculations. J Phys Chem A 111:3387–3393
6. Grabowski SJ (2007) $\pi-H \cdots O$ hydrogen bonds: multicenter covalent $\pi-H$ interaction acts as the proton-donating system. J Phys Chem A 111:13537–13543

7. Pauling L (1960) The nature of the chemical bond, 3rd edn. Cornell University Press, Ithaca
8. Stoyanov ES, Hoffmann SP, Kim K-C, Tham FS, Reed CA (2005) The structure of the H_3O^+ hydronium ion in benzene. J Am Chem Soc 127:7664–7665
9. Rodham DA, Suzuki S, Suenram RD, Lovas FL, Dasgupta S, Goddard WA III, Blake GA (1993) Hydrogen bonding in the benzene-ammonia dimer. Nature 362:735–737
10. Suzuki S, Green PG, Bumgarner RE, Dasgupta S, Goddard WA III, Blake GA (1992) Benzene forms hydrogen bonds with water. Science 257:942–945
11. Balocchi FA, Williams JH, Klemperer W (1983) Molecular beam studies of C_6F_6, $C_6F_3H_3$, and C_6H_6 complexes of HF. The rotational spectrum of C_6H_6-HF. J Phys Chem 87:2079–2084
12. Gotch AJ, Zwier TS (1990) The spectroscopy and dynamics of π hydrogen-bonded complexes: benzene-HCl/DCl and toluene-HCl/DCl. J Chem Phys 93:6977–6986
13. Read WG, Flygare WH (1982) The microwave spectrum and molecular structure of the acetylene-HF complex. J Chem Phys 76:2238–2246
14. Li G, Parr J, Fedorov I, Reisler H (2006) Imaging study of vibrational predissociation of the HCl-acetylene dimer: pair-correlated distributions. Phys Chem Chem Phys 8:2915–2924
15. Douberly GE, Merritt JM, Miller RE (2007) Infrared-infrared double resonance spectroscopy of the isomers of acetylene-HCN and cyanoacetylene-HCN in helium nanodroplets. J Phys Chem A 111:7282–7291
16. Klooster WT, Koetzle TF, Siegbahn PEM, Richardson TB, Crabtree RH (1999) Study of the N–H···H–B dihydrogen bond including the crystal structure of BH_3NH_3 by neutron diffraction. J Am Chem Soc 121:6337–6343
17. Crabtree RH, Siegbahn PEM, Eisenstein O, Rheingold AL, Koetzle TF (1996) A new intermolecular interaction: unconventional hydrogen bonds with element-hydride bonds as proton acceptor. Acc Chem Res 29:348–354
18. Alkorta I, Elguero J, Grabowski SJ (2008) How to determine whether intramolecular·H···H interactions can be classified as dihydrogen bonds. J Phys Chem A 112:2721–2727
19. Alkorta I, Elguero J, Solimannejad M, Grabowski SJ (2011) Dihydrogen bonding vs metal-σ interaction in complexes between H_2 and metal hydride. J Phys Chem A 115:201–210
20. Allen FH, Kennard O (1993) 3D search and research using the Cambridge structural database. Chem Des Autom News 8:31–37
21. Matsuo Y, Isobe H, Tanaka T, Murata Y, Murata M, Komatsu K, Nakamura E (2005) Organic and organometallic derivatives of dihydrogen-encapsulated [60] fullerene. J Am Chem Soc 127:17148–17149
22. Moore DT, Miller RE (2004) Rotationally resolved infrared laser spectroscopy of $(H_2)_n$-HF and $(D_2)_n$-HF (n = 2–6) in helium nanodroplets. J Phys Chem A 108:1930–1937
23. Grabowski SJ, Sokalski WA, Leszczynski J (2006) Can H···σ, π···H^+···σ and σ···H^+···σ interactions be classified as H-bonded? Chem Phys Lett 432:33–39
24. Jena P (2011) Materials for hydrogen storage: past, present and future. J Phys Chem Lett 2:206–211
25. Urban J, Roszak S, Leszczynski J (2001) Shellvation of the ammonium cation by molecular hydrogen: a theoretical study. Chem Phys Lett 346:512–518
26. Szymczak JJ, Grabowski SJ, Roszak S, Leszczynski J (2004) H···σ interactions – ab initio and 'atoms in molecules' study. Chem Phys Lett 393:81–86
27. Pimentel GC, McClellan AL (1960) The hydrogen bond. W. H. Freeman and Company, San Francisco/London
28. Leiserowitz L (1976) Molecular packing modes. Carboxylic acids. Acta Cryst B32:775–802
29. Tsuzuki S, Fuji A (2008) Nature and physical origin of CH/π interaction: significant difference from conventional hydrogen bonds. Phys Chem Chem Phys 10:2584–2594
30. Tsuzuki S, Honda K, Uchimaru T, Mikami M, Tanabe K (2000) The magnitude of the CH/π interaction between benzene and some model hydrocarbons. J Am Chem Soc 122:3746–3753
31. Grabowski SJ (2011) What is the covalency of hydrogen bonding? Chem Rev 111:2597–2625
32. Grabowski SJ, Ugalde JM (2010) Bond paths show preferable interactions: ab initio and QTAIM studies on the X–H···π hydrogen bond. J Phys Chem A 114:7223–7229

33. Grabowski SJ, Sokalski WA, Dyguda E, Leszczynski J (2006) Quantitative classification of covalent and noncovalent H-bonds. J Phys Chem B 110:6444–6446

34. Morita S, Fuji A, Mikami N, Tsuzuki S (2006) Origin of the attraction in aliphatic C–H/π interactions: infrared spectroscopic and theoretical characterization of gas-phase clusters of aromatics with methane. J Phys Chem A 110:10583–10590

35. Morokuma K, Kitaura K (1982) Energy decomposition analysis of molecular interactions. In: Politzer P, Truhlar DG (eds) Chemical applications of atomic and molecular electrostatic potentials. Plenum, New York, pp 215–242

36. Domagała M, Grabowski SJ (2009) X–H$\cdots\pi$ and X–H$\cdots$N hydrogen bonds – acetylene and hydrogen cyanide as proton acceptors. Chem Phys 363:42–48

37. Sokalski WA, Roszak S, Pecul K (1988) An efficient procedure for decomposition of the SCF interaction energy into components with reduced basis set dependence. Chem Phys Lett 153:153–159

38. Grabowski SJ, Sokalski WA, Leszczynski J (2004) Is a $\pi\cdots H^{+}\cdots\pi$ complex hydrogen bonded? J Phys Chem A 108:1806–1812

39. Douberly GE, Ricks AM, Ticknor BW, McKee WC, Schleyer PvR, Duncan MA (2008) Infrared photodissociation spectroscopy of protonated acetylene and its clusters. J Phys Chem A 112:1897–1906

40. Grabowski SJ (2007) Covalent character of three-center, two-electron systems – $C_2H_3^{+}$ and $C_2H_5^{+}$. Chem Phys Lett 436:63–67

41. Bader RFW (1990) *Atoms in molecules*. A quantum theory. Oxford University Press, New York

42. Matta C, Boyd RJ (eds) (2007) The quantum theory of atoms in molecules. From solid state to DNA and drug design. Wiley-VCH, Weinheim

43. Coppens P (1997) *X-Ray charge densities and chemical bonding*, IUCr. Oxford University Press, New York

44. Grabowski SJ, Ugalde JM (2010) High-level ab initio calculations on low barrier hydrogen bonds and proton bound homodimers. Chem Phys Lett 493:37–44

45. Bader RFW (1998) A bond path: a universal indicator of bonded interactions. J Phys Chem A 102:7314–7323

46. Bader RFW (2009) Bond paths are not chemical bonds. J Phys Chem A 113:10391–10396

47. Rozas I, Alkorta I, Elguero J (2000) Behavior of ylides containing N, O, and C atoms as hydrogen bond acceptors. J Am Chem Soc 122:11154–11161

48. Weinhold F, Landis C (2005) Valency and bonding. A natural bond orbital donor – acceptor perspective. Cambridge University Press, Cambridge

Chapter 18
Computational Approaches Towards Modeling Finite Molecular Assemblies: Role of Cation-π, π–π and Hydrogen Bonding Interactions

A. Subha Mahadevi and G. Narahari Sastry

Abstract The current review focuses on theoretical approaches for various kinds of noncovalent interactions such as cation-π, π–π stacking, and hydrogen bonding which govern the formation of finite molecular assemblies. Cation-π interactions were shown to be arguably the strongest of noncovalent interactions through a series of systematic computations and their comparison with experiments. The major factors affecting cation-π interaction, including the role of solvation, nature and size of systems and regioselectivity for cation attack have been discussed using theoretical studies. The mutual dependence of cation-π interactions with the neighboring non bonded interactions, such as stacking and hydrogen bonding has been explained. Cooperativity in systems containing cation-π interactions has been quantified. Relevance of cation-π and π–π interactions in function and structure of biological molecules and materials has also been dealt with. The role of quantum chemical calculations and molecular dynamics simulations in understanding the structure and energetics of nonbonded interactions is explored.

Keywords Noncovalent interactions • Cation-π • π–π stacking • Hydrogen bonding • Molecular clusters • Cooperativity • Biological relevance of metal ion interactions

18.1 Introduction

Chemists have made outstanding progress in understanding bonding principles which primarily explore how molecules make and break bonds thereby accounting for stability and reactivity of molecules. Covalent bonds result from the interaction

A.S. Mahadevi • G.N. Sastry (✉)
Molecular Modeling Group, Indian Institute of Chemical Technology, Tarnaka,
Hyderabad, 500607, India
e-mail: gnsastry@gmail.com

J. Leszczynski and M.K. Shukla (eds.), *Practical Aspects of Computational Chemistry I:*   517
An Overview of the Last Two Decades and Current Trends,
DOI 10.1007/978-94-007-0919-5_18, © Springer Science+Business Media B.V. 2012

of atoms leading to the formation of molecules under certain conditions. The molecules formed have properties completely different from those of the original systems [1]. Noncovalent interactions were first recognized by J. D. van der Waals in the later part of nineteenth century helping him to reformulate the equation of state for real gases [2]. These interactions lead to the formation of molecular clusters while covalent interactions lead to the formation of a classical molecule. Interactions observed between any two entities whose valencies are satisfied form noncovalent interactions. In the last couple of decades the study of material formation from molecules has gained considerable attention. Thus the importance and significance of noncovalent interactions has been widely recognized by material scientists and biologists. Noncovalent interactions are known to act at long distances of several angstroms unlike covalent bonds. Although these interactions are traditionally considered to be weak, their strength covers a substantial range from a few kJ/mol to several hundreds of kJ/mol depending on the type of interaction. Cation-π interactions, hydrogen bonding, $\pi-\pi$ interactions, hydrophobic interactions and van der Waals forces are representative of different kinds of noncovalent interactions.

Nature has adopted noncovalent interactions in its fold for several crucial processes specifically in context of biomolecules. The synergistic interplay between different kinds of nonbonded interactions is relevant to maintain the structure of important bio-macromolecules like DNA and protein and in retaining the fidelity of information processing needed for normal life (Fig. 18.1).

The gecko lizard is a classic example of the effect of dispersion based noncovalent interactions creating an incredible ability of the geckos to climb rapidly up smooth vertical surfaces. Full et al. [3] show how a large animal such as a gecko can fully support its substantial body weight by the noncovalent interactions between the few hundred thousands of keratinous hairs, or setae, on their feet and the surface using a two-dimensional micro-electro-mechanical systems force sensor and a wire as a force gauge. Adhesive force values support the hypothesis that individual seta operate by van der Waals forces. The role of noncovalent interactions in the process of molecular recognition process has attracted substantial interest in recent times [4]. Several early studies have demonstrated how cation-π and $\pi-\pi$ interactions in the active sites of numerous enzymes involved in catalysis, ion channels as well as ligand recognition drive the functional aspects in these molecules [5–12].

Finite functional assemblies evolved over millions of years have been a source of inspiration for scientist's world over. Pioneering studies by J. M. Lehn and coworkers lead to the development of numerous chemical self-organizing systems, furthering the generation of supramolecular architectures from their components assembled through important noncovalent interactions like hydrogen-bonding and ligand-metal ion recognition processes. These supramolecules include double and triple helicates, circular inorganic helices, multi compartmental nanocylinders, grid-type entities, ordered polymetallic arrays etc. [4, 13, 14]. Important studies revealing the implication of self assembly in large complex biological molecules involved generation of a closed circular helicate assembly analogous to viral DNA [13]. Proteins form remarkably intricate structures by component self-assembly, such as the icosahedral framework combining 180 subunits in the cowpea chloritic mottle

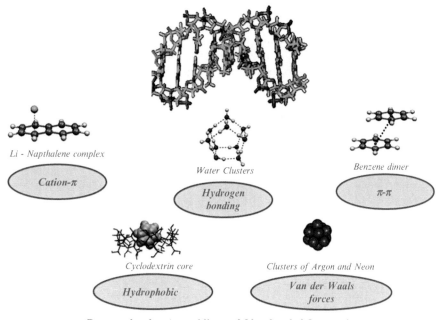

Li - Napthalene complex

Cation-π

Water Clusters

Hydrogen bonding

Benzene dimer

π-π

Cyclodextrin core

Hydrophobic

Clusters of Argon and Neon

Van der Waals forces

Supramolecular Assemblies and Non-bonded Interactions

Fig. 18.1 Representation of important noncovalent interactions observed in small molecules and larger macromolecules

virus [15]. These frameworks create highly specific pockets of chemical space that can induce selective reactivity [16, 17]. The role of weak interactions in controlling the three dimensional structure of macromolecules and supramolecular assemblies has thus been well recognized [18–20].

The experimental determination of noncovalent interactions relies heavily on various spectroscopy methods. Theoretical studies have also played a vital part in understanding different kinds of noncovalent interactions. Stacked nucleic acid base pairs which model DNA have been subjected to ab initio calculations to determine the theoretical methods suitable to generate reliable characteristics [21]. Besides stacking interactions numerous theoretical and experimental studies have been performed on water and other molecular clusters to understand hydrogen bonding and other noncovalent interactions [22, 23]. The interactions of a number of π systems with different ligands including rare gas atoms and complex clusters of water and methanol as model systems was reviewed by Kim et al. These studies reveal how interaction energies of these π-complexes are dependent on both the nature of the ligand and the π-system and establish how in all cases, the repulsive interactions have a vital role in governing the observed geometry [24]. Dougherty's group has established the generality of cation binding to π face of an aromatic structure through gas phase measurements and studies of model receptors in aqueous medium [5–8]. The importance of cation-π interactions in chemistry and biology particularly in proteins have been demonstrated especially with the cationic side chains of basic amino acids and the amino acids phenylalanine, tyrosine and tryptophan [5]. Several prototypical cation-π systems, including structures of relevance to biological receptors and prototypical heterocyclic systems important in medicinal chemistry have also been thoroughly studied and the dominant role played by the electrostatic component in cation-π interaction was clearly delineated with quantum chemical calculations. They demonstrated that the electrostatic potential (ESP) evaluated at a single point above the center of a substituted aryl ring predicts the strength of the cation-π interaction, in particular more negative ESPs indicate stronger interactions [6]. Using the ability to design and evaluate low-molecular weight model systems while surveying several PDB crystal structures alongside Gokel et al. confirm that arenes can serve as π donors for alkali metal cations [25]. Recent studies by Houk et al. reveal π-polarization models of cation-π interactions are flawed and that substituent effects arise primarily from direct through-space interactions with the substituents [26]. Studies modeling not only DNA base pairs but also amino-acid pairs at the CCSD(T)/CBS limit yield accurate interaction energies and demonstrate how wave function theories seem to reliably explain hydrogen bonding and stacking interactions while DFT fails to describe dispersion bound stacking interactions [27]. Sherill et al. performed an assessment of theoretical methods for nonbonded interactions where a comparison of complete basis set limit coupled-cluster potential energy curves for the benzene dimer, methane dimer, benzene-methane, and benzene-H_2S was undertaken with various spin component scaled methods SCS-MP2, SCS-CCSD, DFT methods corrected for dispersion (DFT-D) and meta-generalized-gradient approximation functionals (M05-2X and M06-2X) [28]. They suggest that a combination of general approximations, which significantly reduce computational time and newer approximate electronic structure methods, provide fairly reliable results for the nonbonded interactions making trustworthy computations for much larger chemical systems a possibility. Elaborating on the chemical variety and energy span of numerous kinds of hydrogen bonds the concept of a 'hydrogen bridge' has been suggested recognizing a hydrogen bonding interaction without borders considering variation in its relative covalent, electrostatic, and van der Waals content [29]. Special type of cation-π interactions, metal ligand aromatic cation-π interaction (MLACπ) where ligands are coordinated to a metal interacting with aromatic groups been observed in several metalloproteins, DNA and RNA have been reviewed by Zaric from a computational perspective [30]. The field of noncovalent interactions has thus been a subject of extensive and thorough study motivating analysis from varied perspectives.

An atomic level comprehension of condensed phase structure warrants a clear understanding of the role, strength and relevance of noncovalent interactions and how they mutually influence each other. In the following discussions we present a detailed view on three prominent classes of nonbonded interactions namely cation-π interactions, π–π stacking and hydrogen bonding while emphasizing

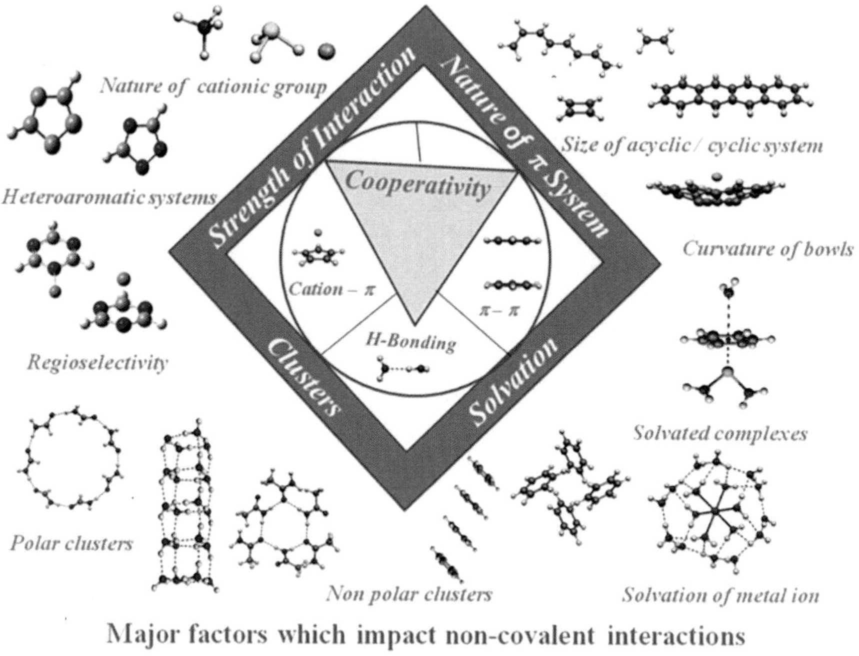

Major factors which impact non-covalent interactions

Fig. 18.2 Major factors which govern the three noncovalent interactions, cation-π, π–π and hydrogen bonding

their contemporary relevance and significance (Fig. 18.2). In our efforts to employ rigorous computations on the medium sized molecules which involve noncovalent interactions several interesting features were unraveled. We have focused on metal ion interactions as well as molecular clusters in chemical and biological systems.

The primary emphasis in this review is on cation-π interactions. The relative preferences of different kinds of alkali and alkaline earth metals for binding to different aromatic systems, the dependence on size of different π systems and solvation have been dealt with in detail. Besides cation-π interactions we present studies on hydrogen bonded clusters and π–π stacked systems of different sizes. Two databases, cation aromatic interaction database and the aromatic-aromatic interaction database have been developed based on our studies. An interesting observation is the mutual enhancement in interaction strength when noncovalent interactions work in concert with each other. The extent of cooperativity in these interactions is analyzed with an emphasis on its effect on controlling structure using different model systems. The role of quantum chemical calculations and molecular dynamics simulations in understanding the structure and energetics of these interactions is studied.

18.2 Noncovalent Interactions

18.2.1 Cation-π Interactions

Cation-π interactions are amongst the strongest noncovalent interactions ranging from 5 to 80 kJ/mol. They include interactions between any aromatic group (may be part of amino acid side chain) and any cation (alkali, alkaline earth metals, transition metals and cations such as NH_4^+, NMe_4^+, SH_3^+, OH_3^+ etc.). We focus primarily on cation-π interactions of alkali and alkaline earth metals in our studies. The bonding observed in cation-π interactions of transition metals are quite different from those seen with alkali and alkaline earth metals due to the presence of d orbitals in these metals. Although they are important in biology they do not fall under the purview of interactions mentioned here and are out of the scope of the current review. Electrostatic interaction and induction are two important components for the metal ion-aromatic interaction.

Benzene, fluorobenzene, anisole, nucleobases and nitrogen heterocycles have been used as model aromatic groups in numerous studies [31–34]. Dougherty et al. have pursued studies on cation-π interactions involving alkali and alkaline earth metal cations that are significant in biological macromolecular structure and function [5–8]. Dunbar et al. calculated the Na^+, K^+ affinities of phenylalanine, tyrosine, and tryptophan using kinetic method [35]. The bond dissociation energies of cation-π complexes of anisole and alkali metal cation has been determined by Rodgers et al. using the collision-induced dissociation technique and also by theoretical studies at MP2(full)/6-311+G** level [36]. They report the energetics of Na^+, K^+ complexes with aromatic amino acids such as phenylalanine, tyrosine, and tryptophan using the guided ion beam tandem mass spectrometry method. Zhu et al. have employed B3LYP/6-311++G** calculations to systematically explore the geometrical multiplicity and binding strength for the alkali and alkaline earth metal complexes with nucleobases (namely adenine, cytosine, guanine, thiamine, and uracil) [37]. HF and MP2 computations on the interaction of mono- and divalent metal ions with nucleobases have been reported by Sponer et al. [38]. Garau et al. [39] have studied the nonbonded interactions of different anions with benzene using a topological analysis of the electron density and molecular interaction potential with polarization (MIP) energy partition scheme calculations. Further quantitative estimation of cation-π and anion-π interactions was carried out, emphasizing the changes in the aromaticity of the ring upon complexation and charge-transfer. Studies on the dependence of basis set quality, electron correlation, and structural variations on the interaction of the alkaline-earth metal divalent cations with benzene were undertaken by Tsuzuki et al. [40]. Ikuta has described the interaction between the monovalent cations (Li^+, Na^+, and K^+) with anthracene and phenanthrene molecules at the hybrid DFT [41]. Thus very subtle factors such as the size of the system and change in the nature of bond formation can substantially modulate the strength of the nonbonded interactions.

In the context of evaluating the major factors which play a role in cation-π interactions we have considered the impact of various kinds of cations such as alkali, alkaline earth metal cations and their site of attack, analyze the size effect of π systems on the nature of cation-π interactions, understand the role of substituents in aromatic rings and importantly gauge the impact of solvent on the behavior of the interaction as well as on individual metal (Fig. 18.2). All the aspects mentioned above have been tackled using different model systems which mentioned below.

1. Two model studies are taken into account while exploring the role of different kinds of cations. First cation interaction with aromatic side chain motifs of four amino acids (viz., phenylalanine, tyrosine, tryptophan and histidine) [42] were investigated followed by a study where the structural and energetic preferences of π, σ and bidentate cation binding to aromatic amines (Ph–$(Ch_2)_n$–NH_2, n = 2–5) is theoretically studied [43].
2. π and σ complexation of various heteroaromatic systems which include mono-, di-, and tri substituted azoles, phospholes, azines and phosphinines with various metal ions, viz. Li^+, Na^+, K^+, Mg^{2+}, and Ca^{2+} [44]
3. Exploring the size dependence of cyclic and acyclic π systems on strength of interaction [45, 46] and
4. Solvation of metal ions as well as cation-π complexes [47–51].

In the following section we provide an overview of the four factors mentioned above, revealing the diverse behavior of cation-π interactions.

18.2.1.1 Computational Details

Stabilization of all the mentioned types of noncovalent complexes is due to favorable energy. It implies that the energy of a complex is lower than the sum of the energies of its separated subsystems if a complex is formed in vacuum. However the situation in presence of a solvent particularly in the aqueous phase is quite different. For our studies on noncovalent interactions, the strength of interaction is measured in terms of interaction energy. For a binary cation-aromatic complexes, the interaction energy (IE) was calculated as the difference of the total energy of the complex and sum of the energies of the aromatic system and the metal taken

$$IE = E_{complex} - (E_{aromatic\ system} + E_{cation}) \qquad (18.1)$$

In case of ternary complexes involving cation-π (IB) and π–π (BB) interactions, the interaction energies in the ternary complex was calculated based on equations mentioned below

$$\Delta E'_{IB} = E_{IBB} - (E_{BB} + E_I) + BSSE \qquad (18.2)$$

$$\Delta E'_{BB} = E_{IBB} - (E_{IB} + E_B) + BSSE \qquad (18.3)$$

Where, E_{IBB}, E_{BB}, E_{IB}, E_B and E_I are the total energies of the ternary, binary and monomeric systems. The interaction energy for molecular clusters was evaluated using the following equation

$$IE = E_{cluster-energy} - n * E_{monomer} \qquad (18.4)$$

Where $E_{cluster-energy}$ – Total energy of water cluster and $E_{monomer}$ – Total energy of a single monomer unit. To evaluate basis set super position error the Boys and Bernadi method has been applied for the purpose of counterpoise correction (CP) in the different systems wherever mentioned [52]. All calculations reported in the model systems have been performed using Gaussian 03 suite of programs [53]. Based on Bader's AIM (Atoms-in Molecule) analysis we have calculated the electron density values at critical points as mentioned in specific cases [54].

18.2.1.2 Impact of Different Cations and Preferential Site of Binding to Aromatic Group

A systematic analysis of cation (M=H^+, Li^+, Na^+, K^+, Mg^{2+}, Ca^{2+}, NH_4^+, NMe_4^+) binding with different aromatic side chains is undertaken [42]. Scheme 18.1 gives a representation of all the model systems considered for study. The regioselectivity aspect of cation binding to aromatic side chain motifs and protons has been investigated. The regioisomers of protonated complexes assess the relative propensity of various sites for proton attachment. The covalent binding of proton to the aromatic ring carbon atoms is contrastingly different to the other metal cations as well as ammonium ions which are found to form cation-π and cation–heteroatom interactions. The NH_4^+ and NMe_4^+ ions have shown N–H$\ldots\pi$ interaction and C–H$\ldots\pi$ interaction with the aromatic motifs. The interaction energies of N–H$\ldots\pi$ and C–H$\ldots\pi$ complexes are higher than hydrogen bonding interactions; thus, the orientation of aromatic side chains in protein is effected in the presence of ammonium ions. However, the regioselectivity of metal ion complexation is controlled by the affinity of the site of attack. In the imidazole unit of histidine the ring nitrogen has much higher metal ion (as well as proton) affinity as compared to the π-face, facilitating the in-plane complexation of the metal ions. The interaction energies increase in the order of benzene-M < toluene-M < para-hydroxy benzene-M < methyl indole-M < methyl imidazole-M for all the metal (M) ions considered. Similarly, the interaction energies with the model systems decrease in the following order: Mg^{2+} > Ca^{2+} > Li^+ > Na^+ > K^+ = NH^{4+} > NMe^{4+}. The bond lengths between ring atoms show a definite increase on formation of cation-π complex while the bonds between ring atom and substituent group are noticeably shortened. An important observation is also that is in presence of an alternative basic group, the covalent interaction appears to overtake the cation-π interaction. Thus, the proton and metal ion complexation in biological systems with aromatic motifs can be substantially different.

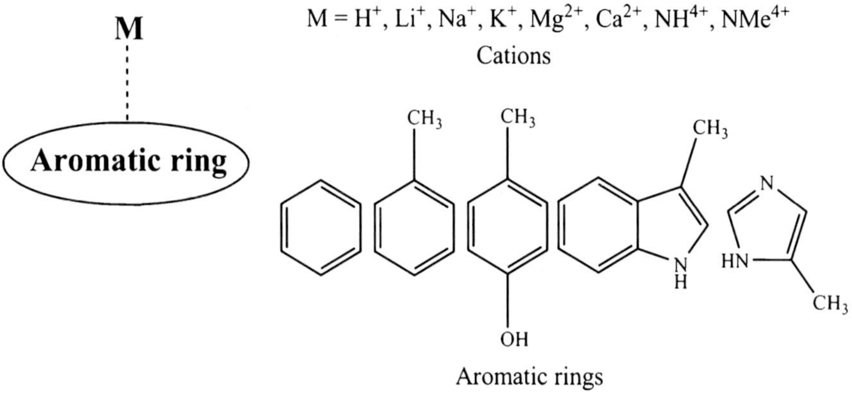

Scheme 18.1

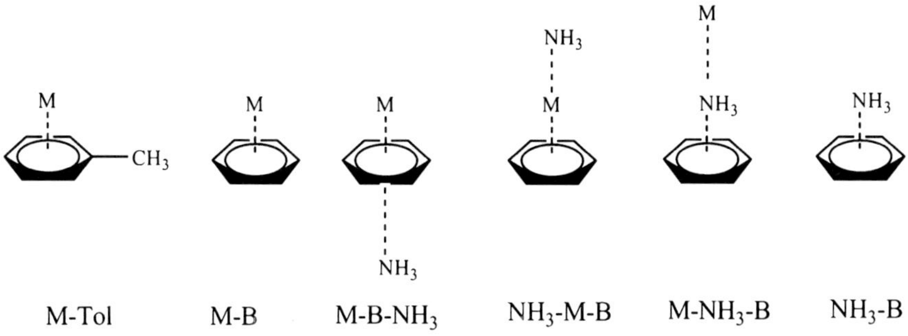

Scheme 18.2

The two principle interactions in proteins with metal ions are cation-π interactions and σ interactions of metal with amines [55–57], besides the coordination of the metal ion with the side chain of acidic residues. Cation-π interactions are competitive with cation-σ interactions within the same molecules and this plays an important role in stabilizing chelating conformations [58]. To probe into this aspect of regioselectivity and preferential binding of the cation in the same molecule MP2/aug-cc-pVTZ calculations were carried out [43]. Three distinct binding preferences, namely, monodentate binding in π and σ fashions to aromatic and amine groups, respectively, and the bidentate mode of binding of Li^+, Na^+ and Mg^{2+} ions with aromatic amines (Ph–$(CH_2)_n$–NH_2, n = 2–5) were considered.

The model systems devised to examine the binding strength of the interactions where the aromatic and amine motifs are not interconnected are shown in Schemes 18.2 and 18.3. The main questions addressed here were regarding the relative preference of Li^+ and Mg^{2+} to bind to an aromatic ring in a π fashion and of amines to bind in a σ fashion, the differences in conformations of complexes involving metal compared with those of protonated ones and the extent of structural reorganization required to achieve bidentate confirmations.

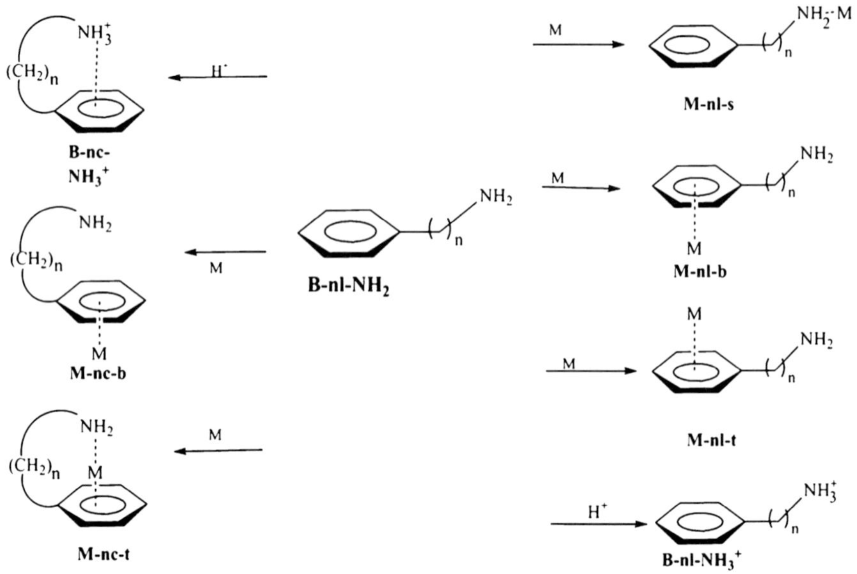

Scheme 18.3

The results obtained in this study reveal how Li^+ and Na^+ have displayed a consistently higher propensity to bind with the amine group compared to the aromatic group. In contrast, Mg^{2+} binds more strongly to the π systems compared to the amine group. From the mono- to bidentate, the chelation gain in the binding energy for Mg^{2+} is about three to four times greater than that of Li^+ and Na^+. Cation-π interactions seem to show a higher dependence on the charge of the metal ion compared to that of the cation interaction with lone-pair-bearing molecules. The monodentate binding of metal ions with $-NH_2$ has a small variation in the interaction energies as the spacer chain length increases. While the binding of Mg^{2+} to an aromatic ring is sensitive to its side-chain orientation and its length, Li^+ and Na^+ complexation is independent of spacer chain length and orientation. Structural reorganization due to Mg^{2+} complexation is slightly higher than that due to Li^+ or Na^+ complexation. Thus, the divalent metal ion complexation leads to a significant variation in the macromolecular structure and the function. The charge on the metal ion depends on the side-chain length and the mode of complexation of metal ions with the aromatic amine (mono or bidentate). Regioselectivity and the nature of cation involved thus play a vital role in determining the effective strength of cation-π interactions.

18.2.1.3 σ vs π **Binding of Cations to Heteroaromatic Systems**

The preferential binding mode of alkali and alkaline earth metal cations to aromatic rings in presence of different substituted aromatic groups has been investigated [44]. While exploring the metal ion binding to the heteroaromatics the following issues were addressed: (a) the relative strength of σ and π binding modes, (b) the regioselectivity of metal ion binding, and (c) all possible minima of metal ion and ring complexes. The presence of several crystal structures which have Li^+ and Na^+ bound to phosphorus systems in CSD (Cambridge Structural Database) and PDB (Protein Data Bank) [59] motivated choosing cation-π complexes of phosphorus containing heteroaromatics, while highlighting the contrasts in the structural and binding affinities between the nitrogen and phosphorus containing heterocyclic systems. The heterocyclic systems studied include mono-, di-, and trisubstituted azoles, phospholes, azines and phosphinines.

Azoles and azines form stronger σ complexes in sharp contrast to the phospholes and phosphinines which form stronger π complexes with the metal ions Li^+, Mg^{2+}, and Ca^{2+}. With Na^+ and K^+ there is little difference between the σ and π complexation energies for phosphorus heteroaromatics. The nitrogen heterocyclic system 4H- [1, 2, 4] triazole and pyridazine formed the strongest σ complex among the five- and six-membered heteroaromatic systems considered. The σ and π complexation energy of azoles and azines is found to decrease as the heteroatom substitution increases in the ring. In contrast, the complexation energies of both phosphole and phosphinines show little dependence on the number of phosphorus atoms in the ring. The complexation energy of a given heteroaromatic system with various metals followed the order $Mg^{2+} > Ca^{2+} > Li^+ > Na^+ > K^+$. Among the chosen metals, Mg^{2+} show a higher propensity to bind to the phosphorus systems while forming a π complex. The phosphinine Mg^{2+} complexes were found to have comparable complexation energy to benzene Mg^{2+} complex. The metal preferred to bind in the bidentate fashion to azoles and azines, whereas for the phospholes and phosphinines, no such binding preference was observed. For both azoles and phospholes, the metal binds away from the electron-deficient heteroatom. Thus, a very important contrast between nitrogen- and phosphorus containing heteroaromatics is revealed through this study.

18.2.1.4 **Cation-Aromatic Database**

Important studies by Gallivan and Dougherty [8] report a quantitative survey of cation-π interactions in high-resolution structures in the Protein Data Bank (PDB). Based on an energy criterion for identifying significant side chain interactions, they conclude how the geometry is biased toward one that would experience a favorable cation-π interaction when a cationic side chain is near an aromatic side chain. Energetically significant cation-π interactions are available from their program CaPTURE [8]. Besides several other studies focus on using different criterion such as angle, distance etc. in order to understand the propensity of different cation-π interactions in the PDB [60]. In most of these studies, the cation is the side chain protonated nitrogen of the basic amino acid residues. An understanding of the metal ion interactions however is rather scarce. Understanding the nature of metal cation-π interactions is key to realize and model various processes in metalloproteins. Towards this end we developed an exhaustive database (Cationic-Aromatic Database) of Metal-aromatic motifs present in the PDB based on relevant geometrical criteria so as to help in screening and developing new methods for identifying and ranking these interactions [61]. Metal-aromatic motifs include both σ as well as π interaction geometries in the database.

A statistical analysis of this database reveals that the aromatic side of the histidine moiety prefers to bind in a σ fashion, while the rest show a propensity to bind in π-fashion. The predominance of σ-type interaction in His moiety may be traced to the presence of electron deficient nitrogen atom. Considering cation–π distance the following trend is observed Trp > Phe > Tyr. Even though most of the cation–aromatic interactions are contributed by basic amino acid residue cations Lys, Arg, and His, metal ions too have significant number of cation–aromatic interactions (cation–σ and cation–π). Coming to cations of basic amino acid residues, Arg cation forms interactions within cation–π interaction distance range. Among all the metal cations studied, Zn followed by Fe show more cation–aromatic interactions. An on-line tool has been incorporated in the database, which furnishes all the cation–π interactions for any new protein. This database is available in the public domain.

18.2.1.5 Size of System

The size of the π-system chosen has important implication on the structural and functional aspects of metal binding. To explore the size effect calculations were performed on the cation–π complexes of Li^+ and Mg^{2+} with the π-face of linear and cyclic unsaturated hydrocarbons [45]. In the case of the acyclic π-systems, we started with the simplest system, e.g.: ethylene followed by buta-1,3-diene, hexa-1,3,5-triene, and octa-1,3,5, 7-tetraene with 2, 3 and 4, conjugated π units, respectively. These linear systems with two and more number of π units can have various conformations wherein the π units can have cis, trans or a combination of both cis and trans orientations. Similarly for cyclic systems cyclobutadiene, benzene, cyclooctateraene, naphthalene, anthracene, phenanthrene and naphthacene have been included. Thus a wide range of sizes for aromatic systems have been covered.

Similar to the earlier case where the role of differing cations was noted, here the impact of the variation in the interaction energy of linear and cyclic conjugated π-systems as a function of the size of the π-system is probed. The interaction energies depend on the size of the π-system, with larger molecules exhibiting higher complexation energy. The increase in the interaction energy can also be correlated to the strain induced in the system upon metal ion complexation and also upon charge transfer. In acyclic systems, which have higher flexibility to reorient the structure upon metal ion complexation, the stabilization energy is higher than in cases where

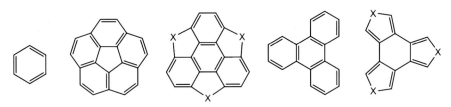

X = O, NH, CH$_2$, BH, S, PH, PH$_3$, Si, SiH$_2$, AlH

Scheme 18.4

the π-system is highly distorted from the idealized planar form. The electrostatic interaction seems to be a major factor and there is some correlation between the interaction energy and the charge transfer. However, this study does not show any quantitative correlation between interaction energy and the cation–π distance. Thus, the size of the π-system which can be estimated as the number of double bonds present (n) shows a dramatic increase when n goes from 1 to 9. This increase is uniform both in cyclic and acyclic systems and thus the number of double bonds in conjugation may be taken as a general signature, to estimate the cation–π binding at least in the gas phase.

In the context of assessing the size effect we also reviewed the impact of curvature of polycyclic systems in terms of their role in cation-π interactions. B3LYP/6-311+G** calculations were performed to assess the effect of curvature and remote electronic perturbations on the cation–π interactions of a large series of aromatic hydrocarbons and their hetero analogs shown in Scheme 18.4 [46].

In all cases, except corannulene, the π-system is structurally and electronically modified aromatic six-membered ring. The metal ions (Li$^+$ and Na$^+$) bind to both the faces of the buckybowls arising to two possibilities for π-complexes; convex face binding is preferred over concave binding in all the cases by about 1–4 kcal/mol. Both the bowl and planar forms yield similar binding energies, indicating that the curvature of the buckybowls has very little effect on the complexation energies. The strength of cation binding to the six-membered ring is mainly controlled by electronic factors, while the curvature plays only a marginal role. Heterosumanene or heterotrindene has a very high complexation energy compared to other compounds when X═NH. The interaction energies observed in this class of compounds exhibit a wide range from 25–59 to 15–43 kcal for Li$^+$ and Na$^+$ ions, respectively. Importantly, the curvature and flexibility of the curved surfaces are virtually undisturbed upon metal ion complexation.

18.2.1.6 Solvation

Biological systems exist in aqueous phase and are influenced by presence of a solvent not only in terms of retaining structure but also from a functional view point. Modeling studies incorporate a solvent to model its impact on strength of different noncovalent interactions. The solvent effect in quantum chemical calculations on model systems is addressed by including an implicit solvent or by explicit addition of water molecules at various positions to gauge the solvent effect on nature of nonbonded interaction. Dougherty et al. have shown that cation-π interactions are frequently found on the surfaces of proteins and exposed to aqueous solvation in a study which projects the importance of solvent on cation-π interactions [62]. An analysis, of 2,878 energetically significant cation-π interactions present in the data set of 593 non homologous proteins shows that 20% of the aromatic amino acid residues have at least 20% of their surface exposed to water, whereas 70% of all cationic amino acids expose more than 20% of their surface to water. Vaden et al. investigated the hydrated Na^+, K^+ ion complexes with benzene and phenol using infrared spectroscopy, wherein hydrated ions interact with the π-system of benzene as well as the phenol oxygen atom [63]. According to their analysis in hydrated environments, cation-π interactions are size-selective toward the hydrated ions. Earlier computational studies on the influence of water molecules on cation-π interactions revealed that the strength of the cation-π interactions gets substantially reduced when solvated with water [64].

An analysis performed on the structures of PDB and CSD clearly demonstrates that a higher number of cation-π interactions exist in the distance range of 3–4 Å from the cation to the centroid of the aromatic system. While structural parameters obtained by X-ray analysis are normally in excellent agreement with computations on single molecules in the gas phase, the role of the environment on the geometric parameters appears to be rather critical for cation-π interactions. A quick look at a cation-aromatic database built reveals that the frequency of cation-π interactions is relatively high in cation-π distances around 3.5–4.5 Å. These bond lengths are well over 1–2 Å longer than the optimized geometries obtained for smaller model systems using reliable quantum chemical methods. To identify possible reasons for this disparity between bond distances from x-ray crystal structures and theoretical studies, the effect of explicit solvation of the cation-π system where the first solvation shell of cations is saturated with water is considered. This provides a realistic description of the first solvation shell, not only in a manner relevant for biomolecules, but also to mimic the saturation of metal ion coordination (Fig. 18.3).

As part of the effort to gauge the relevance of solvation using theoretical methods we initially performed studies on hydrated metal ion (Li^+, Na^+, K^+, Mg^{2+}, Ca^{2+}) complexes with benzene (cation-π) as model systems (Scheme 18.5) [47]. The geometrical parameters and interaction energies of these complexes were evaluated using B3LYP/6-31G (d,p) level of theory.

This study reveals that the strength of interaction of an arene, modeled by benzene, with a fully solvated metal ion is nearly half of the gas-phase complex for K^+, while it reduces to almost one-fifth for the divalent Mg^{2+}. There is a stepwise decrease in the strength of cation-π interactions as the metal ion is solvated and thus the cation-π strength is actually much smaller in the condensed phase. However, all energies are still higher than the interaction energy of the water-benzene complex ($\sim$3 kcal/mol). Further coordination of metal ions with water molecules results in lengthening the cation-π distance. Interestingly, while the potential in the gas phase

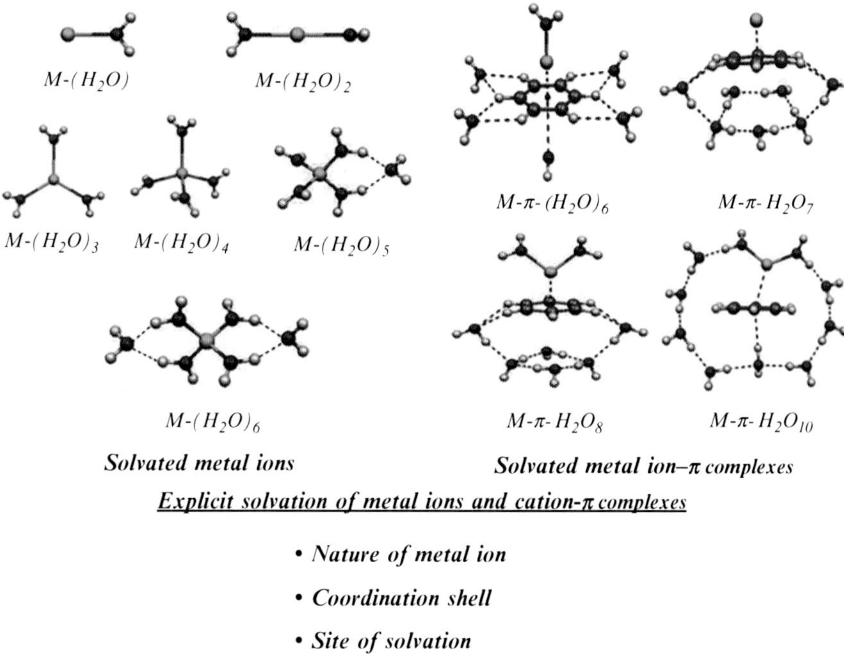

M-(H₂O)

M-(H₂O)₂

M-(H₂O)₃   *M-(H₂O)₄*   *M-(H₂O)₅*

M-(H₂O)₆

Solvated metal ions

M-π-(H₂O)₆   *M-π-H₂O₇*

M-π-H₂O₈   *M-π-H₂O₁₀*

Solvated metal ion–π complexes

Explicit solvation of metal ions and cation-π complexes

- **Nature of metal ion**
- **Coordination shell**
- **Site of solvation**

Fig. 18.3 Impact of solvation on metal, strength of a noncovalent interaction and relevance of site solvated

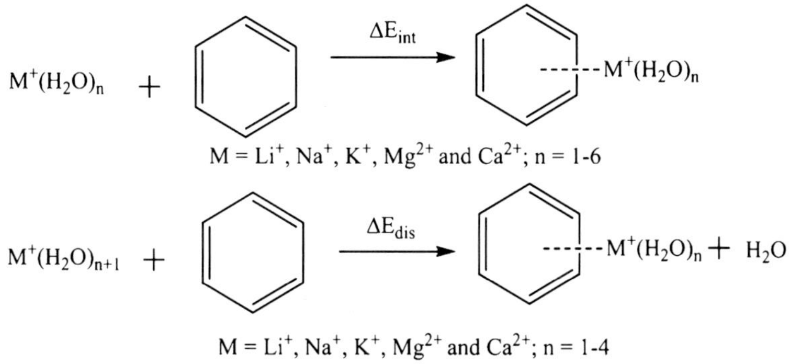

$M^+(H_2O)_n$ $+$ ⬡ $\xrightarrow{\Delta E_{int}}$ ⬡ ----- $M^+(H_2O)_n$

M = Li$^+$, Na$^+$, K$^+$, Mg^{2+} and Ca^{2+}; n = 1-6

$M^+(H_2O)_{n+1}$ $+$ ⬡ $\xrightarrow{\Delta E_{dis}}$ ⬡ ----- $M^+(H_2O)_n$ $+$ H_2O

M = Li$^+$, Na$^+$, K$^+$, Mg^{2+} and Ca^{2+}; n = 1-4

Scheme 18.5

for the cation-π may be very tight, it appears to be loose in the condensed phase. This is precisely the reason for the adaptation of a large span of cation-π distances that are observed in protein databases. Thus there exists a tremendous disparity in the behavior of cation-π interactions in gas and condensed phases.

Sequential attachment of water molecules via the explicit solvation mode to Li$^+$, K$^+$ and Mg^{2+} complexes with benzene was pursued [48]. This study reveals how cation-π interaction energy is sensitive to the site of solvation of cation-π systems, and size and charge of the metal ion. Compared to Mg^{2+} and Li^+, K-π interaction energies are more competitive with metal-water interaction energies. The approach of the water molecules determines how the cation-π strengths are altered upon solvation, the strength is attenuated when water molecules selectively surround the metal ion while it is enhanced upon selectively solvating the π system. Thus, solvation of metal ions lowers the interaction energy and causes lengthening of the cation-π distance, while when water molecules selectively approach the π system a starkly contrasting effect results, i.e., increase of interaction energy and shortening of the cation-π distance. Although the qualitative observation is virtually similar in the three cations (Li^+, K^+, and Mg^{2+}) studied, the solvent-assisted augmentation and attenuation of the cation-π strength depending on the face of metal ion attack is more dramatic in the case of K^+ ion. RVS analysis indicates that the major contributions to the cation-π interaction energy are coming from the POL (polarization) and CT (charge transfer) energy terms of benzene. The topological analysis of electron density distribution within Bader's atoms in molecules theory (AIM) consolidates how depending on the site of solvation, cation-π interaction energy becomes stronger or weaker. The electron density values at BCPs confirm the C–H$\cdots$O bridge type hydrogen bonding between the oxygen atom of water and the C–H group of benzene.

A more recent study done towards understanding impact of solvation on cation-aromatic interaction involved quantum chemical calculations done on the binding of hydrated Li^+, Na^+, K^+, Mg^{2+}, Cu^+, and Zn^{2+} metal ions with biologically relevant heteroaromatics such as imidazole and methylimidazole [49]. This study reveals how alkali, alkaline earth metal, and transition metal ions binding to imidazole and methyl imidazole motifs, which model the histidine side chain, have a strong preference to bind to the lone pair of nitrogen. The water molecules virtually fill the first solvation shell to start the explicit solvation process. The metal ions show higher propensity to bind to the imidazole motif compared to water or benzene. Histidine thus is the most ubiquitous amino acid residue in metal binding sites. The study carefully analyzes hydrogen-bond-donating abilities of the N–H group of histidine upon unsolvated and solvated metal ion binding to its other nitrogen [N(3)] center. The study ensures that the first solvation shell is satisfied for all metal ion complexes. The presence of solvent molecules at the N(1) position of imidazole or methylimidazole enhances the metal ion binding at the N(3) position, and the binding of metal ion at the N(3) position strengthens the hydrogen bonding at the N(1) position. The shift in the vibrational frequencies of N–H indicates that the presence of metal ion at the N(3) position strengthens the hydrogen bond at the N(1) position. The study demonstrates a higher cooperative effect in the hydrogen bonding due to transition metal ions compared to alkali and alkali earth metals.

There is also a high interest in the individual hydration characteristics of metal ions. The study of hydrated metal ions in gas phase provides a connection between the essential chemistry of the isolated ion and that in the solvent. Solvated ions also appear in high concentrations in living organisms, where their presence or absence can fundamentally alter the functions of life. In fact, the structure and dynamics of solvation shells have a large impact on any chemical reaction of metal ions in solution. Solvation of alkali and alkaline earth metal ions in particular has stimulated considerable theoretical interest. Different experimental techniques such as high-pressure mass spectrometry (HPMS), collision induced dissociation (CID) using guided ion beam mass spectrometry, blackbody infrared radiative dissociation (BIRD) kinetics, and electrospray ionization (ESI) with Fourier transform mass spectrometry have been adopted to study water solvation of alkali and alkaline earth metal ions [65–67]. A combined experimental and theoretical investigation on the solvation of Ca^{2+} with water molecules by Armentrout et al. reveals how the sequential binding energies are changed by the addition of each water molecule [68]. While reporting the predominance of electrostatic energies on the binding of alkali metal cations with water molecules Kim et al. high how the sum of induction and dispersion energies are almost canceled out by exchange-repulsion energy [69]. Merrill et al. have evaluated the performance of effective fragment potential method (EFP) with 6-31+G(d) basis set to the description of solvation in simple metal cationic systems [70]. The variation in energetic boundary between the first and the second solvation shell based on size of metal ions was explored using dipole moment and polarizabilities of $M(H_2O)_{1-8}$ (M = Be^{2+}, Mg^{2+}, Ca^{2+}, and Zn^{2+}) clusters by Pavlov et al. [71] using density functional theory. Glendening et al. have reported the binding energies of alkali and alkaline earth metal ions with water molecules using the HF and the MP2 methods [72]. According to their analysis, the HF method provides a reasonable description of cation-water interactions for small (n = 1−3) clusters, whereas it is not adequate for large clusters involving water-water hydrogen bonding. $M(H_2O)_n$ [M = Mg^{2+}, Ca^{2+}, Sr^{2+}, Ba^{2+} and Ra^{2+}] clusters seem to favor structures in which all water molecules directly coordinate to the di-cation in highly symmetric arrangements.

Thus there is a huge amount of current interest in metal ion solvation. We pursued a systematic study wherein water molecules were added to metal ions [M = Li^+, Na^+, K^+, Be^{2+}, Mg^{2+} and Ca^{2+}] and the conformational space of these hydrated metal ion complexes [$M(H_2O)n$; n = 1−6] explored using ab initio and density functional theory methods with a range of basis sets [50]. This benchmark study not only provides insight into the nature of solvation of metal ions but also establishes the method and basis set dependency of this solvation. As hydration of metal ions is a topic of great interest, it is necessary to identify theoretical methods that can satisfactorily reproduce experimental results at the lowest computational cost. Experimental sequential binding energies of hydration of Li^+, Na^+, and K^+ were taken as reference values to evaluate the computational procedures. In those cases where experimental results of sequential binding energies for all of the hydrated complexes involving divalent ions is unavailable high level G3 energies were employed as reference values to assess various levels employed in the study. The following set of results was generated from this benchmarking study.

Triple-ζ basis set with B3LYP or MP2 method seems to serve best for correct identification of the lowest energy conformer of hydrated metal ions involving more than four water molecules. No single level (at B3LYP and MP2) is found to model consistently hydration of all the metal ions chosen for this study. Sequential binding

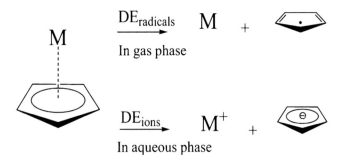

Scheme 18.6

energies at various levels of theory follow a definite trend. For alkali metal cations, MP2 and CCSD(T) perform consistently well with both double- and triple-ζ basis sets ($\sim$6.5 kJ/mol average deviation from experimental results). The performance of MP2 and CCSD(T) seems to be promising with an average deviation of 2–10 kJ/mol for alkaline earth metal ions, and MP2(FULL)/6-31+G(d) is found to be reliable. For both $Be^{2+}(H_2O)_n$ and $Ca^{2+}(H_2O)_n$ complexes, the MP2 method shows much better agreement with G3 compared to B3LYP functional. Considering the fact that experimental values are unavailable, validation of the performance of these routine levels of theory is not straightforward. The variation in the M–O distance is higher for B3LYP than MP2 method upon addition of each water molecule. As the size of metal-water cluster increases, the charge on metal ion decreases monotonically.

Another study on impact of solvation, considers the dissociation preference of metal-cyclopentadienyl (M–Cp) complexes into either radicals or ions in presence and absence of solvent [51]. In this study two plausible dissociation pathways of half sandwich complex of selected main group metallocenes has been considered and nature of these paths as a function of metal ion and solvent has been explored (Scheme 18.6).

The types of systems are considered in this study include a metal (M = Li, Na, K, Mg and Ca) interacts with a single cyclopentadienyl unit to form a half sandwich structure (M–Cp). In order to gauge the impact of presence of solvent on dissociation of M–Cp complex, PCM optimization was performed with water as implicit solvent. Substantially high values of DE (dissociation energy in kcal/mol) in case of gas phase calculation are noted when ions are involved ranging between (171 and 122 kcal/mol for neutral complexes of Li–Cp > Na–Cp > K–Cp; Mg–Cp > Ca–Cp). The corresponding DE in solvent phase calculations for ions shows a total reversal with much lower dissociation energies ranging between (29 and 56 kcal/mol). A drastic decrease in dissociation energy DE_{ions} of the complexes is observed on comparing aqueous and gas phase, with all neutral complexes showing $\sim$90–150 kcal/mol lesser dissociation energy and the +1 complexes of Mg and Ca having $\sim$350–250 kcal/mol lower dissociation energy values in solvent phase. In contrast to this the dissociation energy calculated when radicals are involved DE_{rad}, represented as DE_{rad} (gas) and DE_{rad} (water), do not show such a drastic variation from gas (between ~81 and 35 kcal/mol) to solvent phase (between ~73 and 33 kcal/mol). While all metals show a clear preference to dissociate into ions in solvent phase, the Mg and Ca complexes having +1 charge on the complex show surprisingly identical preference to dissociate into M^{2+} and Cp^- or $M^{+\cdot}$ and $Cp^{\cdot}$ (M = Mg, Ca) Therefore, there appears to be an equal probability in these cases for dissociation by either competing pathways. Thus a marked preference for dissociation of the complex as radicals in gas phase and as ions in solvent phase can be inferred from these PCM studies.

Thus we have explored several aspects of solvation in terms of how they affect cations and cation-π interaction. Although the extent of effect which solvent has on each system is quite case specific, it remains an important factor in reliable quantum chemical calculations of model systems. We demonstrate how even subtle and minor variations in solvation seem to have a significant alteration in the manifestation of the interaction as seen by studies of different model systems.

18.2.2 π–π Interactions

The importance of stacking interactions is made obvious by its ubiquitous presence in DNA, the most important nucleic acid in a majority of life forms. While hydrogen bonding has been considered for a long time as the more predominant interaction while stabilizing nucleic acid structure, the role of π–π stacking is no less important. Jennings et al. reveal how relatively weak intramolecular edge-to-face interactions between aromatic rings can affect or determine the conformation of organic molecules in the solid state and in solution. They show how experimental estimates indicate that these interactions are energetically attractive by 1.5 kcal/mol but disfavored in solution by entropic factors due to the restricted internal mobility. Hence, these interactions are more manifest at low temperature in solution or in crystal structures where conformational entropy effects are negligible [73]. Several experimental and theoretical modeling studies work on stacking interactions primarily in terms of evaluating their role in supramolecular assembly [74–76]. Though the issue of supramolecular stacking leading to stabilization of macromolecules is well grounded in experimentally determined structures, there is not much information available on the energetics of stabilization [12, 77]. The aromatic side chains anchored on the peptide backbone and several CH–π and π–π stacking interactions involving them in the hydrophobic pocket are implied to be functionally important [78, 79]. The benzene dimer has served as a prototype model system to model both CH–π and π–π interactions [80–82]. Several major bottlenecks exist in treating large molecules which are bound through π–π or dispersive interactions. The conventional DFT functionals [83] fail while MP2 calculations are known to overestimate the attraction by dispersion even at the basis set limit [84]. Calculations with CCSD(T) are shown to be more reliable compared to MP2 method particularly for benzene dimer as evidenced from several key studies [85–87]. Although the accuracy of CCSD(T) method is indisputable

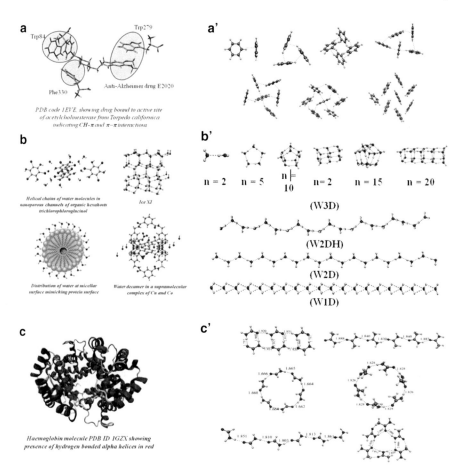

Fig. 18.4 Molecular clusters used as model systems (**a'**, **b'** and **c'**) to study π–π, CH–π and hydrogen bonding interactions observed in several natural systems (**a**, **b**, and **c**)

the prohibitive expense of these calculations on larger systems makes it difficult to perform them. Basis set superimposition error (BSSE) also becomes a stumbling block for obtaining reliable energetics, especially when one includes polarization and diffuse functions in the basis set and at MP2 level. We have however applied MP2 method for larger systems as it is computationally feasible and less expensive when compared to higher levels of other correlated methods, viz., CC, CI, etc. (Fig. 18.4a).

The relative strengths of π–π and CH–π interactions have been modeled using clusters of benzene $(Bz)_n$, n = 2–8, at MP2/6-31++G** level of theory based on a linear scaling method called MTA (Molecular Tailoring Approach) using a divide and conquer approach [88]. Although several lower scaling MP2 methods [89, 90] (which do not involve fragmentation) have been recently applied to large molecular systems we adopt a linear scaling strategy for MP2 through MTA. This work provides a proof of concept for MTA where reasonably accurate energetics for large benzene clusters is evaluated. The final energies and corresponding complexation energies for these clusters indicate that compact structures are more stable than the linear stacked ones. While the existence of more number of interactions in compact structures than in linear ones may seem a pretty obvious conclusion to make we also notice how compact structures reveal a greater predominance of CH–π interactions in preference to π–π interactions. MTA-based calculations also reveal that long-range interactions in benzene rings do not play a significant role in energetics and contributions from important two- and three-body terms (in which the distance between two benzene rings is less than 4 Å) are adequate to produce faithful energy estimates. Although a substantial contribution of BSSE is present even on employing MTA at the MP2/6-31++G** level of theory the results are virtually identical to the standard counterpart in dealing with impact of basis set or BSSE. Also it is noteworthy that the trends of uncorrected and BSSE-corrected energies are similar to each other.

π–π stacking interactions have also been employed as a model system to gauge the impact of basis set superposition error (BSSE) on geometry of the molecules [91]. The fact that BSSE contributes a significant percentage of the interaction energy of the nonbonded interactions has received broad consensus [92–96]. The impact of BSSE on affect the potential energy surface of complexes bound by hydrogen bonding has been elucidated by Dannenburg et al. [97]. Studies on the effect of BSSE on the geometry of the stacked systems however have been relatively scarce, with a few groups having thrown some light on this aspect [98–100]. Hermida-Ramón et al. [100] have carried out a detailed study on the effect of BSSE on the structure, energy and vibrational frequencies of benzene–benzene and benzene–naphthalene dimers. Berski et al. [101] also have examined the effect of BSSE on the stacking interactions of the water dimer in a very systematic manner. Their study reveals that BSSE effect may be substantial on the geometry as well as energy, which appears to depend critically on the orientation and the distance separating the van der Waals complexes. More recently, Hobza et al. present studies on the effect of BSSE on the structure, energetics, and other properties of a large number of noncovalently bound molecules, including those bound exclusively through dispersive interaction [99, 102]. Saeki et al. [103] have found a critical dependence in the structural and energetic aspects, with emphasis on the shifts in frequencies of the spectrum upon using counter poise correction based on their studies on the structures and energies of naphthalene dimers at MP2 level of theory.

With the objective of evaluating how BSSE manifests in the stacking interaction of larger π-systems, we modeled the stacking interactions of bowl shaped molecules, sumanene and corannulene (Fig. 18.5). The contrasts in the crystal packing of corannulene and sumanene, lead us to examine the most approachable way to tackle the problem. While the crystal structure of sumanene is packed in a stacking fashion [104], corannulene structure shows a CH–π type of interaction [105]. A comparison of impact of BSSE correction in smaller stacked systems as against larger buckybowls dimer systems is made to estimate impact of BSSE as a function of size of system. Coming to the benzene dimer the interaction energies

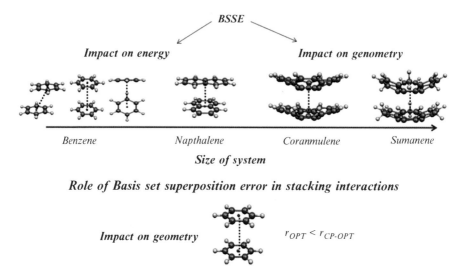

Fig. 18.5 Role of basis set super position error on energy and geometry for π-stacked dimers

obtained are very similar on the geometries obtained with and without employing CP correction, despite the presence of substantial geometric differences.

An analysis of the results in this study crucially reveals how counterpoise optimized geometries at MP2 level with basis sets higher than double ζ quality is in much better agreement with higher level compared with the geometry obtained without employing the counterpoise correction scheme in the optimization. For most of the medium-sized basis set, which require a BSSE correction, need to employ the BSSE correction not only to obtain a reliable energy but also geometry. It is obvious that the best quality basis set needs to be employed for getting reliable structural, energetic, and other properties.

The extent of cation-π interaction seems to dramatically increase as the size of the π-system increases. Subsequently we ventured on exploring BSSE's effect on larger planar aromatic analog like naphthalene and two of the prototypical buckybowls corannulene and sumanene. Consistent with the results of the benzene dimer, we see that for the larger systems also, the π–π distance computed for the CP geometry is longer than that for the uncorrected geometry. Incorporation of the CP correction during the optimization may be crucial to predict more reliable geometrical parameters for the larger aromatic systems, especially as the most practical basis sets on these systems are not sufficiently large. The interaction energies predicted on the CP corrected PES is slightly larger than that predicted on the uncorrected surface. This difference indicates magnitude of error caused due to basis set superposition. A key observation from this study is essentiality of modeling of the noncovalent interactions in the counterpoise corrected PES and that the counter poise correction has significant impact not only on the energetics of the system, but also to obtain reliable intermolecular geometrical parameters. π–π interactions thus have substantial impact in biological molecules especially proteins.

The type of interactions in clusters containing aromatic moieties includes both CH . . . π and π–π stacking. To capture these important noncovalent interactions in biomolecules we have generated a database of all aromatic-aromatic interactions from PDB "Aromatic–Aromatic Interactions Database" (A2ID). This is available in the public domain and it comprises of all types of π–π networks and their connectivity pattern present in proteins [106].

18.2.3   Hydrogen Bonding

One of the most important noncovalent interactions is the hydrogen bond. It plays an essential role in the properties of various materials such as synthetic polymers, biomolecules, and molecular solids and fluids [107, 108]. Hydrogen bonding is also responsible for the conformational preferences of a large number of molecules and produces significant modifications in the kinetics and mechanism of enzymatic reactions. The formation of matter from molecules is governed by the way in which non–covalent interactions operate inter alia. Hydrogen bonding is arguably the most extensively studied among all the non–covalent interactions. The stability of water clusters, based on the arrangement of individual molecules in different phases has been widely explored [109].

Formamide and water are the simplest molecules usually chosen as models for studying biological systems exhibiting the peptide type of bonding as in proteins. Hydrogen bonding complexes of formamide such as formamide-water, formamide-methanol can thus serve as model systems for protein-water and protein-solvent interactions. On account of the simplicity of this model, the characterization of the hydrogen bond interactions between water and formamide has been of considerable interest to experimentalists and theoreticians.

In the current review we focus on two kinds of model systems, first hydrogen bonded molecular clusters of water and the second is hydrogen bonded 1:1 complexes formed between formamide and water molecules. Coming to the case of water clusters we have undertaken a computational analysis of three linear forms of water cluster arrangements to gauge their relative stability and reactivity [110]. Constrained microenvironment of organic and metal organic host lattices give way to multitudes of hydrogen bonded water cluster arrangements [111]. Thus quite exotic structures which are not minima in the gas phase have a high probability to occur in supramolecular architecture and there is a need to look at such structures Theoretical calculations have been performed from dimer to eicosamer of water clusters (Fig. 18.4b) using HF, DFT and MP2 methods. The caged clusters of $(H_2O)_n$; $n = 2-20$ obtained from the Cambridge Structural Database (CSD) [112] have been widely employed in earlier ab initio investigations and we employed them as a standard reference. The impact of various levels of theory and basis sets employed in the calculations along with the effect of basis set superposition error on complexation energy of the water clusters is explored. The cationic $(H_2O)_n^+$ and anionic $(H_2O)_n-$ radical counterparts of the four water cluster arrangements were generated. A comparison of calculated ionization potential (IP) and electron affinity (EA) of the corresponding optimized ionic radicals was used to get an indication of the relative propensity of various water cluster arrangements towards ionization.

The primary aim of this study was to evaluate cooperativity of hydrogen bonding in water clusters, an elaborate discussion of which is made in the subsequent section. Other significant results we obtained from this exhaustive study are summarized in this section. The caged water clusters seen in W3D arrangement are more stable than the other three linear arrangements at HF, B3LYP and MP2 levels of theory. The average electron density values (ρ) obtained at the bond critical points using AIM analysis follow the trend W3D > W2DH > W2D > W1D and exhibit a high correlation with complexation energy of all the arrangements. It is observed that the electron density values at the bond critical point of the hydrogen bonds present towards the centre of the linear chain of W1D, W2D and W2DH arrangements are marginally higher in magnitude when compared to terminal H–bonds. PCM based implicit solvation yields lower complexation energy in the solvated clusters than corresponding gas phase geometries. In all the four arrangements W2DH clusters show highest propensity to form cation and anion radicals. Thus, the neutral and ionized water clusters adopt preferentially caged and helical arrangements respectively.

For the study on 1:1 complexes of formamide and water, systematic calculations using HF, B3LYP, MP2 and CCSD(T) methods have been performed [113]. Three stable structures are considered on the potential energy surface of formamide and water system. The optimized geometric parameters and interaction energies for various isomers at different levels are estimated. The IR frequencies, intensities, and frequency shifts have also been reported. The geometrical parameters of formamide at B3LYP/aug-cc-pVDZ method are in good agreement with experimental studies and other reported values from calculations at higher level of theory. Out of the three stable structures which are considered study we find the cyclic double bonded structure to be most stable while hydrogen bonding to the carbonyl group seems to be slightly more energetically favorable compared to the complex which is hydrogen bonded to the amide group. The geometrical parameters (non-counterpoise corrected) of formamide-water complexes obtained from B3LYP method are close to higher level (MP2) method and experimental values. Similar trend was observed with geometries obtained through CP optimization. Scaled and non scaled harmonic frequencies (non-CP) of formamide-water complexes demonstrate that B3LYP method frequencies are closer to experimental values when compared with HF and MP2 results. Thus the two studies mentioned above give a computational exploration of relative preferences of arrangement of hydrogen bonded molecules.

18.3   Cooperativity

Considering the strong interest we have in cation-π interactions one of our primary objectives is to see how much is the impact of cation-π on neighboring nonbonded interactions. The discussions in the foregoing sections clearly demonstrate that the structure, energy and site and facial preference of cation- π interactions could be dramatically altered by a range of factors. Considering such dramatic changes in the structure and strength of cation-π interactions are due to the neighboring nonbonded interactions we felt that it is appropriate to ask how nonbonded interactions mutually influence each other. A cursory look at the current literature reveals that cooperativity is a widely used term, especially in biology. This cooperative behavior has served as the first clue towards discovery of conformational transition and allosteric interactions in proteins. The best known example in biology for cooperativity is offered by allosteric oxygenation of the haemoglobin molecule [114–116].

Cooperativity however is not limited to ligand binding processes and allosteric proteins. In 1957, Wen et al. were the first to discuss the importance of many-body effects in water in their description of the cooperativity of hydrogen bonds [117]. They postulate that the formation of hydrogen bonds in water is predominantly a cooperative phenomenon, so that, in most cases, when one bond forms several will form, and when one bond breaks then, typically a whole cluster will dissolve. Other early studies consider factors governing the influence of a first hydrogen bond on the formation of a second one by the same molecule or ion [118]. The contribution of many body effects in both aqueous and non aqueous systems have been extensively reviewed [119]. Suhai explored cooperative effects in hydrogen bonding based on the structural and electronic properties of ice and hydrogen-bonded periodic infinite chains of water molecules using first principles quantum mechanical modeling [120]. An enhancement of the binding energy per hydrogen bond by 47% was reported for an infinite polymer compared to water dimer. Dannenberg et al. explored the geometric and vibrational properties of numerous hydrogen bonded model systems including chains of formamide, acetic acid, nitroanilines, urea etc. [121]. The studies performed on these model systems indicate that the electrostatic model for pair wise interactions to be inadequate to describe the major component of cooperativity in case of long chains. Thus quite a few studies have focused on cooperativity observed in hydrogen bonded systems. However studies considering rigorous quantitative estimation of cooperativity or anti cooperativity in systems with cation-π interactions are rather limited.

To explore this vital issue of cooperativity which reflects the nonadditive behavior of noncovalent interactions we have considered different model systems. In the first case we seek to check for presence of cooperativity among cation-π and hydrogen bonding [122] and cation-π and π–π interactions [123]. In the second case we quantify cooperativity in systems with noncovalent interactions other than cation-π interaction, for example in hydrogen bonded clusters of water, formamide and acetamide [110], clusters of π–π stacked benzene [124] etc. (Fig. 18.6).

The common coexistence of M–π and π–π interactions in biology and chemistry has been explored in order to garner a better understanding of how one kind of noncovalent interaction affects the strength of another [82]. This influence is typically described in terms of cooperativity and anti cooperativity in bonding. The occurrence of M–π–π (M = Li$^+$, K$^+$, Na$^+$, Mg^{2+}, and Ca^{2+}) interactions in

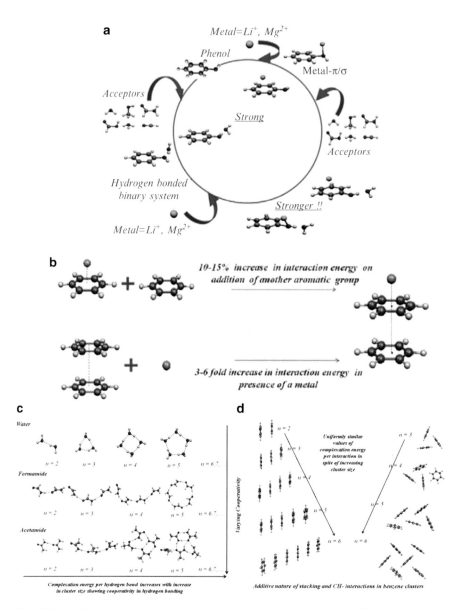

Fig. 18.6 (**a**) Demonstration of cooperativity between cation-π interaction and hydrogen bonding (**b**) Cooperativity between cation-π and π–π interactions working in concert with each other (**c**) Cooperativity in polar clusters of water and formamide (**d**) Additive nature of cooperativity in benzene clusters

the Cambridge Structural Database (CSD, CSD V5.26) and Brookhaven Protein Data Bank (PDB) databases was chosen as subject of study to check for veracity of mutual influence of noncovalent interactions over one other [59]. Different forms of benzene dimers (PD-parallel displaced, S-stacked and T shaped) were also subjected to ab initio calculations as part of this study, in order to establish the relative preference of differently oriented aromatic moieties (represented by benzene) to bind to each other. The PDB was searched for metal ion containing structures with less than 3 Å resolution and R-value <0.3. To remove redundancy, the structures with greater than 90% sequence identity were eliminated from the above dataset. This resulted in a total of 1,941 protein structures containing the 5 metal ions, which were then subjected to analysis. We choose to present the results for 7 Å cutoff, based on results of earlier reported studies. The database analysis reveals the following points: (a) There exists a high occurrence of M–π and M–π–π interactions in chemistry and biology. (b) The prevalence of M–π–π is seen to be higher in most cases or comparable to that of exclusively M–π configurations. (c) The number of M–π–π motifs present is only marginally lower than the number of metal ions available. The non availability of enough metal ions appears to be a reason that a small percentage of π–π interactions are metal ion assisted. (d) In proteins, the aromatic amino acid side chains seem to prefer the PD–M and TB–M orientations

A combined approach of database analysis and computational study was then undertaken. All the initial geometry optimizations for benzene dimer were done at the MP2/6-31G* level. When M = Li$^+$, there is a substantial increase in the π–π interactions for the S and PD type, while the T-shaped stacking has a higher interaction energy only in the TB–M orientation. Essentially the same trends have been noticed for M = Na$^+$ and K$^+$, albeit to a minor extent. Surprisingly, the enhancement of π–π interaction in the presence of metal ion is quite substantial, especially when the metal has a higher charge. This study also reveals that Mg^{2+} ion enhances the strength of the π–π interactions by almost more than fivefold in three configurations of benzene, S–M, PD–M, and TB–M. Thus, with the π–π interaction energy value around 17 kcal/mol, their strength of stabilization is quite substantial. Hence, the strengths of subtle π–π interactions transform to substantial, under the influence of a di-cationic metal. Importantly, the interaction between the two benzene rings in most orientations is enhanced to almost a comparable extent. It indicates that the metal ions stabilize the π–π interactions in most orientations. The influence that the M ion has on the first ring could be the major source of the enhanced interaction. Interestingly, interaction of the metal ion with a single benzene molecule is much lower compared to the interaction of the metal.

The study thus reveals that M–π and π–π interactions work in concert, and the subtle π–π interaction become substantial in the presence of a metal ion. We find the metal ion assisted π–π interaction strengths may become comparable in magnitude to that of the hydrogen bonding interaction. This study triggered our interest to explore further the role of cooperative effect of noncovalent interaction in the 3D aggregation of supramolecular entities and biomolecules.

Quantum chemical calculations were performed to gauge the effect of cation-π and hydrogen bonding interactions on each other, with M-phenol-acceptor (M = Li$^+$ and Mg^{2+}; acceptor (A) = H$_2$O, HCOOH, HCN, CH$_3$OH, HCONH$_2$ and NH$_3$) taken as model ternary systems. (Fig. 18.6a) Direct assembly of the ternary complex from its constituents phenol (P), metal cation (M), and hydrogen bond acceptor (A) proceeds with reaction energy ΔE_{APM}, which can be calculated as the energy difference between the ternary complex and the energies of components P, M, and A. The extent to which the hydrogen bonding and cation-π interaction act in concert in these ternary systems can be deduced quantitatively by comparing the overall reaction energy ΔE_{APM} with the three individual interaction energy terms as in (18.5):

$$E_{coop} = \Delta E_{APM} - \Delta E_{AP} - \Delta E_{MP} - \Delta E_{M-A} \qquad (18.5)$$

The "cooperativity energy" E_{coop} amounts to -2.77 kcal/mol for the example of $M = Li^+$ and $A = H_2O$. The degree of cooperativity may alternatively be quantified by comparing hydrogen bonding energies in the absence of Li^+ ($\Delta E_{AP} = -6.18$ kcal/mol) and in the presence of this cation ($\Delta E_{c,A-PM} = -8.95$ kcal/mol). The complexation of phenol with Li^+ or Mg^{2+} in either π- or σ-fashion strengthens the phenol-H_2O hydrogen bonding interaction energy by about 2 and 7 kcal/mol in Li^+ and Mg^{2+} complexes respectively. Essentially the same trends are found when a range of additional acceptors (HCO_2H, HCN, CH_3OH, $HCONH_2$ and NH_3) are also considered. In all these cases an increase in the hydrogen bonding interactions with phenol can be observed upon complexation with a metal ion. Although the hydrogen bond interaction energy increase is almost 3 kcal/mol with Li^+, it is about 4–9 kcal/mol for Mg^{2+}.

The cooperativity of cation–π and π–π interactions has also been demonstrated taking a series of mono-valent cationic species such as, Li^+, Na^+, K^+, NH_4^+, PH_4^+, OH_3^+ and SH_3^+ where systematic quantum chemical studies were performed to estimate the effects of cation–π and π–π interaction on each other in cation–π–π systems (Fig. 18.6b) [123]. In this study all possible orientations of onium ions to form cation–π complexes have been explored and the most stable conformation has been taken further to evaluate its effect on π–π interaction. The results from this study indicate a notable increase of 2–5 kcal/mol in the π–π interaction energy in the presence of the cations. The cation–π interaction energy is also enhanced in the presence of π–π interaction albeit to a smaller extent. The wide varieties of cations (which include both metal and inorganic ions) employed here underline the generality of the results.

Cooperativity in case of molecular clusters may be defined in terms of energetics as a nonadditive property of interacting monomers, where the interaction energy per monomer depends on the degree of aggregation (Fig. 18.6c). A case in point being the universal example of cooperativity observed in water. Hydrogen bonding here works in a highly cooperative fashion; i.e., the cumulative strength of networks of hydrogen bonds is larger than the sum of the individual bond strengths when they work together. Water has been subject to wide range of studies, where the mode of aggregation of the individual molecules in different cluster arrangements and their relative stabilities have been explored using not only high level *ab* initio calculations but also molecular dynamics approaches [109]. We mentioned in the prior section on hydrogen bonding, the impact of different arrangements of water molecules in different kinds of clusters as a function both of size of cluster (n = 2−20) and method of computation used. The same model systems have also been used to explain cooperativity in hydrogen bonding. The ratio of complexation energy per hydrogen bond from decamer to dimer [$H_2O_{10:2}$] and eicosamer to decamer [$H_2O_{20:10}$] is calculated to quantify the strength of hydrogen bond with increase in cluster size and is used as an indicator of cooperativity that is seen in the water clusters.

Out of a wide variety of basis sets which have been employed for single point calculations on the B3LYP/6-311+G* optimized geometries, aug–cc–pVTZ shows maximum ratio of complexation energy as the cluster size increases from dimer to decamer in case of W1D, W2D and W2DH clusters. In case of W3D, this ratio is maximum with 6-31G** followed by 6-31G* basis set. The complexation energy ratios ([$H_2O_{10:2}$], [$H_2O_{20:10}$]) of BSSE corrected single point calculations with 6-311+G* basis set for W1D (1.33, 1.040), W2D (1.380, 1.050), W2DH (1.352, 1.040) and W3D (1.913, 1.030) shows that the strength of complexation energy increases by 33%, 38%, 35% and 91% respectively as cluster size increases from 2 to 10. However, as the cluster size increases from 10 to 20, the increase in strength of hydrogen bonding is in the range of 4–6% only. This is valid in all the calculations where split valence basis sets have been used. A similar trend is observed when correlation consistent basis sets are used. Thus the non additivity of hydrogen bond strength (cooperativity) is much more evident as cluster size increases from 2 to 10 rather than from 10 to 20, where the augmentation of hydrogen bond strength is only marginal for all the four arrangements. The clear inference from the above mentioned results is the presence of definite amount of cooperativity in hydrogen bonded water clusters where the addition of a subsequent monomer to a hydrogen bonded cluster augments the strength of existing interactions. The mode of arrangement, in particular, how these interactions are arranged also plays a crucial role on the strength of interaction.

Besides water clusters the concept of cooperativity in hydrogen bonding is explored using theoretical calculations in several other molecular clusters including formamide [121, 125], acetamide, N methyl acetamide [126, 127] etc. as test cases mimicking the protein secondary structure. In order to detect its presence and quantify the extent of cooperativity in a set of benzene clusters (Fig. 18.6d), the complexation energy per interaction between a pair of benzene molecules (CE/(n−1), n = number of benzene monomers in the cluster) is compared from dimer to hexamer [111]. The results obtained in this study indicate a near uniform CE/(n−1) value of ∼−3.3 kcal/mol from dimer to hexamer in case of parallel displaced stacked clusters at MP2/6-31G* level of theory. The corresponding value of CE/(n−1) is ∼−6.5 kcal/mol at MP2/6-31++G** level of theory for the same structures. The addition of another benzene molecule does not seem to enhance the strength of each individual interaction. This is quite in contrast to earlier reported studies dealing with cation-π, π–π, and hydrogen bonding interactions. In those clusters where CH–π interactions are more prominent, clusters considered in this instance being tri-05, tet-10, pen-09 and hex-09 a similar absence of cooperativity is observed. The CE per interaction values here are in the range of -2.7 to -4.6 kcal/mol at MP2/6-31G* and consistently about a couple of kcal/mol higher at MP2/6-31++G** level. An inspection of all the clusters and their CEs indicate that the interaction energies are additive irrespective of the number or type (C–H ... π or π–π) of interactions. Hence we may infer that stacking interactions in benzene clusters are essentially additive in nature. The cooperativity observed in clusters of polar molecules is substantially due to induction interactions which play an important role in determining their structure. As electrostatic interactions contribute little in case of benzene clusters, the resultant small induction interactions may be the cause for negligible cooperativity that is observed here.

Thus based on the above mentioned studies we infer a wide variety in the manifestation and extent of cooperativity in the systems analyzed. While hydrogen bonded systems demonstrate significant cooperativity, it a does not appear to contribute much to clusters dominated by CH–π and π–π interactions. However there seems to be quite a substantial enhancement in strength of other noncovalent interactions for instance in the mutual presence of cation-π and hydrogen bonding and even π–π and hydrogen bonding interactions together.

18.4   Correlation and Dispersion

For modeling the nonbonded interactions it is important to employ a method that judiciously accounts for the electrostatic and London dispersion forces. Traditionally the HF method is unsuitable to model noncovalent interactions and methods employing dynamic electron correlation become indispensible. Earlier studies reveal that it is important to employ highly electron correlated methods such as couple-cluster to accurately model nonbonded interactions [128, 129]. However considering the prohibitive cost involved in such calculations in most cases MP2 appears to be the only viable alternative. There are even more economical variants of the MP2 method such as RIMP2 which have proved to be fair in modeling nonbonded interactions. However one of the major bottlenecks in modeling nonbonded interactions is the failure of popular Becke's functional such as B3LYP. Recently the dispersion corrected DFT-functionals have become a good alternative and another possibility is the new Minnesota functional such as M05, M05-X, M05-2X which appears to fair well for modeling nonbonded interactions [130, 131]. The dispersion corrected DFT functionals include those corrected by empirical or exchange-hole derived dispersion coefficients, DFTs parameterized to reproduce dispersion behavior, or DFTs used with dispersion-correcting potentials [83]. The spin-component-scaled MP2 (SCS-MP2) approach developed by Grimme is also a widely used improvement over MP2 for the π-stacked dimers, reducing the binding energy by $\sim$10 kJ/mol [89, 90, 132].

18.5   Materials

Besides biomolecules, the field of material science is an area where noncovalent interactions play a pivotal role, which is actively pursued currently using theoretical and high level computational approaches. Nanomaterials form a class of materials, involving research at the interfaces of chemistry, biology, materials science, physics, and engineering [133]. Computation is extensively employed in nanostructures with implication in nano biotechnology, nanofabrication and self directed-assembly. The potential value of several materials hitherto unrecognized for example zeolites, clatharates etc. for adsorption and storage of CO_2, H_2, and other gases impacting environment based on their ability to form noncovalent interactions have been explored using computational tools. Materials with possible utility in drug delivery including dendrimers [134], funtionalized single walled nanotubes [135] and cyclodextrins [136] etc. are currently being modeled extensively using quantum chemical approaches. The importance of noncovalent interactions in controlling the catalytic activity of the hydrogen oxidation, oxygen reduction and methanol oxidation reactions on platinum surfaces in alkaline electrolytes serve as an example of the current interest in this field [137]. Studies on interactions of small molecules and metal atoms on surface of branched and linear polycyclic aromatic hydrocarbons mimicking grapheme reveal the growing importance of noncovalent interactions in nanomaterials [138]. The feasibility for application of high level ab initio computational methods on large molecules have long been a major bottleneck owing to their large size, conformational flexibility and limitation to model real time conditions Several low scaling methods as well as combined approaches like QM/MM techniques and MD simulations have now evolved as efficient tools to counter this problem.

18.6   Molecular Dynamics

Several early reviews on MD simulations of biomolecules and in chemistry such as those by Karplus et al. [139] and van Gunsteren et al. [140] give an exhaustive coverage of methodology and salient features involved in dynamics calculations while illustrating their utility in understanding functionality of proteins. MD simulations seek to understand the properties of assemblies of molecules in terms of their structure and the microscopic interactions between them. Molecular dynamics simulation consists of the numerical, step-by-step, solution of the classical equations of motion. For this purpose the forces acting on the atoms are calculated, and these are usually derived from a potential energy based on complete set of 3N atomic coordinates. The size of the configurational space that is accessible to the molecular system, and the accuracy of the molecular model or atomic interaction function or force field that is used to model the molecular system are two basic problems one encounters while performing MD calculations. The broadness and level of sophistication of this technique can hardly be understated (Fig. 18.7).

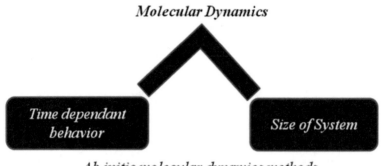

Fig. 18.7 The need for molecular dynamics (MD) calculations

Ab initio molecular dynamics methods which form part of an extension of traditional MD and electronic structure calculations compute the forces acting on the nuclei from electronic structure calculations that are performed on the fly as the molecular dynamics trajectory is generated. The approximation involved in the MD process is shifted from the level of selecting the model potential to the level of selecting a particular approximation for solving the Schrodinger equation. One such approach which includes Car-Parrinello [141] and Born-Oppenheimer molecular dynamics schemes is the CPMD package [142] where the electronic structure approach is dealt by Kohn Sham density functional theory within a plane wave pseudo potential implementation and the Generalized Gradient Approximation [143].

Besides CPMD several other powerful packages such as CASTEP [144], CP-PAW [145], VASP [146] etc. form part of ab initio molecular dynamics programs now available based on a similar idea. The breadth of this technique is demonstrated by the range of fields where it finds applications such as solids, polymers and materials like silicon, surfaces, interfaces, adsorbates, amorphous substances, clusters, fullerenes, nanotubes, zeolites and several biomolecules etc.

18.7 Biological Relevance

The discussions in the preceding sections emphasize the relevance of noncovalent interactions including hydrogen bonding and stacking in stabilizing and retaining supramolecular structure [110, 124]. The role of cation-π interactions in the function and structure of enzymatic proteins also can hardly be understated. In particular metal ions play a pivotal role in the function of metalloproteins. Protein crystal structures from PDB are a great source of the information to understand the structure-activity relationship. Metal histidine complexes are highly abundant in nature, and protein crystal structure data indicates that such complexes are available with essentially most metals. Metal ions not only determine the catalytic activity of proteins but also determine the stability of DNA duplex, depending on the modes of binding and the nature of the metal ions. An extremely important functional and regulatory role is played by metal ions in biomacromolecules [147]. Many histidine-rich sites in proteins bind to transition metal ions thereby controlling protein folding and are important for the enzymatic capabilities, such as hydrolysis, dioxygen transport, and electron transfer. A few examples of metalloproteins, where cation-aromatic interactions are crucial and the metal ions are partially solvated and bound to histidine are given below. These examples show the myriad ways in which metal ions play an important role in biological systems. Metal ions Cu^{2+}, Zn^{2+}, and Fe^{3+} are known to be involved in the oxidation process of methionine of amyloid β peptide by complexing to the imidazole of the same peptide [148]. Carbamoyl phosphate synthetase (CPS) from *Escherichia coli* involved in the catalytic formation of carbamoyl phosphate has two polypeptide chains where the larger chain provides the active site and the smaller chain catalyzes the hydrolysis of glutamine to glutamate and ammonia. In the larger chain, carbamoyl phosphate is synthesized from two molecules of Mg^{2+}ATP, one molecule of bicarbonate, and one molecule of glutamine [149]. Carbonic anhydrases catalyze the hydration and dehydration of carbon dioxide and bicarbonate, respectively. The α-class of anhydrase present mostly in vertebrates, has a Zn ion which is tetravalently coordinated to three imidazole rings and one water molecule in the active site Carbonic anhydrase isozyme IX belongs to the α-class and it has been found that in some cancers over expression of carbonic anhydrase isozyme IX promotes growth and metastatis of the tumor, making it a good target for cancer treatment. In this protein, water acts as the fourth ligand and as a critical component. It has been suggested that the mechanistic pathway for activating the water depends on the identity of the other three ligands coordinated to zinc ion and their spacing [150]. Superoxide dismutases (SODs) are a class of enzymes that catalyze the dismutation of superoxide into oxygen and hydrogen peroxide. They are an important antioxidant defense in nearly all cells exposed to oxygen. Among three major families of SODs copper-zinc superoxide dismutases (CuZnSOD) are most common in eukaryotes. In CuZnSOD the active metal is Cu; i.e., Cu plays an essential role in enzymatic behavior, whereas Zn atom acts as a structural stabilization factor [151]. Thus it should be amply clear that they metal ions play an extremely diversified and important role in determining biological structure and function. Clearly, the effect and role of solvated water in metal ion complexes is profound and is a topic of outstanding importance and relevance in understanding biological processes.

Site directed mutagenesis studies on the structure of a chitinase complex (PDB:1H0G) show that mutation of aminoacids Y240 and W252 to alanine which are greater than 12 Å away from the ligand show a great loss in activity. W470 and F190 too which are more than 5 Å away from the active site mutations exhibit activity loss. Mutation of these residues, results in breaking of the network which could be one of the reasons for the decreased activity [152]. Mutation studies on HIV-1 reverse transcriptase (PDB: 1TKT), at position Y181 and Y188 lead to high-level resistance to first generation non-nucleoside inhibitors [153]. This resistance was mainly attributed to the loss of extensive stacking interactions of the side chains of Y181C and Y188C mutants with the aromatic rings of the inhibitors. These observations however warrant more site directed mutagenesis studies in which the residues are mutated to other aromatic and non-aromatic residues. Thus analysis from studies on aromatic-aromatic interactions in crystal structures from PDB lead us to postulate that a mutation resulting in breaking of π–π networks could result either in destabilizing the protein or in activity loss [106, 124]. Studies on kinases, including choline kinase [154] and p38 kinase [155] which are important targets in treatment of diseases such as cancer, have also helped to reinforce the role and relevance of cation-π interactions in the active site of enzymes. The above mentioned discussion forms a strong base to explain the biological relevance of noncovalent interactions. It is but the tip of the iceberg with a vast number of large biological molecules yet to be explored by rigorous application of quantum chemical studies quantifying the contribution of noncovalent interactions in them.

18.8   Outlook

A multitude of nonbonded interactions working in "concordance or discordance" are usually involved in bestowing the structure, function and dynamics of supramolecular assemblies. Quantum mechanical calculations on x-ray structures play a very important role in understanding the factors responsible for stabilization/destabilization of these assemblies (Fig. 18.8).

Although the nonbonded interactions are synonymous with weak interactions some of them are not weak. The total stabilization achieved by several nonbonded interactions is greater than the sum of individual stabilization. Thus nonbonded interactions have a profound influence on other neighboring interactions and on operating cooperatively generate an amplified effect. It is also important to employ reliable molecular dynamics simulations along with quantum chemical approaches to get picture of time dependant behavior of different model systems studied. In this chapter a detailed explanation of several strategies employed effectively on noncovalent interactions is presented. Subtle variations in different factors are shown to play a key role on the strength of weak interactions. This includes solubility, nature and size of systems considered, computational method applied to

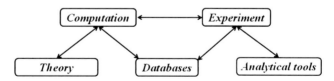

Fig. 18.8 Interplay between different factors is essential for understanding noncovalent interactions

evaluate interaction and so forth. We demonstrate using numerous examples, how weak attractive intramolecular and intermolecular interactions between aromatic rings, cations and in differently bonded clusters can play a significant role in determining the preferred conformation of molecules. The field of noncovalent interactions presents a lot of scope for inter disciplinary research on account of huge complexity inherently present in it and computations emerge as an extremely vital method to tackle this complexity.

Acknowledgements GNS acknowledges financial support from Department of Science and Technology (DST) New Delhi through its Swarnajayanthi fellowship, ASM is grateful to DST for financial support under its Woman Scientist Scheme (WOS-A).

References

1. Special Issue: 90 years of chemical bonding (2007) J Comput Chem 28:1–466, 411–350
2. van der Waals JD (1873) Over de Continuiteit van den Gas- en Vloeistoftoestand (on the continuity of the gas and liquid state) Doctoral dissertation, Leiden, Holland
3. Autumn K, Liang YA, Hsieh ST, Zesch W, Chan WP, Kenny TW, Fearing R, Full RJ (2000) Nature 405:681–685
4. Buhler E, Candau SJ, Schmidt J, Talmon Y, Kolomiets E, Lehn JM (2007) J Polym Sci: Part B: Polym Phys 45:103–115
5. Dougherty DA (1996) Science 271:163–168
6. Mecozzi S, West AP Jr, Dougherty DA (1996) Proc Natl Acad Sci USA 93:10566–10571
7. Ma JC, Dougherty DA (1997) Chem Rev 97:1303–1324
8. Gallivan JP, Dougherty DA (1999) Proc Natl Acad Sci USA 96:9459–9464
9. (a) Bindu PH, Sastry GM, Murty US, Sastry GN (2004) Biochem Biophys Res Commun 319:312–320; (b) Bindu PH, Sastry GM, Sastry GN (2004) Biochem Biophys Res Commun 320:461–467
10. Chourasia M, Sastry GM, Sastry GN (2005) Biochem Biophys Res Commun 336:961–966
11. Zondlo NJ (2010) Nat Chem Biol 6:567–568
12. Brunsveld L, Folmer BJB, Meijer EW, Sijbesma RP (2001) Chem Rev 101:4071–4097
13. (a) Hasenknopf B, Lehn JM, Kneisel BO, Baum G, Fenske D (1996) Angew Chem Int Ed 35:1838–1840; (b) Lehn JM (1995) Supramolecular chemistry: concepts and perspectives. VCH, Weinheim
14. Breuning E, Hanan GS, Romero-Salgeuro FJ, Garcia AM, Baxter PNW, Lehn JM, Wegelius E, Rissanen K, Nierengarten H, van Dorsselaer A (2002) Chem Eur J 8:3458–3466
15. Douglas T, Young M (1998) Nature 393:152–155
16. Meissner RS, Rebek J Jr, de Mendoza J (1995) Science 270:1485–1488
17. Roman M, Cannizzo C, Pinault T, Isare B, Andrioletti B, van der Schoot P, Bouteiller L (2010) J Am Chem Soc 132:16818–16824
18. Hoeben FJM, Jonkheijm P, Meijer EW, Schenning APHJ (2005) Chem Rev 105:1491–1546
19. Davis AV, Yeh RM, Raymond KN (2002) Proc Natl Acad Sci USA 99:4793–4796
20. Schneider H, Strongin RM (2009) Acc Chem Res 42:1489–1500
21. Hobza P, Sponer J (1999) Chem Rev 99:3247–3276
22. Mller-Dethlefs K, Hobza P (2000) Chem Rev 100:143–168
23. Hobza P, Zahradnik R, Mller-Dethlefs K (2006) Collect Czech Chem Commun 71:443–531
24. Kim KS, Tarakeshwar P, Lee JY (2000) Chem Rev 100:4145–4185
25. Gokel GW, De Wall SL, Meadows ES (2000) Eur J Org Chem 2967–2978
26. Wheeler SE, Houk KN (2009) J Am Chem Soc 131:3126–3127

27. Cerny J, Hobza P (2007) Phys Chem Chem Phys 9:5291–5303
28. Sherill CD, Takatani T, Hohenstein EG (2009) J Phys Chem A 113:10146–10159
29. Desiraju GR (2002) Acc Chem Res 35:565–573
30. Zaric SD (2003) Eur J Inorg Chem 2197–2209
31. Hong BH, Bae SC, Lee CW, Jeong S, Kim KS (2001) Science 294:348–351
32. Hong BH, Lee JY, Lee CW, Kim JC, Bae SC, Kim KS (2001) J Am Chem Soc 123:10748–10749
33. Gu J, Leszczynski J (2001) J Phys Chem A 105:10366–10371
34. Gu J, Leszczynski J (2000) J Phys Chem A 104:6308–6313
35. Ryzhov V, Dunbar RC, Cerda B, Wesdemiotis C (2000) J Am Soc Mass Spectrom 11:1037–1046
36. Rodgers MT, Armentrout PB (2004) Acc Chem Res 37:989–998
37. Zhu W, Luo X, Puah CM, Tan X, Shen J, Gu J, Chen K, Jiang H (2004) J Phys Chem A 108:4008–4018
38. Sponer J, Burda JV, Sabat M, Leszczynski J, Hobza P (1998) J Phys Chem A 102:5951–5957
39. Garau C, Frontera A, Quinonero D, Ballester P, Costa A, Deya PM (2004) Chem Phys Lett 392:85–89
40. Tsuzuki S, Honda K, Uchimaru T, Mikami M, Tanabe K (2000) J Am Chem Soc 122:3746–3753
41. Ikuta S (2000) J Mol Struct (THEOCHEM) 530:201–207
42. Reddy AS, Sastry GN (2005) J Phys Chem A 109:8893–8903
43. Rao JS, Sastry GN (2009) J Phys Chem A 113:5446–5454
44. Vijay D, Sastry GN (2006) J Phys Chem A 110:10148–10154
45. Vijay D, Sastry GN (2008) Phys Chem Chem Phys 10:582–590
46. Priyakumar UD, Punnagai M, Krishna Mohan GP, Sastry GN (2004) Tetrahedron 60:3037–3043
47. Reddy AS, Zipse H, Sastry GN (2007) J Phys Chem B 111:11546–11553
48. Rao JS, Zipse H, Sastry GN (2009) J Phys Chem B 113:7225–7236
49. Sharma B, Rao JS, Sastry GN (2011) J Phys Chem A 115:1971–1984
50. Rao JS, Dinadayalane TC, Sastry GN, Leszczynski J (2008) J Phys Chem A 112:12944–12953
51. Mahadevi AS, Sastry GN (2011) J Phys Chem B 115:703–710
52. Boys SF, Bernardi F (1970) Mol Phys 19:553–566
53. Gaussian 03, Revision E.01, Frisch MJ, Trucks GW, Schlegel HB, Scuseria GE, Robb MA, Cheeseman JR, Montgomery JA Jr., Vreven T, Kudin KN, Burant JC, Millam JM, Iyengar SS, Tomasi J, Barone V, Mennucci B, Cossi M, Scalmani G, Rega N, Petersson GA, Nakatsuji H, Hada M, Ehara M, Toyota K, Fukuda R, Hasegawa J, Ishida M, Nakajima T, Honda Y, Kitao O, Nakai H, Klene M, Li X, Knox JE, Hratchian HP, Cross JB, Bakken V, Adamo C, Jaramillo J, Gomperts R, Stratmann RE, Yazyev O, Austin AJ, Cammi R, Pomelli C, Ochterski JW, Ayala PY, Morokuma K, Voth GA, Salvador P, Dannenberg JJ, Zakrzewski VG, Dapprich S, Daniels AD, Strain MC, Farkas O, Malick DK, Rabuck AD, Raghavachari K, Foresman JB, Ortiz JV, Cui Q, Baboul AG, Clifford S, Cioslowski J, Stefanov BB, Liu G, Liashenko A, Piskorz P, Komaromi I, Martin RL, Fox DJ, Keith T, Al-Laham MA, Peng CY, Nanayakkara A, Challacombe M, Gill PMW, Johnson B, Chen W, Wong MW, Gonzalez C, Pople JA (2004) Gaussian Inc., Wallingford
54. Bader RFW (1990) Atoms in molecules: a quantum theory. Clarendon, Oxford
55. Dinadayalane TC, Sastry GN, Leszczynski J (2006) Int J Quant Chem 106:2920–2933
56. Kumar MK, Rao JS, Prabhakar S, Vairamani M, Sastry GN (2005) Chem Commun 1420–1422
57. Rao JS, Sastry GN (2006) Int J Quant Chem 106:1217–1224
58. Siu FM, Ma NL, Tsang CW (2004) Chem Eur J 10:1966–1976
59. (a) Allen FH (2002) Acta Crystallogr Sect B 58:380–388; (b) Berman HM, Westbrook J, Feng Z, Gilliland G, Bhat TN, Weissig H, Shindyalov IN, Bourne PE (2000) Nucleic Acids Res 28:235–242

60. Biot C, Buisine E, Kwasigroch JM, Wintjens R, Rooman M (2002) J Biol Chem 227:40816–40822
61. Reddy AS, Sastry GM, Sastry GN (2007) Proteins Struct Funct Bioinform 67:1179–1184
62. Gallivan JP, Dougherty DA (2000) J Am Chem Soc 122:870–874
63. Vaden TD, Lisy MJ (2004) J Chem Phys 120:721–730
64. Adamo C, Berthier G, Savinelli R (2004) Theor Chem Acc 111:176–181
65. Dzidic I, Kebarle PJ (1970) J Phys Chem 74:1466–1474
66. Rodriguez-Cruz SE, Jockusch RA, Williams ER (1998) J Am Chem Soc 120:5842–5843
67. Rodgers MT, Armentrout PB (1997) J Phys Chem A 101:1238–1249
68. Carl DR, Moision RM, Armentrout PB (2007) Int J Mass Spectrom 265:308–325
69. Lee HM, Tarakeshwar P, Park J, Kołaski MR, Yoon YJ, Yi HB, Kim WY, Kim KS (2004) J Phys Chem A 108:2949–2958
70. Merrill GN, Webb SP, Bivin DB (2003) J Phys Chem A 107:386–396
71. Pavlov M, Siegbahn PEM, Sandstrom M (1998) J Phys Chem A 102:219–228
72. Glendening ED, Feller D (1995) J Phys Chem 99:3060–3067
73. Jennings WB, Farrell BM, Malone JF (2001) Acc Chem Res 34:885–894
74. Pallan PS, Lubini P, Egli M (2007) Chem Commun 1447–1449
75. Mehta G, Sen S, Guru Row TN, Chopra D, Chattopadhyay S (2008) Eur J Org Chem 2008:805–815
76. Sponer J, Riley KE, Hobza P (2008) Phys Chem Chem Phys 10:2595–2610
77. Bradeanu IL, Flesch R, Kosugi N, Pavlychev AA, Rühl E (2006) Phys Chem Chem Phys 8:1906–1913
78. McGaughey GB, Gagne M, Rappe AK (1998) J Biol Chem 273:15458–15463
79. Frank BS, Vardar D, Buckley DA, McKnight CJ (2002) Protein Sci 11:680–687
80. Tsuzuki S, Honda K, Uchimaru T, Mikami M, Tanabe K (2002) J Am Chem Soc 124:104–112
81. Sinnokrot MO, Sherrill CD (2006) J Phys Chem A 110:10656–10668
82. (a) Reddy AS, Vijay D, Sastry GM, Sastry GN (2006) J Phys Chem B 110:2479–2481; (b) Reddy AS, Vijay D, Sastry GM, Sastry GN (2006) J Phys Chem B 110:10206–10207
83. Johnson ER, Mackie ID, Dilabio GA (2009) J Phys Org Chem 22:1127–1135
84. Min SK, Lee EC, Lee HM, Kim DY, Kim D, Kim KS (2008) J Comput Chem 29:1208–1221
85. Hobza P, Selzle HL, Schlag EW (1996) J Phys Chem 100:18790–18794
86. Tsuzuki S, Uchimaru T, Matsumara K, Mikami M, Tanabe K (2000) Chem Phys Lett 319:547–554
87. Sinnokrot MO, Sherrill CD (2004) J Phys Chem A 108:10200–10207
88. Gadre SR, Shirsat RN, Limaye AC (1994) J Phys Chem 98:9165–9169
89. Gerenkamp M, Grimme S (2004) Chem Phys Lett 392:229–235
90. Grimme S (2003) J Chem Phys 20:9095–9102
91. Vijay D, Sakurai H, Sastry GN (2011) Int J Quant Chem 111:1893–1901
92. Liu B, McLean AD (1973) J Chem Phys 59:4557
93. Grabowski SJ, Sokalski WA, Leszczynski J (2006) Chem Phys Lett 432:33–39
94. van Duijneveldt FB, van Duijneveldt-van de Rijdt JGCM, van Lenthe JH (1994) Chem Rev 94:1873–1885
95. Xantheas SS (1996) J Chem Phys 104:8821–8824
96. Rayon VM, Sordo JA (1998) Theor Chem Acc 99:68–70
97. Simon S, Duran M, Dannenberg JJ (1996) J Chem Phys 24:11024–11031
98. Sinnokrot MO, Valeev EF, Sherrill CD (2002) J Am Chem Soc 124:10887–10893
99. Hobza P, Havlas Z (1998) Theor Chem Acc 99:372–377
100. Hermida-Ramón JM, Graña AM (2007) J Comput Chem 28:540–546
101. Berski S, Latajka Z (1996) Comput Chem 21:347–354
102. Kolář M, Hobza P (2007) J Phys Chem A 111:5851–5854
103. Saeki M, Akagi H (2006) J Chem Theor Comput 2:1176–1183
104. Sakurai H, Daiko T, Sakane H, Amaya T (2005) J Am Chem Soc 127:11580–11581
105. Kawase T, Kurata H (2006) Chem Rev 106:5250–5273
106. Chourasia M, Sastry GM, Sastry GN (2011) Int J Biol Macromol 48:540–552

107. Espinosa E, Molins E (2000) J Chem Phys 113:5686–5694
108. Sunita SS, Rohini NK, Kulkarni MG, Nagaraju M, Sastry GN (2006) J Am Chem Soc 128:7752–7753
109. Maheshwary S, Patel P, Sathyamurthy N, Kulkarni AD, Gadre SR (2001) J Phys Chem A 105:10525–10537
110. (a) Neela YI, Mahadevi AS, Sastry GN (2010) J Phys Chem B 114:17162–17171; (b) Mahadevi AS, Neela YI, Sastry GN (2011) J Chem Sci (accepted); (c) Mahadevi AS, Neela YI, Sastry GN (2011) Phys Chem Chem Phys 13:15211–15220
111. Barbour LJ, Orr GW, Atwood JL (1998) Nature 39:671–673
112. Wales DJ, Doye JPK, Dullweber A, Naumkin FY (1997) The Cambridge cluster database. http://brian.ch.cam.ac.uk/CCD.html
113. Nagaraju M, Sastry GN (2010) Int J Quant Chem 110:1994–2003
114. Monod J, Wyman J, Changeux JP (1965) J Mol Biol 12:88–118
115. Ogata RT, McConnell HM (1971) Proc Natl Acad Sci USA 69:335–339
116. Acerenza L, Mizraji E (1997) Biochim et Biophys Avta 1339:155–166
117. Frank HS, Wen WY (1957) Discuss Faraday Soc 24:133
118. Hyskens PL (1977) J Am Chem Soc 99:2578–2582
119. Elrod MJ, Saykally RJ (1994) Chem Rev 94:2578–2582
120. Suhai S (1994) J Chem Phys 101:9766–9782
121. (a) Kobko N, Paraskevas L, Rio E, Dannenberg JJ (2001) 123:4348–4349; (b) Turi L, Dannenberg JJ (1994) J Am Chem Soc 116:8714–8721; (c) Masunov A, Dannenberg JJ (2000) J Phys Chem B 104:806–810; (d) Turi L, Dannenberg JJ (1996) J Phys Chem 100:9638–9648
122. Vijay D, Zipse H, Sastry GN (2008) J Phys Chem B 112:8863–8867
123. Vijay D, Sastry GN (2010) Chem Phys Lett 485:235–242
124. Mahadevi AS, Rahalkar AP, Gadre SR, Sastry GN (2010) J Chem Phys 133:164308
125. Kobko N, Dannenberg JJ (2003) J Phys Chem A 107:10389–10395
126. Jiang X, Sun C, Wang C (2009) J Comp Chem 31:410–1420
127. Esrafili MD, Behzadi H, Hadipour NL (2008) Theor Chem Acc 121:135–146
128. Rappé AK, Bernstein ER (2000) J Phys Chem A 104:6117–6128
129. Steele RP, DiStasio RA, Shao Y, Kong J, Head-Gordon M (2006) J Chem Phys 125:074108–074111
130. Noguera M, Bertran J, Sodupe M (2004) J Phys Chem A 108:333–341
131. Zhao Y, Truhlar DG (2007) J Chem Theor Comput 3:289–300
132. Bachorz RA, Bischoff FA, Höfener S, Klopper W, Ottiger P, Leist R, Frey JA, Leutwyler S (2008) Phys Chem Chem Phys 10:2758–2766
133. Sun S, Zhang G, Geng D, Chen Y, Li R, Cai M, Sun X (2011) Angew Chem Int Ed 50:422–426
134. Peterca M, Percec V, Imam MR, Leowanawat P, Morimitsu K, Heiney PA (2008) J Am Chem Soc 130:14840–14852
135. Balamurugan K, Gopalakrishnan R, Sundar Raman S, Subramanian V (2010) J Phys Chem B 114:14048–14058
136. Nagaraju M, Sastry GN (2009) J Phys Chem A 113:9533–9542
137. Strmcnik D, Kodama K, van der Vliet D, Greeley J, Stamenkovic VR, Markovic NM (2009) Nat Chem 1:466–472
138. Umadevi D, Sastry GN (2011) J Phys Chem C 115:9656–9667
139. Karplus M, McCammon JA (2002) Nat Struct Biol 9:646–651
140. van Gunsteren WF, Berendsen HJC (1990) Angew Chem Int Ed 29:992–1023
141. Car R, Parrinello M (1985) Phys Rev Lett 55:2471–2474
142. CPMD Version 3.3 (1995–1999) developed by Hutter J, Alavi A, Deutsch T, Bernasconi M, Goedecker St, Marx D, Tuckerman M, Parrinello M Max-Planck-Institut für Festkörperforschung and IBM Zurich Research Laboratory
143. Grimme S (2006) J Comput Chem 27:1787–1799

144. Payne MC, Teter MP, Allan DC, Arias TA, Joannopoulos JD (1992) Rev Mod Phys 64:1045–1097
145. Blöchl PE (1994) Phys Rev B 50:17953–17979
146. VASP, Kresse G, Furthmüller J (1996) Phys Rev B 54:11169–11186
147. Dudev T, Lim C (2008) Annu Rev Biophys 37:97–116
148. Barman A, Taves W, Prabhakar R (2009) J Comput Chem 30:1405–1413
149. Thoden JB, Miran SG, Phillips JC, Howard AJ, Raushel FM, Holden HM (1998) Biochemistry 37:8825–8831
150. (a) Genis C, Sippel KH, Case N, Cao W, Avvaru BS, Tartaglia LJ, Govindasamy L, Tu C, McKenna MA, Silverman DN, Rosser CJ, McKenna R (2009) Biochemistry 48:1322–1331; (b) Vallee BL, Auld DS (1990) Proc Natl Acad Sci USA 87:220–224
151. (a) D'Alessandro M, Aschi M, Paci M, Nola AD, Amadei A (2004) J Phys Chem B 108:16255–16260; (b) Pelmenschikov V, Siegbahn EM (2005) Inorg Chem 44:3311–3320
152. Katouno F, Taguchi M, Sakurai K, Uchiyama T, Nikaidou N, Nonaka T, Sugiyama J, Watanabe T (2004) J Biochem 136:163–168
153. Ren J, Nichols C, Bird L, Chamberlain P, Weaver K, Short S, Stuart DI, Stammers DK (2001) J Mol Biol 312:795–805
154. (a) Janardhan S, Srivani P, Sastry GN (2006) Curr Med Chem 13:1169–1186; (b) Janardhan S, Srivani P, Sastry GN (2006) QSAR Comb Sci 25:860–872
155. (a) Badrinarayan P, Sastry GN (2011) J Chem Inf Model 51:115–129; (b) Soliva R, Gelpi JL, Almansa C, Virgoli M, Orozco M (2007) J Med Chem 50:283–293

Chapter 19
Unusual Properties of Usual Molecules. Conformational Analysis of Cyclohexene, Its Derivatives and Heterocyclic Analogues

Oleg V. Shishkin and Svitlana V. Shishkina

Abstract Cyclohexene, its derivatives and heterocyclic analogues usually are considered as well-investigated molecules especially from viewpoint of their molecular structure and conformational properties. However, close inspection of published data indicates that this is wrong opinion. Detailed analysis of conformational characteristics of tetrahydroaromatic rings revealed that these usual molecules possess some very unusual properties. This especially concerns ring inversion process. In many cases potential energy surface has very flattened character in the area of saddle point including also some additional minima. Both $\pi-\pi$ conjugation and $n-\sigma$ and $\pi-\sigma$ hyperconjugation interactions significantly influence conformational properties of ring. All these questions are considered in chapter.

Keywords Tetrahydroaromatic rings • Conformation • Ring inversion

Modern stereochemistry of organic compounds is based on knowledge of conformational properties of set of the simplest and fundamental molecules like ethane, butane, butadiene, cyclohexane etc. In the case of partially hydrogenated ring this set includes cyclopentene, cyclohexene and isomeric cyclohexadienes [1, 2]. Stereochemistry of all other compounds usually may be analyzed in terms of combination of conformational properties of these simplest molecular fragments [3]. Therefore, detailed investigation of conformational characteristics of such basic molecules always represents a cornerstone of conformational analysis. Taking into account numerous studies of structure and properties of such molecules usually they are considered as well-known and well-investigated. However, more detailed inspection of published data clearly indicates that this opinion very often is wrong.

O.V. Shishkin (✉) • S.V. Shishkina
Division of Functional Materials Chemistry, SSI "Institute for Single Crystals",
National Academy of Science of Ukraine, 60 Lenina ave, Kharkiv, 61001 Ukraine
e-mail: shishkin@xray.isc.kharkov.com

J. Leszczynski and M.K. Shukla (eds.), *Practical Aspects of Computational Chemistry I:*   557
An Overview of the Last Two Decades and Current Trends,
DOI 10.1007/978-94-007-0919-5_19, © Springer Science+Business Media B.V. 2012

In particular this concerns energy profile for ring inversion process. According to classical theory transition between two conformers proceeds as movement of molecule along valley of the potential energy surface with smoth increase of energy up to maximal point corresponding to transition state followed by smoth decrease of energy due to move of molecule along other hill of valley to second stable conformer corresponding another minimum of potential energy surface. Such picture of conformational transitions was derived mainly from investigation of rotation around single bonds in aliphatic compounds. Investigation of this process in cyclic molecules is much more complicated because of coupling between torsion angles within ring. Change of one endocyclic torsion angle causes changes of neighboring torsion angles. Therefore reaction coordinate for ring inversion process is quite complicated and represents a combination of at least several torsion angles. This requires an application of some special procedure for accurate investigation of conformational transitions in cyclic molecules instead of simple scan of one torsion angle which is usual method for acyclic molecules. The intrinsic reaction coordinate method (IRC) represents very suitable tool for such studies [4, 5]. This approach allows finding accurate enough pathway with minimal energy in every point from transition state to minima of the potential energy surface. Reaction coordinate includes changes of all geometrical parameters of molecules providing reliable and clear energy profile for ring inversion process.

In this chapter we would like to demonstrate that careful analysis of conformational properties of such simple molecules as cyclohexene, its derivatives and heterocyclic analogues provides new and unexpected information about general background of modern stereochemistry. Especially this concerns character of ring inversion process.

19.1   Conformational Properties of Cyclohexene

Conformational properties of cyclohexene were a subject of numerous experimental and theoretical studies using various methods during almost half of century. First papers concerning stereochemistry of cyclohexene were published in 40th years of last century [6, 7]. It was attempt to determine equilibrium conformation of ring using IR and Raman spectroscopy. Conformation of cyclohexene was described in terms of ring bending and ring twisting angles. It was concluded that ring has intermediate conformation between boat and chair possessing C_2 symmetry. For the first time Barton et al. in 1954 introduced new notation for description of ring conformation of cyclohexene – "half-chair" [8]. Further investigations of equilibrium geometry of cyclohexene using microwave spectroscopy [9, 10], electron diffraction [11, 12], IR [13–15], Raman [13, 16] and NMR [17, 18] spectroscopy and calculations by molecular mechanics methods clearly confirmed half-chair conformation of ring with C_2 symmetry and the $Csp^3–Csp^3–Csp^3–Csp^3$ torsion angle being about $60°$. Finally structure of the cyclohexene was established in the crystal state [19]

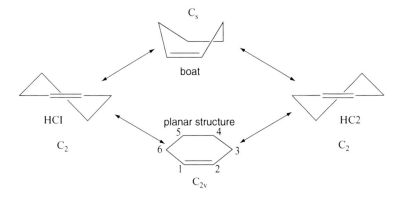

Fig. 19.1 Scheme of possible conformational transitions in cyclohexene

Table 19.1 Values of the C3–C4–C5–C6 torsion angle and ring inversion barrier (E_{inv}, kcal/mol) as determined by different experimental and theoretical methods

Method	Angle, deg.	E_{inv}	Ref.	Method	Angle, deg.	E_{inv}	Ref.
Raman spectroscopy		7.6	[16]	RHF/3-21G		5.51	[33]
IR spectroscopy		7.4	[14]	RHF/6-31G(d)		6.56	[33]
IR spectroscopy		10.3	[15]	MP2/6-31G(d)		5.70	[33]
Electron diffraction	60.2		[11]	MP4SDQ		6.07	[33]
MW spectroscopy	62.6		[9]	QCISD		6.05	[34]
Electron diffraction	61.2		[12]	QCISD(T)		5.87	[34]
NMR spectroscopy		5.3	[17]	MP2/aug-cc-pvdz	64.0	5.19	[35]
MM3		6.6	[31]	MP2/6-31G(2df,2pd)	63.9	5.51	[35]
MM3		7.2	[32]	MP4SDQ/6-31G(d,p)	62.6	6.03	[35]

using X-ray and neutron diffraction methods. Such conformation of ring was also predicted by numerous theoretical studies using molecular mechanics with various force fields [20–32] and quantum chemical methods [33–37].

Half-chair conformation of cyclohexene may be described in terms of the torsion angle formed by the saturated carbon atoms (the C3–C4–C5–C6 torsion angle, see Fig. 19.1). According to experimental and theoretical data value of this angle is within 60–64° (Table 19.1).

Symmetry of ring conformation implies existence of two symmetrical half-chair conformations of cyclohexene with the same energy. Based on IR and Raman data it was suggested that conformational transition from one half-chair conformation (HC1) to another one (HC2) may proceed via either boat conformation with C_s symmetry or planar structure with C_{2v} symmetry (Fig. 19.1).

According to this two possible pathways of ring inversion it was considered two possible values of energetic barriers for these processes called inversion barrier (E_{inv}) and barrier to planarity (E_{plan}) calculated as difference in energy between equilibrium half-chair conformation and boat (E_{inv}) or planar (E_{plan}) conformations as a transition state. Determination of these values by Raman (7.6 and 25.4 kcal/mol, respectively) and IR (7.4 and 16.7 kcal/mol, respectively) spectroscopy [14, 16] clearly demonstrated that ring inversion proceeds via boat conformation as transition state. More recent investigation of ring inversion barrier [15] gave higher value (10.3 kcal/mol).

Theoretical estimations of energetic parameters agree quite reasonably with experimental data (Table 19.1). Value of ring inversion barrier is within 4.2–7.9 kcal/mol for molecular mechanics and 5.2–6.6 kcal/mol for quantum chemical methods.

However, despite of presence of experimental and theoretical data for equilibrium conformation and ring inversion barrier the question about ring inversion pathway was quite controversial during many years. First attempt of analysis of inversion profile was performed by N. Allinger using scan of the C3–C4–C5–C6 torsion angle by molecular mechanics method [25]. Unexpectedly, results of calculations demonstrated that boat conformation corresponds to local minimum on the potential energy surface (PES) connected with two half-chair conformations by symmetric twist-boat conformations as transition states. Difference in energy between boat and twist boat conformations was not negligible (1.6 kcal/mol).

Further calculations by MM2 method [26] did not indicate existence of additional minimum on PES of cyclohexene. Boat conformation was clearly classified as single transition state for ring inversion process. However, it was stressed that potential energy surface around saddle point is rather flat. Variation of the C3–C4–C5–C6 torsion angle within ±20° results in energy change only by 0.08 kcal/mol. Application of next generation of Allinger's force field (MM3) again led to controversial data. According to these calculations [31] boat conformation corresponds to shallow local minimum of PES with differences in energy between boat and twist boats conformations less than 0.1 kcal/mol. This ambiguity was not resolved also by scans of potential energy surface using ab initio quantum chemical methods within Hartree-Fock approximation with small basis sets.

Problem was solved by application of intrinsic reaction coordinate (IRC) procedure in combination with Density Functional Theory and second order Moller-Plesset perturbation theory methods [35]. It was demonstrated that potential energy surface around boat conformation is extremely flat (Fig. 19.2) and contains wide plateau where energy of molecule remains almost the same. Conformation of cyclohexene ring is changed from one twist-boat to symmetric another twist-boat via boat conformation representing a central point of plateau. Width of plateau in terms of value of the C3–C4–C5–C6 torsion angle slightly depends on method of calculations (Table 19.2) and is varied within ±30–40°.

Thus, the ring inversion process in cyclohexene includes three stage. The first stage involves the transition from an equilibrium half-chair conformation to a

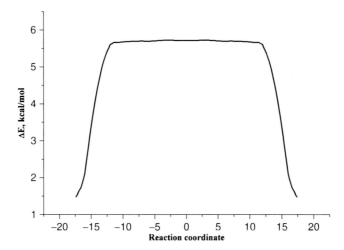

Fig. 19.2 Cyclohexene ring inversion profile calculated by MP2/6-31G(d,p) method using intrinsic reaction coordinate

Table 19.2 Values of the C3–C4–C5–C6 torsion angle where difference in energy between boat and twist-boat conformations is 0.5 kcal/mol (in according to ref [35])

Method	Torsion angle, deg.
B3LYP/6-31G(d,p)	30.3
B3LYP/6-311G(d,p)	29.9
B3LYP/D95(d,p)	31.6
B3LYP/cc-pvdz	32.0
B3LYP/aug-cc-pvdz	32.4
MP2/6-31G(d,p)	37.8
MP2/6-311G(d,p)	36.0
MP2/D95(d,p)	39.0
MP2/cc-pvdz	40.6
MP2/aug-cc-pvdz	36.0

twist-boat conformation with a significant increase in energy. The second stage may be described as a very easy transformation of one twist-boat conformation to another via boat almost without any changes in energy. The third stage includes the transition from a twist-boat conformation to a half-chair conformation accompanied by a significant decrease in energy. Taking into account the extremely top-flattened character of the potential energy profile for the cyclohexene ring inversion, it is possible to conclude that question concerning the conformation in saddle point is meaningless. The difference in energy between the twist-boat and the boat conformation is less than 0.01 kcal/mol. Therefore, we can consider any of these conformations as a saddle point for the ring inversion from a physical viewpoint. In this case we should say about the multitude of saddle points within the plateau on the potential energy surface rather than about one specific point. The boat conformation is just central point of this plateau and it is only the most convenient point for a description of the saddle point for the ring inversion. However, an interpretation of the character of this conformation in terms of eigenvalues of Hessian is meaningless.

Detailed investigation of experimental vibrational spectra of cylohexene confirmed theoretical data [36]. Analysis of ring-twisting potential indicated good agreement between theoretical and experimental data including plateau around boat conformation.

Thus, ring inversion profile in cyclohexene strongly differs from classical form due to its top-flattened shape. It is possible to suggest that this difference is caused by correlation between changes of geometrical parameters of ring during conformational transition.

19.2 Derivatives of Cyclohexene Containing Exocyclic Double Bond

It is quite clear that unusual character of ring inversion profile in cyclohexene is determined by subtle balance of intramolecular interactions. Therefore replacement of one methylene group by exocyclic double bond should lead to some changes of these interactions accompanied by change of equilibrium conformation as well as ring inversion pathway.

All derivatives of cyclohexene containing exocyclic double bond may be divided in two groups depending on position of this bond with respect to endocyclic C=C bond. Replacement of methylene group next to endocyclic C=C bond (molecules **2-8**) leads to formation of butadiene (X=CH$_2$) or heterobutadiene fragment with two conjugated double bonds. It is possible to suggest that $\pi-\pi$ conjugation should significantly influence conformational properties of ring in these molecules.

Conjugation is absent in derivatives where two double bonds are separated by methylene group. Therefore one can expect that conformational characteristics of cyclohexene ring in these molecules should be similar to unsubstituted cyclohexene. Taking into account different character of intramolecular interactions in these two types of derivatives of cyclohexene we will consider their conformational properties separately.

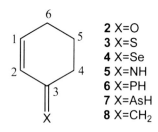

2 X=O
3 X=S
4 X=Se
5 X=NH
6 X=PH
7 X=AsH
8 X=CH$_2$

Table 19.3 Values of puckering parameters (puckering degree S and polar angles Θ and Ψ) [53] for ring conformation, barrier or ring inversion (E_{inv}, kcal/mol) of cyclohexene **1** and its derivatives **2–8** calculated by MP2/6-311G(d,p) method [49]

Molecule	X	S	Θ	Ψ	E_{inv}
1	H_2	0.78	36	30	5.55
2	O	0.74	38	14	8.25
3	S	0.75	38	19	8.11
4	Se	0.75	38	17	8.13
5	NH (trans)	0.75	38	15	8.38
5	NH (cis)	0.75	37	18	8.04
6	PH (trans)	0.76	38	20	8.02
6	PH (cis)	0.77	37	23	7.67
7	AsH (trans)	0.75	38	18	8.20
7	AsH (cis)	0.76	38	22	7.88
8	CH_2	0.76	37	19	8.03

Parameters for ideal conformations are following: half-chair – $\Theta = 45°$. $\Psi = 30°$; sofa – $\Theta = 45°$, $\Psi = 0°$

19.2.1 Derivatives of Cyclohexene with Conjugated Double Bonds

Among molecules **2–8** the oxo derivative **2** is the best studied by experimental and theoretical methods. It was established that cyclohexene ring in molecule **2** adopts a sofa conformation according to NMR [38, 39], microwave [40], Raman [41] and IR [42–45] spectroscopic studies. The same results were obtained from theoretical calculations by molecular mechanics [46, 47] and quantum chemical HF/6-31G(d,p) and MP2/6-31G(d,p) methods [48]. More detailed analysis of cyclohexenone conformation by MP2/6-311G(d,p) method indicated that its conformation may be described as slightly distorted sofa [49]. Similar conclusion was made based on results of X-ray diffraction study of inclusion complex [50] containing molecule **2**.

According to theoretical data [49] decrease of polarity of exocyclic double bond leads to shift of equilibrium conformation of cyclohexene ring from sofa to half-chair (Table 19.3). Unfortunately experimental data for molecules **3–7** are absent. In the case of methylene derivative **8** IR spectroscopic study [45] suggested a sofa conformation of cyclohexene ring similar to cyclohexenone **2**. However, quantum chemical calculations by MP2/6-311G(d,p) method indicated slightly asymmetric half-chair equilibrium conformation of ring in this molecule.

Thus, appearance of conjugated system in molecules **2–8** results in flattening of part of ring containing π-system as compared to cyclohexene. Therefore it is possible to suppose that strength of conjugation between double bonds should significantly influence conformational properties of ring.

According to common understanding of conjugation effects in organic chemistry presence of highly electronegative terminal heteroatom in butadiene fragment should lead to strengthening of the π–π interactions between double bonds due to

Table 19.4 Length (Å) of the C2–C3 bond and energy of π–π and n–σ^* (E(2), kcal/mol) interactions derived from NBO theory in molecules **2–8** [49]

Molecule	X	C2–C3	E(2), π–π	E(2), n–σ^*
2	O	1.484	25.9	20.8
3	S	1.464	33.9	13.1
4	Se	1.458	38.0	10.8
5	NH (trans)	1.472	27.4	1.2
5	NH (cis)	1.475	27.5	12.4
6	PH (trans)	1.462	34.4	–
6	PH (cis)	1.460	33.8	7.5
7	AsH (trans)	1.456	38.2	–
7	AsH (cis)	1.455	37.5	6.5
8	CH_2	1.466	29.6	–

Fig. 19.3 Scheme of the n–σ^* hyperconjugation in derivatives of cyclohexene contained exocyclic double bond with heteroatom

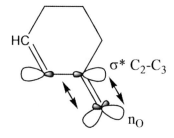

shift of π-electrons from the C=C bond to the C=X bond. This effect should be accompanied by increase of bond order of single C–C bond causing its shortening. From this viewpoint the strongest conjugation should be observed in oxo and imino derivatives **2** and **5** leading to the most significant flattening of ring.

However, comparison of length of the C2–C3 bond in molecules under consideration led to unexpected results [49]. This bond is considerably longer in oxo and imino derivatives **2** and **5** as compared to methylene derivative **8** and it is slightly shorter in molecules containing heavier heteroatoms (Table 19.4). In common sense such trend should be interpreted as weakening of conjugation between double bonds in molecules **2** and **5** as compared to **8** containing non-polar butadiene fragment. The same conclusion was made from analysis of electron density distribution and character of intramolecular interactions using Bader's "Atoms in molecules" (AIM) [51] and Natural Bonding Orbitals (NBO) [52] theories [49].

So unexpected results leads to obvious question – what is wrong: modern ab initio quantum chemistry or theoretical background of organic chemistry. However, more detailed analysis of intramolecular interactions in these molecules using NBO theory results in another answer. It was demonstrated that length of single bond between two double bonds is affected by hyperconjugation interaction between lone pair of terminal heteroatom and antibonding orbital of the C2–C3 bond (Fig. 19.3). Strength of this interaction depends on size of heteroatom determining energy of

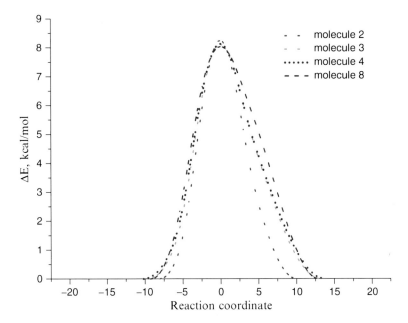

Fig. 19.4 Energy profile for ring inversion in molecules **2, 3, 4** and **8** according to calculations by MP2/6-311G(d,p) method

non-bonding orbital of lone pair and length of the C═X bond. The smallest size of the oxygen and the nitrogen atoms and the shortest C═X bond lead to the strongest n–σ* hyperconjugation in molecules **2** and **5**. Population of anti-bonding orbital of the C2–C3 bond results in weakening of this bond and its elongation (Table 19.4). Decrease of hyperconjugation in other derivatives of cyclohexene causes shortening of the C–C bond.

It should be noted that presence of n–σ* hyperconjugation does not influence equilibrium conformation of cyclohexene ring. This may be clearly determined using puckering parameters for quantitative description of ring conformation [53]. Some flattening of ring in oxo and imino derivatives reflects strengthening of the π–π conjugation despite of presence of strong enough hyperconjugation.

Process of ring inversion in molecules **2–8** was subject of several experimental studies by NMR [38, 39] and IR [42–45] spectroscopy. Value of inversion barrier was determined within 4–5 kcal/mol. It was suggested that ring inversion proceeds via boat or twist boat conformation as saddle point.

More detailed theoretical investigation of ring inversion process by MP2/6-311G(d,p) method results in considerably higher value of barrier (Table 19.3). Moreover, presence of conjugated system in molecules **2–8** results in drastic change of topology of potential energy surface around transition state (Fig. 19.4). In all cases it was found absence of top-flattened character of PES near a boat conformation as transition state of ring inversion. It is interesting that change of strength of the π–π conjugation only slightly influence inversion barrier (Table 19.3).

19.2.2   Derivatives of Cyclohexene with Non-conjugated Double Bonds

Derivatives of cyclohexene containing exocyclic double bond separated to endocyclic C$=$C bond by methylene group (molecule **9–15**) were investigated considerably less as compared to their isomeric counterparts with conjugated double bonds. Experimental data are known only for cyclohexenone **9** and methylenecyclohexene **15**. Investigation of these compounds by infrared [44, 45] and microwave [54, 55] spectroscopy led to conclusion that equilibrium geometry of molecules may be described as half-chair or boat which may exist in some equilibrium. However, more detailed quantum-chemical investigation by MP2/6-311G(d,p) method demonstrated [49] that these molecules adopt only one equilibrium conformation which is a half-chair in **15** and intermediate conformation between sofa and half-chair in carbonyl derivative **9**.

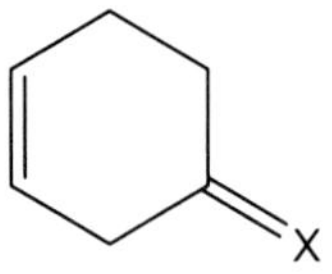

9 X=O
10 X=S
11 X=Se
12 X=NH
13 X=PH
14 X=AsH
15 X=CH$_2$

Comparison of puckering parameters of ring in molecules **9–15** indicates systematic changes of cyclohexene ring conformation from sofa to half-chair in the rows O–S–Se and NH–PH–AsH (Table 19.5). This is very similar to molecules **2–8** where conformation of ring depends on conjugation between double bonds. In the case of compounds **9–15** conjugation is absent. However, two double bonds can interact to each other using hyperconjugation mechanism [49]. Analysis of intramolecular interactions indicated that only the n–σ* interactions between lone pair of heteroatom and neighboring C–C bonds become systematically weaker with increase of atomic number of heteroatom while the π–σ hyperconjugation between π-system of double bonds and neighboring methylene groups remains almost the same in all molecules. Therefore it is possible to suggest that the n–σ* hyperconjugation is responsible for flattening of ring in carbonyl and imino derivatives. In the case of absence of such interactions in methylene derivative **15** cyclohexene ring adopts almost ideal half-chair conformation (Table 19.5).

Investigation of ring inversion by MP2/6-311G(d,p) method demonstrated that this process is much more complex in molecules **9–15** as compared to their isomers **2–8** [49]. Energy profile for ring inversion is highly asymmetric for all molecules (Figs. 19.5–19.7) and it is more or less similar to cyclohexene. Moreover in the case of thione derivative **10** and trans isomers of the N, P and As containing molecules potential energy surface contains additional minimum corresponding to boat or twist boat conformation. Energy of this conformer is considerably higher than energy of

Table 19.5 Values of puckering parameters (puckering degree S and polar angles Θ and Ψ) [53] for ring conformation, barrier of ring inversion (E_{inv}, kcal/mol) of molecules **9–15** calculated by MP2/6-311G(d,p) method [49]

Molecule	X	S	Θ	Ψ	E_{inv}
9	O	0.74	39	10	3.96
10	S	0.80	37	22	4.88
11	Se	0.79	37	22	4.90
12	NH (trans)	0.77	37	18	4.62
12	NH (cis)	0.76	37	17	4.86
13	PH (trans)	0.79	37	25	5.36
13	PH (cis)	0.79	37	25	5.60
14	AsH (trans)	0.80	37	26	5.33
14	AsH (cis)	0.85	37	28	5.56
15	CH_2	0.78	37	23	5.36

Parameters for ideal conformations are following: half-chair – $\Theta = 45°$. $\Psi = 30°$; sofa – $\Theta = 45°$, $\Psi = 0°$

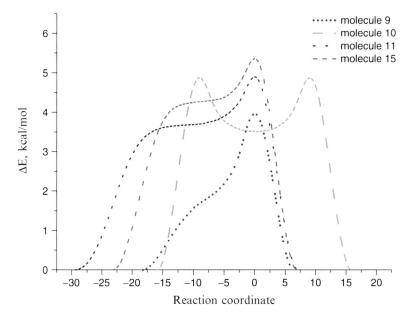

Fig. 19.5 Energy profile for ring inversion in molecules **9–11, 15** according to calculations by MP2/6-311G(d,p) method

half-chair by 3.0–4.5 kcal/mol. Nevertheless ring inversion is a multi-stage process going via two transition states (Fig. 19.8).

Comparison of ring inversion profile in cis and trans isomers of molecules **12–14** (Figs. 19.6 and 19.7) allows suggesting possible reasons for appearance of high-lying conformer of cyclohexene ring. It was found that change of orientation of the X–H bond at the heteroatom results in significant changes in character of potential energy surface around boat conformation of ring. Additional minimum on PES is observed in the case of trans orientation of the X–H bond with respect to the

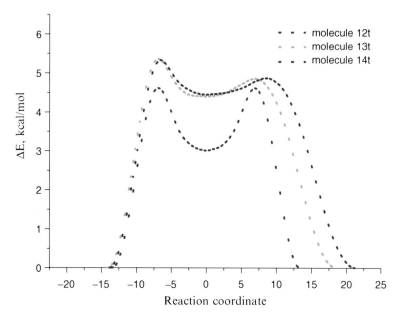

Fig. 19.6 Energy profile for ring inversion in trans isomers of molecules **12–14** according to calculations by MP2/6-311G(d,p) method

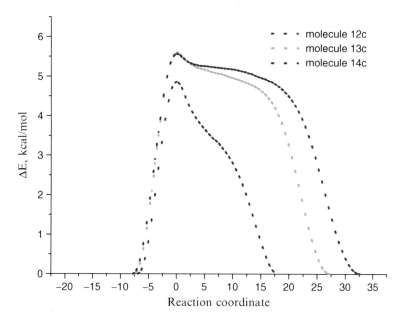

Fig. 19.7 Energy profile for ring inversion in cis isomers of molecules **12–14** according to calculations by MP2/6-311G(d,p) method

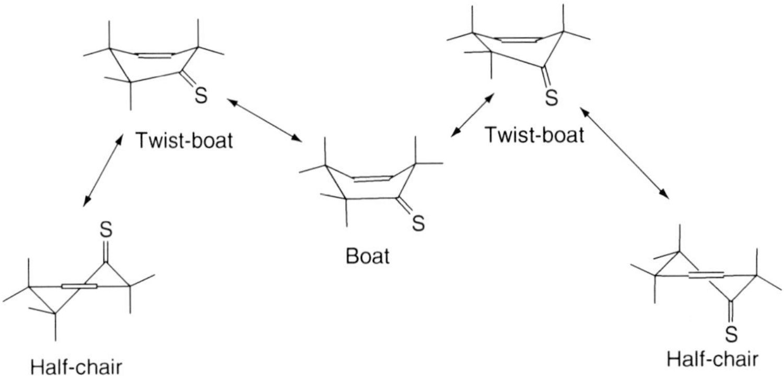

Fig. 19.8 Scheme of ring inversion process in molecule **10**

C3–C4 bond of ring and it disappears in cis isomers. Difference between these two isomers is only in character of the n–σ* hyperconjugation between lone pair of heteroatom and neighboring C–C bonds. Existence of such interactions with the C4–C5 bond leads to appearance of minimum in trans isomers. Probably, this effect may explain presence of high-lying minimum in thione derivative **10**. However, reasons of absence of such minimum in oxo- and selenone derivatives **9** and **11** remain unclear.

Values of ring inversion barriers in non-conjugated molecules **9–15** are considerably lower that in conjugated isomers **2–8** (Table 19.5). This clearly indicates that π–π conjugation significantly influence dynamical properties of ring in the contrary to equilibrium conformation. However, general trend of changes of ring inversion barriers in **9–15** completely differs from conjugated isomers. These values become systematically smaller with increase of atomic number of heteroatom. Reasons of this trend are not clear. However it is possible to assume that ring inversion barrier depends on total energy of hyperconjugation interaction. One of the highest values was observed for methylene derivative **15** where hyperconjugation is the smallest [49].

19.3 Tetrahydroheterocycles

Another way for change of character of intramolecular interactions in cyclohexene ring is replacement of one methylene group by heteroatom. Like derivatives with endocyclic double bond all tetrahydroheterocyclic rings may be divided in two groups depending on location of heteroatom. Replacement of the nearest to the C=C double bond methylene group creates suitable conditions for n–π conjugation between lone pair of heteroatom and π-system of double bond. It is quite clear that this interaction should significantly influence conformational properties of ring. Separation of double bond and heteroatom by methylene group results in considerably weaker perturbation of intramolecular interactions. Therefore it is possible to expect that properties of such rings should be similar to cyclohexene. However, the n–π conjugation considerably weaker influences conformational properties of ring than π–π interactions. Much more differences are caused by possibility of existence of pnictogen heteroatoms (N, P, As) in two different configurations with planar and pyramidal (N) or two pyramidal (P, As) geometry. Careful analysis of conformational properties of tetrahydroaromatic heterocycles indicates that presence or absence of n–π conjugation does not provide so many differences as presence of two configurations of three-coordinated heteroatom.

19.3.1   Tetrahydroheterocycles Containing Chalcogen

Among heterocycles with conjugated heteroatom and double bond 3,4-dihydro-2H-pyrane **17** is the most well studied. Investigation by IR [14, 56–58], Raman [16] and NMR [59] spectroscopy led to conclusion that equilibrium conformation of ring is a half-chair. The same results were obtained from theoretical calculations by molecular mechanics [60, 61] and quantum-chemical [37, 62] methods. Ring inversion barrier estimated as difference in energy between half-chair and boat conformations is within 6.6–10.0 kcal/mol according to experimental [56, 58, 59] as well as theoretical [60–62] data. It is considerably lower than planarity barrier (15.4–17.6 kcal/mol) estimated from IR spectroscopic data [56, 57]. This is additional confirmation that ring inversion process proceeds via boat conformation like cyclohexene.

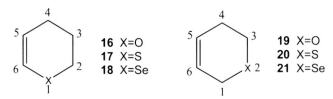

Some experimental data are also available for thiopyrane **17**. Infrared spectroscopic study of this compound [63] led to conclusion about half-chair equilibrium conformation of ring with inversion barrier being slightly smaller than for oxygen analogue (4.3 kcal/mol).

Systematic theoretical investigation of tetrahydroaromatic heterocycles with conjugated heteroatom and endocyclic double bond demonstrated [62] that equilibrium geometry of ring is asymmetric half-chair which characteristics strongly depend on n–π conjugation. Weakening of interaction between lone pair of heteroatom and π-system of the C=C double bond results in an increase of puckering degree of ring especially in the area of the C=C–X–C fragment (Table 19.6). Decrease of strength of n–π conjugation results in increase of puckering degree of heterocycle.

Table 19.6 Values of puckering parameters (puckering degree S and polar angles Θ and Ψ) [53] for ring conformation, barrier of ring inversion (E_{inv}, kcal/mol) of molecules **16–21** calculated by MP2/6-311G(d,p) method [62]

Molecule	X	S	Θ	Ψ	E_{inv}
16	O	0.78	39	28	7.53
17	S	0.83	39	25	4.14
18	Se	0.83	39	16	3.85
19	O	0.84	38	26	6.40
20	S	0.83	35	25	5.97
21	Se	0.82	34	25	5.38

Parameters for ideal conformations are following: half-chair – $\Theta = 45°$. $\Psi = 30°$; sofa – $\Theta = 45°$, $\Psi = 0°$

In the case of non-conjugated heterocycles **19–21** change of heteroatom almost does not influence equilibrium conformation of ring (Table 19.6). Only slight decrease of ring puckering is observed. This agrees well with known experimental data for 5,6-dihydro-4H-pyrane **19** [57, 58, 61] and its thio analogue **20** [64, 65] obtained by IR and microwave spectroscopy.

The main differences in conformational properties of heterocycles were found for ring inversion profile. It should be noted that character of ring inversion does not depend on presence or absence of n–π conjugation and it is tuned only by nature of heteroatom. In the case of dihydropyranes **16** and **19** energy profile for this conformational transition is similar to derivatives of cyclohexene with conjugated endo- and exocyclic double bonds (Fig. 19.9). It is observed only slight top-flattened character in the area of boat or twist-boat conformation serving as transition state. However, for sulphur and selenium containing rings character of ring inversion is completely different. Investigation of conformational transition between two half-chair conformation by intrinsic reaction coordinate method (Fig. 19.10) revealed existence of additional minimum on the potential energy surface corresponding to boat (molecules **20, 21**) or twist-boat (molecules **17, 18**) conformation. Energy of this additional minimum is considerably higher (by 2.8–3.6 kcal/mol) as compared to half-chair conformations except molecule **21** where difference in energy is less than 1 kcal/mol.

Thus ring inversion process in sulphur- and selenium-containing molecules is a multi-stage one similar to derivatives of cyclohexene with non-conjugated exocyclic double bonds. However, thiopyranes **17** and **20** have additional feature (Fig. 19.10). Energies of two transition states between half-chair and boat or twist-boat conformers considerably differ (ΔE is up to 3 kcal/mol) as calculated by different methods [62].

It is interesting that general trend of change of ring inversion barriers is the same for both types of heterocycles with conjugated and non-conjugated heteroatom and the C==C double bond (Table 19.6). Increase of size of heteroatom results in decrease of barrier. This fact clearly indicates that both n–π conjugation and n–σ* hyperconjugation strongly influence properties of rings. Systematic decrease of conjugation in molecules **16–18** or hyperconjugation in molecules **19–21** results in decrease of ring inversion barriers.

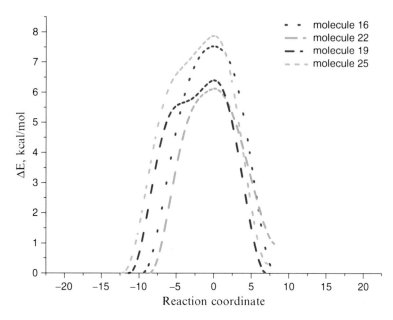

Fig. 19.9 Energy profile for ring inversion in cis isomers of molecules **16, 19, 22, 25** according to calculations by MP2/6-311G(d,p) method

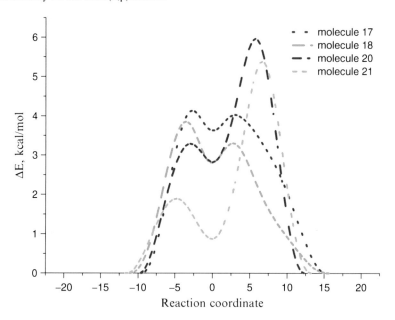

Fig. 19.10 Energy profile for ring inversion in cis isomers of molecules **17, 18, 20, 21** according to calculations by MP2/6-311G(d,p) method

Table 19.7 Values of puckering parameters (puckering degree S and polar angles Θ and Ψ) [53] for ring conformation, relative stability of conformers with different configuration of heteroatom (ΔE_{conf}, kcal/mol) and barrier of ring inversion (E_{inv}, kcal/mol) of molecules **22–27** calculated by MP2/6-311G(d,p) method [62]

Molecule	X	Configuration	S	Θ	Ψ	ΔE_{conf}	E_{inv}
22	NH	Planar	0.73	41	12	0.94	6.12
	NH	Pyramidal	0.81	35	22	0	
23	PH	Pyramidal	0.88	34	26	0.75	4.85
	PH	Pyramidal	0.80	45	12	0	
24	AsH	Pyramidal	0.88	35	29	0.97	4.99
	AsH	Pyramidal	0.81	46	7	0	
25	NH	Pyramidal	0.77	37	27	0	7.87
	NH	Pyramidal	0.86	37	29	0.30	
26	PH	Pyramidal	0.74	35	20	0	5.89
	PH	Pyramidal	0.87	34	22	1.47	
27	AsH	Pyramidal	0.75	35	19	0	4.78
	AsH	Pyramidal	0.87	34	22	1.26	

Parameters for ideal conformations are following: half-chair – $\Theta = 45°$. $\Psi = 30°$; sofa – $\Theta = 45°$, $\Psi = 0°$

19.3.2 Tetrahydroheterocycles Containing Pnictogen

Experimental data about structure and conformational properties of tetrahydro-heterocycles containing pnictogens (N, P, As) are absent. According to quantum-chemical calculations [62, 66] the main feature of pnictogen-containing tetrahy-droaromatic heterocycles is a possibility of heteroatom to adopt two different configurations. However, it should be noted that equilibrium conformation of ring is a half-chair in all cases (Table 19.7).

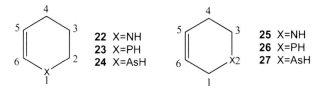

In the case of tetrahydropyridine **22** the nitrogen atom may exist in two configu-rations (Fig. 19.11) with almost planar and pyramidal geometry of the imino group. Flattening of configuration of the nitrogen atom results in considerable decrease of puckering of heterocycle. However such geometry is slightly less favorable than conformation with pyramidal nitrogen ($\Delta E = 0.94$ kcal/mol). Transition between conformers with different configuration of the nitrogen atom requires 2.0 kcal/mol.

More complex situation is observed for tetrahydrophosphine and tetrahydroar-sine. In these molecules heteroatom exists only in pyramidal configuration. How-ever, orientation of the hydrogen atom with respect to ring may be different leading also to formation of two conformers for each half-chair conformation of ring

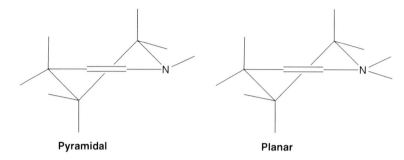

Pyramidal   **Planar**

Fig. 19.11 Two conformation of 1,2,3,4-tetrahydropyridine **22** differing in configuration of the nitrogen atom

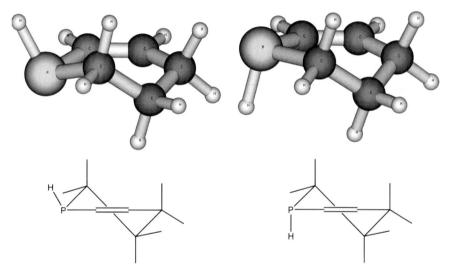

Fig. 19.12 Two conformers of tetrahydrophosphine **23** with different orientation of the hydrogen atom of the PH group at the same conformation of ring (MP2/6-311G(d,p) data)

(Fig. 19.12). In both conformers heteroatom has highly pyramidal configuration (sum of bond angles centered at heteroatom is 292–294° for **23** and 286–288° for **24**) however energy of them is almost equal (ΔE is less than 1 kcal/mol). In contrast to tetrahydropyridine **22** inversion of heteroatom in molecules **23, 24** is almost impossible due to very high barriers (more than 10 kcal/mol).

Similar results were obtained for non-conjugated heterocycles **25–27**. Only in the case of tetrahydropyridine **25** the nitrogen atom adopts two pyramidal configurations differing in orientation of the N–H bond with respect to heterocycle (axial and equatorial). Transition between these two conformers proceeds via planar configuration as transition state and it has also considerably higher barriers (6.7 kcal/mol) as compared to isomeric molecule **22**.

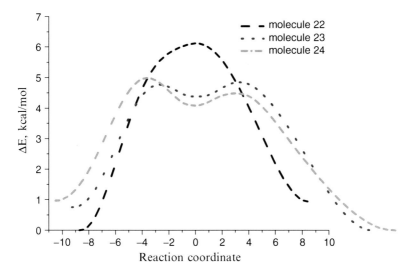

Fig. 19.13 Ring inversion profile for molecules **22–24** according to calculations by MP2/6-311G(d,p) method

Existence of two conformers with the same conformation of ring but with different configuration of the heteroatom makes ring inversion process more complex. In tetrahydropyridines **22** and **25** ring inversion profile is very similar to isomeric pyranes **16, 19** (Fig. 19.9). More detailed analysis of potential energy surface revealed [62] that ring inversion is strongly coupled with change of configuration of the nitrogen atom. Transition of heterocycle from one to another half-chair conformation is accompanied by change of configuration of the nitrogen atom. Probably this is caused by smaller barrier of change of configuration of heteroatom as compared to ring inversion.

High value of heteroatom inversion barrier in tetrahydrophosphine and tetrahydroarsine results in fixation of configuration of the P and As atom during ring inversion. Moreover, pyramidality degree of heteroatom remains almost unchanged during conformational transition between two half-chair conformations.

Ring inversion profiles in molecules **22–27** are very similar to heterocycles containing chalkogens. Tetrahydropyridines **22** and **25** have slightly top-flattened profiles (Fig. 19.9) with boat conformation as a transition state. Tetrahydrophosphines and tetrahydroarsines possess highly assimetric energy profiles with additional minimum corresponding to a twist-boat conformation (Figs. 19.13 and 19.14). However relative energy of these stable conformers considerably differs for heterocycles with conjugated and non-conjugated heteroatoms. In the case of molecules **23, 24** twist-boat conformer lyes more than 4 kcal/mol above the equilibrium half-chair conformation. Difference in energy becomes significantly smaller for tetrahydrophosphine **26** (2.5 kcal/mol) and especially for tetrahydroarsine **27** (0.5 kcal/mol). It should be noted that these additional minima are considerably shallower as compared to thio- and selenopyranes.

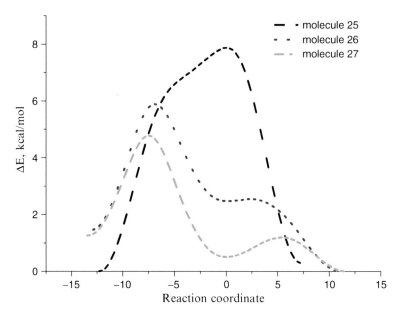

Fig. 19.14 Ring inversion profile for molecules **25–27** according to calculations by MP2/6-311G(d,p) method

19.4   Conclusions

Detailed investigation of cyclohexene, its analogues and derivatives clearly demonstrated that conformational characteristics of cyclic systems are much more sensitive
to influence of relatively weak intramolecular interactions. This especially concerns
ring inversion process. Energy profile, structure of transition state and topology
of the potential energy surface around saddle point are strongly affected not
only classic relatively strong interactions like π–π, n–π conjugation and steric
repulsion but also much weaker n–σ* and π–σ* hyperconjugation. This leads
to appearance of structural unusual properties of ring like plateau of transition
states for cyclohexene inversion, elongation of single C–C bond in heterobutadiene
fragment, coupling between inversion of the nitrogen configuration and ring conformation in tetrahydropyridines, fixation of pyramidal configuration of heteroatom in
heterocycles with heavier pnictogens (PH, AsH) etc. These results clearly indicate
that development of new very reliable theoretical methods of quantum chemistry
promotes re-investigation of conformational characteristics of small molecules
which are considered as fundamental in stereochemistry of organic compounds.

References

1. Anet FAL (1987) In: Rabideau PW (ed) The conformational analysis of cyclohexenes, cyclo-hexadienes, and related hydroaromatic compounds. VCH Publishers, New York, Chapter 1
2. Vereshchagin AN (ed) (1990) Conformational analysis of hydrocarbons and its derivatives. Science, Moscow
3. Eliel EL, Wilen SH (1994) Stereochemistry of organic compounds. Wiley, New York
4. Gonzalez C, Schlegel HB (1989) J Chem Phys 90:2154
5. Gonzalez C, Schlegel HB (1990) J Chem Phys 94:5523
6. Lister MW (1941) J Am Chem Soc 63:143
7. Beckett CW, Freeman NK, Pitzer KS (1948) J Am Chem Soc 70:4227
8. Barton DHR, Cookson RC, Klyne W, Shoppee CW (1954) Chem Ind (London) 21
9. Scharpen LH, Wollrab JE, Ames DP (1968) J Chem Phys 49:2368
10. Ogata T, Kozima K (1969) Bull Chem Soc Jpn 42:1263
11. Chiang JF, Bauer SH (1969) J Am Chem Soc 91:1898
12. Geise HJ, Buys HR (1970) Rec Trav Chim Pays-Bas 89:1147
13. Neto N, Di Lauro C, Castellucci E, Califano S (1967) Spectrochim Acta A 63:1763
14. Smithson TK, Wieser H (1980) J Chem Phys 72:2340
15. Rivera-Gaines VE, Leibowitz SJ, Laane J (1991) J Am Chem Soc 113:9735
16. Durig JR, Carter RO, Carreira LA (1974) J Chem Phys 60:3098
17. Anet FA, Haq MZ (1965) J Am Chem Soc 87:3147
18. Suarez C, Tafazzoli M, True NS (1992) J Mol Struct 271:89
19. Ibberson RM, Telling MTF, Parsons S (2008) Cryst Growth Des 8:512
20. Dashevsky VG, Lugovskoy AA (1972) J Mol Struct 12:39
21. Saebo S, Boggs JE (1981) J Mol Struct 73:137
22. Bucourt R, Hainaut D (1964) C R Acad Sci Paris Sect C 258:3305
23. Bucourt R, Hainaut D (1965) Bull Soc Chim Fr 5:1366
24. Bucourt R, Hainaut D (1967) Bull Soc Chim Fr 7:4562
25. Allinger NL, Hirsch JA, Miller MA, Tyminski IJ (1968) J Am Chem Soc 90:5773
26. Favini G, Buemi G, Raimondi M (1968) J Mol Struct 2:137
27. Allinger NL, Sprague JT (1972) J Am Chem Soc 94:5734
28. Anet FAL, Yavari I (1978) Tetrahedron 34:2879
29. Ermer O (1981) Aspecte von Kraftfeldrechnungen. W. Baur Verlag, Munchen
30. Vanhee P, Tavernier D, Baass JMA, van der Graaf B (1981) Bull Soc Chim Belg 90:697
31. Allinger NL, Li F, Yan L (1990) J Comput Chem 11:848
32. Laane J, Choo J (1994) J Am Chem Soc 116:3889
33. Anet FAL, Freedberg DI, Storer JW, Houk KN (1992) J Am Chem Soc 114:10969
34. Pople JA, Head-Gordon M, Raghavachari K (1987) J Chem Phys 87:5968
35. Shishkina SV, Shishkin OV, Leszczynski J (2002) Chem Phys Lett 354:428
36. Lespade L (2008) J Mol Struct 891:370
37. Ocola EJ, Brito T, McCann K, Laane J (2010) J Mol Struct 978:74
38. Jensen FR, Beck BH (1968) J Am Chem Soc 90:1066
39. Anet FAL, Chmurny GN, Krane J (1973) J Am Chem Soc 95:4423
40. Manley SA, Tyler JK (1970) J Chem Soc Chem Commun 382
41. Carreira LA, Towns TG, Malloy TB Jr (1979) J Chem Phys 70:2273
42. Duckett JA, Smithson TL, Wieser H (1979) J Mol Spectrosc 78:407
43. Smithson TL, Wieser H (1980) J Chem Phys 73:2518
44. Wieser H, Smithson TL, Krueger PJ (1982) J Mol Spectrosc 96:368
45. Smithson TL, Ibrahim N, Wieser H (1983) Can J Chem 61:442
46. Allinger NL, Tribble MT, Miller MA (1972) Tetrahedron 28:1173
47. Ohta Y, Jaime C, Osawa E, Iitata Y, Shimizu N, Nishihara S, Ohsaka T, Hori H, Shibata T, Inayama S (1985) Chem Farm Bull 33:400
48. Organ MG, Froese RDJ, Goddard JD, Taylor NJ, Lange GL (1994) J Am Chem Soc 116:3312

49. Shishkina SV, Shishkin OV, Desenko SM, Leszczynski J (2008) J Phys Chem A 112:7080
50. Tanaka K, Mizutani H, Miyahara I, Hirotsu K, Toda F (1999) Cryst Eng Comm 1:8
51. Bader RFW (1990) Atoms in molecules. Clarendon, Oxford
52. Weinhold F (1998) In: Schleyer PvR, Allinger NL, Clark T, Gasteiger J, Kollman PA, Schaefer HF III, Schreiner PR (eds) Encyclopedia of computational chemistry, vol 3. Wiley, Chichester, p 1792
53. Zefirov NS, Palyulin VA, Dashevskaya EE (1990) J Phys Org Chem 3:147
54. Cervellati R, Damiani D, Dore L, Lister DG (1990) J Mol Spectrosc 135:22
55. Cervellati R, Damiani D, Dore L, Lister DG (1990) J Mol Spectrosc 139:328
56. Lord RC, Rounds TC, Ueda T (1972) J Chem Phys 57:2572
57. Dixon EA, King GSS, Smithson TL, Wieser H (1981) J Mol Struct 71:97
58. Tecklenburg MMJ, Laane J (1989) J Am Chem Soc 111:6920
59. Bushweller CH, O'Neil JW (1969) Tetrahedron Lett 10:4713
60. Dodziuk H, von Voithenberg H, Allinger NL (1982) Tetrahedron 38:2811
61. Choo J, Lee S-N, Lee K-H (1996) Bull Korean Chem Soc 17:7
62. Shishkina SV, Shishkin OV, Desenko SM, Leszczynski J (2007) J Phys Chem A 111:2368
63. Choo J, Meinander NT, Villarreal JR, Laane J (1995) J Chem Phys 102:9506
64. Leal LA, Lister DG, Alonso JL, Tecklenburg MMJ, Villarreal JR, Laane J (1994) J Chem Soc Faraday Trans 90:2849
65. Tecklenburg MMJ, Villarreal JR, Laane J (1989) J Chem Phys 91:2771
66. Tran T, Malloy TB Jr (2010) J Mol Struct 970:66

Chapter 20
Molecular Models of the Stabilization of Bivalent Metal Cations in Zeolite Catalysts

G.M. Zhidomirov, A.A. Shubin, A.V. Larin, S.E. Malykhin, and A.A. Rybakov

Abstract A review of quantum chemical modeling of bivalent metal ion stabilization in zeolites is presented. Location of single metal ions in zeolite cationic positions and formation of polynuclear metal-oxo clusters are considered. Special attention is paid to the stabilization of single bivalent metal ions in the cationic positions with distant separation of the two lattice Al ions forming these exchange positions. It is shown that such a type of cation trapping generates increased number of the Lewis acid sites. Comparative stability and catalytic reactivity of different forms of cation species in zeolites are discussed on the example of Zn/HMFI. Dehydrogenation catalytic reaction of ethane molecule on the single Zn(II) and polynuclear zinc-oxo ions is considered. It is found that binuclear metal-oxo ions can be the effective traps of molecular oxygen and so they can direct the way of oxidation catalysis. This is demonstrated by the theoretical treatment of CO oxidation on alkaline earth zeolites. The CO_2 molecule can be activated by the binuclear metal-oxo ions with possibility of further reaction functionalization. The last part of this review is devoted to discussion of the structure of single Fe(II) active sites in Fe/HZSM-5 zeolites. This system has attracted great attention as selective oxidation of hydrocarbons with N_2O but up to now the understanding of the active structures remains challenging.

Keywords Zeolites • Bivalent metal cation stabilization • ZnZSM-5 • FeZSM-5 • Ethane dehydrogenation • Benzene oxidation • DFT cluster and periodical calculations • Carbon oxide • Carbon dioxide • Carbonate • Carbonilation

G.M. Zhidomirov (✉) • A.A. Shubin • S.E. Malykhin
Boreskov Institute of Catalysis, Siberian Branch of the Russian Academy of Sciences, Pr. Akad. Lavrentieva 5, Novosibirsk 630090, Russia

G.M. Zhidomirov • A.V. Larin • A.A. Rybakov
Chemistry Department, Lomonosov Moscow State University, Leninskiye Gory 1-3, Moscow, GSP-2, 119992, Russia
e-mail: zhidomirov@mail.ru; nasgo@yandex.ru; rybakovy@mail.ru

J. Leszczynski and M.K. Shukla (eds.), *Practical Aspects of Computational Chemistry I: An Overview of the Last Two Decades and Current Trends*,
DOI 10.1007/978-94-007-0919-5_20, © Springer Science+Business Media B.V. 2012

20.1   Introduction

Zeolites relate to the structural class of ion-exchangeable systems. Their ion-exchange properties are created by the inclusion of Al atoms in the lattice positions of silicates. Diversification of zeolite systems is associated with wide range of crystal types, possible change of zeolite module (Si/Al relation), and character of the distribution of Al over the lattice. Due to the variation of exchanged cations together with unique porous structure, zeolites are perspective heterogeneous catalysts. H-forms of zeolites revealing Brönsted acid properties are applied as catalysts in petroleum treatment. The catalytic activity is associated with acidic bridged hydroxyl groups Si–(OH)–Al where the proton appears as a counterion. The recent state of practical applications and basic researches of the acidic zeolite catalysis can be found in two reviews [1, 2]. Well-defined zeolite structures and comparative simplicity of acid sites composition were the basis for the molecular modeling of the sites and quantum chemical calculations of their properties. Cluster model of a small fragment of the silicate chain with included Al atom $((H^*O)_3Si–(OH)–Al(OH^*)_3)$ was found to be sufficient to describe Brönsted acid properties of the site. Boundary H* atoms were used to saturate dangling bonds of the cluster. The evident way of an expanding of the cluster was the advantage of such approach. Similar cluster structure $[(H^*O)_3Si–O^{(1)}–Al(OH^*)_2–O^{(2)}–Si(OH^*)3]^-[M^+]$ was used to model the trap of volumetric exchange ion M^+: NH_4^+, monovalent metal cation, etc. At first, various semiempirical quantum chemical methods were applied for the theoretical treatment of the electronic structure, acid properties and adsorption activity of the sites. The common approach proved to be rather fruitful and a number of qualitative and semiquantitative conclusions have been made. For example, the appearance of Brönsted acidity has been explained by the contact of silicon hydroxyl group with aluminum Lewis acid site and the relationship of acid power of the sites and Al distribution over the lattice has been analyzed. Moreover, the mechanism of isotope H/D exchange of water and small alkane molecules with bridged hydroxyl group was proposed. Outcome of this initial period of the quantum chemical modeling of the Brönsted acidity of zeolites have been summarized in [3, 4]. The next period was associated with preferable application of nonempirical quantum chemical methods and the consideration of large volume fragments of zeolite structure. Different approaches used in these studies were described in the review literatures [5, 6].

In contrast to the stabilization of monovalent exchange cations in zeolites the structural aspects of stabilization of multivalent metal exchange cations are not quite clear. For example, the ion exchange position for bivalent metal cations should be formed by two lattice Al atoms. According to the traditional point of view the aluminum distribution over zeolite lattice is predominantly stochastic. It creates a variety of mutual localization of two nearest lattice Al atoms and results in a number of possible ion exchange structures for the bivalent cation stabilization. On the other hand, it is evident that structural peculiarities of bivalent metal cations stabilization influence the adsorption ability and catalytic reactivity of the cation.

In principle there is a possibility to stabilize bivalent metal cation with a ligand, which transforms bivalent cations to monovalent species, for example, $(ZnOH)^+$. It is reasonable to expect appearance of such species as a result of dissociative adsorption of water and it will be discussed below. However, the most interesting catalytic processes proceed at high temperatures when the above-mentioned species became unstable. It should be noted that all the above considerations are applicable to zeolites with rather high zeolite module Si/Al ≥ 5. The theoretical treatment of traps of exchange metal cations in zeolites with lower module requires different approach. The stabilization of bivalent metal cations in zeolites is considered in the Sect. 20.2 of this review which discusses the example of Zn/ZSM-5 zeolites. These zeolites are active in the catalytic process of small alkane dehydrogenation. Thus, we will explore the influence of the stabilization forms of zinc cation on different stages of this catalytic reaction.

Discussed structural problems in the stabilization of bivalent metal cations in zeolites occur during application of traditional techniques of the insertion of active component into zeolite matrix (such as witness impregnation or ion exchange). In both cases it is very difficult to reach sufficiently full exchange. At the same time the quantity of inserted active component can be essentially larger than the exchange capacity of zeolite determined by the quantity of Al atoms in the lattice positions. This forces to the search of other forms of the bivalent metal cation stabilization besides a trapping of single metal ion in cationic positions. The most evident possibility is the stabilization of oxo-binuclear complexes $(M–O–M)^{2+}$ especially for high silica zeolites. Simultaneously, it is probable to expect trapping of neutral or charged form of oxide-hydroxide metal clusters inside the zeolite pores. Indeed, the experimental studies prove the formation of such clusters. But there is only limited information about their location, composition, structure, charge, and the mechanisms of formation and immobilization. The questions of the structures and some interesting chemical properties of oxo-binuclear complexes are discussed in the Sect. 20.3. From the chemical point of view these complexes are interesting due to the chemical peculiarities of bridged oxygen. We would like to highlight the possibility of such structures to accommodate additional oxygen atoms as is illustrated on the example of alkaline-earth metal oxo-binuclear complexes. Such complexes can transform oxo-binuclear metal structures into effective sites for oxidation, for example, by molecular oxygen. Another interesting chemical activity of such complexes is their ability to capture CO_2 molecule as the first step of their activation.

The formation of oxide-hydroxide metal clusters is considered in Sect. 20.4. Various places of their localization are discussed. One of them is associated with accommodation of metal oxide species in the cationic position of zeolites. The $(Zn_3O_2)^{2+}$ cluster formation was studied and its activity in the dehydrogenation of ethane was calculated. Besides the condensation of polynuclear oxide species in ion-exchange positions a possibility of the grafting of small metal oxide clusters to zeolite or to pure silica lattice was considered on the example of immobilization of ZnO, $(ZnO)_2$ and $(ZnO)_3$ species.

The Sect. 20.5 of the review contains a consideration of probable molecular models and description of the electronic structure of single Fe(II) active sites in the Fe/HZSM-5 zeolites. Recently, these catalysts attracted great attention due to their unique ability to provide very selective oxidation for hydrocarbons by N_2O.

Cluster model scheme and periodical method were used in the molecular model calculations of active sites of zeolite catalysts; results of both approaches are presented and discussed in this review. In cluster models of zeolite structures hydrogen boundary atoms (H*) were used to saturate dangling bonds of the Si and Al atoms. Definite restrictions were imposed on the optimization of positions of these boundary H* atoms. In the optimization, the geometry of an appropriate fragment of zeolite lattice was taken from the experimental X-ray diffraction data [7]. Only Si–H* and Al–H* bond distances were optimized, while the positions of other atoms (except M), as well as directions of O–H* bonds, were kept frozen. The M ion was allowed to move freely in the structure.

Subsequently the second step of optimization was performed with the fixed positions of H* atoms. Positions of these atoms obtained after the first step of this procedure were frozen and used further in the restricted geometry optimizations of H-forms and metal ion exchanged zeolite structures. Described technique was applied in the majority of studies. At the same time the procedures using the experimental X-ray diffraction data and subsequent restricted geometry optimization (with fixed boundary H* atoms) can lead to some inaccuracy. The most important factors contributing to the inaccuracy are the following: (1) the use of an 'average' experimental X-ray diffraction structures, because there is no possibility to know from experiment at which positions of the ring Si atoms are isomorphic substituted by Al, that can result in some artificial tensions and deformations in the vicinity of Al atoms; (2) basis set dependence of difference between experimental and optimized bond lengths and angles; (3) the use of fixed positions of boundary atoms H* obtained from the Zn-form of the ring may influence the calculated value of the energy difference between cationic and H-form because it can result in some stress when such positions are used for the corresponding H-form. Nevertheless, this scheme can help to estimate the role of real lattice geometry of zeolite rings.

20.2   Structural Forms of the Stabilization of Single Bivalent Metal Ions in Cationic Positions of Zeolites

Structures of cationic positions in zeolites are determined by location of aluminum atoms in the lattice. There are two factors which can disturb the stochastic distribution over the lattice. The first one is the availability of preferential lattice points for the location of Al atoms in zeolite. It can be important especially in the stabilization due to exchange of monovalent metal cations. The second one is mutual interaction of lattice Al atoms, which, for example, is the reason of direct adjacency impossibility for the placing of Al in the lattice (Löwenstein's rule).

It can influence distribution of Al atom over zeolites lattice, which is important for the stabilization of bivalent metal cations. The problem of deviation of Al atoms distribution over zeolite lattice from stochastic model was considered with regard to the stabilization of exchange metal cation [8]. But the question arises whether it is so important to take into account the above-mentioned factors of the deviation from stochasticity during the molecular modeling of stabilization of exchange metal cation in zeolites? Certainly, in some cases it is desirable. In fact the above discussed factors are based on energy estimation of the resulting structures (the energy preferences are comparatively small) [8]. Herewith the kinetic peculiarities of the zeolite synthesis are not taken into account which can be very important. Below, we will propose the stochastic location of Al atoms over zeolite lattice. It is evident that the most effective capture of bivalent metal cations should be provided by zeolite structures with nearest-neighbor location of two lattice Al. Thus, one can suggest that reasonable molecular models for the cationic positions of bivalent metal ions in zeolites are the 6-, 5- and 4-membered rings containing two Al ions. Such models have been actively used in quantum chemical calculations since 2000 [9–13]. This approach was used in [14] for the simulation of preferential localization of Co(II) in pentasil zeolites with MOR, FER and MFI topology by comparative analysis of the UV-VIS-NIR spectra of Co(II). Three preferential positions referred as α, β and γ have been identified. The α position – an effective six-membered ring on the wall of direct channel, formed by two interconnecting five-membered rings, and readily available to reagents – seems to be of special interest with respect to catalytic action.

The quantity of sites with necessary location of two lattice Al atom depends on the Si/Al ratio. For a rough estimation at given Si/Al ratio one can use the fraction of Al ions with the next nearest neighbors (NNN). The necessary NNN is rather a marked value even for high silica zeolites. So, it was found that according to various computational approaches [11, 13] 5–20% of Al atoms comply with the above condition at Si/Al = 50. Comparison of the energetic stability of various cation species is a key task for the theoretical simulation of cation localization in zeolites. Comparing the stability of M(II) located in different cationic positions $M(II)/Z_i$, it is reasonable to consider corresponding H-forms $2H^+/Z_i$ and the ion-exchanged reaction

$$2H^+/Z_i + M \rightarrow M\,(II)\,/Z_i + H_2 \qquad (20.1)$$

The energy of the reaction (20.1) is denoted as stabilization energy (E_{st}). The negative E_{st} value indicates that metal cation form is more stable than H-form. E_{st} can be considered as a generic index of reaction ability of the trapped metal cation where higher stability corresponds to lower reactivity. As an example let us consider the stabilization of Zn(II) cation in Zn/HZSM-5 zeolite [10] in connection with the activity of this catalyst in the process of the dehydrogenation of small alkanes. The dehydrogenation of ethane molecule was studied. Two five membered rings and the α position of ZSM-5 zeolite were chosen as probable active sites

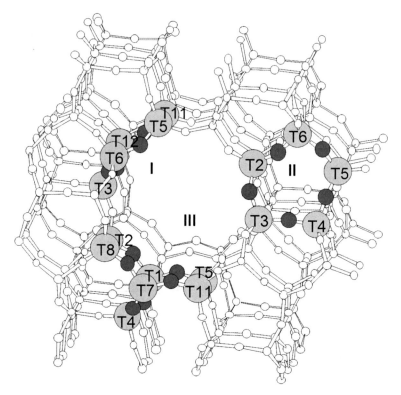

Fig. 20.1 Possible cationic positions in the crystal structure of ZSM 5 zeolite: a five ring in the straight channel (**I**), a five ring in the sinusoidal channel (**II**), and α site (**III**) (Reproduced from Ref. [10] with kind permission of The American Chemical Society)

(see Fig. 20.1). The cluster models of these sites were constructed according of the crystallographic data. For comparison a four membered ring was considered with full optimization geometry because of enough rigid geometry of such structure. All the calculations were carried out using the GAUSSIAN 98 program [15]. The level of the calculations is shown in Table 20.1.

The stabilization energies of Zn^{2+} ions in various probable cationic positions of ZSM–5 zeolite were calculated according to the following reaction

$$2H^+/Z + Zn^0 \rightarrow Zn^{2+}/Z + H_2, \quad E_{st} \tag{20.2}$$

It is generally accepted that dehydrogenation reaction of alkanes starts with the heterolytic dissociation of ones through the "alkyl" route

$$Zn^{2+}/Z + C_2H_6 \rightarrow H^+/Z + C_2H_5^-Zn^{2+}/Z, \quad E_{diss}(C_2H_6) \tag{20.3}$$

Table 20.1 Calculated (B3LYP/6-311G**) energies of reactions (kcal/mol, energy is positive for endothermic reactions)

Structure	E_{st}	$E_{diss}(C_2H_6)$	$E_{des}(C_2H_4)$	$E_{des}(H_2)$	Comments
(1) 5-ring in the straight channel of ZSM-5, 4-coordinated Zn^{2+}	−25.5	14.2	29.6	−0.8	(a) Geometry optimisation with imposed restrictions according to 5-membered ring formed by T3, T5, T6, T11 and T12 positions in the straight channel of ZSM-5. Aluminium atoms are in T5 and T12 positions. Hydrogen atoms in H–form are near O5 and O11 positions. (b) 4-coordinated Zn in ($H^-Zn^{2+}/Z + H^+/Z$) and ($C_2H_5{}^-Zn^{2+}/Z + H^+/Z$) (c) There is hydrogen atom near O11 for all forms except Zn^{2+}/Z
(2) 5-ring in the sinusoidal channel of ZSM-5, 4-coordinated Zn^{2+}	−35.7	12.9	29.1	1.1	(a) Geometry optimisation with imposed restrictions according to 5-membered ring formed by T2, T3, T4, T5, and T6 positions in the sinusoidal channel of ZSM-5. Aluminium atoms are in T3 and T5 positions. Hydrogen atoms in H–form are near O2 and O5 (b) 5-coordinated Zn in ($H^-Zn^{2+}/Z + H^+/Z$) and ($C_2H_5{}^-Zn^{2+}/Z + H^+/Z$) (c) There is hydrogen atom near O5 for all forms except Zn^{2+}/Z
(3) 4-ring, 4-coordinated Zn^{2+}	−16.3	8.7	30.2	4.1	(a) Full geometry optimisation (b) 4-coordinated Zn in ($H^-Zn^{2+}/Z + H^+/Z$) and ($C_2H_5{}^-Zn^{2+}/Z + H^+/Z$)
(4) α-site in ZSM-5 (effective 6-ring), 4-coordinated Zn^{2+}	−19.4	9.9	31.9	1.2	(a) Geometry optimisation with imposed restrictions according to α-site formed in the straight channel by T1, T2, T5, T7, T8, T11, and T4 positions. Aluminium atoms are in T2 and T11 positions. Hydrogen atoms in H–form are near O1 and O22 (b) 3-coordinated Zn in ($H^-Zn^{2+}/Z + H^+/Z$) and ($C_2H_5{}^-Zn^{2+}/Z + H^+/Z$) (c) There is hydrogen atom near O1 for all forms except Zn^{2+}/Z

Reproduced from Ref. [10] with kind permission of The American Chemical Society

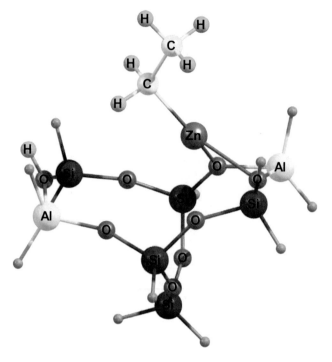

Fig. 20.2 Decomposition of ethane on α-site in ZSM–5 zeolite

The next step is formation and desorption of alkene

$$C_2H_5{}^-Zn^{2+}/Z \rightarrow H^-Zn^{2+}/Z + C_2H_4, \quad E_{des}\,(C_2H_4) \qquad (20.4)$$

and the elimination of H_2

$$H^-Zn^{2+}/Z + H^+/Z \rightarrow Zn^{2+}/Z + H_2, \quad E_{des}\,(H_2) \qquad (20.5)$$

with the closure of the catalytic cycle.

Another possibility for the dehydrogenation reaction is the "carbenium" $(Zn^{2+}/Z + C_2H_6 \rightarrow H^-Zn^{2+}/Z + C_2H_5{}^+/Z)$ route. Early intermediates and transition states for "alkyl" and "carbenium" mechanisms were compared for Zn(II), stabilized in four-membered zeolite ring [16]. The activation energies for the heterolytic dissociation of ethane in these mechanisms were evaluated to be 18.6 and 53 kcal/mol, respectively. The conclusion was made that the "alkyl" route is more preferential for the ethane activation. The reaction energies of processes described by Eqs. 20.2–20.5 are present in Table 20.1. Cluster model of the first step (described by Eq. 20.3) of the reaction on the α-site is presented in Fig. 20.2. The calculations showed that the stabilization of Zn^{2+} in five-membered rings is significantly stronger than in four-membered rings. This correlates with their activity in the ethane heterolytic dissociation. The charge polarization (positive charge on zinc cation) decreases with decreasing of the cation stability. It is interesting that E_{st} and $E_{diss}(C_2H_6)$ for Zn^{2+} in the α-site are quite similar to those obtained for the four-membered ring. Peculiarities of the active site structures notably affect the steps described by Eqs. 20.3 and 20.5 in contrast to Eq. 20.4 (see Table 20.1). The E_{des} value (C_2H_4) is mostly determined by endothermic nature of the ethane dehydrogenation reaction. The result of this study shows that we can expect definite hindrance in the first step of the reaction for considered type of active sites, whereas step described by Eq. 20.5 can proceed quite easy.

With an increase of the Si/Al ratio, a role of zeolite structures (where two nearest neighbour lattice Al atoms located in different zeolite rings) is also increased. This issue was studied in [9] where the possibility of bivalent cation stabilization was firstly analyzed. These sites were classified as MZ_d in contrast to MZ_s (related to the sites, where both Al atoms are located in one small zeolite ring). In [9] stabilization of Zn^{2+} in the zeolite fragment consisting of two connected five-membered rings was considered. This cluster model included possibility to describe both ZnZ_s and ZnZ_d sites. A diminishing stability, higher Lewis acidy and reactivity of ZnZ_d in comparison with ZnZ_s were found. Implementation of a ZnZ_d sites allowed to explain [17, 18] an abnormal low frequency shifts (220–230) cm^{-1} of H_2 [19–21] and 110 cm^{-1} of symmetric (A_1) stretching band of the CH_4 molecules adsorbed on ZnZSM-5 zeolite [19, 22]. It is very difficult to explain such large shifts by interaction of H_2 [23] and CH_4 [24] with traditional forms of zinc ion species: various ZnZ_s sites, binuclear or polynuclear zinc oxide species. Such large frequency shifts were suggested to be due to the decrease in stability and corresponding increase of Lewis acid power and adsorption ability.

Two connected five-rings on the wall of straight channel of ZSM-5 zeolite have been suggested as a possible cationic site for the zinc ion, with Al ions placed in T_{12} and T_8 lattice positions. The cluster $ZnAl_2Si_6O_9 H^*_{14}$ was chosen as model of the ZnZ_d site. Quantum chemical calculations have been carried out within the gradient-corrected density functional theory in conjunction with the B3LYP functional using the GAUSSIAN-98 program [15]. The 6-31G basis set was used for zinc and the 6-311G basis set for zeolite oxygen atoms, and hydrogen atoms of bridging hydroxyl groups. Zeolitic Al and Si atoms, as well as H* hydrogen atoms saturating the broken Si–O bonds at the edges of the cluster, were calculated using the D95-Dunning/Huzinaga basis set.

The cluster structure of Zn^{2+} ion stabilized in the ZnZ_d cationic position with distant placing of two lattice Al atoms is shown in Fig. 20.3 (structure I). The distance between two Al is equal to 7.78 Å. It is important that in this case zinc ion directly interacts with oxygen surrounding only one Al ion in contrast to the ZnZ_s site, where the interacting oxygen is surrounded by both Al ions. Molecular adsorption of dihydrogen molecule on ZnZ_d is shown in Fig. 20.4 [17]. The calculated adsorption energy of H_2 was found to be 7.7 kcal/mol, which is a reasonable value for the interaction with strong Lewis acid sites.

The stretching frequency shift of H_2 induced by the adsorption was evaluated as 254 cm^{-1} and this number agrees well with the experimental data [20]. Scaling

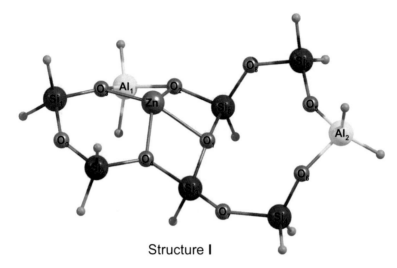

Structure I

Fig. 20.3 Cluster structure (ZnZ$_d$) of Zn^{2+} ion stabilized in the cationic position with the distant placing of two Al ions

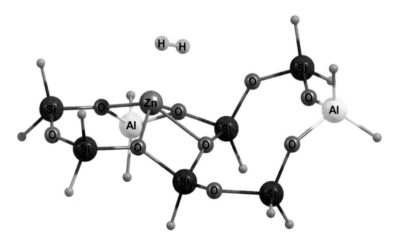

Fig. 20.4 H$_2$ adsorption on ZnZ$_d$ cluster in ZnZSM–5 zeolite

factor of 0.94 was used to correct calculated vibration frequencies. It was evaluated from calculation of the stretching vibration band of the free H$_2$.

The molecular adsorption of the methane molecule on ZnZ$_d$ is shown in Fig. 20.5 [18]. Calculated adsorption energy is equal to 11 kcal/mol. The C–H bond lengths in adsorbed CH$_4$, see Table 20.2, are changed slightly in comparison with the bond length in free methane molecule (1.091 Å). The adsorption of CH$_4$ also initiates slight increase of Zn–O bond length, Table 20.2. Changes of the atomic charges under methane adsorption indicate electron transfer (-0.1) from the methane molecule

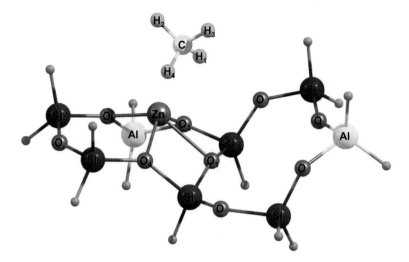

Fig. 20.5 Methane adsorption on ZnZ_d cluster in ZnZSM–5 zeolite

Table 20.2 Bonds lengths (in angstroms) for ZnZ_d active site

	ZnZ_d	CH_4/ZnZ_d
Zn–O_1	2.081	2.122
Zn–O_2	1.973	2.013
Zn–O_3	1.916	1.947
Zn–O_4	2.019	2.104
Zn–C		2.366
C–H_1		1.101
C–H_2		1.088
C–H_3		1.090
C–H_4		1.106

Reproduced from Ref. [8] with kind permission of Elsevier

to the zinc cation and to the neighboring oxygen ions according to the Lewis acidity of zinc in the cationic position. Calculated vibrational frequencies for free and adsorbed CH_4 are presented in the Table 20.3. The most intensive (originated from IR forbidden A_1 band of free methane molecule) the C–H stretching band of adsorbed CH_4 is strongly shifted to lower frequencies (-107 cm^{-1}). This is in a reasonable agreement with the experiment (-111 cm^{-1}) [22]. High frequency T_2 asymmetric C–H stretching band of free CH_4 is found to split due to adsorption. Further, its components are also revealed to be shifted. The intensity of this band is rather low while it is the most intensive for free methane molecule.

Increasing temperature resulted in heterolytic dissociation of the adsorbed H_2 [20, 21] and CH_4 [22] with formation of bridged hydroxyl group, Zn–H or Zn–CH_3, correspondingly and it was confirmed by the IR spectra. Evaluated energies of dissociative adsorption of H_2 and CH_4 were found to be 44 and 36 kcal/mol,

Table 20.3 Calculated IR frequencies (in cm^{-1}) and IR intensities (in km mol^{-1}) for free methane and methane adsorbed on ZnZSM 5 zeolite[a]

CH$_4$ (exp.)[b]	CH$_4$ (theor.)		CH$_4$/ZnZ$_d$		CH$_4$/ZnZ$_d$	
	IR freq.	IR int.	IR freq.	IR int.	IR freq.	IR int.
1306 T$_2$	1293 T$_2$	51	1206 ($-$87)	60	1270 ($-$23)	57
			1306 ($+$13)	16	1299 ($+$6)	21
			1356 ($+$63)	23	1323 ($+$30)	21
1534 E	1504 E	0	1493 ($-$11)	7	1499 ($-$5)	2
			1522 ($+$18)	3	1518 ($+$14)	3
2917 A$_1$	2917 A$_1$	0	2810 ($-$107)	46	2847 ($-$70)	34
3019 T$_2$	3019 T$_2$	84	2906 ($-$113)	4	2963 ($-$56)	3
			2992 ($-$27)	2	3017 ($-$2)	1
			3055 ($+$36)	4	3028 ($+$9)	3

Reproduced from Ref. [18] with kind permission of John Wiley and Sons

[a]Scaling factor equal to 0.964 was used to minimize systematic errors in calculated frequencies. For the case of comparison it was chosen so that after scaling computed A$_1$ frequency for free methane was equal to the experimental value of 2917 cm^{-1}

[b]CRC Handbook of Chemistry and Physics 73rd edition (Editor-in-Chief D.R. Lide), 1992–1993, CRC Press, Inc., Boca Raton, US

respectively. They are significantly higher than the energies of the molecular adsorption. It is not found that much for ZnZ$_s$ sites and this is a principle difference for these two sites (ZnZ$_d$ and ZnZ$_s$) in their reaction ability. This will be shown on the example of dehydrogenation reaction of ethane molecule. Discussed processes of dissociative adsorption of H$_2$ and CH$_4$ include migration of H atom to oxygen surrounding of the distant Al atom. The path of this migration for H$_2$ dissociative adsorption was considered in [17]. Some activation of H$_2$ in the adsorption state (Fig. 20.4) is already apparent from the elongation of H–H bond length (0.762 Å) in comparison with the bond length of free molecule (0.744 Å). Such activation is the stimulating factor for further dissociation of H$_2$ on the site. The first step of the dissociation process involves migration of one hydrogen atom to the one of the nearest oxygen ions of the zeolite structure and formation of the Zn–H chemical bond with the second hydrogen atom. Based on the results of the computational analysis one of the possible targets for hydrogen atom migration is zeolite lattice O$_{(8)}$ oxygen ion. The structure of the corresponding transition state is shown in Fig. 20.6. The H–H distance increases to 1.025 Å. The Mulliken atomic charges on the hydrogen atoms are +0.332 (H$_{(1)}$) and $-$0.229 (H$_{(2)}$), which is in agreement with the chemical meaning of the dissociation path. The resulting structure is characterized by an increase of positive atomic charge on H$_{(1)}$ and considerable decrease of negative atomic charge on H$_{(2)}$. This is not the final position for the hydrogen migration. The most stable localization of H$_{(1)}$ should be near the one of two oxygen ions connected with Al. The next possible step of H$_{(1)}$ migration is complex. It includes the "rotation" of O$_{(8)}$–H$_{(1)}$ bond around Si$_5$–O$_{(8)}$–Si$_6$ quasilinear fragment in order to form intermediate transition state for the subsequent step of H$_{(1)}$ proton migration from O$_{(8)}$ to O$_{(5)}$. The resulting structure

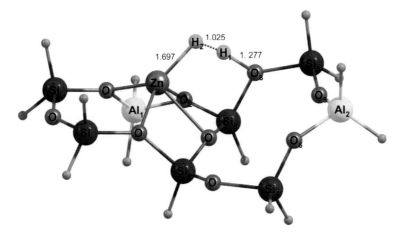

Fig. 20.6 Transition state for H_2 dissociation on ZnZ_d cluster

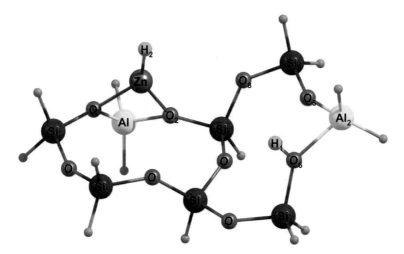

Fig. 20.7 Final structure for $H_{(1)}$ proton migration through ZnZ_d cluster

for the final step of the migration is shown in Fig. 20.7. The energy path for $H_{(1)}$ migration is shown in Fig. 20.8. It can be concluded that H_2 dissociation followed by migration of H atom to the second Al ion is not characterized by a serious energy hindrance.

Dehydrogenation of the ethane molecule on ZnZ_d was studied in [25]. The calculation was initiated by direct experimental IR spectroscopy observation of two stages of the reaction described by Eq. 20.3 and 20.4 [26].

Peculiarities of the inclusion technique of zinc active component into HZSM-5 zeolite and some experimental data lead to conclusion that the reaction occurred on ZnZ_d sites. The same cluster model and level of the quantum chemical approach

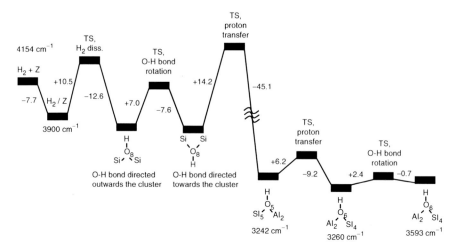

Fig. 20.8 Energy path for $H_{(1)}$ proton migration through ZnZ_d cluster (Reproduced from Ref. [17] with kind permission of Springer)

as in [17, 18] were applied. Several forms of ethane adsorption were found with very close adsorption energies. They can be approximately converted to each other by rotation of ethane molecules around the Zn–C bond. More preferred structure with adsorption energy of 13.4 kcal/mol is shown in Fig. 20.9a (structure **II**). After ethane adsorption, the fourfold coordination of Zn by the lattice oxygen atoms remains unchanged; however, all four Zn–O distances became somewhat longer. Effective transfer of electron density from ethane to the adsorption site is equal to 0.11 electron that is in agreement with a slight decrease in the positive charge of Zn ion. This indicates significant Lewis acidity of the considered active site. Ethane adsorption also resulted in strong polarization of adsorbed molecule and in some change in charge distribution resulting from interaction with the zinc cations. The calculated C–H stretching frequencies of adsorbed ethane and intensities of corresponding IR bands are indicated in Fig. 20.9b. The solid lines show the simulated IR spectra carried out using scaled by 0.964 calculated frequencies; the corresponding IR bands are slightly broadened. For comparison purposes, scaled in the same manner, the IR spectrum of free ethane molecules (shown by the dashed line) was also calculated at the same DFT level. The most remarkable features of the calculated spectra of the molecular form of ethane adsorption are very intense lines at 2720 cm^{-1} that are very strongly red shifted in comparison with the spectrum of free ethane. Positions and intensities of these lines are in good agreement with the most strongly red-shifted IR band at 2727 cm^{-1} in the experimentally observed spectrum of ethane adsorption reported in [26]. Several much weaker C–H stretching bands that were observed in [26] in the region of 2800–3000 cm^{-1} are also associated with this form of adsorption. According to the calculations their scaled frequencies occur at 2849, 2943, 2983, 3012, and 3017 cm^{-1} (see Fig. 20.9b).

a

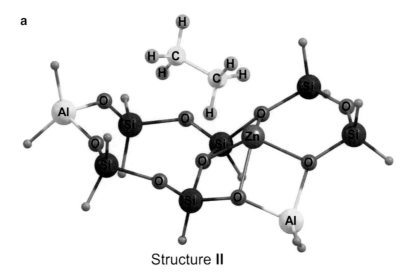

Structure II

b

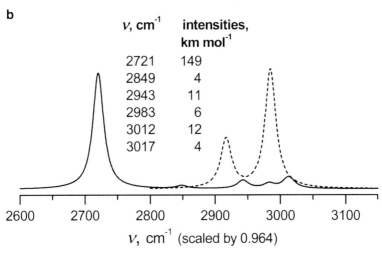

ν, cm^{-1}	intensities, km mol^{-1}
2721	149
2849	4
2943	11
2983	6
3012	12
3017	4

2600   2700   2800   2900   3000   3100

ν, cm^{-1} (scaled by 0.964)

Fig. 20.9 (**a**) Adsorption of ethane molecule on ZnZ$_d$ cluster (**b**) Calculated IR frequencies with their intensities and simulated IR spectrum. IR spectrum of free C$_2$H$_6$ molecule calculated at the same DFT level is shown for comparison purposes by the *dashed line* (Reproduced from Ref. [25] with kind permission of Springer)

Heating of ZnZSM-5 zeolite with adsorbed ethane at 423 K results in heterolytic dissociation of ethane with the formation of a Zn–ethyl fragment and that the bridging hydroxyl group. Prolonged evacuation of the sample with adsorbed ethane at 523 K results in the formation of ethylene and an appearance of a Zn–H bond characterized by the stretching vibration at 1934 cm^{-1} [26]. The reactions described by Eqs. 20.3 and 20.4 were studied and structures of **I–VI** were calculated. These

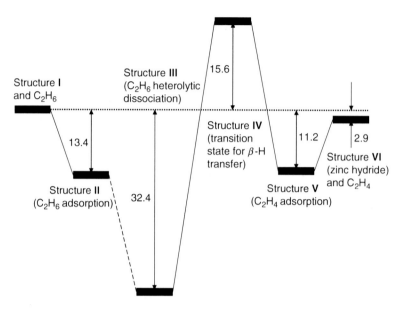

Fig. 20.10 Reaction energy diagram of ethylene formation. Reaction path (shown in *dash*) between structure **II** and structure **III** includes several additional steps (not shown) corresponding to heterolytic dissociation of C_2H_6 and subsequent proton migration similar to those considered for H_2 (Reproduced from Ref. [25] with kind permission of Springer)

structures are characterized by the energy diagram, see Fig. 20.10. Structure **V** with adsorbed ethylene molecule, which was formed after β-H transfer to the Zn ion, is presented in Fig. 20.11. This structure contains zinc hydride and adsorbed ethylene coordinated by the π- bonding with the zinc ion. Calculated Zn–H stretch frequency of 1881 cm^{-1} is in qualitative agreement with the experimental value of 1934 cm^{-1}. The activation energy of 15.6 kcal/mol for the β-H transfer seems to be reasonable. Desorption of ethylene results in structure **VI**, which contains the bridging hydroxyl group and the zinc hydride group. The calculated desorption energy of ethylene is equal to 8.3 kcal/mol. It is quite remarkable that desorption of ethylene decreased the Zn–H stretching frequency to 1788 cm^{-1}.

Recently comparative DFT study of ethane molecule dehydrogenation on ZnZ_s and ZnZ_d sites have been carried out [27]. A new mechanism of transformation of Zn-alkyl species was proposed to consist of direct consistent formation and elimination of H_2 and C_2H_4 instead of stages defined by Eqs. 20.4 and 20.5. The conclusions have been made that ZnZ_d site is more active than ZnZ_s and the proposed direct mechanism requires lower activation energy.

However, comprehensive analysis of the alkane dehydration mechanism on Zn zeolite catalysis is not the main aim of this review. We would like to mention that the observation of Zn–H hydride form [26] give an evidence of a competitive role of β-H transfer mechanism described by Eqs. 20.3–20.5.

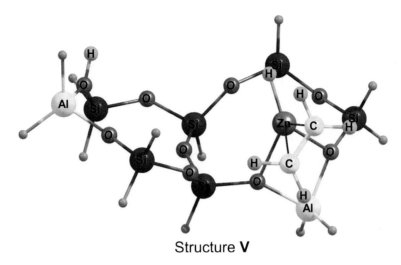

Structure V

Fig. 20.11 The product (which contains zinc hydride and adsorbed ethylene) of the reaction of β-H transfer to the Zn ion

Conception of ZnZ_s and ZnZ_d sites was exploited in [28], in which dehydrogenation ability of gallium cation species stabilized in Ga-exchanged HZSM-5 zeolites was considered. It was proposed that $[GaH]^{2+}$ bivalent structures were the active species in dehydrogenation of light alkanes. The mechanism of the dehydrogenation was considered to involve stages given by Eqs. 20.3 and 20.5. From the previous discussion it is clear that these stages occur differently on ZnZ_s and ZnZ_d sites: step described by Eq. 20.3 is difficult and step (Eq. 20.5) is facilitated for the ZnZ_s sites but is slower for ZnZ_d. The next question is if it is possible to find such distance between lattice Al atoms in order to provide equal effectiveness of both steps. These active sites were found on the base of ZSM-5 zeolite structure [28].

We have till now discussed the structure and chemical properties of ZnZ_d sites using the cluster models. But the possibility of cluster approach to model ZnZ_d sites is strongly limited especially for the sites with large distance between lattice Al atoms. Application of periodical quantum chemical methods seems to be more favourable. Such approach was used firstly to compare stabilization energy of Zn^{2+} cation in ZnZ_d sites with different distance between Al atoms [29]. Periodical calculations were performed using the Vienna Ab Intio Simulation Package [30] for high-silica ferrierite. It was found that the decrease of the stabilization energy at large distances between Al ions (more than $10\,\text{Å}$) is about of 2 eV in comparison with the zeolite lattice (where two Al ions are in the nearest possible positions). It only slightly depends on the following increase of the Al—Al distance. Main changes in the stabilization energy occur within $3\,\text{Å}$ of these distances. Only for the localizations of both Al ions in one zeolite ring zinc cationic form is more stable than the hydrogen form. For example, energy of substitution for α-position is equal to -4.5 kcal/mol. It is interesting to note that the DFT cluster model studies predicted some larger Zn^{2+} stabilization energy for α-cationic position [10], see Table 20.1, which is associated with difference in the quantum chemical methods of these two calculation schemes. Similar difference between the results obtained from the periodical [31] and cluster calculation [17, 18, 25] was reveled from the study of H_2 dissociation on ZnZ_d sites.

Important part of chemistry of bivalent metal exchanged cations in zeolites is their reactions with water. Particularly the water molecule can dissociate on Me^{2+}/Z with formation of Brönsted groups and hydroxyl groups linked to the cation:

$$Me^{2+} + H_2O + [-Si - O - Al-]^- \rightarrow MeOH^+ + [-Si - O\,(H) - Al-] \quad (20.6)$$

This process is often proposed as an essential step in a tentative mechanism of acidic catalysis on metal ion-exchanged forms of zeolites, alkene protonation, alkene oxidation and alkene isomerization [32]. The dissociative adsorption of water molecule on divalent metal cations in zeolites has been studied only in few papers compared with the molecular adsorption. Particularly, Zn ion-exchange zeolites attracted more attention due to their activity as catalysts of the dehydrogenation reaction of small alkanes [18, 33]. So, the calculations of the interaction of Zn^{2+}, localized in the six-membered cycle of Y zeolite with a water molecule showed that the energy of molecular adsorption is quite significant and is equal to 29 kcal/mol [18]. No structure corresponding to the heterolytic dissociation of water (Eq. 20.6) was found [18]. All optimization resulted in molecular adsorption regardless of the starting point. In contrast to Y zeolite the structures corresponding to the water heterolytic dissociation on Zn^{2+} in various five-membered zeolite cycles were revealed. The process (20.6) was calculated as exothermic (26–28 kcal/mol) in relation to H_2O molecule in the gas phase. However, there was no comparison of the energies of the molecular and dissociative adsorption in this work. It was provided in [33] for Zn^{2+} in five-membered zeolite cycle and in α-position of ZSM-5. It was shown that for these cluster models molecular adsorption of H_2O on Zn^{2+} is more stable than dissociative one (Eq. 20.6) by 11.94 kcal/mol. It is reasonable to propose that zinc ion in the cation positions with distant Al will be more effective in the heterolytic decomposition of water as compared with that discussed above for ZnZ_s zinc sites. Indeed, the calculations of the molecular and dissociative adsorption of H_2O molecule on ZnZ_d site (Fig. 20.3) resulted in the adsorption energies of 42.7 and 50.3 kcal/mol. It can be suggested that experimental observation of the process (Eq. 20.6) [32] is associated probably with the dissociation of water on ZnZ_d sites.

20.3   OxoBinuclear Structures of the Alkaline Earth (AE) Metal Cations

Stabilization of multivalent metal ions in zeolites can be achieved by the incorporation of extra lattice oxygen (ELO) into the structure of bi- or multinuclear complexes, wherein each cation interacts efficiently only with the zeolite fragment containing a single Al ion. Although such complexes can form even in low-silica zeolites, they seem to be of particular importance in high-silica zeolites, where the super exchange phenomenon can be related to them.

The bridge $MeOMe^{2+}$ species in zeolites have been discussed in literature for Me$=$Zn [9, 34], Cu [35–39], and Fe [40–42]. For the latter, experimental evidences were first found by Boudart and co-workers [43] who observed redox pairs of iron ions in FeY. Afterwards Mössbauer [44] and IR spectroscopy [45] confirmed the bridging position of oxygen. Some homologues in the series, such as MeO_2Me species, were observed experimentally as a part of organic complexes for Me$=$Ni and Co [46], V [47], Fe [48], or modeled in zeolites for Me$=$Cu [49, 50].

Very interesting question arises about the possibility of the formation of the MeO_XMe ($X = 2, 3, 4, \ldots$) complexes, trapped in the zeolite exchange ion positions. We have recently considered this problem for the AE form of mordenite (MOR) and faujasite (FAU) zeolites at the cluster [51–54] and periodic [53, 54] levels. It was found that CaO_XCa moieties can accumulate the oxygen atoms up to $X = 4$ conserving their ground singlet states. We will focus mainly on the Ca forms with the accent on the $X = 2$, 3, and 4. Such species can explain the experimental data on the desorption of singlet dioxygen from various cation exchanged forms of the MOR and FAU type zeolites under heating [55, 56]. Such oxygen accumulation process can be described by the reaction

$$^1MeO_{X-1}Me + 1/2\,^3O_2 \rightarrow\,^1MeO_XMe \qquad (20.7)$$

This reaction should be considered as the formal scheme that is used for the thermodynamic evaluations of heat (ΔU), where negative value allows the formation of the $^1MeO_XMe$ species. The mechanism possibly involves the intermediate triplet state that converts to the singlet one. The experimental facts prove that AE zeolite easily trap and accumulate oxygen [31, 57], which can oxidize organics under thermal stimulation without additional source of oxygen. The authors concluded [57] that the oxidation of propane takes place only after a preliminary treatment in air/oxygen atmosphere. The oxidation on the Ca forms with the accent on the $X = 2$ and 3 will be described in more details.

The experimental facts whose interpretation demonstrates the presence of the oxo-binuclear structures are the followings: (1) The AE carbonate formation in the zeolites [58]. The $MeCO_3Me$ carbonates are fixed in the zeolite framework and do not move so that AE carbonates can appear when a MeO_XMe binuclear cluster exists already in the zeolite. The immobility of the AE carbonates was confirmed from comparison of electron densities at the critical Me$\cdots$O bond points with the oxygen atoms of the carbonates or of the framework [54]; (2) The CO oxidation processes under thermal activation, for which the mechanism with a charge transfer stabilized CO–O_2 complex cannot be applied. The present interpretation arrives to a reasonable barrier values around of 15 kcal/mol [53, 54]; (3) The formation of two different hydroxyl groups at 3643 and 3533 cm^{-1} after addition of dried HCl (or HCN) to totally dehydrated MgY [58]. The first band can be interpreted due to the barrierless transformation of MeOMe to ClMe–O(H)–Me upon HCl addition. The second band probably corresponds to acid Si–O(H)–Al, but could also be caused by presence of different MeOMe types in the same structure [54].

Now, we will pay more attention to the question concerning the stability of higher MeO_xMe species in mordenite. Their role in the oxidation of CO and application of obtained AE carbonates as the activated forms of carbon dioxide in the reaction with methanol will be discussed.

Computational details: The geometry of a supercell with two primitive cells of Zn-form of mordenite (MOR) with Al in the position T4 on the opposite sides of the large channel was fully optimized using the GULP code [59] and Catlow force field (FF) [60, 61]. One of the 8R windows that opens the side pockets in ZnMOR was isolated from the 3D structure and after that the second Si atom was replaced by Al. Keeping the Zn cation near the 8R window and capping the ruptured T–O bonds of all T atoms (T = Si, Al) by H atoms, a neutral fragment $ZnAl_2Si_6O_8H_{16}$ was obtained. In this fragment Zn was replaced by Ca to get $CaAl_2Si_6O_8H_{16}$. The T–H lengths were further optimized using GAUSSIAN03 [15] at the B3LYP/6-31G* level with fixed O–T–H and T'–O–T–H angles and fixed positions of all other atoms. To determine the reaction coordinate for the CO oxidation, we applied the QST3 algorithm as implemented with GAUSSIAN03 [15]. The influence of the spatial restraints on the resulting geometry and energy for a larger fragment $(CaO_xCa)Al_2Si_6O_{24}H_{16}$ (obtained by addition of a layer of 16 O atoms to the 8R window T atoms) was checked by replacing the capping hydrogens. To add oxygens, the optimized small cluster was embedded in the initial crystallographic position. It was achieved *via* a series of rotations of the small cluster as a whole, *i.e.*, with the fixed relative optimized coordinates. It was performed in order to reach minimal distances between the O atoms in the crystallographic positions and optimized H atoms. The hydrogens in the optimized positions were replaced by the nearest oxygens in crystallographic positions while additional H atoms were added to the new O atoms in the direction of the nearest T atoms. Then, the optimization procedure of the H coordinates was repeated for the smaller cluster. For the FAU, we repeated the same procedure as used for MOR but starting from the $Na_2Al_2Si_6O_9H_{14}$ fragment of the X-ray diffraction (XRD) model of the NaFAU zeolite (Si/Al = 3) including 6R and 4R windows [7]. The Na cations, corresponding to the II and III types, were then replaced by the Ca cations.

In the periodic approach, we optimized the cell parameters and structure of the $MeO_xMe(MOR)$ and $MeO_xMe(PHI)$ moieties (Me=Mg, Ca, with $X = 1$–4) using VASP [30]. The projected-augmented wave (PAW) method [62] was used to describe the electron–ion interactions, and a plane wave basis set was employed for the valence electrons. The plane-wave cutoff was set to 500 eV. Results were obtained with the PW91 generalized gradient approximation (GGA) functional [63]. The Brillouin zone sampling was restricted to the Γ–point. The chemical composition of the unit cell was $MeAl_2Si_{46}O_{96}$ for MeMOR and $Me_2Al_4Si_{12}O_{32}$ for MePHI. The composition of the zeolites with MeO_xMe corresponds to the replacement of Me cations by these species. The cell parameters are given in Table 20.1 of ref. [54]. After optimization of the $MeO_xMe(MOR)$ moiety, we added

Table 20.4 Relative ΔU (kcal/mol) energies of the $Ca_2O_X(8R)$ in MOR and $Ca_2O_X(6R+4R)$ in FAU formation for $X = 2-4$ at singlet states according to reaction (20.7), $Ca\cdots Ca$ and O–O distances (Å), calculated at the B3LYP/6-31G* level and PDFT (PW91) level in Ca_2O_XMOR

X	Type	ΔU		$Ca\cdots Ca$		O–O	
		8R	6R+4R	8R	6R+4R	8R	6R+4R
1	CLU	–	–	3.746	3.946	–	–
	PDFT	–	–	3.849	–	–	–
2	CLU	−24.8, −27.9[a]	−26.5	3.754	3.904	1.556	1.572
	PDFT	−14.0	–	3.762	–	1.537	–
3	CLU	−6.1, −12.7[a]	−8.6	3.661	3.801	1.493, 1.495	1.496, 1.498
	PDFT	1.4	–	3.691	–	1.499, 1.504	–
4	CLU	5.2[b]	−1.7	3.940	3.913	1.340, 1.359, 1.998, 2.498	1.358, 1.362, 1.928, 2.514
	PDFT	−10.2	–	3.819	–	1.359, 1.363, 2.381, 2.112	–

Negative ΔU values of reaction (20.7) correspond to an exothermic effect

[a] At the MP2/6-31G* level; total electronic energy of O_2 is −149.944087 a.u

[b] Slight exothermic effect of −1.1 kcal/mol is observed according to the $Ca_2O_2(8R) + O_2 \rightarrow Ca_2O_4(8R)$ formation

CO_2 and performed the optimization of the entire systems. Vibrational frequencies were calculated using the finite difference method as implemented in VASP. We scaled all frequencies by fitting one band position ν_{cal} to the experimental ν_{exp} value and then re-calculating the positions of other bands by multiplying using a factor of (ν_{exp}/ν_{cal}). Because the frequencies obtained with B3LYP are very similar to the experimental ones, we re-scaled only the frequencies calculated at the periodic level with PW91. Visualization of the results obtained by both cluster and periodic theoretical approaches was realized with the MOLDRAW code [64].

Me$_2$O: The geometries of the CaOCa species are shown in Tables 20.4 and 20.5. Comparing the Ca–O bond lengths between the optimized cluster and periodic models we can confirm that the Ca–O distances are shortened in the periodic model (Table 20.4). A trend was revealed, *i.e.*, a denser attachment of the CaO_XCa species to the framework conserves for all X in the periodic model.

Me$_2$O$_2$: The cluster geometries at the traditional B3LYP level have been discussed in details in refs. [51, 52]. The heat ΔU of the reaction (20.7) is negative for both Mg and Ca clusters allowing their formations. For the discussed question of the formation and properties of MeO_XMe species the finding of the favored multiplicity of each MeO_XMe species is very important. The last one is determined by the sign of the ΔU_{ST} difference between the total energies of the singlet and triplet states ($\Delta U_{ST} = U_S - U_T$), a positive value corresponds to a favored triplet state (Table 20.6). The sign of ΔU_{ST} coincides at both MP2 and B3LYP cluster levels as shown earlier for CaO_XCa species [51].

Table 20.5 Mulliken charges $Q_0^0(e^-)$, bond lengths (Å), and bond angles (°) between Ca atoms and O atoms of the O_X group for $Ca_2O_X(8R)$ in MOR and $Ca_2O_X(6R+4R)$ in FAU calculated at the B3LYP/6-31G* level and PDFT (PW91) level in CaMOR[a]

X	Q_0^0(Ca1), Q_0^0(Ca2)		Ca–O–Ca		Ca–O	
	8R	6R+4R	8R	6R+4R	8R	6R+4R
1	0.913, 0.954	0.777, 0.987	124.48	137.35	2.115, 2.118	2.089, 2.147
1[a]	–	–	151.52	–	1.970, 2.001	–
2	1.069, 1.099	0.910, 1.144	111.27, 111.74	121.4, 121.2	2.244, 2.245, 2.247, 2.236	2.200, 2.275, 2.206, 2.275
2[a]	–	–	121.68, 121.71	–	2.148, 2.160 2.142, 2.165	–
3	1.090, 1.122	0.916, 1.163	103.82, 94.16, 103.96	108.99, 102.37, 109.40	2.312, 2.513, 2.324, 2.336, 2.484, 2.328	2.374, 2.294, 2.503, 2.374, 2.377, 2.280
3[a]	–	–	109.23, 104.02, 110.18	–	2.243, 2.284 2.328, 2.355 2.235, 2.266	–
4	1.145, 1.157	0.949, 1.153	108.33, 100.79, 121.15, 109.38	111.81, 89.13, 111.72, 89.04	2.442, 2.583, 2.274, 2.445, 2.417, 2.530, 2.249, 2.382	2.383, 2.401, 3.142[b], 3.150[b], 2.345, 2.381, 2.327, 2.375
4[a]	–	–	108.86, 103.70, 117.30, 110.68	–	2.326, 2.368 2.405, 2.451, 2.235, 2.237 2.307, 2.336	–

[a] PDFT (PW91) level in CaMOR

[b] The bond is not shown in Fig. 20.15b

Table 20.6 Relative energies (eV) of the singlet (U_S) and triplet (U_T) states of the O_2 molecule as well as of the MOR unit cells including the CaO_XCa species, singlet-triplet energy differences ($\Delta U_{ST} = U_S - U_T$, kcal/mol), ΔU_{ST}^{CL} (kcal/mol) at the B3LYP and PW91 levels of the isolated 8R cluster [51–54], and heats ΔU of the reaction (20.7)[a] (kcal/mol) between singlet states using VASP (PW91). Reproduced from Ref. [54] with kind permission of Elsevier

Species	ΔU_{ST} PW91	ΔU_{ST} PBE	ΔU_{ST}^{CL} PW91	ΔU_{ST}^{CL} B3LYP	ΔU
O_2	23.65	25.14	41.37	33.12	–
CaO_2Ca	−0.37	−0.27	−26.2	−31.2, −41.9[b]	−14.04
CaO_3Ca	0.46	0.31	−21.1	−14.2	1.38
$CaOCa + O_2$	8.62	10.00	–	–	19.67
CaO_4Ca	0.02	0.06	−8.4	–[c]	−10.23

[a] The steps for reaction (20.7) at $X = 2-4$ for Ca are illustrated in Fig. 20.13
[b] MP2/6-31G* level
[c] No SCF convergence while using B3LYP/6-31G* for triplet states

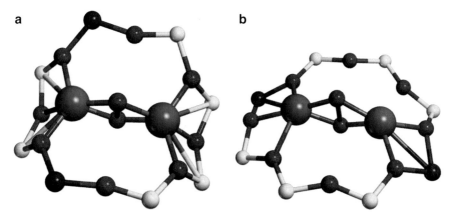

Fig. 20.12 Optimized structures of the singlet MgO_2Mg moiety in (**a**) PHI and (**b**) 8R moiety of MOR as calculated at the B3LYP/6-31G* level. The color code is: O in *red*, Si in *green* (*smaller spheres*), Al in *yellow*, H in *grey*, Mg in *dark cyan*

The difference between the singlet and triplet geometries of $CaO_2Ca(8R)$ resembles the geometries of the CuO_2Cu models [49, 50] with one O–O bond present in (μ-$\eta2$:$\eta2$-peroxo)dicopper. Their ΔU_{ST}^{CL} values (absolute) remain large for $X = 2$ (Table 20.6) and decrease for the expanded $CaO_2Ca(8R)$ cluster. Hence, we consider only small periodic ΔU_{ST} values as reliable ones for the evaluation of close relative stabilities between the singlet and triplet states.

For comparison, we present the optimized geometries of the MgO_2Mg species in the PHI zeolite and in the 8R cluster of MOR in Fig. 20.12. Even if the 8R window looks more distorted in PHI *versus* the one in 8R, the angles and distances in the MgO_2Mg moiety are almost the same for both $MgO_2Mg(PHI)$ and $MgO_2Mg(8R)$

Table 20.7 Relative and absolute energies (eV) of the singlet (U_S) and triplet (U_T) states of the MOR unit cell including the MgO_XMg species, singlet-triplet energy differences ($\Delta U_{ST} = U_S - U_T$, kcal/mol), and ΔU heats for reaction (20.7) (kcal/mol) between singlet states using VASP (PW91) level. Reproduced from Ref. [54] with kind permission of Elsevier

Species	U_S	U_T	ΔU_{ST}	ΔU
MgOMg	−1157.017	−1157.014	−0.07	–
MgO_2Mg	−1162.393	−1162.392	−0.02	−14.04

complexes. Details about their geometries are given in ref. [53] where it was shown that MgO_XMg clusters possess ground singlet states for $X = 1$–2 and ground triplet states for $X = 3$–4 at the level of the isolated 8R cluster. At the periodic level the Ca–O distances shorten in the $CaO_2Ca(MOR)$ species while the Ca$\cdots$Ca distance are enlarged. It corresponds to a tighter contact between the CaO_2Ca species and the MOR framework in the periodic model. The CaO_2Ca is the unique case, for which the O–O distance is shorter in the opposite direction in the periodic model than for the upper clusters with $X = 3$ and 4. We got a negative value ($\Delta U = −9.01$ kcal/mol/cell[1]) for $X = 2$. This suggests the thermodynamic possibility of the $MgO_2Mg(PHI)$ formation at the periodic level *via* the same reaction equation as described by Eq. 20.7. The ΔU_{ST} difference is indeed very small for MgOMg and MgO_2Mg suggesting clearly that the MgO_XMg species in singlet states can interact with CO_2 and form carbonates. All the carbonate species studied herein possess a singlet electronic ground state. Hence, their formation is allowed owing to the change of the total spin in the course of the interaction between MgO_XMg and CO_2. The results of periodic calculations (Tables 20.6 and 20.7) show that the MgO_3Mg and MgO_4Mg moieties are characterized by positive ΔU values, *i.e.*, 21.10 and 21.68 kcal/mol/cell, respectively. This means that they are not thermodynamically feasible at standard conditions. Moreover, both MgO_3Mg and MgO_4Mg species are less stable than the $MgO_XMg + O_2$ pairs for $X = 1$ or 2, respectively, and thus they have to decompose. The oxidation activity of Mg-zeolites can be related with the MgO_2Mg formation.

Me_2O_3: Compared with the MeOMe and MeO_2Me clusters of transition metals in zeolites (already discussed in the literature [34–42]), the MeO_3Me cluster is of a new type, for which the relation to the problem of $^1O_2$ desorption is straightforward. The reaction (20.7) for Me $=$ Ca is exothermic for all X values ($\Delta U < 0$) with the exception of $X = 3$. For $X = 3$ an endothermic effect of 1.38 kcal/mol (Table 20.6 and Fig. 20.13) was obtained. For the reaction $^1CaOCa(MOR) + {}^3O_2 \rightarrow {}^1CaO_3Ca(MOR)$, we observed an exothermic heat of $(−14.04 + 1.38) = −12.66$ kcal/mol (Fig. 20.12). In opposite to the Ca case, the

---

[1]−4.51 kcal/mol (Table 3.3) or −9.01 kcal/mol/cell $= (−413.71 + 403.543 + 9.766) \times 23.05$ regarding two MgO_XMg species per cell.

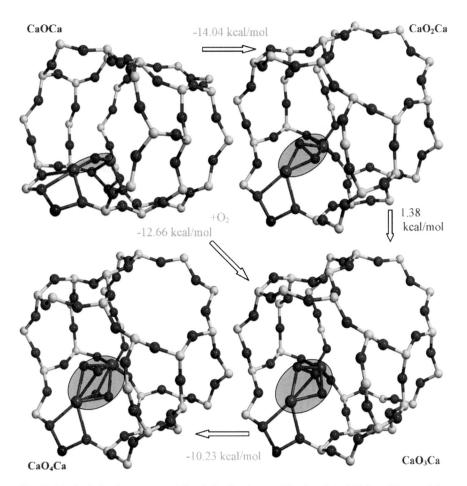

Fig. 20.13 Optimized structures of the CaO$_X$Ca clusters ($X = 1 - 4$) in MOR and heats of the CaO$_{X-1}$Ca + ½ ^{3}O$_2$ → CaO$_X$Ca reactions (kcal/mol, shown near *arrows*) as calculated at the PDFT/PW91 level with VASP. The CaO$_X$Ca moieties are shown by the ellipses. The color code is the same as in Fig. 20.12, Ca in *dark cyan*

heats ΔU of the reaction (20.7) at $X = 3$ are negative for the Me$=$Sr and Ba clusters. The ozone like O$_3$ part of the Ca$_2$O$_3$ cluster has an angular O–O–O moiety of A-letter shape with the close O–O distances (1.493 Å, 1.495 Å) and O–O–O angle of 105.56° calculated at the B3LYP/6-31G* level. The A-shape O$_3$ part for Ca$_2$O$_3$ is turned perpendicularly to the Ca$\cdots$Ca axis and symmetrically coordinated to both Ca cations (Fig. 20.14a). Two additional "inverted" and "rotated" isomers with local minimum for Ca$_2$O$_3$ were observed (Fig. 20.14b, c) but they are less stable than symmetric form (Fig. 20.14a). The O$_3$ moiety keeps its O–O bonds and O–O–O angle in the inverted form (less stable by 17.7 kcal/mol at the B3LYP/6-31G* level, both Ca atoms are symmetrically oriented towards two O atoms of one O–O side of the O$_3$ part, Fig. 20.6c) and the "rotated" form (less stable by 9.7 kcal/mol,

Fig. 20.14 The CaO$_3$Ca
structures in MOR optimized
at the B3LYP/6-31G* level.
The most (**a**) stable, (**b**)
"inverted", and (**c**) "rotated"
isomers. The color code is the
same as in Fig. 20.13

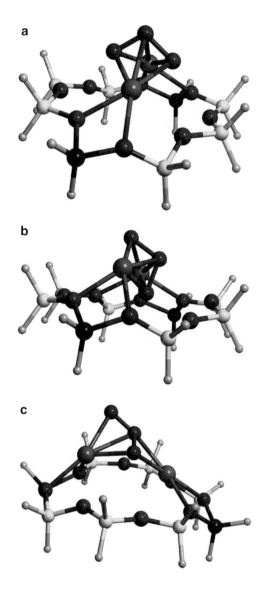

Fig. 20.14b). The rotated Ca$_2$O$_3$ conformer with the O–O distances of 1.497 and
1.504 Å and O–O–O angle of 105.34° corresponds to an approximate rotation by
90° of the symmetric form (Fig. 20.14a) around the Ca$\cdots$Ca axis. Different initial
models were tested and all led to the same conformers. The invariance of the O–O
distances and O–O–O angle of the O$_3$ geometry confirms the stability of the Ca$_2$O$_3$
cluster.

Small periodic ΔU_{ST} values were obtained for Ca$_2$O$_3$ (Table 20.6). A decrease of
the energy differences ΔU_{ST} between the singlet and triplet complexes containing
O$_2$ species was already discussed in the literature [55, 65]. A precise value of small

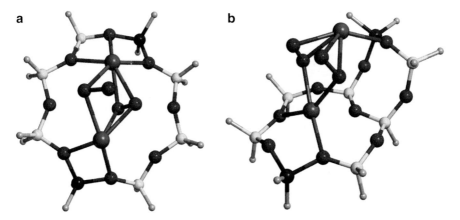

Fig. 20.15 The CaO$_4$Ca structures in (**a**) MOR (8R) and (**b**) FAU (6R+4R) optimized at the B3LYP/6-31G* level. The color code is the same as in Fig. 20.13

ΔU_{ST} separation calculated using the MR-CISD method for the Li$^+$–O$_2$ complex is difficult to evaluate from the illustration in ref. [55]. The molecular HO$_2$$^+$ cation is much better characterized by the CCSD(T) level and a complex combination of accurate basis sets up to 5Z quality. This gives a ΔU_{ST} value of 4.45 ± 0.15 kcal/mol for the equilibrium geometry [65]. It is much smaller compared with the one for the O$_2$ gas state (from 23.65 to 41.37 kcal/mol in the upper row of Table 20.6). Decrease of the barrier between the singlet and triplet O$_2$ states to 4.8 kcal/mol was calculated over the BaO surface [66]. One should note that the sizes of both cluster and periodic models are large enough compared with the Li$^+$–O$_2$ and HO$_2$$^+$ complexes. Hence, the main contribution to the ΔU_{ST} value comes from the electron correlation [55, 65].

Ca$_2$O$_4$. Stable geometries for the Ca$_2$O$_4$ cluster were found for both 8R in MOR and 6R+4R in FAU fragments (Fig. 20.15). The reaction (20.7) for Me$=$Ca is exothermic for $X = 4$ with the heat of -10.23 kcal/mol. The O$_4$ moiety presents a plain trapezium form based on two sides of the O$_2$ molecules with bond lengths of 1.359 and 1.360 Å in MOR, and 1.358 and 1.362 Å in FAU. They are strongly elongated *versus* the experimental gas value $r_e = 1.201$ Å [67] or theoretical value 1.214 Å obtained at the B3LYP/6-31G* level. The lengths of the parallel sides of the trapezium equal to 1.998 and 2.498 Å for MOR, and to 1.928 and 2.514 Å for FAU. For MOR, the O–O axes are located almost in a same plane perpendicular to the Ca$\cdots$Ca axis. The plane is slightly shifted toward Ca$_2$ resulting in an average difference between the Ca$_1$–O and Ca$_2$–O distances of 0.03–0.06 Å (Fig. 20.15a). For FAU, the shift of the Ca$_2$O$_4$ plane towards the upper Ca cation is similar for two lower O atoms, *i.e.*, 0.03–0.07 Å. But it is more emphasized, *i.e.*, 0.8–0.9 Å, for two upper O atoms (Fig. 20.15b). Distances of 1.998 and 1.928 Å of the central O–O bond (upper parallel side of the O$_4$ trapezium for 8R and 6R+4R, respectively (Table 20.4)), are smaller than the sum of the covalent or ionic radii for

O, *i.e.*, 1.21 and 1.35 Å [68]. This short length leads to unusual O–O–O–O chains stabilized in the zeolite space. Characteristic vibrations of the $Ca_2O_4(8R)$ at 1129.4, 1032.9, 556.8, and 431.9 cm^{-1} computed at the B3LYP/6-31G* level also clearly manifest a unique Ca_2O_4 group and not two separate O_2 species with the O–O frequency of 1657 cm^{-1} at the same B3LYP/6-31G* level. The upper symmetric valence vibration at 1129.4 and 1032.9 cm^{-1} can be compared to 1254, 1224, 1112, and 1067 cm^{-1} of the isolated tetragonal puckered O_4 structure computed at the CISD/DZ+P level [69]. The vibrations with the lower frequencies (351 cm^{-1} for the isolated O_4 species [69]) are strongly coupled to the framework ones. Finally, only four-member O_4 ring as Ca_2O_4 was optimized. Together with cyclic O_4 species [69], the larger O_8 [70] molecule has been discussed as an analogy to known sulfur rings. The relatively small 8R window size can be the reason for the nonobservance of larger Ca_2O_8 ring.

No qualitative deviations between the cluster and periodic models were calculated for the Ca_2O_4 species in the MOR zeolite (Table 20.4). Compared to smaller species at $X = 1$, 2, and 3 the Ca$\cdots$Ca distance decreases from the cluster model to periodic one in an opposite trend. The "periodic" Ca_2O_4 structure is less distorted, distances are closer to each other (O–O bond lengths are elongated compared to the cluster distances (Table 20.4)), with one exception of the longest O–O distance. Another permanent trend observed for the Ca_2O_4 cluster and all lower homologues is shortening of the Ca–O bond lengths for the periodic model showing a closer coordination to zeolite. These variations are essential (greater than or equal to 0.1 Å) so that it can lead to change of the reactivity. But understanding of this phenomenon requires more detail investigation.

20.3.1   Reactivity of the Me_2O_X Clusters in the AE Zeolites

20.3.1.1   CO Oxidation

In this Chapter the thermal activation of the reagents for large variety of the oxidation reactions observed over zeolites will be discussed . There are two reasons for the selection of CO oxidation process for testing our Me_2O_X system was chosen:

$$MeO_XMe + CO \rightarrow MeCO_{X+1}Me \tag{20.8}$$

First, the CO oxidation in exhaust gases remains an important topic from application and theoretical point of view. Second, such CO oxidation over the AE forms was observed experimentally [71]. The oxidation activity of the AE forms take an intermediate position between that of the transition metal forms and rare earth metal forms on one hand, and alkali metal forms, on the other hand (if one discusses thermally stimulated oxidation only).

It is important to note that the reactions (20.7 and 20.8) keep the entity of the MeOMe chain (both Me–O bonds). This happens also during the reactions with

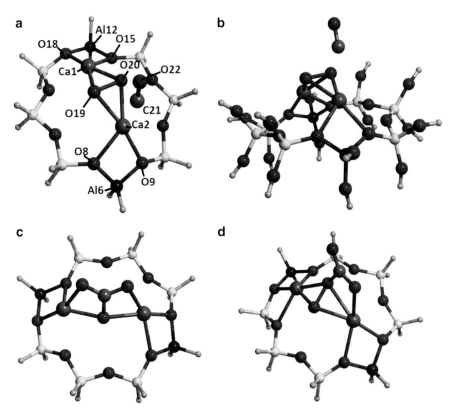

Fig. 20.16 Optimized structures of the (**a, b**) reagents and (**c, d**) products of CO oxidation over the CaO$_X$Ca(8R) moieties in MOR, if (**a, c**) $X = 2$ and (**b, d**) $X = 3$, as calculated at the B3LYP/6-31G* level. The color code is the same as in Fig. 20.13, C in *brown*

water [54] and carbon dioxide [55]. It illustrates the stability of this MeOMe unit and possibility of inverse process for each of the reactions (it shows especially upon heating for CO_2 (see below) and H_2O).

Ca$_2$O$_2$(8R): Ca$_2$O$_2$(8R) in MOR holds its geometry in the reaction complex (Fig. 20.16a). This explains the moderate heat of adsorption for physisorbed CO (Table 20.8). The reaction coordinate corresponds to the vibrations of CO towards one O atom of the Ca$_2$O$_2$ group. This motion is easily partitioned due to the large reduced mass, *i.e.*, 14 a.m.u., *versus* other 15 imaginary frequencies whose reduced masses range between 1 and 9 a.m.u. and describe the degrees of freedom of the fixed H atoms. Imaginary frequencies calculated at the B3LYP/6-31G* level along the reaction coordinates were $674i$, $695i$, and $750i$ cm^{-1} for the –Al–O–(Si–O)$_m$–Al–($m = 1$–3) 8R window, respectively. The computations with larger 8R model (Fig. 20.16b) slightly shift the activation barrier (*i.e.*, 33.2 kcal/mol instead of 35.7 kcal/mol in Table 20.9). This results in minor changes of the cluster geometry [51].

Table 20.8 $\Delta E^{\#}$ activation barriers of the $Ca_2O_X(6R+4R) + CO \rightarrow Ca_2CO_{X+1}(6R+4R)$ oxidation reaction, imaginary frequencies ω, ΔE_{CO} heat of CO adsorption, ΔU heat of oxidation reaction, and ΔE_{CO2} heat of CO_2 adsorption, on $Ca_2O_X(6R+4R)$ clusters of FAU calculated at the B3LYP/6-31G* level. Reproduced from Ref. [52] with kind permission of Wiley

Values	X = 2	X = 3
$\Delta E^{\#}$	36.4[a]	34.2[a]
$i\omega$	−756	−833, −793[b], −962[c]
ΔE_{CO}	15.2	13.6
ΔU	101.4	72.4
ΔE_{CO2}	72.6	24.1

All energy values are in kcal/mol, frequencies are in cm^{-1}
[a]Related to CO molecule adsorbed at the upper Ca cation
[b]B3P86
[c]PW91PW91

Table 20.9 $\Delta E^{\#}$ activation barriers of the $Ca_2O_X(8R) + CO \rightarrow Ca_2CO_{X+1}$ (8R) oxidation reaction, ΔE_{CO} heat of CO adsorption, ΔU heat of oxidation reaction, and ΔE_{CO2} heat of CO_2 adsorption, on small (S) or large (L) $Ca_2O_X(8R)$ clusters of MOR calculated at the B3LYP level and 6-31G* basis sets for all framework atoms, and 6-31G* or LANL2DZ for Ca cations. Reproduced from Ref. [52] with kind permission of Wiley

Energy values	X = 2			X = 3	
	S/ps	S/6-31G*	L/6-31G*	S/ps	S/6-31G*
$\Delta E^{\#}$	31.5	35.7	33.2	29.4	32.4, 19.7[a]
ΔE_{CO}	12.7	13.4	13.0	11.5	12.4, 15.1[b]
ΔU	108.1	106.9	108.2	73.5	72.3, 78.9[a], 68.9[b]
ΔE_{CO2}	74.4	74.5	75.1	22.4	20.4, 25.7[b]

All values are in kcal/mol
[a]At the PW91PW91/6-31G* level
[b]at the MP2/6-31G* level

Detailed analyses of the geometrical parameter and atomic charge variations for the CO and cluster atoms were performed for three steps of the CO oxidation (Table 20.10). Comparison of the bond lengths reveals a slight shortening of the Al–O and Si–O bonds, around 0.03–0.04 Å and 0.01–0.02 Å, respectively. Variations of the Si–O–Al angles do not exceed 1 degree. A small window relaxation seems to correlate with small structural changes of the adsorption complex. The maximal variation of the torsional Ca–O–O–Ca angle is 5.5°, while the Ca–O bonds vary by less than 0.01–0.02 Å, with the exception of 0.06 Å for the Ca_1–O_{18} case.

It is interesting to trace the charge redistribution of the $CO \cdots Ca_2O_2(8R)$ reactive part along the REA $\rightarrow$ TS $\rightarrow$ PRO sequence (Table 20.10). We will consider the values related to the large model as the most accurate one (even though difference *versus* data for the small model is minor). The total charge of the two Ca cations rises by $0.07\,e^-$, while the total charge of CO and two O19 and O20 atoms (which form together the CO_3^{8-} ion), decreases by the same $0.07\,e^-$ value. We would like to mention that this variation is by $0.07\,e^-$ smaller than the changes of −0.134 and

Table 20.10 Mulliken atomic charges Q_0^0 (e^-) and geometrical parameters of the reagents (*REA*), transition states (*TS*), and products (*PRO*) of CO oxidation reaction over the small and large $Ca_2O_2(8R)$ complex of the MOR type (distances in Å, angles in degrees) calculated at the B3LYP/6-31 G* level (atomic labels are shown in Fig. 20.16a). Reproduced from Ref. [52] with kind permission of Wiley

Parameter	REA		TS		PRO	
	Large	Small	Large	Small	Large	Small
$Q_0^0(Ca1)$	1.101	1.114	0.995	1.021	1.036	1.058
$Q_0^0(Ca2)$	0.958	0.965	1.027	1.033	1.090	1.088
$Q_0^0(O19)$	−0.567	−0.572	−0.594	−0.601	−0.632	−0.629
$Q_0^0(O20)$	−0.569	−0.575	−0.363	−0.361	−0.755	−0.768
$Q_0^0(C21)$	0.280	0.278	0.349	0.338	0.839	0.854
$Q_0^0(O22)$	−0.243	−0.220	−0.482	−0.460	−0.630	−0.631
Ca2–O9	2.349	2.368	2.318	2.338	2.300	2.300
Ca2–C21	2.686	2.695	2.665	2.629	2.692	2.687
Ca2–O8	2.397	2.394	2.437	2.414	2.420	2.408
Ca2–O19	2.265	2.254	2.251	2.252	3.814	3.923
Ca2–O20	2.274	2.268	2.964	2.970	2.259	2.259
Ca2–Ca1	3.671	3.754	3.847	3.936	4.195	4.302
O19–O20	1.552	1.556	1.857	1.851	2.237	2.239
Ca1–O20	2.263	2.258	2.428	2.405	2.253	2.259
Ca1–O19	2.263	2.252	2.216	2.217	2.376	2.348
C21–O20	2.702	2.781	1.297	1.304	1.359	1.363
C21–O22	1.139	1.136	1.228	1.226	1.275	1.275
C21–Ca1	4.635	4.687	3.564	3.527	2.677	2.681
C21–O19	2.583	2.657	2.605	2.603	1.276	1.273
Ca1–O15	2.339	2.339	2.323	2.321	2.354	2.347
Ca1–O18	2.393	2.455	2.354	2.410	2.333	2.389
Ca1–O14	2.546	2.534	2.572	2.552	2.755	2.663
Al6–O8	1.744	1.782	1.745	1.785	1.745	1.785
Al6–O9	1.744	1.778	1.745	1.778	1.745	1.778
Al12–O18	1.763	1.793	1.764	1.796	1.760	1.819
Al12–O15	1.780	1.814	1.787	1.821	1.785	1.795
Si4–O8	1.612	1.625	1.617	1.629	1.620	1.632
Si5–O9	1.601	1.621	1.601	1.622	1.598	1.619
Si13–O18	1.592	1.608	1.595	1.612	1.596	1.612
Si11–O15	1.611	1.628	1.617	1.634	1.616	1.632
Ca2–C21–O22	162.1	161.3	141.8	140.7	61.6	60.6
O20–Ca2–O19	40.0	40.2	38.8	38.5	29.3	29.2
O20–Ca1–O19	40.1	40.3	46.9	47.0	57.8	58.1
Al6–O8–Si4	130.6	131.2	129.4	130.1	129.3	129.6
Al6–O9–Si5	135.1	134.4	135.7	135.2	136.7	136.4
Ca1–O20–O19–Ca2	119.1	124.6	119.8	124.7	130.6	139.3

0.045 e^- of the Ca charges obtained due to the CO adsorption (the reagent complex). Absolute values of the O charges rise from the REA to the PRO stages, *i.e.*, by 0.4 e^- for O22. However, the total charge δ of the CO_3^{8-} ion is −1.099, −1.090, and

$-1.178\,e^-$ for the REA, TS, and PRO stages, respectively, for the large model. The total charge of two Ca cations also varies only a little, *i.e.*, 2.069, 2.022, and 2.096 e^- in the REA, TS, and PRO reaction steps. Even in the case of more essential charge variation of Ca2 (*i.e.*, 0.958, 1.027, and 1.090 e^-) the corresponding bond lengths and angles change insignificantly relative to the ones of Ca1. These minor variations of the total charge of the $CO \cdots Ca_2O_2$ group between the reaction steps suggest a moderate or small influence of the EP on the stabilization of the TS relative to the stabilization of the reagents and products. However, the redistribution of the charge values over the O atoms signifies a possible effect of the electric field (EF) and its gradient. This effect is of lower order than the usual stabilization effect due to the EP.

The reaction products of the CO oxidation over $Mg_2O_2(8R)$ can be optimized as the "co-existent" physisorbed CO_2 and chemisorbed CO_3^{2-} species. The CO_3^{2-} species are found to be more stable by 28.80 kcal/mol. For other AE forms (Ca, Sr, Ba) no physisorbed CO_2 was stabilized with the cluster approach. This is probably related to the absence of reaction barrier.

$Ca_2O_3(8R)$: The favored TS geometry for the $CO \cdots Ca_2O_3(8R)$ reaction complex in MOR corresponds to the CO coordination near Ca (Fig. 20.16a). The most intriguing part, while searching the reaction coordinate for the CO oxidation over $Ca_2O_3(8R)$, was the choice between the extreme or central O atom of the Ca_2O_3 group to form CO_2. Such problem does not exist for $Ca_2O_2(8R)$, which have two oxygens in approximately symmetric positions. Finally, either of the two extreme O atoms of the O_3 group can be involved into the CO_2 formation. The TS geometry turns out to be very similar to the case of CO oxidation over $Ca_2O_2(8R)$.

Different geometries of the products of the CO oxidation over $Ca_2O_2(8R)$ and $Ca_2O_3(8R)$ determine the larger heats of reaction and of CO_2 adsorption for $Ca_2O(8R)$ than for $Ca_2O_2(8R)$. We assigned this increase to the strong tetra-dentate CO_2 chemisorption over $Ca_2O(8R)$ resulting in the CO_3^{2-} anion (Fig. 20.15c). Spatial restrictions do not allow the same CO_2 coordination for $Ca_2O_2(8R)$ (Fig. 20.16d). These observations reveal different heats for each reaction step of the CO_2 desorption. The $\Delta E^\#$, ΔE_{CO}, ΔE_{CO2}, and ΔU variations (Table 20.8) follow the same trends as observed for the CO oxidation over $Ca_2O_2(8R)$ (Table 20.8). The MP2 computations confirm the ΔE_{CO}, ΔE_{CO2}, and ΔU values with the same basis set (Table 20.9). The data obtained at the MP2 level coincide with those at the periodic level with respect to the absence of the transition state for the CO oxidation over $Ca_2O_3(8R)$. The respective tri-dentate $CaCO_4Ca(8R)$ product of the modeling at the cluster level possesses the frequencies that do not correspond to the experimental ones. We suppose that this coincidence between the MP2 and periodic solution denotes more complex mechanism of the reaction between CO and Ca_2O_3.

$Ca_2O_2(6R+4R)$ and $Ca_2O_3(6R+4R)$: The data for the CO oxidation over two ($X = 2$ and 3) complexes in CaFAU are presented together due to their similarities relative to the data obtained for the $Ca_2O_X(8R)$ of CaMOR. A difference appears relative to the non-equivalent Ca cationic positions in all $Ca_2O_X(6R+4R)$ clusters. The favored CO location at the upper Ca cation is accompanied by an energy gain of 6.2 and 5.7 kcal/mol for $X = 2$ (Fig. 20.17a) and $X = 3$ (Fig. 20.17b), respectively. The same effect of different mono- or bi-dentate CO_2 chemisorption appears in

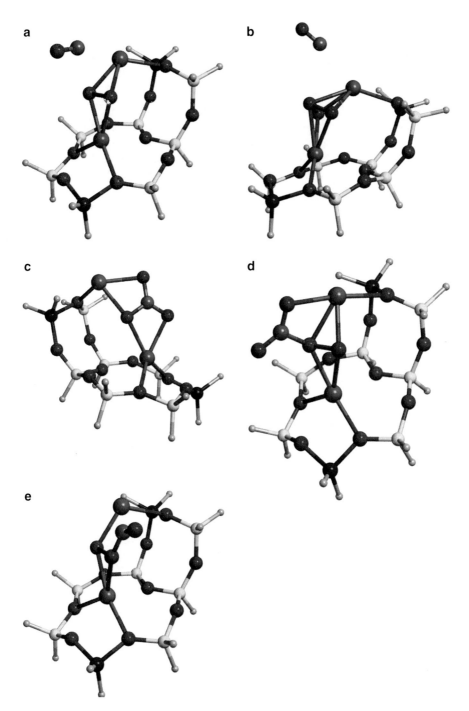

Fig. 20.17 Optimized structures of the (**a, b**) reagents, (**c, d**) products, and (**e**) TS of CO oxidation over the CaO$_X$Ca(6R+4R) moieties in FAU, if (**a, c**) $X = 2$ and (**b, d, e**) $X = 3$, as calculated at the B3LYP/6-31G* level. The color code is the same as in Fig. 20.16

Table 20.11 Distances (Å) and angles ($^\circ$) for the reaction complex (*REA*), transition state (*TS*) complex, and products (*PRO*) of CO oxidation over Mg_2O_2 (Eq. 20.8) calculated with PDFT (GGA) in MgPHI and MgMOR zeolites and with the cluster approach (*CLU*) at the B3LYP/6-31G* level in MgMOR (using the 8R window). Reproduced from Ref. [53] with kind permission of Elsevier

Parameter	REA			TS			PRO		
	PHI	MOR	CLU	PHI	MOR	CLU	PHI	MOR	CLU
$Mg_{42}\cdots Mg_{104}$	3.805	3.517	3.473	3.654	3.680	3.673	3.893	3.937	3.950
$Mg_{42}-O_{96}$	1.949	1.914	1.909	1.909	1.914	1.905	1.956	1.965	1.975
$Mg_{42}-O_{98}$	**2.570**	**1.974**	**1.934**	2.600	2.606	2.608	3.626	3.609	3.615
$Mg_{104}-O_{96}$	1.892	1.910	1.896	1.862	1.885	1.865	1.940	1.980	1.978
$Mg_{104}-O_{98}$	1.976	1.932	1.917	1.966	2.031	2.055	2.015	2.011	2.005
$O_{96}-O_{98}$	**1.528**	**1.592**	**1.608**	1.852	1.881	1.864	2.237	2.242	2.225
$Mg_{42}-C_2$	**2.201**	**2.331**	**2.417**	2.284	2.230	2.254	2.355	2.345	2.355
C_2-O_{40}	**1.210**	**1.139**	**1.131**	1.233	1.219	1.214	1.282	1.280	1.270
$C_2\cdots O_{98}$	**1.482**	**2.555**	**3.220**	1.302	1.351	1.322	1.278	1.274	1.269
$Mg_{42}-O_{96}-Mg_{104}$	**164.2**	**133.80**	**131.8**	151.4	151.27	153.8	176.7	172.67	175.3
$Mg_{42}-O_{98}-Mg_{104}$	113.0	128.47	128.8	105.4	104.40	103.3	81.9	83.78	84.0
$O_{96}-Mg_{42}-O_{98}$	36.3	48.30	49.5	45.4	46.12	45.7	32.5	33.26	32.5
$O_{96}-Mg_{104}-O_{98}$	46.5	48.95	49.9	57.8	57.28	56.6	68.9	68.36	68.8

Atomic numbers are shown in Fig. 20.19d

FAU between the products of CO oxidation over the $Ca_2O_2(6R+4R)$ (Fig. 20.17c) and $Ca_2O_3(6R+4R)$ (Fig. 20.17d) moieties. Higher heat of reaction for $X = 2$ is also assigned to the tetra-dentate CO_2 chemisorption ($\Delta E_{CO2} = 72.6$ kcal/mol, Fig. 20.17c) *versus* the tri-dentate one ($\Delta E_{CO2} = 24.1$ kcal/mol, Fig. 20.17d), respectively (Table 20.8). Vibration coordinates for the TS calculated with B3LYP/LANL2DZ(Ca)/6-31G*(Al, Si, O, H) correspond to the vibrations of CO and one O atom with imaginary frequencies of $668i$ and $802i$ cm^{-1} for $X = 2$ (Fig. 20.17e) and $X = 3$, respectively. Smaller number of 4–6 imaginary frequencies of the isolated $Ca_2O_X(6R+4R)$ cluster suggests a less important perturbation of the initially optimized T-H lengths in the course of the optimization as compared to the one for $Ca_2O_X(8R)$ that possesses 15 or 16 imaginary frequencies. The vibration coordinates for TS when $X = 3$ are very similar obtained using the B3LYP, B3P86, and PW91PW91 functionals and 6-31G* basis set, even if the frequencies differ substantially (*i.e.*, $833i$, $793i$, and $962i$ cm^{-1}, respectively (Table 20.8)).

We computed the reaction pathway for the reaction (20.8) using both the periodic and cluster approach. Reagent, transition state (TS), and product geometries are given in Table 20.11. Results obtained with GGA are similar to those obtained with LDA. Thus, only results obtained with GGA are given in the Table 20.11. The computations led to a fact that the atoms of the whole $MgCO_3Mg$ complex were located almost in one plane during the reaction. The TS geometry is very close to the one of the reagent. The main difference between both geometries is the $O_{96}-O_{98}$ distance. This O–O distance is an important parameter for the oxidation activity of the MgO_2Mg moiety; it varies between 1.528 Å (PDFT) and 1.608 Å

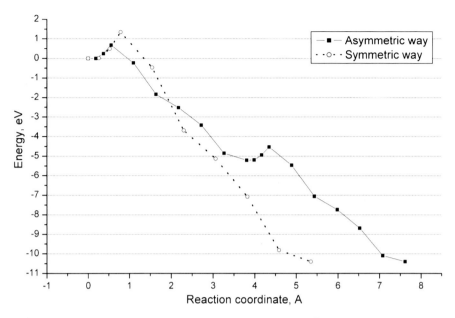

Fig. 20.18 Energy profiles (eV) along the reaction coordinate (Å) for CO oxidation at the MgO$_2$Mg cluster in MgPHI calculated at the PDFT/GGA level, 1 eV = 23.05 kcal/mol, for symmetric P2$_1$/m (*circles*) and asymmetric P1 (*squares*) cell groups. Reproduced from Ref. [53] with kind permission of Elsevier

(CLU) for the reaction complex and between 1.852 Å (PDFT) and 1.864 Å (CLU) for the transition state.

The heat of reaction (20.8), calculated at the PDFT level, is 121.6 (LDA) and 119.9 kcal/mol (GGA). It is larger by 10% *versus* the heat of reaction (109.6 and 111.0 kcal/mol) calculated with GAUSSIAN03 for the extended MgMOR fragment at the B3LYP/6-31G* and 6-311G** levels. For MgY, the heat of reaction was similar, *i.e.*, 105.1 kcal/mol. The evaluation of the CO$_2$ chemisorption energy gives −48.0 kcal/mol per CO$_2$ at the cluster level. One also has to emphasize difference between the carbonate formation over the CaO$_X$Ca and MgO$_X$Mg species. The CO$_2$ product participates in a barrierless formation of a mono-coordinated carbonate anion at the CaO$_2$Ca species, while CO$_2$ remains physisorbed over one of Mg atom of MgO$_2$Mg. For MeOMe, a tetra-dentate carbonate anion appears for both Me=Ca and Mg cations. Finally, the respective coordination number should be tetra to partition the products of CO$_2$ reaction with the CaO$_2$Ca and CaO$_3$Ca species. In such a way, the number of the Ca$\cdots$O bonds counts (not the number of the cations coordinated). The coordination number 2 has to be used when two O atoms of carbonate anion are coordinated to one cation. But this case will not be considered here.

Two parallel reactions take place simultaneously in one PHI cell with P2$_1$/m cell symmetry (Fig. 20.18, circles). The CO oxidation activation energy obtained with PDFT (*i.e.*, 15.3 and 15.6 kcal/mol per one MgO$_2$Mg moiety with LDA and GGA, respectively), is much smaller than the cluster result of 35.5 kcal/mol calculated

Table 20.12 Bader type atomic charges (QB, e) calculated with PDFT (GGA) in MgPHI and Mulliken charges (QM, e) for the extended $CaO_2Ca(8R)$ cluster at the B3LYP/6-31G* level [52] for the reaction complex (*REA*) and symmetric transition state (*TS*) complex of CO oxidation (20.8). Reproduced from Ref. [53] with kind permission of Elsevier

Atom	Me = Mg [53]			Me=Ca [52]		
	QB(REA)	QB(TS)	QB(TS)-QB(REA)	QM(REA)	QM(TS)	QM(TS)-QM(REA)
C_2	0.906	1.192	0.286	0.280	0.349	0.069
O_{40}	−1.189	−1.111	**0.078**	−0.243	−0.482	**−0.239**
O_{98}	−0.660	−0.889	**−0.229**	−0.569	−0.363	**0.206**
O_{96}	−0.738	−0.834	−0.096	−0.567	−0.594	−0.027
Me_{42}	1.650	1.651	0.001	0.958	1.027	0.069
Me_{104}	1.681	1.681	0.000	1.101	0.995	−0.106

Atom labels are shown in Fig. 20.19d. Reproduced from Table 20.5 in [53] with kind permission of Elsevier

for the extended MgMOR fragment and 34.9 kcal/mol for the MgY one (6R+4R), with B3LYP and 6-31G* or 6-311G** basis sets. While without any symmetry (Fig. 20.18, squares), *i.e.*, P1, we observed nearly the same activation barriers, *i.e.*, one half of the barrier value for the symmetric reactions, for the two consequent reactions in one MgPHI cell containing two MgO_2Mg species. It shows a negligible synergism between the two reactions in one PHI cell, the variation of the activation energy being around 0.13 kcal/mol. The imaginary frequencies were estimated as $638i$ and $704i$ cm^{-1} for the first and second TS along with the asymmetric reaction coordinates. Similar values around $750i$ and $756i$ cm^{-1} were estimated as well for the isolated $CaO_2Ca(8R)$ and $CaO_2Ca(6R+4R)$ cluster models [51].

The role of electrostatic effects was elucidated with the charge analyses. In Table 20.12, we compared the atomic charge variations calculated with the Mulliken scheme for the cluster approach (Table 20.10) and Bader scheme for the periodic approach. It shows that the charge redistribution between the reaction complex and TS is the reason of a drastic fall of the activation energy obtained with the periodic approach. However, even if the total charges of the TS complexes remain small, the charge distributions differ with respect to the O_{98} and O_{40} atoms (Table 20.12, bold values), *i.e.*, the parts of the "forming" CO_2 molecule. The charge variations of CO_2 fragment (0.135 e) and O_{96} atom (−0.096 e) correspond to more ionic TS complex for MgO_2Mg(PHI) than respective changes of 0.036 e and −0.027 e in $CaO_2Ca(8R)$. The same less ionic TS type was calculated using Mulliken scheme for $CaO_2Ca(6R+4R)$ with the minor changes of 0.029 e for the CO_2 fragment and of −0.047 e for O_{96} atom (not shown in Table 20.12). These charge variations lead to more polar TS complex. They can not be reflected in the energy variations between the REA and TS by the cluster model without long-range terms. This difference is the reason of different activation energies, *i.e.* 35 kcal/mol with the cluster and 15.3 kcal/mol with periodic models.

A coherence between most of the geometric parameters obtained by the isolated cluster calculations (Fig. 20.19a–c) and PDFT (Fig. 20.19d–f) can be noted for each reaction step (Table 20.11). To understand the reasons of the lowest activation

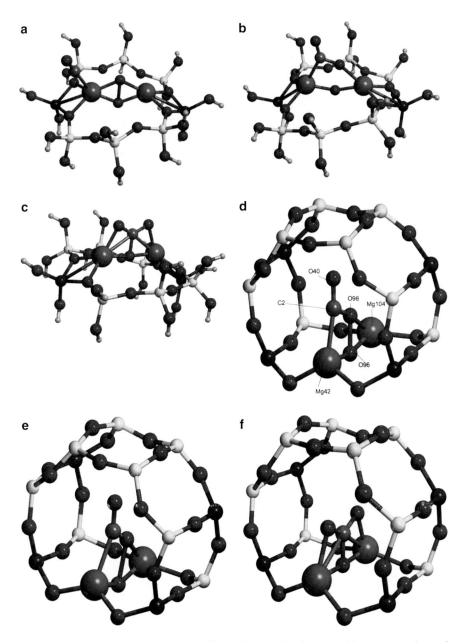

Fig. 20.19 Optimized structures of the (**a**, **d**) reaction complex, (**b**, **e**) transition state complex, and (**c**, **f**) products of CO oxidation calculated at the (**a**–**c**) B3LYP/6-31G* cluster level for MgMOR and (**d**–**f**) PDFT/GGA level for MgPHI. The color code is as in Fig. 20.12, C in *brown*

energies, the geometry analysis of the reaction complex was performed (the values are given in bold in Table 20.11). The C_2–Mg_{42} distance is the shortest in the PHI case, even if the CO is attached solely to the Mg_{42} cation in two other reaction complexes (MOR and CLU). For the PHI model the C term of the molecule is coordinated to both O_{98} and Mg_{42} atoms (so that one expects longer C_2–Mg_{42} bond length). The O_{96}–O_{98} distance increases while the C_2–O_{40} one decreases in the sequence PHI $\rightarrow$ MOR $\rightarrow$ CLU. This shows the extent of the highest CO activation in the MgO_2Mg(PHI) model (Table 20.11). One should compare the O_{96}–O_{98} distance with 1.613 Å in Mg_2O_2(8R) and with drastic growth from 1.528 Å in the reaction complex to 1.852 Å in the TS complex for PHI. All the Mg–O distances do not vary between the geometries of the reaction and TS complexes for the PHI case. This denies their importance for the REA $\rightarrow$ TS transformation. For the latter step, two distances O_{96}–O_{98} and C_2–O_{40} are important. The variation of the O_{96}–O_{98} and C_2–O_{40} are maximal for the PHI model compared to MOR and CLU. The C_2–O_{40} distance for the adsorbed CO molecule (REA case) is strongly elongated by 0.08 Å in the periodic PHI model. It shows a strong chemisorption state of the oxidation reaction that cannot be achieved for the Mg-form as well as for the Ca-form within the cluster model. Hence, this difference in the C_2–O_{40} distance between the reaction and TS complexes is the main reason of lower activation barrier.

The periodic solution does not lead necessarily to the chemisorption complex as revealed by the CO oxidation over MgO_2Mg(MOR) (Table 20.11) with the activation energy of 23.0 kcal/mol. The barrier is larger than 15.3 kcal/mol for the periodic phillipsite (PHI) model and smaller than 35 kcal/mol for the MgO_2Mg(8R) cluster. All of the shown geometry parameters for the CO$\cdots MgO_2Mg$ species are similar between the TS in three demonstrated systems, i.e., MgO_2Mg(PHI), MgO_2Mg(MOR), and MgO_2Mg(8R). On the other hand, the parameters for the chemisorbed complex in MgO_2Mg(PHI) (shown in bold, Table 20.11) differ relative to two other cases. It signifies that the periodic solution in MOR stabilizes the TS in higher extent than periodic solution of the reagents and products. The lowest barrier for the oxidation over the MgO_2Mg(PHI) and not over the MgO_2Mg(MOR) confirms importance of the framework type for the optimal chemisorbed complex.

20.3.1.2 CO_2 Interaction with the MeO_XMe(MOR) Species

It has been demonstrated that weak interactions between CO_2 and O atom of zeolite cannot be the reason of carbonate formation [72] while the CO_2 interaction with the MeO_XMe(MOR) species is barrierless [55]. In order to model the carbonate formation, we needed to determine the ground state of each of the reagents and products. All of the carbonate species studied herein possess a singlet electronic ground state. The closeness between the triplet and singlet states (see Part 3.1) of the $MeCO_XMe$ reagents allows the CO_2 addition reaction and formation of carbonate in the singlet (ground) state.

The geometries of the carbonate anions regarding CO_2 addition to the MeO_XMe(MOR) reagents (Me=Mg, Ca) were calculated at the periodic level using

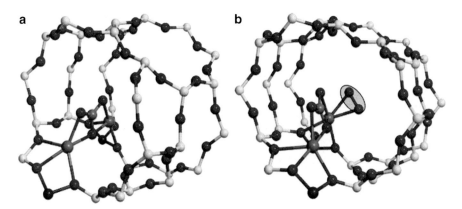

Fig. 20.20 Optimized structures of the $CaCO_3Ca$ moieties in $Ca(MOR)$ obtained for the (**a**) CaOCa and (**b**) CaO_3Ca species as calculated at the PDFT/PW91 level with VASP. The O_2 molecule is shown by the ellipse. The color code is the same as in Fig. 20.12, C in *brown*

VASP [30] for the Me forms of mordenite (MeMOR) and the Me forms of phillipsite (MePHI). The 3D optimized structures are illustrated for the $CaCO_XCa(MOR)$ for $X = 1$ (Fig. 20.20a) and $X = 3$ (Fig. 20.20b). The oxidation reaction sequence regarding the $CaO_XCa(MOR)$ reagent structures from $X = 1–4$ as optimized at the periodic level is presented in Fig. 20.13 and Table 20.6. The MeO_XMe geometries obtained at the periodic level are pretty similar to the cluster ones obtained with B3LYP/6-31G* [51–55].

Regarding the CO_2 adsorption over the $CaOCa(MOR)$ and $CaCO_3Ca(MOR)$ species, we obtained stable 3D structures with two frequencies having low absolute values, 1587 and 1255 cm^{-1} (relative to 1649 and 1347 cm^{-1} for the 8R cluster model of MOR). We re-scaled the calculated periodic PW91 values as the most strongly displaced ones relative to the experimental data. Fitting the symmetric vibration of 1255 cm^{-1} to the "cluster" value of 1347 cm^{-1}, instead of 1587 cm^{-1}, we got a re-scaled[2] asymmetric vibration of 1703 cm^{-1}. For the 8R cluster value of 1649 cm^{-1} was obtained for MOR and value of 1664 cm^{-1} for the (6R+4R) cluster for FAU (Table 20.13).

Considering the interaction between the MeO_2Me homologues and CO_2 at the periodic level, we did not succeed to obtain tri-dentate carbonate species as the ones optimized at the cluster level for MOR (Fig. 20.16d) or FAU (Fig. 20.17d) with B3LYP. At the periodic level, no product of interaction between MeO_2Me and CO_2 neither for MOR, nor for PHI was obtained. Remarkably, we did not get this product at the MP2 level for the cluster model as well. We consider this as a confirmation that tri-dentate species do not appear in the AE forms. This is also confirmed by the

---

[2]$(1347/1255) \times 1587 = 1703$ cm^{-1}.

Table 20.13 Band positions (cm^{-1}) of the asymmetric (ν_{asym}) and symmetric (ν_{sym}) vibrations as well as the intensity ratios (I_{asym}/I_{sym}) for the carbonate species in the $MeO_XMe(6R+4R)$ cluster (Me=Mg, Ca, Sr, Ba) calculated at the B3LYP/6-31G* level compared to the experimental data for $X = 1$ [73] and $X = 2$ [74]. Reproduced from Ref. [54] with kind permission of Elsevier

	Mg	Ca	Sr	Ba	Experiment
$X = 1$					
ν_{asym}	1646	1664	1663	1651	1700, 1665, 1625[a]
ν_{sym}	1330	1334	1336	1343	1365, 1390, 1440[a]
I_{asym}/I_{sym}	1.24	1.62	1.37	1.37	1.45–1.55[a]
$X = 2$					
ν_{asym}	1912	1960	1875	1869	1850[b]
ν_{sym}	1260	1268	1279	1304	1180[b]
I_{asym}/I_{sym}	1.16	1.31	1.47	2.50	–

[a]Upper ν_{asym} and lower ν_{sym} values correspond to the three bands in CaX zeolite [73]
[b]For alumina [74]

absence of an upper frequency branch for the chemisorbed species (*i.e.*, as calculated between 1860 and 1960 cm^{-1} ($X = 2$ in Table 20.13), in the experimental spectra of AE form zeolites).

Similarly, we did not get any tri-dentate carbonate species *via* the periodic study for the interaction between CO_2 and MeO_2Me homologues. With respect to the periodic model of CO_2 adsorption in $CaO_3Ca(MOR)$, the latter confirms the results of the cluster computations. The periodic model also led to the formation of an O_2 fragment separated from $CaCO_3Ca(MOR)$ with carbonate bands at 1517 and 1226 cm^{-1} (Fig. 20.20b). Assigning the periodic value of 1226–1347 cm^{-1} of the "cluster" one gets a re-scaled value[3] of 1669 cm^{-1}. This value has to be compared to 1649 cm^{-1} for the 8R cluster of MOR or to 1664 cm^{-1} for the (6R+4R) cluster of FAU (Table 20.13).

20.3.1.3   AE Carbonate Interaction with Methanol

The carbonates can participate in many catalytic processes, which include CO_2 binding and reactions. We studied methanol interaction with CO_2 and found a similar activity for CaY as compared to CuY at elevated temperatures [104]. Proton transfer towards CO_2 in the gas phase can be proposed as initial reaction step with an endothermic heat evaluated as 11.5 kcal/mol at the B3LYP/6-31G* level, which was applied in our calculations discussed above. This step results in the activation

---

[3]$(1347/1226) \times 1517 = 1669\ cm^{-1}$.

energy of 42.3 kcal/mol and the imaginary frequency is $1699i\,cm^{-1}$. The favored orientation of CO_2 and CH_3OH is determined by the electrostatic interactions between CO_2 quadrupole and CH_3OH dipole.

We repeated the same reaction with methanol introducing the $CaCO_3Ca(8R)$ carbonate instead of the CO_2 molecule. The relevance of the Ca carbonate for the carbonylation reaction can be confirmed by a slightly exothermic effect of -3.7 kcal/mol, if one estimates the energy of the reagents involving the initial structure I (Fig. 20.21a) and the energy of final product IV (Fig. 20.21d). The energy of the reagents is the sum of the energy of the optimized structure I (Fig. 20.21a) and the energy of CO_2 molecule. The product IV (Fig. 20.21d) is formed in the barrierless reaction between CO_2 and dimethylcarbonate. The reaction restores the $CaCO_3Ca(8R)$ carbonate and completes the catalytic cycle. We present herein the first step of the reaction that leads to the $CH_3CO_3^-$ anion formation as the part of the complex III:

$$CaCO_3Ca\,(8R) + 2CH_3OH \rightarrow Ca\,(CH_3O \ldots CO_3H)\,Ca\,(8R) \ldots HOCH_3\,(II, V) \rightarrow$$

$$\rightarrow HO - Ca\,(CH_3CO_3)\,Ca\,(8R) \ldots HOCH_3\,(III) \rightarrow CH_3HCO_3$$

$$+ CaOCa\,(8R) \ldots HOCH_3 \qquad\qquad (20.9)$$

For all of the compounds (Fig. 20.21a–c, e) with the same chemical composition (with the exception of the product IV that contains additional CO_2 molecule and has a higher energy, Fig. 20.21d), the energy values are given in Figures. The endothermic effect of the one-step reaction is smaller (9.8 kcal/mol between the structures I and III (Fig. 20.21c) relative to 11.5 kcal/mol in the gas phase). We obtained two intermediate II (Fig. 20.21b) and V (Fig. 20.21e) states with energies that are less than 40 kcal/mol and smaller than the activation energy in the gas phase (42.3 kcal/mol). It is very interesting that the carbon geometry in state II corresponds to sp^3 hybridization while it is usually sp^2 in the initial and final CO_3 states. Both the II and V states can serve as a step for TS *via* the following mechanism as the first step of dimethylcarbonate formation (20.9). The state II is especially promising as an initial point for the proton transfer. However, the problem of the TS localization at the cluster level met an obstacle created by bending vibrations of capping hydrogens in the 8R cluster. Due to the restrictions imposed on the cluster, their frequencies are also imaginary and hinder the search of TS with a close frequency less than $400i\,cm^{-1}$. The periodic calculations, which are under progress, do not have such problems in the TS search. Nevertheless, the initial modeling step shows a perspective of carbonate inclusion into the reaction schemes for CO_2 binding and activation.

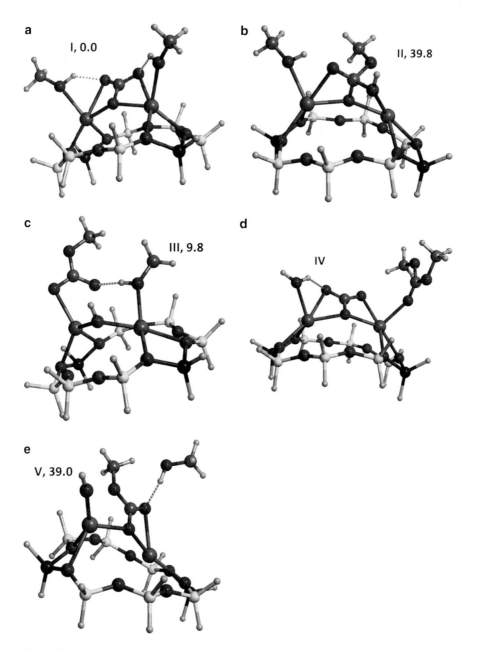

Fig. 20.21 Optimized structures of the (**a**) reagents, (**b, c, e**) intermediate states, and (**d**) products for the reaction between CH_3OH and $CaCO_3Ca(8R)$ moieties at the B3LYP/6-31G* level. The relative energies (kcal/mol) are shown for the different steps of the same brutto chemical composition. Hydrogen bonds are shown by *dotted lines*. The color code is the same as in Fig. 20.12, C in brown

20.4   On the Feasible Ways of the Polynuclear Metal-Oxo Clusters Formation

Rather often Zn-exchanged catalysts are prepared with the amount of included active component that exceeds exchange capacity of the zeolite. In these cases one can expect a formation of zinc oxide cluster species, entrapped in the zeolite matrix. But there exists evidence that it is difficult to avoid formation of small oxide clusters in e.g. ZSM-5 zeolite channels even at low loading of active component under conventional methods of preparation [75, 76]. It was found that there is a strong interaction between framework oxygen atoms of zeolite and ZnO clusters [75]. This interaction can be the reason of increased stability of encapsulated clusters in comparison with the isolated ones. A question arises concerning the zeolite structures that can retain metal oxide clusters. We consider the cation positions as a probable variant of such places. Indeed, binuclear metal-oxo species discussed in Sect. 20.3 may be an example of similar cluster formation as a result of ZnO molecule adsorption on Zn^{2+} ion located in the cation positions. Generation of the optimal structures, analyzed above, need definite geometrical configuration of the position. At the same time, one can suggest formation of different condensed structures with lower stability. The computational details of the calculations presented below are the same as they were in the Sect. 20.2 for the study of ZnZ_d sites. Let us present an illustrative example of Zn^{2+} ion in the cation position modeled by the four-membered ring, see Fig. 20.22. Attachment of ZnO molecule leads to the structure presented in Fig. 20.23. This structure is similar to that considered earlier binuclear metal-oxo species but because of small size of the ring its formation is associated with the certain structural tension. Despite this fact, the energy of ZnO molecule addition is quite substantial and has the value of 106.4 kcal/mol. Addition of a second ZnO molecule results in the structure shown in

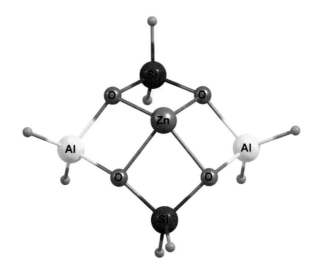

Fig. 20.22 Zn^{2+} ion stabilized in the cation position modeled by four-membered ring

Fig. 20.23 The structure
formed by attachment of one
ZnO molecule to Zn^{2+} ion in
the cation position
(Fig. 20.22)

Fig. 20.24 The structure
formed by attachment of two
ZnO molecules to Zn^{2+} ion
in the cation position
(Fig. 20.22)

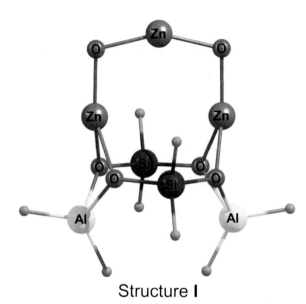

Structure I

Fig. 20.24; the gain of energy is 107.5 kcal/mol. Evidently it is possible to provide subsequent condensation of the cluster by this manner until the steric hindrance of zeolite channel will prevent this process. The experimental study [76] showed that small zinc oxide clusters in ZSM-5 zeolites are also active in the dehydrogenation of light alkanes as well as the zinc ion in the cation position (see Sect. 20.2). The mechanism of this process for the ethane molecule on $(Zn-O-Zn)^{2+}$ in ZSN-5 zeolite was studied in [10] (see also [27]). It was shown that the first stage described by Eq. 20.3 is very exothermic but the last stage (Eq. 20.5) turned out to be very endothermic. In the case of $(Zn_3O_2)^{2+}$ cluster (Fig. 20.24) the modeling of this stage leads to the same conclusion. The desorption energy of H_2 is more than 60 kcal/mol.

Fig. 20.25 The result of H_2 dissociation on the Structure **I** (Fig. 20.24)

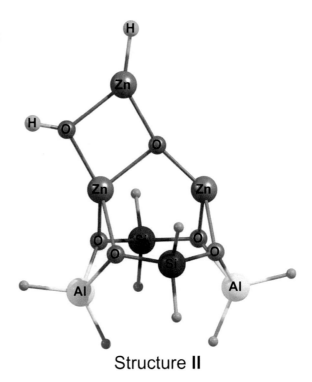

Structure II

The agreement between experiment and theory can be reached by consideration of the hydride forms of the clusters. In the reductive conditions of dehydrogenation reactions this proposition is quite reasonable. Dissociative addition of dihydrogen to the structure I in the Fig. 20.24 convert it to the structure II (see Fig. 20.25) with reconstruction of the cluster. The result of ethane heterolytic dissociation on this cluster is shown in the Fig. 20.26, structure III. The activation energy of this stage was evaluated to be 23.8 kcal/mol. This process is suggested to proceed: (a) through β hydrogen transfer with the formation of Zn–H bond as well as adsorbed or eliminated ethylene molecule, (b) through the direct formation of H_2 and of adsorbed ethylene. The transition states for these stages are shown in the Figs. 20.27 and 20.28 with the activation energy of 54.8 and 51.1 kcal/mol, respectively. The final product of path (a) is shown in Fig. 20.29, subsequent desorption of H_2 regenerate structure II with the closure of the catalytic cycle. Activation energy of this stage is 27.3 kcal/mol. Resulting product of path (b) is shown in Fig. 20.30 where energy of ethylene adsorption is 20.9 kcal/mol. Reaction energy diagram of both variants of ethane dehydrogenation on hydride $(Zn_3O_2)^{2+}$ is presented in Fig. 20.31. It is interesting to note that energetic profile of the process is similar to the one for the reaction on the ZnZ_s site are discussed in Sect. 20.2.

While considering the structural problem of a stabilization of zinc oxide clusters in zeolite pores, it is very important to pay attention to the process of incorporation

Fig. 20.26 Heterolytic
dissociation of ethane on the
hydride form of the model
cluster (Fig. 20.25)

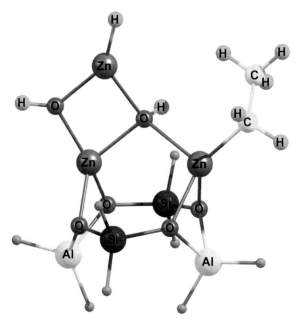

Structure **III**

Fig. 20.27 The transition
state for β-hydrogen transfer
and Zn–H bond formation

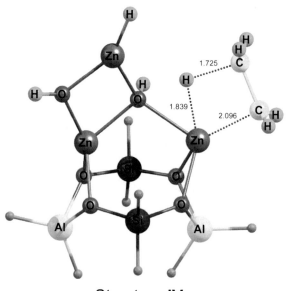

Structure **IV**

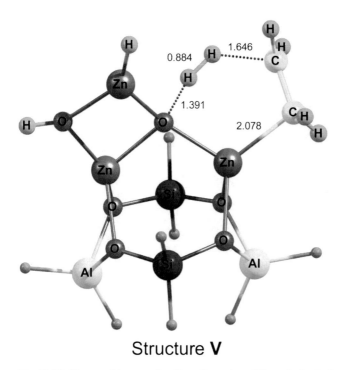

Structure V

Fig. 20.28 The transition state for direct formation of H_2 and adsorbed ethylene

Fig. 20.29 The final product of path (a) which includes β–hydrogen transfer

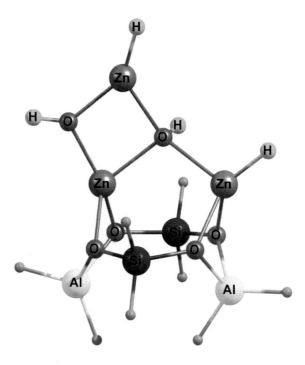

Fig. 20.30 The final product
of path (b) which includes
direct formation of H$_2$ and
adsorbed ethylene

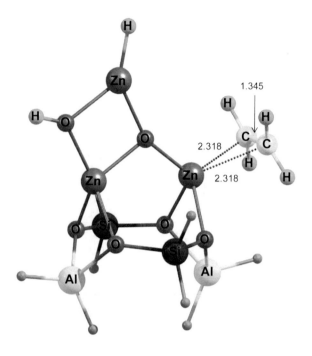

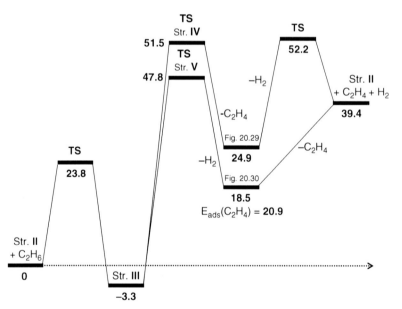

Fig. 20.31 Reaction energy diagram of both variants of ethane dehydrogenation

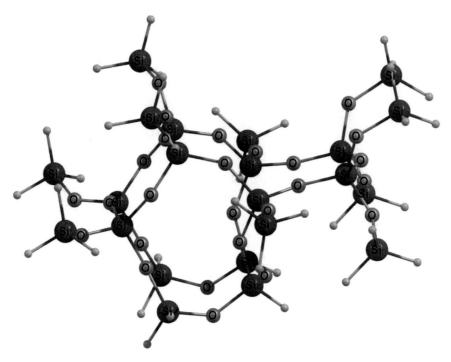

Fig. 20.32 Model cluster simulating a fragment of silicalite lattice near the intersection of straight and zig-zag channels

of zinc active component into silicalite [76]. In [76] stabilized zinc oxide species were studied both in HZSM-5 zeolite and in silicalite with the same MFI structure. It was found that stability of these oxide species and mechanism of their interactions with propane molecules are rather similar. In silicalites ion exchange positions do not exist and different ways to stabilize the zinc oxide clusters should be investigated.

Our test calculations showed that free small zinc oxide clusters are easily broken under the interaction with light alkanes, in this case the confinement effect of silicalite pores is not reliable to account for zinc oxide species stabilization. Thus, we propose the existence of other ways of incorporation of zinc oxide species into the silicate framework. It was confirmed by the observation of zinc incorporation into the silica mesoporous molecular sieves [77]. As an example we considered a grafting of $(ZnO)_n$, $n = 1$–3, clusters on the silicalite framework. The cluster simulating a fragment of silicalite lattice is shown in the Fig. 20.32. Grafting of ZnO molecule leads to the structure presented in Fig. 20.33. The energy gain for such incorporation of ZnO into the silicalite lattice is 76 kcal/mol. The grafting of $(ZnO)_2$ and $(ZnO)_3$ are shown in the Fig. 20.34 and in the Fig. 20.35, respectively. The energy gain of the addition of second ZnO molecule is 132 kcal/mol and for the

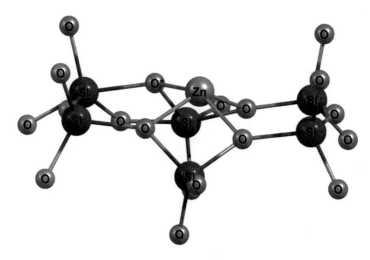

Fig. 20.33 The structure resulting from grafting of single ZnO molecules on the silicalite cluster shown in Fig. 20.32

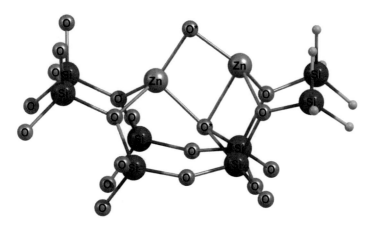

Fig. 20.34 The structure resulting from grafting of two ZnO molecules on the silicalite cluster shown in Fig. 20.32

third one is 112 kcal/mol. Certainly, the principle point of such model consideration of the grating of zinc component is zinc compounds, which probably interact with silicate lattice. It should depend on the loaded species and on the technique of the insertion. We used the ZnO molecules as a reasonable choice to insert an active component within the framework of solid state ion exchange technique.

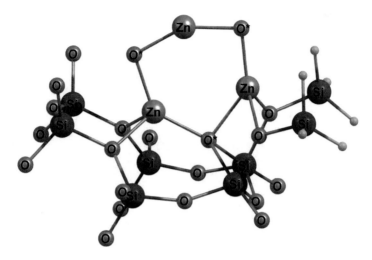

Fig. 20.35 The structure resulting from grafting of three ZnO molecules on the silicalite cluster shown in Fig. 20.32

20.5 On Possible Forms of Single Fe(II) Ion Stabilization in Fe/HZSM-5 Selective Oxidation Catalyst

The Fe/HZSM-5 system is a unique catalyst for selective oxidation of hydrocarbons using N_2O as an oxidative agent [78, 79]. The oxidation of benzene to phenol according to reaction

$$C_6H_6 + N_2O \rightarrow C_6H_5OH + N_2 \qquad (20.10)$$

attracts a lot of attention [79]. This reaction proceeds with high selectivity (over 95%) and satisfactory conversion. Numerous studies were performed on this system to determine the oxidative state of active iron species [80–87]. There is general agreement that the active species are associated with Fe^{2+} [80]. However, the coordination environment of these cations is unknown. Low active iron content (below 0.5% wt, 10^{19} sites/g) and severe conditions of the catalyst preparation (such as steaming, high temperature treatment, etc.) are the main problems of the investigation.

Capturing of the Fe^{2+} ions in the zeolite α-cation-exchange positions was first considered theoretically in [81]. The Fe^{2+} ion "grafted" to the zeolite framework, or captured by a zeolite lattice defect $\equiv$Si–O–Fe–O–Si$\equiv$ was proposed in [82] as an active center with low coordination of Fe^{2+}. Such structure could be emerged as the result of iron immobilization on vicinal hydroxyl zeolite groups. Also, it was found that addition of trimethylaluminum to Fe-silicalite drastically improves the catalyst activity in the process (20.10). Based on these data a conclusion about the formation of $FeAlO_x$ active species was made [83].

Cluster models for each kind of possible Fe^{2+} single ion stabilization in zeolite are shown in the Fig. 20.36. The first one is a four-membered zeolite ring with two lattice Al^{3+}. Silica defect position is simulated as the $Fe(OSi)_2O(OH)_4$ structure. $FeAlO_x$ extraframework iron species pseudoshpinel structure $FeAl_2O_4$ was suggested.

Decomposition of N_2O on the active site under mild heating (200°C) results in the formation of oxidative $[FeO]^{2+}$ center:

$$N_2O + Fe^{2+} \rightarrow [FeO]^{2+} + N_2 \qquad (20.11)$$

The most intriguing feature of the catalytic system under discussion is possibility to separate the oxidative agent (N_2O) loading phase and the oxidation reaction phase (hydrocarbon loading) [79, 80, 84, 85]. The N_2O molecules decompose on the Fe^{2+} sites of Fe/ZSM-5 and form highly reactive iron-oxygen species. They are rapidly able to oxidize methane to methanol and benzene to phenol at low temperature and with very high selectivity. Without reagents these oxidative species are rather stable. This form of oxygen stabilization is called *alpha-oxygen* (denoted here as FeO^α). The reactivity of this center resembles one of the active iron species of enzymes [86]. This analogy is widely discussed in the literature [87, 88]. Understanding of the electronic structure of the FeO^α species and reasons of high oxidation activity is important.

20.5.1   The Electronic Structure of FeO^α Centers

The FeO^α center is bonded to the zeolite support through bridged oxygen or hydroxyl groups. The geometry of the FeO^α surrounding is of distorted tetrahedron. It is widely accepted that the formal oxidation state of the iron center in (hydr)oxide is determined by the number of OH^- and O^{2-} ligands in its first coordination sphere. For these reasons the $OFe(OH)_2$ system has been adopted as the simplest neutral model of the FeO^α [89].

One may describe the electronic state of $[FeO]^{2+}$ unit in terms of the following valence bond schemes: $Fe^{IV}=O$ and $Fe^{III}-O\bullet$. It should be emphasized that $Fe^{IV}=O$ and $Fe^{III}-O\bullet$ are not simple resonance structures of each other, but are different states of the $[FeO]^{2+}$ unit (one may be the ground state and the second one can be the first excited state). Which one is actually present under the catalytic conditions remains unresolved [79, 89–91]. It is not clear whether they differ enough to be distinguished via some spectroscopic methods. However, some attempts have been made. For example, Pirngruber et al. [92] using resonant inelastic x-ray spectroscopy (RIXS) provided an experimental evidence of the $Fe^{III}-O\bullet$ structure of the FeO^α site. The RIXS experiment was designed to obtain unbiased "yes/no" information about FeO^α electronic state. However, it is unknown, which electronic state ($Fe^{IV}=O$ or $Fe^{III}-O\bullet$) drives an oxidative reaction. Under reaction conditions

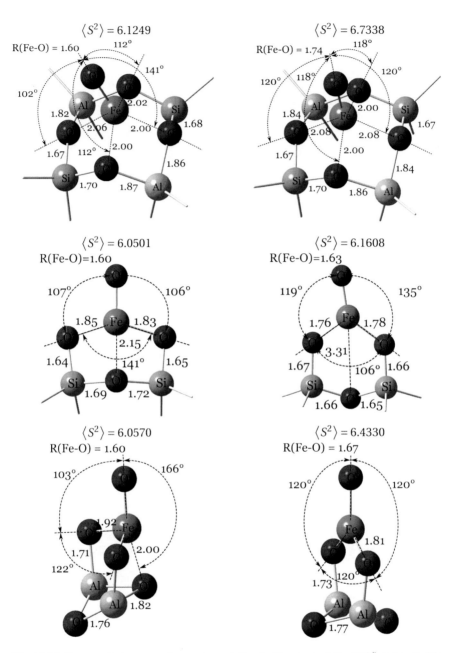

Fig. 20.36 From *top* to *bottom*: cation-exchanged, "grafted" and spinel-like FeO$^\alpha$ center models at the Fe^{IV} $=$O (*left*) and Fe^{III}–O• (*right*) electronic states. Optimal geometries were calculated on UB3LYP/6-31G(d) level of theory

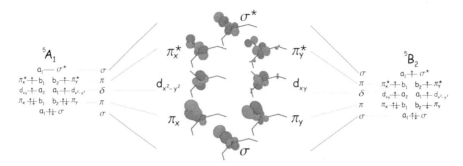

Fig. 20.37 CASSCF natural orbitals of OFe(OH)$_2$, *left* – molecular orbitals occupation scheme for $^5A_1$ solution, *right* – $^5B_2$. *Red* color marks split π-electron pair

Table 20.14 Properties of OFe(OH)$_2$ optimized at the different level of theory

State	Energy, kcal/mol	Geometry		
		Fe–O	Fe–OH	(H)O–Fe–O(H)
B3LYP/6-311g(d) theory				
$^5A_1$	0	1.60	1.79	149°
$^5B_2$	−6	1.66	1.78	111°
ISA-MRMP/6-311g(d) theory				
$^5A_1$	0	1.62	1.78	146°
$^5B_2$	0	1.68	1.77	118°

The energies are given with respect to the $^5A_1$ state

the energetic order of states may vary (for example when a hydrocarbon molecule approaches the active center). The OFe(OH)$_2$ molecule has the C_{2v} point group symmetry. According to the high-level quantum-chemical method (ISA-MRMP [93]) there is a close coincidence between the Fe^{IV}=O scheme and $^5A_1$ solution, and between Fe^{III}–O• and $^5B_2$ solution [89]. The $^5A_1$ solution has π^* highest occupied molecular orbital (HOMO) and σ^* lowest unoccupied molecular orbital (LUMO), while $^5B_2$ has σ^* HOMO and π^* LUMO, see Fig. 20.37. The character of the molecular orbitals resembles those of molecular oxygen. There are no occupied d-orbitals in the O$_2$ case. This analogy was first recognized by Shaik et al. [94], and was confirmed later by Baerends et al. [91].

The properties of the $^5A_1$ and $^5B_2$ states listed in Table 20.14 are not very different. By changing the angle (H)O–Fe–O(H) or twisting dihedral O–Fe–O–H angle it is possible to switch $^5A_1$ to $^5B_2$ or $^5B_2$ to $^5A_1$ to be the ground state [89]. The global energy minimum (the conformation with zero dihedral O–Fe–O–H angle) corresponds with the $^5A_1$ solution, but this value for $^5B_2$ is only 11 kcal/mol higher.

There is a general tendency of the DFT methods to overestimate the electron density delocalization. This happens due to self-interaction error [95]. The Fe^{III}–O•/FeIV=O competition may also originate from this problem. What is more favorable, charge delocalization over double Fe=O bond or charge separation,

Table 20.15 The OFe(OH)$_2$ properties predicted via different quantum-chemical approaches: Fe–O bond length, vibrational frequency, multiplet splitting and the heat of oxygen desorption, see Eq. 20.12

Method	BLYP	B3LYP	B2PLYP	ISA-MRMP
R(Fe–O), Å	1.62	1.60	1.61	1.62
ω (Fe–O), cm^{-1}	918	969	749	–
$\Delta E_{5 \rightarrow 7}$, eV	2.85	2.35	1.73	1.08
ΔH, kcal/mol	39	22	1	–

resulting in breaking the electron pair and subsequent appearing of the radical structure Fe^{III}–O•? The answer depends on the DFT functional used under investigation. One of the ways to overcome DFT self-interaction and improve the accuracy of predictions is hybrid (Hartree-Fock plus DFT) approach [95]. There are no self-interaction in the Hartree-Fock theory. Therefore, the admixture of HF to pure DFT decreases the self-interaction effect. The most popular hybrid density functional B3LYP contains 20% of HF. There are also double-hybrid DFT functionals [96], where the HF fraction is higher (50% or even more). In that case second-order perturbation theory is used to compensate partial correlation energy loss due to higher amount of HF. The B2PLYP functional is the most accurate but also computationally demanding method [96].

Table 20.15 illustrates the dependence of the OFe(OH)$_2$ properties on the DFT functionals. The Fe–O bond length as a first-order property changes slightly, while second-order properties, such as vibrational frequency, bond energy, multiplet splitting, definitely correlate with the used method. Panov at el. [79] claimed that Fe$^\alpha$ centers are stable against the action of molecular oxygen since the Fe–O bond formation can not energetically compensate the double bond breaking of the molecular oxygen.

$$FeO^\alpha \rightarrow Fe^\alpha + {}^1/_2O_2 \qquad (20.12)$$

Calorimetric measurements performed for the reaction (20.12) have shown that oxygen resulting from N$_2$O decomposition on Fe$^\alpha$ has low bond energy (1–6 ± 3.5 kcal/mol) [97]. The B2PLYP estimation (1 kcal/mol) agrees well with the experiment, while BLYP (39 kcal/mol) overestimates FeO$^\alpha$ stability, see Table 20.15. The presence of low-lying excited state is a characteristic feature of chemically active species. From this point of view in the case of OFe(OH)$_2$ high-level electron correlation methods predict greater activity than pure DFT method (BLYP).

The chemical activity of OFe(OH)$_2$ is close to that of a free OH• radical which can be judged on a basis of estimated Mulliken electronegativities (see Table 20.16). They are found to be approximately equal. It is interesting to note that in an adjacent field of the Fenton reaction chemistry similar [FeO]$^{2+}$ aqua ion was suggested to be the key intermediate. However, there is still a competition between [FeO]$^{2+}$ and

Table 20.16 Ionization potentials, electron affinities and Mulliken electronegativities for the typical oxidative agents compared to $OFe(OH)_2$ estimated via ump4/6-311g(df,p) quantum-chemical method

	OH•	O_2	F_2	O=Fe=O	$OFe(OH)_2$
IP	13.02	12.07	15.70	9.50	11.59
EA	1.83	0.45	3.00	2.36	3.35
χ	7.42	6.26	9.35	5.93	7.47

All units are in electron-volts

aqua OH• radical as candidates to drive the oxidation reaction [91]. More realistic extra-framework iron site models (Fig. 20.36) may not possess the C_{2v} symmetry as $OFe(OH)_2$. This is the reason why it becomes difficult to differentiate solutions $^5A_1$ from $^5B_2$ and to reveal the radical character of the $[FeO]^{2+}$ moiety. However, it is possible to recognize the difference (in quantum-chemical simulation) between the Fe^{III}–O• and Fe^{IV}=O structures and even to obtain both states for the same kind of iron site using difference between the HOMO/LUMO occupation pattern. The oxygen-radical-like structure Fe^{III}–O• has σ^* HOMO and π^* LUMO, while Fe^{IV}=O structure has an opposite order: π^* HOMO before σ^* LUMO.

The second criterion used to detect Fe^{III}–O• bonding is the difference of $<S^{**}2> - S_z(S_z + 1)$ [89, 98]. The ground spin state for the $[FeO]^{2+}$ moiety is quintet, $S = 2$. Under low crystal field of the zeolite environment Fe^{3+} ions possess high spin $S = 5/2$ state. Radical oxygen $S = 1/2$ is bonded to Fe^{3+} ion antiferromagneticaly – ↑↓. As it was proposed by Noodleman [99], difference between actual $<S^{**}2>$ and expected $<S^{**}2> = S_z(S_z + 1)$ values can serve as an indicator of such antiferromagnetic coupling. Using this criteria we obtain energetic minima for Fe^{III}–O• and $Fe^{IV} = O$ like structures of various FeO^α models. For the cation-exchanged iron center the energy difference is negligible (less than 0.5 kcal/mol). "Grafted" iron center prefers Fe^{IV}=O bonding (it is 4 kcal/mol lower than Fe^{III}–O•) while for the pseudospinel center the opposite order has been obtained (Fe^{III}–O• is 3 kcal/mol lower). One can see that different kinds of possible FeO^α structures are always at two states. At least there is no considerable preference for some kind of bonding. Important questions are whether the radical character of oxygen matters or not and what the influence of the radical character of oxygen in the $[FeO]^{2+}$ moiety is on the oxidative reaction?

20.5.2   The Impact of the Radical Electronic State of FeO^α on Oxidative Reactions

In the case of ordinary electrophilic or nucleophilic reactions electrons move in pairs. Therefore, the main process of an iron ion reduction is $Fe^{4+} + 2e = Fe^{2+}$. A radical reaction step is a single electron transfer, such as $Fe^{4+} + e = Fe^{3+}$.

An intermediate (or final product) of hydrocarbon oxidation in the presence of Fe^{3+} is a product of a radical reaction step. If there are no Fe^{3+} species, the reaction can not be considered as radical. For example:

$$2FeO^{\alpha} + CH_4 = Fe^{3+}OH + Fe^{3+}OCH_3 \tag{20.13}$$

$$FeO^{\alpha} + CH_4 = Fe^{\alpha}CH_3OH \tag{20.14}$$

The reaction (20.13) occurs via hydrogen atom transfer step followed by the $CH_3\bullet$ addition to the next available FeO^{α} site. The stoichiometry of the FeO^{α} reaction with methane was measured and it was found to be very close to 2:1 [100]. Therefore, FeO^{α} behavior is similar to oxygen radical [80].

Benzene to phenol oxidation is somewhat different at first glance. The characteristic feature of HAT is pronounced kinetic isotope effect [87]. But, its absence in the benzene oxidation [84, 87] reactions suggests the lack of steps regarding the breaking of the C–H bond (HAT). Therefore, it is not clear how does radical reaction occur in this case. However, the Fe^{3+} intermediate indicates single electron transfer that means radical reaction. Benzene oxide (see Fig. 20.38:2a) as the first step intermediate was first predicted by Kachurovskaya et al. [81, 101]. This one is not a Fe^{3+} intermediate. Therefore, the reaction pathway proposed by these authors [81, 101] is not based on radical oxygen hypothesis. The mechanism proposed later by Fellah et al. [102] includes one-electron step resulting in benzene σ-complex formation (Fig. 20.38) as the first intermediate. It is expected that this intermediate contains the Fe^{3+} ion.

Our quantum-chemical simulation of benzene oxidation reaction based on pseudospinel iron center (see Fig. 20.36, bottom) reveals the same structure. The characteristic feature of such intermediate is the presence of $C(sp^3)$–H bond. The presence of the $C(sp^3)$–H bond intermediate was confirmed by *in-situ* IR experiment of Panov et al. [84]. The IR band at 2874 cm^{-1} appeared immediately after benzene was fed to the FeO^{α} catalyst. At the same time no phenol signals were detected. Heating of the sample resulted in complete disappearance of this band. According to our quantum-chemical simulation only the σ-complex structure has the characteristic of this IR band. For benzene oxide, which also has two $C(sp^3)$–H bonds, this band is not present, since all of the vibrational frequencies are within narrow range of 3182–3218 cm^{-1}. In the case of the benzene σ-complex the calculated IR frequency for the $C(sp^3)$–H vibration is 2930 cm^{-1}, while the other C–H vibrations are within 3178–3215 cm^{-1}. Applying anharmonic scaling factor $f = 0.96$ one may obtain quite reasonable agreement: 2813 cm^{-1} and 3050–3086 cm^{-1} (theory estimation) versus 3037–3090 cm^{-1} and 2874 cm^{-1} (experimental data).

The benzene σ-complex may isomerize to the keton (Fig. 20.38:3a), where two hydrogen atoms are equivalent. This makes 1,2-isotope shift possible when 1,3,5-D-substituted benzene is used. This effect was first recognized on the benzene

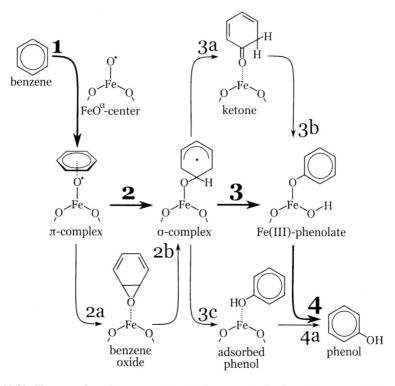

Fig. 20.38 The route from benzene to phenol: the energetically favorable path (1-2-3-4) and alternative branches predicted *via* B3LYP/6-31G(d) simulation

Table 20.17 Thermochemical data for the benzene to phenol reaction (see Fig. 20.38) predicted theoretically using b3lyp/6-31g(d) theory

Reaction	0	1	2	2a	2b	3	3a	3b	3c	4	4a
$\Delta_r H^\circ_{T=0}$	−33.8	−7.6	−1.1	+7.1	−8.2	−53.1	−28.8	−24.4	−30.2	+44.4	+21.5
E_a	−	−	9.5	17.4	−	7.6	12.1	0.5	18.6	−	−

oxidation via P450 enzyme (so called "NIH-shift" described in [86]). Similar behavior predicted for FeO^α is an evidence of its biomimetic feature.

The most stable product of benzene oxidation is not the adsorbed phenol as predicted in [102], but the $Fe^{3+}OC_6H_5$ form with hydrogen located nearby. To desorb the phenol molecule and close up catalytic cycle requires 44 kcal/mol (see Table 20.17). This energy can be obtained from subsequent N_2O decomposition:

$$Fe^\alpha(OH)C_6H_5 + N_2O \rightarrow FeO^\alpha + C_6H_5OH + N_2 \qquad (20.15)$$

The limiting step of benzene to phenol transformation is suggested to be the reaction between zeolite adsorbed N_2O and catalyst active sites bound with phenol

[103]. This conclusion is based on the fact that activation energy measured for the N_2O decomposition on the unoccupied Fe^α sites is 42 kJ/mol. This is lower than the activation energy of overall benzene to phenol reaction (105–126 kJ/mol). However, the limiting step is still the N_2O decomposition.

20.6 Outlook

Considered problem of the bivalent metal cations stabilization in zeolites is a good example of the usefulness of quantum chemical model approach in the treatment of complex catalytic systems. Even in the case of comparatively strongly defined lattice structure of a substrate we can observe a wide number of various active sites of the metal supported catalysts. The computational modeling effectively helps to examine diversification of the structures. Regarding the topics of this review we would like to emphasize the non-traditional variants of the bivalent metal cations stabilization in zeolites. This is interesting especially in the case of the distant placing of lattice Al atoms (Sect. 20.2) and in the theoretical prediction of possibility of the grafting of metal oxide molecules on a silicalite structures (Sect. 20.4). The metal cation species in zeolite matrix are very significant family of the catalytic systems. Recently, high-silica zeolites attracted great attention as one of the most effective supports for the active components. A possibility of a wide range of conditions for stabilization of active species is an important factor promoting the use of these systems as catalysts. The synthesis problems should be taken into account during theoretical modeling of the aluminum distribution.

The cluster model approach with various constraints imposed by the lattice appeared to be effective in the studies of the structure and reactivity of active sites. Nevertheless, the need to apply embedding scheme and especially periodical calculations is quite evident. Unfortunately, the cluster and periodical approaches can lead to disagreement in the calculated results. Understanding of the reasons and removal of this shortcoming is a task for the future studies.

The treatment and application of the alkaline earth (AE) zeolites could become an important problem in the nearest future because they are the products of water softening during washing process. Taking into account that zeolites take 15% w. of washing powder, the depositions of the partially exchangeable AE zeolites will present a real threat for environment. But the chemistry and reactivity of the AE zeolites are not indeed well understood.

Regarding the cluster and periodic studies of the AE cationic Me_2O_X moieties, Me=Mg and Ca, we observed two important consequences from their comparison. Both approaches showed more or less similar results with respect to the heat ΔU of oxidation reactions (20.7) by triplet oxygen. Different evaluations of the singlet-triplet energy differences were obtained with the periodic (ΔU_{ST}) or cluster (ΔU_{ST}^{CL}) methods. The fact that the absolute ΔU_{ST}^{CL} values decrease with the expansion of the cluster model at both the B3LYP and MP2 levels supports the results of the periodic models. Both these results show the possibility of the reactions involving the Me_2O_X moieties, for which the spin restriction presents a minor obstacle. The experimentally known reactions of the CO oxidation over the AE form zeolites were interpreted due to the Me_2O_2 and Me_2O_3 species. The products of the CO oxidation form tri- or tetra-dentate AE carbonates (regarding the number of the Me–O coordination bonds). The spectra of the tetra-dentate carbonates are in good agreement with known experimental data, while the spectra for tri-dentate ones were not observed on zeolites but over oxides. This last feature limits possible carbonates by the tetra-dentate carbonates. In the same way, the spectroscopic data restricts possibility of reactions between the CO_2 and the Me_2O_X species. The tetra-dentate $CaCO_3Ca$ carbonates were tested as the activated form of CO_2 in the course of dimethylcarbonate formation from methanol. It was demonstrated that various intermediate states can be considered as good initial points for search of the transition states for reactions between carbonates and methanol under moderate temperatures.

Unusual form of the bivalent iron cation stabilization in the Fe/HZSM-5 zeolite is specific for zeolite environment. Despite many attempts these centers were unable to be obtained outside of the zeolite channels. Thus, zeolites are not only high surface area catalytic supports, but they also provide some kind of additional stabilization, possibly similar to enzyme biological systems. In turn, this iron center can be loaded with oxygen (originating from N_2O decomposition) to make highly reactive and at the same time stable oxygen form, so called *alpha-oxygen*. Its behavior is similar to the radical oxygen species, such as O^-, OH•, for example. However, it is still unclear whether alpha-oxygen is Fe^{III}–O• radical structure from the beginning, or its stable form is $Fe^{IV}=O$ (which can be changed to Fe^{III}–O• at some stage of the hydrocarbon oxidation reaction). Therefore, the quantum-chemical simulations can significantly help to understand these problems and help the new experiments planning.

In summary we would like to note that a number of problems in the theoretical interpretation of catalytic activity of polyvalent cations in high-silica zeolites is of particular interest and the field is open to further studies using the theoretical methods.

References

1. Stöcker M (2005) Gas phase catalysis by zeolites. Microp Mesop Mater 82:257
2. Caeiro G, Carvalho RH, Wang X, Lemos MANDA, Lemos F, Guisnet M, Ribeiro FR (2006) Activation of C_2–C_4 alkanes over acid and bifunctional zeolite catalysts. J Mol Catal A Chem 255:131
3. Zhidomirov GM, Kazansky VB (1986) Quantum chemical cluster models of acid-base sites of oxide catalysts. Adv Catal 34:131
4. Zamaraev KI, Zhidomirov GM (1986) Active sites and the role of the medium in homogeneous, heterogeneous and enzymatic catalysis similarities and differences. In: Proceedings of the fifth international symposium on relations between homogeneous and heterogeneous catalysis, Novosibirsk, pp 23–74
5. Sauer J (1989) Molecular models in ab initio studies of solids and surfaces: from ionic crystal to catalysis. Chem Rev 89:199

6. van Santen RA, Kramer GJ (1995) Reactivity theory of zeolite Brönsted acid sites. Chem Rev 95:637
7. Uytterhoeven L, Dompas D, Mortier WJ (1992) Theoretical investigations on the interaction of benzene with faujasite. J Chem Soc Faraday Trans 88:2753
8. Zhidomirov GM, Shubin AA, van Santen RA (2004) Structure and reactivity of metal ion species in high silica zeolites. In: Catlow CRA, van Santen RA, Smit B (eds) Computer modelling of microporous materials. Elsevier, London, pp 201–241
9. Yakovlev AL, Shubin AA, Zhidomirov GM, van Santen RA (2000) DFT study of oxygen-bridged Zn^{2+} ion pairs in Zn/ZSM-5 zeolites. Catal Lett 70:175
10. Shubin AA, Zhidomirov GM, Yakovlev AL, van Santen RA (2001) Comparative quantum chemical study of stabilization energies of Zn^{2+} ions in different zeolite structures. J Phys Chem B 105:4928
11. Rice MJ, Chakraborty AK, Bell AT (2000) Site availability and competive siting of divalent metal cation in ZSM-5. J Catal 194:278
12. Rice MJ, Chakraborty AK, Bell AT (2000) Theoretical studies of the coordination and stability of divalent cations in ZSM-5. J Phys Chem B 104:9987
13. Bell AT (2001) Siting and stability of metal cations in zeolites for catalysis. In: Centi G, Wichterlova B, Bell AT (eds) Catalysis by unique metal ion structures in solid matrices, vol 13, NATO science series II. Kluwer, Dordrecht, pp 55–73
14. Wichterlova B, Dedecek J, Sobalik Z (2001) Single metal ions in host zeolite matrices. Structure-activity-selectivity-relationships. In: Centi G, Wichterlova B, Bell AT (eds) Catalysis by unique metal ion structures in solid matrices, vol 13, NATO science series II. Kluwer, Dordrecht, pp 31–54
15. Frisch MJ, Trucks GW, Schlegel HB, Scuseria GE, Robb MA, Cheeseman JR, Zakrzewski VG, Montgomery JA Jr., Stratmann RE, Burant JC, Dapprich S, Millam JM, Daniels AD, Kudin KN, Strain MC, Farkas O, Tomasi J, Barone V, Cossi M, Cammi R, Mennucci B, Pomelli C, Adamo C, Clifford S, Ochterski J, Petersson GA, Ayala PY, Cui Q, Morokuma K, Salvador P, Dannenberg JJ, Malick DK, Rabuck AD, Raghavachari K, Foresman JB, Cioslowski J, Ortiz JV, Baboul AG, Stefanov BB, Liu G, Liashenko A, Piskorz P, Komaromi I, Gomperts R, Martin RL, Fox DJ, Keith T, Al-Laham MA, Peng CY, Nanayakkara A, Challacombe M, Gill PMW, Johnson B, Chen W, Wong MW, Andres JL, Gonzalez C, Head-Gordon M, Replogle ES, Pople JA (2001) Gaussian 98, Revision A.11. Gaussian Inc., Pittsburgh
16. Frash M, van Santen RA (2000) Activation of ethane on Zn-exchanged zeolites: a theoretical study. Phys Chem Chem Phys 2:1085
17. Shubin AA, Zhidomirov GM, Kazansky VB, van Santen RA (2003) DFT cluster modeling of molecular and dissociative hydrogen adsorption on Zn^{2+} ions with distant placing of aluminium in the framework of high – silica zeolits. Catal Lett 90:137
18. Zhidomirov GM, Shubin AA, Kazansky VB, van Santen RA (2004) Possible molecular structure of promoted Lewis acidity sites in ZnZSM-5. Int J Quantum Chem 100:489
19. Kazansky VB, Kustov LM, Khodakov AM (1989) On the nature of active sites for dehydrogenation of saturated hydrocarbons in HZSM-5 zeolites modified by zinc and gallium oxides. In: Jacobs PA, van Santen RA (eds) Proceedings of the international conference on zeolities. Zeolites: facts, figures, future. Studies in surface science and catalysis, vol 49, Part B. Elsevier, Amsterdam, pp 1173–1182
20. Kazansky VB, Borovkov VYu, Serikh AI, van Santen RA, Anderson BG (2000) Nature of the sites of dissociative adsorption of dihydrogen and light paraffins in ZnHZSM-5 zeolite prepared by incipient wetness impregnation. Catal Lett 66:39
21. Kazansky VB, Serykh AI, Anderson BG, van Santen RA (2003) The sites of molecular and dissociative hydrogen adsorption in high-silica zeolites modified with zinc ions. III DRIFT study of H_2 adsorption by the zeolites with different zinc content and Si/Al ratios in the framework. Catal Lett 88:211
22. Kazansky VB, Serykh AI, Pidko EA (2004) DRIFT study of molecular and dissociative adsorption of light paraffins by HZSM-5 zeolite modified with zinc ions: methane adsorption. J Catal 225:369

23. Barbosa LAMM, Zhidomirov GM, van Santen RA (2001) Theoretical study of the molecular hydrogen adsorption and dissociation on different Zn(II) active sites of zeolites. Catal Lett 77:55

24. Barbosa LAMM, Zhidomirov GM, van Santen RA (2000) Theoretical study of methane adsorption on Zn(II) zeolites. Phys Chem Chem Phys 2:3909

25. Zhidomirov GM, Shubin AA, Kazansky VB, van Santen RA (2005) Spectroscopic identification of adsorption properties of Zn^{2+} ions at catalytic positions of high-silica zeolites with distant placing of aluminium ions. Theor Chem Acc 114:90

26. Kazansky VB, Pidko EA (2005) Intensities of IR stretching bands as a criterion of polarization and initial chemical activation of adsorbed molecules in acid catalysis. Ethane adsorption and dehydrogenation by zinc ions in ZnZSM-5 zeolite. J Phys Chem B 109:2103

27. Pidko EA, van Santen RA (2007) Activation of light alkanes over zinc species stabilized in ZSM-5 zeolite: a comprehensive DFT study. J Phys Chem C 111:2643

28. Joshi YV, Thomas KT (2005) The role of gallium hydride and Brönsted acidity in light alkane dehydrogenation mechanism using Ga-exchanged HZSM-5 catalyst: a DFT pathway analysis. Catal Today 105:106

29. Kachurovskaya NA, Zhidomirov GM, van Santen RA (2004) Comparative energies of Zn(II) cation localization as a function of the distance between two forming cation position aluminium ions in high-silica zeolites. Res Chem Intermed 30:99

30. (a) Kresse G, Hafner J (1993) Ab initio molecular dynamics for liquid metals. Phys Rev B 47:558; (b) Kresse G, Furthmüller J (1996) Efficient iterative schemes for ab initio total-energy calculations using a plane–wave basis set. Phys Rev B 54:11169

31. Benco L, Bucko T, Hafner J, Toulhoat H (2005) Periodic DFT calculation of the stabilization of Al/Si substitutions and extraframework Zn^{2+} cation in mordenate and reaction pathway for the dissociation of H_2 and CH_4. J Phys Chem B 109:20361

32. Kao H-M, Grey CP, Pitchumani K, Lakshminarasimhan PH, Ramamurthy V (1998) Activation conditions play a key role in the acidity of zeolite CaY: NMR and product studies of Brönsted acidity. J Phys Chem A 102:5627

33. Aleksandrov HA, Vayssilov GN, Rösch N (2006) Heterolytic dissociation and recombination of H2 over Zn, H-ZSM-5 zeolites – a density functional model study. J Mol Catal A Chem 256:149

34. Barbosa LAMM, van Santen RA (2007) The activation of H2 by zeolitic Zn(II) cations. J Phys Chem C 111:8337

35. Lei GD, Adelman BJ, Sarkany J, Sachtler WMH (1995) Identification of copper(II) and copper(I) and their interconversion in Cu/ZSM-5 De-NOx catalysts. Appl Catal B 5:245

36. Sayle DC, Catlow CRA, Gale JD, Perrin MA, Nortier P (1997) Computer modeling of the active-site configurations within the NO decomposition catalyst Cu-ZSM-5. J Phys Chem A 101:3331

37. Teraishi K, Ishida M, Irisawa J, Kume M, Takahashi Y, Nakano T, Nakamura H, Miyamoto A (1997) Active site structure of Cu/ZSM-5: computational study. J Phys Chem B 101:8079

38. Catlow CRA, Bell RG, Gale JD, Lewis DW, Sayle DC, Sinclair PE (1998) In: Derouane EG et al (eds) Catalytic activation and functionalization of light alkanes. Kluwer Academic, Dordrecht, pp 189–214

39. Goodman BR, Schneider WF, Hass KC, Adams JB (1998) Theoretical analysis of oxygen-bridged Cu pairs in Cu-exchanged zeolites. Catal Lett 56:183

40. El-Malki El-M, van Santen RA, Sachtler WMH (2000) Active sites in Fe/MFI catalysts for NOx reduction and oscillating N2O decomposition. J Catal 196:212

41. Yakovlev AL, Zhidomirov GM, van Santen RA (2001) DFT calculations on N2O decomposition by binuclear Fe complexes in Fe/ZSM-5. J Phys Chem B 105:12297

42. Hansen N, Heyden A, Bell AT, Keil FJ (2007) A reaction mechanism for the nitrous oxide decomposition on binuclear oxygen bridged iron sites in Fe-ZSM-5. J Phys Chem C 111:2092

43. Delgass WN, Garten RL, Boudart M (1969) Dehydration and adsorbate interactions of Fe-Y zeolite by Mossbauer spectroscopy. J Phys Chem 73:2970

44. Garten RL, Delgass WN, Boudart M (1970) A Mössbauer spectroscopic study of the reversible oxidation of ferrous ions in Y zeolite. J Catal 18:90

45. Dalla Betta RA, Garten RL, Boudart M (1976) Infrared examination of the reversible oxidation of ferrous ions in Y zeolite. J Catal 41:40

46. Hikichi S, Yoshizawa M, Sasakura Y, Akita M, Moro-oka Y (1998) First synthesis and structural characterization of dinuclear M(III) Bis(*i*-oxo) complexes of nickel and cobalt with hydrotris(pyrazolyl)borate ligand. J Am Chem Soc 120:10567

47. Duan Z, Schmidt M, Young VG Jr, Xie X, McCarley RE, Verkade JG (1996) The novel Bis(oxo-bridged) dinuclear vanadium(IV) complex {(*i*-O)2V2[N(SiMe3)2]4}: an unexpected reaction product. J Am Chem Soc 118:5302

48. Costas M, Rohde J-U, Stubna A, Ho RYN, Quaroni L, Munck E, Que L Jr (2001) A synthetic model for the putative FeIV2O2 diamond core of methane monooxygenase intermediate Q. J Am Chem Soc 123:12931

49. Groothaert MH, Smeets PJ, Sels BF, Jacobs PA, Schoonheydt RA (2005) Selective oxidation of methane by the Bis(*i*-oxo)dicopper core stabilized on ZSM-5 and mordenite zeolites. J Am Chem Soc 127:1394

50. Groothaert MH, van Bokhoven JA, Battiston AA, Weckhuysen BM, Schoonheydt RA (2003) Bis(*i*-oxo)dicopper in Cu-ZSM-5 and its role in the decomposition of NO: a combined in situ XAFS, UV-Vis-near-IR, and kinetic study. J Am Chem Soc 125:7629

51. Zhidomirov GM, Larin AV, Trubnikov DN, Vercauteren DP (2009) Ion-exchanged binuclear clusters as active sites of selective oxidation over zeolites. J Phys Chem C 113:8258

52. Larin AV, Zhidomirov GM, Trubnikov DN, Vercauteren DP (2010) Ion-exchanged binuclear Ca2OX clusters, X = 1–4, as active sites of selective oxidation over MOR and FAU zeolites. J Comput Chem 31:421

53. Rybakov AA, Larin AV, Zhidomirov GM, Trubnikov DN, Vercauteren DP (2011) DFT investigation of mechanism of CO oxidation over Mg exchanged zeolite. Comput Theor Chem 964:108

54. Larin AV, Rybakov AA, Zhidomirov GM, Mace A, Laaksonen A, Vercauteren DP (2011) Oxide clusters as source of the third oxygen atom for the formation of carbonates in alkaline earth dehydrated zeolites. J Catal 281:212

55. Udalova OV, Khaula EV, Bykhovski MY, Rufov YN, Romanov AN (2003) Thermal generation of singlet oxygen in zeolites and its role in the alkene oxidation. Russ J Phys Chem 77:912

56. Udalova OV, Khaula EV, Bykhovski MY, Rufov YN (2007) The heterogeneous catalytic oxidation of propylene in the presence of singlet oxygen in the reaction mixture. Russ J Phys Chem A 81:1511

57. Xu J, Mojet BL, van Ommen JG, Lefferts L (2005) Formation of M2+(O2)(C3H8) species in alkaline-earth-exchanged Y zeolite during propane selective oxidation. J Phys Chem B 109:18361

58. Angel I CL, Schaffer PC (1965) Infrared spectroscopic investigations of zeolites and adsorbed molecules. I. Structural OH groups. J Phys Chem 69:3463

59. Gale JD (1992/1994) GULP 1.3. Royal Institution/Imperial College, London

60. Schröder KP, Sauer J, Leslie M, Catlow CRA, Thomas JM (1992) Bridging hydrodyl groups in zeolitic catalysts: a computer simulation of their structure, vibrational properties and acidity in protonated faujasites (H-Y zeolites). Chem Phys Lett 188:320

61. Gale JD, Henson NJ (1994) Derivation of interatomic potentials for microporous aluminophosphates from the structure and properties of berlinite. J Chem Soc Faraday Trans 90:3175

62. Kresse G, Joubert J (1999) From ultrasoft pseudopotentials to the projector augmented-wave method. Phys Rev B 59:1758

63. Perdew JP, Chevary JA, Vosko SH, Jackson KA, Pederson MR, Singh DJ, Fiolhais C (1992) Atoms, molecules, solids, and surfaces: applications of the generalized gradient approximation for exchange and correlation. Phys Rev B 46:6671; (b) Perdew JP, Burke K, Ernzerhof M (1996) Generalized gradient approximation made simple. Phys Rev Lett 77:3865

64. Ugliengo P, Viterbo D, Chiari G (1993) MOLDRAW: molecular graphics on a personal computer. Z Kristall 207:9

65. Huang X, Lee TJ (2008) A procedure for computing accurate *ab initio* quartic force fields: application to HO_2^+ and H_2O. J Chem Phys 129:044312

66. Lu N-X, Fu G, Xu X, Wan H-L (2008) Mechanisms for O_2 dissociation over the BaO(100) surface. J Chem Phys 128:034702-1

67. Huber KP, Herzberg G (1979) Molecular spectra and molecular structure, vol. 4. Constants of diatomic molecules. Van Nostrand, New York

68. Shannon RD (1976) Revised effective ionic radii and systematic studies of interatomic distances in halides and chaleogenides. Acta Cryst A 32:751

69. Seidl ET, Schaefer HF III (1988) Theoretical studies of oxygen rings: cyclotetraoxygen O_4. J Chem Phys 88:7043

70. Kim KS, Jang JH, Kim S, Mhin B-J, Schaefer HF III (1990) Potential new high density energy materials: cyclooctaoxygen O_8, including comparison with well-known cyclo-S_8 molecule. J Chem Phys 92:1887

71. Förster H, Frede W, Peters G, Schumann M, Witten U (1981) Zeolite-catalysed oxidation of carbon monoxide at unusually low temperatures. J Chem Soc Chem Comm 1081:1064–1065; (b) Shete BS, Kamble VS, Gupta NM, Kartha VB (1998) Fourier transform infrared study on the encapsulation of CO in zeolite Y under the moderate temperature and pressure conditions. J Phys Chem B 102:5581

72. Garrone E, Bonelli B, Lamberti C, Civalleri B, Rocchia M, Roy P, Otero Areán C (2002) Coupling of framework modes and adsorbate vibrations for CO_2 molecularly adsorbed on alkali ZSM-5 zeolites: mid- and far-infrared spectroscopy and *ab initio* modeling. J Chem Phys 117:10274

73. Jacobs PA, van Cauwelaert FH, Vansant EF (1973) Surface probing of synthetic faujasites by adsorption of carbon dioxide. Part I.-infra-red study of carbon dioxide adsorbed on Na–Ca–Y and Na–Mg–Y zeolites. J Chem Soc Faraday Trans I 69:2130

74. Parkyns ND (1969) The surface properties of metal oxides. Part II. An infrared study of the adsorption of carbon dioxide on γ-alumina. J Chem Soc A4: 10–417

75. Chen J, Feng Z, Ying P, Li C (2004) ZnO cluster encapsulated inside micropores of zeolite studied by UV Raman and laser induced luminescence spectroscopies. J Phys Chem B 108:12669

76. Kolyagin YuG, Ordomscy VV, Knimyak YZ, Rebrov AI, Fajula F, Ivanova II (2006) Initial stages of propane activation over Zn/MFI catalyst studied by in situ NMR and IR spectroscopic techniques. J Catal 238:123

77. Kowalak S, Stawinski K, Mackoviak A (2001) Incorporation of zinc into silica mesoporous molecular sieves. Microp Mesop Mater 44–45:283

78. Panov GI (2000) Advances in oxidation catalysis; oxidation of benzene to phenol by nitrous oxide. CATTECH 4:18

79. Panov GI, Uriarte AK, Rodkin MA, Sobolev VI (1998) Generation of active oxygen species on solid surfaces. Opportunity for novel oxidation technologies over zeolites. Catal Today 41:365

80. Panov GI, Dubkov KA, Starokon EV (2006) Active oxygen in selective oxidation catalysis. Catal Today 117:148

81. Kachurovskaya NA, Zhidomirov GM, Hensen EJM, van Santen RA (2003) Cluster model DFT study of the intermediates of benzene to phenol oxidation by N_2O on FeZSM-5 zeolites. Catal Lett 86:25

82. Berlier G, Bonino F, Zecchina A, Bordiga S, Lamberti C (2003) Anchoring Fe ions to amorphous and crystalline oxides: a means to tune the degree of Fe coordination. Chemphyschem 4:1073

83. Hensen EJM, Zhu Q, van Santen RA (2003) Extraframework Fe–Al–O species occluded in MFI zeolite as the active species in the oxidation of benzene to phenol with nitrous oxide. J Catal 220:260

84. Panov GI, Dubkov KA, Paukshtis EA (2001) Identification of active oxygen species over Fe complexes in zeolites. In: Centi G et al (eds) Catalysis by unique metal Ion structures in solid matrices, vol 13. Kluwer, Dordrecht, p 149

85. Starokon EV, Parfenov MV, Pirutko LV, Abornev SI, Panov GI (2011) Room-temperature oxidation of methane by alpha-oxygen and extraction of products from the FeZSM-5 surface. J Phys Chem C 115:2155

86. de Montellano PRO (ed) (2003) Cytochrome P450: structure, mechanisms and biochemistry. Plenum, New York

87. Dubkov KA, Sobolev VI, Talsi EP, Rodkin MA, Watkins NH, Shteinman AA, Panov GI (1997) Kinetic isotope effects and mechanism of biomimetic oxidation of methane and benzene on FeZSM-5 zeolite. J Mol Catal A Chem 123:155

88. Zecchina A, Rivallan M, Berlier G, Lamberti C, Ricchiardi G (2007) Structure and nuclearity of active sites in Fe-zeolites: comparison with iron sites in enzymes and homogeneous catalysts. PCCP 9:3483

89. Malykhin S, Zilberberg I, Zhidomirov GM (2005) Electron structure of oxygen complexes of ferrous ion center. Chem Phys Lett 414:434

90. Jia J, Sun Q, Wen B, Chen LX, Sachtler WMH (2002) Identification of highly active iron sites in N_2O-activated Fe/MFI. Catal Lett 82:7

91. Louwerse MJ, Baerends EJ (2007) Oxidative properties of FeO^{2+}: electronic structure and solvation effects. PCCP 9:156

92. Pirngruber GD, Grunwaldt J-D, Roy PK, van Bokhoven JA, Safonova O, Glatzel P (2007) The nature of the active site in the Fe-ZSM-5/N_2O system studied by (resonant) inelastic x-ray scattering. Catal Today 126:127

93. Witek HA, Choe Y, Finley JP, Hirao K (2002) Intruder state avoidance multireference Möller-Plesset perturbation theory. J Comput Chem 23:957

94. Fiedler A, Shaik S, Schröeder D, Schwarz H (1994) Electronic structures and gas-phase reactivities of cationic late-transition-metal oxides. J Am Chem Soc 116:10734

95. Fiolhais C, Nogueira F, Marques M (eds) (2003) Density functionals for non-relativistic coulomb systems in the new century. Springer, Berlin/Heidelberg/New York

96. Schwabe T, Grimme S (2008) Theoretical thermodynamics for large molecules: walking the thin line between accuracy and computational cost. Acc Chem Res 41:569

97. Sobolev VI, Kovalenko ON, Kharitonov AS, Pankrat'ev YD, Panov GI (1991) Anomalously low bond energy of surface oxygen on FeZSM-5 zeolite. Mendeleev Commun 1:29

98. Malykhin S, Zilberberg I, Ruzankin SPh, Zhidomirov GM (2004) Unrestricted density functional theory of the bonding between NO radical and ferrous ion. Chem Phys Lett 394:392

99. Noodleman L (1981) Valence bond description of antiferromagnetic coupling in transition metal dimers. J Chem Phys 74:5737

100. Dubkov KA, Paukshtis EA, Panov GI (2001) Stoichiometry of oxidation reactions involving α–oxygen on FeZSM-5 zeolite. Kinet Catal 42:205

101. Kachurovskaya NA, Zhidomirov GM, van Santen RA (2004) Computational study of benzene to phenol oxidation catalyzed by N_2O on iron-exchanged ferrierite. J Phys Chem B 108:5944

102. Fellah MF, van Santen RA, Onal I (2009) Oxidation of benzene to phenol by N_2O on an Fe^{2+} ZSM–5 cluster: a density functional theory study. J Phys Chem C 113:15307

103. Ivanov AA, Chernyavsky VS, Gross MJ, Kharitonov AS, Uriarte AK, Panov GI (2003) Kinetics of benzene to phenol oxidation over Fe-ZSM-5 catalyst. Appl Catal A Gen 249:327

104. Rebmann G, Keller V, Ledoux MJ, Keller N (2008) Cu–Y zeolite supported on silicon carbide for the vapour phase oxidative carbonylation of methanol to dimethyl carbonate. Green Chem 10:207

Chapter 21
Towards Involvement of Interactions of Nucleic Acid Bases with Minerals in the Origin of Life: Quantum Chemical Approach

Andrea Michalkova and Jerzy Leszczynski

Abstract This review describes the results of studies of interactions of nucleic acid bases with water, cations and minerals. It focuses on three areas investigated computationally: (1) carbon fixation cycle; (2) 1-methylcytosine tautomerism due to presence of water and cation; (3) adsorption of thymine and uracil on clay minerals. It reveals how the individual reactions responsible for the generation of acetic acid on iron-nickel sulfide surfaces as catalysts could have operated to produce carboxylic acids from carbon oxide and water. The importance of these results in terms of a primordial chemistry on iron-nickel sulfide surfaces is discussed. The interaction with the Na^+ cation has a dramatic effect on equilibrium between *oxo-amino* and *oxo-imino* tautomers of methylcytosine and can lead to a decrease in concentration of *imino-oxo* tautomer (present in the cell during DNA synthesis) responsible for the point mutations in DNA during formation of a mismatched base pair. Thymine and uracil interact quite strongly with the kaolinite mineral surfaces but uracil is slightly better stabilized than thymine on these clays. The adsorption leads to their structural changes, which are induced by the deposition way and are driven by the surface potential. Explicit addition of water molecule changes only slightly its adsorption properties, comparing the presence of the sodium cation.

Keywords Nucleobase • Kaolinite • DFT • Adsorption • Tautomerism • Carbon fixation cycle

A. Michalkova • J. Leszczynski (✉)
Interdisciplinary Nanotoxicity Center, Jackson State University, Jackson, MS, 39217, USA
e-mail: jerzy@icnanotox.org

J. Leszczynski and M.K. Shukla (eds.), *Practical Aspects of Computational Chemistry I:*
An Overview of the Last Two Decades and Current Trends,
DOI 10.1007/978-94-007-0919-5_21, © Springer Science+Business Media B.V. 2012

21.1   Introduction

Two dominating hypotheses concerning the origin of life are the replication-first and metabolism-first theory (for a review see ref. [1]). Currently, there are two detailed evolutionary scenarios representing the "metabolism first" concept (see ref. [2, 3] for recent surveys). Wächtershäuser's key concept proposes envisioned "two-dimensional" primordial metabolic cycles driven by oxidation of iron monosulfide (FeS) into iron disulfide (FeS_2, pyrite) and confined to the Ni-Fe-S and Fe-S mineral surfaces at the sea floor near deep submarine hydrothermal vents [4–11]. Common Ni-Fe-S mineral found at the sea floor submarine hydrothermal vents is pentlandite. However, it was stated that greigite and violarite are more likely to have been the catalysts [12]. Violarite was generated and Huber's and Wächtershäuser's generation of acetic acid [9] with this catalyst was successfully repeated. Moreover, it was concluded that violarite nanocrystals may have the potential to act as a carbon monoxide dehydrogenase (in place of pentlandite) because in this sulfide mineral the sulfurs on the surface are likely to be hydrogenated, leaving the nickel and iron sites open [13]. Violarite possesses an empirical $Fe^{2+}Ni_2S_4$ formula and it is of hydrothermal origin. It is a normal thiospinel with extensively delocalized valence electrons [14]. In this mineral nickel occurs in both the tetrahedral and octahedral sites, whereas the iron is most likely restricted to the octahedral site and occurs in the low-spin state.

The involvement of surface chemistry on clays and other minerals in the prebiotic chemical evolution that culminated in the origin of life was suggested in several studies [15, 16]. Among proposed models and theories the first one is Cairns-Smith's "clay world theory" [17] according to which all the major control structures (such as genes, membranes and catalysts) were at first inorganic and in a second step had been replaced by more evolved organic structures. This model seems to have several weaknesses since the genetic takeover, leading from the clay world to the biological world, has not been proven. However, the role of clay minerals in prebiotic chemistry as catalysts has been shown extensively. The adsorption on clay minerals and their catalytic properties might have had a significant role in the binding the organic molecules from the surrounding water. This process significantly increases their concentration and aggregates them into biologically relevant structures. Such structures act as naturally occurring environment for polymerization of large molecules, including self-replicating informational molecules [17–22]. This theory was supported by other studies (for review see ref. [23]) where it was revealed that the genetic material retains its integrity and functionality upon interaction with clays while maintaining its biological activity.

Clay minerals with their charged aluminosilicate layered structure were assumed to have the appropriate characteristics to harbor precursor organic molecules for the synthesis of important biomolecules. Minerals with positively charged surfaces were suggested to catalyze the formation of sugar phosphates [24, 25] while clays with negatively charged surfaces [26] could have bound and concentrated organic molecules from the surrounding water [17–19, 22]. The suggestions of a surface-mediated origin of life were reinforced by several experiments. The replication of DNA analogs on templates adsorbed on solid phase was demonstrated [27] with a theory for the assembly of biopolymers on silica-rich minerals [28, 29]. On montmorillonite the polymerization of oligonucleotides, up to the length of small ribozymes was also demonstrated [30]. Moreover, since clays exist everywhere it is highly likely that they were important in protecting materials from degradation by adsorption on active sites [31].

Mineral surfaces may protect nucleic acids, by shielding certain bonds from chemical hydrolysis [32, 33], enzymatic hydrolysis [34], or UV radiation [35]. They may also catalytically degrade them by providing Lewis or other catalytic sites in proximity to labile bonds [36–38]. Whether catalytic degradation or preservation predominates may be somewhat environmentally idiosyncratic, but both effects are likely dependent on the affinity of nucleic acids for mineral surfaces.

Studies on the characteristics of nucleic acid molecules adsorbed on the clay minerals such as montmorillonite and kaolinite have indicated that the formation of clay-nucleic acid complexes could have been an important step in the preservation of genetic material in primeval habitats [39]. Specifically, the nucleic acid adsorption on clay minerals was investigated as early as 1952 by Goring and Bartholomew [40]. Since then DNA/RNA adsorption on clay minerals has been studied in several works [27]. In 1996 Ferris et al. [30] demonstrated the polymerization of oligonucleotides on clay. Nucleic acids adsorbed and bound on clay minerals are partially protected against degradation by nucleases and other degradative enzymes [34, 41–46]. Two hypotheses have been suggested to describe the adsorption and binding of nucleic acids onto clays. According to Khanna et al. [47] one end of DNA is bound to the edges of the clay with a fraction of the DNA bound on the planar surface. Another model of the DNA-clay complexes has been described by Paget and Simonet [48] where DNA is partially adsorbed on soil having a part ("train") interacting with the soil particles and a remaining moiety ("tail") not involved in the interaction. Some aspects of this work, which relate to the effect of differing molecular masses of DNA and their adsorption to clays still, remain controversial [49]. In clay-nucleic acid complexes, the nucleic acid is adsorbed on the surface of the clay minerals through binding of the substrate by electrostatic and/or hydrogen bonds [39, 50]. The lengths and shape of DNA molecules affects the adsorption on minerals [51]. In the adsorption process, the DNA likely changes its configuration to have a more compact length and shape to allow better adsorption on clay. In practical applications, clays can incorporate the DNA molecules to provide a remedial for the gene therapy of leukemia [52], carriers for the gastrointestinal release of selected cationic drugs [53], and chemotherapeutic treatment of colorectal cancer [54].

The adsorption and binding of DNA by clay minerals can be affected by many different factors [34, 48, 55] such as for example pH [43, 56], presence of water [57], the clay type [58], the nature of exchangeable interlayer cations [56, 59], and also the DNA structure [44, 51]. However, the obtained results are in conflict with each other and the interactions of DNA with clay minerals have not been clarified yet except the predictions that the adsorption could occur also due to the Coulombic attraction, van der Waals interactions [26] and hydrogen bonding [43].

Furthermore, the results helped to characterize how nucleic acid molecules are adsorbed on clay minerals including kaolinite [28, 29]. For example, double stranded DNA molecules that differ in their guanine-cytosine content were adsorbed in equal amounts by clays including kaolinite [49]. Linear DNA was revealed to be adsorbed on illite and kaolinite to a greater extent than on montmorillonite [60]. This emphasizes the influence that positive charges of the lattice edges and the microorganization of clay particles have on the mechanism of DNA adsorption.

The minerals of the kaolinite group can adsorb polar organic molecules relatively easily. In the ideal case, the layers of these minerals are not charged and their interlayer space does not contain exchange cations or intercalated molecules. Dickite and kaolinite are typical representatives of minerals of the kaolinite group with a dioctahedral 1:1 layer structure consisting of an octahedral aluminum hydroxide sheet and a tetrahedral silica sheet [61, 62]. The tetrahedral side of a layer is characterized by ditrigonal cavities. The layers are kept together via hydrogen bridges between the surface hydroxyl groups of the octahedral side and the basal oxygen atoms of the tetrahedral side. Therefore, the active sites for adsorption on the clay minerals are hydroxyl groups of the octahedral side and oxygen atoms of the Si-O-Si groups on the tetrahedral side.

The isomorphic substitution of some Al^{3+} for Si^{4+} in the tetrahedral sheet and some Mg^{2+} for Al^{3+} in the octahedral sheet (or to some local partial charges on Si-O-Si surfaces) was suggested to occur in the minerals of the kaolin group [63], which can lead to the generation of the surface cations and the presence of local charge defects [64]. The basal siloxane surfaces of kaolinite are believed by many [65–68] to carry a constant structural charge due to the isomorphous substitution of Si^{4+} by Al^{3+} [69]. Natural kaolinites do contain very small amount of Mg [70] and Mg-rich kaolinite was synthesized [71].

In particular, the stability and conformational integrity of nucleic acids are largely controlled by the interactions with surrounding water molecules [72–74]. The water molecules interact with uracil through two hydrogen bonds [75] and four uracil-water minima were found [76]. Hydration involving the oxygen of uracil is energetically more favorable than hydration involving nitrogen [77, 78]. Moreover, all uracil + $n_w H_2O$ complexes retained a planar geometry of uracil [79]. Depending on the nature of tautomers these complexes have cyclic or open structures [80]. Several theoretical works investigated tautomerism of uracil [81] and thymine base pairs [82].

Metal ions were shown to be vitally important in the origin of life [49, 56, 59, 60]. For example, selective adsorption of the molecules onto mineral surfaces promotes concentration and polymerization of various activated monomers. Such absorption, in fact can be considered as the interaction of biologically active species with completely or partially hydrated metal cations including alkali cations. They are necessary for stabilization of biologically active structures of nucleic acids [83–89]. Among a number of different metal cations, which are found to be important for the interaction with DNA bases [83, 87], the alkali cations play a very special role. At low concentration, the alkali cations interact with the negatively charged phosphate groups of the nucleic acid chain to provide charge neutralization [90–92].

At higher concentrations they also interact with the nucleic acid bases so that they can even destroy the base pair hydrogen bonding, and compromise the structural integrity of the nucleic acid polymers [90–92].

21.2 Computational Details

The calculations of the systems of nucleic acid bases with water, cations and clay minerals were performed using the density functional theory (DFT) [93] as implemented in the Gaussian03 and Gaussian09 program packages [94]. Firstly, several DFT functionals (B3LYP (Becke, three-parameter, Lee-Yang-Parr)), BLYP [95, 96] and M05-2X [97] were applied. An application of B3LYP functional in studies of large systems has become extremely popular in the area of adsorption on the clay minerals but this functional has unsatisfactory performance issues such as underestimation of interaction energies for weak non-covalent interactions [98]. Therefore, the interaction energies were also calculated using the M05-2X functional. M05-2X is a hybrid meta exchange-correlation functional [97] derived from the M05 functional [99], which adds a kinetic energy component to the exchange-correlation function.

Several different basis sets, including pseudopotential LANL2DZ [100] as well as two Aldrich's valence split basis sets (triple zeta (TZVP) and quadruple zeta (QZVP) valence quality) [101] were used to calculate modeled reactions involving small FeS clusters. Employing an electron core potential (ECP) basis set such as LANL2DZ (Los Alamos National Laboratory 2 double ζ for transition metals) has become popular in computations on a transition-metal-containing systems. Due to large size of calculated models in the case of the adsorption of nucleic acid bases on the clay mineral surfaces the 6-31G(d) basis set [102] was applied.

The values of the interaction energy (Eint) of studied systems are obtained as difference between the energy of the complex and the sum of the energy of the adsorbate and adsorbent subsystems. The Eint value was corrected by the basis set superposition error (BSSE) using the counterpoise method [103] (the corrected interaction energies are denoted Ecorr). The topological characteristics of the electron density distribution were obtained following Bader's "Atoms in Molecules" approach (AIM) [104]. This method reveals insightful information on the nature of bonds. A $(3,-1)$ critical point of the electron density located between two atomic centers denotes the presence of stabilizing interaction, in particular case the presence of a bond [105, 106]. Charge density at such a point is referred as $\rho(r)$. Typically, a closed-shell interaction of electrons (ionic, van der Waals, or hydrogen bonds) is identified with a small $\rho(r)$, and a large and positive the Laplacian of the electron density $(\nabla^2\rho(r))$. The maps of electrostatic potential (MEPs) of nucleobases adsorbed on clay minerals were also calculated using the Molekel program package [107].

The solvation of the system was modeled in two different ways. Firstly the supermolecular approximation was applied, which involves the explicit consideration of water molecules (microsolvation). Second approach is based on replacement of the explicit solvent molecules by a dielectric continuum with a permittivity ε, surrounding the solute molecules outside of a molecular cavity (Conductor-like Screening Model, COSMO) [63] (water was used as solvent with the relative dielectric constant $\epsilon_r = 78.39$). In the case of study of thermodynamics, the values of interaction enthalpy (ΔH), Gibbs free energy (ΔG) and entropy $S(T)$ have been calculated using rigid rotor – harmonic oscillator – ideal gas approximation based on the vibrational frequencies and optimized structures of studied systems.

To estimate the rate constants of proton transfer reactions, the approximate instanton approach has been employed as implemented in the DOIT 1.2 program [108]. The tunneling rate constant for the G*C* → GC proton transfer has been calculated using the expression:

$$k_r(T) = (\Omega_0^t/2\pi)e^{-S_I(T)} \tag{21.1}$$

where Ω_0^t is the effective tunneling frequency in the equilibrium configuration of G*C*, and $S_I(T)$ is the multidimensional instanton action. The rate constants for the proton transfer processes were calculated from $k_f(T) = K(T) k_r(T)$, where $K(T)$ is the equilibrium constant calculated using the Boltzmann's formula:

$$K(T) = e^{-\frac{\Delta G}{RT}} \tag{21.2}$$

The models of kaolinite mineral fragments were prepared using its crystal structure data [109]. Both types of mineral surfaces (tetrahedral and octahedral) were mimicked. These models consist of tetrahedral or octahedral rings and the borders of the clusters were saturated by the hydrogen atoms. Several different initial positions of adsorbate were tested to investigate the most advantageous orientation (parallel, perpendicular orientations, cation-π and cation-heteroatom interactions) of the interacting species. Substitution of Si^{4+} by Al^{3+} in the tetrahedral and Al^{3+} by Mg^{2+} in octahedral fragment was modeled. After the substitution the model was kept electroneutral by addition of sodium 1+ cation (Na^+).

Models of Ni-Fe sulfide were constructed as cut offs from the crystal structure of violarite [110] with space group Fd-3m and cubical unit cell with the parameters $a = b = c = 9.465\,\text{Å}$. Due to total structural changes caused by the optimization the original model was modified, its size was decreased by removal of the ambiguous part. The new model consists of one Fe, one Ni atom, and six sulfur atoms. Dangling bonds that occur on four sulfur atoms were saturated by the hydrogen atoms to ensure the electroneutrality of the entire system.

21.3 Experimental Studies of Interactions of Minerals with DNA

The prebiotic biochemistry is predicted to begin when two volatiles met at a hydrothermal vent rich in metal sulfides [111]. In this 'hydrothermal reactor' hypothesis, a primitive, inorganically catalyzed analogue of the exergonic acetyl-CoA pathway, an application of H_2 as the initial electron donor and CO_2 as the initial acceptor, was instrumental in the synthesis of organic precursors to fuel primordial biochemical reactions. Primordial biochemistry is suggested to house in an acetate-producing hydrothermal reactor that retained reduced carbon compounds produced within its naturally forming inorganic confines. A few papers were published on investigation of the mechanism of the redox system FeS/H_2S and its properties. For these studies a novel method to produce and immobilize FeS nanoparticles has been developed [112]. The parageneses thermodynamic analysis of components of chemoautotrophic CO_2 archaic fixation (CAF) cycle, their stability and self-organization in hydrothermal systems was studied by Marakushev and Belonogova [113]. The authors used thermodynamic data for aqueous organic compounds to construct the chemical potential diagrams. The tentative integrated system of CAF was developed by combining acetyl-CoA pathway, 3-hydroxypropionate cycle (3-HPC) and reductive citric acid cycle (RCC) containing a succinate–fumarate core, capable of switching electron flow in forward or reverse direction depending on the redox potential of the geochemical environment.

Despite continuing interest in the metabolism-first theory, this approach possesses many fundamental problems that have not been sufficiently addressed and a generally accepted proposal is still lacking [1]. In paper by Kalapos [114] the operation of a surface-bound system and presentation in the original version of the archaic RCC is criticized, particularly from an energetic point of view (an aspect not addressed in recent reviews by neither Anet [1] nor Cody [115]). The most significant differences between the Wächtershäuser hypothesis [6] and new proposal is that Wächtershäuser did not consider individual reactions in his calculations. Kalapos [114] disputed involvement of a considerable number of sulfur-containing organic intermediates as building blocks and questioned the involvement of seven molecules of pyrite, which does not emerge as a direct consequence of the chemical reactions presented in the archaic RCC. The new scheme of the cycle was proposed, according to which less free energy is released than hypothesized by Wächtershäuser. The advantages of this new scheme are that the number of pyrite molecules involved in the cycle is reduced, the free energy changes for the individual reactions can be calculated, and fewer sulfur-containing intermediates are required for the cycle to operate [116].

It is of high interest to gain a more fundamental understanding of the adsorption behavior of nucleic acids (NA) on minerals and metal oxides in general. Thus, many experimental studies on the interactions of mineral surfaces with DNA were published, which are summarized in the Introduction section. Here we will describe only a few, which relate directly to the results of theoretical studies discussed in the next section.

It was shown that crystallographically different surfaces presented by rock-forming minerals [117, 118] may have markedly different affinities for NAs. The adsorption of NAs to mineral surfaces was proposed to occur through ionic interactions between the negatively charged phosphate groups of the DNA backbone and positively charged surface groups. Various low molecular weight phosphate compounds were shown to desorb nucleic acids from mineral surfaces [40]. At low pH the functional groups of the NA bases may also contribute to adsorption via ionic interactions if the mineral surface is negatively charged [119].

While interactions of clays with nucleotides, nucleic acids and nucleoproteins have been investigated [40], only a few papers were published about reactions of nucleic acid bases with inorganic soil constituents. Both organic and inorganic soil components were active in adsorbing of NA, and thus the role of the latter could be more important than had been assumed [120]. An increase in the concentration of cations and/or a decrease in the pH favor adsorption of NA and nucleotides by reducing the electrostatic repulsion between the negatively charged polyanions and the clay surface [121]. In addition, adsorption can also be promoted by the presence of polyvalent cations [56, 122]. The chemical and molecular structure of nucleic acid bases may also influence their adsorption on clay minerals. In the presence of cations, the purines adsorb more excessively to the clay surface than pyrimidines [123] and NAs differing in their composition could represent different affinities for mineral surfaces.

Particularly the study of the adsorption of adenine, guanine, cytosine, thymine and uracil on clays (montmorillonite, illite and kaolinite), Fe- and Al-oxides (goethite, hematite and gibbsite) revealed that both inorganic and organic soil constituents adsorb NA [124]. Portions of the clays and soil were saturated with H^+, Fe^{3+} and Ca^{2+}. The adsorption was strongly affected by pH, with a tendency to decrease with increase in pH. The main adsorption mechanism at pH 4 appeared to be cation exchange whereas at pH 8 complex formation between the nucleic acid bases and cations on inorganic surfaces seemed to occur. The adsorption of nucleic acid bases by the clays was proportional to their exchange capacities, but the nature of the dominant cation had only minor effects. On goethite and gibbsite the amount of NA was lower than that on clays, while adsorption of NA bases on soils was slightly lower than that on oxides.

Formamide (NH_2CHO) is one of the simplest molecules containing the four most common elements of the universe (H, C, O and N). Several experimental works investigated formamide as a highly versatile building block for the syntheses of precursor nucleic bases in the presence of minerals as catalysts. The authors reported [38, 125–129] the synthesis of purine, N9-formylpurine, adenine, N9, N6-diformyladenine, hypoxanthine, cytosine, hydroxypyrimidine, 4(3H)-pyrimidinone, uracil, 5, 6-dihydrouracil, thymine, 5-hydroxymethyluracil, AICA (5-aminoimidazole-4-carboxamide), FAICA (5-formylaminoimidazole-4-carboxamide), and urea from formamide. Conditions were temperature between 90°C and 160°C and the presence of simple inorganic catalysts such as silica, alumina, zeolites, $CaCO_3$, TiO_2, common clays, kaolin, montmorillonites, olivines, and cosmic dust analogues of olivines. Several compounds were simultaneously formed in all cases and for each catalyst (see ref. [129] for summary). Each catalyst yields a specific panel of compounds, and the chemical processes involved have been revealed in details [130–133]. The highest amount of products was found in the presence of clays, TiO_2, and cosmic dust analogues. The presence of guanine was not detected. Formamide was also revealed to be an efficient activator of nucleoside transphosphorylation [134–136]. This finding supports possible relevance of formamide in the prebiotic polymerization reactions that did lead to pregenetic informational macromolecules.

Due to above described findings the adsorption behavior of single-stranded DNA homo-oligomers of adenine and thymine (including the monomers, dimers, tetramers, hexamers, octomers, and decamers) with sulfide (pyrite, FeS_2), a silicate (olivine, Mg_2SiO_4), a carbonate (calcite, $CaCO_3$), and two metal oxides (rutile and hematite, TiO_2 and Fe_2O_3, respectively) was also investigated experimentally [137]. These minerals were selected based on their considerably different surface properties, which create potential to be highly variable sites of interaction with nucleic acids. It was found that adenine-containing molecules tend to bind much more strongly than thymine-containing molecules. The number of moles adsorbed at saturation tends to fall with increasing chain length, while adsorption affinity tends to rise. Oligomer length appears to affect adsorption more than the mineral type.

21.4 Theoretical Studies of Nucleic Acid Bases and Their Interactions with Water, Cations and Minerals

21.4.1 Wächterhäuser Experiment

All the above mentioned studies have been performed experimentally and they have not been supported by ab initio techniques capable to determine physico-chemical conditions favorable (thermodynamically and kinetically) for the formation of components of the primordial pathway catalyzed by the transition metal sulfides using the carbon fixation cycle. Therefore, recently the investigations of thermodynamic aspects of this cycle using DFT approaches and simple models of Ni-Fe sulfide as catalysts at 373 K temperature were carried out [138] (this temperature was shown at hot oceanic vents on the early Earth). The viability of this cycle and its possible influence on the origin of low-molecular bioorganic compounds of primary archaic metabolism was examined in this theoretical study.

Wächtershäuser [5] proposed that the crucial point for Earth's first life occurrence is the formation of pyrite (FeS_2) from hydrogen sulfide (H_2S) and iron monosulfide (FeS), which could have provided a viable energy source. This specific reaction can be expressed as follows:

$$FeS\,(\alpha) + H_2S\,(aq) \rightarrow FeS_2\,(pyrite) + H_2 \qquad (21.3)$$

The reduction of CO_2 alone to form formic acid was found to be endergonic [5]:

$$CO_2\,(aq) + H_2 \rightarrow HCOOH\,(aq) \quad \Delta Gr = +7.2 kcal/mol \qquad (21.4)$$

However, combining the formation of pyrite (reaction 21.3) with the reduction of CO_2 (reaction 21.4) yields the reaction 21.5 presented below, which is exergonic:

$$FeS\,(\alpha)\ + H_2S\,(aq)\ + CO_2\,(aq)\ \rightarrow FeS_2\,(pyrite)$$

$$+ HCOOH\,(aq)\ \Delta Gr = -2.0 kcal/mol \tag{21.5}$$

The 21.3 reaction was assumed to be the energy source of the primordial initiation reaction for a chemoautotrophic origin of life. Huber and Wächtershäuser [9] assumed that this reaction is initiated by the reductive formation of methyl mercaptan (CH_3SH), its subsequent carbonylation to thioacetic acid (CH_3COSH) and hydrolysis to acetic acid. The transition metal (Fe, Ni, Co) sulfides were assumed to be catalysts of formation of acetic acid by carbon fixation.

The formation of acetic acid from CO and CH_3SH, which can be summarized by net Eq. 21.6, was also modeled by the computational study:

$$CH_3SH + CO + H_2O \rightarrow CH_3COOH + H_2S \tag{21.6}$$

This equation can be divided into two separate equations as follows:

$$CH_3SH + CO \rightarrow CH_3COSH \tag{21.7}$$

$$CH_3COSH + H_2O \rightarrow CH_3COOH + H_2S \tag{21.8}$$

We modeled the sequence of transformations of adsorbed reactants on the NiS-FeS surface [138]. The thermodynamics of each step of this mechanism was calculated. Figure 21.1 shows the ΔGr values for each step of the proposed mechanism of carbon fixation pathway obtained at the B3LYP/TZVP level of theory. In the first step carbon monoxide was added into the Fe-Ni-S model by binding with the iron center (formation of the Fe-CO center). This step is characterized by positive ΔGr value equal to 4.3 kcal/mol. This shows that the above reaction will most likely not proceed in a gas phase at 373 K temperature.

Similarly, in the next step the CH_3SH molecule was added into the system with endergonic Gibbs free energy (13.1 kcal/mol). A thiol group interacts with the hydrogen atom from the SH ligand forming H_2S molecule, which is removed from the system. The remaining methyl group is bonded to the sulfur ligand of both central atoms (ΔGr is equal to −2.8 kcal/mol). Then the CH_3 group migrates to a surface iron. However, this step is highly endergonic (41.5 kcal/mol) to be driven by a naturally occurring energy source. Total ΔGr value of all previously described reactions (expressed by the Eq. 21.7) is positive (37.7 kcal/mol). This means the reaction is not feasible in modeled situation. Thus, it is not clear if the cycle can operate at suitable rates and if it could have participated in the chemistry leading to life, therefore further kinetic studies are needed.

Most of the further steps of the calculated cycle are characterized by negative ΔGr values. At first the CH_3 group is detached from the iron center and is

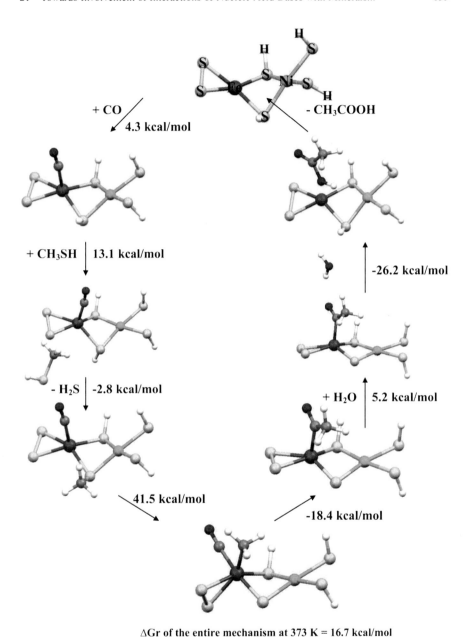

ΔGr of the entire mechanism at 373 K = 16.7 kcal/mol

Fig. 21.1 Modeling of the carbon fixation pathway catalyzed by the Fe–Ni–S model. The ΔGr values (kcal/mol) are calculated at the B3LYP/TZVP level of theory

bonded to the carbonyl group anchored to the iron center through the carbon atom
(-18.4 kcal/mol). Secondly, addition of water into the system is not thermodynamically feasible (ΔGr of this step is equal to 5.2 kcal/mol). Finally, the water
molecule dissociates into a hydroxide ion and a hydrogen cation. The hydrogen
cation binds with the sulfur ligand of both iron and nickel center. The hydroxide ion
is attached to the CH_3-CO (acetyl) ligand of iron and forms acetic acid. The results
indicate that this structure is not stable and desorption of acetic acid is favored.
Such a step is characterized by ΔGr equal to -26.2 kcal/mol. If one considers
only the reaction described by the Eq. 21.8 (formation of acetic acid and hydrogen
sulfide from thioacetic acid and water), this process is exergonic (-21.0 kcal/mol).
However, the ΔGr value that characterizes all above discussed reactions of the entire
mechanism (summarized by the Eq. 21.6) is positive (16.7 kcal/mol). This means
that the proposed cycle will not proceed spontaneously under modeled conditions.

21.4.2   Interactions of Nucleic Acid Bases with Sodium Cation and Water

Before analysis of the interactions of the nucleic acid bases with the clay minerals in
the presence of water and cation one needs to understand the individual interactions
of NAs with isolated water and with a cation. Such theoretical study was performed
for 1-methylcytosine (MeC) [139]. The study revealed influence of water and cation
in the proton transfer for this compound. This leads to the formation of *imino-oxo*
(MeC*) tautomer. Topology of the proton transfer potential surface and thermodynamics of stepwise hydration of $MeCNa^+$ and MeC^*Na^+ complexes is further
discussed. The one dimensional potential energy profile for this process followed
by the proton transfer with the formation of hydrated MeC^*Na^+ is presented
in Fig. 21.2. One-dimensional potential energy profile for *amino-imino* proton
transfer in monohydrated N1-methylcytosine (this represents the situation when
tautomerization is promoted by a single water molecule without the influence of
Na^+ cation) and for the case of pure intramolecular proton transfer (tautomerization
is not assisted by any internal interactions) is also included. The most important
features of this profile do not depend on the presence or absence of Na^+ cation. All
the potential energy curves have local minima corresponding to MeC and MeC*.
However, the significant difference is observed in the relative position of local
minima and transition state, which results in a different thermodynamic and kinetic
behavior for all presented cases (see Fig. 21.2).

The thermodynamic parameters related to local minima and the transition state,
which are responsible for the interaction between $MeCNa^+$ (MeC) and water
molecules, are given in Tables 21.1 and 21.2. Due to the well known nature of
compensation effects the values of relative Gibbs free energy of interaction are much
smaller than corresponding relative enthalpies. It is also expected that interaction of
$MeCNa^+$ and MeC^*Na^+ with the first water molecule is larger than the interaction

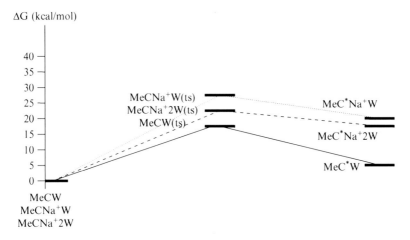

ΔG (kcal/mol)

MeCNa⁺W(ts)
MeCNa⁺2W(ts)
MeCW(ts)

MeC*Na⁺W

MeC*Na⁺2W

MeC*W

MeCW
MeCNa⁺W
MeCNa⁺2W

Fig. 21.2 The one dimensional potential energy profile for the stepwise hydration of the MeC complex followed by the proton transfer with the formation of hydrated MeC* with and without presence of the Na⁺ cation. The ΔG value differences between the reactant and transition state correspond with the $\Delta G^{\neq}$ values in Table 21.1 and the ΔG value differences between the reactant and product correspond with the $\Delta G_{taut}{}^{b}$ values in Table 21.1

with the MeC and MeC* species. New in this study is a conclusion that the rare form of N1-methylcytosine in the presence of the sodium cation interacts more strongly with the water molecules compared to the canonic forms. To explain the reason for this finding we have compared the differences $\Delta\Delta H_{int}$ and $\Delta\Delta G_{int}$ for species having the same number of coordinated water molecules-W (for example, in the case of MeCNa⁺W the $\Delta\Delta H_{int} = -17.3$ kcal/mol $- (-16.3$ kcal/mol$) = -1.0$ kcal/mol and the $\Delta\Delta G_{int} = -6.2$ kcal/mol $- (-5.5)$ kcal/mol $= -0.7$ kcal/mol). We concluded that this phenomenon is due to the enthalpy contribution because those differences have very similar values for all species considered.

Since we expect that the thermodynamic parameters of the stepwise hydration of MeC, MeC*, MeCNa⁺, MeC*Na⁺ are calculated with the accuracy of experimental measurements [140, 141], it is convenient to convert the ΔG_{int} values into the equilibrium constants (K). An analysis of the equilibrium constant values presented in Table 21.2 shows that difference in K considerably increases when MeC and MeC* are coordinated by Na⁺ and stepwise hydrated. Moreover, it suggests that interaction of isolated MeC and MeC* with the water molecules results in the presence of a significant amount of non-bonded MeC and MeC* in an equilibrium mixture. In contrast, the stepwise hydration of the MeCNa⁺ and MeC*Na⁺ species (up to two water molecules) leads to a virtually complete binding of MeC and MeC*.

Another important thermodynamic characteristic is represented by the free energy difference between *amino-* and *imino-* tautomers of methylcytosine. These parameters describe a relative stability of tautomers or in other words the relative concentration of tautomers at the equilibrium state. Moreover, this energy difference is very often discussed in regards to the contribution of this proton transition to the

Table 21.1 Values of enthalpies and Gibbs free energies (kcal/mol, ΔH_{int}, ΔG_{int} – enthalpy and Gibbs free energy of interactions of MeC with water and cation, $\Delta H^{\#}$, $\Delta G^{\#}$ – difference between the total enthalpies and Gibbs free energies of reactant and transition state, $\Delta H_{taut}{}^{b}$, $\Delta G_{taut}{}^{b}$ – difference between the total enthalpies and Gibbs free energies of reactant and product, $\Delta H'_{int}$, $\Delta G'_{int}$ – enthalpy and Gibbs free energy of the interaction of MeC* with water and cation) for the processes presented in Fig. 21.2 calculated at the MP2 level and extrapolated to complete basis set

	ΔH_{int}	$\Delta H^{\neq}$	$\Delta H_{taut}{}^{a}$	$\Delta H'_{int}$
MeC	–	40.3	1.3	–
MeCW	-10.4^{b}	13.8	2.1	–8.1
MeCNa^{+}W	–16.3	22.3	17.7	–17.3
MeCNa^{+}2W	–13.2	21.0	15.5	–14.9
	ΔG_{int}	$\Delta G^{\neq}$	$\Delta G_{taut}{}^{a}$	$\Delta G'_{int}$
MeC	–	40.3	2.4	–
MeCW	-0.9^{b}	15.6	2.1	+0.8
MeCNa^{+}W	–5.5	25.2	19.0	–6.2
MeCNa^{+}2W	–5.2	23.0	16.5	–7.5

[a]The data obtained at MP2/cc-pVDZ//B3LYP/cc-pVDZ
[b]The data obtained using basis set free optimization technique

Table 21.2 Values of equilibrium constant (K_{taut}), rate constants (k_f (forward) and k_r (reverse)) for the interactions of MeC with water and cation and equilibrium constant (K') for the interactions of MeC* with cation and water presented in Fig. 21.2 calculated at the MP2 level and extrapolated to complete basis set

	$k_f \ sec^{-1}$	$k_r \ sec^{-1}$	K_{taut}	K'
MeC	1.2×10^{-13}	7.1×10^{-12}	1.7×10^{-2}	–
MeCW	7.9×10^{5}	2.7×10^{7}	2.9×10^{-2}	2.6×10^{-1}
MeCNa^{+}W	3.2×10^{-16}	2.7×10^{-2}	1.2×10^{-14}	3.5×10^{4}
MeCNa^{+}2W	3.8×10^{-17}	4.5×10^{-5}	8.4×10^{-13}	3.1×10^{5}

frequency of spontaneous point mutations [142, 143]. The previous studies [144, 145], which have been performed for cytosine at the MP2 and CCDT levels of theory, have revealed the energy difference (uncorrected by ZPE) of 1.5–1.8 kcal/mol. This is in very reasonable correspondence with the value obtained in the above discussed study for a similar reaction in MeC (see ΔH_{taut} MeC in Table 21.1). As it follows from data presented in Tables 21.1 and 21.2, just the interaction with water molecule does not significantly change the equilibrium between MeC and MeC*. However, the interaction with Na^{+}-cation has a dramatic effect on changing the equilibrium constant (by 11–12 orders of magnitude) and making the concentration of MeC* coordinated by hydrated Na^{+} absolutely negligible.

21.4.3   Interactions of Nucleic Acid Bases with Clay Minerals

According to our best knowledge only three theoretical papers investigated the interactions of nucleobases with clay minerals. They include calculations of adsorption of RNA/DNA nucleobases on the external surfaces of Na^+-montmorillonite by using periodic plane wave calculations based on the PBE functional [146]. The authors have considered different orientations of the nucleobases (parallel and orthogonal to the surface plane) at the side comprising the Na^+ counterion or on the opposite side, where only siloxane bonds are present. For guanine and cytosine the metal cation interacts with two basic centers (N and O). It was indicated that guanine and cytosine are the ones with larger adsorption energies (about -27 kcal/mol). The remaining three bases present smaller adsorption energies (about -21.0 kcal/mol). On the other hand, adsorption of the nucleobase on the surface free from Na^+, either in a face-to-face or orthogonal orientation, is sizable for all bases (adsorption energies amount from -3.7 to -11.3 kcal/mol), due to the stabilizing effect of dispersion interactions.

21.4.3.1   Geometrical Parameters and Charges of Thymine and Uracil Adsorbed on Minerals of Kaolinite Group

Two works were devoted to the ab initio cluster calculations of the interactions between minerals of the kaolinite group with thymine (TH) and uracil (U). The key purpose of such studies was to determine: (i) the equilibrium adsorption of selected nucleic acids, differing in chemical, molecular structure and functions (DNA, RNA) on specific clay mineral surfaces; and (ii) the nature of the interaction between nucleic acids and clay using computational chemistry methods and modeling. An additional objective was to assess the effect of presence of water and sodium cation commonly occurring in soils on the process of adsorption.

Firstly certain types of adsorption of thymine and uracil with hydrated and non-hydrated dickite surface fragments [147] were investigated. This preliminary study has shown that the target molecules are less stable on the tetrahedral surface (denoted as D(t)) than on the octahedral surface (denoted as D(o)). The most energetically favorable adsorption on non-hydrated and hydrated (addition of one water molecule, D(t)w and D(o)w) surfaces was predicted. Thymine and uracil are bounded to the octahedral mineral fragments and hydrated tetrahedral mineral fragments by their highly effective sites (N3-H groups and O2 and O4 atoms) similarly as they are bounded in A:T and A:U base pairs [148]. These centers play a key role in the stabilization of studied NAs on clays. The presence of proton donors (OH groups of the octahedral fragment or water) and proton-acceptors (oxygen atoms of the octahedral fragment or water) governs the adsorption. The presence of only proton-acceptors in the case of tetrahedral systems causes that the N1-H group (proton-donor) surrounded by other proton-donors (the C-H groups) is

Table 21.3 Geometric parameters of Isolated thymine (TH) and thymine adsorbed on the tetrahedral (D(t), D(t)W) and octahedral (D(o), D(o)W) surface of Dickite obtained at the B3LYP/6-31G(d) level of theory

System/parameters[a]	D(t)–TH	D(t)w–TH	D(o)–TH	D(o)W–TH	TH
D(N1–C2)	1.386	1.393	1.378	1.357	1.390
D(N1–H)	1.013	1.012	1.024	1.040	1.010
D(C2=O)	1.219	1.218	1.223	1.253	1.217
D(C2–N3)	1.388	1.386	1.391	1.369	1.386
D(N3–C4)	1.408	1.396	1.393	1.405	1.408
D(C4=O4)	1.224	1.236	1.249	1.239	1.222
D(C4–C5)	1.464	1.461	1.449	1.446	1.468
D(C5–C(–H$_3$))[b]	1.502	1.502	1.513	1.502	1.501
α(N1–C2–O2)	123.8	122.9	125.9	125.3	123.2
α(N1–C2–N3)	112.4	112.8	112.6	114.7	112.4
α(C2–N3–C4)	128.4	127.3	126.3	125.7	128.3
α(N3–C4–O4)	120.2	120.5	118.3	118.2	120.4
α(N3–C4–C5)	114.3	115.7	114.3	114.9	114.5
α(C4–C5–C(–H$_3$))[b]	118.1	118.4	121.1	118.8	117.8
γ(N1–C2–O2–N3)	179.9	179.5	178.8	179.5	−179.9
γ(C2–N3–C4–O4)	179.6	174.9	−151.3	−157.8	179.9
γ(N3–C4–O4–C5)	179.9	−179.4	178.04	176.1	−179.9

[a]Bond lengths (D) are in Å and angles (α) and dihedral angles (γ) are in degrees [b]Average value

the main factor in the intermolecular binding. Experimental evidence of stronger hydrogen bonds in RNA for A:U than in DNA for A:T base pairs has been reported [149, 150].

The main geometrical features of thymine adsorbed on the tetrahedral and octahedral surface of dickite are presented in Table 21.3. Generally, the bond lengths of the molecule are not modified by the adsorption on the tetrahedral surface but interactions with the octahedral surface change the structural characteristics. The largest changes were found for the D(o)w-TH complex in which the strongest intermolecular interactions are revealed. The difference in these bond lengths is about 0.035 Å. Changes of bond lengths in the non-hydrated octahedral systems are less significant than previously mentioned difference. The formation of weak intermolecular interactions between thymine, uracil and the mineral surface results in small changes in angles of adsorbed thymine and uracil comparing to the isolated target molecule. The changes in the octahedral systems are not large but they are more evident than in the systems with the tetrahedral fragment. On the other hand, a significant change of the C2-N3-C4-O4 dihedral angle (about 30°) was revealed in the D(o)-TH, D(o)w-TH, D(o)-U, and D(o)w-U systems. This means that the interactions of thymine and uracil with the non-hydrated and hydrated octahedral mineral fragment largely affect planarity of the target molecule.

Table 21.4 presents the atomic charges of adsorbed and isolated thymine obtained by the fitting of the electrostatic potential. For uracil and thymine the adsorption

Table 21.4 ESP atomic charges (e) of isolated thymine (TH) and thymine adsorbed on the tetrahedral (D(t), D(t)W) and octahedral (D(o), D(o)W) surface of Dickite obtained at the B3LYP/6-31G(d) level of theory

System	D(t)–TH	D(t)W–TH	D(o)–TH	D(o)W–TH	TH
N1	−0.535	−0.627	−0.483	−0.525	−0.453
C2	0.814	0.785	0.815	0.898	0.707
O2	−0.579	−0.543	−0.577	−0.670	−0.546
N3	−0.681	−0.581	−0.659	−0.679	−0.578
C4	0.651	0.679	0.745	0.725	0.629
O4	−0.524	−0.568	−0.604	−0.547	−0.513
C5	−0.077	−0.041	−0.269	−0.190	−0.079
C6	0.115	−0.020	0.158	0.155	−0.014

leads to the same trend in changes of charges. The largest changes were found in the D(o)-TH and D(o)-U systems for the N3, C4, O4, C5 and C6 atoms. The modification is more significant in the complex containing thymine than uracil. In the D(o)w-TH and D(o)-U systems the changes are the largest for C2 and N3 atoms. In the D(t)w-TH and D(t)w-U complexes only the charge of the N1 atom was changed significantly since only N1-H group of the target molecule is involved in the H-binding with the surface. The adsorption on the octahedral fragment causes larger changes in charges than the adsorption on the tetrahedral surface. The presence of water on the mineral surfaces increases the variations of the atomic charges.

21.4.3.2 Interactions of Thymine and Uracil with Hydrated Surface of Minerals of the Kaolinite Group

Second theoretical study on the adsorption of thymine and uracil on the kaolinite minerals is much more comprehensive and extended to include the interactions with the negative (1−) charged mineral fragment. In addition, different level of theory was used with two different ways of calculation of the hydration effect [151]. The structure and interactions of studied complexes were also investigated using much larger models with different termination. The optimized structure of the most stable thymine adsorbed on large model of hydrated tetrahedral and octahedral kaolinite mineral surface with sodium cation is displayed in Fig. 21.3. These systems are denoted K(3t)NaW-TH and K(3o)NaW-TH, respectively. The analysis of the topological characteristics for these complexes is given in Table 21.5.

The same type of adsorption was revealed to be favorable after addition of water compared to the non-hydrated complexes. It can be described as physisorption. Two H-bonds (denoted HB1 and HB2 in Fig. 21.3 and Table 21.5) between the N1-H1 group and the basal oxygen atom directly bonded with the substituted aluminum cation from outside of the ditrigonal cavity and between C6-H6 and surface oxygen are created in the tetrahedral system. Three H-bonds, in which the molecular O2,

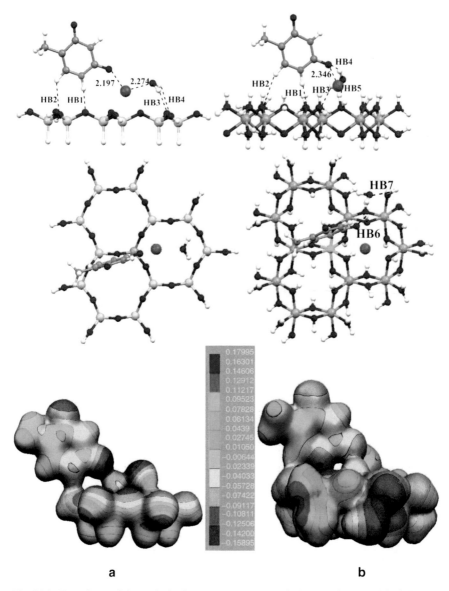

Fig. 21.3 Two views of the optimized structure and maps of electrostatic potential of thymine adsorbed on hydrated tetrahedral and octahedral surface of kaolinite mineral (**a** − K(3t)NaW-TH, **b** – K(3o)NaW-TH) obtained at the B3LYP/6-31G(d) level of theory

N1-H1 and C6-H6 groups are involved with two different surface oxygen (Os) atoms and one hydroxyl group are formed in the octahedral systems.

A water molecule on both tetrahedral and octahedral clay surfaces is strongly attracted to a sodium cation in a few different positions. In first position (the most stable on the octahedral surface) the water monomer interacts mainly via two

Table 21.5 H···Y and X···Y (in parentheses) distances (Å) and X—H···Y angles (°) and electron density characteristics (ρ (au) and $\nabla^2\rho$ (au)) for hydrated complexes

	K(3t)NaW–U				K(3t)NaW–TH			
	H···Y (X···Y)	X–H···Y	ρ	$\nabla^2\rho$	H···Y (X···Y)	X–H···Y	ρ	$\nabla^2\rho$
Na···O2	2.194	–	0.025250	0.180895	2.197	–	0.025216	0.179667
HB1	1.902 (2.930)	175.5	0.029890	0.085760	1.901 (2.927)	174.1	0.029991	0.086217
HB2	2.410 (3.294)	137.8	0.011084	0.037527	2.450 (3.348)	139.3	0.010177	0.034650
Na···Ow	2.272	–	0.024196	0.149032	2.274	–	0.024090	0.148102
HB3	2.032 (2.918)	150.1	0.021836	0.069389	2.025 (2.908)	149.7	0.022195	0.070493
HB4	2.575 (3.185)	121.0	0.007524	0.031356	2.600 (3.211)	121.2	0.007103	0.029942
	K(3o)NaW–U				K(3o)NaW–TH			
Na···O2	2.342	–	0.019959	0.117361	2.346	–	0.019802	0.115947
HB1	1.692 (2.741)	170.8	0.050392	0.140470	1.690 (2.739)	171.0	0.050573	0.141400
HB2	2.001 (2.940)	141.3	0.028783	0.077192	2.037 (2.970)	140.9	0.026894	0.071701
HB3	1.926 (2.842)	149.9	0.032872	0.087553	1.949 (2.865)	150.2	0.031454	0.083967
HB4	2.015 (2.884)	146.8	0.022184	0.072299	2.011 (2.883)	147.2	0.022370	0.072633
HB5	1.852 (2.781)	156.7	0.036149	0.105428	1.876 (2.799)	155.6	0.034290	0.100386
HB6	2.599 (3.256)	124.4	0.007526	0.031092	2.573 (3.239)	125.2	0.007872	0.032178
HB7	2.082 (2.736)	122.7	0.020689	0.068300	2.047 (2.711)	123.4	0.022278	0.072789

H-bridges with the surface OH groups (HB5 and HB7 in Fig. 21.3 and Table 21.5), one Ow-Hw...Os H-bond (HB3 in Fig. 21.3 and Table 21.5) and one O2...H-Os H-bond (HB4 in Fig. 21.3 and Table 21.5). In second configuration (dominant on the tetrahedral fragments) water remains interacting with Na^+ via the oxygen atom. The Na...Ow distance, which amounts to ~ 2.27 Å in K(3t)NaW, is fairly independent on the initial orientation of the water molecule. This finding agrees well with the results of theoretical study of hydration of Na^+ in a montmorillonite model [152] where the cation was shown to coordinate to water molecules as well as to the surface oxygen atoms. In the case of Na-smectite surface water is adsorbed through one H-bond with the surface oxygen atom next to the place of substitution [153, 154]. On the other hand, thymine and uracil form two hydrogen bonds (Ow–Hw...O and N–H...Ow) with isolated water [79, 155] with Ow and Hw being co-planar with the thymine molecule, whereas the second hydrogen atom points out of the plane. The experimental spectra obtained for $Li^+(H_2O)$-thymine and uracil also show similar structural features in terms of bonding between the base and Li^+, as well as for solvation [156].

The nucleic acid bases do not interact directly with water in the K(3o)NaW systems as it is observed for isolated thymine and uracil. These isolated bases interact with water molecules through two H-bridges [79, 155]. Then if cation is added into the system one ionic bond (bond length 2.1 Å) is formed between uracil and cation [157]. It can be concluded that besides the substitution also the surface oxygen atoms govern the binding with the adsorbent, cation and water. This is shown to be true also for the basal oxygen atoms, which help to stabilize the water molecule via the formation of hydrogen bonds [154].

The characteristics of adsorption binding of thymine and uracil were confirmed by the maps of electrostatic potential drawn for all complexes. The isomorphic substitution and the presence of the cation in both fragments affect the electron redistribution and this influence varies based on the surface type.

21.4.3.3 Energetics of Thymine and Uracil Adsorbed on Hydrated Surface of Minerals of the Kaolinite Group

In older studies adsorption of nucleobases on clay minerals was shown to be promoted by the presence of polyvalent cations [56, 122]. This finding corresponds well with our results [151]. Periodic plane wave calculations based on the PBE functional [146] revealed sizeable adsorption of thymine and uracil on the external surfaces of Na-montmorillonite in the case of surface free from Na^+ (from -6 to -11 kcal/mol), due to the stabilizing effect of dispersion interactions. As one can see from the comparison of these interaction energies with the Ecorr values obtained for K(3t)Na and K(3t)NaW systems with thymine and uracil (-28, -25 kcal/mol for uracil and -27, -24 for thymine as given in Table 21.6) and for D(t), D(t)W without cation (from -1 to -9 kcal/mol) [147], the addition of Na^+ leads to a significant stabilization of the tetrahedral systems. The same is true for K(3o)Na systems, for which interaction energies increase (in absolute value) from -30 to -36 for uracil

Table 21.6 BSSE corrected interaction energies (kcal/mol, BSSE values are in parentheses) of uracil and thymine systems

	M05–2X	B3LYP(large)	COSMO	M05–2X	B3LYP(large)	COSMO
	Uracil			Thymine		
Tet	−31.0 (5.1)	−28.2 (5.9)	−6.7 (5.9)	−29.9 (5.1)	−26.9 (6.0)	−5.3 (6.0)
Tet–w	−28.5 (5.1)	−25.1 (6.0)	–	−27.0 (5.1)	−24.0 (6.0)	–
Oct	−46.1 (7.3)	−35.9 (8.5)	−7.5 (8.5)	−44.4 (7.4)	−34.3 (8.5)	−5.8 (8.5)
Oct–w	−43.7 (7.5)	−35.4 (8.7)	–	−43.4 (7.6)	−36.0 (7.7)	–

and from −21 to −34 kcal/mol for thymine. It is due to a fact that Na^+ forms strong interaction with the adsorbate. This agrees well with a statement about the effect of positive charges of the lattice edges on the DNA adsorption [60]. Ecorr values for K(3t)Na are larger than those obtained for thymine and uracil binding with Na^+ placed orthogonally or in a parallel way on the external surface of montmorillonite (from −16.2 to −21.7 kcal/mol) [146]. This difference corresponds with the nature of the adsorption (see Sect. 21.4.3.2 for more details). Also the binding energy of pyridine on dry surface of Na-smectite (−17.2 kcal/mol) [154] and on clay cluster with Mg^{2+} substituent (−15 kcal/mol) [158] is smaller than that obtained in the study for thymine and uracil.

The calculations of solvated (hydrated) complexes simulated by explicit inclusion of water molecule shows the energy loss only for the tetrahedral systems, which is due to the structural changes of the substrate induced by water adsorption [151]. It amounts to about 3 kcal/mol. In the case of octahedral systems presence of water does not increase the Ecorr value for K(3o)NaW-U but it stabilizes thymine by 1.7 kcal/mol. Isolated uracil is more stable in the water-cation environment than in the microcosmic environment with only water [157]. In all K(3o)NaW systems orientation by O2 towards Na^+ remains to be the most favorable. On the other hand, in study of hydrated Li^+-thymine and Li^+-uracil complexes binding of $(H_2O)Li^+$ to the O4 site is stabilized by 1.7–2.9 kcal/mol over the binding to the O2 site [156].

The nucleobases interact more strongly with the octahedral than with tetrahedral surface (the energy difference is 6–7 kcal/mol for non-hydrated surface and 9–12 kcal/mol for hydrated systems) [151]. It is in line with the nature of the adsorption, the binding energies of K(3o)Na and K(3o)NaW are in the order of strength of additional hydrogen bonds between the adsorbate and surface. The same trend was revealed also for the monoaromatic hydrocarbons on kaolinite, which also form more stable interactions with the hydroxylated surface [159]. Moreover, the interaction energy of water over hydroxylated kaolinite surface is equal to −13.1 kcal/mol while the silicon surface has an interaction energy of −4.1 kcal/mol [160]. The K(3o)NaW-TH and K(3o)Na-U systems are the most stable (Ecorr are about −36 kcal/mol) among all studied complexes.

The solvation of the surface was also calculated applying the COSMO model (see Table 21.6) [151]. In contrast to the changes caused by an explicit addition of a water molecule (the K(3t)NaW and K(3o)NaW systems), which hardly affects the adsorption strength, COSMO approach leads to significantly lower interaction energy. These values are more than two times lower than those obtained for K(3t)NaW and K(3o)NaW. This suggests an importance of performing both types of calculation instead applying only the explicit addition of water. The general trend in the stabilization is also changed compared with the results obtained using the supermolecular approach. Uracil appears the most strongly adsorbed on the non-hydrated octahedral surface followed by uracil adsorbed on non-hydrated tetrahedral fragment. The Ecorr values of K(3o)Na-TH and K(3o)Na-U from the COSMO calculation vary by about 1.7 kcal/mol. This number agrees well with the energy difference from the microsolvation.

The interaction energies of the most stable uracil and thymine adsorbed on large mineral fragments (K(3o)Na, K(3t)Na, K(3t)NaW and K(3o)NaW) were also calculated at the B3LYP/6-31G(d)//M05-2X/6-31G(d) level of theory and the results are presented in Table 21.6 [151]. The M05-2X results confirm predictions obtained using the B3LYP functional. Comparison of the Ecorr values for the most stable uracil and thymine systems obtained at B3LYP and M05-2X levels (presented in Table 21.6) reveals that dispersion term amounts to about 3 kcal/mol for K(3t)Na, K(3t)NaW and 7–10 kcal/mol for K(3o)Na and K(3o)NaW. Such difference in the energy reveals similar trend in the stability of thymine and uracil with exception of K(3o)NaW-TH (it exhibits smaller adsorption strength than K(3o)NaW found the most stable at the B3LYP level).

21.4.3.4 Implication to Origin of Life

The adsorption strength of clay fragments depends on their structure and physico-chemical properties since DNA bases possess different binding affinities and slightly different conformations when adsorbed to various kaolinite mineral surfaces. This was found to be true also for DNA molecules adsorbed on different layered materials [161]. The substitution in the mineral sheet and the presence of the counterion affect the adsorption of the nucleobases the most significantly. Such a conclusion is in agreement with statement of Paget and co-workers [59] that DNA interaction with clay minerals is a charge-dependent process. Structural and chemical differences of specific layered silicate mineral surfaces cause varying adsorption of DNA bases. Higher affinity was predicted for the octahedral surface, which induces the changes in the conformation of DNA bases. The clays ability to organize the nucleic acids (based on their orientation towards the surface and presence of the interlayer cation and water) could indicate that specifically designed clay mineral surfaces may posses a catalytic potential when used as substrates for biomolecules or to activate and concentrate the monomer building blocks in the prebiotic chemistry. Furthermore, the isolated mineral fragments with well defined edges may have also played an important role in the adsorption of DNA bases and their derivatives on the early earth.

21.5   Conclusions

This chapter summarizes the results of experimental and theoretical studies on the interactions of nucleic acid bases with minerals, water and sodium cation. The computational study to some extent supported experimentally proposed Wächtershäuser's cycle (the production of acetic acid from CO and CH_3SH) [9]. It was shown that Fe-Ni-S surface model as catalyst can partially catalyze this reaction through the creation of different coordination complexes [138]. But synthesis of formic acid from CO_2 and H_2S in the presence of pyrite was shown to be endergonic under modeled conditions. The simulations show that this reaction pathway does not lead to sufficient amount of the product in isolated systems and the cycle can possibly operate at low rates. But to make the final conclusion about the rates, the kinetic study based on reaction rates needs to be performed and inclusion of conditions close to those that occurred on the early Earth, are required to confirm feasibility of studied prebiotic reactions.

The thermodynamic and kinetic parameters of the stepwise hydration of 1-methylcytosine and its *imino-oxo* tautomer in the presence of the Na^+ cation have been investigated [139]. Hydration of 1-methylcytosine by one water molecule leads to an increase of the concentration of its *imino-oxo* tautomer in the equilibrium mixture and decrease of the barrier of the tatutomer formation (to 15.6 kcal/mol). If the sodium cation is present the tautomeric form is much less favored and tautomerization barrier increases to 25.2 kcal/mol. The computationally predicted values of the rate constants suggest that the tautomerization of 1-methylcytosine to its *imino-oxo* form proceeds mainly due to a presence of the hydrated (MeCW) species. Based on the kinetic analysis of the tautomerization process in hydrated MeC in the presence of sodium ions it was concluded that complexes of hydrated MeC with Na^+ are unlikely to contribute to the frequency of DNA point mutations caused by the tautomers. This is due to the fact that the interactions with Na^+ lead to a decrease of both the rate and the equilibrium constants of the tautomerization reactions in hydrated 1-methylcytosine.

The calculations of the adsorption of thymine and uracil on the octahedral and tetrahedral surface of clay minerals of the kaolinite group in the presence of sodium cation and solution show that thymine and uracil interact in a very similar way with the mineral surface [147, 151]. Studied clay mineral fragments posses high sorption affinity for nucleic acid bases. The presence of Na^+ leads to a significant stabilization of these molecules on both surfaces due to the formation of strong interaction between Na^+ and the molecular oxygen atoms. Explicit addition of water molecule has only small influence on the stabilization of the nucleic acid bases in the presence of the Na^+. The binding strength of thymine with the clay surface is slightly larger than of uracil, which can be explained in terms of their chemical nature. This attractive interaction contributes the most to the adsorption strength, which depends on several other factors such as the type of the surface, its chemistry, position and orientation of the target molecule towards the surface. Such a large effect is suggested to be an indication of the catalytic properties that these materials may have shown also when used as substrates for biomolecules in the prebiotic chemistry.

The above summarized studies of the interaction of nucleic acid bases with water, cations, clay minerals and clay-based materials are only the first step in attempt to understand such adsorption. To provide full details necessary for the understanding of the fate of nucleobases when interacting with soil components, simulations need to include the effects of different type of soil and nucleobase, the chemical environment (the pH of the system), and of external physical conditions (temperature). Therefore, future studies should concentrate on other types of layered minerals with considering above mentioned factors.

The reviewed data strongly suggest that computational approaches are able to support studies devoted to the origin of life. We believe that due to fast enhancements of the computer power the contribution of such approaches will certainly increase in near future.

Acknowledgments This work was facilitated by the support of the Origin of Life project funded by the National Science Foundation (NSF) through the Georgia Tech University grant no. CHE-0739189.

References

1. Anet FAL (2004) Curr Opin Chem Biol 8:654
2. Kauffman S (2007) Orig Life Evol Biosph 37:315
3. Trefil J, Morowitz HJ, Smith E (2009) Am Sci 97:206
4. Wächtershäuser G (1988) Microbiol Rev 52:452
5. Wächtershäuser G (1988) Syst Appl Microbiol 10:207
6. Wächtershäuser G (1990) Proc Natl Acad Sci USA 87:200
7. Wächtershäuser G (1992) Prog Biophys Mol Biol 58:85
8. Wächtershäuser G (1994) Proc Natl Acad Sci USA 91:4283
9. Huber C, Wächtershäuser G (1997) Science 276:245
10. Wächtershäuser G (2006) Philos Trans R Soc Lond B Biol Sci 361:1787
11. Wächtershäuser G (2007) Chem Biodivers 4:584
12. Russell MJ, Daia DE, Hall AJ (1998) In: Wiegel J, Adams MWW (eds) Thermophiles: the keys to molecular evolution and the origin of life. Taylor & Francis, Washington, DC, pp 77–126
13. Qui D, Kumar M, Ragsdale SW, Spiro TG (1994) Science 264:817
14. Vaughan DJ, Craig JR (1978) Mineral chemistry of metal sulphides. Cambridge University Press, Cambridge
15. Anders E (1989) Nature 342:255
16. Ehrenfreund P (1999) Science 283:1123
17. Cairns SG (1986) Clay minerals and the origin of life. Cambridge University Press, Cambridge
18. Bernal JD (1951) The physical basis of life. Routledge and Kegan Paul, London
19. Rao M, Odom DG, Orò JJ (1980) J Mol Evol 15:317
20. Ferris JP, Ertem G, Agarwal VK (1989) Orig Life Evol Biosph 19:153
21. Ferris JP, Ertem G (1992) Orig Life Evol Biosph 22:369

22. Ferris JP (1993) Orig Life Evol Biosph 23:307
23. Stotzky JV, Gallori E, Khanna M (1996) In: Akkermans ADL, van Elsas YD, Brujin FJ (eds) Transformation in soil, molecular microbial ecology manual. Kluwer, Dordrecht, pp 1–28
24. Pitsch S, Eschenmoser A, Gedulin B, Hui S, Arrhenius G (1995) Orig Life Evol Biosph 25:297
25. Krishnamurthy R, Pitsch S, Arrhenius G (1999) Orig Life Evol Biosph 29:139
26. Mortland MM (1986) Mechanisms of adsorption of nonhumic organic species by clays. In: Huang PM, Schnitzer M (eds) Interactions of soil minerals with natural organics and microbes, Soil Science Society of America (Special publication no. 17), Madison, WI
27. Luther A, Brandsch R, von Kiedrowsky G (1998) Nature 396:245
28. Smith JV (1998) Proc Natl Acad Sci USA 95:3370
29. Smith JV, Frederick PA Jr, Parsons I, Lee MR (1999) Proc Natl Acad Sci USA 96:3479
30. Ferris JP, Hill AR, Lin R, Orgel LE (1996) Nature 381:59
31. Cairns-Smith AG (1982) Genetic takeover and the origin of life. Cambridge University Press, New York, p 477
32. Keil RG, Montlucon DB, Prahl FG, Hedges JI (1994) Nature 370:549
33. Ferris JP (2005) Elements 1:145
34. Lorenz MG, Wackernagel W (1994) Microbiol Rev 58:563
35. Scappini F, Casadei F, Zamboni R, Franchi M, Gallori E, Monti S (2004) Int J Astrobiol 3:17
36. Baldwin DS, Beattie JK, Coleman LM, Jones DR (1995) Environ Sci Technol 29:1706
37. Cohn CA, Mueller S, Wimmer E, Leifer N, Greenbaum S, Strongin DR, Schoonen MA (2006) Geochem Trans 7:3
38. Saladino R, Neri V, Crestini C, Costanzo G, Graciotti M, Di Mauro E (2008) J Am Chem Soc 130:15512
39. Franchi M, Bramanti L, Morassi Bonzi L, Orioli L, Vettori PL, Gallori E (1999) Orig Life Evol Biosph 29:297
40. Goring CAI, Bartholomew WV (1952) Soil Sci 74:149
41. Arderma BW, Lorenz MG, Krumbein WE (1983) Appl Environ Microbiol 46:417
42. Lorenz MG, Wackernagel W (1987) Appl Environ Microbiol 53:2948
43. Khanna M, Stotzky G (1992) Appl Environ Microbiol 58:1930
44. Gallori E, Bazzicalupo M, Dal Canto L, Fani R, Nannipieri P, Vattori C, Stotzky G (1994) FEMS Microbiol Ecol 15:119
45. Vettori C, Paffetti D, Pietramellara G, Stotzky G, Gallori E (1996) FEMS Microbiol Ecol 20:251
46. Crecchio C, Stotzky G (1998) Soil Biol Biochem 30:1061
47. Khanna M, Yoder L, Calamai L, Stotzky G (1998) Soil Sci 3:1
48. Paget E, Simonet P (1994) FEMS Microbiol Ecol 15:109
49. Pietramellara G, Franchi M, Gallori E, Nannipieri P (2001) Biol Fertil Soils 33:402
50. Mao Y, Daniel N, Whittaker N, Saffiotti U (1994) Environ Health Perspect 102:165
51. Ogram AV, Mathot ML, Harsh JB, Boyle J, Pettigrew CA (1994) Appl Environ Microbiol 60:393
52. Choy JH, Park JS, Kwak SY, Jeong YJ, Han YS (2000) Mol Crystals Liquid Crystals 341:1229
53. Fejer I, Kata M, Eros I, Berkesi O, Dekany I (2001) Colloid Polym Sci 279:1177
54. Lin FH, Lee YH, Wong JM, Shieh MJ, Wang CY (2002) Biomaterials 23:1981
55. Stotzky G (1989) In: Levy SB, Miller RV (eds) Gene transfer in the environment. McGraw-Hill, New York, pp 165–222
56. Greaves MP, Wilson MJ (1969) Soil Biol Biochem 1:317
57. Pietramellara G, Dal Canto L, Vettori C, Gallori E, Nannipieri P (1997) Soil Biol Biochem 29:55
58. Lorenz MG, Wackernagel W (1992) In: Gauthier MJ (ed) Gene transfer and environment. Springer, Berlin, pp 103–113
59. Paget E, Jocteur Monrozier L, Simonet P (1992) FEMS Microbiol Lett 97:31
60. Poly F, Chenu C, Simonet P, Rouiller J, Monrozier LJ (2000) Langmuir 16:1233

61. Newman ACD (1987) The chemistry of clays and clay minerals (Mineralogical Society Monograph No. 6). Longman Scientific & Technical, London
62. Bailey SW (1980) In: Brindley GW, Brown G (eds) Crystal structures of clay minerals and their x-ray identification. Mineralogical Society, London, pp 6–28
63. Essington ME (2004) Soil and water chemistry: an integrative approach. Taylor & Francis e-Library/CRC Press LLC, Boca Raton
64. Sayed Hassan M, Villieras F, Razafitianamaharavo A, Michot LJ (2005) Langmuir 21:12283
65. van Olphen H (1977) Clay colloid chemistry, 2nd edn. Wiley, New York, p 318
66. McBride MB (1976) Clays Clay Miner 24:88
67. Rand B, Melton IE (1977) J Colloid Interface Sci 60:308
68. Williams DJA, Williams KR (1978) J Colloid Interface Sci 65:79
69. Schroeder PA, Pruett RJ (1996) Am Miner 81:26
70. Huertas FJ, Fiore S, Linares J (2004) Clay Miner 39(4):423
71. Bentabol M, Ruiz Cruz MD, Huertas FJ, Linares J (2006) Clays Clay Miner 54:667
72. Texter J (1978) Prog Biophys Mol Biol 33:83
73. Weshof E (1988) Annu Rev Biophys Biophys Chem 17:125
74. Saenger W (1987) Annu Rev Biophys Biophys Chem 16:93
75. Chandra AK, Nguyen MT, Zeegers-Huyskens TJ (1998) J Phys Chem A 102(29):6010
76. Gaigeot MP, Ghomi M (2001) J Phys Chem B 105(21):5007
77. Kryachko ES, Nguyen MT, Zeegers-Huyskens TJ (2001) J Phys Chem A 105(10):1934
78. van Mourik T, Price SL, Clary DC (1999) J Phys Chem A 103(11):1611
79. Dolgounitcheva O, Zakrzewski VG, Ortiz JV (1999) J Phys Chem A 103(39):7912
80. Bao X, Sun H, Wong NB, Gu J (2006) J Phys Chem B 110:5865
81. Leszczynski J (1992) J Phys Chem 96:1649
82. Gorb L, Podolyan Y, Dziekonski P, Sokalski WA, Leszczynski J (2004) J Am Chem Soc 126:10119
83. Saenger W (1984) Principles of nucleic acid structure. Springer, New York
84. Eichhorn GL (1981) Adv Inorg Biochem 3:1
85. Martin RB (1993) Acc Chem Res 22:255
86. Sigel H (1993) Chem Soc Rev 22:255
87. Sigel A, Sigel H (1996) Interactions of metal ions with nucleotides, nucleic acids and their constituents, vol 32, Metal ions in biological systems. Marcel Dekker, New York
88. Sigel A, Sigel H (1996) Probing of nucleic acids by metal ion complexes of small molecules, vol 33, Metal ions in biological systems. Marcel Dekker, New York
89. Nakano S, Fujimote M, Hara H, Sugimoto N (1999) Nucleic Acids Res 27:2957
90. Lippard SJ, Berg JM (1994) Principle of bioinorganic chemistry. University Science Books, Mill Valley
91. Kaim W, Schwedersky B (1994) Bioinorganic chemistry: inorganic elements in the chemistry of life. Wiley, Chichester
92. Loeb LA, Zakour AR (1980) In: Spiro TG (ed) Nucleic acid-metal ion interactions. Wiley, New York, pp 115–144
93. Parr RG, Yang W (1989) Density-functional theory of atoms and molecules. Oxford University Press, New York
94. Gaussian 09, Revision A1, Frisch MJ, Trucks GW, Schlegel HB, Scuseria GE, Robb MA, Cheeseman JR, Scalmani G, Barone V, Mennucci B, Petersson GA, Nakatsuji H, Caricato M, Li X, Hratchian HP, Izmaylov AF, Bloino J, Zheng G, Sonnenberg JL, Hada M, Ehara M, Toyota K, Fukuda R, Hasegawa J, Ishida M, Nakajima T, Honda Y, Kitao O, Nakai H, Vreven T, Montgomery JA Jr, Peralt JE, Ogliaro F, Bearpark M, Heyd JJ, Brothers E, Kudin KN, Staroverov VN, Kobayashi R, Normand J, Raghavachari K, Rendell A, Burant JC, Iyengar SS, Tomasi J, Cossi M, Rega N, Millam NJ, Klene M, Knox JE, Cross JB, Bakken V, Adamo C, Jaramillo J, Gomperts R, Stratmann RE, Yazyev O, Austin AJ, Cammi R, Pomelli C, Ochterski JW, Martin RL, Morokuma K, Zakrzewski VG, Voth GA, Salvador P, Dannenberg JJ, Dapprich S, Daniels AD, Farkas Ö, Foresman JB, Ortiz JV, Cioslowski J, Fox DJ (2009) Gaussian Inc., Wallingford

95. Becke AD (1993) J Chem Phys 98:5648
96. Lee C, Yang W, Parr RG (1988) Phys Rev B 37:785
97. Zhao Y, Schultz NE, Truhlar DG (2006) J Chem Theory Comput 2(2):364
98. Zhao Y, Truhlar DG (2008) Theor Chem Acc 120(1–3):215
99. Zhao Y, Gonzalez-Garcia N, Truhlar DG (2005) J Phys Chem A 109(9):2012
100. Hay PJ, Wadt WR (1985) J Chem Phys 82:270, 284, 299
101. Weigend F, Ahlrichs R (2005) Phys Chem Chem Phys 7:3297
102. Frisch MJ, Pople JA, Binkley JS (1984) J Chem Phys 80:3265
103. Boys SF, Bernardi F (1970) Mol Phys 19:553
104. Bader RWF (1990) Atoms in molecules: a quantum theory. Oxford University Press, Oxford
105. Koch U, Popelier PLA (1995) J Phys Chem 99:9747
106. Popelier PLA (1998) J Phys Chem A 102:1873
107. Molekel 4.0, Flükiger P, Lüthi HP, Portmann S (2000) Weber J, Swiss National Supercomputing Centre CSCS, Manno (Switzerland)
108. Smedarchina Z, Fernandez-Ramos A, Zgierski MZ, Siebrand W (2000) DOIT 1.2, A computer program to calculate hydrogen tunneling rate constants and splittings, National Research Council of Canada
109. Joswig W, Drits VA (1986) N Jb Miner Mh 19:1986
110. Vaughan DJ, Craig JR (1985) Am Miner 70:1036
111. Russell MJ, Martin W (2004) Trends Biochem Sci 29(7):358
112. Dörr M, Alpermann T, Weigand W (2007) Orig Life Evol Biosph 37:329
113. Marakushev SA, Belonogova OV (2009) J Theor Biol 257:588
114. Kalapos MP (2007) J Theor Biol 248:251
115. Cody GD (2004) Annu Rev Earth Planet Sci 32:569
116. Kalapos MP (1997) J Theor Biol 188:201
117. Hazen RM (2004) Chiral crystal faces of common rock-forming minerals. In: Palyi G, Zucchi C, Cagglioti L (eds) Progress in biological chirality. Elsevier, New York, pp 137–151
118. Hazen RM, Papineau D, Bleeker W, Downs RT, Ferry JM, McCoy TJ, Sverjensky DA, Yang HX (2009) Am Miner 93:1693
119. Dawson RMC, Elliott DC, Elliott WH, Jones KM (1986) Data for biochemical research, 3rd edn. Oxford University Press, Oxford
120. Cortez J, Schnitzer M (1979) Can J Soil Sci 59:277
121. Theng BKG (1982) Clays Clay Miner 30:1
122. Hesselink FT (1983) Adsorption form solution at the solid/liquid interface. Academic, London, p 377
123. Lahav N, Chang S (1976) J Mol Evol 8:357
124. Cortez J, Schnitzer M (1981) Soil Biol Biochem 13:173
125. Saladino R, Crestini C, Ciciriello F, Di Mauro E, Costanzo G (2006) J Biol Chem 281(9):5790
126. Saladino R, Crestini C, Costanzo G, Negri R, Di Mauro E (2001) Bioorg Med Chem 9:1249
127. Saladino R, Ciambecchini U, Crestini C, Costanzo G, Negri R, Di Mauro E (2003) Chembiochem 4:514
128. Saladino R, Crestini C, Ciambecchini U, Ciciriello F, Costanzo G, Di Mauro E (2004) Chembiochem 5:1558
129. Saladino R, Crestini C, Neri V, Brucato JR, Colangeli L, Ciciriello F, Di Mauro E, Costanzo G (2005) Chembiochem 6:1368
130. Saladino R, Crestini C, Neri V, Ciciriello F, Costanzo G, Di Mauro E (2006) Chembiochem 7:1707
131. Saladino R, Crestini C, Costanzo G, Di Mauro E (2005) In: Walde P (ed) Prebiotic chemistry, vol 259, Topics in current chemistry. Springer, New York, pp 29–68
132. Saladino R, Crestini C, Ciciriello F, Costanzo G, Di Mauro E (2007) Helv Chim Acta 4:694
133. Saladino R, Crestini C, Ciciriello F, Costanzo G, di Mauro E (2006) Orig Life Evol Biosph 36(5–6):523
134. Schoffstall AM (1976) Orig Life 7:399
135. Schoffstall AM, Barto RJ, Ramos DL (1982) Orig Life 12:143

136. Schoffstall AM, Mahone SM (1988) Orig Life Evol Biosph 18:389
137. Kopstein M, Sverjensky DA, Hazen RM, Cleaves HJ (2009) American Geophysical Union, Fall Meeting, abstract #B43A-0362 http://adsabs.harvard.edu/abs/2009AGUFM.B43A0362K
138. Michalkova A, Kholod Y, Kosenkov D, Gorb L, Leszczynski J (2011) Geochim Cosmochim Acta 75:1933
139. Michalkova A, Kosenkov D, Gorb L, Leszczynski J (2008) J Phys Chem B 112:8624
140. Isayev O, Gorb L, Leszczynski J (2007) J Comput Chem 28:1598
141. Isayev O, Furmanchuk A, Gorb L, Leszczynski J (2008) Chem Phys Lett 451:147
142. Gorb L, Kaczmarek A, Gorb A, Sadlej AJ, Leszczynski J (2005) J Phys Chem B 109:13770
143. Podolyan Y, Gorb L, Leszczynski J (2003) Int J Mol Sci 4:410
144. Fogarasi G, Kobayashi R (1998) J Phys Chem 102:110813
145. Gorb L, Podolyan Y, Leszczynski J (1999) J Mol Struct Theochem 487:47
146. Mignon P, Ugliengo P, Sodupe M (2009) J Phys Chem C 113(31):13741
147. Robinson TL, Michalkova A, Gorb L, Leszczynski J (2007) J Mol Struct 844–845:48
148. Asensio A, Kobko N, Dannenberg JJ (2003) J Phys Chem A 107:6441
149. Vakonakis I, LiWang AC (2004) J Am Chem Soc 126:5688
150. Fonesca GC, Bickelhaupt FM, Snijders JG, Baerends EJ (2000) J Am Chem Soc 122:4117
151. Michalkova A, Robinson TL, Leszczynski J (2011) Phys Chem Chem Phys 13:7862
152. Mignon P, Ugliengo P, Sodupe M, Hernandez ER (2010) Phys Chem Chem Phys 12:688
153. Clausen P, Andreoni W, Curioni A, Hughes E, Plummer CJG (2009) J Phys Chem C 113(34):15218
154. Clausen P, Andreoni W, Curioni A, Hughes E, Plummer CJG (2009) J Phys Chem C 113:12293
155. Frigato T, Svozil D, Jungwirth P (2006) J Phys Chem A 110(9):2916
156. Gillis EAL, Rajabi K, Fridgen TD (2009) J Phys Chem A 113:824
157. Hu X, Li H, Zhang L, Han S (2007) J Phys Chem B 111(31):9347
158. Chatterjee A, Ebina T, Iwasaki T (2001) J Phys Chem A 105(47):10694
159. Castro EAS, Martins JBL (2005) J Comput Aided Mater Des 12:121
160. Tunega D, Benco L, Haberhauer G, Gerzabek MH, Lischka H (2002) J Phys Chem B 106:11515
161. Antognozzi M, Wotherspoon A, Hayes JM, Miles MJ, Szczelkun MD, Valdre G (2006) Nanotechnology 17(15):3897

Index

A

ab initio model potential (AIMP), 218, 223, 226, 238, 244, 245, 247

ab initio quantum chemistry, 70, 256, 281, 466

Absorbance, 369, 455

Adsorption, 241, 242, 370, 379, 380, 547, 580, 581, 587–590, 593, 607–610, 617, 618, 621, 623, 646–649, 651–653, 659–661, 664–668

AgH, 136–139

AIM. *See* Atoms in molecules (AIM)

Algebraic approximation, 35, 39, 124

Alpha helix, 15–19

Alpha keratin, 17

Alpha-oxygen, 630, 638

Anderson, P., 19, 256

Angle contraction in Si-O-N linkages, 490

Anion-radical, 540

Arm-chair, 342, 356, 358–360

Atoms in molecules (AIM), 197, 505–510, 524, 532, 540, 564, 649

AuH, 136–139

Avery, O.T., 13, 14

B

Babbage, C., 34

Backdoor, 41

Backdoor intruder states, 41, 42

Bartell, L.S., 10

Basis-free method, 168

Basis set superposition error (BSSE), 241, 256, 501, 503, 506, 512, 524, 536–538, 545, 649, 663

Becke-3-Lee-Yang-Parr (B3LYP), 130, 149–151, 154, 155, 322, 428, 429, 443, 445, 464, 465, 468, 522, 529, 530, 533, 534, 540, 545, 546, 561, 585, 587, 598–601, 603–609, 611–615, 617, 618, 620, 632, 633, 636, 637, 649, 653, 654, 658, 660–662, 665, 666

Benzene-ammonia complex, 499

Bernal, J.D., 14

Beta keratin, 17

Bishop–Kirtman perturbation theory (BKPT), 135–139, 159, 163

Blackbody infrared radiative dissociation, 533

B3LYP. *See* Becke-3-Lee-Yang-Parr (B3LYP)

Bohr, N., 5, 34, 188

Bond critical point (BCP), 506–510, 512–514, 532, 540

Bond path (BPs), 506–514

Born and Oppenheimer, 34, 136, 266, 352, 461, 548

Born, M., 21, 22

Boron hydride (BH), 87, 94, 95

B3PW91, 355

Bragg, W.H., 11

Bragg, W.L., 11

Brandow, B.H., 43

Breit–Pauli Hamiltonian, 304

Brillouin–Wigner, 38, 40, 44–51, 57, 64

Brillouin–Wigner coupled cluster singles and doubles theory (BW-CCSD), 43

Brillouin–Wigner coupled cluster theory, 43, 52

Brillouin–Wigner full configuration interaction (BW-FCI), 42

Brillouin–Wigner methods, many-body, 33–64

Brillouin–Wigner perturbation theory, 35, 39–44, 46, 48, 51–53

Brillouin–Wigner theory, 33–64

Brownian motion, 34, 259

Brown, R., 34

Brown–Ravenhall disease, 116
Brueckner, K.A., 38, 39
BSSE. *See* Basis set superposition error (BSSE)

C

CADPAC, 221, 238, 466
Cambridge Structural Database (CSD), 486, 500, 527, 530, 539, 542
CaPTURE, 526
Carbon fixation cycle, 653
CASCCD, 74–78
CASPT2, 107, 112, 113, 135–140, 163, 224, 242
CASSCF, 74, 81, 94, 95, 99, 112, 113, 135, 136, 138–140, 224, 241, 242, 273, 388–392, 395, 410, 632
Cation-π interaction, 339, 340, 517–551
cc-pVDZ, 94, 95, 221, 241, 389–392, 428, 540, 559, 561
CCSD(T), 57, 81, 82, 105, 109, 130, 132–134, 136–140, 143–146, 162, 163, 241, 242, 502, 520, 534, 535, 540, 605
CdS, 131–134
Charge density distribution, 168
Charge transfer numbers, 422, 444
CHARMM27, 393, 396, 400
Chebyshev–Davidson method, 181, 182, 186, 188
Chebyshev filtered subspace iteration, 167–188
Chebyshev–Jackson approximation, 181
Chirality, 374, 451–472
Chlorotrinitromethane Cl–C(NO$_2$)$_3$, 482, 484–488
Classification of hydrogen bonds, 497–501
Clay, kaolinite, 647, 648, 652, 659, 667
Closed-shell interactions, 507, 649
CMOS. *See* Complementary metal-oxide semiconductor (CMOS)
Coester, F., 38
Coinage metal hydrides, 136–139
Collaborative research, 37, 55
Collaborative virtual environment (CVE), 37, 38, 52–57
Collectivity number, 422, 423, 428, 430, 433–436, 439, 443
Collision induced dissociation (CID), 522, 533
Complementary metal-oxide semiconductor (CMOS), 351, 362
Computer and communications technology, 35
Computing, high performance, 35, 39
Conductor-like screening model (COSMO), 459, 467, 650, 665, 666

Configuration interaction (CI), 38, 39, 42, 43, 52, 57, 58, 70, 75, 78, 88, 153, 163, 192, 232–234, 237, 240, 256, 263, 265, 270, 275, 416, 422, 425, 427, 431, 432, 438, 441–443, 536
Conical intersection, 388, 390, 391, 395, 411
Contemporary computers, power of, 35
Cooperativity, 521, 540–546
Core electron binding energy (CEBE), 111–114, 243, 244
Core orbital, 211, 214–216, 218, 232, 244, 273, 274
Core-valence separability, 211
Corey, R., 17
Correlating functions, 221, 238, 241, 244, 245
COSMO. *See* Conductor-like screening model (COSMO)
Coulson, C., 20, 21, 205
Counterpoise method (CP), 241, 649
Coupled cluster (CC), 33, 39, 43, 52, 57, 70–75, 78, 79, 83–88, 99, 100, 133, 134, 192, 193, 256, 270, 273, 282, 294, 416, 462, 518
Coupled cluster singles and doubles theory (CCSD), 57, 70–77, 88, 416
Coupled cluster (CC) theory, 42, 43, 58, 70, 71, 77, 79, 82, 256
Covalent interactions, 321, 507, 518, 524
CPD. *See* Cyclobutane pyrimidine dimer (CPD)
Crick, F., 14, 287
CuH, 136–139, 314, 315
Current trends in computational chemistry (CCTCC), 11
Cusp correction, 303, 309–315
CVE. *See* Collaborative virtual environment (CVE)
Cyclobutane pyrimidine dimer (CPD), 386–388, 402
Cyclohexene, 557–576
Cyclopropenone, 140–142
Cyclopropenethione, 140–142

D

Dale, H., 14
Das, T.P., 39
Democritos's maxim, 2
deMon2K, 238
Density functional theory (DFT), 23, 130, 168, 192, 215, 256, 320, 352, 416, 458, 520, 592, 649
Density of state (DOS), 184, 196, 197, 205, 331, 336–338, 355

Deoxyribonucleic acid, 14
Desoxyribose, 14
Dewar, M., 23, 24, 444
DFT. *See* Density functional theory (DFT)
DFT catastrophe, 149
DFT-GF formalism, 355
Diagonalization, 119, 120, 169, 171, 173–176,
    181, 185–186, 194
Diagrammatic formalism, 38
Diagrammatic techniques, 38
Dickenson, R., 15
Diffusion Monte Carlo (DMC), 256–268, 277,
    282, 283, 286–288, 294, 298–301, 311,
    315
1,2-Dihydro fullerenes, 149, 152
Dihydrogen bond, 479, 498, 500
Dirac density matrix, 419
Dirac Hamiltonian, 115, 116, 118–121, 123,
    303, 304
Dirac, P.A.M., 34, 103
Dissociative processes, 39
Divide and conquer approach, 536
DKHn approach, 122
DNA, 14, 210, 243, 287, 341–344, 356, 361,
    385–392, 396–400, 405, 410, 411, 454,
    518–520, 535, 549, 647, 648, 650–653,
    659, 660, 665–667
DNA sequencing, 319–344
Double-helix model, 14, 386, 394
Duplex, 387, 388, 392, 395, 398, 399,
    403–405, 549
1 D vibrational Schroedinger equation, 262
Dynamic correlation, 126, 224, 226, 230, 234,
    245, 247, 270

E
Effective core potential (ECP), 158, 209,
    211–215, 218, 219, 226, 241, 243, 244,
    246, 266, 268, 274–275, 283, 353, 649
Effective Hamiltonian, 43, 160, 214, 275, 300,
    328
Effectively unpaired electrons, 425, 426, 441
Eigenproblem, 176, 181
Einstein, A., 3, 34, 259
Electron affinity, 105, 233, 540
Electron correlation, 39, 53, 55, 69–71, 81,
    112, 113, 132–134, 140, 151, 212, 215,
    226, 238, 241–243, 247, 256, 267–269,
    275–277, 285, 293, 294, 301, 315, 316,
    354, 355, 464, 470, 522, 546, 605, 633
Electron correlation energy, 464
Electron density distribution, 7, 13, 25, 532,
    564, 649

Electron diffraction, 12, 13, 558, 559
Electronic structure, 20, 70, 82, 83, 111, 169,
    174, 181, 182, 191–206, 210, 230, 242,
    244, 255–288, 331, 386, 464–465, 520,
    548, 580, 582, 630–634
Electronic structure theory, 83
Electron-nucleus coalescence condition,
    302–303
Electron transport, 327, 329–330, 333, 335,
    342, 355, 376, 379
Electrostatic potentials, 349, 364, 480–485,
    488, 493, 520, 649, 660, 662, 664
Encyclopedia of computational chemistry, 58
Endohedral fullerenes, 149–159
European metalaboratory for multireference,
    52
Ewald, P.P., 393
Excitation indices, 423, 426, 438

F
FCI. *See* Full configuration interaction (FCI)
FDTD simulations. *See* Finite-difference
    time-domain (FDTD) simulations
$Fe^{IV}=O$, 631–634, 638
Fermi–Dirac distribution function, 328
Feynman, R.P., 273
Fine-structure splitting, 222
Finite basis set expansion, 35, 39
Finite difference method, 172, 599
Finite-difference time-domain (FDTD)
    simulations, 323–325
First-order RA (FORA), 117, 305, 306
Fixed-node approximation (FNA), 261, 262,
    267, 300
Fixed-node diffusion Monte Carlo, 311
Fluorine hydride (FH), 94–99
Fock space Brillouin–Wigner methods, 59–64
Fock, V.A., 23, 25, 39
FORA. *See* First-order RA (FORA)
Fragment molecular orbital method (FMO),
    221, 243
Friedrich, W., 11
Full configuration interaction (FCI), 42, 79, 81,
    82, 88, 94, 99, 265, 270, 442, 443
Fullerenes, 149–159, 548

G
GAMESS-US, 221, 232, 242, 243, 245
Gamma helix, 18
Gauge invariant atomic orbitals (GIAO), 463,
    465
GAUSSIAN, 352, 466, 584, 587, 598, 613

Gaussian function, 39, 126, 262, 353
Gay–Lussac, J.L., 27
Generalized bond index, 426, 427
Generalized Van Vleck perturbation theory,
    161
GENIP, 355, 357, 358, 363, 378
Gillespie, R.J., 7, 10
GNR. *See* Graphene nanoribbon (GNR)
Goldstone, J., 38, 39, 43
Google sites, 54, 56
Graphene, 335, 341, 343, 356–360, 363–379
Graphene nanoribbon (GNR), 320, 335–339,
    341–344, 368, 369
Green function (GF) theory, 352, 374
Green's function, 256, 262, 299, 300, 328, 332,
    336
Growth of the internet, 35
Guanine, 287, 388, 406–408, 410, 411, 522,
    648, 652, 659

H
Hairpin, 388, 395–402, 404, 406, 410
Harmonic vibrational analysis, 242
Hartree, D.R., 23, 25
Hartree–Fock (HF), 39, 70, 72, 90, 93, 95, 131,
    145, 150, 155, 156, 158, 192, 193, 211,
    214, 215, 223, 235, 238, 239, 247, 256,
    261, 262, 266, 268–271, 294, 301, 303,
    312, 313, 315, 353–354, 396, 416–419,
    421, 427, 441, 464, 560, 633
H_2 as the proton acceptor, 504
Hauptman, H., 11
Heitler and London, 15, 34
Heitler, W., 5, 15, 34
Hellman–Feynman theorem, 266
Hemoglobin, 16
Heredity, 14
HERZBERG package, 88
Hessian matrix, 463
FH–H_2 complex, 500, 514
HgS, 131–134
High performance computing (HPC), 39, 286,
    287
High-pressure mass spectrometry (HPMS),
    533
HLG. *See* HOMO-LUMO gap (HLG)
H_2 molecule, 498, 500, 501
Hoffmann, R., 20
Hohenberg–Kohn, 354, 443, 483
π-Hole interactions, 482
σ-Hole interactions, 479, 482, 492, 494
Homochirality, 454
HOMO-LUMO gap (HLG), 350, 356–359, 377

Hubač, I., 33, 43, 52
Hubbard, J., 38
Hugenholtz, N.M., 38
Human-human communication, 37, 55
Human-machine communication, 37, 38
Hund, F., 20–22
Hund–Mulliken theory, 22
Hybrid DFT, 241, 522, 633
Hydration, 243, 532, 533, 549, 648, 656, 657,
    661, 664, 667
Hydrogen bonding, 16, 17, 340, 361, 454, 459,
    466, 468–470, 479, 498, 499, 502, 503,
    507, 509, 512, 517, 551, 647, 649
Hydrophobic interaction, 518
Hyperpolarizabilities, 64, 108, 130–132, 134,
    135, 139–153, 155, 156, 162

I
Infinite order two-component approach
    (IOTC), 106, 112, 113, 118–122, 125,
    126
Infinite-order regular approximation (IORA),
    117, 305, 306, 311
Insulin, 14
Intel, 35, 186
Intermediate coupling (IC), 229, 233, 235
Internet, growth of the, 35
Internet users, number of, 35, 36
Intramolecular interactions, 479–495, 505,
    562, 564, 566, 569, 570, 576
Intrinsic reaction coordinate (IRC), 558, 560,
    561, 571
Intruder state problem, 38, 40, 41, 44, 52, 54
Intruder states, 40–42, 57
Inversion in NH_3, 159–163

J
Jahn–Teller, 198, 205, 238
Jastrow correlation factor, 301, 302
Jastrow–Slater wave function, 268, 301, 309,
    311
j-j coupling, 238

K
Kaldor, U., 39
Karle, J., 11
Kelly, H.P., 39
Kepler, J., 2, 3
Ketones, 129, 140–142
Kinetics, 121, 160, 163, 193, 246, 247, 259,
    262, 297, 300, 302, 304–308, 352, 507,

522, 533, 539, 583, 635, 649, 654, 656,
    667
Kitaura–Morokuma decomposition, 503
Knipping, P., 11
Kohn–Sham, 167–188, 192–194, 205, 215,
    261, 303, 354, 443, 548
Kohn, W., 23, 193
Krylov subspace, 185
Kumar, M., 34
Kümmel, H., 38

L

Landauer–Büttiker formalism, 327, 328
Large amplitude motions, 159
Laue, M., 10
LEVEL package, 92
Lewis, G.N., 5
Li@C_{60}, 154–159
[Li@C $_{60}$]$^+$, 154, 155, 157
Linear scaling, 53, 256, 276, 283–285, 536
Linked diagram theorem, 38, 39, 51
Lipscomb, W.N., 20
Lithium hydride (LiH), 87, 90–93, 130, 143,
    144, 294
Local energy, 257, 258, 261, 264, 278–280,
    282–285, 294, 295, 297, 298, 301–303,
    306–307, 309, 311, 313–315
Localization operator, 423, 425
Locked nucleic acid, 388–405
London dispersion, 546
London, F.W., 5, 15, 34
L-S coupling, 235, 238

M

Magic structures, 329
Magnetoresitance (MR), 57, 74, 335, 338, 344,
    605
Many-body Brillouin–Wigner methods, 33, 35,
    37
Many-body methods, 33, 38, 42, 52
Many-body perturbation theory (MBPT), 39,
    43, 51, 57, 70, 256
Mášik, J., 43
Massively parallel, 172
Matrix-isolation VCD, 468
Mavridis, A., 52
McWeeny formula, 419
Medal, C., 14
Mehra, J., 34
Meissner, L., 52
Metal-cycopentadienyl (M–Cp), 534
Metal hydrides, 136–139, 500

Metal-oxide semiconductor, 362
Metal-sigma interactions, 500
Metropolis method, 257, 296
Mirsky, A., 16
Model core potential, 209–247
Molar extinction coefficient, 456
MOLCAS, 118, 238
Molecular biology, 13–15, 482
Molecular electrostatic potential (MEPs), 347,
    349–350, 360, 361, 363–365, 367, 374,
    379
Molecular engineering, 347–380
Molecular mechanics, 3–4, 558–560, 563
Molecular orbital (MO) theory, 5, 15, 22, 294,
    350, 352–353, 426
Molecular tailoring approach (MTA), 194, 536,
    537
Moore, G.E., 35
Moore's law, 35
MO theory. *See* Molecular orbital (MO) theory
MRCC. *See* Multireference coupled cluster
    (MRCC)
Mulliken, R.S., 1, 5, 20–22, 27, 590, 600, 609,
    614, 633, 634
Multicenter proton acceptors, 495–514
Multiconfiguration quasi-degenerate
    perturbation theory, 229
Multi-reference, 35, 37–42, 48–54, 57–59,
    69–100, 238, 240
Multi-reference Brillouin–Wigner perturbation
    theory, 42, 52
Multireference coupled cluster (MRCC), 52,
    71, 72, 82, 88
Multi-reference functions, 39, 48, 49, 58
Multi-reference Rayleigh–Schrödinger
    perturbation theory, 40–42, 52
Mutation, 385, 386, 549, 550, 658, 667
M05-2X, 355, 520, 546, 649, 666
M06-2X, 444, 520

N

NAMD, 393
Nanocrystal, 186, 187, 646
Nano-scale phenomena, 320
Natural spin-orbitals, 421
Natural spinors, 235, 238, 421
NEGF. *See* Non-equilibrium Green's function
    (NEGF)
Neogrády, P., 43
NH_4^+-$(H_2)_n$ clusters, 504
Ni-Fe Sulfide, 645, 653
Nobel Prize, 5, 11, 22, 23, 27
Nodeless orbitals, 212–214

Non-bonded interaction, 8–10, 355, 440, 518, 520, 522, 530, 537, 541, 546, 550

Noncovalent interaction, 479–482, 518–543, 546, 547, 550, 649

Non-equilibrium Green's function (NEGF), 320, 327, 331, 342

Nonlinear eigenproblem, 170, 194

Non-linear optical properties, 129–163

Non-relativistic cusp condition, 308

Non-relativistic local energy, 294

Nuclear physics, 38, 59

Nucleic acid base, 645–668

Number of internet users, 35, 36

Numerov–Cooley (NC), 135–139, 485

Nyholm, R.J., 7

O

Oliphant–Adamowicz coupled cluster model, 72

Oppenheimer, J.R., 34, 136, 266, 352, 461, 548

Organic nanostructures, 320

Organoxenon complexes, 241

Origin of life, 645–668

Orthogonalization, 181, 186

P

Paldus, J., 38, 39, 82, 91, 92

Parr, R.G., 23, 24

PARSEC. *See* Pseudopotential algorithm for real-space electronic calculations (PARSEC)

Pauli exclusion principle, 6

Pepsin, 14

Perturbation theory (PT), 38, 39, 51, 57, 58, 69, 70, 112, 117, 161, 229, 230, 233–237, 240, 247, 633

Perturbation theory, a posteriori corrections to, 43, 52, 57

Perturbation theory, Brillouin–Wigner, 39–44, 48, 52, 53

Perturbation theory, many-body, 39, 43, 51, 256

Perturbation theory, multi-reference, 40–42, 52, 53

Perturbation theory, Rayleigh–Schrödinger, 40–42, 52, 53

Perutz, M., 15

Phenylalanine, 520, 522, 523

Photoadduct, 386, 405

6-4 Photoadduct, 386

Photodimerization, 385–411, 437

Pittner, J., 40, 52

Plasmonic, 367–368, 370–374, 379, 380

Polanyi, M., 1, 28

Polarizabilities, 64, 108, 109, 123, 124, 130, 132, 133, 145, 154–156, 162, 533

Polarizable continuum model (PCMs), 467, 468, 534, 535, 540

Polarization functions, 465

Polycyclic aromatic hydrocarbons (PAHs), 428, 436

Pople, J., 1, 11, 23

Posteriori adjustments, 38

Posteriori corrections to Brillouin–Wigner perturbation theory, 43, 51–52

Powell, H.M., 7

Power of contemporary computers, 35

Protein data bank (PDB), 520, 527, 528, 530, 539, 542, 543, 548, 550

Protein folding, 549

Pseudopotential, 160, 168, 169, 186, 205, 209–212, 215, 218, 222, 223, 234, 244, 265, 269, 311, 649

Pseudopotential algorithm for real-space electronic calculations (PARSEC), 169, 172, 181, 184–187

Pseudo-valence orbital, 211, 212, 244

Pure vibrational contribution, 132, 134–137, 142, 145, 147, 148

Pyrimidine dimer, 385–387

Pyrite, 646, 651, 653, 654, 667

Pyrrole, 145–149, 340

Q

Quantum chemical methods, 34, 37, 38, 52, 54, 470, 530, 559, 560, 580, 595, 596, 632, 634

Quantum field theory, 31

Quantum many-body theories, 35

Quantum Monte Carlo (QMC), 255–288, 293–316

Quantum theory of atoms in molecules (QTAIM), 504–510, 512

Quantum yield, 388, 392, 394–402, 404–411

R

Rate constants, 650, 658, 667

Rayleigh–Ritz method, 181

Rayleigh–Schrödinger, 38, 40–47, 51–53, 64, 231

Rayleigh–Schrödinger formalism, 40

Rayleigh–Schrödinger perturbation theory, 40–43, 45, 51–53, 231

Real-space method, 168, 169, 172

Rechenberg, H., 34
Reduced density matrix (RDM), 419–421, 423–425, 442, 443
Relativistic Brillouin–Wigner methods, 59
Relativistic cusp correction, 309, 314
Relativistic effect, 103–126, 131, 136, 138, 139, 195, 211, 222–238, 245, 247, 294, 309, 316
Relativistic effects, 103–126, 131, 136, 138, 139, 195, 211, 222–238, 245, 247, 294, 309, 316
Relativistic Hamiltonian, 105, 115, 117, 247, 294, 298, 300, 307
Relativistic many-body Brillouin–Wigner methods, 59
Relativistic QMC method, 295
Renn, J., 34
Roberts, J.M., 20, 21, 35
Rockefeller, 14
Rotatory strength, 456, 461, 464, 465, 470
R4QMC program, 311, 314
Rydberg–Klein–Reese (RKR) approach, 96–99

S
Sayre, D., 11, 12, 14
Scalar mass velocity (MV), 106
Scalar-relativistic effects, 211, 222–226, 245
Scanning tunneling microscope (STM), 200, 361
Schrödinger, E., 15
SCS-CCSD, 520
SCS-MP2, 520, 546
Selenazofurin, 492, 494
Selenazoles, 493
Self consistent field, 74, 169, 270, 329, 464
Shavitt, I., 39
Sidgwick, N.V., 7
Silver, D.M., 39
Single crystal, 12, 196
Single-reference CC theory, 70
Singular value decomposition, 421
Si-PETN, 490, 491
Snowflakes, 2, 3
Solid state physics, 38, 464
Sommerfeld, A., 10, 15, 104
Spectrum slicing, 181–184
Spin-forbidden radiative transitions, 222
Spin orbital, 70, 72, 73, 75, 80, 81, 83, 125, 126, 268, 352, 417, 418, 421, 426
Spin-orbit coupling (SOC), 205, 211, 212, 215, 219, 222, 223, 226–235, 237, 238, 244–246
Spin-orbit effect, 212, 219

Spin-orbit operator, 219, 304
Spintronic, 319–344
SSMRCC. See State specific MRCC (SSMRCC)
Stacking interaction, 321, 343, 519, 520, 535, 537, 546, 550
Staemmler, V., 52
State specific, 35, 37, 40, 41, 47, 48, 51–53, 69–100, 237
State specific MRCC (SSMRCC), 71–77, 79
State-specific multi-reference correlation problem, 35
Statistical overlap, 417, 425
Steiner, M.M., 41
Stereochemistry, 3, 456, 557, 558, 576
Strong orthogonality constraint, 211, 214
Structural chemistry, 1–28, 422, 438
Super-magnetoresistance (SMR), 320, 335, 338
Super-refraction, 320
Surface-plasmon excitation, 323

T
Tautomerism, 648
Teller, E., 19
Theory of resonance, 15–19
Thermodynamics, 597, 602, 650, 651, 653, 654, 656, 657, 667
Thiazoles, 493
Thiones, 140, 141, 566, 569
Thymine, 287, 385–411, 438, 648, 652, 653, 659, 667
Thymine-thymine dimer (TT dimer), 386–388, 394, 396, 398–400, 406–411
Tiazofurin, 492–494
Topochemical rules, 387
$n\pi^*$-Transition, 430–432
$\pi\pi^*$-Transition, 431
Transition metal systems, 200, 220, 221, 223, 230, 239, 240, 246, 266, 288, 500, 522, 532, 549, 606, 649, 653, 654
Trinucleotide, 388, 406–411
Tryptophan, 520, 522, 523
Tsipis, C., 52
TT dimer. See Thymine-thymine dimer (TT dimer)
TT dimerization, 388–392, 396–399, 402
Tyrosine, 520, 522, 523

U
Ultrafast DNA sequencing, 320, 344
Unlinked diagrams, 38, 58

V

Valence-bond theory (VB theory), 15, 20, 271
Valence orbital, 124, 210–212, 214–216, 219,
    226, 232, 244, 245, 274
Valence photoisomerization, 438
Valence shell electron pair repulsion (VSEPR),
    5–8
van der Waals interaction, 465, 647
Variance minimization technique, 278, 279,
    307, 311
Variational Monte Carlo (VMC), 256–258,
    260, 263, 266, 267, 277, 282, 294, 295,
    297–298, 307, 311, 313–315
VB theory. *See* Valence-bond theory (VB
    theory)
Vibrational contributions, 131–137, 140–143,
    145–149, 153, 154, 158, 161, 163
Vibrational electronics (Vibronics), 348–349,
    362, 363, 365, 367–370, 374, 379, 380
Violarite, 646, 650
Viral DNA, 518

W

Walker, 258, 260–266, 274, 275, 277–279,
    282, 283, 286, 297, 299–302, 307
Watson, R.E., 14, 287
Weakly coupled subsystems, 433–434

Weaver, W., 14
Well-tempered basis sets (WTBS), 220, 231,
    232
Wenzel, W., 41
Westheimer, F., 3–4
Wheland, G., 15, 16
Wiberg bond indices, 417, 427
Wigner–Witmer rules, 28
William Astbury, 14, 17
Wilson, S., 33–64

X

X-ray crystallography, 10–12, 14, 15, 17, 387

Z

Zeroth order regular approximation (ZORA),
    117, 295, 304–309, 311, 312, 314, 315
Zigzag, 335, 336, 356, 358–360, 373, 627
Ziman, J., 37
ZnS, 131–134
ZORA. *See* Zeroth order regular approximation
    (ZORA)
ZORA local energy, 307, 315
ZORA pseudo local energy, 307
ZPVA contribution, 134, 135, 163

Printed by Publishers' Graphics LLC
CAMZ140302.20.06.344